THE ROUTLEDGE HANDBOOK ON SPACES OF URBAN POLITICS

The Routledge Handbook on Spaces of Urban Politics provides a comprehensive statement and reference point for urban politics. The scope of this handbook's coverage and contributions engages with and reflects upon the most important, innovative and recent critical developments to the interdisciplinary field of urban politics, drawing upon a range of examples from within and across the Global North and Global South.

This handbook is organized into nine interrelated sections, with an introductory chapter setting out the rationale, aims and structure of the *Handbook*, and short introductory commentaries at the beginning of each. It questions the eliding of 'urban politics' into the 'politics of the city', reconsidering the usefulness of the distinction between 'old' and 'new' urban politics, considering issues of 'class', 'gender', 'race' and the ways in which they intersect, appear and reappear in matters of urban politics, how best to theorize the roles of capital, the state and other actors, such as social movements, in the production of the city and, finally, issues of doing urban political research. The various chapters explore the issues of urban politics of economic development, environment and nature in the city, governance and planning, the politics of labour as well as living spaces. The concluding sections of the *Handbook* examine the politics over alternative visions of cities of the future and provide concluding discussions and reflections, particularly on the futures for urban politics in an increasingly 'global' and multidisciplinary context.

With over forty-five contributions from leading international scholars in the field, this handbook provides critical reviews and appraisals of current conceptual and theoretical approaches and future developments in urban politics. It is a key reference for all researchers and policy-makers with an interest in urban politics.

Kevin Ward is Professor of Human Geography in the School of Environment, Education and Development and Director of the Manchester Urban Institute (www.mui.manchester.ac.uk) at the University of Manchester, UK.

Andrew E.G. Jonas is Professor of Human Geography at the University of Hull, UK.

Byron Miller is Professor and the Coordinator of the Urban Studies Interdisciplinary Program, University of Calgary, Canada.

David Wilson is Professor of Geography, Urban Planning, and the Unit for Criticism and Interpretive Theory at the University of Illinois at Urbana-Champaign, USA.

THE ROUTLEDGE HANDBOOK ON SPACES OF URBAN POLITICS

Edited by Kevin Ward, Andrew E.G. Jonas, Byron Miller
and David Wilson

LONDON AND NEW YORK

First published 2018
by Routledge
2 Park Square, Milton Park, Abingdon, Oxon OX14 4RN

and by Routledge
711 Third Avenue, New York, NY 10017

Routledge is an imprint of the Taylor & Francis Group, an informa business

© 2018 selection and editorial matter, Kevin Ward, Andrew E.G. Jonas, Byron Miller and David Wilson; individual chapters, the contributors

The right of Kevin Ward, Andrew E.G. Jonas, Byron Miller and David Wilson to be identified as the authors of the editorial matter, and of the authors for their individual chapters, has been asserted in accordance with sections 77 and 78 of the Copyright, Designs and Patents Act 1988.

All rights reserved. No part of this book may be reprinted or reproduced or utilised in any form or by any electronic, mechanical, or other means, now known or hereafter invented, including photocopying and recording, or in any information storage or retrieval system, without permission in writing from the publishers.

Trademark notice: Product or corporate names may be trademarks or registered trademarks, and are used only for identification and explanation without intent to infringe.

British Library Cataloguing-in-Publication Data
A catalogue record for this book is available from the British Library

Library of Congress Cataloging-in-Publication Data
Names: Ward, Kevin, 1969– editor.
Title: The Routledge handbook on spaces of urban politics / edited by Kevin Ward, Andrew EG Jonas, Byron Miller and David Wilson.
Other titles: Handbook on spaces of urban politics
Description: Abingdon, Oxon ; New York, NY : Routledge, 2018. | Includes bibliographical references and index.
Identifiers: LCCN 2017054392| ISBN 9781138890329 (hardback : alk. paper) | ISBN 9781315712468 (ebook)
Subjects: LCSH: Urban policy–Cross-cultural studies. | Urbanization–Political aspects–Cross-cultural studies. | City planning–Cross-cultural studies. | Urban renewal–Cross-cultural studies. | Municipal government–Cross-cultural studies.
Classification: LCC HT151 .R664 2018 | DDC 307.76–dc23
LC record available at https://lccn.loc.gov/2017054392

ISBN: 978-1-138-89032-9 (hbk)
ISBN: 978-1-315-71246-8 (ebk)

Typeset in Bembo
by Wearset Ltd, Boldon, Tyne and Wear

Printed and bound in Great Britain by
TJ International Ltd, Padstow, Cornwall

CONTENTS

Lists of figures   xi
List of tables   xiii
Notes on contributors   xiv
Acknowledgements   xxii
Glossary   xxiii

1 Spaces of urban politics: an introduction   1
   Andrew E.G. Jonas, Byron Miller, Kevin Ward and David Wilson

PART I
Approaching the space(s) of urban politics   11
Andrew E.G. Jonas and Byron Miller

2 Here, there and everywhere: rethinking the urban of urban politics   14
   Allan Cochrane

3 Place-based or place-positioned? Framing and making the spaces of urban politics   26
   Deborah G. Martin

4 Ambivalence of the urban commons   35
   Theresa Enright and Ugo Rossi

5 The smart state as utopian space for urban politics   47
   Yonn Dierwechter

PART II
Spaces of economic development — 59
David Wilson and Andrew E.G. Jonas

6 Pro-growth urban politics and the inner workings of public-private partnerships — 62
Christopher Mele and Matthew H. McLeskey

7 The urban politics of strategic coupling in global production networks — 70
Rachel Bok and Neil M. Coe

8 The sky is not the limit: negotiating height and density in Toronto's condominium boom — 85
Ute Lehrer and Peter Pantalone

9 Digital technologies and reconfiguration of urban space — 96
Barney Warf

PART III
Spaces of the environment and nature — 107
Byron Miller and Andrew E.G. Jonas

10 Climate science and the city: consensus, calculation and security in Seattle, Washington — 110
Jennifer L. Rice

11 Democratizing the production of urban environments: working in, against and beyond the state, from Durban to Berlin — 122
James Angel and Alex Loftus

12 Politics of urban gardening — 134
Marit Rosol

13 Just green spaces of urban politics: a pragmatist approach — 146
Ryan Holifield

14 From sustainability to resilience: the hidden costs of recent socio-environmental change in cities of the Global North — 157
David Saurí

15 Transforming Rainey Street: the decoupling of equity from environment in Austin's smart growth agenda — 167
Eliot Tretter and Elizabeth J. Mueller

PART IV
Spaces of governing and planning 181
David Wilson and Byron Miller

16 Cities on a grand scale: instant urbanism at the start of the twenty-first century 184
   Martin J. Murray

17 Urbanization, planning and the possibility of being post-growth 197
   Jason Hackworth

18 Troubled buildings, distressed markets: the urban governance of the US foreclosure crisis 206
   Philip Ashton

19 Housing the banlieue in global times: French public housing and spaces between neoliberalization and hybridization 217
   David Giband

PART V
Spaces of labour 229
Kevin Ward and Andrew E.G. Jonas

20 Roll-against neoliberalism and labour organizing in the post-2008 crisis 232
   Luis L.M. Aguiar and Yanick Noiseux

21 Urbanization as a bordering process: non-citizen labour and precarious construction work in the Greater Toronto Area 245
   Michelle Buckley and Emily Reid-Musson

22 Organizing the ruins: the thin institutional geography of labour in the US Midwest 259
   Marc Doussard

23 Mobilities and moralities of domestic work in Indonesian cities 271
   David Jordhus-Lier and Debbie Prabawati

24 Street work as a key site of urban politics 282
   Ilda Lindell

25 Urban informality and the new politics of precarity: day labourer activism in the USA 293
   Nik Theodore

PART VI
Spaces of living 305
David Wilson and Kevin Ward

26 The political spaces of urban poverty management 307
    Joshua Evans and Geoff DeVerteuil

27 Urban community gardens as new spaces of living 320
    Rina Ghose and Margaret Pettygrove

28 Envisioning liveability and do-it-together urban development 336
    Helen Jarvis

29 Infrastructural citizenship: spaces of living in Cape Town, South Africa 350
    Charlotte Lemanski

30 The politics of urban agriculture: sustainability, governance and contestation 361
    Nathan McClintock, Christiana Miewald and Eugene McCann

31 Retroactive utopia: class and the urbanization of self-management in Poland 375
    Kacper Pobłocki

PART VII
Spaces of circulation 389
Andrew E.G. Jonas and Kevin Ward

32 Circulating risks: coastal cities and the spectre of climate change risk 393
    Kevin Fox Gotham and Clare Cannon

33 The logics and politics of circulation: exploring the urban and non-urban spaces of *Amazon.com* 404
    Markus Hesse

34 Circulating experiments: urban living labs and the politics of sustainability 416
    James Evans, Harriet Bulkeley, Yuliya Voytenko, Kes McCormick and Steven Curtis

35 Assembling and re-assembling Asian carp: the Chicago Area Waterways System as a space of urban politics 426
    Julie Cidell

36 Google buses and Uber cars: the politics of tech mobility and the future of urban liveability 439
Jason Henderson

37 Making multi-racial counter-publics: towards egalitarian spaces in urban politics 451
Helga Leitner and Samuel Nowak

PART VIII
Spaces of identity 465
Byron Miller and David Wilson

38 A city of migrants: migration and urban identity politics 468
Virginie Mamadouh

39 Class, territory and politics in the American city 479
Kevin R. Cox

40 Urban middle-class shifting sensibilities in neoliberal Buenos Aires 492
Carolina Sternberg

41 Compassionate capitalism: tax breaks, tech companies and the transformation of San Francisco 504
Lauren Alfrey and France Winddance Twine

42 Gendering urban protest: politics, bodies and space 518
Fran Tonkiss

43 Queering urban politics and ecologies 528
Will McKeithen, Larry Knopp and Michael Brown

PART IX
Spaces of utopia and dystopia 539
Andrew E.G. Jonas and David Wilson

44 Dystopian dynamics at work: the creative validation of urban space 542
Ulf Strohmayer

45 Deconstructing modern utopias: sustainable urbanism, participation and profit in the 'European City' 555
Samuel Mössner and Rob Krueger

46 Dystopian spaces and Roma imaginaries: the case of young Roma in
    Slovenia and Romania   565
    Stuart C. Aitken and Jasmine Arpagian

47 Mobile futures: urban revitalization and the aesthetics of transportation 577
    Theresa Enright

48 Reimagining the urban as a dystopic resilient space: scalar materialities in
    climate knowledge, planning and politics   589
    Andrew Kythreotis

Index   601

FIGURES

3.1	Image of City Hall, Worcester, MA	28
7.1	Urban politics and the strategic coupling process	73
10.1	Information about climate impacts from departmental briefing memo	118
10.2	Reproduction of 'risk tolerance' guidelines provided by the city of Seattle to department heads	118
15.1	Smart growth zones, Austin	170
15.2	Revitalization and East Austin	173
26.1	Image of Skid Row	308
27.1	Harambee neighbourhood	323
27.2	Urban agriculture in Milwaukee, Wisconsin	325
27.3	Networks of association for urban agriculture in Milwaukee, Wisconsin	328
28.1	Let the children play: street closure for Big Lunch street party; children 'playing out', Newcastle upon Tyne, 2008	343
28.2	Let the adults play: street closure for Big Lunch street party, Newcastle upon Tyne, 2008	344
28.3	LA Ecovillage, view of White House Place Learning Garden	346
28.4	Open-air curb-side 'lending library' in front of LA Ecovillage	347
30.1	A front yard garden in a gentrifying neighbourhood of Portland, Oregon	368
30.2	Sole Food Street Farms, a large social enterprise located in downtown Vancouver, British Columbia	370
31.1	New construction in Poland, 1950–2014	377
31.2	Spatial distribution of *My-Poznaniacy* support	382
31.3	The layout for Strzeszyn	383
33.1	*Amazon.com* locations in Europe	409
35.1	The Chicago Diversion as a result of reversing the flow of the Chicago River	431
35.2	Asian carp leaping from the Wabash River in Ohio	432
44.1	The *Ceinture* near the Rue de Crimée	545
44.2	Studios below the *Ceinture* near the Rue de l'Argonne	546
44.3	Urban gardening adjacent to the *Ceinture* below the Rue du Ruisseau	547
44.4	Squatting along the *Ceinture* in the Parc des Buttes Chaumont	548

Figures

46.1	Study sites at Maribor and Cluj-Napoca	566
48.1	The linear model of (climate) expertise	594
48.2	The multilinear model of (climate) expertise	595
48.3	Urban dystopias: the short-circuiting of knowledge flows from the city to the international scale	597

TABLES

7.1	Key types of strategic coupling between city-regions and global production networks	76
10.1	Climate science and governance in Seattle, WA	114
27.1	History of Milwaukee food projects	324
34.1	Strategic urban living lab example	421
34.2	Civic/municipal urban living lab example	421
34.3	Grassroots urban living lab example	422
41.1	2010 racial demographics for the Tenderloin vs San Francisco	508

CONTRIBUTORS

Luis L.M. Aguiar is Associate Professor of Sociology at the University of British Columbia, Okanagan Campus. He has published on low wage workers and organizing in cities across the Global North. He has been especially interested in the migration of the Justice for Janitors model to global cities for implementation in organizing workers.

Stuart C. Aitken is Professor of Geography and June Burnett Chair at San Diego State University. His research interests include critical social theory, qualitative methods, children, youth and families, film and masculinities.

Lauren Alfrey is Assistant Professor of Sociology at the University of Portland. She studies the cultural sociology of inequality, race, class and gender, the sociology of elites, and race and residential segregation. She is the co-author, with France Winddance Twine, of *Geek Girls: Race, Gender, and Sexuality in the Tech Industry*, forthcoming from Cambridge University Press.

James Angel is a PhD candidate in Geography at King's College London. His work focuses upon the democratization of urban infrastructure networks, with a particular focus on energy. As well as publishing academic articles and reports on this theme, James participates in activist networks developing ideas and practices of 'energy democracy'.

Jasmine Arpagian is a PhD student in Geography at San Diego State University. Her research interests include urban geography, disadvantaged families in Romania and displacement.

Philip Ashton is Associate Professor of Urban Planning and Policy at the University of Illinois at Chicago. His research interests are the restructuring of retail finance in US cities, the relationship between financialization and urban governance and progressive finance sector reform.

Rachel Bok is a doctoral student in Geography at the University of British Columbia. Her research interests lie in the area of urban political economy, especially in relation to topics of urban policy, politics and governance.

Michael Brown is Professor of Geography at the University of Washington in Seattle. His interests are in urban political geography, sexuality and health.

Contributors

Michelle Buckley is Assistant Professor of Human Geography at the University of Toronto Scarborough. She has written a number of articles on the transnational labour geographies of construction work in cities in the United Arab Emirates, Canada and the United Kingdom. Her work is broadly concerned with the intersectional and colonial politics of gender, citizenship and race that sustain contemporary processes of urbanization and the production of urban built environments.

Harriet Bulkeley is Professor of Geography, Durham University. Her research focuses on environmental governance and the politics of climate change, energy and sustainable cities.

Clare Cannon is Assistant Professor in the Department of Human Ecology at the University of California, Davis. Her areas of research interests include gender and the environment, global and urban sustainability, gender inequality and critical social theory.

Julie Cidell is Associate Professor of Geography and GIS at the University of Illinois. She studies how local governments and individual actors matter in struggles over large-scale infrastructure and policy development, and the corresponding urban environments that are produced.

Allan Cochrane is Emeritus Professor of Urban Studies at the Open University and the focus of his work is on cities and regions, their politics and policies. It combines theoretical contributions on the nature of urban and local politics and the making and remaking of regions with extensive empirically based research and an interest in qualitative research methods.

Neil M. Coe is Professor of Economic Geography and Co-Director of the Global Production Networks Research Centre at the National University of Singapore. His research interests are global production networks and local economic development, labour geographies and the geographies of innovation. He explores these interests through the study of globalizing service sectors such as retailing, staffing and logistics.

Kevin R. Cox is Distinguished Emeritus Professor of Geography at the Ohio State University. He has been a Guggenheim Fellow. His latest book is *The Politics of Urban and Regional Development and the American Exception* (Syracuse University Press, 2016).

Steven Curtis is a PhD student at Lund University in Sweden. His research focuses on platforms and organizational structures that support sustainable consumption and urban innovation for sustainability.

Geoff DeVerteuil is Senior Lecturer of Social and Urban Geography at Cardiff University, School of Geography and Planning. He has published extensively on urban poverty, homelessness, resilience, gentrification and inequality.

Yonn Dierwechter is Professor of Urban Studies at the University of Washington, Tacoma. His current research focuses on comparative regional planning systems, smart growth, smart city regions and the politics, policies and spaces of metropolitan sustainability, in the USA, Western Europe and South Africa.

Marc Doussard is Assistant Professor of Urban and Regional Planning at the University of Illinois at Urbana-Champaign. His research covers economic development and the problem of low-wage work.

Contributors

Theresa Enright is Assistant Professor of Political Science and Senior Fellow at the Global Cities Institute at the University of Toronto. Her research focuses on critical theory, metropolitan politics and urban political economy.

James Evans is Professor of Geography at the School of Environment, Education and Development, University of Manchester. His current research projects focus on urban living labs, smart cities and resilience.

Joshua Evans is Assistant Professor of Human Geography at the University of Alberta. He studies geographies of exclusion and inclusion. His most recent research focuses on spaces of care, home and work and their role in shaping the lived experiences of socially marginalized and vulnerable individuals, and spaces of policy development and implementation and their role in the creation of healthy, enabling and equitable urban environments.

Kevin Fox Gotham is Professor of Sociology and Associate Dean in the School of Liberal Arts (SLA) at Tulane University in New Orleans. His research interests are in the areas of real estate and housing policy, political economy of tourism and post-disaster recovery and rebuilding.

Rina Ghose is Professor of Geography at the University of Wisconsin-Milwaukee. Her research interests include urban political economy and GIS. She has published many academic articles and book chapters. Currently, she is working on the use of open GIS in analysing urban deprivation and hunger.

David Giband is Professor of Urban Planning at the University of Perpignan. He is the director of the research centre UMR ART-DEV (Universities of Montpellier and Perpignan). He is the author of numerous articles, book chapters and book about urban inequalities. His current researches deal with neoliberal policies in the education and housing fields.

Jason Hackworth is Professor of Geography and Planning at the University of Toronto. He has written extensively about neoliberalism, housing and land abandonment. He is currently writing a book on the role that ethno-racial conflict plays in urban decline.

Jason Henderson is Professor in Geography and Environment at San Francisco State University. His research examines how culture, economics and politics shape urban transportation policy and the geography of cities.

Markus Hesse is Professor of Urban Studies at the University of Luxembourg, Faculty of Humanities. As an urban and economic geographer, his research interest is situated at the intersection of places and flows (of all kinds).

Ryan Holifield is Associate Professor of Geography at the University of Wisconsin-Milwaukee. His research interests include environmental justice policy and practice, social and political dimensions of urban environmental change, and stakeholder participation in environmental governance.

Helen Jarvis is Reader in Social Geography at Newcastle University. Her publications include three books on cities, gender, work/life balance and social reproduction. Her current research focuses on sustainable degrowth and social phenomenon of cooperation in collaborative community innovations.

Andrew E.G. Jonas is Professor of Human Geography in the School of Environmental Sciences at the University of Hull. His research interests include city-regionalism and the politics of urban and regional development in the USA, Europe and Asia.

David Jordhus-Lier is Professor of Human Geography at the University of Oslo. His research focuses on civil society mobilization and labour organization in South Africa, Norway, Indonesia and the Democratic Republic of Congo.

Larry Knopp is Professor of Interdisciplinary Arts and Sciences at the University of Washington, Seattle and Adjunct Professor of Geography and Gender, Women and Sexuality Studies at the University of Washington, Seattle.

Rob Krueger is Associate Professor of Geography at Worcester Polytechnic Institute in the USA and Guest Professor of Sustainable Spatial Planning at the University of Luxembourg. His current research interests lie at the intersection of urban development, alternative economies and the environment.

Andrew Kythreotis is an associate professor in Environmental Studies, Policy and Management at the Universitie Brunei Darussalam. His research interests are trans-disciplinary and interdisciplinary, spanning geography, political and environmental science and cognate disciplines.

Ute Lehrer is an associate professor at the Faculty of Environmental Studies, York University. Taking a critical and comparative approach, her research interests include urban design and land use; housing, gentrification and the condominium boom; discourse and mega-projects; as well as the social construction of public space.

Helga Leitner is Professor of Geography at the University of California, Los Angeles. She has published three books and has written numerous articles and book chapters on the politics of immigration and citizenship, urban governance and social movements, global urbanism and socio-spatial theory.

Charlotte Lemanski is Senior Lecturer in Human Geography at the University of Cambridge. She is also a fellow of Robinson College, Cambridge. Her research focuses on everyday inequalities in the Southern city, with a specific focus on housing and infrastructure in post-apartheid South Africa.

Ilda Lindell is an associate professor in the Department of Human Geography, Stockholm University. Her interest is on the politics of informal work in urban Africa, with a focus on street work. She has edited *Africa's Informal Workers: Collective Agency, Alliances and Transnational Organizing in Urban Africa* (Zed Books, 2010) and several special issues in journals.

Alex Loftus is Senior Lecturer in Environment and Development at King's College London. He is the author of *Everyday Environmentalism: Creating an Urban Political Ecology* (University of Minnesota Press, 2012) and co-editor of *Gramsci: Space Nature Politics* (Wiley-Blackwell, 2012) and *The Right to Water: Politics, Governance and Social Struggles* (Earthscan, 2012).

Contributors

Virginie Mamadouh is Associate Professor of Political and Cultural Geography at the University of Amsterdam and affiliated to the Centre for Urban Studies and ACCESS EUROPE. Her research interests pertain to (critical) geopolitics, political culture, European integration, (urban) social movements, transnationalism and multilingualism.

Deborah G. Martin is Professor of Geography at Clark University, Worcester, Massachusetts. She is a qualitative urban geographer with research interests in place and place identity, neighbourhood activism and urban politics, law and geography, and socio-political dimensions of environmental issues and policies.

Eugene McCann is University Professor of Geography at Simon Fraser University. An urban political geographer, he researches policy mobilities, urban policy making, development and planning. He has co-edited and co-authored three books and numerous articles and book chapters. He is managing editor of *Environment & Planning C: Politics & Space*.

Nathan McClintock is Associate Professor of Urban Studies and Planning at Portland State University. An urban political ecologist, he researches the co-production of nature, society and the built environment. His theoretical and applied work on urban agriculture and food systems has appeared in a range of geography, planning and food studies journals.

Kes McCormick is an associate professor at Lund University in Sweden. His research focuses on the implementation of renewable energy technologies, sustainable urban transformation and education for sustainability.

Will McKeithen is a doctoral student and instructor of Geography at the University of Washington. He has published academic articles on the political ecology of intimacy under late capitalism. His current work examines issues of gender, embodiment and health in US incarceration.

Matthew H. McLeskey is a doctoral student in the University at Buffalo Department of Sociology and is interested in sociology of law, urban sociology and social theory. His dissertation focuses on how tenants navigate the legal system when landlords fail to comply with lead poisoning abatement protocols for older urban housing units.

Christopher Mele is Professor of Sociology and Adjunct Professor of Geography at the University at Buffalo-SUNY. He is the author of *Race and the Politics of Deception: The Making of an American City* (NYU Press, 2017). His current research focuses on issues of race and the political economy of urban development in North American cities.

Christiana Miewald is an adjunct professor in the Department of Geography at Simon Fraser University. Her recent research has focused on the intersections of food security and housing, harm reduction and health. She has also worked on issues of local food production such as community gardening and urban farming in Vancouver, BC.

Byron Miller is Professor of Geography and Coordinator of the Urban Studies Program at the University of Calgary. His recent work focuses on the spatial constitution of social movements, urban governance and governmentality, neighbourhood change and inequality, the politics of urban/regional sustainability, and the social and environmental implications of smart cities.

Samuel Mössner is a professor at the University of Münster, Institut für Geographie. His research has a focus on post-politics of urban sustainability and social justice in European cities.

Elizabeth J. Mueller is Associate Professor of Community and Regional Planning and Social Work at the University of Texas at Austin. Her work focuses on the tensions between urban sustainability planning and social equity.

Martin J. Murray is Professor of Urban Planning at Taubman College of Architecture and Urban Planning, and Adjunct Professor, Department of Afro-American and African Studies, University of Michigan. His most recent book is *The Urbanism of Extremes: City Building in the 21st Century* (Cambridge University Press, 2016).

Yanick Noiseux is Professor of Sociology at the Université de Montréal and the lead investigator at Groupe de recherche interuniversitaire et interdisciplinaire sur l'emploi, la pauvreté et la protection sociale (www.gireps.org). His work focuses on labour transformation, union renewal and social policies in the context of globalization.

Samuel Nowak is a PhD student in the Department of Geography at the University of California, Los Angeles. An urban geographer, his research is broadly concerned with the relationship between mobility and inequality and currently examines the governance of public transit in Los Angeles.

Peter Pantalone is a graduate of York University's Masters of Environmental Studies Planning Program. He is a Toronto-based land use planning practitioner engaged in cultural heritage conservation and adaptive reuse projects.

Margaret Pettygrove has a doctorate in geography and is currently a GIS analyst with the US Forest Service. Her research centres on urban food systems, political ecology and health geographies. She has published articles on these topics in a variety of peer-reviewed journals.

Kacper Pobłocki is Assistant Professor of Anthropology and Urban Studies at the Warsaw University. He just published *Spatial Origins of Capitalism* (in Polish), and writes about class, space and uneven development. He was also an activist and led the Alliance of Urban Movement that ran in 2014 in municipal elections in eleven Polish cities.

Debbie Prabawati is a researcher at the Faculty of Social and Political Sciences of Universitas Gadjah Mada (UGM) in Yogyakarta, Indonesia. Her main interests relate to the urban poor and women's rights, and her research has been part of the Power, Welfare and Democracy (PWD) project at UGM.

Emily Reid-Musson is a human geographer and postdoctoral fellow in the area of work and health at the University of Waterloo, Canada. Her current research focuses on the impacts of exceptions to labour regulation and law on worker well-being.

Jennifer L. Rice is an associate professor in the Department of Geography at the University of Georgia. Her research is in the areas of urban political ecology, carbon and climate governance, and science and technology studies.

Contributors

Marit Rosol is Associate Professor of Geography and Canada Research Chair in "Global Urban Studies" at the University of Calgary. She holds a PhD in Geography from Humboldt-Universitaet zu Berlin. She works on urban food movements, housing, urban gardening, participation and governance.

Ugo Rossi is Senior Researcher in Economic and Political Geography at the University of Turin, where he teaches at both undergraduate and postgraduate levels. His research focuses on the politics of urban and regional development and on critical theory.

David Saurí is Professor of Human Geography in the Universitat Autònoma de Barcelona. His research has focused on natural hazards, water management and, more recently, on tourism and climate change. He is currently working on the social, political and territorial aspects of water reuse.

Carolina Sternberg is Assistant Professor of Latin American and Latino Studies at DePaul University. She has published academic articles on neoliberal urban governance and its sociophysical impacts on racialized communities in both the US and Latin American urban settings.

Ulf Strohmayer teaches Geography at the National University of Ireland in Galway, where he is currently a professor in the School of Geography and Archaeology. His interests are rooted in social philosophies, historical geographies of modernity and are connected to urban planning, with a particular emphasis on the geographies and histories of European metropolitan areas.

Nik Theodore is Professor of Urban Planning and Policy and Associate Dean for Research and Faculty Affairs in the College of Urban Planning and Public Affairs, University of Illinois at Chicago. His current research pursuits include the study of urban informal economies, low-wage labour markets and worker organizing.

Fran Tonkiss is Professor of Sociology at the London School of Economics. Her work focuses on urban inequalities, urban development and design, and social and spatial divisions in the city. Her books include *Space, the City and Social Theory* (Polity 2005) and *Cities by Design: The Social Life of Urban Form* (Polity, 2013).

Eliot Tretter is an assistant professor in the Department of Geography and the Urban Studies Program at the University of Calgary. He is the author of *Shadows of a Sunbelt City: The Environment, Racism, and the Knowledge Economy in Austin* (University of Georgia Press, 2015).

Yuliya Voytenko is an assistant professor at Lund University in Sweden. She works in the areas of sustainability, urban governance, innovation and sustainable consumption.

Kevin Ward is Professor of Human Geography and Director of the Manchester Urban Institute at the University of Manchester. An urban economic and political geographer, he researches urban finance, governance and redevelopment policy and their mobilities. Currently, he is the co-editor of *Urban Geography*.

Barney Warf is Professor of Geography at the University of Kansas. He has published over a hundred articles in human geography, primarily emphasizing services, telecommunications and the Internet.

Contributors

David Wilson is Professor of Geography at the University of Illinois at Urbana-Champaign. His current projects examine the politics of urban growth regimes in US cities, the politics of competing discursive formations that generate gentrified neighbourhoods and poverty communities, and the racializing of the contemporary urban issues of crime and city growth.

France Winddance Twine is Professor of Sociology and a documentary filmmaker at the University of California, Santa Barbara. She has published ten books and numerous articles on intersectionality, inequality and assisted reproductive technologies.

ACKNOWLEDGEMENTS

Andy thanks Kevin Cox for helping him to understand the importance of seeing urban space not as a passive backdrop but as an active political construction. More prosaically, he thanks the Regional Studies Association for funding support through a Fellowship Research Grant.

Byron thanks the many friends and colleagues who have stimulated, challenged and encouraged him over the years. It's a long list that begins with Justin Beaumont, Lawrence Berg, Kevin Cox, Andy Jonas, Helga Leitner, Deb Martin, Roger Miller, Samuel Mössner, Walter Nicholls, Mike Pasqualetti, Eric Sheppard, Theano Terkenli, Kevin Ward, David Wilson and Elvin Wyly.

Kevin thanks the following who over two decades have provided intellectual and social nourishment on urban political matters: Josh Akers, Neil Brenner, Allan Cochrane, Kevin Cox, Jonathan Darling, Mark Davidson, Jason Hackworth, Nik Heynen, Kurt Iverson, Andy Jonas, Martin Jones, Roger Keil, Bill Kutz, Ute Lehrer, Gordon MacLeod, Eugene McCann, Colin McFarlane, Byron Miller, Jamie Peck, Scott Prudham, Mary Beth Tegan, Cristina Temenos, Nik Theodore, Rachel Weber, David Wilson, Helen Wilson, Raegan Wilson and Andy Wood. Closer to home, he thanks Colette and Jack, for whom this book is yet another example of his work being a long way from what he should be doing, were he a 'real geographer'!

David thanks the following friends and colleagues who have proved so helpful and nourishing in the last two years: Matt Anderson, Dennis Baron, Bill Cope, Phil Kalantzis Cope, Marc Doussard, Tamsyn Gilbert, Barbara Hahn, Ulrike Gerhard, David Green, Brian Jefferson, Andy Jonas, Mary Kalantzis, Roger Keil, James Kilgore, Andy Merrifield, Byron Miller, Faranak Miraftab, Ken Salo, Carolina Sternberg and Kevin Ward.

Collectively we thank Andrew Mould and Egle Zigaite at Routledge for the commissioning of this book and their support during its editing.

GLOSSARY

Accumulation by dispossession – a predatory tendency of capitalist expansion that works to obliterate common or public assets through enclosure, appropriation and privatization (of, for example, land, natural resources or lifeworlds). Although sometimes associated solely with primitive stages of capitalist evolution, recent scholarship focuses on dispossession as a continually renewed and iterated dynamic of accumulation, one particularly prominent within regimes of financialized neoliberal governance.

Age-friendly city – this concept considers ageing in relation to place. It encourages active ageing by optimizing opportunities for health, participation and security in order to enhance quality of life as people age. In 2006, the World Health Organization (WHO) developed the Global Age-Friendly Cities Network. Member cities measure age-friendliness across eight interconnected themes. Age-friendly policies can be regarded as beneficial to all ages.

Allochtonous/authochonous – in the Dutch context, the category allochtonous was used for people born outside the Netherlands (more exactly outside the European part of the Kingdom) or when one of a person's parents was born abroad – regardless of citizenship. The category was introduced as a 'neutral' category by the Dutch statistical office CBS to replace the prejudices associated with 'immigrant' and has been used in policy documents and public debate since the 1980s. The Dutch distinguish between first and second generation, and between Western and Non-Western, allochthones. After years of contestation, the CBS decided in early 2017 to replace it. The categories have been relabelled *inwoners met een migratieachtergrond* and *inwoners met een Nederlandse achtergrond* (literally resident with a migration background and resident with a Dutch background), arguably not solving any of the problems associated with the connotations of allochtonous as 'not really Dutch'.

Alt-labour organizing – a set of organizing strategies that depart from formal collective bargaining campaigns established by the New Deal industrial relations regime. Rather than setting out to collect union cards and win an election (and contract) governed by the National Labor Relations Board, alt-labour organizing approaches use worker centres, social media, public protest and relationships with electoral officials to pursue labour causes.

Ambivalence (capitalist) – ambivalence in this context refers to the internal tensions within capitalist processes of value production that generate exploitative and liberatory potentials. In particular neoliberalism is marked by an ambivalent dynamic of 'negative' forces: predatory capitalism, dispossession and state retrenchment; and 'affirmative' forces: the mobilization of pro-growth imaginaries, stimulus of prosperity and creation of new subjectivities.

American pragmatism – a philosophical tradition with the following broadly shared characteristics: the idea that truth and meaning are to be assessed through practical consequences; a conception of knowledge as social; the premise that chance and contingency are central in the emergence and uptake of ideas; commitment to fallibilism and experimentation; and embrace of epistemological pluralism.

Antonio Gramsci – an Italian communist, jailed under the Fascist government of Benito Mussolini in the 1920s and 1930s. Gramsci developed a philosophy of praxis that has influenced studies of ideology and state-society relations.

Apartheid – the political and legal framework and ideology that governed South Africa from 1948 to 1994 through a system which spatially and socially separated, and discriminated against, different racial groups based on white supremacy.

Asian carp – a collective term used to describe several species of carp, which were imported from China into the USA in the 1970s to filter out plankton from aquaculture and retention ponds in the southern US states but which subsequently found their way into major rivers and, eventually, urban waterways.

Assemblage – developed out of work by Deleuze, Guattari and DeLanda, among others, the concept of assemblage is related to complex systems in that it explores heterogeneity and complexity, where the whole is something other than the sum of its parts. The concept concedes much more space for indeterminacy, emergence, becoming, processuality and the socio-materiality of phenomena (McFarlane 2011). An assemblage can be thought of as an arrangement held together through 'relations of exteriority' between heterogeneous entities whose individual capacities (or potential) cannot be known in advance.

Banlieue – references autonomous suburban communities outside major cities in France that have recently become disproportionally populated by near-poor and poor immigrants. Housing and living conditions here tend to be squalid and problematic to residents.

Bare life – Italian geo-philosopher Giorgio Agamben's notion of being stripped of legal and political representation and entitlement to essentials that are less than human, as opposed to a politically embedded life.

Baugruppen (German for homeowner groups) – self-organized and largely self-financed groups of individuals involved in developing and constructing buildings to live in. *Baugruppen* usually consist of ten or more individuals that gather together during an early phase of planning before starting to develop and construct the buildings. They promote alternatives to traditional models of housing development because of their focus on saving costs by developing co-operations with fellow owners, who share a common vision for living together and, as a consequence, establish a community before the construction of the building starts.

Biopolitical production – forms of power that work to create modes of subjectivity, embodiment and social relations; in other words, to administer life itself. The metropolis is a crucial site of these dynamics where knowledges, affects, languages and lifestyles are formed and regulated. Owing to the productive aspects of power, however, the biopolitical metropolis is also a terrain of resistance where people live together, share resources, communicate, exchange and create in ways that are autonomous from state and capital.

Bordering – while 'borders' have traditionally been conceived as the territorial limits of nation-states, the term 'bordering' helps to re-conceptualize borders as mobile, fluid, shifting, contradictory and processual rather than universal, homogeneous and static. In geography and cognate disciplines, recent scholarship considers borders as more complex socio-spatial phenomena which are being reworked by trade liberalization, national security and war-making agendas, and migration and refugee crises. On the one hand borders are being reformulated to ensure the free flow of commodities and capital, and yet borders are also increasingly 'hardening' to restrict the flow of people, particularly migrants and refugees.

Children's Rights – from the United Nations Convention on the Rights of the Child (UNCRC), pertaining to the rights of young people in terms of provision, protection and participation in policies that affect them.

Circulating risks – the risks, such as vulnerability to **climate change**, that are associated with circulations of energy, people, policies and financing through cities.

Circulation – the flow of materials, resources, knowledge and/or virtual goods from origin to destination as part of the networked economy, including the politics that provide the associated infrastructure and regulation.

Citizenship – the relationship between the individual and the state (not necessarily nation-state). Citizenship is increasingly recognized as both a legal act (e.g. the rights that citizens receive from the state), as well as a practice (e.g. the ways in which citizens exercise and demonstrate those rights and responsibilities).

Citizenship/non-citizenship – citizenship refers to a fabric of statutes, norms and terms that determine political membership within polities, in which membership is typically associated with a set of rights and responsibilities that simultaneously 'produce' both a sovereign territory and its citizens. Modern states invest considerable energy creating non-citizen categories, such as 'asylum seekers' and 'temporary foreign workers'. For this reason, many scholars now agree that citizenship cannot be conceptualized on ideal, a priori terms. Rather, they argue, citizenship norms – and by extension state sovereignty itself – are dynamically constituted through non-citizenship formations.

Class monopoly rent – the rent that accrues to developers and to lenders as a result of attributes exclusive to particular neighbourhoods in a metropolitan housing market. Lenders gain when they restrict mortgage loans to certain parts of the city, as in the practice of redlining. Developers appropriate class monopoly rent through the novelty of their developments; subsequent to sale to homeowners, the homeowner imputed rent is siphoned off in the form of mortgage interest.

Climate adaptation – the adjustment in natural or human systems in response to actual or expected climatic stimuli or their effects (i.e. **climate change**), which moderates harm or exploits beneficial opportunities.

Climate change – recent trends identified by scientists indicate a rise in global mean temperatures, which in turn is contributing to melting ice sheets and rising sea levels. Global climate change represents a particular threat to low-lying coastal cities but it has also trigged a range of political interventions around climate adaptation and mitigation in cities throughout the world.

Climate governance – is the process of steering societies towards prevention, mitigation or adaptation of the risks posed by global **climate change**. Either in response to, or in the absence of, national state intervention urban authorities have assumed greater responsibility for climate governance in recent years.

Climate reductionism – a form of analysis and prediction in which climate is extracted from, and analysed independently of, the matrix of interdependencies that shape human life within the physical world.

Climate resilience – the capacity for a given system to absorb stresses and maintain function in the face of external stresses imposed upon it by **climate change** and to reorganize and evolve into more desirable (urban) configurations that improve the sustainability of the overall human settlement system, leaving it better prepared for future climate-change impacts.

Collaboration/co-production – this is a working practice or process of two or more people or organizations, working together to realize shared goals. Collaboration may involve asynchronous (time-shifted) interaction, but place-making typically entails real-time meetings in shared space. It is very similar to cooperation, and both are opposite to competition. Collaboration research involves the co-production of knowledge: rather than to conduct research *on* people, research is participatory *with* people.

Collective consumption – is a term used by Manuel Castells to describe those aspects of social life which are not subject to pricing through the market, yet are necessary for the maintenance of everyday life (so, this might include schools, personal social services, health services, social housing, public space, even transport infrastructure). This sphere of activity was held to be the space of a quintessentially urban politics, capable of generating urban social movements which might challenge existing relations of political and economic power. While the notion was helpful in highlighting some aspects of the popular politics of urban protest, the rise of privatization and the fragmentation of welfare states in Europe and North America has undermined its foundations.

Commons or common – both terms are used to refer to a range of collaborative institutions for managing natural and human resources. Most frequently associated with solidaristic organizations and practices aiming to produce a more cooperative and egalitarian society, the notion of the commons invokes non-private property relations and non-market economies in environmental, cultural, intellectual or digital goods. This notion has, however, also been re-appropriated into the capitalist mainstream through, for example, the proliferation of entrepreneurial discourses and the rise of the so-called 'sharing economy'.

Community benefits – facilities and services that may include public art, streetscape improvements, park land or park improvements, cultural/community/institutional facilities, heritage preservation, rental housing replacement and transit improvements.

Community-labour coalition – a strategic political alliance between local labour unions and community-based organizations. Community-labour coalitions developed after the decision of US unions to localize politics and organizing in the late 1990s. Community-labour coalitions play a central role in current efforts to pass legislation addressing economic inequality.

Community unionism – an organizational construct uniting in partnership unions and social movements organizations such as student bodies and/or neighbourhood organizations.

Compassionate capitalism – is characterized by contradictions of extreme income inequality, neoliberalism and race-evasion. It refers to a type of late capitalism that consists of both symbolic and discursive manoeuvres including the philanthropic commitments that accompany corporate redevelopment efforts in underserved neighbourhoods (Alfrey and Twine 2017).

Condo boom – the proliferation of high-rise residential intensification throughout the Greater Toronto Area since the early 2000s engendered by provincial smart-growth policies, resulting in a shift from the production of low-density residential development to predominantly mid- and high-density condominium development.

Condominium – is a legal arrangement of land ownership where a multi-unit building, usually in the form of a mid- to high-rise structure, is divided into individually owned units and undivided shares of collective areas (such as hallways, elevators, amenities, roofs, etc.), that are under a collective governance structure. This legal instrument is an innovative form of multiplying finite land into vertical land ownership.

Contingent work – a general term that covers a range of employment arrangements in which workers are hired into jobs that are understood to be temporary, seasonal or otherwise and not permanent. These arrangements include employment through temporary staffing agencies, freelance work and various on-call positions, such as day labour. In addition to the lack of employment security, many forms of contingent work do not offer employment benefits such as the provision of sick leave or vacation time.

Counter-publics – according to Fraser (1990: 67) counter-publics constitute "parallel discursive arenas where members of subordinated social groups invent and circulate counter-discourses". Progressive counter-publics are alternative material and discursive arenas constructed to challenge undemocratic forms of authority and discrimination, and to enact egalitarian political spaces.

Creative destruction – the tendency embodied by capitalist modes of production periodically to invalidate past investment. Key in this process is the proliferation of innovations that render past production and related social and cultural practices redundant. It is of interest here because it fundamentally contributes to the shaping of urban environments within capitalist economies.

Day labour – a form of on-demand employment arrangement in which a worker is hired and paid one day at a time, with no promise of future employment. Some day labourers are dispatched to workplaces by temporary staffing agencies, often in the manufacturing and warehousing sectors. Other day labourers look for work at informal hiring sites located in public spaces. These workers typically are employed in the construction and landscaping industries.

Density bonus – the provision of 'bonus' height or density beyond what would typically be allowed under the in-force zoning regulations in exchange for the provision of community benefits to offset the planning impacts of the increased magnitude of development.

Direct action – the use of tactics of protest, demonstration, boycott, strike, occupation, etc. in pursuit of social or political aims.

Dispositif – a strategic response to a historical problem. Foucault (1980: 194–195) employed the concept apparatus to draw attention to a network of relations formed between discourses, institutions and practices, at a given historical moment, in response to an urgent need. Apparatuses are invested with strategy, meaning they are structured by explicit objectives, inscribed in specific webs of power relations and linked to specific types of knowledge.

Distributed agency – a reference to the difficulties of assigning causal responsibility to a single agency or individual operating within a complex system or **assemblage**.

Distribution or fulfilment centres – facilities for the storage, consolidation and sorting of consignments to prepare them for delivery to the customer.

Domestic worker – a person who performs paid work in a household. While 'domestic work' refers to all paid and unpaid work performed in or for a household or households, a 'domestic worker' is someone who performs such work within an employment relationship. Many domestic workers are not recognized as workers, but rather referred to as maids, helpers or servants.

Dystopia – broadly the antithesis of **utopia**, dystopia is an imaginary undesirable or hellish place.

E-commerce – the trading of commodities of all kind (e.g. consumer goods, music files, films for download, airline tickets, used goods) through digital platforms between different peers, most notably corporate agents and private households, but also governments and municipalities.

Eco-gentrification – a particular type of gentrification that is characterized by a 'green' aesthetic and is abetted by design and planning agendas that value green spaces and 'green' infrastructure, including spaces for urban agriculture of one sort or another. This largely class- and race-specific ecological agenda, or environmental ethic, facilitates the exclusion and displacement of low-income and vulnerable populations from neighbourhoods where it is implemented in favour of their replacement by largely white middle-class residents.

Ecological modernization – a managerial approach originating in developed Europe to solve environmental challenges (especially pollution) through the application of technological and economic instruments without questioning the logic of capitalist growth.

Ecovillage – a human scale settlement which connects communal living with environmental concerns, the ecovillage concept usually conjures up a rural retreat, away from mainstream society. Contemporary ecovillages take many forms and include urban as well as remote locations. They are one form of 'intentional community' whereby pioneer residents intentionally create a place in which to practise four dimensions of sustainability: ecological, cultural/spiritual, social and economic. Local variations on the ecovillage idea share common goals and values as part of a wider organization called the Global Ecovillage Network (GEN).

Energy poverty – A situation by which a household is deprived of adequate energy services due to insufficient economic means. While in the developed world energy poverty is linked to affordability, in the developing world, it is linked to accessibility as well as to affordability.

Entrepreneurial governance – a term that references the drive by city coalitions of powerful state and civic institutions to regulate cities in a more business-like fashion. This typically means privileging government power and resources to cultivate and retain businesses, turning over civic affairs to the private sector, and encouraging a social ethos of personal entrepreneurialism and efficiency.

Erasure – the removal of minorities (e.g. in the case of Slovenia, Croats, Serbs, Roma, Bosnians) from a country's permanent residential register, resulting in the loss of legal status.

Ethnicity – refers to identification with a category of people who identify with each other based on common identity features (common history, language, religion, culture, etc.).

Ethnic minority – a minority group defined on ethnic grounds. In certain cases the term implicitly suggests that members of the majority (i.e. the dominant group) have no ethnic identity and that the minority occupies a marginalized social position with inferior status.

European City model – as a strong planning discourse, the European City model emerged as at the end of the 1980s in response to efforts to impose North American models of urban development on European cities. Central aspects of the European City model are a compact urban structure, a historical dependency that is embedded in its architectural form, and an intrinsic belief in the ability to reconcile market-logics with social cohesion. As a model of sustainable urbanism it has influenced political processes at the international scale, in particular the Aalborg Charter of European municipalities and the Leipzig Charter adopted by the European Union (EU).

Eventful subjectivity – focuses on activism at the ground level, so to speak, and the long-lasting impact of this experience on participants.

Eviction – the legal (and sometimes illegal) displacement of an individual, family or community from their homes.

Exclusion – the process or mechanism by which certain people (e.g. racial or ethnic minorities) are kept out of particular areas of a city, region or nation, as well as political arenas or discourses within those spaces.

Expat – the word expat or expatriate comes from the Latin terms ex ('out of') and patria ('country, fatherland') and refers to a person residing in another country than that of their citizenship. The term has a cultural stigma and is associated with the international mobility of middle-class and upper-class people, particularly from Western countries. Expat communities refer to areas in post-colonial cities where international migrants live and can access separate social networks and luxury consumer services.

Experimental city – a condition where the urban is both an arena for and formed through processes of experimentation. One version of experimentation that is growing rapidly in popularity in cities is the **urban living lab** (oratory).

Familial-vocational privatism – a term coined by the German sociologist Jürgen Habermas and referring to a motivational structure emphasizing family consumption and leisure on the one hand and career achievement on the other. It overlaps in its family consumption aspects with the 'symmetrical family' of Young and Willmott. It is useful in understanding living place meanings.

Fast-track urbanism – a process where the social relations, physical textures and spaces of cities are quickly re-made to facilitate capital accumulation. Supposedly successful urbanization schemes are often replicated in this undertaking.

Fight for $15 – a campaign across North American cities to raise (mainly) fast food workers' wage to $15 dollars per hour of work.

Figuration – the movement of bodies in social space understood as a form of communication and interaction.

Food justice – views food as a central component to creating an equitable society. Food justice programmes seek to address disparities in access to affordable, healthy, culturally appropriate food that disproportionately impact people based on race and class. Food justice does more than increase access to healthy food; it works to dismantle the structural inequalities that cause food insecurity and social marginalization.

Freiburg Charter for Sustainable Urbanism – in 2010, the City of Freiburg, Germany, supported by the Institute of Urbanism, developed a Charter of Sustainable Urbanism that generalizes Freiburg-specific experiences into a manual of urban development. The charter shows significant parallels to the findings of best-practice research, projects them onto Freiburg's development and, by doing so, can claim leadership in this realm.

French pragmatic sociology – a sociological approach that analyses how ordinary actors challenge perceived injustices by critiquing others and justifying their own claims, drawing on various orders of worth – or logics of the common good – in situated processes of argumentation and negotiation.

Gender inequality – a mode of social stratification that describes the uneven distribution of resources, wealth and power between men and women – usually to the detriment of the latter.

Geopolitics of capitalism – in the original development of the idea, David Harvey applied it to countries and the international, including the competition to attract flows of value through respective spaces. It is a class politics with a necessarily territorial form. The same argument can also be applied to metropolitan areas and the struggles between central city and suburbs, older and newer suburbs, and one suburb and another.

Global production networks – organizational arrangements, coordinated by powerful transnational corporate actors known as lead firms, which produce goods and services across multiple geographical locations for global markets. As well as describing this empirical phenomenon, the term also denotes a particular theoretical framework emerging from economic geography research.

Google Bus – moniker for privately run, third-party-contracted commuter buses that transport tech workers in the San Francisco Bay Area; often large, over-the-road coaches outfitted with wireless service, ample seating and legroom, laptop plug-ins, and accessed via employer-provided digital IDs and smart phones.

Governance – the political management of social behaviour indicating how power and politics operate not only through formal government institutions, but also via interactions between the state, society and the economy. The term captures an increasingly complex and coordinated set of socio-economic relationships involved in contemporary forms of regulation and governing between public, private and non-profit sectors. An example would be how to manage poverty in ways that quell social disorder and assimilate low-income communities into the existing regime of social relations, all while maintaining state legitimacy.

Governmentality – a concept developed by philosopher Michel Foucault to describe the dispersed forms of power that allow states to govern a population. It emphasizes the role of knowledge and discourse in producing self-governing individuals that generally act in accordance with the goals of the state.

Group home – a residential facility in which services, such as life skills or health resources, are available to residents.

Habitus – Pierre Bourdieu defines habitus as 'structuring structures': the dispositions and tastes that are cultivated primarily through one's class position and interactions with other members of one's class. Although these dispositions are often unconscious, they are practised in a way that constructs and reproduces class difference in everyday life (Bourdieu 1984).

Hope – the curtailment of despair and the feeling that what is wanted or desired can be achieved in practice.

Identity Politics – sometimes called identitarian politics, foregrounds identity features in political mobilizations stressing that people's politics are shaped (or – in a normative variant – should be shaped) by their identification with a cultural group and their affiliation with identity-based organizations, rather than economic interests or civic loyalties, for example.

Inclusion – the process or mechanism that allows or accepts diverse peoples into areas, regions, nations, political arenas and discourses (see **exclusion**).

Informal economy – a varied set of economic activities, business enterprises and jobs that are largely unregulated and unprotected by government enforcement agencies. Activities include some forms of street vending, various domestic work arrangements through which caregivers and housekeepers are hired, and **day labour** in the construction sector where workers are hired into short-term positions that are filled on an as-needed basis.

Informal urbanism – a contested term which is typically associated with the so-called developing world. It generally refers to socio-economic marginality and a set of processes (including do-it-yourself house-building and trade or exchange) that occur outside of formal legal structures and processes. In Western place-making, the term usefully describes the active agency and alternative social and economic practices of community-led development. There are compelling arguments for regarding informal and formal urbanism as part of a continuum. Informal urbanism may appear chaotic but it reflects locally embedded knowledge and its own internal logic of resilience.

Information and communication technologies (ICT) – programmes (e.g. software) and devices (e.g. servers, computers, mobile phones) with which the processing of information can be steered, displayed and used.

Infrastructural citizenship – how citizens' everyday access to, and use of, infrastructure in the city affect, and are affected by, their citizenship identity and practice. This explores connections between the material and civic nature of state-society relations in ways that do not privilege radical expressions (e.g. protest). The concept provides the foundations for initiating critical connections between the two scholarly fields of urban infrastructure and citizenship.

Instant cities – a process whereby actors and institutions quickly and efficiently build cities often simulating the morphology and architecture of other cities. This process is especially evident in African and Asian countries in recent years.

Intelligent cities – a term that identifies the drive by city governances to re-create their economies around high technology services and 'smart' work forces (e.g. micro-chip, computing, fibre optics production). The desire is to replenish innovation and growth both economically and civically that are seen as necessary for cities to be globally competitive in the current international economy.

Intersectionality – a term used to describe the ways in which different modes of identity or social divisions – including gender, race, class, religion, sexuality – overlap and shape each other, and interact in producing patterns of discrimination and disadvantage.

Jurisdictional fragmentation – a reality across metropolitan environments where municipalities administratively establish themselves as separate and distinctive political entities. Such autonomy means political coordination across nearby municipalities becomes considerably more difficult.

Just City – the concept of the Just City has been widely promoted by intellectuals and planners, such as Susan Fainstein, who highlights the relationship between democratic processes and just and equitable urban outcomes. An extension of the concept of the Just City is the idea of Just Sustainability, which recognizes the universal right to live in and enjoy clean and healthy urban environments.

Knowledge co-creation – the process of involving groups of people who are the subjects of research in designing, executing and analysing the research.

Labour standards – a general term encompassing a range of protections for employees in the workplace. These include wage and hour laws (e.g. minimum wage provisions and regulations governing overtime pay), anti-discrimination policies, as well as health and safety rules. Labour standards are applied broadly so that employers in affected industries operate on a level playing field and so that employees in these industries receive fair treatment on the job.

Late neoliberalism – a phase of neoliberalism – roughly periodized in the lead-up to and aftermath of the 2008 financial crisis – that has intensified the contradictions of capitalist accumulation. Late neoliberalism is characterized both by persistent limits to growth and resilience and also by austerity and prosperity measures resulting simultaneously in the dismantling of public welfare and the mobilization of innovative strategies to boost livelihoods through value production.

Leipzig Charter on Sustainable European Cities – in 2007, the European Union (EU) developed the Charter for Sustainable European Cities under German presidency. Under the Charter, EU ministers responsible for urban development commit themselves to develop policies aimed at the integration of sustainable principles at all levels of governance including the national, regional and the local. The aim is to achieve balanced territorial development based on a polycentric European urban structure. The Charter follows a neoliberal logic of reconciling economic growth and sustainable urbanism with social and ecological outcomes, which in turn are subordinated to economic growth. The Leipzig Charter has played an important role in shaping urban policies in cities throughout Europe.

LGBTIQ – Acronym meant to include all sexual minorities, as opposed to the heteronormative majority; LGBTIQ stands for Lesbian, Gay, Bisexual, Transsexual, Intersex and Queer (or Questioning).

Liveability – place 'liveability' (or attractiveness) is typically defined by the extent to which a place can retain its population and assets; one with a good quality of life that makes provision for children and older people, clean air, green spaces, social and spiritual belonging, heritage and cultural sites. While the precise formula of liveability remains elusive and relatively intangible, competition for 'most liveable city' is typically determined by narrow metrics of formal planning (e.g. visitor numbers, green spaces, bicycle routes, transit-oriented development, etc.). Planned or designed urban liveability in the USA is associated with smart growth or the new urbanism, and in Europe with compact cities.

Logistics – the process of planning, implementing and controlling the efficient, cost-effective and synchronized flow and storage of raw materials, inventory, finished goods and related information, from point of origin to point of consumption and vice versa.

Low carbon urbanism – is an approach to designing and developing cities that focuses on reducing their carbon emissions.

Mass social housing policy – a term that describes a central governmental programme designed to stimulate the production of affordable housing for all city residents to have decent shelter. This central programme follows from the belief that there is too little affordable housing for many residents and the failings of the private sector to stimulate housing investment requires a strong governmental response.

Method of equality – emerging from Rancière's understanding of radical democracy, the method of equality presumes an egalitarianism that is realized through the ongoing practice of being together in difference. Equality is a normative assumption from which to proceed not the endpoint. Such a politics of equality has no single, proper place and time; it emerges through multiple practices and spatialities.

Metromobility – refers to the physical infrastructures of mass urban transit; to the social, economic, cultural and political systems which they compose; and to the interdependence of these systems with the production of the metropolis. Whereas automobility was dominant in the twentieth century, in the twenty-first century metromobility is one of a number of mobility systems that is emerging alongside and often in contention with automobility.

Microelectronics revolution – a term that references the dramatic rise of microelectronics as a distinctive and powerful economic sector in advanced capitalist societies. Such producers are seen to be economically propulsive and in need of being courted to boost city economies.

Migrant – someone who is migrating or has migrated. Migrants are emigrants from the perspective of their country and immigrants from the perspective of the country of destination. Immigrants are increasingly called migrants. Immigrants are often categorized according to their motives and status. Terms such as expats, guest-workers, *braceros*, knowledge workers, international students, refugees, asylum seekers, illegals, sexual slaves, etc., refer to different trajectories, opportunities and constraints, and forms of prejudice.

Migrant labour – refers to workforces who travel relatively long distances to secure wages, both within nation-state borders (in the case of rural-to-urban migration) and transnationally (in the case of cross-border labour migration), and whose residency within their destination locale is, for a diversity of reasons, temporary in some way. Migrant labourers face significant exclusions from social, political and civic rights in host states (e.g. migrants may be restricted to particular labour market sectors). There is high demand for migrant labour among employers in host states, who frequently view migrant labour as a highly pliable or cheaper workforce. At the same time, researchers are increasingly focusing attention on migrant workers' own identities and agency.

Migration – physical movement from one area to another. Although migration does not necessarily imply crossing a state border, 'migration' is often used for international migration (domestic migration refers to migration between different areas inside the territory of the same state). Migration can also be categorized according to its temporality (permanent versus temporary migration, definitive versus circular migration), its motives (forced versus free migration, labour migration versus family reunification), its procedural characteristics (guest-worker schemes versus spontaneous migration, documented versus undocumented border crossings, legal versus illegal residence and work permits), etc.

Migration-urbanization nexus – conventional definitions frame urbanization as a phenomenon comprising population growth within a given metropolitan area, or in other words, as a process of territorial-demographic accretion. Earlier migration research on rural-to-urban migration considered how migrants to cities contributed to the expansion of the territorial borders and/or populations of municipal or metropolitan regions. These interests have shifted to consider how migrants shape various aspects of urbanization, including labour markets, capital accumulation, citizenship, place-making and belonging.

Mobility – the capability of moving or being moved from a place to another. In human geography often related to the 'mobilities turn' in the 1990s, where social phenomena were studied as flows and movements rather than static entities. International migration often involves both 'spatial mobility' (movement between locations) and 'social mobility' (movement between social strata).

Moral geography – the idea that certain people, things and practices belong in certain spaces and places and not in others. Through symbols, displays and visible borders landscapes are often seen to accommodate particular identities while explicitly distancing themselves from others. Politics of migration and citizenship are often viewed through the lens of moral geographies.

Multi-racial organizing – the practice of social justice organizing across racial and ethnic lines of difference.

Neighbourhood – a district or area within a city – typically but not only in the inner city – with predominantly residential land use.

Neoliberalism – an ideology and policy model that emphasizes individual responsibility and the value of free market competition. It is commonly associated with nineteenth-century laissez-faire economics. As a policy model, neoliberalism emphasizes privatization, austerity, deregulation and free trade.

Neoliberal redevelopment governance – refers to a set of policies, programmes and procedures, and an associated assemblage of institutions (builders, developers, financial institutions, the local state and auxiliary institutions), unified around a common vision of urban redevelopment.

Networked utopia – the notion of a networked utopia suggests that the ideal city today is not a closed and autonomous territory as in traditional utopian imaginings (Plato's *Kallipolis* or Thomas More's *Utopia*) but comprises a non-topographical set of relations, connections and flows. Utopian cities in this sense are exemplified by fantastic infrastructures, such as those of mass transportation. This gives a new spin on the traditional interpretation of utopia as being both the 'good-place' (*eu-topia*) and 'no-place' (*ou-topia*).

New Urbanism – a movement of a variety of urban practitioners primarily associated with urban design. In particular, the movement advocates redesigning cities to be more pedestrian-oriented by changing land-use and transportation practices. The movement began in the 1980s and in 1993 the Congress for the New Urbanism was founded. Its core principles were codified in the Charter of the New Urbanism. Architects Peter Calthorpe and Andrés Duany are two of the movement's most notable advocates.

NIMBY – an acronym for Not in My Back Yard, a residents' movement triggered by a protest of local residents against an impending development or siting of an unwanted facility near their place of residence. Although usually NIMBY movements are defensive and aimed at defending a local, parochial interest, there are also examples when such movements eventually embrace a broader, more inclusive political agenda.

Nuisance law – references a legal mandate that prohibits acts from damaging the 'right of quiet enjoyment' by people. As one of the oldest kinds of legal mandates preventing impingement on human activities, cases framed in nuisance go back almost to the beginning of recorded case law.

Obsolessencialization – the process of making the social protection system outdated and incapable of covering people engaged in new working arrangements such as non-standard jobs.

Ontology – the philosophy and study of existence or being; in social science and social theory, the term is often used to describe objects of study or inquiry that are claimed, metaphysically, to exist.

Participation and the culture of communicative planning – participation sometimes refers to formal public consultation processes, which are institutionalized in building codes and planning laws, and tend to be restricted to the process of informing aggrieved parties. More often than not it is used in a broader sense to refer to efforts to find agreement and shared opinions at the beginning of the planning process. Participation is also used synonymously with various forms of democratic empowerment. In the 1990s, informal ways of participation became more important in urban planning, suggesting the emergence of a new communicative planning culture in which citizens become partners in, rather than subjects of, urban planning.

Place framing or place frames – in collective action, references to specific features and uses of a geographical area in terms of norms or grievances of people in or using the area; including articulations about how to protect or change those features and uses.

Place positioning – references to a geographical area or place which situate or describe it in relation to other places. Urban politics might be place based but are at the same often constructed in relation to processes operating across several places, i.e. place positioned.

Planetary urbanization – a concept often associated with Neil Brenner and Christian Schmid which rejects agglomeration as the defining characteristic of the urban, instead suggesting that urbanization is a "[P]lanetary phenomenon" that is an "increasingly worldwide, if unevenly woven, fabric" (Brenner and Schmid 2014: 751), whose "sociospatial dimensions … are polymorphic, variable and dynamic" (Brenner and Schmid 2014: 750) which generate and are incorporated into thick webs of relationships. Or, as Andy Merrifield puts it: "The urban is shapeless, formless and apparently boundless, riven with new contradictions and tensions that make it hard to tell where borders reside and what's inside and what's outside" (Merrifield 2013: 910).

Post-apartheid – the social, spatial, political, economic and cultural transition in South Africa after the end of the apartheid era can technically be dated from South Africa's first democratic elections in 1994, although apartheid was in demise from the late 1980s/early 1990s. The post-apartheid era can be defined as the drive to establish an integrated country in the context of ongoing spatial, social and economic inequality and division.

Post growth – a planning concept that suggests cities cannot indefinitely grow and that other techniques must be planned to increase human well-being. Here economic growth can generate benefits up to a point but cannot be relied on as a sole policy and planning goal into the future.

Precarious legal status – refers to the ways that state immigration and citizenship policies can restrict those deemed 'non-citizens' from accessing rights, benefits or resources associated with citizenship, including, for example, rights to labour market mobility or to social security entitlements like housing or education. While precarious legal status is related to government policies and regulations, recently researchers have documented how employers and other intermediaries shape precarious legal status, including teachers, employers, landlords, doctors and labour recruiters.

Privatized urbanism – a logical consequence of the processes of neoliberalization that has dominated cities since the 1980s. At one level, privatized urbanism refers to the proliferation of private and often controlled-access developments, such as gated communities and shopping malls, along with non-democratically accountable agencies responsible for strategic urban development, such as Business Improvement Districts (BIDs). At a subtler level, it is the creeping hegemony of private corporations and capital in urban planning processes. The loss of public spaces, dismantling of welfare structures and social fragmentation and polarization are the particular consequences of privatized urbanism.

Problematization – the notion of problematization featured prominently in Foucault's (2007, 2008) writings on governmentality. He described governmentality as a patterned style of political reasoning (encompassing procedures, calculations, reflections, critiques) within which problems of life "take on their intensity" and "assume the form of a challenge" (Foucault 2008: 317). Problematizations occur when a domain of experience loses its familiarity, becomes uncertain, and registers as a practical difficulty or problem.

Pro-growth urban politics – a mix of city rules, regulations and mandates designed to expand a municipality's ability to attract and create jobs, investment and an enhanced entrepreneurial business climate. Changing physical forms and social relations are deemed important outcomes in offering this distinctive politics.

Property capital – that branch of capital responsible for the supply of physical premises. In its current form it is highly speculative and competes both through cost, continual innovation in design features, and an artful arrangement of land uses within a project to enhance rental streams further. By the way it makes existing premises and their sites obsolete, it also creates its own demand.

Property-led development partnerships – a notion that references the organizing of an institutional coalition designed to stimulate city growth and development around the drive to enhance land and property values. Land and property are also identified as things in need of policy and programmatic attention.

Public-private partnerships – a cooperative arrangement between one or more public and private sectors designed to enhance the effectiveness and efficiency of resource provision in a city or place. Governments have used such a mix of public and private endeavours throughout history but this has become more pronounced after 1980 across the globe.

Publics – groups of people participating in collective discourses or sharing common concerns; the idea challenges a unitary notion of 'the public' as a sphere of debate, opinion and general interest including all people on equal terms.

Queer ecology – a loose set of interdisciplinary practices and theories that aim to disrupt heterosexist discursive and material formulations of nature-society relations and to reimagine our more-than-human worlds in light of queer theory.

Race – a classification system that imbues physical differences, such as eye colour, skin colour, hair texture, nose size and shape with social meaning. These differences in appearance (sometimes called phenotype) become linked to differences in power, access to resources, identities and interests that are economic, political, social and cultural.

Railways and rail infrastructure – part of the infrastructural make-up of cities in the form of tracks and stations, rail-related forms of transportation have come to shape urban development since the 1830s. Railways and rail infrastructure are used to transport goods and people across space, with speeds and container-based standardization having increased tremendously since the Second World War.

Glossary

Remunicipalization – a process gaining significant momentum since the early 2000s in which local authorities have resumed ownership and control over basic services, thereby reversing an earlier trend towards privatization.

Rentier class – a term from Marxist theory that identifies a group of people who generate profits for themselves through control and management of land and property but provide no useful contribution to people and society as producers of goods and services. Rentiers, as Marx's parasitic class, often actively lobby to shape programmes and policies that reproduce their current privileged position.

REO properties – a term that identifies real estate owned properties, also known as foreclosures, which are bank-owned. Banks become owners of properties when mortgage holders fall critically delinquent in their payment of monthly mortgage charges.

Resilience – a concept with multiple interpretations and meanings. In the mainstream environmental literature and in a growing number of international organizations it is used to indicate the ability to adapt to and survive ongoing environmental threats.

Resilient cities – a resilient city is a place that has developed capacities to help absorb future climate shocks and stresses to its social, economic and technical systems and infrastructures in order to maintain essentially the same functions, structures, systems and identity.

Right to the City – on the one hand it is, as Henri Lefebvre put it, a 'cry and a demand' to rebuild the city in a completely new image. On the other hand, it can also be a daily, material practice of various groups of residents who claim the right to use certain spaces and demand the right to be democratically involved in shaping urban processes. This practice in turn can become a fulcrum of making broader political claims and reinvigorating political institutions such as national citizenship.

Right to stay put – the right to not be evicted or moved; to remain in one place. Chester Hartman, writing in 1984, was one of the first to argue that property 'ownership' is more than a simply economic matter but is also about the right to an abode.

Right-to-Work legislation – State-level US legislation prohibiting mandatory union membership and dues payment in a unionized workplace. Right-to-Work laws create a free-rider problem, in which unions are obligated to provide resources and services to individuals who do not pay for them. Because this arrangement makes unions prohibitively difficult to fund, the US states with Right-to-Work laws have extremely low rates of union membership.

Roll-against neoliberalism – this is the resisting of neoliberalism in the aftermath of the Great Recession of 2008.

Roma – a so-called travelling people (although many live sedentary lives) who purportedly originated in northern India over 2,000 years ago and are now found primarily in Europe and the Americas.

Ruin – the material remainder of social and economic change in the city; often temporary in nature unless embedded in processes of heritage construction or safeguarded through the incorporation into new materialities.

Satellite cities – a concept in urban planning that refers to smaller metropolitan areas which are located somewhat near to, but are mostly independent of larger metropolitan areas. Despite political independence from larger nearby cities, satellite centres are often economically organized in order to serve the needs of both their current residents and those in the nearby city.

Section 37 – Section 37 of the Ontario *Planning Act* that states a municipal council may permit increases in height or density beyond that allowed under prevailing zoning restrictions in exchange for the provision, by the developer, of community benefits to the municipality.

Glossary

Self-construction – a process by which residents claim their 'right to the city' by engaging with space in daily, mundane activities such as building urban infrastructure or amenities. The term comes from the Brazilian *autoconstrução* but in many other parts of the world it goes by a different local name. In the case for Poland, for example, it is denoted by *czyn społeczny* – a voluntary work in the construction of local schools or transport infrastructure in the times of state socialism.

Self-driving car (or driverless car) – automobile operated by computers with microscopic sensors, cameras, remote sensing technology, three-dimensional map navigation tools and other nanotechnologies.

Sexuality – a set of discourses and practices – particularly in the 'West' – that produce and reproduce desire, embodied forms of erotic and other pleasure, and (in some contexts) identities.

Smart cities – urban authorities that use **ICT** to improve the efficiency of their services, as well as the well-being and satisfaction of their residents.

Smart growth – smart growth is a regional planning strategy. The term was first used to characterize a group of state-centred planning efforts to limit urban sprawl. Increasingly, the term is applied to a range of regional planning efforts that combine state and market-driven approaches to both limit sprawl and revitalize older urbanized areas. The term became popular when it was adopted as a state-wide planning practice by Maryland in the late 1990s under Governor Parris Glendenning. Glendenning used a variety of state government policies to incentivize the private sector open-space preservation and to redirect investment away from suburbs.

Smart state – an extension of the idea of the **smart city** such that the city draws upon the powers of the state to establish programmes, policies, projects and imaginaries which valorize technological competences in order to achieve social preferences.

Social class aesthetic – A highly organized form of suburban landscape consisting of single family homes on winding streets, cul-de-sacs, manicured grass verges and copious tree plantings. One effect is to underline the character of the residential neighbourhood as a contrast to the workplace: a consumptive relation to nature and a street layout that is antithetical to workplace rationality.

Social reproduction – consists of the mechanisms and processes that contribute to the intergenerational transmission of social advantage or disadvantage. Social reproduction theory highlights those barriers that constrain, without completely blocking, the capacity of the lower and working class to access the upper reaches of the class structure.

Social service agency – a government department or non-profit organization providing social support services to clients, such as health or counselling.

Spatial rights – the right of people to create, recreate, produce and reproduce their own spaces and thereby also themselves.

Spatial scale – a concept sometimes used to describe the different levels in an organized territorial hierarchy, such as organization of the state around national, regional and local jurisdictions. It can also refer to the spatial scope and extent of human or physical processes. Examples of such processes include global climate change, international migration or regional devolution.

Starchitecture – a portmanteau ('star architecture') used to describe iconic works by celebrity architects. Since at least the 1990s, it has become common for post-industrial cities to seek flagship works of starchitecture (especially in the form of art galleries and museums) by prestigious designers in order to attract visitors and investors and boost local development. Starchitecture is a key aspect of the strategy of culture-led urban regeneration made famous in cities like Bilbao, Spain.

State-of-exception – Italian geo-philosopher Giorgio Agamben's notion of an exception to legal citizenship (and also the creation of a space for that exception) that creates a person or persons who are not recognized by the law. Refers to a context wherein a group of people are placed outside the protection of national laws.

Structural coupling – a specific mode of strategic coupling in which city-regions connect into **global production networks** in a somewhat dependent and vulnerable manner that is dictated by the requirements of extra-local corporate actors.

Suburbs – conventionally understood to be residential areas on the outskirts of the city, occupied by those seeking to avoid the intensity of urban living, while wage or salary earners commute back into the city for employment and child raising and domestic labour is conducted at home – often associated with a clear gendered division of labour. This set of understandings was always inadequate, but the vital role of the suburbs in defining the urban is increasingly clear, both as new economic spaces are generated and as the everyday of urban life is refracted through a suburban lens.

Surplus populations – an extremely vulnerable class dispossessed through economic restructuring and abandonment within a post-welfare state configuration.

Sustainability fix – a set of political discourses, practices and institutional capacities that partially mitigate economic and ecological crises while assuaging related popular concerns and opposition in order to allow development to proceed for the benefit of hegemonic interests. This particular 'fix' appeals to mainstream notions of ecological stewardship and concerns about environmental futures and tends to propose solutions that are technical and certainly not radical (see While *et al.* 2004).

Sustainable Urban Development (SUD) and sustainable urbanism – both represent a new planning paradigm that incorporates sustainability principles into all aspects of urban planning, design and development. SUD is, however, a contested notion due to its attempt to reconcile growth with sustainability, which often involves compromises and hence acceptance of the status quo with respect to economic development and consumption. Sustainable urbanism refers not just to the policies and practices that shape settlement patterns, buildings and urban infrastructure but also the ideas that inform these policies and practices.

Tech mobility – fusing smart phone apps, Global Positioning Systems (GPS), and nanotechnology, this is a highly automated and surveilled transport regime shaped by pricing and markets, emphasizing privatized and deregulated access to mobility services, with less emphasis on the public realm, and intensifying precariousness in labour and austere governance.

Technocratic governance – based on the idea that key policy decisions should be based on scientific and technological knowledge, possessed by an elite class of experts, rather than political deliberation and debate. It has become closely connected to the phenomenon of post-politics.

Telegraphy – a notion that references the practice of using or constructing communications systems to transmit or reproduce information. These communications involve the long-distance transmission of textual or symbolic (as opposed to verbal or audio) messages without exchanging the object that bears the message.

Territorial state – has arguably functioned as the basic geopolitical unit of the capitalist world economy since at least the sixteenth century. Today, however, the traditional economic and political function of the territorial state is being eroded by globalization, the rise of city-regions as geopolitical actors, and new city-based social and political movements that transcend national borders.

Throwntogetherness of place – a phrase introduced by Doreen Massey to capture some of the mundane and apparently haphazard ways in which urban social existence is put together and the extent to which cities are the sites in which a range of different political outcomes may be explored and struggled over. From this perspective what matters are the sets of relations which come together to construct the urban experience, and the urban is understood as the place where sets of relationships overlap, settle and combine with a particular intensity. There is, in other words, no foundational or essential urban, but cities are the places "where different stories meet up" (Massey 1999: 134).

Transit-led urbanism – a planning approach that situates mass transit networks – especially metro and light rail systems – at the heart of urban development initiatives. Based on the idea that contemporary urban life is defined by flows and connectivity, transit-led urbanism aims to optimize patterns of movement. It purports that the creation of polycentric mobility hubs will intensify land use and facilitate the production of compact, dense, sustainable, attractive, mixed-use and, ultimately, more 'liveable' urban communities.

Transportation Network Companies or TNCs – pre-arranged, on-demand car services such as Uber, that use global positioning systems for navigation and tracking, and require smart phones to connect drivers to potential customers using an 'app' that allows 'e-hailing', cashless payment and driver ratings.

Uneven development – the process by which ebbs and flows of capital differentially shape the built environment over time. The investment of capital is concentrated in certain areas, while other areas are neglected. As rates of profit fall over time, however, owners of capital seek new spaces where a higher return on investment may be possible. These new sites of investment tend to be those areas previously neglected by capital, where the costs of land and labour are likely to be lower. Capital thus 'seesaws' back and forth between such spaces, both within and between cities and regions (see Smith 2008). Urban agriculture often arises in devalued, neglected spaces, preceding the return of capital. Increasingly, however, it actually accompanies such investment.

Uplift – the increased value of land resulting from public or private investments on neighbouring or nearby land.

Urban agriculture – consists of a variety of food growing practices and spaces – including household and community gardening, urban farming, orchards, animal husbandry (e.g. chickens, goats) and aviaries – found in urban and suburban areas. Unlike traditional agriculture, urban agriculture usually operates at a relatively small scale and can be motivated by other purposes, outside of strictly food production, such as community building, civic-engagement and environmental sustainability.

Urban citizenship – the rights and duties given to individuals or groups through their membership in an urban community. Processes of political globalization have led many geographers to examine cities as spaces where rights are practised and contested, in ways that complement or supplement the nation-state. Socio-economic inequality and cultural exclusion are seen as key challenges to extending urban citizenship in cities across the world.

Urban commoning – the processes of collective (not necessarily open-access), cooperative, non-commodified creation, maintenance, protection and transformation of urban spaces. There are no commons without 'commoning'.

Urban commons – can create and promote alternatives to capitalist social relations by providing a mechanism for redistribution, by challenging the hegemonic social order of organization and planning of urban space, as well as by challenging the market and profit logic that dominates it. Commons challenge the exchange-value and (abstract) private property logic of how urban space is produced and used. They can be produced through the de-commodification of formerly private commodified spaces.

Urban experimentation – involves deploying social and/or technical alternatives in cities in order to test new solutions to urban problems and learn from them.

Urban infrastructure – the physical manifestation of the city but also the means through which the urban is able to function. Urban infrastructure is both visible (e.g. roads, bridges, water) but also invisible (e.g. pipes, cables, gas, electricity). Such examples of infrastructure are technologies that facilitate urban exchanges by linking bodies and processes across space; as technological advances to enduring urban problems, these often embody a mix of longevity and periodic re-imaginings which quickly become taken-for-granted within the fabric of urban life while remaining tied into processes of **creative destruction**. Contemporary urban scholarship has experienced an infrastructural 'turn', conceptualizing infrastructure as a physical representation of socio-political processes.

Urban land nexus – summed up by Allen Scott and Michael Storper as "an interacting set of land uses expressing the ways in which the social and economic activities of the city condense out into a differentiated, polarised, locational mosaic" and as "the essential fabric of intra-urban space" (Scott and Storper 2015: 8). This is associated with the agglomeration process which characterizes the urban and "has powerful feedback effects not only on economic development, but also on society as a whole" (Scott and Storper 2015: 6).

Urban living labs – can be considered both as an arena (geographically or institutionally bounded spaces) and as an approach designed to enable intentional collaborative experimentation of researchers, citizens, companies and local governments in cities.

Urban planning – discourses and practices that aim to shape the concrete material make-up of cities. These involve questions of land use, design, urban politics and infrastructural arrangements within a forward-orientated, community-embedded frame of reference.

Urban political ecology – a subfield of geographic research that analyses: (1) the political, economic, social and cultural processes that generate environmental change within cities or associated with urbanization; and (2) the inequalities, relations of power and social reconfigurations associated both with these processes and their outcomes and effects.

Urban poverty management – an approach to managing extreme poverty in cities, through a mix of caring, curing and controlling institutions operated at the local level by urban elites, voluntary sector organizations and the state to 'regulate' urban poverty and more generally maintain civil order in the city.

Urban revanchism – a term coined by Neil Smith (1996) in analysing the race/class insecurity and fear of New York City's middle class during the 1990s. It is characterized by a discourse of revenge against minorities, the working class, feminists, environmental activists and immigrants.

User engagement – refers to the practice of working with groups of people who use a service, place or technology in order to understand their needs and preferences.

Utopia – attributed to the writings of Thomas More, utopia is an imaginary ideal place or a kind of 'heaven on earth'.

Validation of space – the process of transforming nominally Euclidian space into land that attracts or embodies value. Most often in the city the latter materializes in the form of land values or 'real-estate' prices; equally important, however, are processes of social, cultural or environmental validation which can enter into competitive arrangements with one another in concrete historical settings. Hence the need for urban politics, broadly construed.

Visible minority – a statistical category created by Statistics Canada for "persons, other than aboriginal peoples, who are non-Caucasian in race or non-white in colour". Crafted as a 'neutral' demographic category, the term has been a subject of debate regarding both its contrived nature and its role in reproducing racial hierarchies in Canadian society.

Wage theft – the practice of employers not paying workers their full wage, by violating minimum wage laws, stealing tips, withholding overtime pay, and forcing workers to clock out and continue working.

Water poverty – a situation by which a household is deprived of adequate water services due to insufficient economic means. As in the case of energy, while in the developed world water poverty is linked to affordability, in the developing world, it is linked to accessibility as well as to affordability.

Worker centre – not-for-profit, community-based organizations that advocate on behalf of workers, typically those in low-wage industries where violations of labour standards are most common. These grassroots organizations are involved in worker organizing, redressing wage theft and other workplace violations, and advocating for pro-worker public policies especially for immigrants and people of colour. There are approximately 200 worker centres in the USA, covering a range of industries including domestic work, construction, restaurants, and warehousing and logistics.

Youth activism – the mobilization (on the streets and on the Internet) of young people on behalf of causes for which they are politically motivated to work for and rebel against.

References

Alfrey, L. and Winddance Twine, F. (2017). Gender-Fluid Geek Girls Negotiating Inequality Regimes in the Tech Industry. *Gender & Society*, 31(1), pp. 28–50.

Bourdieu, P. (1984). *Distinction: A Social Critique of the Judgement of Taste*. Boston: Harvard University.

Brenner, N. and Schmid, C. (2014). The Urban Age in Question. *International Journal of Urban and Regional Research*, 38(3), pp. 731–755.

Foucault, M. (1980). *Power/Knowledge: Selected Interviews and Other Writings, 1972–1977*. London: Pantheon Books.

Foucault, M. (2007). *Security, Territory, Population: Lectures at the College de France, 1977–1978*, trans. G. Burchell. Basingstoke: Palgrave Macmillan.

Foucault, M. (2008). *The Birth of Biopolitics: Lectures at the College de France, 1978–1979*, trans. G. Burchell. Basingstoke: Palgrave Macmillan.

Fraser, N. (1990). Rethinking the Public Sphere: A Contribution to the Critique of Actually Existing Democracy. *Social Text*, 25/26, pp. 56–80.

Hartman, C. (1984). *The Transformation of San Francisco*. Totwa, NJ: Rowman & Littlefield.

Massey, D. (1999). Cities in the World. In: D. Massey, J. Allen and S. Pile, eds. *City Worlds*. London: Routledge, pp. 99–156.

McFarlane, C. (2011). Assemblage and Critical Urbanism. *City: Analysis of Urban Trends, Culture, Theory, Policy, Action*, 15(2), pp. 202–224.

Merrifield, A. (2013). *The Politics of the Encounter: Urban Theory and Protest under Planetary Urbanization*. Athens: University of Georgia Press.

Scott, A. and Storper, M. (2015). The Nature of Cities: The Scope and Limits of Urban Theory. *International Journal of Urban and Regional Research*, 39(1), pp. 1–15.

Smith, N. (1996). *The New Urban Frontier: Gentrification and the Revanchist City*. New York: Routledge.

Smith, N. (2008). On the Eviction of Critical Perspectives. *International Journal of Urban and Regional Research*, 32(1), pp. 195–197.

While, A.H., Jonas, A.E.G. and Gibbs, D.C. (2004). The Environment and the Entrepreneurial City: The 'Sustainability Fix' in Leeds and Manchester. *International Journal of Urban and Regional Research*, 28(3), pp. 549–569.

1
SPACES OF URBAN POLITICS
An introduction

Andrew E.G. Jonas, Byron Miller, Kevin Ward and David Wilson

> The urban politics of the twenty-first century will be both a local politics and a global politics: the challenge to be faced by those seeking to analyze it effectively will be to hold both aspects together at the same time, without allowing either dominate as a matter of principle.
>
> *(Cochrane 1999: 123)*

Introduction

The above quote from Allan Cochrane has proven highly prescient, probably more than Cochrane himself could have guessed at the time. Ignoring for the moment important disciplinary and theoretical differences in how urban politics have been examined, the last decade has seen an intellectual consensus – of sorts – emerge on the importance of understanding and analysing urban politics in a wider relational or 'global' context. This consensus has grown out of critiques of approaches which, in their efforts to identify the urban's specificity, often restricted the analysis of urban politics to the jurisdictional territory of the city (i.e. municipal or local government). Yet the tensions and political conflicts that frequently punctuate capitalist society are not so easily contained within local territorial boundaries. Instead,

> [t]he political processes at work in civil society are much broader and deeper than local government's particular compass ... Its boundaries do not coincide with the fluid zones of urban labor and commodity markets or infrastructural formation; and their adjustment through annexation, local government reorganization, and metropolitan-wide cooperation is cumbersome, though often of great long-run significance.
>
> *(Harvey 1985: 153)*

Relational perspectives strive to move beyond such overly 'territorialist' readings of urban politics and examine the broader societal processes which shape the production, consumption, restructuring and political contestation of urban space.

Inspiration for relational thinking in geography comes from formative statements by the likes of Harvey (as per the above quote), Ed Soja and Doreen Massey, who have persuasively argued that wider social processes do not simply occur in space; instead, space makes a difference to how those processes operate (see, e.g. Massey 1994, 2005). However, given the variety of ontological positions from which the 'difference that space makes' has been explored, what this means for urban theory remains open to further interpretation. For example, some urbanists claim that urban politics are best understood as being the contingent effect of wider contradictions and conflicts in capitalist society; in other words, urban space *does* make a difference but only insofar as it alters the spatial configuration of wider social and political processes (see, e.g. Saunders 1981). For others – not least those who continue to draw inspiration from historical materialism – urban space plays a much more fundamental role in capitalist society. This is due to the ways in which actors with different material interests derive profits and income from the production and circulation of value through the built environment – the accumulation strategies of property capitals, for example, necessarily depend upon the production of particular urban spatial forms (Cox 2012). For these scholars, space – especially urban space – is of causal significance rather than merely contingent. In summary, within the broad category of 'relational' thinking, starkly different ways exist with which to approach the analysis of urban politics in respect of its necessary spatial attributes.

Whilst scholars continue to debate urban theory, urban social conditions have evolved and changed significantly. All sorts of new urban social movements and political struggles, ranging from the Arab Spring and Occupy to struggles for environmental justice and anti-austerity movements, have sprung up within cities across the world. The urban has become a vibrant and highly contested political arena for social struggle around issues of democracy, citizenship, identity and human rights. As much as we need to rethink the urban from a relational perspective, we must also recognize the importance of using evidence drawn from different cities to investigate how the territorial landscape of democratic politics is being transformed on an international stage through more or less globalized social and economic processes.

This *Handbook* aims to contribute to this wider intellectual project by investigating the diversity of spaces of urban politics within and between which socio-spatial processes of a more essential or 'global' nature do their work. As we embark upon this challenging and exciting task, we set out some general observations about the changing socio-spatial form and scope of urban politics. Thus, in this introductory chapter we examine three questions: What are urban politics? Where are such politics to be found? And what are some of the significant, newly emerging spaces of urban politics?

In addressing the first question, we suggest that the analysis of urban politics should continue to be about (1) struggles around the socio-distributional effects of capitalist urban development and (2) struggles for the recognition and inclusion of diverse citizen voices in urban political processes. Such struggles go to the heart of how global capitalism, operating in conjunction with neoliberal state interventions, deeply penetrate into many aspects of urban life to the point that our received understanding of the city as a democratic polity is being questioned. To turn a phrase from Agamben (1995), we want to restore some concrete substance to the 'bare life' of urban politics, which all-too-often has been abstracted from its spatial contexts and stripped of its contextual meanings and significance. One of the aims of this collection is to use a variety of examples from cities around the world to reveal the richness of urban politics in the current phase of globalization.

Second, the book makes a case for examining urban politics through a territorial lens in order better to grasp the place-specific effects of wider societal pressures and state restructurings. Much has been written of late that we are living in a post-national, post-territorial world in which globalization and urbanization have coalesced to produce a planetary space of neoliberal capitalism (Brenner 2013; Brenner and Schmid 2015); a space, moreover, which arguably functions at the expense of local and regional political differences and identities. However, we caution against a perspective that ignores the role of urban politics in differentiating processes of globalization and demarcating territory. Many of the chapters in this book demonstrate that urban political processes continue to shape territory as much as they connect it through globally extended social relations and networks.

This brings us to the third, and perhaps most important, objective. Our overall aim in this book is not to make a claim for the specificity of the urban as *the* space of politics. We do not suggest that society is in an era in which all political struggles are essentially urban struggles and that, consequently, struggles for democracy, social justice and equality are essentially struggles for what Lefebvre (1968) and his followers describe as the 'right to the city'. Just as capitalist society is constructed in and around many different spaces and scales, so there are corresponding spaces and scales of politics. Some urban political struggles intensify processes of territorial differentiation and localization. Others, however, are more about globalization and the establishment of new scales of connection and flow between urban places. And still others are about the multitude of ways cities shape interaction and political mobilization, even when the grievance is not specific to the city (Miller and Nicholls 2013). In other words, urban political processes are always socially constructed and contested in and through a variety of places, spaces and scales, and they generate spaces of urban difference as well as those of flow and connection. The overarching aim of this collection is to provide some substance to the claim that struggles around and through different kinds of urban spatial formations generate a great variety of territorial politics.

What are urban politics?

Before we examine the socio-spatial characteristics of the 'urban' more closely, it is important to note that there has been much discussion amongst critical urban scholars over the years about the changing substance of urban politics. For Marxist scholars, the main concern might be summarized as follows: what do the tensions and conflicts that occasionally erupt inside cities say about capitalist society in general and the capitalist state in particular? Early answers to these questions concluded that the study of urban politics helps to expose the failure on the part of the state to support two things, the production of cities as optimal sites for industrial accumulation and the social reproduction of the working class in capitalism. This first conclusion was reached by David Gordon (1978) and Michael Stone (1978), for example, who assessed the ever relentless quest by capital and the state to create more entrepreneurial city morphologies. The second conclusion was reached particularly by Manuel Castells, who examined urban social movements in the 1970s and early 1980s (Castells 1977, 1978, 1983). While the first thematic focus in urban theory continues today, by the end of the 1990s Castells had abandoned his position and instead made a case for seeing the world of global capitalism as comprised of networks of interconnected urban agglomerations (Castells 2009). Nonetheless a consistent theme running through both foci is the central theoretical position of the 'urban' within wider social processes. The writings of critical urban theorists, including Castells and Lefebvre, both of whom have made a strong case for thinking seriously about the essential or causal attributes of urban space, continue to inspire scholars of the contemporary urban condition (Brenner and Elden 2009; Merrifield 2013; Ward and McCann 2006).

Acknowledging the problem of drawing territorial boundaries around otherwise unbounded social and political processes, urbanists eventually abandoned the collective consumption concept. If cities did – and still do – manifest all sorts of social tensions and struggles around housing, education, health and social provision, it became increasingly unrealistic to argue that such social considerations could be separated out from national or global political and economic processes. This did not mean that urbanization followed a single global trajectory since researchers quickly discovered that urban social conditions and the politics of welfare and development vary considerably from one country to another (Cockburn 1977; Dickens et al. 1985; Piven and Friedland 1984). Knowledge of differences in the spatial configuration of state institutions for urban development and collective consumption, respectively, remain important for explaining differences in the ways in which struggles around distribution and recognition emerge in particular urban contexts in capitalist democracies (Cox and Jonas 1993; Young 1990). And even if urban politics has altered its spatial appearance, perhaps its substance might not have changed quite as much as has been implied in discussions that were largely insensitive to geographical differences in state structures and political processes.

Nonetheless, towards the end of the 1980s urban scholars had reached the broad conclusion that the role of urban politics within wider distributional struggles in capitalist democracies was undergoing a decisive shift: promoting economic development through place entrepreneurialism had dramatically subsumed the once powerful drive to ensure social redistribution through welfare policy (Harvey 1989). The rise of this 'New Urban Politics' (Cox 1993) could further be explained in terms of the search for new forms of consumption and profitable urban development within a context of enhanced capital mobility on the part of finance capital and some branches of manufacturing industry. In seeking to attract such mobile investments, moreover, urban land-based elites behaved increasingly like a 'growth machine' (Logan and Molotch 1987), competing with each other to attract corporate headquarters, prestigious office buildings, international sporting and cultural events, creative industries and so forth, in their collective efforts to capture exchange value. At the same time, the jurisdictional space of local government was fast becoming occupied by new forms of local governance, including special assessment districts, business improvement districts and urban redevelopment authorities, which channelled scarce public resources towards increasingly controversial private-sector urban development projects (Cochrane 1993). In light of the relatively recent global proliferation and spread of such arrangements (Ward 2012), scholars are inclined to the view that urban politics in this global age of neoliberal-*cum*-hypermobile capitalism has become decidedly elitist, undemocratic, socially exclusive, and a threat to basic human rights and dignity (Harvey 2012).

At the same time, the very notion of urban politics itself began to move to a different conceptual terrain: to the world of the discursive. Important work by Beauregard (2003) and Fischer (2003) revealed that the notion of urban politics could fruitfully be extended to the arena of realities constructed and how urban issues were imaginatively set up for public consumption. Here the world of the discursive becomes a vital contested terrain for determining urban outcomes, i.e. who can control and manage how issues are understood is as important an act of politics as how the state, labour, and other actors and institutions regulate and respond to issues (Lovering 1995; Wilson 2004, 2007). In the process, discourses are identified as new, vital objects of investigation: as sources of power and persuasion, they 'take over' the world and define how it is understood by offering symbolic universes to produce realities for public consumption. It follows that to know, for example, how public resources are being distributed, how cities are being restructured and who is being provided state subsidies, one must know how competing interests build their best case for understanding these issues.

Nevertheless, there are still plenty of urban researchers who remain sceptical of claims that the substance of urban politics has irreversibly changed or, to the extent that it *has* changed, the precise direction. These critics broadly fall into two camps. First, there are those who argue for the value of examining a greater variety of spatial contexts in which urban politics operate, especially examples that draw upon evidence from cities in the Global South which are experiencing new social and spatial patterns of growth along with all sorts of newly emergent social and political tensions (Robinson and Roy 2016; Roy 2011). Second, there are those who suggest that too much theoretical emphasis has been placed on urban development trajectories in a few select cases, especially cities located higher up in the global urban hierarchy (Williams and Pendras 2013). Consequently, urban scholarship might be missing out on cases where alternative forms of urban development politics have emerged, such as cities where conditions of austerity have undermined political institutions formerly dominated by established growth elites or where left-populist political movements have gained some sort of a foothold in local government, as has occurred of late in Spanish cities like Barcelona. Such alternative forms of urban politics might be characterized as less growth-oriented and more redistributive and social democratic in intent (if not necessarily in effect).

So if the substance of urban politics has not changed quite as much as might have been assumed by a select group of influential urban scholars, *where* should one begin to look for further clues? This brings us to the second theme informing this collection.

Where are urban politics?

We believe that one can only get only so far in understanding the changing nature of urban politics by looking at how it varies from one city to another. Indeed, the broader insights to be garnered from traditional forms of urban comparison – for example, studies that compare and contrast urban political institutions in case city A with those in case city B – are fruitfully questioned by those who argue for the intellectual benefits of producing generalizable theoretical abstractions about capitalist urban processes (Dear 2002; Scott and Storper 2015). In any case, the opposition between empirical specificity and theoretical generalization may be a red herring. What seems to be even more important today is understanding how larger processes of accumulation and social restructuring work through *all* cities to produce, in some cases, broadly similar urban political outcomes such as the tendency for urban elites to pursue growth at whatever social or political cost to the urban citizenry (Jessop *et al.* 1999) and, in other cases, quite different political outcomes connected through common processes such as the capital investment (in one place) and capital flight (from another place) that underpins uneven development (Markusen 1989; Savitch and Kantor 1995). In other words, when looking to answer the question 'where is urban politics?' theoretical attention has seemingly embraced the 'global' at the expense of the 'local'. But should the 'local' be completely ignored? In some respects, the answer to this depends upon one's view as to whether the 'local' represents an ontological position or instead refers to a particular space – or set of spaces – where 'urban' politics happen to converge.

In terms of ontological positioning within the social sciences, it is now commonplace to read how the 'global' is in the 'local' (or 'urban') or how the 'local' (urban) is in the 'global'. Nonetheless different disciplines have approached this relationship in different ways and, in doing so, have drawn upon different theoretical traditions, some of which downplay the importance of space and spatial difference. In human geography, which has taken its intellectual departure from debates about uneven development and the restructuring of the nation state, work on the 'local' has emphasized how urban political decision-making occupies a particular scale (or set of scales) within a wider hierarchy of state territorial structures and governance processes, which range from the neighbourhood and district scale to the regional, national and the global scales. Moreover, there has been growing interest in the political convergence of formerly separate scales as can be illustrated by efforts by national governments to promote city regionalism as an answer to various ongoing distributional and political challenges arising from uneven development (Jonas and Moisio 2016).

In political science, by way of comparison, the Multi-Level Governance (MLG) theoretical framework (see, e.g. Pierre and Peters 2000) has emerged as a way to capture how urban politics function as a distinct level in wider governance systems but are increasingly shaped by agreements, frameworks, guidelines and governance processes located at other governance levels (e.g. at the European Union or international levels). Unlike the aforementioned geographical approaches, the relationships between each level in the governance hierarchy is assumed to be determined by a relatively fixed set of political criteria and subject to pre-given rules of negotiation, allowing political science to retain a fairly consistent emphasis on urban government (rather than governance) as the main site of political decision-making in cities (compare, for example, Davies and Imbroscio 2009; Judge et al. 1995). Despite these disciplinary differences, there is an across-the-board acceptance of the role that something called the 'global' plays on something called the 'urban', together with a somewhat more begrudging acknowledgement that something called the 'urban' shapes or at the very least is 'holding down' something called the 'global' (Amin and Thrift 1994).

Some scholars question whether ontologically urban politics has a spatial status which is separate from its necessary social structures and conditions. In other words, in isolating the 'urban' as a separate political *and* spatial category operating inside the 'global', there is a danger of eliding spatial form and underlying causal mechanisms since "what is commonly referred to as 'urban politics' is typically quite heterogeneous and by no means referable to struggles within, or among, the agents structured by some set of social relations corresponding ambiguously to the urban" (Cox 2001: 756). Cox's solution is to separate out in thought the spaces of dependence of various economic and political interests from their scales of political engagement, which in turn may or may not be limited to the 'urban' scale, at least in a formal jurisdictional sense. In other words, there is no necessary territorial correspondence between the 'urban', considered as a particular space of dependence, and political mobilization by various urban interests and actors (Cox 1998). Rather, the focus turns to understanding how "local alliances attempt … to create and realize new powers to intervene in processes of geographical restructuring" (Cox and Mair 1991: 208). Such powers might be located at other (i.e. non-urban) scales of the state or they could involve creating new state spaces inside the political jurisdiction of local government. For Cox, it is important to unpack the different kinds of spaces that underpin the making of 'urban' politics.

Whilst suggesting more fruitful lines of analysis, this approach still raises more questions than it answers. Cox's critics suggest that a thorough reframing of the question might help. Perhaps holding the 'local' and the 'global' together involves a comprehensive re-thinking of spatial relations around the question of '*where* are urban politics?' As Rodgers et al. (2014) and others have argued, there are always elements of elsewhere that go into the making up or assembling of 'urban politics' in particular contexts. This in turn involves a different approach to urban comparison, one which examines the interconnections and relationships between different places rather than generalizes about their differences (Ward 2011). Such an approach has been widely adopted of late, not only in studies of urban politics (McCann 2011; McCann and Ward 2011) but also in related fields of inquiry, such as urban planning, which in some respects has led the way (see Healey and Upton 2010).

Reframing the question around the 'where' of urban politics would seem to mean two things conceptually: first, the nature of the social relations among and between urban localities warrant further study. For example, which agents are involved in the movement – borrowing, evaluating, learning, etc. – of urban policies from one place to another and how are the processes of comparison and learning performed? Second, instead of the idea that urban politics differ among places, it should be possible to show how material and territorial outcomes in one place affect what happens in another place. Urban localities are implicated in each other's pasts, presents and futures in ways that challenge more traditional ways of comparing and approaching urban politics. Both territorial and network relations are intertwined in the urban politics of here and there (Rodgers et al. 2014). Moreover, the where of 'here' and 'there', along with the causal influence of the 'global' on the 'local', must be treated as processes and relations to be investigated rather than taken as given. Rising to these challenges – theoretically and methodologically – should be at the centre of any serious study of contemporary urban politics since they invite different ways of thinking about urban space and its attendant politics.

What are some significant emerging spaces of urban politics?

Although there is a growing consensus that boundaries cannot be imposed a priori around the urban political arena, the full implications of this have yet to be properly explored with respect to identifying the variety of spaces in and through which the politics of the 'urban' are constructed and contested today. In fact, there has been very little agreement over how best to 'hold', in Cochrane's words, the 'local' and the 'global' together. This has proved an especially thorny issue, not least because it speaks to a wider concern in the social sciences about what is meant by the 'urban'. For an interdisciplinary endeavour calling itself the study of 'urban politics', this poses fundamental questions which are not resolved simply by saying that "it is not possible to separate out neatly the local from national or even international politics as each one affects the other" (John 2009: 17). Such assertions are often repeated by urban scholars. Yet they only get this line of inquiry so far.

Although there are good reasons why the analysis of urban politics should be rid of its 'localness', this does not imply that questions of place, space and scale are incidental to the discussion. To be sure, the territorial scope of urban political analysis can no longer be restricted to city limits, especially if those limits are reduced to a matter of formal jurisdictional boundaries. Nonetheless, we can still learn a lot by looking out how struggles around economic development and collective consumption assume place-specific forms. Many such struggles do assume an urban form by virtue of the fact that the production of the built environment offers all sorts of opportunities for a select few to accumulate wealth and power at the same time as it denies many access to the basic necessities of life, such as decent housing. Struggles for access to different kinds of urban space have for many urban dwellers become essential to their material well-being and, indeed, very survival. At the same time, urban agglomerations are getting larger with some coalescing into city-regions and mega-urban agglomerations. Yet questions of scale are not simply about the growing size of urban areas. Instead, new scales of urban governance provide alternative spaces of engagement for different urban political actors, both 'above' and 'below' the scale of the municipality. All of this suggests that the concepts of place, space and scale that inform the analysis of urban politics need to be examined carefully and critically and, in so doing, expanded and deepened.

The scope of the task is established in Part I, the introductory section of the book, which reflects upon some of the different ways in which scholars have approached the 'urban' as a space of politics. Chapters in this section move on to consider the 'where' of urban politics by drawing attention to how concepts such as globalization, place, space, scale, territory, commons and nation-state have been put to work in urban theory. This introductory part is followed by eight substantive sections (Parts II–IX), each of which systematically examines the spaces of urban politics from particular thematic vantage points.

Part II examines urban politics through the lens of spaces of economic development. Drawing on examples of such spaces, ranging from glitzy high tech production zones and declining industrial areas to retail districts and gentrifying neighbourhoods, the chapters in this section collectively illuminate the underlying politics of their production. Part III sets out to problematize the tendency to separate nature from knowledge about the production and contestation of urban space and to integrate nature into our understanding of urban development politics. The six chapters in this section explore a range of examples, exposing the politics and power relationships underpinning the urban, understood as an intertwined set of social, political and ecological processes. The foci of Part IV are spaces of governance and planning, respectively. Whereas planning establishes a template for how cities are formally divided into zones and districts, governance involves the negotiation of both formal and informal rules that guide urban development. Reflecting wider social concerns and expectations about spatial realities and outcomes, the study of the spaces of governance and planning can reveal more precisely how the urban development process is often profoundly contested.

Part V focuses on spaces of labour. It explores the types of work done in cities around the world, the conditions under which such work takes place, the spaces in which it occurs and the politics around its representation. Given changing patterns of work and remuneration in the city, Part VI notes an intensification of existing tensions around urban living: rising housing costs; lack of housing alternatives; creaking public services; persistently low real wages; and declining quality of life. Chapters in this section examine some of the issues that are fought out through the contemporary politics of the urban living place. Part VII considers a crucial yet hitherto under-investigated dimension of urban politics, namely, how cities and the spaces within them become interconnected through all sorts of circulatory systems, ranging from flows of materials and transport to circuits of nature and knowledge. Each chapter highlights the importance of seeing cities not as fixed political jurisdictions but instead as comprised of intertwined spaces of flows, negotiations, experimentations and contestations. Part VIII explores a series of issues related to identity and urban politics. Questioning received categories of urban politics that refer to divisions around, for instance, class, gender, race and sexuality, chapters in this section outline the ways in which different forms of identity become manifested in the everyday politics of the city and beyond.

The final set of chapters in Part IX explores the interplay of utopian and dystopian urban imaginaries through examples of urban abandonment and ruin, ideas of the utopian model city, the hopes and fears associated with immigration, tensions between mobility and immobility in urban political imaginaries, and the pressing global political issue of climate change. These concluding chapters reinforce the overarching aim of the *Handbook on Spaces of Urban Politics*, which argues that struggles around and through different urban spatial formations and imaginaries generate a great variety of urban politics. Just as there is no single space of urban politics, so can there be no all-encompassing theory of urban politics.

References

Agamben, G. (1995). *Homo Sacer: Sovereign Power and Bare Life*. Stanford, CA: Stanford University Press.
Amin, A. and Thrift, N. (1994). Holding Down the Global. In: A. Amin and N. Thrift, eds. *Globalization, Institutions, and Regional Development in Europe*. Oxford: Oxford University Press, pp. 257–260.
Beauregard, R.A. (2003). *Voices of Decline: The Postwar Fate of US Cities*. 2nd ed., New York: Routledge.
Brenner, N. (2013). Theses on Urbanization. *Public Culture*, 25(1), pp. 85–114.
Brenner, N. and Elden, S., eds. (2009). *State, Space, World: Selected Essays, Henri Lefebvre*. Minneapolis: University of Minnesota Press.
Brenner, N. and Schmid, C. (2015). Towards a New Epistemology of the Urban? *City*, 19(2–3), pp. 151–182.
Castells, M. (1977). *The Urban Question*. London: Edward Arnold.
Castells, M. (1978). *City, Class and Power*. London: Macmillan.

Castells, M. (1983). *The City and the Grassroots: Cross-cultural Theory of Urban Social Movements*. London: Edward Arnold.
Castells, M. (2009). *The Rise of the Network Society: The Information Age: Economy, Society, and Culture, Volume I*. 2nd ed., Oxford: Wiley.
Cochrane, A. (1993). *Whatever Happened to Local Government?* Buckingham: Open University Press.
Cochrane, A. (1999). Redefining Urban Politics for the Twenty First Century. In: A.E.G. Jonas and D. Wilson, eds. *The Urban Growth Machine: Two Decades Later*. Albany, NY: State University of New York, pp. 109–124.
Cockburn, C. (1977). *The Local State: Management of Cities and People*. London: Pluto.
Cox, K.R. (1993). The Local and the Global in the New Urban Politics: A Critical View. *Environment and Planning D: Society and Space*, 11, pp. 433–448.
Cox, K.R. (1998). Spaces of Dependence, Spaces of Engagements and the Politics of Scale, Or: Looking for Local Politics. *Political Geography*, 17, pp. 1–23.
Cox, K.R. (2001). Territoriality, Politics and the 'Urban'. *Political Geography*, 20, pp. 745–762.
Cox, K.R. (2012). Marxism, Space and the Urban Question. In A.E.G. Jonas and A. Wood, eds. *Territory, the State and Urban Politics: A Critical Appreciation of the Selected Writings of Kevin R. Cox*. Farnham: Ashgate, pp. 55–73.
Cox, K.R. and Jonas, A.E.G. (1993). Urban Development, Collective Consumption and the Politics of Metropolitan Fragmentation. *Political Geography*, 12, pp. 8–37.
Cox K.R. and Mair, A.J. (1991). From Localised Social Structures to Localities as Agents. *Environment and Planning A*, 23, pp. 197–213.
Davies, J. and Imbroscio, D., eds. (2009). *Theories of Urban Politics*. 2nd ed., London: Sage.
Dear, M., ed. (2002). *From Chicago to LA: Making Sense of Urban Theory*. Thousand Oaks, CA: Sage.
Dickens, P., Duncan, S., Goodwin, M. and Gray, F. (1985). *Housing, States and Localities*. London: Methuen.
Fischer, F. (2003). *Reframing Public Policy: Discursive Politics and Deliberative Practices*. Oxford: Oxford University Press.
Gordon, D. (1978). Capitalist Development and the History of American Cities. In: W. Tabb and L. Sawers, eds. *Marxism and the Metropolis*. New York: Oxford University Press, pp. 25–63.
Harvey, D. (1985). *The Urbanization of Capital*. Baltimore: Johns Hopkins University Press.
Harvey, D. (1989). From Managerialism to Entrepreneurialism: The Transformation of Urban Governance in Late Capitalism. *Geografiska Annaler*, 71B, pp. 3–17.
Harvey, D. (2012). *Rebel Cities: From the Right to the City to the Urban Revolution*. London: Verso.
Healey, P. and Upton, R., eds. (2010). *Crossing Borders: International Exchange and Planning Practices*. London: Routledge.
Jessop, B., Peck, J.A. and Tickell, A. (1999). Retooling the Machine: Economic Crisis, State Restructuring, and Urban Politics. In: A.E.G. Jonas and D. Wilson, eds. *The Urban Growth Machine: Critical Perspectives Two Decades Later*. Albany, NY: State University Press of New York, pp. 141–159.
John, P. (2009). Why Study Urban Politics? In: J. Davies and D. Imbroscio, eds. *Theories of Urban Politics*. 2nd ed., London: Sage, pp. 17–24.
Jonas, A.E.G. and Moisio, S. (2016). City Regionalism as Geopolitical Processes: A New Framework for Analysis. *Progress in Human Geography*. Available online: DOI: 10.1177/0309132516679897.
Judge, D., Stoker, G. and Wolman, H., eds. (1995). *Theories of Urban Politics*. London: Sage.
Lefebvre, H. (1968). *Le Droit à la Ville* [*The Right to the City*]. Paris: Anthropos.
Logan, J. and Molotch, H. (1987). *Urban Fortunes: The Political Economy of Place*. Berkeley: University of California Press.
Lovering, J. (1995). Creating Discourses Rather than Jobs: The Crisis in the Cities and the Transition Fantasies of Intellectuals and Policy Makers. In: P. Healey, S. Cameron, S. Davoudi, S. Graham and A. Madani-Pour, eds. *Managing Cities: The New Urban Context*. Chichester, John Wiley, pp. 109–126.
Markusen, A.R. (1989). Industrial Restructuring and Regional Politics. In: R. Beauregard, ed. *Economic Restructuring and Political Response*. Newbury Park, CA and London: Sage, pp. 115–147.
Massey, D. (1994). *Space, Place and Gender*. Cambridge: Polity Press.
Massey, D. (2005). *For Space*. London: Sage.
McCann, E. (2011). Urban Policy Mobilities and Global Circuits of Knowledge: Towards a Research Agenda. *Annals of the Association of American Geographers*, 101(1), pp. 107–130.
McCann, E. and Ward, K., eds. (2011). *Mobile Urbanism: Cities and Policymaking in the Global Age*. Minneapolis: University of Minnesota Press.

Merrifield, A. (2013). The Urban Question Under Planetary Urbanization. *International Journal of Urban and Regional Research*, 37(3), pp. 909–922.

Miller, B. and Nicholls, W. (2013). Social Movements in Urban Society: The City as a Space of Politicization. *Urban Geography*, 34(4), pp. 452–473.

Pierre, J. and Peters, B.G. (2000). *Governance, Politics and the State*. Houndsmills: Macmillan Press.

Piven, F.F. and Friedland, R. (1984). Public Choice and Private Power: A Theory of the Urban Fiscal Crisis. In: A. Kirby, P. Knox and S. Pinch, eds. *Public Service Provision and Urban Development*. New York: Croom Helm/St Martin's, pp. 390–420.

Robinson, J. and Roy A. (2016). Global Urbanisms and the Nature of Urban Theory. *International Journal of Urban and Regional Research*, 40(1), pp. 181–186.

Rodgers, S., Barnett, C. and Cochrane, A. (2014). Where is Urban Politics? *International Journal of Urban and Regional Research*, 38(5), pp. 1551–1560.

Roy, A. (2011). Slumdog Cities: Rethinking Subaltern Urbanism. *International Journal of Urban and Regional Research*, 35(2), pp. 223–238.

Saunders, P. (1981). *Social Theory and the Urban Question*. London: Hutchison.

Savitch, H.V. and Kantor, P. (1995). City Business: An International Perspective on Marketplace Politics. *International Journal of Urban and Regional Research*, 19(4), pp. 495–512.

Scott, A. and Storper, M. (2015). The Nature of Cities: The Scope and Limits of Urban Theory. *International Journal of Urban and Regional Research*, 39(1), pp. 1–15.

Stone, M. (1978). Housing, Mortgage Lending and the Contradictions of Capitalism. In: W. Tabb and L. Sawers, eds. *Marxism and the Metropolis*. New York: Oxford University Press, pp. 179–207.

Ward, K. (2011). Urban Politics as a Politics of Comparison. *International Journal of Urban and Regional Research*, 35, pp. 864–865.

Ward, K. (2012). Policy Transfer in Space: Entrepreneurial Urbanism and the Making Up of 'Urban' Politics. In: A.E.G. Jonas and A. Wood, eds. *Territory, the State and Urban Politics: A Critical Appreciation of the Selected Writings of Kevin R. Cox*. Farnham: Ashgate, pp. 131–149.

Ward, K. and McCann, E. (2006). 'The New Path to a New City'? Introduction to a Debate on Urban Politics, Social Movements and the Legacies of Manuel Castells' *The City and the Grassroots*. *International Journal of Urban and Regional Research*, 30(1), pp. 189–193.

Williams, C. and Pendras, M. (2013). Urban Stasis and the Politics of Alternative Development in the United States. *Urban Geography*, 34, pp. 289–304.

Wilson, D. (2004). Toward a Contingent Urban Neoliberalism. *Urban Geography*, 25(8), pp. 771–783.

Wilson, D. (2007). *Cities and Race: America's New Black Ghetto*. London: Routledge.

Young, I.M. (1990). *Justice and the Politics of Difference*. Princeton, NJ: Princeton University Press.

PART I

Approaching the space(s) of urban politics

Andrew E.G. Jonas and Byron Miller

The introductory chapters in this section set out some general conceptual themes for exploring the urban as a particular space – or, more accurately, a multiplicity of spaces – of politics. Such spaces are not restricted to those within the city viewed solely as a legal or jurisdictional entity; nor can such spaces be separated so easily from knowledge of wider social relations and processes, including those that operate around the state, as has often been the case in utopian political imaginaries. Rather, it is the premise of this collection of chapters that there is a great variety of spaces in and through which 'urban' politics takes place – the 'where' of urban politics is not necessarily located exclusively inside the city itself (Rodgers *et al.* 2014). Conceptually, such spaces can be examined from the perspective of the hierarchical spatial scales, such neighbourhood, metropolitan, national or international, around which social life, the economy and the state tend to be organized (Jonas 2017). But they can also refer to the specific locations, sites and practices inside the city where more or less globalized social and economic processes and policies circulate and converge (McCann and Ward 2011).

Not only are different scales of socio-spatial organization always present in urban politics but also the material and social substance of urban politics itself varies from place to place. In some places urban politics are fought around matters of collective consumption, in others economic development, and in many more besides a combination of both processes operating in stark tension (Cox and Jonas 1993). In these global times, urban politics is often infused with contentious debates about inequality, democracy, civil rights, and environmental and social justice (Harvey 2012). Accordingly, the chapters featured in this section examine the variety of ways in which urban politics are spatially framed and, for that matter also, the different ways in which the 'urban' might be seen to be a politically contested and contentious space in its own right. In particular, the chapters consider whether and how different spaces of urban politics afford opportunities for more or less progressive visions of social life and, ultimately, to what extent the spaces of urban politics open up opportunities to imagine more hopeful existence for all irrespective of differences in class, race, gender or identity.

Allan Cochrane begins with the proposition that hitherto there has been little debate about the urban of urban politics. Rather it has been conventionally understood as a relatively straightforward question. From this perspective urban politics is the politics that takes place within those areas defined as urban. In contrast to the received wisdom, Cochrane explores some of the issues that arise when defining the urban of urban politics is itself rather more uncertain. He does this by reflecting on four distinctive approaches.

The first has a relatively matter-of-fact starting point, looking at the ways in which the urban is framed and reframed through the behaviour and claims-making of political actors. The second and third approaches open up very different ways of thinking, but both set out to develop more precise and consistent (possibly even scientific) definitions: one starts from the material importance of agglomeration to highlight the significance of land and property as foundational to the urban; while the other sees urbanization as a more or less universal social and economic process – of planetary urbanization – in which settlement patterns are of secondary significance.

The chapter builds on these arguments and the work of Doreen Massey to reflect, fourthly, on the many spaces of urban politics, moving beyond the narrow confines of assumed territorial geographies, while maintaining a profound acknowledgement of the significance of urban place. Holding these two aspects of the urban together makes it possible to explore and understand both the nature of actually existing urban politics and the possibilities incorporated within it.

The next chapter by Deborah Martin considers the role of place in urban politics. In the urban context, the politics of place is often conceptualized as 'local' politics, with local usually connoting either a particular municipality or, especially in cities, a 'neighbourhood'. Place politics involves explicit and implicit references to geographical norms or grievances in the form of 'place-frames'. Such place politics may seem parochial and limited in potential transformative scope, given the often-restricted scope of territory enrolled in any given place conflict. Relational conceptualizations of place, however, emphasize the spatially complex and disparate networks and processes producing any particular locality (city or neighbourhood). These, in turn, necessitate that considerations of politics reach beyond the locality.

Martin argues that researchers need to pay close attention to the politics of place-framing because this helps to reveal the processes at work in shaping the dynamics of place politics, where the territory of contention, the stakeholders and claims-makers, can all be points of contestation and deliberation. The politics of place, then, is politics over what place, and whose place, and how much place are explicitly enrolled in contention and potential solutions.

Contemporary critical representations of urban space increasingly refer to it as a 'commons' – a space to which all the inhabitants of the city should have equal access. However, the commons (or common) is always contested and such struggles seem to have intensified under the onslaught of neoliberalism. The chapter by Theresa Enright and Ugo Rossi considers the concept of the commons from the perspective of critical urban theory. In particular, it identifies two predominant approaches – neo-institutionalist and neo-Marxist – used to understand the commons and their relation to contemporary urban life. The chapter explores the political stakes of these approaches, particularly in light of the current stage of what the authors define as late neoliberalism: a phase of neoliberalism characterized by the tension between the fear of a permanent crisis (or secular stagnation) and the search for a reinvention of capitalist narrative and culture. The chapter questions to what extent the notion of the commons is a necessary addition to established concepts of social justice and citizenship within the urban political lexicon.

Seen as a specific socio-spatial conjuncture of power relations, the state should be an essential ingredient in the menu of topics examined in a volume about the spaces of urban politics. Yet as Yonn Dierwechter argues, utopian thought across diverse traditions has often rendered 'the state' invisible, unhelpful or dangerous, even as it has concomitantly elevated specific kinds of places as more worthy of our normative-political aspirations. Stateness per se can be understood as the presence of socially legitimate institutions that articulate political democracy and help to puzzle out problems. As such, the state remains resilient in the face of constant attacks and legitimate critiques, particularly as a bulwark against corporatized economies that, untethered from popular control, have repeatedly generated social and ecological turmoil.

At the same time, the role of the state in reconstructing politically an ideal of the good society has returned to prominence. Despite this, the macro-dynamics of globalization, urbanization and city-regionalism have shifted the ways in which the state can be mobilized in utopian thinking. In short, the state now has to be smart.

Dierwechter's chapter argues the case for examining urban politics through the rubric of the smart state as a possible utopian space. The discussion highlights the tense intellectual-theoretical relationship between utopianism and the (territorial or centralized) state. It then shifts to the rising but still open and contested role of cities in global, 'post-Westphalian' state space. Finally, the chapter explores specificities of urban-regional politics across Greater Seattle, which is often used as an exemplar of the so-called smart turn in global urbanism. Here Dierwechter establishes the tone for the remainder of the *Handbook* by using a place-specific example to examine more abstract propositions about the spaces of urban politics.

References

Cox, K.R. and Jonas, A.E.G. (1993). Urban Development, Collective Consumption and the Politics of Metropolitan Fragmentation. *Political Geography*, 12, pp. 8–37.

Harvey, D. (2012). *Rebel Cities: From the Right to the City to the Urban Revolution*. London: Verso.

Jonas, A.E.G. (2017). Politics. In: M. Jayne and K. Ward, eds. *Urban Theory: New Critical Perspectives*. London: Routledge, pp. 231–242.

McCann, E. and Ward, K., eds. (2011). *Mobile Urbanism: Cities and Policymaking in the Global Age*. Minneapolis: University of Minnesota Press.

Rodgers, S., Barnett, C. and Cochrane, A. (2014). Where is Urban Politics? *International Journal of Urban and Regional Research*, 38(5), pp. 1551–1560.

2
HERE, THERE AND EVERYWHERE
Rethinking the urban of urban politics

Allan Cochrane

Introduction

Reflecting on the where of urban politics may at first sight seem a slightly odd thing to be doing, if only because the where seems implicit (maybe even explicit) in the label itself. After all, urban politics is presumably simply the politics that takes place in, is shaped by and shapes the experience of those living in urban areas. In other words, urban politics is the politics pursued in those areas that everybody knows to be urban – the politics of what goes on in cities or city neighbourhoods. In practice, however, attempting to define the urban of urban politics as an object of study is by no means as straightforward as its easy use in everyday language – or academic writing – might suggest.

In practice, the administrative borders of local government have often provided the frames within which politics are explored (Davies and Imbroscio 2009; Judge *et al.* 1995). And it is, perhaps, hardly surprising that existing institutional frameworks become the routes through which this is done. City governments and local governments provide an easy and acceptable way into empirical research, delivering both political institutions and a whole set of physical and social infrastructures that are oriented towards them and the resources they have to offer. There is an understandable tendency to start from existing structures, to utilize existing territorial framings of the urban (or the local), even if those structures and framings are themselves sometimes subject to change as borders are redrawn or new institutions created (or abolished). Tight territorial boundaries may fit uneasily with some forms of political practice, but that has not diminished their significance as meaningful entities through which institutional actors – from councils to mayors, chief executives to social workers, town planners to local community associations – define their day-to-day political practice.

But there is another (sometimes overlapping) approach which emphasizes the importance of the urban as a site across which more radical forms of politics may emerge – the residential equivalent of the factory as a place of organization (see, e.g. Hardt and Negri 2009). From this perspective what matters are the spaces of the city, which allow and enable various populations to come together and identify themselves as collective political actors. Sometimes these are literally public spaces – the squares and streets of demonstrations and riots or uprisings – but they may equally be expressed in the language of community or particular populations imagined as quintessentially urban – Wacquant's 'urban outcasts' for example (Wacquant 2008).

But what if the urban itself becomes an uncertain or contested category? What if the where of urban politics itself becomes an issue (see Rodgers *et al.* 2014). Elsewhere I and John Allen have suggested that:

> There is a nagging excess to the urban, where the political relations that construct the city as an urban political arena exceed the boundaries drawn. There is, in other words, a tension between the city understood as a political arena or figuration and the city as a bounded political space.
>
> *(Allen and Cochrane 2014: 1610)*

At its simplest, it is not always clear where the politics of urban territory begins and ends, how it is bounded and for what purposes, and it often seems equally difficult to identify what the other spaces of politics might be – the non-urban politics. Of course, it might be argued that this is a scalar question (global, national or regional scales might be contrasted with the urban scale) but it is not long before it is apparent that each of those also has an urban aspect, and other spatial categories (such as rural, suburban or regional, again) are equally problematic in framing any other distinctive form of politics, as borders become porous and newly invented terms are introduced (from polycentric region and exopolis to mega region and city region).

There is an uneasy relationship between the almost careless way in which urban is often mobilized in popular and academic speech – we all know where it is – and the search for more specific – even scientific – definitions. And it is on that relationship that this chapter is focused. The point here is not to come up with a final definitive answer but, rather, to highlight a range of possibilities, some of which overlap, some of which are indicative and some of which may simply be incompatible.

I want to start by looking at some of the ways in which the urban of urban politics has been reimagined in recent years, before moving on to discuss some of the attempts that have been made to conceptualize the urban more precisely as a distinctive analytical category. There is a long tradition of doing the latter. So, for example, famously the Chicago School and Louis Wirth, in particular, specified a set of criteria which constituted 'urbanism as a way of life'. More recent attempts to specify the urban and the politics associated with it include Manuel Castells' identification of collective consumption as the fundamental axis around which urban social movements might develop; while a focus on land and development as underpinning the urban growth machine explored by John Logan and Harvey Molotch has been generalized by Allen Scott and Michael Storper for whom the urban land nexus is identified as the key defining characteristic around which urban politics is mobilized; and, building on the work of Henri Lefebvre, Neil Brenner and Christian Schmid go beyond any bounded conceptions of the city to emphasize what they call a process of planetary urbanization. The different theoretical approaches adopted may be quite distinctive, but in effect each seeks – as Storper and Scott argue in their review of urban theory – to develop a "foundational concept of the urban" (Storper and Scott 2016: 1115).

Bordering and rebordering the urban

One response to the issues around the urban has simply been to generate a range of imaginations, drawing on the changing empirical patterns of political action within cities and then mobilizing different descriptors of those patterns to suit whatever the current concern happens to be. Over the years the focus has moved easily (and often without any explicit reflection on the shifts in emphasis) between neighbourhood, community, inner city and locality, as well as moving beyond the city to the edge city, city region, the metropolitan region or mega region. This has not been a linear direction of change, and each of these ways of thinking about the spaces of urban politics has survived alongside the others, even as the emphasis moved around.

In a sense the injunction has been to follow where the politics leads rather than to worry too much about definitional questions. The stress has been placed on the political and policy actors, and following what they do is what has defined the nature of urban politics. That might include the politics of the streets – the riots and uprisings – community action or gentrification stories as well as those of urban regimes or major development projects. The politics may still somehow be bounded within places understood to be urban, but the boundaries and frames change with the political processes being explored. Even where the starting point is given by the pattern of government institutions, the focus of attention shifts in ways that also imply changing interpretations of the urban (for example, in moving from policing and community development to neighbourhood plans and urban development corporations to the invention of city regions).

Taking such an open approach to urban politics as an active process is not to be dismissed. It makes it possible to reflect on emergent and uncertain political formations without having to claim that they are somehow necessary outcomes of some pre-existing model. So, for example, I have recently been seeking to explore the ways in which new political forms have begun to emerge on the edges of England's largest urban region – London and the South East. There has been some reluctance to identify this as a city region (London is often treated as an exception) but it is hard to disagree with the conclusions of Ian Gordon, who notes that: "In this regionalized version of London, outer areas now substantially contribute to its agglomeration economies, as well as continuing to benefit from those rooted in central London" (Gordon 2004: 41). For Hall and Pain (2006) there can be no doubt that this is what they call a 'polycentric metropolitan region'. It is the spread of activities rooted in and dependent on the London economy, coupled with a transport network that focuses on London, that help define the city region.

Those seeking to govern London and the South East have historically found it difficult to contain dynamic growth within the formal structures of territorial governance. It is in this context that recent attempts to formalize and develop arrangements are particularly interesting. They confirm that the process is an uncertain and ambiguous one, which fits uneasily with traditional hierarchies of government. For instance, a more or less explicit 'growth region' strategy was developed in the early years of the twenty-first century, supported by sets of partnerships, urban development corporations and targets for housing – even the regeneration plans associated with the 2012 Olympics could be incorporated into this vision. In practice, the market based assumptions underpinning the strategy were unsustainable and targets for new housing were never achieved – the policy model failed (Cochrane et al. 2013).

However, that does not mean that the underlying tensions have disappeared. On the contrary, a whole range of different spatial imaginaries continues to be generated as different actors seek to come to terms with the protean shifts of urban life. So, for example, in 2016 in the area on which my research (funded by the ESRC ES/I038632/1) has focused, initiatives have included: the launch of the Midlands Engine (which incorporates a range of infrastructural and other schemes in a newly imagined region that stretches across a large swathe of central England from Birmingham down to Luton); the associated Midlands Connect, which is focused on transport connections and infrastructure; the formation of England's Economic Heartland, which brings together the counties of Buckinghamshire, Oxfordshire and Northamptonshire across the northern part of London's city region, or the Greater South East, with the claim that they are 'At the Heart of Science and Technology Innovation'; the identification of an Oxford to Cambridge Arc, around which growth is to be clustered with the help of infrastructural investment, particularly transport related investment; and the continuing presence of a set of Local Enterprise Partnerships at a different level, but each seeking to position itself within the broader frameworks – in one case planning to merge to reflect these shifts.

Of course, the politics associated with these initiatives might be understood to be corporatist in the sense that they have little direct connection with the popular politics of the regions across which they are playing – they represent the politics of development; a politics within which state and private interests are deeply intertwined to deliver new opportunities and open up new spaces for profitable investment, often couched in the relative anodyne language of growth and regional competitiveness. And these particular labels (or regional brands) may turn out to be temporary as governmental and business strategies change. Meanwhile, for many of those living in these areas, however much their activities stretch across space for employment and consumption, the focus remains on more immediate localities; and the different understandings of social space often come into conflict over particular development projects and infrastructural or transport investments (Cochrane *et al.* 2015a).

It is important and necessary to explore the detailed practices of urban politics. But, even doing so along the lines I have indicated begins to suggest that such an approach is not quite enough, because the detailed practices may themselves reflect wider shifts, different ways of thinking through the nature of cities and urban forms. In this case, for example, it may mean that it is necessary to question conventional conceptualizations of the suburbs.

Traditionally (most clearly reflected in the models of the Chicago School back in the 1920s and 1930s) the suburbs have been allocated a secondary status – a consequence of urban development around a core or development node. More recently, however, they have been given a more active role in the urban imaginary, the importance of the suburbs as a central, rather than peripheral or secondary, aspect of contemporary urbanism and the making of global urban regions. The outer suburbs are increasingly being reinterpreted in public policy and political discourse both as sources of economic dynamism – capable of generating economic growth – and as places whose purpose is to provide the housing for the labour force required to feed the insatiable demands of the wider urban economy.

This shift in emphasis can be illustrated with the help of examples from many places. It is perhaps particularly apparent in the work of those who have sought to use the experience of Los Angeles and Southern California to question the presumed dominance of the Chicago School (Dear 2000, 2002), but it has also found an expression in the work of those seeking to highlight the experience of development in places that have too often been off the global map of urban modelling, in Africa or South Asia, where new urban developments often owe little to the older urban cores, and what may look like suburbs are better understood as quite distinctive spaces of urban living (Simone 2004). At the same time a focus on the messy, crowded urbanization of Western Europe has drawn attention to the emergence of what has been called the *Zwischenstadt* (or 'in between city') which cannot easily be characterized in the spatial language of concentric rings (Sieverts 2003), and which also seems to be identifiable in other parts of the urbanized world (see, e.g. Young and Keil 2014 on the Toronto experience).

New social and economic spaces are emerging, even if they are ambiguous and uncertain, and in the English context something significant *is* happening on the edge of the South East, in ways that require a reimagination of how we interpret the urban and the suburban. It is increasingly necessary to move beyond a notion of the urban as having some sort of defined core, of the drivers of growth being concentrated in the centre and somehow rolling out from there. This is being made up not just through the everyday lives of its residents, but also through the practices of the (public and private sector) policy professionals as they develop their plans, which are realized in built form in housing developments and material infrastructure. Instead of being understood as little more than a secondary consequence of urban growth out from a central place it becomes necessary to recognize that the suburbs themselves play an increasingly active part in defining the urban as a lived experience. The distinctions between city and suburb are becoming less clear-cut, even as some residents call on rural and market town imaginaries to distance themselves from the perceived encroachment of suburbia, as much as the urban (Cochrane *et al.* 2015b).

In other words, a starting point rooted in the empirical identification and investigation of the politics of an urbanizing suburbia and a suburbanizing urbanism, inexorably leads towards the search for a wider theoretical framing with the help of which such empirical observations can be better understood.

Defining the urban

If one approach has been to respond to and explore what might be described as actually existing urban politics, a second seeks to develop more or less scientific definitions, through which forms of urban politics may in turn be explored. Nor is this a recent phenomenon. So, for example, back in the 1930s Louis Wirth identified the city as: "a relatively large, dense and permanent settlement of socially heterogeneous individuals" (Wirth 1938: 8). Of course this could simply be seen as a common sense interpretation of "what we all mean by the city", but it has significant implications. A series of more or less logical consequences were seen to follow from the four characteristics that Wirth identified.

The large population size of cities was held to mean that contact between urban residents would generally be impersonal, superficial and transitory, since no one person could know all the others in the city. According to Wirth, the size of cities meant that contact between individuals was usually restricted to those occasions in which there was some mutual advantage to be gained. As a result, he argued, human relationships were segmentalized according to the services that each individual might perform for another. Similarly, the density of population was said to produce increased differentiation between individuals, an increasingly complex social structure, social segregation and a dissociation of workplace from residence. It fostered a spirit of competition, aggrandizement and mutual exploitation, requiring complex rules to manage the resulting friction. It was the permanence of cities that was assumed to create the spaces within which the social processes identified by Wirth worked themselves out, as, to echo the title of Wirth's seminal article, urbanism was 'a way of life'. Meanwhile social heterogeneity meant that,

> The city has ... historically been the melting pot of races, peoples and cultures. It has not only tolerated but rewarded individual differences. It has brought together people from the ends of the earth because they are different and thus useful to one another, rather than because they are homogeneous and like-minded.
>
> *(Wirth 1938: 8)*

No single group has the full loyalty of individuals, since different groups are important for different segments of each individual's personality.

Although Wirth was a sociologist, his understanding of the city seeks to follow through the logics of an urban space economy – logics that flow from the clustering/concentration of people which generates social processes in place. And similar understandings have been reflected in a whole series of interventions since then, even if the conclusions that have been drawn have not always been the same. Jane Jacobs, for example, was the poet of the city as social organism, describing the ways in which city streets helped to generate what she called a "sidewalk ballet

… in which the individual dancers and ensembles all have distinctive parts which miraculously reinforce each other and compose an orderly whole" (Jacobs 1961: 50).

But this was more than just a romantic vision, because for Jacobs it was of direct practical significance. As she went on to argue:

> Although it is hard to believe, while looking at dull grey areas or at housing projects or at civic centres, the fact is that big cities *are* natural generators of diversity and prolific incubators of new enterprises and ideas of all kinds … The diversity, of whatever kind, that is generated by cities rests on the fact that in cities so many people are so close together, and among them contain so many different tastes, skills, needs, supplies, and bees in their bonnets.
>
> *(Jacobs 1961: 145 and 148)*

In contrast to Wirth, Jacobs emphasized the extent to which the complexity and heterogeneity of the urban experience might create order out of apparent chaos and, above all, generate dynamism and innovation. And others, too, have stressed the extent to which urban social relations may be characterized by a process that involves learning to live together and even generate collective political action (for example, in the form of urban social movements), rather than the sort of anomic vision identified by Wirth (see, for example, Castells 1983; Hamel *et al.* 2000; Hardt and Negri 2009; Harvey 2012; Neal *et al.* 2013).

A related but distinct set of approaches has clustered around questions of land and development as the key aspects in defining the urban and the politics associated with it. John Logan and Harvey Molotch point to the centrality of what they call an "urban growth machine" (Logan and Molotch 1987) and develop the argument that whatever else is going on, the driving force of urban politics is focused around land and property values, as politicians and developers seek to foster forms of growth that will increase those values and generate returns for those owning and developing the land.

Scott and Storper take this emphasis on land further in developing their own model, stressing that just because something happens in the city, that does not necessarily make it 'urban' – as they put it, "Political outcomes in the city … [need to] distinguish the specifically urban from what is merely contingently so" (Storper and Scott 2016: 1117). This means that the task is to identify the "fundamental common genetic factors underlying urban patterns, and a robust set of conceptual categories within which urbanisation processes and urban experiences can be analysed, wherever they may occur in the world" (Storper and Scott 2016: 1115). And they find this in the process of agglomeration which constitutes the urban land nexus, that is "an interacting set of land uses expressing the ways in which the social and economic activities of the city condense out into a differentiated, polarised, locational mosaic" as "the essential fabric of intraurban space" (Scott and Storper 2015: 8). The agglomeration process, they argue, "has powerful feedback effects not only on economic development, but also on society as a whole" (Scott and Storper 2015: 6).

Questions of urban politics and urban governance, therefore, flow from:

> The tensions created by competition for land uses, the urge to secure access to positive externalities and to avoid the effects of negative externalities, the rent-seeking behaviour of property owners and the need to protect or enhance certain kinds of urban commons (such as agglomeration economies), among other frictions, all create constantly shifting circles of urban social collisions.
>
> *(Storper and Scott 2016: 1117)*

This generates a theoretical framing, which allows for some uncertainty – the defining principle with its attendant (genetic) logics is clear, but there is "no rigid line that separates urban land nexus definitively from the rest of geographic space, but rather a series of spatial gradations in which we move from one to the other", so that in practice a "pragmatic rule of thumb", may be used "to locate the line of division in some more or less workable way relative to available data" (Storper and Scott 2016: 1130). The strength of this approach lies in the way that it manages to build on a widely agreed aspect of urbanism and the urban economy to highlight some key features of urban politics – the disputes, tensions and possibilities associated with land, its uses and markets, as well as conflicts over the ways in which the social costs (as well the private benefits) of development (and decline) are managed.

The thesis developed by Scott and Storper is in key respects a response to writers from a post-colonial perspective who have persuasively questioned the extent to which it is possible to identify some universal process of urbanization which holds across time and space, arguing that agglomeration does not necessarily have the same implications everywhere and always and that not all urban experiences can be captured under the heading of agglomeration (see, e.g. Roy 2009, 2011). The extent to which the arguments of Scott and Storper effectively deal with this broader critique of geographical thinking remains uncertain (see, e.g. Robinson and Roy 2016; Roy 2016), but an equally fundamental issue may relate to the underlying ambition neatly to delimit what counts as the 'urban' and therefore what counts as urban politics. Even within the 'urban', Storper and Scott identify what they call "ontologically distinctive scale[s] of urban space" – each of which "*poses uniquely problematical scientific and political questions* deriving from its modes of operation" (Storper and Scott 2016: 1129, emphasis in original) – and which apparently include neighbourhoods, slums, industrial quarters, central business districts and suburbs. This highlights the underlying weakness of attempts to identify what turn out to be a series of structurally determined boxes into which everything that matters is expected to fit.

Scott and Storper appear to promise an approach rooted in the familiar spaces of the urban but the elision between agglomeration and the urban land nexus generates a narrower definition of urban politics, excluding many of those political movements popularly understood as urban (for example around welfare or poverty or class or ethnic mobilization). They may exist, but from this perspective they are not quintessentially urban. But then the danger is that definitional debates become substitutes for attempts either to explain actually existing political movements that do not fit into the typology or to contribute to the development of those policies. In other words, instead of the city being "a distinctive, concrete social phenomenon" (Scott and Storper 2015: 12) it becomes something that only exists through their necessarily narrow definition of it. The changing frames of urban politics in practice – as understood by those who engage in them – simply become category mistakes, either reducible to the workings of the urban land nexus, or excluded from the urban. In one sense, of course, all this is perfectly legitimate (the urban is defined and we know what is or is not urban) but the danger is that exploring some of the other political dynamics that tend to be focused on and in cities for whatever reason is left to others to consider.

Thinking without boundaries: the urban as a social phenomenon

In that context, it is interesting to look at attempts that seek to think the urban differently, not in terms of space and place, but rather through a focus on the urban as social process. In his early writing on urban social movements (before *The City and the Grassroots*) Manuel Castells developed a particular understanding of urban politics defined through collective consumption. Urban politics was said to be the politics of collective consumption (that is those activities consumed collectively through forms of public provision, rather than through market relations or individual consumption) wherever it took place (Castells 1977, 1978). Urban social relations were to be found wherever such activities took place rather than simply in those places popularly understood to be urban, defined through agglomeration or the concentration of population. And urban social movements were focused on conflicts and campaigns around collective consumption as the post-welfare state sought to roll back provision and urban populations sought to maintain and even extend it.

The clarity underpinning this approach was very persuasive and picked up on the experience of some key aspects of the political movements and tensions of the 1960s and 1970s and into the 1980s, with the rise of the new Right and the emergence of a series of movements seeking to challenge the implications of the new politics, soon to attract the soubriquet of neoliberalism. However, as Andy Merrifield (2014: 19) has noted, it fitted less easily with the ways in which the old welfare regimes were actively restructured across the period, through privatization and restructuring. And, if Scott and Storper manage to sideline the urban politics of social welfare, then Castells' approach managed to ignore the importance of the politics of growth and development – the rise of urban competiveness as a political focus (Cochrane 1999).

More recently, the argument that the urban should be defined otherwise than through agglomeration has been taken further by Brenner and Schmid (2014), who have sought to develop the notion of planetary urbanization as an alternative way of thinking. In other words, instead of imagining urban expansion from a central core (or central cores), they focus attention on the emergence of an urban society ("urbanism as a way of life", but without Wirth's morphological criteria) in which agglomeration is only one possible expression of urbanization. In some respects, however, the starting point is similar to that of Scott and Storper. As Brenner and Schmid put it, without some "reflexive theoretical perspective, the concept of the urban will remain an empty abstraction devoid of substantive analytical content" (Brenner and Schmid 2014: 748). The urban and urbanization, they argue, are "theoretical categories" (Brenner 2013: 96; Brenner and Schmid 2014: 749; 2015: 163). But, for them, the "sociospatial dimensions of urbanization are polymorphic, variable and dynamic" (Brenner and Schmid 2014: 750) in forms of extended urbanization, generating and incorporated into thick webs of relationships. Urbanization is, they say, a "[P]lanetary phenomenon" that is an "increasingly worldwide, if unevenly woven, fabric" (Brenner and Schmid 2014: 751). Merrifield makes a similar point in arguing that "The urban is shapeless, formless and apparently boundless, riven with new contradictions and tensions that make it hard to tell where borders reside and what's inside and what's inside" (Merrifield 2013: 910). The complexity of the process/es involved is captured in the notion of implosion/explosion as a reflection of the ways in which socio-spatial organization is restructured (Brenner 2014).

Brenner and Schmid construct their arguments through sets of theses (Brenner 2013; Brenner and Schmid 2015) which highlight their dismissal of any attempts to define the urban through a binary (urban/non-urban) distinction, questioning any focus on settlement types. Instead they argue that it is the variegated forms of urbanization that need to be understood and explored (concentrated, extended and differential). In other words, they recognize the importance of uneven development and even concentrated development (in what might be called cities, although they resist using that term) but they do not see these as definitional criteria, because of what is identified as the "dialectically intertwined moments" of concentration and extension which they see as central to urbanization (Brenner 2013). In this context, the politics of the urban might be understood to emerge from the "collective project in which the potentials generated through urbanization are appropriated and extended" (Brenner and Schmid 2015: 176).

The power of this vision of urbanization is unmistakable, both because of the ways in which it deals with some of the uncertainties (the nagging excess) of the urban that cannot easily be captured through a focus on agglomeration and because of the way in which it opens up the possibility of a planetary politics rooted in urban social relations. It brings to the fore what populations have in common, rather than looking for ways of separating them out through the identification of particular spatial patterns and the implicit boundaries constructed through them. It begins to suggest possibilities of collective action across urban space that is capable (in principle at least) to challenge what is understood to be a neoliberal world – a world of capital. Andy Merrifield calls on the science fiction writer Isaac Asimov's dystopian/utopian description of the "thoroughly urbanised" planet Trantor as a metaphor through which it might be possible to imagine a new politics (Merrifield 2013: 909–910). The logic of existing urban social relations is given a more coherent form – science fiction helps to give a material expression to an abstract phenomenon – at the same time as providing a vision of what might be politically possible in the context of planetary urbanization.

And yet the confidence with which it appears to dismiss formulations that start from actually existing urban politics also raises questions for this approach. For all its emphasis on differentiation this is a universalizing vision as well as teleological one. In other words, its undoubted strengths also have associated weaknesses. Some critiques have emphasized the importance of incorporating more explicitly the differences within urbanization – even if one can agree that the process is planetary, what may matter is precisely the variation and difference – as Ananya Roy puts it, while we may be able to agree that "urbanism is a worldwide process … that does not necessarily mean that such urban transformations can be understood as the universalization of a singular and basic urban form" (Roy 2016: 204). The theoretical starting point makes it more difficult to acknowledge the importance of the urban as spaced and placed, even if attempts are made to incorporate such points in specific moments (of concentration and extension).

A curiously antiseptic form of politics emerges – just as surely as it does from the urban land nexus approach. Politics as an active process is difficult to identify, and it is even difficult to see how it would emerge. This critique is at the heart of Richard Walker's arguments as he expresses his frustration at the lack of any clear linkages between the theoretical abstractions however well expressed and actually existing urban politics (Walker 2015) and, from a slightly different angle, Kate Shaw similarly expresses doubts about the efficacy of the approach as guide to political action (Shaw 2015). The planetary urbanization thesis may point to possibilities (through its mobilization of various metaphors, including those drawn from science fiction) but it is less convincing in identifying the ways in which they might be realized through political practice. What emerges from these critiques (which find an echo in the arguments of Scott and Storper) is that the city or the urban has to be understood as a material space (as well as a theoretical category). In other words, urbanization may be planetary, but the urban is also placed and it is this which makes it possible to conclude (as Mark Davidson and Kurt Iveson do) that the city remains a category that can productively be mobilized in developing and understanding urban politics (Davidson and Iveson 2015).

Thinking differently about space, place and the urban

It is probably not helpful to start by setting up false distinctions between the different approaches discussed above and dismissing one or the other as a result. All of them highlight important aspects of contemporary urbanization and urban politics. Instead, a more productive way forward is, it seems to me, to think through the lived experiences of the urban in ways that reflect what Doreen Massey has called the "throwntogetherness of place" (Massey 2005: 149), which highlights the more mundane ways in which urban social existence is put together and the extent to which cities continue to be the sites where a range of different political outcomes may be explored and struggled over. For Massey, in other words, space has to be understood "as an open and ongoing production" (Massey 2005: 55) so that places are defined through the contingent juxtaposition of communities, groups and individuals linked out into wider networks yet actively engaged in the co-construction of neighbourhoods, cities and regions.

From this perspective what matters are the sets of relations which help to construct those experiences, so that the urban is understood as the space within which sets of relationships overlap, settle and come together, with a particular intensity. If, as Massey and others argue (Massey et al. 1999), the urban experience is defined through its intensity and the juxtaposition of difference, because of the ways in which cities bring people together from a range of different backgrounds, requiring them to interact in mundane as well as sometimes more active ways, then it is that which provides a way into an understanding of urban politics. Cities, in other words, are the places "where different stories meet up" (Massey 1999: 134).

Massey's work requires us to explore the ways in which places are defined and define themselves, are made and remade, imagined and reimagined. As a consequence, it is necessary to approach urban politics in relational terms, that is, to explore the linkages through which politics is constructed in practice. In this context, we can interpret the urban as a node around which sets of relationships come together. Some of the relations that matter here will be stretched across space but be very close in other respects. Such an approach offers a way into a less narrowly territorial – or bounded – but still fundamentally 'placed' urban politics which, as John Allen and I have argued, needs to be understood as assembled (Allen and Cochrane 2010), and in ways that reflect the extent to which places are the product of relations that stretch across space, far beyond the administrative or other boundaries that help to define them, reflecting and helping to construct a "global sense of place" (Massey 1994: 146–156).

This implies that urban politics – in particular places – will also reflect interests and conflicts apparently drawn from elsewhere, while even the most apparently local of political actors may actively reach out to draw in policy lessons and political understandings developed in quite other contexts. From this standpoint, in other words, the urban and urban politics are assembled and put together in place, yet are shaped by the nature of their connections to elsewhere rather than being limited by the territorial boundaries of particular urban spaces. Indeed, as Massey (2007) has argued in her discussion of London as a World city, the political identity, legitimacy and responsibility of large, globalizing cities in particular is stretched across national and transnational spaces.

Rather than reflecting any desire to escape from the recognition that urbanization is a globalized (or planetary) phenomenon, this approach implies the need to explore the active processes by which it is made up, bringing places together at the same time as separating them from each other. Circulating policy knowledges illustrate how urban politics and policy cannot, if they ever could, be understood by studying activities conceived as taking place within particular scales or places; fundamentally, it takes place through knowledge of and making reference to 'elsewhere' (Cochrane and Ward 2012; McCann 2011; McCann and Ward 2011). Global urban policies are always defined in place. From this perspective urban politics becomes a process of assemblage – an active and continuing process in which the spaces of urban politics are defined and redefined, becoming more or less settled in practice. And the realities of urban agglomeration as well as the drivers associated with the urban land nexus will also be important in this process, even if urban politics cannot be reduced to those drivers.

And, of course, in this context territory may still provide an important basis for mobilization, alongside the overlapping administrative hierarchies of government and state. They are still meaningful political entities through which institutional actors – from councils to mayors, chief executives to social workers, town planners to local community activists – define their day-to-day political practice. Urban space is actively constructed – always in the process of being invented, so struggles over urban space continue to have a strong political resonance, even as they are being made and unmade. As Scott Rodgers, Clive Barnett and I (Rodgers et al. 2014) argued in seeking to identify the where of urban politics, it is necessary to work across the troubled and troubling borderlines in which sets of territorial definitions and the practices defined as urban, and which define the urban, are in tension with each other. And it is that which I have sought to do in this chapter.

References

Allen, J. and Cochrane, A. (2010). Assemblages of State Power: Topological Shifts in the Organization of Government and Politics. *Antipode*, 42(5), pp. 1071–1089.
Allen, J. and Cochrane, A. (2014). The Urban Unbound: London's Politics and the 2012 Olympic Games. *International Journal of Urban and Regional Research*, 38(5), pp. 1609–1624.
Brenner, N. (2013). Theses on Urbanization. *Public Culture*, 25(1), pp. 85–114.
Brenner, N., ed. (2014) *Implosions/Explosions: Towards a Study of Planetary Urbanization*. Berlin: Jovis.
Brenner, N. and Schmid, C. (2014). The 'Urban Age' in Question. *International Journal of Urban and Regional Research*, 38(3), pp. 731–755.
Brenner, N. and Schmid, C. (2015). Towards a New Epistemology of the Urban? *City*, 19(2–3), pp. 151–182.
Castells, M. (1977). *The Urban Question*. London: Edward Arnold.
Castells, M. (1978). *City, Class and Power*. London: Macmillan.
Castells, M. (1983). *The City and the Grassroots: Cross-cultural Theory of Urban Social Movements*. London: Edward Arnold.
Cochrane, A. (1999). Redefining Urban Politics for the Twenty First Century. In: A.E.G. Jonas and D. Wilson, eds. *The Urban Growth Machine: Critical Perspectives Two Decades Later*. Albany, NY: State University of New York, pp. 109–124.
Cochrane, A. and Ward, K. (2012). Researching the Geographies of Policy Mobility: Confronting the Methodological Challenges. *Environment and Planning A*, 44(1), pp. 5–12.
Cochrane, A., Colenutt, B. and Field, M. (2013). Developing a Sub-regional Growth Strategy: Reflections on Recent English Experience. *Local Economy*, 28(7–8), pp. 786–800.
Cochrane, A., Colenutt, B. and Field, M. (2015a). Governing the Ungovernable: Spatial Policy, Markets and Volume House-building in a Growth Region. *Policy and Politics*, 43(4), pp. 527–544.
Cochrane, A., Colenutt, B. and Field, M. (2015b). Living on the Edge: Building a Sub/urban Region. *Built Environment*, 41(4), pp. 567–578.
Davidson, M. and Iveson, K. (2015). Beyond City Limits. *City*, 19(5), pp. 646–664.
Davies, J. and Imbroscio, D., eds. (2009). *Theories of Urban Politics*, 2nd ed. London: Sage.
Dear, M. (2000). *The Postmodern Urban Condition*. Malden, MA: Blackwell.
Dear, M., ed. (2002). *From Chicago to LA: Making Sense of Urban Theory*. Thousand Oaks, CA: Sage.
Gordon, I. (2004). A Disjointed Dynamo. The Greater South East and Inter-regional Relationships. *New Economy*, 11, pp. 40–44.
Hall, P. and Pain, K., eds. (2006). *The Polycentric Metropolitan Region: Learning from Mega-City Regions in Europe*. London: Earthscan.
Hamel, P., Lustiger-Thaler, H. and Mayer, M., eds. (2000). *Urban Movements in a Globalizing World*. London: Routledge.
Hardt, M. and Negri, A. (2009). *Commonwealth*. Cambridge, MA: Belknap Press of Harvard University Press.
Harvey, D. (2012). *Rebel Cities: From the Right to the City to the Urban Revolution*. London: Verso.
Jacobs, J. (1961). *The Death and Life of Great American Cities: The Failure of Town Planning*. New York: Random House/Harmondsworth: Penguin.
Judge, D., Stoker, G. and Wolman, H., eds. (1995). *Theories of Urban Politics*. London: Sage.
Logan, J. and Molotch, H. (1987). *Urban Fortunes: The Political Economy of Place*. Berkeley: University of California Press.

Massey, D. (1994). *Space, Place and Gender*. Cambridge: Policy Press.
Massey, D. (1999). Cities in the World. In: D. Massey, J. Allen and S. Pile, eds. *City Worlds*. London: Routledge, pp. 99–156.
Massey, D. (2005). *For Space*. London: Sage.
Massey, D. (2007). *World City*. Cambridge: Polity.
Massey, D., Allen, J. and Pile, S., eds. (1999). *City Worlds*. London: Routledge.
McCann, E. (2011). Urban Policy Mobilities and Global Circuits of Knowledge: Towards a Research Agenda. *Annals of the Association of American Geographers*, 101(1), pp. 107–130.
McCann, E. and Ward, K., eds. (2011). *Mobile Urbanism: Cities and Policymaking in the Global Age*. Minneapolis: University of Minnesota Press.
Merrifield, A. (2013). The Urban Question Under Planetary Urbanization. *International Journal of Urban and Regional Research*, 37(3), pp. 909–922.
Merrifield, A. (2014). *The New Urban Question*. London: Pluto.
Neal, S., Bennett, K., Cochrane, A. and Mohan, G. (2013). Living Multiculture: Understanding the New Spatial and Social Relations of Ethnicity and Multiculture in England. *Environment and Planning C: Government and Policy*, 31(2), pp. 308–323.
Robinson, J and Roy, A. (2016). Global Urbanisms and the Nature of Urban Theory. *International Journal of Urban and Regional Research*, 40(1), pp. 181–186.
Rodgers, S., Barnett, C. and Cochrane, A. (2014). Where is Urban Politics? *International Journal of Urban and Regional Research*, 38(5), pp. 1551–1560.
Roy, A. (2009). The 21st-century Metropolis: New Geographies of Theory. *Regional Studies*, 43(6), pp. 819–830.
Roy, A. (2011). Slumdog Cities: Rethinking Subaltern Urbanism. *International Journal of Urban and Regional Research*, 35(2), pp. 223–238.
Roy, A. (2016). Who's Afraid of Postcolonial Theory? *International Journal of Urban and Regional Research*, 40(1), pp. 2000–2009.
Scott, A. and Storper, M. (2015). The Nature of Cities: The Scope and Limits of Urban Theory. *International Journal of Urban and Regional Research*, 39(1), pp. 1–15.
Shaw, K. (2015). The Intelligent Woman's Guide to the Urban Question. *City*, 19(6), pp. 781–800.
Sieverts, T. (2003). *Cities Without Cities: An Interpretation of the Zwischenstadt*. London: Spon.
Simone, A.M. (2004). *For the City Yet to Come. Changing African Life in Four Cities*. Durham, NC: Duke University Press.
Storper, M. and Scott, A. (2016). Current Debates in Urban Theory: A Critical Assessment. *Urban Studies*, 53(6), pp. 1114–1136.
Wacquant, L. (2008). *Urban Outcasts: A Comparative Sociology of Advanced Marginality*. Cambridge: Polity Press.
Walker, R. (2015). Building a Better Theory of the Urban: A Response to 'Towards a New Epistemology of the Urban?' *City*, 19(2–3), pp. 183–191.
Wirth, L. (1938). Urbanism as a Way of Life. *The American Journal of Sociology*, 44, pp. 1–24.
Young, D. and Keil, R. (2014). Locating the Urban In-between: Tracking the Urban Politics of Infrastructure in Toronto. *International Journal of Urban and Regional Research*, 38(5), pp. 1589–1608.

3

PLACE-BASED OR PLACE-POSITIONED?

Framing and making the spaces of urban politics

Deborah G. Martin

Introduction

A famous saying in the United States usually attributed to 'Tip' O'Neill, a former Massachusetts politician and Speaker in the US House of Representatives, is that "all politics is local", and certainly few politics are more local than 'urban politics', which by definition are based in some urban area or another. Politics can mean many things to people, but one way to think about urban politics is to think about how decisions in cities get made that affect all or most residents in one way or another. There are many spaces of politics within cities. 'Local' urban politics are those that shape or affect a particular part of a city, such as a business district or a neighbourhood.

Focusing on different spaces of urban politics, this chapter explores the dynamics of place politics research. Place politics involve explicit and implicit references to geographical norms or grievances in the form of 'place-frames'. Such place politics may seem parochial and limited in potential transformative scope, given the often-restricted scope of urban territory enrolled in any given place conflict. Relational conceptualizations of place, however, which emphasize the spatially complex and disparate networks and processes producing any particular locality (city or neighbourhood), necessitate that considerations of politics reach beyond the urban locality. Attention to the politics of place-framing help to attend to the processes at work in shaping the dynamics of place politics, where the territory of contention, the stakeholders and claims-makers, can all be points of contestation and deliberation. The politics of place, then, are politics over what place, and whose place, and how much place are explicitly enrolled in contention and potential solutions.

Spaces of urban politics

There are at least three different kinds of politics that occur in urban areas and in two types of spaces. First, there are the 'official' decisions made in formal government locations like city hall. In this case, it is usually elected officials who are the key political actors, but appointed and city staff members of agencies also are involved in making policies in cities that affect people locally, such as whether the parks have playground equipment for kids, the traffic flow on city streets, and how often garbage is collected, and by whom (city employees or contracted private companies, for example). A second realm of urban politics, which extends from the first, involves the people who live in cities. These might include residents who come to city hall or write or call to complain or inquire about an issue, such as why there is no crosswalk near a busy park, how much the city charges for parking, or why the police budget is bigger than the school budget (perhaps). Sometimes people come to city hall to speak at a city council meeting or public forum on some issue. These sorts of actions comprise part of the 'formal' realm of government and occur, at least in part, in official settings of urban government, although the decisions that are made influence all sorts of spaces in a city.

A third way that people practise urban politics is more informal, occurring away from, or overtly in opposition to, the formal realms and spaces of government; this latter form falls under popular notions of protest. Protests can be planned marches for some cause, or relatively unplanned and spontaneous gatherings. Protest politics may occur at formal spaces of government, such as city hall, but they occur more frequently in the public open spaces of a city rather than inside government offices. As such, they are more public and informal than the first two types of urban politics. A form of protest that has occurred more and more frequently in some North American cities are calls against police brutality and the seeming indiscriminate killing of black people associated with a 'shoot first' mentality of policing.

In all of these examples, urban politics is mostly about the city as a whole, or focuses on some constituent 'local' part of a city. Yet, the case of police brutality highlights that urban politics sometimes invoke national politics in that they can highlight issues that affect people in cities and elsewhere. But because cities are places where lots of people live together in relatively dense environments, the issues might be particularly evident in urban areas. Cities – or metropolitan areas comprised of cities, suburbs and sometimes even rural areas – are places where difference is manifest, and where access to jobs and services, such as schooling or housing, differentiate people and spaces within an urban area. For example, not everyone fears police, because not everyone is viewed by police in the same way. The relations between urban residents and formal systems of government such as policing are differently experienced; expectations on both sides, those of police and those of everyday individuals, are shaped in part on where they occur and who is involved (Duneier 1999). These differences can be especially stark where there are people in relative close proximity to one another yet experience the same service – policing or transportation, for example – in different ways.

Places of urban politics

Just as cities can be places of difference, so also urban politics vary from place to place across the city. Politics change in different urban spaces, and much of the decision-making and contestations about urban politics have an underlying geography to them. The physical setting of municipal governments, most obviously evident at a place like 'city hall' (Figure 3.1), provides a clear signal in the built environment of the formal, official role of government in providing the resources and structure for daily life in that urban place. The politics that happen at city halls have an order and set of expectations governing their conduct at that specific place.

Likewise, other spaces have an expected order or a set of activities that are more or less regulated: sitting and playing in a park; driving by car along highways; cycling on city streets; walking on sidewalks, etc. When people protest some aspect of city policies or current events, they seek to disrupt the order of spaces such as these so that their actions and importantly, their speech, is noticeable as objecting to something. Tim Cresswell (1996) noted that activism relies upon a day-to-day ordinariness of, and expectations about, the regulation of urban landscapes in order for such regulation to be contested or disrupted. For example, people marching in the street or discursively objecting to auto-oriented transit change the dynamic of the street and

Figure 3.1 Image of City Hall, Worcester, MA
Source: photo: Deborah Martin.

assumptions about transit space (for examples, see Henderson 2015 and Van Neste 2015 in Cidell and Prytherch 2015). It means something to disrupt spaces such as streets, whether physically or discursively because the disruption changes (however temporarily) the space or assumptions about space. Similarly, people camping in a park that usually closes at night are changing the use of that space, a strategy used by the Occupy movement in 2011. By pitching tents in public green spaces such as Zuccotti Park in New York, members of Occupy challenged the day-to-day understandings of, and rules about, how such space was meant to be used.

Urban politics, therefore, can be understood in respect of not only the city examined as a single territorial entity but also the diverse spaces *within* the city. Indeed, an important type of urban politics is the politics *of place* itself, or conflicts about the purpose and physical look of various spaces in the city. When the City of Seattle set out to redesign and rebuild several of its parks, for example, it faced vociferous opposition from youth who objected to the loss of a beloved skateboarding park facility in the planned renovations (Carr 2012). Subsequent to protests by skateboarding groups, the city tried to respond by incorporating new skate park spaces in its renovation plans, but then faced opposition from people who objected to the skateboarding and wanted parks oriented to other sorts of activities such as playground equipment for young children (Carr 2012). Thus, two visions of urban space emerged: one which celebrated skateboarding as a distinctive part of Seattle's youth culture and identity; and one which sought more green space and a quieter and less trafficked city park environment. This particular conflict underlies the idea that different spaces in the city have different meanings for different groups of people; meanings that are signalled by design (layout, physical structure, etc.), by function or by use, and imposed differently by the people who live or spend time in an area.

Place imaginaries can be very powerful in motivating place-based conflict: Mark Purcell (2001) found that residents of part of Los Angeles had a 'suburban ideal' that was contradicted by the reality that some of their neighbours were transforming garages into residences for rental or use by relatives, such as parents-in-law. For these residents, and many people who live in other cities, the ideal of a residential neighbourhood means that the landscape should be dominated by single family homes that are used for residences exclusively, not offices or stores or even uses such as multi-family residential units. The predominant American suburban residential imaginary creates a landscape norm that people may measure landscapes against when looking at their own or other neighbourhoods (Purcell 2001).

Urban politics and place frames

These imagined and actual conflicts over place, which have at their core competing place imaginaries, highlight the role of 'place frames' in urban politics (Martin 2003). In many cities, people who live in the same residential areas often have little else in common; they may speak different languages, practice different religions (or no religion) and have very different household types. Yet, when they live in the same areas, they share spaces in those areas – sidewalks, streets, trees, parks – that foster shared concerns about their neighbourhoods. Shared concerns may include, for example, whether kids and youth in a neighbourhood have safe places to play and interact, whether the streets are clear of garbage, and whether trees are present and healthy along the sidewalks. These concerns may prompt organization among residents and, subsequently, claims for services from the city. Such claims are often articulated in terms of the particular geographical features of the neighbourhood itself, such descriptions of its green spaces, trees, or the characteristics of the residents or the architecture of many of their homes and apartment buildings. I have called these descriptions that reference the people, places and experiences of being in a given neighbourhood place frames because they describe the landscape ideals, or neighbourhood practices, that people seek or want to preserve in their shared spaces (Martin 2003).

Place frames capture a particular kind of urban politics in which the conflict itself is over different place imaginaries; and people collectively act to try to shape the outcome of the place conflict to satisfy their imaginaries and expectations. Residents mobilizing themselves and others to act with and to challenge city agencies for urban development and renewal use place-framing to describe the kind of neighbourhood they want to have (Martin 2003). Such descriptions can highlight elements such as historic housing, politically engaged residents, children looking for better and more parks, while also characterizing some of the threats to safety such as traffic or lack of programmes and jobs for youth in the area. Place frames can also be evident in what some people call NIMBY, or not-in-my-backyard, activism. When residents challenge new land uses that conflict with their ideals of a neighbourhood's residential character, for example, or even residential land uses but which might serve a different group of residents (Duncan and Duncan 2001; Martin 2013a), they use place frames to characterize their place ideals and how they want the place to look and be experienced.

These examples of place frames suggest that they typify a type of urban politics which are exclusively about local neighbourhood conditions. Yet in any given case of neighbourhood politics, the underlying issues are not bound solely in the local place. The imaginaries that drive NIMBY politics are based in much broader ideologies about cities and neighbourhoods, and especially, residential neighbourhoods, as the case in Los Angeles investigated by Mark Purcell (2001) illustrates. Furthermore, it is not only imaginaries that underlie place politics and their discursive place frames. Pierce *et al.* (2011) argue that investigation of place frames can point to and highlight the underlying broader geographies that shape places and engage place imaginaries. These multiple relations highlight that more than being place-based, such urban politics are place-positioned. That is, they situate (or position, via framing) places in relation to imaginaries, and to other places, both explicitly and implicitly. They provide the example of a hospital expansion in an Athens, Georgia, neighbourhood to illustrate the simultaneous local, regional and even global dynamics of place frames and place conflicts.

In the Athens case, a regional hospital seeking to position itself as the premier health care provider in its region sought expansion of its physical site, with plans to increase the hospital building footprint, supporting physical plant and surrounding surface parking (for further details, see Martin 2004). The hospital is located along a major commercial street, but also abuts the primarily single family residential land uses of a neighbourhood of Athens dubbed 'Normaltown', which derives its name from its proximity to the former 'Normal' teaching school, later used as a Navy Supply Corps School and, eventually, as a satellite campus – the Health Sciences campus – of the University of Georgia (also located in Athens). The residents of the areas around the hospital responded in a typically locally defensive style, objecting to the change in character of some of their neighbourhood to a more institutional landscape, and less residential one (Martin 2004). Their place frames offered discourse and imagery focused on the immediate locale, referencing the mostly modest, bungalow-style single-family homes in the area, the importance of pedestrian access, safety and daily practices, and the disproportionate size and scope of the proposed buildings and associated parking in the hospital's proposed plan.

At first glance, the place politics at play were about who gets to define and shape the future production of a neighbourhood. But the questions of shaping (a local) place embed politics of broader spatial scales. The economics of health care in the United States, and in particular in north-east Georgia, shaped the hospital's decision-making and expectations of its board (Pierce *et al.* 2011). The relatively non-urban character of the surrounding region and potential customer base shaped the planning and design of the hospital's expansion, plans that imagined more suburban and rural landscaping and building layout than the more urban, albeit still relatively low-density, yet-pedestrian-friendly, landscape of the neighbourhood. Indeed, the town-feel of the neighbourhood also shaped an expectation of distance from commercial land uses, even as residents criticized the sprawling plan of the hospital's building committee. All of the stakeholders engaged on both sides of the conflict employed discourses about the character of the city of Athens, and drew upon their networks with local and regional politicians, and in some cases, celebrities, to draw attention to and leverage pressure on the other side. The place politics were simultaneously about an approximately one-and-a-half square mile area of a neighbourhood, and about the economics shaping growth in general in the city. The residents pushed to negotiate the smallest detail of the hospital's plan, seeking to preserve every individual house that they could and questioning the large scale of land use change, while the hospital officials made claims about the economic imperatives of growth and the need for Athens to preserve its health care industry. These claims prioritized competing territorial definitions of the conflict, which in turn shape the potential scale of conflict resolution, and the definition of who has power to decide the outcome.

A relational perspective on these place claims, however, would seek to connect even the most local framings to the underlying relations and dynamics that favour such a perspective. For residents of the neighbourhood where the hospital was located, their daily life experiences in the neighbourhood, the layout of the streets, presence of sidewalks and landscape of single family homes were what they prioritized. It is easy to read this framing as only about the local, and not attentive to regional economic needs and forces. Yet as Purcell (2001) noted in his analysis of homeowners in Los Angeles, ideals about residential neighbourhoods do not arise spontaneously within such neighbourhoods. In the United States, a suburban ideal in particular, which prioritizes single family residences, a house with a yard, is pervasive in the popular imaginary (Jackson 1985). Connected to the ideal of a residential neighbourhood with single family homes – even in an urban neighbourhood of Athens, Georgia – there is an underlying economic market which values some residential landscapes more than others. The residents of Athens' Normaltown neighbourhood have literally and figuratively invested in the lifestyle and imaginary that their one- and two-storey bungalow homes, nestled among small businesses and a large well-regarded hospital, enable and foster. They value their homes for the experiences that they have there, but they also value the investment that their homes represent. For them, a changed, larger, more sprawling hospital threatened the balance that enabled the neighbourhood to be thought of as a quiet, well-kept, residential area.

The hospital's references to national trends in the health care industry and the need to grow in order to survive regionally were explicitly relational, and beyond the immediate local. Yet the underlying economic forces and pressures were in fact, represented in both imaginaries and claims. Further, as residents in the conflict pointed out (Martin 2004), the hospital's particular understanding of itself conceptualized the local landscape very differently from the residents – both groups had very specific local imaginaries, or place frames, even as they invoked broader processes differently (or unevenly). For the hospital, the design needs focused on creating a 'campus' which had easy automobile access, large amounts of green space and engineered infrastructure – such as retaining water ponds – which tended to view the hospital's site in isolation from the surrounding built environment. In other words, compatibility and aesthetic integration with the surrounding low-density and small-scale residential and retail landscapes simply were not part of the place imaginary in the initial hospital designs. The hospital, in its own way, had a very local, single-campus site mentality in its plans. The most explicit evidence of this single-site orientation, rather than one attending to the larger neighbourhood built environment, was the plan to build a storm water retention pond (on land where houses would have been razed to clear it). In the conflict, residents had opposed such an approach to storm water management as primarily a rural land use approach, rather than compatible with urban-oriented infrastructure (Soto 1999).

The urban politics in this case involved a host of variables, then, around who gets to define a site or place, and how that territory is imagined, and by whom. In the hospital–neighbourhood conflict, the principal self-defined stakeholders were the local residents, and the board of directors and management of the hospital. At the same time, local politicians and other leaders took sides, alternatively debating the economic imperatives of the region having a top-notch medical facility, and of the city to shape its neighbourhoods and built environment according to competing visions of livability. This case was ultimately an explicitly tangible, material, landscape-oriented conflict, but one with other ideals about urban space and the broader urban and regional economy, embedded within it.

Place frames and social norms

Many cases of urban politics engage explicit debates about the urban landscape, and what land uses are appropriate in which locations. Underlying these important place imaginaries and tangible landscape experiences, however, are also pernicious social contestations and tensions. Some urban politics engage more explicitly social relations, where the place politics are not so clear, despite being oriented around land use disputes. In my research in Massachusetts, for example (Martin 2013a, 2013b), I found that conflicts around the siting of social service agencies highlighted the fact that many people prefer to think of addiction treatment or other life skills training as incompatible land uses within their residential neighbourhoods. Yet, for social service agencies and their clients, residential neighbourhoods are often the ideal locations for group homes for people struggling to overcome addiction or other life challenges which have sometimes led to or compounded problems of homelessness. These differing opinions about appropriate residential land uses are quintessential NIMBY disputes, and they centre quite explicitly on place, often expressed overtly very materially through land use.

Underlying the social discomforts of NIMBY are very much the same sort of place imaginaries as those which motivated the residents opposed to the hospital expansion in Athens that I have discussed above; namely, a place imaginary about residential neighbourhoods. But where the conflict with the hospital pitted a large institution with a significant physical infrastructure against mostly single family homes on modest yards, group-home-based social services occur in the same sort of single-family residences as typically exist in many American residential neighbourhoods. The physical infrastructure is not significantly different, and the underlying discomforts which generate a 'not here' type of reaction are related to social norms and perceptions. A group home might have more cars in the driveway at night, and a larger amount of garbage cans than the houses nearby on trash collection day, but the difference between it and neighbouring land uses isn't really in the look of the building or how it fits on the street. The core difference between a social service residence and any other residence is the imputed differences in the residents themselves – differences imputed by the people who object to the social services.

Opposition to residential social services highlights the exclusionary aspect of any place imaginary; notions about what goes where also embed and draw upon ideas about who goes where (Cresswell 1996). As much as social exclusion may not be an intentional goal of someone engaged in a land use conflict, the underlying discomforts and concerns that drive the conflict are about difference, whether in land use or people, or both. One reason people worry about difference is the reliance that many homeowners have on the economic value of their property. In the United States, homeownership functions as a source of financial value as well as being a site for living; in other words, as a site for both exchange and use value (Logan and Molotch 1987). Since NIMBY is about the place politics of land use, it is also about expectations of reliability and predictability in land use that inhere in property values over time. When people express concerns about new or different land uses in their neighbourhoods, they sometimes explicitly reference property value. In the case of social services in Worcester and Framingham, Massachusetts, for example, residents explicitly identified concerns about a decline in their property values if social service group homes were located nearby (Martin 2013b).

As a result of the intensely local focus on property value and compatible or incompatible land uses for a particular area, NIMBY politics seem narrowly constructed around neighbourhood. But like the case of the hospital expansion in Athens, Georgia, social conflicts about place and belonging connect much broader processes and tensions to any 'place-based' concern. The reliance of individual households on their investments in housing situates them in the context of a broader set of social relations in which homeownership is a primary means of investment in the United States. Such a position in the economy – as a property owner – simultaneously situates at least a major part of their economic outlook at a local scale; homeowners attend to the very local land use dynamics of their neighbourhood because that is the setting of their property investment. At the same time, however, broader social and economic forces shape the process by which the combined challenges of addiction and homelessness are met through social service group housing – at least, in some states where laws and funding structures, such as those in Massachusetts, make such social services practical and feasible (Pierce and Martin 2017). A host of very complicated, multi-sited and multi-scalar processes combine to enmesh individuals into drug or alcohol reliance and addiction, while a web of government and non-profit agencies exist to assist families and individuals with such personal, health and social challenges. These individuals, their problems and services to address them extend far beyond any one local or locally situated process, and yet, in the situation of a social service siting, they meet and manifest in the form of a local land use conflict.

Conclusions

The broader social, political and economic forces that occur across a wide range of places and neighbourhoods rarely enter into the debate about social service siting, except sometimes in the explicit call to elsewhere that the 'not-here' (put it someplace else) politics of NIMBY invokes. Urban politics, then, are always constituted through multiple places and spatial processes, and thus are really more 'place-positioned' than place-based.

Place-positioned politics, on the other hand, are urban politics that invoke or occur in specific sites, yet cannot truly be anchored to or fully solved within those sites, because the processes that produced them extend geographically within, beyond and through the places of the conflict. In their negotiation and resolutions, however, they often sit, fixed – however temporarily – into place-based situations in which they were made manifest through explicit conflict. The place-based conflicts of a neighbourhood where a hospital will, through one plan or another, expand its physical footprint, or where a social service will find a house in which to provide life skills and support to its resident-clients, are always place-positioned in particular sites, while simultaneously drawing upon relations and processes beyond those sites. Indeed, they draw upon formal and informal spaces of urban politics within the localities in which they are positioned: the formal sites of government; the informal spaces of negotiation; and the landscapes, buildings and settings of place-based conflicts. They may also draw upon, use or invoke other spaces, such as place imaginaries or material sites near to and far from the place of conflict.

Urban politics, then, invoke and engage multiple spaces. Conceptualizing place-based politics as place-positioned suggests a need to unravel the various and complicated spatial relations and processes that helped to situate a particular conflict in space, to expose multiple underlying spatialities. In doing so, alternative sorts of conflicts, relations and potential politics may emerge to form new sorts of urban politics, which can make explicit the place-positioning – or indeed, place-framing – underlying these old and new politics.

References

Carr, J. (2012). Public Input/Elite Privilege: The Use of Participatory Planning to Reinforce Urban Geographies of Power in Seattle. *Urban Geography*, 33, pp. 420–441.
Cidell, J. and Prytherch, D., eds. (2015). *Transport, Mobility and the Production of Urban Space*. New York: Routledge.
Cresswell, T. (1996). *In Place/Out of Place: Geography, Ideology and Transgression*. Minneapolis: University of Minnesota Press.
Duncan, J. and Duncan, N. (2001). Sense of Place as a Positional Good: Locating Bedford in Space and Time. In: P. Adams, S. Hoelscher and K. Till, eds. *Textures of Place: Exploring Humanist Geographies*. Minneapolis: University of Minnesota Press, pp. 41–54.
Duneier, M. (1999). *Sidewalk*. New York: Farrar, Straus and Giroux.
Henderson, J. (2015). From Climate Fight to Street Fight: The Politics of Mobility and the Right to the City. In: J. Cidell and D. Prytherch, eds. *Transport, Mobility and the Production of Urban Space*. New York: Routledge, pp. 154–179.
Jackson, K. (1985). *Crabgrass Frontier: The Suburbanization of the United States*. New York: Oxford University Press.

Logan, J. and Molotch, H. (1987). *Urban Fortunes: The Political Economy of Place*. Berkeley and Los Angeles: University of California Press.

Martin, D.G. (2003). 'Place-framing' as Place-making: Constituting a Neighborhood for Organizing and Activism. *Annals of the Association of American Geographers*, 93, pp. 730–750.

Martin, D.G. (2004). Reconstructing Urban Politics: Neighborhood Activism in Land Use Change. *Urban Affairs Review*, 39, pp. 589–612.

Martin, D.G. (2013a). Place Frames: Analyzing Practice and Production of Place in Contentious Politics. In: W. Nicholls, J. Beaumont and B. Miller, eds. *Spaces of Contention: Places, Scales and Networks of Social Movements*. Farnham: Ashgate, pp. 85–99.

Martin, D.G. (2013b). Up Against the Law: Legal Structuring of Political Opportunities in Neighborhood Opposition to Group Home Siting in Massachusetts. *Urban Geography*, 34, pp. 523–540.

Pierce, J. and Martin, D. (2017). The Law is Not Enough: Seeking the Theoretical 'Frontier of Urban Justice' Via Legal Tools. *Urban Studies*, 54(2), pp. 456–465.

Pierce, J., Martin, D.G. and Murphy, J. (2011). Relational Place-making: The Networked Politics of Place. *Transactions of the Institute of British Geographers*, 36, pp. 54–70.

Purcell, M. (2001). Neighborhood Activism Among Homeowners as a Politics of Space. *The Professional Geographer*, 53, pp. 178–194.

Soto, J. (1999). ARMC Plan Opposed: Residents Continue Fight Against Expansion. *Athens Banner-Herald*, 29 April. Available at: http://onlineathens.com/stories/042999/new_0429990001.shtml#.Vz3YaORRRb0 [accessed 19 May 2016].

Van Neste, S. (2015). Place-framing and the Regulation of Mobility Flows in Metropolitan 'In-betweens'. In: J. Cidell and D. Prytherch, eds. *Transport, Mobility and the Production of Urban Space*. New York: Routledge, pp. 245–262.

4
AMBIVALENCE OF THE URBAN COMMONS

Theresa Enright and Ugo Rossi

Introduction

The notion of the commons or, as some authors put it, the common (singular) is at the centre of recent debates concerned with how societies manage natural and human resources. Specifically, the commons is conceived in contrast to the profit-driven arrangements of marketization and privatization which are hegemonic within contemporary neoliberalized societies. This chapter will interrogate the relevance of this notion from the perspective of urban studies and critical theory, showing how its different and even contrasting meanings are illustrative of the political, cultural and economic intricacies of capitalist societies in (late) neoliberal times. Our analysis considers the commons as a contested terrain in contemporary urban political economy: as a site of experimentation with post-capitalist cooperative relations; as a site of an anti-capitalist practice of resistance; and/or as a site of capitalist re-appropriation. This chapter will argue that the politics of the urban commons sheds light not only on multiple and even competing understandings and uses of the notion of the common(s), but also on the more general ambivalence of contemporary capitalism in its urban manifestation (Virno 1996).

In conceptual terms, we identify two main approaches to the commons: (1) a neo-institutionalist strand inspired by the seminal work of Elinor Ostrom, which has been very influential within the public sphere and within mainstream collaborative economies; (2) a neo-Marxist strand, looking at the both the defence of the commons against iterated processes of 'accumulation by dispossession' (Harvey 2003) and the production of alternative communal economies outside of capitalism as such. Across each of these formulations, today's capitalist city acts as an important mediator of economic, political, social and ecological dynamics. Notably, contemporary cities and the spaces within offer a rich repertoire of practices illustrative of the co-existence of capitalist appropriations of common goods, on the one hand, and commons-oriented grassroots initiatives, on the other hand. Moreover, the interest in cities as privileged sites of the politics of the common(s) stems from an idea of the urban as a source of a substantially plural, democratic, open and collective way of 'seeing' (Magnusson 2011) that dovetails with the normative project of producing robust and just common life.

An analysis of the variegated politics of the commons reveals a stratification of city life that enables us to better locate the urban political (Enright and Rossi, 2018) – that is, the city understood as an ontological machine producing subjectivities and forms of life within biopolitical capitalism (Rossi, 2017) – in what we call "the ambivalent spaces of late neoliberalism" (Enright and Rossi forthcoming), where the adjective 'late' refers to the crisis-prone character of contemporary capitalism. In this context, we observe how anti-capitalist movements, grassroots communities and neo-capitalist forces differently commit to performing and imagining new worlds, mobilizing the idea of the commons. Through such an analysis a number of important questions and points of controversy come to the fore. Notably: What is the commons? Is the commons a particular type of 'thing' or 'good' or is it a discourse, a cultural dispositif that can be appropriated for different and even opposing purposes? Is there one common or many commons? Through what institutional procedures are the commons instituted? As a discursive formation increasingly appropriated by market forces does the commons exhaust its more revolutionary meanings and interpretations?

This chapter elaborates these inquiries, drawing out their implications for understanding spaces of urban politics as follows: in the first section, we will provide an overview of the different theorizations of the commons; the second section looks at the growing centrality of cities and the urban within current debates on the commons; the third section looks more empirically at the urban commons, highlighting its ambivalent uses within contemporary capitalist cities. We conclude with final remarks on the commons as a crucial terrain of political and intellectual struggle in the current urban age.

Theorizing the common(s)

Within academic debates, the publication of Elinor Ostrom's *Governing the Commons* in 1990 sparked a renewed interest in the notion of common goods (Ostrom 1990). In this book, Ostrom criticized the individualistic understanding of common property originally theorized by Garrett Hardin in his controversial essay *The Tragedy of the Commons*, published in *Science* in 1968 (Hardin 1968). In particular, Ostrom repudiated Hardin's claims associating the failure of the commons and the degradation of the environment with the fact that self-interested individuals inevitably compete for collectively held scarce resources. Ostrom also critically engaged with Mancur Olson's utilitarian conception – spelled out in his influential *The Logic of Collective Action* (Olson 1965) – that in the absence of specific obligations rational and utility-maximizing individuals tend not to pursue group interests. Against Hardin and Olson's contention that natural common property resources (CPR) such as land, air and minerals cannot be effectively used and managed based on the fulfilment of social needs, Ostrom contends that such resources are, in fact, often organized through networks of mutual aid and solidarity and through non-market arrangements of democratic rule.

In explicit contrast to the individualistic understandings of CPR which had been widely adopted as foundations for public policy, Ostrom centred her theorization on processes of self-organization and self-governance in the management of a "natural or man-made common-pool resource-system" (Ostrom 1990: 30), understanding them as an alternative to both the market and the state. In her view, both 'privatizers' and 'centralizers' fail to take adequately into account the role of intangible institutions such as networks of information and reciprocity, which characterize cooperative forms of human interaction (Ostrom 1990). Ostrom explicitly positioned herself within the wider neo-institutionalist interest in ideas of social capital, trustworthiness and community which had gained wide currency within the social sciences in the 1990s (Fukuyama 1995; Putnam 2000). In Ostrom's view, theories of social capital provide clues to what she defines "second-generation theories of collective action" (Ostrom and Ahn 2003: 2), contesting the assumption in both state- and market-centred policy prescriptions that rational, atomized and selfish individuals are best organized through impersonal and homogenous structures. On the contrary, Ostrom and Ahn argued that:

Self-governing systems in any arena of social interactions tend to be more efficient and stable not because of any magical effects of grassroots participation itself but because of the social capital in the form of effective working rules those systems are more likely to develop and preserve, the networks that the participants have created, and the norms they have adopted.

(2003: 11)

The award of the Nobel Prize in Economic Sciences to Elinor Ostrom in 2009 (shared with Oscar Williamson, another prominent figure in neo-institutionalist economics) further increased the popularity of the concept of the commons and ushered it into the social science mainstream.

A second influential contribution to the theorization of the commons – or, in this instance, the common – is found in Hardt and Negri's (2009) *Commonwealth*. Their point of departure is similar to that of Elinor Ostrom, namely the contestation of "the seemingly exclusive alternative between the private and the public" and, as they elaborate, the "equally pernicious political alternative between capitalism and socialism" (Hardt and Negri 2009: ix). For Hardt and Negri, however, unlike Ostrom, the "political project of instituting the common" cannot be contained within existing institutional apparatuses, but is linked with the coming into being of a communist future (2009: ix). For these thinkers, the common is not something to be inserted into liberal republican structures, but is tied etymologically and historically to the political vocabulary of communism. As they note, "what the private is to capitalism and what the public is to socialism, the common is to communism" (2009: 273).

The radical orientation of Hardt and Negri's work is linked to their understanding of the common as having a key function within capitalist accumulation. Whereas Ostrom's work emerged from the challenges of accounting for the role of civil-society associations, local governments and informal actors in resource management within a post-Fordist and increasingly interdependent world, Hardt and Negri base their understanding of the common on a heterodox Marxist interpretation of contemporary knowledge-based capitalism. In doing so, these authors criticize any inert conception of static common goods, arguing that contemporary capitalism relies on communal forms of value production, namely on the exploitation of what they call "biopolitical labour": the vast array of knowledge, affects and social relations that are external to capital but that capital appropriates (Hardt and Negri 2009: 133–142). In addition to the earth and natural common, then, commonwealth also and even primarily refers to collectively produced and used human resources such as ideas, language, information and affects. For Hardt and Negri (2009: 139), "this form of the common does not lend itself to a logic of scarcity as does the first". The creation and management of the common – as an always excessive product – defines accumulation processes and class struggle today. As such, the common is part of a new political vocabulary appropriate to the decentred sovereignty of Empire and its plural forms of multitudinous resistance.

In the strand of theorization inspired by politically engaged Marxist positions, the commons is understood primarily in relation to collective struggles. This stands in marked contrast to the kinds of collaborative practices of maintenance found in the institutionalist scholarship, which is influenced by the work of Elinor Ostrom. For their part, feminists have offered a more radical version of the pragmatic understanding of the commons offered by the institutionalists. For example, Silvia Federici draws attention on issues relating to the reproduction of everyday life, customarily overlooked in major strands of Marxist theory, looking at women's efforts to collective reproductive labour as a way to protect each other from poverty and the violence of both the state and men (Federici 2010). In her study of a manufactured housing community,

Elsa Noterman (2016) has put forward the notion of 'differential commoning', where the recognition of difference among members of communities managing collective resources is a way of ensuring greater resilience for these communities.

Another important source of inspiration for radical feminist scholarship on grassroots economic practices as privileged sites of the commons is Gibson-Graham's diverse economies research. In their work non-capitalist forms of subjectivity, practice and politics exist alongside and often in conjunction with dominant capitalist dynamics (Gibson-Graham 1996, 2006). Their project to chronicle and imagine economic alternatives and experiments attempts to foment regimes of accumulation that entail solidaristic modes of organizing collective life: from cooperative enterprises to environmental and agricultural organizations; and from local currencies and non-profits to informal markets. In Gibson-Graham's view, identifying alternative economies is part of a post-capitalist project aimed at taking on the task of "how we might perform new economic worlds, starting with an ontology of economic difference" (2006: 3).

The commons, however, is not a merely academic abstraction; it has been widely used and discussed not only by institutionalist social scientists and radical scholars but by social activists and political militants alike. Indeed, beyond academia, a significant impetus for a return to the commons has arisen from movements themselves. As Silvia Federici (2010) points out, the commons has taken on new life as a political concept at least since the Zapatista uprisings in 1994 and has been elaborated more fully through the new era of resistance against neoliberal practices around the world. From open source software promotion, to landless peasants' movements, to the conservation and defence of natural resources, to anti-eviction campaigns, the commons fits a non-statist model of contentious politics and struggle against what David Harvey (2003) calls "accumulation by dispossession".

Along these lines, Pierre Dardot and Christian Laval also conceive the common – like Hardt and Negri they favour the singular form – as an element of political praxis, and as "a principle of action" (Dardot and Laval 2014: 168) emerging from social movements resisting the neoliberalization of contemporary societies. In their book *Commun: essai sur la revolution au XXIème siècle*, Dardot and Laval build on their earlier analysis of neoliberal governmentality (Dardot and Laval 2010) to elaborate an institutional understanding of the common as a set of activities and liberation practices. For these authors, the common cannot be conceived as a pre-existing object or good, but is rather the dynamic relationship between a thing (e.g. a resource, a place, a value) and its communal institutions of management (Dardot and Laval 2010). The common in this sense is fundamentally rooted in praxis: it is not a fixed entity but a political principle on the basis of which we must construct collective goods, defend them and extend them. The common is thus another name for the shared activity of co-responsibility, reciprocity, solidarity and democracy.

While the feminists draw attention on the differential situatedness of commoning processes, radical historians suggest embracing a trans-historical perspective. In his influential *Magna Carta Manifesto*, Peter Linebaugh (2008: 19) speaks of a "millennium of privatization, enclosure and utilitarianism", in which the current phase of neoliberal globalization appears to be the culmination of capitalism's historical tendency to treat everything as a commodity. In doing so, Linebaugh reconnects the spirit of the Magna Carta of 1215 to social struggles across the world spanning over a millennium: from the neo-Zapatista insurgence in Chiapas in the mid-1990s claiming the rights of campesinos to the Nigerian women seizing the Chevron Oil Terminal preventing them from obtaining wood and water from the forest in 2003; from the native Americans dispossessed of their common resources by the conservation movement of the 1880s to the national upsurge of 1919–1920 in India against the colonial government of Britain taking over the community lands or to the campaigns contrasting the incessant enclosure movement in the Amazon in Brazil from the 1960s until today. Like Harvey, Linebaugh and other libertarian Marxists think that contemporary neoliberal capitalism has stabilized forms of accumulation conventionally associated with the primitive stages of capitalist development: land encroachment; extraction of natural resources (particularly gas and petroleum products); and displacement of peasant populations. In this perspective, social struggles, on the one hand, resist this perpetuation of primitive accumulation over time; on the other hand, in their productive dimension, they constantly renew the sense of the commons through the redefinition of norms, values and measures of things being held in common, thus re-creating an alternative realm in which material and social life can be re-produced outside capital (De Angelis 2007). In transitional times like those following the crisis of 2008, these processes can be understood as a "repair or replacement of broken infrastructure" extending the limits of sociality beyond received modes of belonging (Berlant 2016: 393).

This process of relentless (re)production of value around the commons is not limited to anti-capitalist movements and grassroots communities of commoners, but has reached also the dominant rationality of government: neoliberal regimes have started recognizing the economic value of the commons, allowing a re-appropriation of the commons from within capitalism. In both its 'roll back' and 'roll out' forms (Peck and Tickell 2002) neoliberal reason entails not only the corporatization of governance and the privatization of the public sector, but also a larger entrepreneurialization of society and the self (Gordon 1991) predicated on the increasingly unfettered liberalization of capitalist economies. In this context, the commons is mobilized as part of a new global imaginary of enterprise society (Lazzarato 2009) in which everyone is being asked to embark on purportedly community-oriented business experiments as a way to compensate the shrinkage of labour markets and the impoverishment of the middle class. The contemporary capitalist city is a key site for observing this multifaceted, ultimately contradictory and ambivalent, mobilization of the commons.

Urbanizing the commons

As we have seen, the original formulation of the commons within the social sciences has referred largely to non-urbanized environments: from Elinor Ostrom famously basing her analysis on forest preservation and fisheries to critics of neoliberalism focusing on acts of enclosure typically involving rural areas. Marx himself – whose thoughts about property and common wealth are generally known for being city-centred – was inspired by the rural phenomenon of woodstealing and its widespread prosecution in Prussia and other pre-German states (Bensaïd 2007; Linebaugh 1976). Recent treatments of progressive commoning also frequently see the production of the commons as a process of 'rurbanization' (Federici 2010), through, for example, the regeneration of swidden forests (Tsing 2005) or the planting of foodstuffs in abandoned urban landscapes (Tornaghi 2014). Amanda Huron (2015) relates the implicit anti-urban ideology within conceptualizations of the commons to the fact that urban environments are conventionally regarded as places bringing strangers together, thus originating an experience of fear and suspicion rather than of community and solidarity. In recent times, this anti-urban prejudice has been fostered by resurrecting neo-ruralist ideologies of authenticity that have gained new ground after the economic-*cum*-urban crisis of 2008. Despite this bias, recent years have seen a lively body of literature investigating the commons through urban lenses (Bresnihan and Byrne 2015; Kohn 2016; Noterman 2016). Why is this happening? And how might it help to reshape our thinking about spaces of urban politics and the urban political?

Significantly, the crisis of 2008 and its aftermath have reignited interest in the geographies of social justice and inequality, and in cities as strategic spaces to observe the conflictual dynamics of capitalism, particularly those rooted in the financialized housing market. This urban perspective, moreover, has opened new vantage points from which to view struggles over the commons.

The recognition of the 'urban roots' (Harvey 2012, chapter 2) of the crisis is the most obvious response to this question, but there are two other sets of explanations for this urbanization of the commons. First, and well before the housing and financial crisis of 2008, general scholarship in human geography and other critical social sciences had insisted on how neoliberalism should be understood as an inherently urban phenomenon. This understanding is founded on Henri Lefebvre's and David Harvey's analyses of the secondary circuit of real estate capital, which highlight how the exploitation of the built environment, as well as the production of urban space more generally, play central roles in the dynamics of capitalism. The appropriation of rent, in other words, which is a major form of value production today, is fundamentally about the creation and appropriation of the commons. Moreover, as a result of post-industrial restructurings, cities have been at the forefront of the process of entrepreneurialization of societal governance, shedding light on the structural dimension of the privatization of collective-consumption goods, particularly of public services such as education, water supply, transit, waste disposal. These privatizations have particularly intensified through the recurring rounds of austerity urbanism (Peck 2012). In this perspective, urban environments are viewed as contra-cyclical regulators of economic development in a context of capitalism, particularly through the exploitation of the rent gap in the built environment, the privatization of public services, the expansion of consumerism and the commodification of social relations. While Garrett Hardin conceived the act of enclosure and private management of common resources in a positive sense, on the contrary critics of neoliberalism, and particularly of its urban form, see enclosure as instrumental in the reproduction of capitalist forms of privatization, leading to the dismantling of the welfare state and the displacement and fragmentation of the working class (Jeffrey et al. 2012). In this perspective, struggles over the commons are viewed as a response to the processes of capitalist expansion, particularly within the framework of contemporary 'planetary urbanization' (Brenner and Schmid 2014). Within inner-city areas, protection of the commons – to the point of forming defensive enclaves – is a means to escape the commodification of society and the annihilation of social and civil rights (Bresnihan and Byrne 2015). This is especially pertinent for low-income groups and racial minorities that are disproportionately affected by neoliberal urbanisms.

Second, it is not only the relationship of mutual dependence between cities and neoliberalism that requires attention to the urban manifestation of the commons. The recent post-recession transition in the United States and elsewhere in the Western world has seen cities becoming increasingly central to knowledge-intensive and tech-driven capitalism. Urban economies are being deeply reshaped by the advent of a wide range of experiential economies, in which conventional boundaries between production and consumption, between labour time and leisure time, are increasingly blurred. As a result, a significant part of today's interest in the commons looks at these technology- and life-oriented economies, particularly within the Global North but increasingly also in other regions of the planet. In the heterodox Marxist perspective of Hardt and Negri, urban environments – particularly in knowledge-intensive economies characterized by dense relations of cooperation – provide capitalist production with an "artificial common" created by affective labour:

> In the biopolitical economy, there is an increasingly intense and direct relation between the production process and the common that constitutes the city. The city, of course, is not just a built environment consisting of buildings and streets and subways and

parks and waste systems and communications cables but also a living dynamic of cultural practices, intellectual circuits, affective networks, and social institutions.

(Hardt and Negri 2009: 153–154)

Third, there is a resurging interest in cities as informal sites of institutional experimentation, which has significant convergences with the notion of the commons. This perspective is particularly associated with the work of urban scholars looking at the social fabric of cities in the Global South (e.g. Robinson 2006; Roy 2009). In this scholarship, the experience of urban informality, which was once merely associated with a condition of exclusion, is also seen to be generative of mobile and provisional infrastructures of collaboration (see also Simone 2004). In a more explicitly vitalist and neo-humanist vein, marginalized urban environments are also celebrated as sites of community and solidarity through the "making of micro-collectives of the poor around shared infrastructures" (Amin 2014: 157). The same spaces stimulate the fantasies of a wide range of actors such as civil-society organizations, government agencies, multinationals, aiming to exploit the commonwealth of knowledge and creativity attached to the idea of the 'entrepreneurial slum' (McFarlane 2012).

Thus, within dominant capitalist countries the crisis of the late 2000s and its aftermath have particularly illuminated the dual role of cities as sites of capitalist contradiction, due to the concentration of financialized real estate, and also as central spaces within post-recession trajectories of austerity and economic regeneration. At the same time, the increasingly more advanced globalization of the urban phenomenon has started provincializing conventional understandings of the commons as a Western grassroots politics of experimentation, while it has attracted the attention of pro-market forces interested in extracting the cognitive capital of socially dense urban spaces like the slums of Southern megacities. It is no surprise, therefore, that in this historical conjuncture cities have acquired a growing centrality within contemporary debates over the commons. In brief, these different perspectives share an understanding of the urban as a space where an affirmative production of the commons arises from affective relations of cooperation within a both capitalist and non-capitalist framework. While critics of urban neoliberalism direct attention towards the dispossessing logic behind extended forms of urbanization such as the so-called 'planetary urbanization', scholars with a poststructuralist or post-colonial emphasis tend to highlight the production of cooperative forms of life associated with different expressions of urban social practice.

The multiple valences of the commons

The original rehabilitation of the commons was premised on the idea that human societies are institutionally dense environments in which relations of trustworthiness and knowledge exchange allow the optimization of 'common-pool resources' (Ostrom 1990). This view was particularly influential in the 1990s and the early 2000s within the framework of communitarian and neo-institutionalist conceptualizations of associational forms of economic governance drawing on social capital, trust, face-to-face interactions and other informal institutions and conventions (Cooke and Morgan 1998). In this context, the rediscovery of the local scale reinvigorated long-standing ideals of associative democracy, which advocate the centrality of self-governing associations performing public functions as a response to the failure of both the state and the market (Hirst 1994). Anthony Giddens's 'Third Way' and its adoption by the New Labour, asserting the need to provide socially responsible answers to the challenges of the 'new capitalism', constituted the mainstream version of a larger academic and political stream that sought out a new route between the Keynesian social state and market-driven economic reform (Giddens 1998).

The financial crash of 2008 and Elinor Ostrom's receipt of the Nobel Prize in 2009 had the combined effect of re-creating a legitimate discursive space for associationalist tendencies that had been long present within public debates but had always remained peripheral. An unintended outcome of this ascendancy was the fact that commons-oriented initiatives started to be increasingly incorporated into mainstream urban economic development. Even in the context of intensifying global capitalism, a wide range of community-based collaborative experiments such as housing cooperatives, citizens' networks, solidarity purchase groups and other experiments of local community management and co-ownership have persisted and even proliferated in recent times (see especially Imbroscio 2010). In addition to the more traditional arena of the workplace, housing and food are key sectors in which cooperative economies have taken shape. In some cases, these experiments have become strongly institutionalized, such as the co-housing projects adopted in Great Britain within the framework of the 'sustainable communities' initiative embraced by the New Labour government since 2003 (Williams 2005). In other cases, co-housing projects have resulted from the collaboration between different non-state actors, as has been observed in Tokyo, in Japan, where non-profit organizations have established partnerships with real estate developers and other for-profit housing firms (Fromm 2012). However, co-housing projects still arise from genuinely citizen-led initiatives of cooperation beyond a market logic, some drawing inspiration from the pioneering projects of collective housing in the Scandinavian countries, and in Holland and Germany in the 1970s and the 1980s based on the philosophy of self-work (Vestbro 2000). In recent years, projects of collective self-organization – variously called 'purpose-built cohousing' (Jarvis 2011), 'limited-equity housing cooperatives' (Huron 2015) and the like – have acquired new significance against the background of the housing crisis affecting cities in Europe and North America. In the domain of food, Italy – which has been at the forefront in the slow food movement that has sparked critical consciousness about the risks associated with mainstream food business at the global level – has seen the proliferation of self-organized solidarity purchase groups in cities and towns across the country (Grasseni 2014).

These collaborative initiatives are intended to foster local social capital and civic participation within close-knit social groups at the micro-scale of the neighbourhood. However, there is often a characteristic ambivalence in these initiatives, particularly in those taking place within affluent societal contexts. On the one hand, they are intended to nurture a collaborative ethos within a revitalized civil society in which the 'active citizen' contributes to the regeneration of the sense of belonging to the urban community through involvement in neighbourhood-based social projects (Amin 2005). On the other hand, this active citizenship formally endorses diversity, but in reality encourages the deployment of disciplinary measures aimed at monitoring the moral and physical integrity of the members of the community (Raco 2007). Albeit rife with contradictions, as they are also nourished by a 'genuine' commitment to social interaction at the neighbourhood level, these phenomena shed light on the increasingly moral characterization of urban government within advanced liberal societies. In this context, classic goals of socio-economic emancipation and justice disappear from the official urban-policy agenda and solidarity, reciprocity and mutuality are replaced by facile references to trust and mutual benefit. The institutionalist discourse around the commons is deeply informed by the 'moral turn' that has characterized contemporary cities and their public realms after the advent of neoliberal governmentalities.

The high-tech boom witnessed by several cities in the United States and across the world in the aftermath of the 2008 crisis has capitalized on this moralization of city life and capitalist societies more generally. Technology-based, highly commodified 'sharing economies' have subsumed ideas of collaboration that were once a distinctive trait of grassroots initiatives. Jeremy

Rifkin, for example, a lauded business guru for his visionary thinking on the third industrial revolution, has drawn on the work of Elinor Ostrom prophesying the advent of a 'zero marginal cost society' in which the collaborative commons created within the Internet (e.g. Linux, Wikipedia, Napster, YouTube) are presented as inspirational models for a larger re-configuration of social relations based on gratuitous exchange (Rifkin 2014). Other examples of so-called social enterprises (philanthropic or for-profit) mobilizing a commons-oriented discourse can be typically found in the privatized housing sector. In this highly exploited but also deeply culturalized sector where new lifestyles are being experimented, not only traditional real-estate developers but also technology companies have invested in recent years. Amongst other internet-based offerings, start-up technology companies have set up charming for-profit co-living spaces in trendy neighbourhoods such as Williamsburg and Crown Heights in Brooklyn, New York City, but also in less established areas like Spring Garden in Philadelphia and North Acton in West London (Kasperkevic 2016; Kaysen 2015). One of these companies explicitly appropriates the notion of the common in its brand, promising to offer "shared housing for those who live life in common" (www.hicommon.com/ [accessed 26 May 2017]; see also Cutler 2015). The communication style of this company closely reproduces that of Airbnb, a champion of the home sharing economy that is being increasingly criticized within the wider public owing to its unfair trade practices and its gentrification effects on urban environments – both of which contribute to housing scarcity and unaffordability. Another company has recently launched a co-living business called 'The Collective', headquartered in London, a city known for its overheated housing market (www.thecollective.co.uk/ [accessed 26 May 2017]). These technology-based 'social enterprises' which trade in the language of the collective provide a powerful illustration of what Nicole Aschoff has recently termed the "new prophets of capital": a new generation of "elite storytellers" promoting "solutions to society's problems that can be found within the logic of existing profit-driven structures of production and consumption" and that "reinforce the logic and structure of accumulation" (Aschoff 2015: 11–12).

The incorporation of communal forms of life within the hegemonic discourse on capitalist innovation is symptomatic of the risk of normalization threatening the commons understood as a political ideal and a social practice. Far from being accidental, this normalization is inscribed within the general tendency of capitalism towards the real subsumption of society, in Hardt and Negri's terms, or the commodification of social relations and the gentrification of the urban experience, in more classic Marxist terms.

Conclusion

This chapter started by offering an overview of the differentiated reception of the concept of the commons, showing its multifaceted nature and its applicability to different perspectives and modes of action. Eliner Ostrom's take on the commons, building on a neo-institutionalist framework of analysis, proves attentive to the ways in which contemporary capitalism becomes embedded within collaborative social relations and institutional arrangements. Marxist-oriented approaches interpret the politics of the commons in either 'negative' or 'affirmative' ways: as a response to the colonizing logic of dispossession prevailing in contemporary neoliberal globalization; or as a post-capitalist potential immanent in the biopolitical production of affects, knowledge and relationality within the metropolis. Finally, feminist authors draw attention on practices of commoning as a way of acknowledging the value of difference as a solution to the 'tragedy of the commons', while urban scholars studying cities in the South invite to pay attention to the production of mutable and provisional commons under socially thick conditions of urban informality.

Against the background of this plurality of conceptual sensibilities, this chapter has shown how we should refute essentialistic understandings of the commons. Rather the commons is a politically and discursively contested terrain within the intricate economies of exploitation and reinvention characterizing global capitalism in the neoliberal era. Multiple uses of the commons can be recognized within this diversified politics of the commons, revealing the ultimately ambivalent use of this notion: the commons is at one and the same time a space of resistance to neoliberal accumulation by dispossession and of experimentation with post-capitalist economic practices, but also a site of societal subsumption and commodification within the knowledge-intensive economies of biopolitical capitalism.

In turning its attention to the city and to a number of issues inherently connected to contemporary urbanism, the chapter furthermore considered what is distinctive about the *urban commons* as well as the extent to which the urban context generates novel questions of the commons. Through this it has shown that attempts to urbanize the concept of the commons contribute to its ambiguity but also open up new ways of seeing urban space and its associated politics. In particular debates and discussions of the antagonisms inherent in the urban commons reflect and reveal the larger ambivalence of contemporary global capitalism, where social relations of collaboration and even social projects of post-capitalist transition are incessantly captured within profit-driven logics of economic valorization. This ambivalence, it is argued, is a crucial point of departure for understanding the forms of life, rationalities and political practices that comprise late neoliberal societies.

References

Amin, A. (2005). Local Community on Trial. *Economy and Society*, 34(4), pp. 612–633.
Amin, A. (2014). Lively Infrastructure. *Theory, Culture & Society*, 31(7–8), pp. 137–161.
Aschoff, N. (2015). *The New Prophets of Capital*. London: Verso.
Bensaïd, D. (2007). *Les Dépossedés. Karl Marx, les Voleurs de Dois et le Droit des Pauvres*. Paris: La fabrique.
Berlant, L. (2016). The Commons: Infrastructures for Troubling Times. *Environment and Planning D: Society and Space*, 34(3), pp. 393–419.
Brenner, N. and Schmid, C. (2014). The 'Urban Age' in Question. *International Journal of Urban and Regional Research*, 38(3), pp. 731–755.
Bresnihan, P. and Byrne, M. (2015). Escape into the City: Everyday Practices of Commoning and the Production of Urban Space in Dublin. *Antipode*, 47(1), pp. 37–54.
Cooke, P. and Morgan, K. (1998). *The Associational Economy: Firms, Regions and Innovation*. Oxford: Oxford University Press.
Cutler, K.M. (2015). With $7.35M Raise for Common, General Assembly Co-founder Gets into the 'Co-living' Movement. *Tech Crunch*, 16 July.
Dardot, P. and Laval, C. (2010). *La nouvelle raison du monde. Essai sur la société néolibérale*. Paris: La Découverte.
Dardot, P. and Laval, C. (2014). *Commun. Essai sur la Révolution au XXIe Siècle*. Paris: La Découverte.
De Angelis, M. (2007). *The Beginning of History: Value Struggles and Global Capital*. London: Pluto Press.
Enright, T. and Rossi, U., eds. (2018). *The Urban Political: Ambivalent Spaces of Late Neoliberalism*. New York: Palgrave Macmillan.
Federici, S. (2010). Feminism and the Politics of the Commons in an Era of Primitive Accumulation. In: Team Colors Collective, ed. *Uses of a Whirlwind: Movement, Movements, and Contemporary Radical Currents in the United States*. Chico, CA: AK Press, pp. 283–293.
Fromm, D. (2012). Seeding Community: Collaborative Housing as a Strategy for Social and Neighbourhood Repair. *Built Environment*, 38(3), pp. 364–394.
Fukuyama, F. (1995). *Trust: The Social Virtues and the Creation of Prosperity*. New York: Free Press.
Gibson-Graham, J.K. (1996). *The End of Capitalism (as We Knew it): A Feminist Critique of Political Economy*. Minneapolis: University of Minnesota Press.
Gibson-Graham, J.K. (2006). *A Post-capitalist Politics*. Minneapolis: University of Minnesota Press.

Giddens, A. (1998). *The Third Way: The Renewal of Social Democracy*. Cambridge: Polity.
Gordon, C. (1991). Governmental Rationality: An Introduction. In: G. Burchell, C. Gordon and P. Miller, eds. *Studies in Governmentality*. Chicago: The University of Chicago Press, pp. 1–52.
Grasseni, C. (2014). Seeds of Trust. Italy's Gruppi di Acquisto Solidale (Solidarity Purchase Groups). *Journal of Political Ecology*, 21, pp. 178–192.
Hardin, G. (1968). The Tragedy of the Commons. *Science*, 162(3859), pp. 1243–1248.
Hardt, M. and Negri, A. (2009). *Commonwealth*. Cambridge, MA: Belknap Press of Harvard University Press.
Harvey, D. (2003). *The New Imperialism*. Oxford: Oxford University Press.
Harvey, D. (2012). *Rebel Cities: From the Right to the City to the Urban Revolution*. London: Verso.
Hirst, P. (1994). *Associative Democracy: New Forms of Economic and Societal Governance*. Cambridge: Polity.
Huron, A. (2015). Working with Strangers in Saturated Space: Reclaiming and Maintaining the Urban Commons. *Antipode*, 47(4), pp. 963–979.
Imbroscio, D. (2010). *Urban America Reconsidered: Alternatives for Governance and Policy*. Ithaca, NY: Cornell University Press.
Jarvis, H. (2011). Saving Space, Sharing Time: Integrated Infrastructures of Daily Life in Cohousing. *Environment and Planning A*, 43, pp. 560–577.
Jeffrey, A., McFarlane, C. and Vasudevan, A. (2012). Rethinking Enclosure: Space, Subjectivity and the Commons. *International Journal of Urban and Regional Research*, 44(4), pp. 1247–1267.
Kasperkevic, J. (2016). Co-living – The Companies Reinventing the Idea of Roommates. *Guardian*, 20 March. Available at: www.theguardian.com/business/2016/mar/20/co-living-companies-reinventing-roommates-open-door-common- [accessed 26 May 2017].
Kaysen, R. (2015). The Millennial Commune. *New York Times*, 31 July. Available at: www.nytimes.com/2015/08/02/realestate/the-millennial-commune.html?_r=0 [accessed 26 May 2017].
Kohn, M. (2016). *The Death and Life of the Urban Commonwealth*. Oxford: Oxford University Press.
Lazzarato, M. (2009). Neoliberalism in Action. Inequality, Insecurity and the Reconstitution of the Social. *Theory, Culture & Society*, 26(6), pp. 109–133.
Linebaugh, P. (1976). Karl Marx, the Theft of Wood, and Working Class Composition: A Contribution to the Current Debate. *Crime and Social Justice*, 6 (Fall–Winter), pp. 5–16.
Linebaugh, P. (2008). *The Magna Carta Manifesto: Liberties and Commons for All*. Berkeley: University of California Press.
Magnusson, W. (2011). *Politics of Urbanism: Seeing Like a City*. London and New York: Routledge.
McFarlane, C. (2012). The Entrepreneurial Slum: Civil Society, Mobility and the Co-production of Urban Development. *Urban Studies*, 49(13), pp. 2795–2816.
Noterman, E. (2016). Beyond Tragedy: Differential Commoning in a Manufactured Housing Cooperative. *Antipode*, 48(2), pp. 433–452.
Olson, M. (1965). *The Logic of Collective Action: Public Goods and the Theory of Groups*. Cambridge, MA: Harvard University Press.
Ostrom, E. (1990). *Governing the Commons: The Evolution of Institutions for Collective Action*. Cambridge: Cambridge University Press.
Ostrom, E. and Ahn, T.K. (2003). Introduction. In: E. Ostrom and T.K. Ahn, eds. *Foundations of Social Capital*. Cheltenham: Edward Elgar, pp. 1–24.
Peck, J. (2012). Austerity Urbanism. *City*, 16(6), pp. 626–655.
Peck, J. and Tickell, A. (2002). Neoliberalizing Space. *Antipode*, 34(3), pp. 380–404.
Putnam, R. (2000). *Bowling Alone: The Collapse and Revival of American Community*. New York: Simon & Schuster.
Raco, M. (2007). Securing Sustainable Communities: Citizenship, Safety and Sustainability in the New Urban Planning. *European Urban and Regional Studies*, 14(4), pp. 305–320.
Rifkin, J. (2014). *The Zero Marginal Cost Society: The Internet of Things, the Collaborative Commons, and the Eclipse of Capitalism*. Basingstoke: Palgrave Macmillan.
Robinson, J. (2006). *Ordinary Cities: Between Modernity and Development*. London: Routledge.
Rossi, U. (2017). *Cities in Global Capitalism*. Cambridge: Polity.
Roy, A. (2009). The 21st-century Metropolis: New Geographies of Theory. *Regional Studies*, 43(6), pp. 819–830.
Simone, A. (2004). People as Infrastructure: Intersecting Fragments in Johannesburg. *Public Culture*, 16(3), pp. 407–429.
Tornaghi, C. (2014). Critical Geography of Urban Agriculture. *Progress in Human Geography*, 38(4), pp. 551–567.

Tsing, A. (2005). *Friction: An Ethnography of Global Connections*. Princeton, NJ: Princeton University Press.

Vestbro, D.U. (2000). From Collective Housing to Cohousing: A Summary of Research. *Journal of Architectural and Planning Research*, 17(2), pp. 164–178.

Virno, P. (1996). The Ambivalence of Disenchantment. In: M. Hardt and P. Virno, eds. *Radical Thought in Italy: A Potential Politics*. Minneapolis: Minnesota University Press, pp. 13–36.

Williams, J. (2005). Designing Neighbourhoods for Social Interaction: The Case of Cohousing. *Journal of Urban Design*, 10(2), pp. 195–227.

5
THE SMART STATE AS UTOPIAN SPACE FOR URBAN POLITICS

Yonn Dierwechter

Introduction

Seen as a specific socio-spatial conjuncture of power relations, the state should be an essential ingredient in the menu of topics examined in a volume about the spaces of urban politics. Yet utopian thought across diverse traditions has often rendered 'the state' invisible, unhelpful or dangerous, even as it has concomitantly elevated specific kinds of places as more worthy of our normative-political aspirations. Nonetheless stateness per se – understood here as the presence of socially legitimate institutions that articulate political democracy and help collectives to puzzle through shared problems – remains theoretically relevant in the face of constant attacks and otherwise legitimate critiques, particularly as a bulwark against corporatized economies that, untethered from popular control, have repeatedly generated social and ecological turmoil. The role of the state in reconstructing politically an ideal of the good society has returned to prominence. But the macro-dynamics of globalization, urbanization and especially city-regionalism have shifted the ways in which the state is mobilized in utopian thinking. The state now has to be smart: namely, authoritative but not authoritarian; directive but not bureaucratic; technical but not technocratic; market-shaping but not market-dominated. Or at least, that is the dream.

This chapter develops themes under the synoptic rubric of the smart state as a possible utopian space for urban politics. The discussion highlights the tense intellectual-theoretical relationship between utopianism and the (territorial or centralized) state. It then shifts to the rising but still open and contested role of cities in global, 'post-Westphalian' space, i.e. the world dominated by nation-states since the Peace of Westphalia in 1648, which ended a long period of warfare in Europe and eventually provided the taken-for-granted concepts of nationalism, sovereignty and citizenship. Finally, the chapter briefly explores urban-regional politics across Greater Seattle, a US exemplar of the so-called 'smart turn' in global urbanism (Rossi 2016). Here I highlight the possibilities of a new politics emanating into (and out of) the city-region, in part by 'drawing down' a new kind of state, one that helps to *territorialize* the city-region as an improved space through programmes, policies, projects and imaginaries that valorize competence over control; here, that is, I see "territory as a bounded portion of relational space", as dell'Agnese puts it (2013: 122), wherein, quoting Doreen Massey, "the old notion of territory needs to be reconceptualised, but may turn out to be useful again [and … indeed] can be accepted as 'a way of giving order to space' " (dell'Agnese 2013: 122) (cf. Jonas 2013a). The state is feared, but also needed; the aspiration of smartness has emerged in recent years, I argue, to negotiate this long-standing contradiction in utopian political thought.

Utopianism and the state

The state has not fared well in utopian political thought. Two of the major political ideologies of the nineteenth century – liberalism and socialism – had little place for the state in their renderings of future orders at any territorial scale, save perhaps the often-romanticized locality. Liberals dreamed of integrating diverse societies though unfettered, self-organizing markets. Socialists agitated for uniting the world's industrial workers in an after-state labourite internationalism (Mazower 2012). One of two things would surely happen. Either the state would shrink appropriately to service only the voluntary transactions of the more dynamic free market or it would wither away *in toto* with the liberated consciousness of the now-revolutionary working classes.

With the post-war expansion of social welfarism, Robert Nozick (1974) further reimagined utopia as constituted by a (very) smallish night watchman state. But even his brand of 'minarchism' – i.e. a minimal state that enforces only a framework of natural and legal rights for a free marketplace – was too much for anarcho-capitalists like Murray Rothbard (1970). Rothbard's central project was to eliminate theoretically the necessity of the state *in toto*; markets, he imagined, could even provide so-called 'public' services like law and defence (Chamberlain 1976 [1959]; Friedman 1962; Hayek 1944).

Such appeals to the utopia of pure markets – and the profound reduction of human geohistory to economic calculation – were made easier by many of the tragic political experiments undertaken over the course of the twentieth century; the grotesque statisms of Stalinism, Nazism, apartheid, Pol-Potism, Maoism, Baathism, Salazarism and so forth, that provided the bloody instruments of dystopia. In James Scott's (1998) elegant view, 'seeing' like these sorts of states – but social-democratic states too, from early German to late Nordic – meant oversimplification, mechanical codification, hubris and the erasure of heterodox peoples, places and natures. This is distinctive from but also not widely dissimilar to many post-structuralist concerns with normalization and governmentality in some Foucauldian work on the urban state (e.g. Mackinnon 2000; Rutland and Aylett 2008). With the peaceful denouement of the Cold War, Francis Fukuyama (1992) finally 'ended' history itself, pronouncing liberal-democratic capitalism as good as it gets, however flawed, effectively closing down additional renderings of alternative state worlds. The market liberals of Richard Cobden's day, who helped to repeal the British Corn Laws, champion free trade and support laissez-faire economics, and who together drew on Hegel's notion of the final liberal state, were basically right all along.

In place of the state, utopian thought has imagined ideal places. Notions of the perfect (urban) community are ancient, dating at least to Plato's *Republic*. Bartolomeo Coradini painted his well-known 'ideal city' as long ago as the mid-fifteenth century. To look forward, though, he glanced backwards – a common theme in utopian studies of place that suggests a syncretic admixture of novelty through nostalgia and rupture through restoration. Coradini's aesthetic debt to Greco-Roman forms included the triumphal arch, the agora and the amphitheatre. Likewise in the twentieth century, Lewis Mumford (1961) frequently looked backward for future hope – not like Coradini to the ancient urbanisms of Rome and Greece but to the European medieval city. Inspired by Patrick Geddes' evolutionary theory of an urbanizing history and geography, Mumford conceptualized the medieval socio-spatial form – or at least the European variant of that form (for there were many others, e.g. Islamic, Aztec, Chinese, Zimbabwean) – as the 'eotechnic' period of urbanism, community and society (Mumford 1938). This was the organic world of local wood and stone, not the synthesized commodity of crude oil and plastic culture; of eco-regional bioviability, not global unsustainability.

Mumford simply redirected a specific tradition of American utopianism that had exploded across the landscape in the spatially experimental and culturally creative nineteenth century. Some of these American communities and social movements borrowed and extended Owenist and Fourierist ideals from Britain and France, for example. But many others were legitimately novel New World departures of anarcho-religious-alternative communes set apart from the industrializing society and the centralizing capitalist state that Mumford and many others on the political-cultural left so loathed. This includes the Shakers, the Oneida Perfectionists (also known as the 'Bible communists'), the Harmony Society, and at one time the spiritually more successful Mormons (Nordhoff 1960 [1875]). In short, the diversely imagined and/or partially practised spaces of utopia – e.g. liberated markets, revolutionary workers, recovered urban forms, Bible-based 'communistic societies' – were invariably imagined outside or beyond rather than through the territorial, regulatory or even democratic reach of a modern state apparatus. Utopia has long been a differently imagined place where the overbearing, centralizing, clumsy, inauthentic, capitalist, rapacious state could not unduly contaminate the purity of better worlds.

Still, in *The Great Transformation* Karl Polanyi (1944) detonated the utopia of free markets (Block 2001: xxv). Contra Frederich Hayek's (1944) own meticulous defence in *The Road to Serfdom*, Polanyi had concluded from his study of nineteenth and twentieth century economic history that free markets do not 'self-regulate' for very long without producing profound social and ecological damage – and profound inequality (Lacher 1999). Amongst other themes, Polanyi was trying to understand the economic origins of the First World War, European fascism and the Second World War – at a time when European liberal democracy had either 'failed' (Mazower 1998: 403), or come very close to it. Hayek had referred to communism and fascism as a union of anti-capitalist radical and conservative socialisms, respectively (Block 2001). Polanyi was both anti-fascist and anti-capitalist; but he was not a Marxist utopian either, rejecting the labour theory of value (Dale 2010). Impressed historically with markets, he nonetheless dismissed "the 'economistic prejudice' found in both the market liberalism of Ludwig von Mises and the communism of Karl Marx" (Carlson 2006: 32). To him, the regression to fascism emerged not because of economic planning, as Hayek had claimed, but because of efforts to build a world around market self-regulation. Unleashed from strong democratic controls, Polanyi claimed, the overliberalized market would lead to chaos. Polanyi therefore felt that the problem was not the *existence* of markets per se, but their *social management* within broader political-economic systems at various territorial scales, though he said relatively little of how urbanization impacted all this. As Fred Block (2001: xxxv) notes:

> The key step [for Polanyi] was to overturn the belief that social life should be subordinated to the market mechanism. Once free of this 'obsolete market mentality', the path would be open to subordinate both national economies and the global economy to democratic politics. Polanyi saw Roosevelt's New Deal as a model of these future possibilities. Roosevelt's reforms meant that the US economy continued to be organized around markets and market activity but a new set of regulatory mechanisms now made it possible to buffer both human beings and nature from the pressures of markets.

The global economic catastrophe in 2008 – the deepest since 1929 – and the recent publication of Thomas Picketty's (2014) *Capital in the 21st Century* arguably provided fresh evidence for

Polanyi's core insights about the chaos of under-regulated markets and the nature of capitalism. In the wake of the crisis, Picketty specifically showed that income inequality does not eventually decline as capitalism matures, a view Kuznets (1953) had originally propounded in the 1950s with his famous 'Kuznets Curve'. While politics are treated as exogenous shocks to core economic dynamics, rather than constitutive of these dynamics, Picketty's book nonetheless suggests in the spirit, if not the letter, of Polanyi the importance of re-democratized if multi-scalar states in counteracting capitalism's tendencies towards inequality (Hopkins 2014). In this sense, the role of the state in reconstructing politically the good society has refused to die. However, the macro-dynamics of globalization, urbanization and city-regionalism have shifted the ways in which the state can be (re)mobilized in utopian political thought.

Global space, rising cities, utopian smartness

A key leitmotif in urban studies for a generation or more has been the impact of globalization *on* cities (Olds and Yeung 2004; Robinson 2014; Sassen 1991; Scott 2001). Yet as Toly (2008) suggests, it is no longer a question of what globalization is doing to cities, but increasingly of what cities are doing to globalization. An expanding array of actors who operate and think globally – non-governmental organizations, multinational firms, science networks and indeed cities – are challenging the territorial centrality of the nation-state (Bulkeley and Betsill 2005; Bulkeley and Moser 2007; Engelke 2013; Rosenzweig *et al.* 2010). Novel city-networks like the C40 club, the ICLIE Charter, the Climate Alliance and many other regional, national and transnational movements (Keiner and Kim 2007), in particular, have constituted at least some cities as new sites for international relations (Amen *et al.* 2012). Relatedly, a recent stream of work has traced what Allen Scott (2001) famously called the "rise of city-regions", which he saw as a "new spatial grammar" steadily restructuring how we must increasingly read and collectively talk about the political organization of global space.

Much of the work on city-regions is empirical and explanatory, chasing down and sometimes conceptualizing new patterns in the more urbanized human geographies that have emerged with such quantitative force in recent decades of world history (Clark and Christopherson 2009; Dierwechter 2013a; Jonas 2013b; Pezzoli *et al.* 2006; Scott 2007; Ward and Jonas 2004). But I also want to suggest here that, in some if not all of this new work, there is a strong *normative* element that evokes and even reproduces some of the utopian traditions just recounted, especially as interlaced with the inchoate and seemingly post-political discourses of smartness, efficiency, creativity and agility in confronting and solving classic collective action problems like climate change, participatory decision-making and so forth. Two recent books help to illustrate this last claim: Ben Barber's (2013) scholarly *When Mayors Rule the World: Dysfunctional Nations, Rising Cities*, and Anthony Townsend's (2013) more colloquial *Smart Cities: Big Data, Civic Hackers, and the Quest for a New Utopia*.

Ben Barber, a political theorist, argues that democracy as a meaningful form of active participation is now in deep crisis. The conventional nation-state paradigm distilled in Europe – the 'Westphalian' paradigm of global-territorial politics – has become dysfunctional and outdated. In particular, he suggests, nation-states are now manifestly unable to engage interdependently in cross-border collaboration to solve or even measurably ameliorate severe cross-border problems like weapons, trade, climate change, cultural exchange, crime, drugs, transportation, public health, immigration and technology. Yet his book is optimistic. The solution "stands [right] before us, obviously but largely uncharted: let cities … do what states cannot. Let mayors rule the world" (Barber 2013: 4). Conceived as incubators for problem solving, cities are politically poised to meet the challenges of contemporary globalization, while national governments fail.

Barber's global cartography is populated by novel, mutually beneficial networks of local-yet-global urban action. Indeed, he likens this to a new 'Hanseatic League' of policy activity and learning, evoking the famous network of cities that, in the Middle Ages, fielded its own militaries, constructed an effective exchange mechanism and tariff system, and sometimes met as a Parliament, known as the Hansetage, albeit infrequently.

Barber's new urban networks are, as he sees it, the most robust civic entities in the crowded playground of international politics: e.g. United Cities and Local Governments, the International Union of Local Authorities, Metropolis, ICLEI and C40. Ultimately, Barber concludes, these new entities may even provide a workable prototype for his biggest idea: a global parliament of mayors. "If mayors ruled the world", he entreats,

> the more than 3.5 billion people (over half the world's population) who are urban dwellers and the many more in the exurban neighborhoods beyond could participate locally and cooperate globally at the same time – a miracle of civic 'glocality' promising pragmatism instead of politics, innovation rather than ideology and solutions in place of sovereignty.
>
> *(Barber 2013: 5)*

Promising miracles is utopian. By this I do not mean in the pejorative sense that Barber is Panglossian. It is lazy to dismiss creative ideas like a global parliament of mayors as unattainable or "nutty", a word he enjoins to anticipate pragmatic critics focused only on Tuesday's meeting of the local planning body. Rather I mean that his ideas reflect a well-worn scepticism of the territorial state as necessarily constitutive of utopia, even in part. The state appears profoundly dysfunctional in Barber's book in part because its territorial form is now sclerotic and thus insufficiently supple: too small to manage globalization, too big to express democratic life. Second, I mean that along with the likes of Coradini, Mumford, etc., Barber accordingly looks to a specific kind of place – the once great, long dormant, but now apparently reawakened polis of democratic propinquity, familiarity, accessibility and dense discursive transactions – for global political salvation. His book reflects novelty through nostalgia (for ancient forms of democratic life) and rupture through restoration (of the polis before it was effaced by territorial states now incapable of global interdependency).

In his book, Anthony Townsend, formerly the research director at the imposingly named Institute for the Future, shares Barber's enthusiasm for democratic localism (Townsend 2013). But he is focused on politically contending versions of utopia spawned by the new obsession with smart cities (Wiig 2016). In simple terms, one vision of the smart city is imagined and marketed by large and powerful corporations like IBM, Cisco, Siemens, Oracle, Microsoft and Intel, which see unending business opportunities and profitability in selling to the world's some 500,000 municipalities (and mayors) what Hill elsewhere calls a new "urban intelligence industrial complex" (cited in Hollands 2015: 68). In this not-so future world that these corporations appear to promise, a comprehensive embedding of digital information in the urban fabric – technically urbanizing the already expanding 'internet of things' – will help compute away seemingly intractable urban problems like climate change, traffic congestion, workforce training and declining public health. The second, more desirable utopia for Townsend redirects the purpose of Big Data through local activists, entrepreneurs and hackers – the valorized agents who will/are push(ing) Barber's mayors from the streets, grassroots and garages to avoid the corporate snow job.

In consequence, this is about much more than hooking up traffic lights with sensors that help to reduce unnecessary idling (saving the planet) or analysing location-aware Foursquare apps on the phones of ambulant hipsters who "mash up restaurant reviews with health-inspection data" (Moses 2013: 299). Instead Townsend analyses the contemporary struggle over what the utopia of smart cities can and should mean by referencing past mêlées between towering figures in utopian planning, notably Ebenezer Howard and Patrick Geddes.

Geddes is the unlikely hero in Townsend's text. Howard developed a universal but static place into which desirable change appears: the garden city. Geddes (1950 [1915]) developed an organic, bottom-up theory of "civics", by which he meant a kind rolling citizen engagement in the creative production of new forms of place knowledge about how to improve the human environment, many very small and "surgical" rather than comprehensive and totalizing. Geddes was, if one will, a hacktivist *avant la lettre*, reinventing ways of knowing and engaging the world to improve the now urban human condition. As Townsend (2013: 283) glows, "Geddes would no doubt approve of how today's smart city-builders are applying technology to urban challenges and seeking to develop a new, rigorous empirical science of cities. But he also understood the limits of science."

Townsend appears to accept that Big Data – huge in volume; high in velocity; expansive in scope; fine-grained in resolution, etc. – are "never raw" but "precooked" by someone (Kitchin 2014). He accepts that, as Rob Kitchen explicitly argues, smart urbanism is not about the more efficient mastery of real-time data. But Townsend seems to place considerable faith in the political shrewdness and organizational capacity of (highly educated, mostly male, disproportionately white, invariably healthy and young) data activists, tech-entrepreneurs and cyber hackers – acting in the great Geddesian ethos – to transform through a new digital civics the smart-technocratic forms of urban governance that otherwise "only enable the more efficient management of the manifestations of problems" (Townsend 2013: 9). He shares Barber's utopian optimism in a reanimated, smarter, innovative and intimately known polis as the truly appropriate space for a truly transformative global politics.

Urban politics in city-regional Seattle

If these ideas are going to matter, they are going matter in places like Seattle, Washington, USA. The population of Seattle is, by one measure, the smartest in the United States today (Paton 2009: 15). Washington, DC is a more important "global political city" (Calder and Freytas 2009). San Francisco is bigger, and still richer. But by 2009 Seattle had more workers with college degrees (55.2 per cent) than either of these two otherwise well-educated cities (48.2 and 51.2 per cent, respectively). The main reasons are clear enough: local firms often seen as part and parcel of the smart economy – corporate giants like Microsoft and Amazon along with their growing ecosystems of start-ups, spinoffs and suppliers (Mayer 2013) – continue the IT-led transformation of the city-region's high-tech manufacturing economy associated historically with Boeing (Gray *et al.* 1996).

Seattle long has been a progressive city, though not one without growing contradictions that reverberate across the city-region's geopolitical-economy as a whole (Balk 2014; Gibson 2004; Minard 2014). In his review of the city's political culture, the labour historian James Gregory (2015: 64) has argued that, like San Francisco and a few other US cities, Seattle is characterized by a "left coast formula" of urban politics. Central to this formula is "a set of institutions and expectations that keep radicalism alive while allowing political elites identified as liberal or progressive to stay in power pretty consistently" (Gregory 2015: 65). Two Seattles co-exist: one focused since the World Trade Organization riots in 1999 on 'resurgent radicalism' (e.g. livable wages, immigrant and LGBT rights, a sustainability ethos); the other on 'radical re-urbanization', especially as this pertains to densifying, corporatizing neighbourhoods like South Lake Union, the anchor home of Amazon, where a novel kind of work-life space for some 30,000 engineers, programmers, designers and headquarter staff has been steadily emerging over the past ten years. Seattle in 2015 had an immigrant-born, Trotskyite city councillor and an openly gay mayor; a former mayor, Greg Nichols, while not exactly 'ruling the world', co-founded the Mayors Climate Protection Agreement (Dierwechter 2010). In the popular press, such idiosyncrasies help to make Seattle the "prototype city of the future ... where creative people want to come" (Paton 2009: 15). It is placed to be progressively rich – or richly progressive. And indeed, King County, the home of Seattle, already has some 70,000 millionaires.

Yet the urban politics of resurgent radicalism have not been able to stem the city-regionalized effects of radical reurbanization. Seattle's fastest-changing neighbourhoods in recent years – South Lake Union, Ballard and Capitol Hill – receive the most attention. But historically African American neighbourhoods like the Central District have undergone no less change. By 2010, the Central District – which was 80 per cent black in 1970 – had become majority white (Balk 2014), a reversal of former concerns in US urban affairs with white flight. In 2000, the medium household income of $45,200 for all races in Seattle was 7.5 per cent higher but comparable with the national average of $42,800. By 2013, the median figure of $70,200 for Seattle was 34 per cent higher than the national figure of $53,000. As Seattle's African American population has declined, it has also become poorer. In 2000, medium household income for African Americans was $32,000. By 2013, it had plummeted to $25,700 (Balk 2014). The digitization of the economy, the erosion of standardized labour, and the rise of cognitive-cultural workers like those that populate Microsoft and Amazon are remaking interurban structures of production and work, and concomitantly core patterns of regional development (Scott 2011).

New forms of stratification are pressuring local municipalities in new ways. It is not only that a now heavily regionalized and unevenly digitized social space is being "radically altered by gentrification, i.e. the colonization of former blue collar neighborhoods in inner-city neighborhoods by members of the cognitariat" (Scott 2011: 855). It is also that other cities within the wider metro area, like Tacoma, so far have experienced little such change. With Pierce County, Tacoma may act more as a cheaper consumption zone for the production demands of Seattle. One local joke is that Seattle's affordable housing policy is simple: Pierce County. In fact, the second-tier city of Tacoma located 30 miles south of Seattle, has struggled to encourage private-sector investment in 'upzoned' areas, even in its wealthier neighbourhoods, including its downtown. Like historically blue-collar cities elsewhere, it has thus paradoxically struggled to finance basic social infrastructure outlays for collective consumption (roads, schools, etc.), which has of course suppressed private sector interest. Growing at barely 1 per cent per annum, the 'regional problem' of underperforming Tacoma is neither shrinkage (as has happened in US cities like Detroit, Buffalo and Youngstown) nor gentrification (e.g. Seattle, San Francisco and Washington, DC) but what Williams and Pendras (2013: 289) have called an urban political economy of "stasis". Neither traditional pro-growth coalitions nor alternative-progressive regimes have really captured the local political agenda.

Utopian conclusions: the smart state as urban political space

Seattle is a smart city with a relatively progressive local political culture. Yet the real territorial challenge is how it is supposed to envision itself as a cohesive city-*region* better able to manage its contradictions, tensions and uneven spatialities (Morandi et al. 2015). The same might be said not only of 'left coast cities' in the USA but other high-growth cities reshaped in recent decades by high-tech strategies of accumulation (Tretter 2016). In terms of rethinking spaces of urban politics, what does this mean and who will benefit?

One possibility is for fragmented municipalities across metropolitan space to download the digitized 'everyware' of corporatized utopias on constant offer from the global purveyors of the urban intelligence industrial complex. Here, of course, we simply take our normative cue from the liberal dream of how private markets can and should solve seemingly intractable public problems through efficiency-seeking capitalist innovation.

A second possibility is to reject radically the urban-capitalist fantasy of smartness. Instead we draw utopian sustenance from Marxian traditions leavened with ecological ideals. John Barry (2012: 141), for example, envisions instead a "post-growth, anti-capitalist" paradigm that transcends rather than reshapes current political-economies and institutional matrices of power. "In short", he writes,

> the common green critique of orthodox economics must become a clearer critique of capitalism itself, and relatedly its long-standing and evidence-based critique of economic growth must become a critique of capital accumulation. ... Carbon-based capitalism is destroying the planet's life-support systems and is systematically liquidating them and calling it 'economic growth'.
>
> *(Barry 2012: 141)*

Still a third possibility is to recover Polanyi's older (and to some extent Picketty's newer) desire to embed markets back into a socially legitimated and more deeply democratized politics of space and society. In my reading, this equally utopian agenda explicitly involves the state in the institutional search for greater social equity and ecological resiliency. But as I argued earlier in this chapter, such a recovery – itself a nostalgic framing – paradoxically rubs up against an abiding scepticism of the territorial state within utopian studies, especially in the United States. Thus a serious conundrum: not only for political practice, but also for utopian urban theory.

The geo-political rise of city-regions; the normative celebration of mayors and networked cities as 'glocal' problem-solvers; the invocation of a Geddesian civics pushing digital activism from below – all these recent propositions in the urban literature point collectively to a faint but decipherable utopian space for contemporary politics, one that, I conclude here, seeks to imagine a resolution of this core conundrum: the *smart* state. By smart I mean a state that is neither overbearing nor invisible; neither exactly neoliberal-competitive nor overtly redistributive or rigidly territorial. Accordingly, a new state *space* that is: authoritative in knowledge production, but not authoritarian in requirement; professional in discourse, but not bureaucratic in disposition; market-respectful, but not market-dominated or wholly 'captured'; technocratic in assistance and capable in advice, yet somehow also collaborative, discursive and resourceful; directive, yet open; pro-growth, yet holistically developmental, protecting resilient ecosystems through improved social cohesion. Here the state is now nimble *and relational* in support of a theoretically valorized *territorial* form: the locally competitive yet justly sustainable global city-region. Or so goes the dream.

The smart state so rendered is profoundly syncretic; but so are most ideas of consequence (Dierwechter 2013b) – including most religions. In fact, the smart state may not be wholly imaginary or entirely 'unreal' either, without any relationship whatsoever to the actually existing practices of urban and regional politics. For "practitioners theorize, too" (Whittemore 2015: 76). The utopia of pure markets, for example, has never really captured any society *in toto*, despite the efforts of actors like Hayek or Rothbard; yet as Polanyi showed, the market liberal project has been massively consequential.

Long-range regional planning visions are similarly infused with utopian thinking. The Puget Sound Regional Council (PSRC), the key planning body in the Seattle area, projects utopian themes in its most important documents, e.g. *Vision 2040* (Puget Sound Regional Council

2009). Mumfordian eco-regionalism is discernible, for example, as are related claims associated with how new growth can be made 'smarter' if transport, housing, food and economic policies blend local land use visions with federal-scale funding streams and priorities (e.g. neighbourhood integration, urban sustainability, carbon control, regional transit alternatives). The state is imagined *as smart*, in other words, if city-regional bodies like the PSRC over time can actually articulate – or bridge – a very distant federal-central realm with local-scale spaces that are concomitantly stripped of excessive parochialism. The 'utopia of the smart state' is thus multi-level, downscaled from the feared centre but upscaled from the parochial community. It *is* a utopia, then, but one that both critiques and accepts 'the state' by reimagining a new kind of regional home for the preferred socio-natural developments of future urban life and spatial politics.

References

Amen, M., Toly, N., McCarney, P. and Segbers, K., eds. (2012). *Cities and Global Governance: New Sites for International Relations*. Farnham, UK: Ashgate.

Balk, G. (2014). As Seattle Gets Richer, the City's Black Households Get Poorer. *Seattle Times*, 12 November. Available at: http://blogs.seattletimes.com/fyi-guy/2014/11/12/as-seattle-gets-richer-the-citys-black-households-get-poorer/ [accessed 26 May 2017].

Barber, B. (2013). *If Mayors Ruled the World: Dysfunctional Nations, Rising Cities*. New Haven, CT: Yale University Press.

Barry, J. (2012). Climate Change, the Cancer Stage of Capitalism, and the Return of Limits to Growth. In: M. Mark Pelling, D. Manuel-Navarrete and M. Redclift, eds. *Climate Change and the Crisis of Capitalism: A Chance to Reclaim Self*. London: Routledge, pp. 129–142.

Block, F. (2001). Introduction. *Karl Polanyi. The Great Transformation*. Boston: Beacon Press.

Bulkeley, H. and Betsill, M.M. (2005). Rethinking Sustainable Cities: Multilevel Governance and the 'Urban' Politics of Climate Change. *Environmental Politics*, 14(1), pp. 42–63.

Bulkeley, H. and Moser, S.C. (2007). Responding to Climate Change: Governance and Social Action beyond Kyoto. *Global Environmental Politics*, 7(2), pp. 1–10.

Calder, K. and Freytas, M. (2009). Global Political Cities as Actors in Twenty-First Century International Affairs. *SAIS Review of International Affairs*, 29(2), pp. 79–97.

Carlson, A. (2006). The Problem of Karl Polanyi. *Intercollegiate Review*, Spring, pp. 32–39.

Chamberlain, J. (1976 [1959]). *The Roots of Capitalism*. Indianapolis: Liberty Press.

Clark, J. and Christopherson, S. (2009). Integrating Investment and Equity: A Critical Regionalist Agenda for a Progressive Regionalism. *Journal of Planning Education and Research*, 28(3), pp. 341–354.

Dale, G. (2010). *Karl Polanyi: The Limits of the Market*. Cambridge: Polity Press.

dell'Agnese, E. (2013). The Political Challenge of Relational Territory. In: D. Featherstone and D. Massey, eds. *Spatial Politics: Essays for Doreen Massey*. London: John Wiley & Sons, pp. 115–124.

Dierwechter, Y. (2010). Metropolitan Geographies of US Climate Action: Cities, Suburbs and the Local Divide in Global Responsibilities. *Journal of Environmental Policy and Planning*, 12(1), pp. 59–82.

Dierwechter, Y. (2013a). Smart City-Regionalism Across Seattle: Progressing Transit Nodes in Labor Space? *Geoforum*, 49, pp. 139–149.

Dierwechter, Y. (2013b). Smart Growth and State Territoriality. *Urban Studies*, 50(11), pp. 2275–2292.

Engelke, P. (2013). *Foreign Policy for an Urban World: Global Governance and the Rise of Cities*. Washington, DC: Atlantic Council.

Friedman, M. (1962). *Capitalism and Freedom*. Chicago: University of Chicago Press.

Fukuyama, F. (1992). *The End of History and the Last Man*. New York: Maxwell Macmillan International.

Geddes, P. (1950 [1915]). *Cities in Evolution*. New York: Oxford University Press.

Gibson, T.A. (2004). *Securing the Spectacular City: The Politics of Revitalization and Homelessness in Downtown Seattle*. Lanham, MD: Lexington Books.

Gray, M., Golob, E. and Markusen, A. (1996). Big Firms, Long Arms, Wide Shoulders: The 'Hub-and-Spoke' Industrial District in the Seattle Region. *Regional Studies*, 30(7), pp. 651–666.

Gregory, J. (2015). Seattle's Left Coast Formula. *Dissent*, 62(1), pp. 64–70.

Hayek, F. (1944). *The Road to Serfdom*. Chicago: University of Chicago Press.

Hollands, R.G. (2015). Critical Interventions into the Corporate Smart City. *Cambridge Journal of Regions, Economy and Society*, 8(1), pp. 61–77.

Hopkins, J. (2014). The Politics of Piketty: What Political Science Can Learn From, and Contribute to, the Debate on *Capital in the Twenty-First Century*. *British Journal of Sociology*, 65(4), pp. 678–695.

Jonas, A.E.G. (2013a). Place and Region III: Alternative Regionalisms. *Progress in Human Geography*, 37(6), pp. 822–828.

Jonas, A.E.G. (2013b). City-Regionalism as a Contingent 'Geopolitics of Capitalism'. *Geopolitics*, 18(2), pp. 284–298.

Keiner, M. and Kim, A. (2007). Transnational City Networks for Sustainability. *European Planning Studies*, 15(10), pp. 1369–1395.

Kitchin, R. (2014). The Real-time City? Big Data and Smart Urbanism. *GeoJournal*, 79(1), pp. 1–14.

Kuznets, S. (1953). *Economic Change: Selected Essays in Business Cycles, National Income, and Economic Growth*. New York: Norton.

Lacher, H. (1999). The Politics of the Market: Re-reading Karl Polanyi. *Global Society*, 13(3), pp. 313–326.

Mackinnon, D. (2000). Managerialism, Governmentality and the State: A Neo-Foucauldian Approach to Local Economic Governance. *Political Geography*, 19(3), pp. 293–314.

Mayer, H. (2013). Entrepreneurship in a Hub-and-Spoke Industrial District: Firm Survey Evidence from Seattle's Technology Industry. *Regional Studies*, 47(10), pp. 1715–1733.

Mazower, M. (1998). *Dark Continent: Europe's 20th Century*. New York: Knopf.

Mazower, M. (2012). *Governing the World: The History of an Idea, 1815 to the Present*. New York: The Penguin Press.

Minard, A. (2014). Developers Sue to Make Seattle More Developer-Friendly. *The Stranger*, 29 January. Available at: www.thestranger.com/seattle/developers-sue-to-make-seattle-more-developer-friendly/Content?oid=18781207 [accessed 26 May 2017].

Morandi, C., Rolando, A. and Di Vita, S. (2015). *From Smart City to Smart Region: Digital Services for an Internet of Places*. Cham: Springer.

Mumford, L. (1938). *The Culture of Cities*. New York: Harcourt, Brace and Co.

Mumford, L. (1961). *The City in History: Its Origins, Its Transformations, and Its Prospects*. New York: Harcourt, Brace & World.

Nordhoff, C. (1960 [1875]). *The Communistic Societies of the United States From Personal Visit and Observation*. New York: Hillary House.

Nozick, R. (1974). *Anarchy, State, and Utopia*. New York: Basic Books.

Olds, K. and Yeung, H.W.-C. (2004). Pathways to Global City Formation: A View from the Developmental City-State of Singapore. *Review of International Political Economy*, 11(3), pp. 489–521.

Paton, D. (2009). New Economy Cities: A Seattle Slew of Advantages. *Christian Science Monitor*, 20 November. Available at: www.csmonitor.com/Business/2009/1120/new-economy-cities-a-seattle-slew-of-advantages [accessed 26 May 2017].

Pezzoli, K., Hibbard, M. and Huntoon, L. (2006). Introduction to Symposium: Is Progressive Regionalism an Actionable Framework for Critical Planning Theory and Practice? *Journal of Planning Education and Research*, 25(4), pp. 449–457.

Piketty, T. (2014). *Capital in the Twenty-first Century*. Cambridge, MA: The Belknap Press of Harvard University Press.

Polanyi, K. (1944). *The Great Transformation*. Boston: Beacon Press.

Puget Sound Regional Council. (2009). *VISION 2040*. Seattle: Puget Sound Regional Council.

Robinson, J. (2014). New Geographies of Theorizing the Urban: Putting Comparison to Work for Global Urban Studies. In: S. Parnell and S. Oldfield, eds. *The Routledge Handbook of Cities of the Global South*. New York: Routledge, pp. 57–70.

Rosenzweig, C., Solecki, W., Hammer, S.A. and Mehrotra, S. (2010). Cities Lead the Way in Climate-change Action. *Nature*, 467(7318), pp. 909–911.

Rossi, U. (2016). The Variegated Economics and the Potential Politics of the Smart City. *Territory, Politics, Governance*, 4(3), pp. 337–353.

Rothbard, M. (1970). *Man, Economy, and State: A Treatise on Economic Principles*. Los Angeles: Nash Pub.

Rutland, T. and Aylett, A. (2008). The Work of Policy: Actor Networks, Governmentality, and Local Action on Climate Change in Portland, Oregon. *Environment and Planning D: Society and Space*, 26(4), pp. 627–646.

Sassen, S. (1991). *The Global City: New York, London, Tokyo*. Princeton, NJ: Princeton University Press.

Scott, A. (2001). Globalization and the Rise of City Regions. *European Planning Studies*, 9(7), pp. 813–826.

Scott, A. (2011). A World in Emergence: Notes Toward a Resynthesis of Urban-Economic Geography for the 21st Century. *Urban Geography*, 36(2), pp. 845–870.

Scott, J.C. (1998). *Seeing Like a State: How Certain Schemes to Improve the Human Condition Have Failed*. New Haven, CT: Yale University Press.

Scott, J.W. (2007). Smart Growth as Urban Reform: A Pragmatic 'Recoding' of the New Regionalism. *Urban Studies*, 44(1), pp. 15–35.

Toly, N. (2008). Transnational Municipal Networks in Climate Politics: From Global Governance to Global Politics. *Globalizations*, 5(3), pp. 341–356.

Townsend, A.M. (2013). *Smart Cities: Big Data, Civic Hackers, and the Quest for a New Utopia*. New York: W.W. Norton & Company.

Tretter, E. (2016). *Shadows of a Sunbelt City*. Athens: University of Georgia Press.

Ward, K. and Jonas, A.E.G. (2004). Competitive City-Regionalism as a Politics of Space: A Critical Reinterpretation of the New Regionalism. *Environment and Planning A*, 36(12), pp. 2119–2139.

Whittemore, A.H. (2015). Practitioners Theorize, Too: Reaffirming Planning Theory in a Survey of Practitioners' Theories. *Journal of Planning Education and Research*, 35(1), pp. 76–85.

Wiig, A. (2016). The Empty Rhetoric of the Smart City: From Digital Inclusion to Economic Promotion in Philadelphia. *Urban Geography*, 37(4), pp. 535–553.

Williams, C. and Pendras, M. (2013). Urban Stasis and the Politics of Alternative Development in the United States. *Urban Geography*, 34(3), pp. 289–304.

PART II

Spaces of economic development

David Wilson and Andrew E.G. Jonas

The topic of this section, spaces of economic development, persists as a dominant kind of physical form in cities across the globe. High tech production zones, new conspicuous consumption retail districts, greying industrial zones and neighbourhoods comprising spatial divisions of labour are the most visible imprint of these spaces. The city is fundamentally an economic apparatus, David Harvey repetitiously tells us, and its most visible manifestation in the production of spaces that give form, life and texture to the economic functioning of cities. In these spaces conflict and contentiousness often bubble to the surface or lie lurking. As powerful exchange value advocates typically must navigate the reality of use value fervency, no other kind of space in cities seems to engender more conflict and controversy. The urban economy and its social relations ultimately carve out spatialities via claims and counter-claims about best cities as functioning economic units whose resolution is beyond simple. Economics, politics and space in current cities, as always, dynamically interconnect.

The chapters in this section collectively illuminate the complexities and multi-faceted nature of these ever-evolving spaces. Each in its own way critiques foundational notions of urban economic spaces to offer fresh insight into what is now emerging as cities evolve in neoliberal, transnational and austerity days. Cities here are simultaneously spatial products, complex and shifting terrains of politics, evolving places of meaning, and fundamentally unstable. The 'authorial standpoint', always up for grabs, stands as the elusive resource (the seizing of a powerful political position) that begins the politics of producing economic spaces. From here, the essays delimit, the activities of institutions involved in producing this space must begin.

Christopher Mele and Matthew H. McLeskey begin by interrogating an age-old city power broker, pro-growth coalitions, and crack them open to reveal a murky and complex world of competing claims, aspirations and styles. Their novel excavating of this entity reveals an unstable, heterogeneous political alliance that must toil to present a veneer of uniform interests and values. In reality, a mix of different organizational cultures, divergent goal and value orientations, and differing operational styles becomes occluded from public view under an elaborate presentation of a unified organizational grouping. Mele and McLeskey effectively address 'the myth of the unity', a common organizational trope (that academics too often fail to recognize) in cities today, that enables the likes of pro-growth coalitions to cast into the shadows their undergirding of political turbulence and tumultuousness. This bold unfolding ultimately unearths the shadowy world of seamy engagements, conflicting social relations and oppositional world views that are too often missed in our studies of growth coalitions.

Rachel Bok and Neil M. Coe take on an equally important issue, the spaces in cities today that are emerging from the intersections of urban politics and global production networks. They advance the notion that the popular analytic framework from economic geography – the global production networks (GPN) framework (now the 'world economy's backbone and central nervous system') – can be used to reveal much about the core content, possibilities and limitations of urban politics in the current era. With literatures on both urban political and global production networks being marked by sensitivities to multi-scalar operations, relationalness, and space and territorialization as crucial outcomes and resources, merging these insights makes sense. The authors reveal a mutually constitutive, synergistic set of relations between these spheres and the spaces they create. Urban politics is shown to be a building ingredient, reflection and future agent of change in global production networks that explains the unfolding of current political spaces. Bok and Coe skilfully display a dialectical, ever-evolving and processural city and world which defines a multi-textured relationship between these two domains.

Ute Lehrer and Peter Pantalone turn to another central aspect of cities today, current growth and redevelopment. They examine the long tradition of 'let's make a deal' planning in Canadian cities, and excavate the latest controversial trend in this, the provision of density bonuses by planners to developers. Their analysis shows an ever-enterprising state apparatus, always sensitive to latest trends, technologies and political openings, which fundamentally support an invigorated power broker in these cities, real-estate capital. Yet, as the authors show, this is a quid pro quo arrangement, one where the state expects to reap a central benefit – economic growth – in exchange for this provision of political support. Here a latest urban politics that enables the production of affluent residential and cultural spaces is about the state meeting its need and aspirations through a managing of capital. Lehrer and Pantalone, in this bold foray, ask us to reconsider the widely embraced notion of the state apparatus as essentially a passive prop controlled by ever-enterprising and conspiring real-estate capital.

Barney Warf focuses on a crucial prerequisite to cities striving to become globally competitive: the ability to acquire, process and transmit vast quantities of information. Current economic spaces in North American cites, to Warf, follow logically from this. Warf shows the latest wrinkles in this crucial prerequisite as it helps define smart cities, telework, intelligent transportation systems and e-government (the use of the Internet to deliver public goods). At the same time, the chapter reveals how interactive websites have gradually democratized the process of urban planning, making the drive to 'go global' more participatory even as the dilemma of urban digital divides continues to exist. Warf skilfully ties information production, dissemination and flows to an evolving urban political economy that has needs, requirements and resources. Production of urban economic space, he contends, must always be linked to the evolving informational economy and society that platforms this production. Without the current microelectronics revolution, digital technologies, new operations and styles of the Internet, Warf concludes, spaces of economic development in cities would be extremely different.

Much differs in these chapters, but all form a unity in their common desire to display the multiplicity of ways that cities come to create economic spaces that drive their economic functioning. Cities, these papers realize, are many things (e.g. a lived arena, a terrain for symbolism and meaning extraction, an imagined discursive landscape), but they are also fundamentally economic terrains that produce differential levels of wealth for city dwellers and people and institutions in surrounding regions, states, and nations. The economic may not be the grand determinant of all things in the city, these essays recognize, but they are one dominant buttress for what cities have become and are becoming. Institutions in cities today negotiate a welter of powerful forces – globalization, speeded-up transnational realities, deepening neoliberalism, hegemonic structures of austerity – that are now beginning to reshape the economic fabric of cities in ways that we are just beginning to study. It is in this spirit of seeking to learn more about this that we offer this focus on economic spaces of the city.

6
PRO-GROWTH URBAN POLITICS AND THE INNER WORKINGS OF PUBLIC-PRIVATE PARTNERSHIPS

Christopher Mele and Matthew H. McLeskey

Urban politics in North American and, to some extent, Western European cities continues to be defined by tenacious efforts to capture footloose and highly competitive private investment to stimulate local economic development. In older cities especially, public investment in property-led urban redevelopment, in the shape of mega-sports stadiums, entertainment venues, mixed-used residential-hotel-retail projects and educational-medical research complexes, is now rendered valid and reasonable by the much-promised increase in commerce, tax revenues, employment and other forms of economic growth (Altshuler and Luberoff 2003; Orueta and Fainstein 2008). Economic growth is now so fully normalized as the priority for urban development that competing models, such as sustainable community development, are considered exceptional and, frequently, idealistic. Pro-growth urban politics is equated with an entrepreneurialism and efficiency with local governments expected to work with the private sector (Erie *et al.* 2010; Sagalyn 2007; Stadtler 2014). A pro-growth politics of economic development relocates actual governance of urban development – building of stadiums, revitalized waterfronts, and so on – from traditional (and publicly accountable) government (i.e. city councils, planning departments and housing authorities) to collaborative organizational arrangements between public and private sectors. The central mechanism of pro-growth governance of urban development is the public-private partnership (Dewulf *et al.* 2011; Erie *et al.* 2010; Grimsey and Lewis 2004; Osborne 2000).

Scholarly literature on this subject examines the origins of such partnerships and the consequences of their outcomes, but the process of and internal dynamics within the public partnership itself merits interrogation. In this chapter, we examine the governance of partnerships as an evolving rather than static process. We argue that varying levels of coordination, collaboration, flexibility and adaptability within public-private partnerships influence their effectiveness and, therefore, their impact upon the range of social groups who inhabit areas within cities targeted for contemporary redevelopment.

Much of urban studies scholarship concludes public-private partnerships offer obvious organizational advantages to a pro-growth politics of urban development. Their very design is meant to overcome barriers – ideological, legal and practical – between government and the business sector to maximize economic growth (Reynaers and De Graaf 2014; Salet *et al.* 2013; Stoker 1998). For large urban developments in particular, partnerships are long term, incorporate joint decision-making (but not necessarily power sharing) and allocate negotiated risk between public and private sectors. For theories of pro-growth urban politics, the importance of public-private partnerships pivots around the underlying reasons why the two sectors would cooperate in the first place (motivations) and what each expects to achieve from working together on urban development projects (outcome). That is, pro-growth public and private partnerships matter for urban governance in not only what they do but also why they form and with what consequences, especially for the governed (Guarneros-Meza and Geddes 2010).

Theories of pro-growth urban politics approach questions of partnership motivations and outcomes from two perspectives: those that view partnerships as a means of collaborative governance and those that see them as the dominance of private capital and elites over the public sector. With increasingly mobile investment, cities stand to gain if they can convince the private sector to participate as formal stakeholders in local governance; partnerships secure private sector 'buy in' as collaborators in local economic development. Businesses see collaboration with local governments as not only profitable but beneficial to creating an urban environment conducive to recruiting and retaining skilled labour. Partnerships institutionalize (if only temporarily) an arrangement of mutual dependence of public and private sectors. Pro-growth governance benefits from private sector experience and expertise, including putting deals together, finding investment capital and creatively targeting new markets; businesses, in turn, benefit from favourable connections to city government, bypass bureaucracies, receive exemptions and cost-sharing that lower expenses and risks and enhance return on investment. Regime theorists contend enhanced interdependence of the public and private sectors increases a city's chances at successful economic development. Public-private partnerships bring together the resources and best features of each sector; they are the governance mechanisms of choice for a pro-growth politics of coalition building (Brenner and Theodore 2005; Goldstein and Mele 2016; Grodach and Ehrenfeucht 2015).

The public accountability of partnerships is of minor concern to a theoretical perspective that emphasizes resource sharing between public and private sectors. If partnerships are interdependent and not the sphere of undue influence of elites, then local governments have not forfeited their responsibilities to local citizens. Alternative theoretical approaches aligned with larger critiques of neoliberal governance have a different, more critical take on public-private partnerships. Pro-growth urban politics is defined by the withdrawal of traditional government priorities towards those governed, especially economically vulnerable urban populations. Partnerships are mechanisms of entrepreneurial or pro-market governance in which responsibilities towards community are forfeited to private sector management and control of the 'public good' of economic development. Partnerships are less opportunities for public-private sharing and collaboration than they are venues for governance drift or shift to private sector priorities (Brenner and Theodore 2005; Codecasa and Ponzini 2011). Partnerships serve "as a comprehensive tool for remaking government in the market's image" (Linder 2000: 28). At the core of this critique of public-private partnerships is the concern with accountability of pro-growth urban politics, namely the degree to which local governments have sidelined obligations to their citizenry by contracting out considerable decision-making, control and oversight over urban development to the private sector (Hackworth 2007; Healey 2006; Jessop 2002).

Each of the above perspectives employs its own ideologically tinged understanding of urban economic development to explain why partnerships form and what their outcomes will be. That is, economic development is conceived as either a public good that modern cities should strive for or as another iteration of urban inequality. Both perspectives focus on partners' motivations and the partnership outcome – and not much on what happens in between – and this has its limitations. Regardless of differences in ideological standpoints, both view partnerships as inherently favourable to urban economic development. Pro-growth politics and capitalism seem to be agreeably conjoined: both the public and private sectors share the same objective, each has a stake (political, economic or both) in a successful outcome. As a result, each is motivated to cooperate and collaborate with the other in order to achieve success. The reality, of course, is much more complicated. If public-private partnerships play a dominant role in pro-growth urban politics, a more complete understanding relies on how they operate or, or in other words, their inner workings. Clearly, partnerships do not function on the ground as clear-cut cases of either inter-sectoral sharing and collaboration or elite-dominated governance of urban development (Kimelberg 2011). Instead, a view of the inner workings of public-private partnerships may provide useful insights to theories of pro-growth urban politics.

Inner workings of public-private partnerships

Understandings of public-private partnerships in theories of urban politics may benefit and learn from research in other disciplines (namely, business management, public policy and public administration), where the theoretical underpinnings of partnership governance and its social consequences are not of paramount concern. Rather, such studies are evaluative and tend to examine in detail the performance and internal workings of partnerships (Brinkerhoff and Brinkerhoff 2011; Grimsey and Lewis 2002; Petersen 2011; Reynaers and Grimmelkhuijsen 2015; Stadtler 2014; Velotti et al. 2012; Zarco-Jasso 2005). While the criteria for evaluating partnership performance are many, they may be grouped into two broad categories that are applicable to pro-growth urban politics: degree of coordination and collaboration among partners and the level of organizational flexibility and adaptability. Both are process-oriented, meaning the determination of each is partly a function of how long partnerships exist. These two criteria, then, are significant to economic development partnerships, due to the length of time (typically, several years) involved in planning, implementing and completing projects, such as stadiums or downtown entertainment districts. The focus on the internal operation of partnerships means the evaluation of residents' needs and desires are not taken into consideration.

Partnerships prove to be ideal organizations for contemporary urban economic development because the "rigid division of sectoral responsibilities" of governments and private interests is replaced by an "intersectoral relationship" characterized by the collaboration and even the merging of economic development interests of the public and private sectors (Linder and Vaillancourt Rosenau 2000: 8). Yet, in practice, public-private partnerships bring together each sector's different organizational cultures, potentially divergent expectations and the unique operational styles. Organizational and operational compromise is normally achieved at the formation of a partnership with agreement about a set of precise objectives and protocols. Yet, for some partnerships the rules and norms of internal governance may not be initially decided upon (Hajer 2003; Swyngedouw 2005). In either case, property-led development partnerships are not static over the course of their life cycle, hence their composition and operations are susceptible to changes from both internal and external sources – with the latter being far less predictable and difficult to manage (Tasan-Kok 2010: 127). Given the near certainty of external situations and unforeseen challenges over the life cycle of urban development, a working definition of public-private partnerships must incorporate elements of process and contingency. Examining the inner workings and dynamic nature of partnerships may prove useful to theories of urban politics and future research agendas.

Partnership coordination and collaboration

Urban public-private partnerships are reputedly entrepreneurial enterprises in which the virtues of efficiency and pragmatism result from high quality collaborations among partners (Roberts and Siemiatycki 2015). Such quality is dependent upon the mutuality that exists between public and private sector actors (Kort and Klijn 2011; MacDonald 2012). Mutuality, in this respect, is not simply defined by the character of partners' interpersonal relationships. Mutuality is shaped by the coordination of goals-specific competencies of partners – how well each fits the objectives of the project. Coordination, then, requires collective acknowledgement of partners' capabilities but also how "responsibilities seek to maximize benefits for each party, subject to limits posed by the expediency of meeting joint objectives" (Brinkerhoff 2002: 217). Coordination among partners is not a pregiven attribute of partnership formation; its levels and quality are influenced by the development (or dissolution) of trust among partners over time (Austin and McCaffrey 2002: 42–43). As scholars of organizational dynamics point out, high trust levels foster an environment supportive of innovation, ideas and new approaches; low trust levels entail greater scrutiny, insecurity and lagging commitment to goals and objectives. The unfolding process of collaboration and trust is the major differences between modern partnerships and traditional contract relations between government and private sector actors, in which the division of responsibilities and formalized interactions between actors are stipulated out front. As DiGaetano and Strom (2003: 376) write

> the basic principle in the formation of [public-private] partnerships is to bring together government officials and civic leaders (business, community, etc.), first to break down intersectoral barriers of mistrust and misunderstanding and then to formulate and implement local policies that will address the challenge of governing cities.

During the life cycle of partnerships, interactions between local elites and local governments are reproduced and subject to change. The presumption that the interests of private sector partners are somehow already aligned from the start (and remain so) suggests elites behave as "an undifferentiated, monolithic bloc and fail to capture the full richness and realities of business sector participation in public-private partnerships" (Austin and McCaffrey 2002: 36). Coordinated decision-making is difficult to achieve when each sector is predisposed to its own strategies (Klijn and Teisman 2005). Even the most basic levels of coordination and collaboration among disparate public and private partners evolves through communication and routine interactions – the internal politics of which can neither be assumed nor simply deduced from partners' shared interest in a favourable (i.e. profitable) outcome of urban development. Research that adopts a neoliberal perspective, for example, would benefit from interrogating how a correspondence of interests tilted in favour of the private sector actually occurs and transpires. Such research on the role of agency would benefit a perspective that otherwise emphasizes structures of political economy.

The point here is simple: theories of pro-growth politics cannot assume a degree of coordination and collaboration among partners within public-private partnerships. Regime theory's emphasis upon horizontal power arrangements between public and private sectors may be stipulated in the formation of a partnership but may be tested over time. Partnerships with equally aligned risks and responsibilities between sectors may tilt towards one over the other, giving private developers, for example, effective control over development for many decades. Neoliberal theories, too, cannot assume an imbalance of power and interests automatically tilted in favour of the private sector is an intrinsic aspect of urban development partnerships. Either is possible but neither is inevitable; coordination and collaboration are social processes subject to change over time. With urban development partnerships lasting years (even decades) the reproduction of mutuality and trust is required in order to keep partners engaged and projects on track, especially with likely change in business leaders and city administrations (Austin and McCaffrey 2002: 37).

Changes in partnership coordination and collaboration resulting from unforeseen situations may either enhance or decrease a partnership's public accountability and transparency. Responses to internal tensions and external factors can easily alter the perceived trade-off for local governments between short-term economic gain and long-term public interest. Research into the inner workings of partnerships may demystify decisions frequently assumed as the product of collaboration among partners. For example, Coghill and Woodward (2005: 89) describe how "political decisions" can be presented as "technical decisions", because

> negotiations between government and private sector partners are typically conducted in private under the mantra of 'commercial in confidence' such that the public has little or no input into such negotiations but are presented with the outcome as a *fait accompli*.
>
> *(Coghill and Woodward 2005: 89)*

Partnership flexibility and adaptability

The inner workings of public-private partnerships also depend upon the ability to adapt to changing circumstances throughout the course of the collaboration. Greater levels of trust and coordination among partners, for example, would assist in weathering obstacles and challenges. The internal dynamics of partnerships are prone to changes caused especially by externalities, "as it is impossible to anticipate everything in advance" (Klijn and Teisman 2005: 96). Externalities may be project-specific, such as cost overruns or the collapse of financing, or contextual, such as a change in government or a global economic recession. A partnership's ability to adapt to unanticipated situations is a subject of empirical inquiry. But organizational components point to the likelihood (or not) of adaptation, including an organization's culture of communication among partners, the receptivity to new ideas and the speed and flexibility of corrective action to fix problems that arise. The task ahead for theories of urban politics is to integrate a process approach which captures and accounts for how partnerships, as the mechanisms of pro-growth governance, respond and adapt to unforeseen problems. Otherwise, partnerships are naively understood as fixed, stable and crisis free in practice.

A partnership's response to unanticipated circumstances typically throws the details of its inner governance into sharp relief. The results can reveal power asymmetries, weaknesses in communication and information sharing, and the idiosyncrasies of coordination. Partnerships with an agreed upon and acknowledged distribution of specific roles and responsibilities among partners will respond to external situations very differently than those in which the assignment of responsibilities is informal or voluntary. The assumed entrepreneurial and innovative quality of partnerships may be readily tested.

Responses to external situations are not uniformly predictable for either the private or public sector actors. Flexible government partners may be legally restricted or constrained by administrative protocols or budgetary obligations. The private sector has its own set of constraints linked to profit-driven motives. While the public sector may have limited capacity to engage in strategic and innovative problem solving, private actors' profit desires may narrow the range of possible responses to changing circumstances. Policy studies evaluate partnerships on their level of tolerance to known and unanticipated risks (Edelenbos and Teisman 2008; Estache and Serebrisky 2004; Schaeffer and Loveridge 2002). Private sector members are less willing to accept risks to profit returns than the public sector, where local governments can safely accommodate losses to public coffers, as long as this is legitimated by public support. Ysa (2014) addresses how partnership dynamics respond to contingencies to risk over time. As partnerships evolve, they assume 'variable geometries' in internal arrangements. Symbolic partnerships tend to be more hierarchical and less flexible in the face of change; instrumental partnerships respond to unforeseen situations by automatically deferring to market rules; organic partnerships are the most flexible because their internal governance is network-based and partner-inclusive.

The degree to which public-private partnerships are flexible and can adapt to unforeseen challenges is relevant to theories of urban politics. In partnerships where the local government's concerns over the expectations (and accountability) of the citizenry are paramount and fixed, we are likely to lose the participation of private sector actors if unanticipated risks of continued participation are too high. Conversely, for those partnerships which lean towards the private sector, the likelihood of adapting to changes tends to be greater – however, specific aspects of the project itself may need to change (e.g. fewer low- to moderate-income units in a housing development). In their study of a public-private partnership overseeing urban development in Detroit during the 1980s and 1990s, Austin and McCaffrey (2002: 49) note that external circumstances revealed "functional asymmetries or constraints on effectiveness" forcing a lengthy process in which the partnership's mission and purpose were completely redefined. The extent to which cooperative governance mechanisms are deployed in public-private partnerships are influenced by various political, social, ideological and legal factors (Lauermann and Davidson 2013; MacLeod and Jones 2011).

Conclusion

We began this chapter noting that amid public-private partnerships receiving much attention in theories of urban politics, there remain important gaps in the understanding of their inner workings. Theories of urban politics assume public-private partnerships form around specific projects and are comprised of institutions, agencies and individuals who join together, pool their resources, expertise, skills and plans and form a coalition for action based on the sharing of risks, responsibilities and potential gains (Pierre 2011). Critical perspectives on urban politics have problematized partnerships, question whose purposes they serve and whether their goals truly help the city (understood as containing diverse and divergent populations). Consequently, few urban scholars have questioned the efficiency and tight coordination of current public-private partnerships. Our concern is that this understanding is defined without due consideration to how a partnership's internal governance can have implications for theories of urban politics. We chronicled how variations in coordination and collaboration and adaptability and flexibility influence partnership performance and project outcomes. In addition, we emphasized that public-private partnerships are not static but ever evolving and subject to pressures brought on by unforeseen situations. The governance of public-private partnerships is a grounded social process subject to internal fracturing and pressures from external and often unanticipated situations. Future research should develop a richer understanding of the organizational interactions between private and public actors and explore how risk tolerance influences responses to external situations. These proposed research avenues can deepen and nuance our understanding of current city governances and the economic politics that follow.

References

Altshuler, A. and Luberoff, D. (2003). *Mega-Projects: The Changing Politics of Urban Public Investment*. Washington, DC: Brookings Institution Press.

Austin, J. and McCaffrey, A. (2002). Business Leadership Coalitions and Public-Private Partnerships in American Cities: A Business Perspective on Regime Theory. *Journal of Urban Affairs*, 24(1), pp. 35–54.

Brenner, N. and Theodore, N. (2005). Neoliberalism and the Urban Condition. *City*, 9(1), pp. 101–107.

Brinkerhoff, D.W. and Brinkerhoff, J.M. (2011). Public–Private Partnerships: Perspectives on Purposes, Publicness and Good Governance. *Public Administration and Development*, 31(1), pp. 2–14.

Brinkerhoff, J.M. (2002). Assessing and Improving Partnership Relationships and Outcomes: A Proposed Framework. *Evaluation and Program Planning*, 25, pp. 215–231.

Codecasa, G. and Ponzini, D. (2011). Public–Private Partnership: A Delusion for Urban Regeneration? Evidence From Italy. *European Planning Studies*, 19(4), pp. 647–667.

Coghill, K. and Woodward, D. (2005). Political Issues of Public-Private Partnerships. In: G. Hodge and C. Greve, eds. *The Challenge of Public-Private Partnerships*. Northampton, MA: Edward Elgar, pp. 81–94.

Dewulf, G., Blanken, A. and Bult-Spiering, M. (2011). *Strategic Issues in Public-Private Partnerships*. New York: John Wiley & Sons.

DiGaetano, A. and Strom, E. (2003). Comparative Urban Governance: An Integrated Approach. *Urban Affairs Review*, 38(3), pp. 356–395.

Edelenbos, J. and Teisman, G.R. (2008). Public-Private Partnership: On the Edge of Project and Process Management: Insights From Dutch Practice: The Sijtwende Spatial Development Project. *Environment and Planning C: Government and Policy*, 26(3), pp. 614–626.

Erie, S.P., Kogan, V. and MacKenzie, S.A. (2010). Redevelopment San Diego Style: The Limits of Public-Private Partnerships. *Urban Affairs Review*, 45(5), pp. 644–678.

Estache, A. and Serebrisky, T. (2004). Where Do We Stand on Transport Infrastructure Deregulation and Public-Private Partnerships? *World Bank Policy Research Working Paper 3356*, Washington, DC: The World Bank.

Goldstein, B. and Mele, C. (2016). Governance Within Public Private Partnerships and the Politics of Urban Development. *Space and Polity*, 20(2), pp. 194–211.

Grimsey, D. and Lewis, M.K. (2002). Evaluating the Risks of Public-Private Partnerships For Infrastructure Projects. *International Journal of Project Management*, 20(2), pp. 107–118.

Grimsey, D. and Lewis, M.K. (2004). *Public-Private Partnership: The World Wide Revolution in Infrastructure Provision and Project Finance*. Cheltenham: Edward Elgar.

Grodach, C. and Ehrenfeucht, R. (2015). *Urban Revitalization: Remaking Cities in a Changing World*. New York: Routledge.

Guarneros-Meza, V. and Geddes, M. (2010). Local Governance and Participation Under Neoliberalism: Comparative Perspectives. *International Journal of Urban and Regional Research*, 34(1), pp. 115–129.

Hackworth, J. (2007). *The Neoliberal City: Governance, Ideology and Development in American Urbanism*. Ithaca, NY: Cornell University.

Hajer, M. (2003). Policy Without Polity? Policy Analysis and the Institutional Void, *Policy Sciences*, 36(2), pp. 175–195.

Healey, P. (2006). Transforming Governance: Challenges of Institutional Adaptation and A New Politics of Space. *European Planning Studies*, 14(3), pp. 299–320.

Jessop, B. (2002). Liberalism, Neoliberalism, and Urban Governance: A State-Theoretical Perspective. *Antipode*, 34(3), pp. 452–472.

Kimelberg, S.M. (2011). Inside the Growth Machine: Real Estate Professionals On the Perceived Challenges of Urban Development. *City & Community*, 10(1), pp. 76–99.

Klijn, E.H. and Teisman, G.R. (2005). Public-Private Partnerships as the Management of Co-Production: Strategic and Institutional Obstacles in a Difficult Marriage. In: G. Hodge and C. Greve, eds. *The Challenge of Public-Private Partnerships*. Northampton, MA: Edward Elgar, pp. 95–116.

Kort, M. and Klijn, E-H. (2011). Public-Private Partnerships in Urban Regeneration Projects: Organizational Form or Managerial Capacity? *Public Administration Review*, 71(4), pp. 618–626.

Lauermann, J. and Davidson, M. (2013). Negotiating Particularity in Neoliberalism Studies: Tracing Development Strategies Across Neoliberal Urban Governance Projects. *Antipode*, 45(5), pp. 1277–1297.

Linder, S.H. (2000). Coming to Terms With the Public-Private Partnership. In: P. Vaillancourt Rosenau, ed. *Public-Private Policy Partnerships*. Cambridge, MA: MIT, pp. 19–36.

Linder, S.H. and Vaillancourt Rosenau, P. (2000). Mapping the Terrain of the Public-Private Policy Partnership. In: P. Vaillancourt Rosenau, ed. *Public-Private Policy Partnerships*. Cambridge, MA: MIT, pp. 1–18.

MacDonald, R. (2012). Pinning Down the Moving Target: The Nature of Public-Private Relationships. *Public Performance & Management Review*, 35(4), pp. 578–594.

MacLeod, G. and Jones, M. (2011). Renewing Urban Politics. *Urban Studies*, 48(12), pp. 2443–2472.

Osborne, S.P. (2000). Introduction: Understanding Public-Private Partnerships in International Perspective: Globally Convergent or Nationally Divergent Phenomena? In: S.P. Osborne, ed. *Public-Private Partnerships: Theory and Practice in International Perspective*. London, Routledge, pp. 1–5.

Orueta, F.D. and Fainstein, S.S. (2008). The New Mega-Projects: Genesis and Impacts, *International Journal of Urban and Regional Research*, 32(4), pp. 759–767.

Petersen, O.H. (2011). Multi-Level Governance of Public-Private Partnerships: An Analysis of the Irish Case. *Administrative Culture*, 12(2), pp. 162–188.

Pierre, J. (2011). *The Politics of Urban Governance*. New York: Palgrave Macmillan.

Reynaers, A.M. and De Graaf, G. (2014). Public Values in Public-Private Partnerships. *International Journal of Public Administration*, 37(2), pp. 120–128.

Reynaers, A.M. and Grimmelkhuijsen, S. (2015). Transparency in Public-Private Partnerships: Not So Bad After All? *Public Administration*, 93(3), pp. 609–626.

Roberts, D.J. and Siemiatycki, M. (2015). Fostering Meaningful Partnerships in Public-Private Partnerships: Innovations in Partnership Design and Process Management to Create Value. *Environment and Planning C*, abstract: 0–0.

Sagalyn, L.B. (2007). Public/Private Development: Lessons From History, Research, and Practice. *Journal of American Planning Association*, 73(1), pp. 7–22.

Salet, W., Bertolini, L. and Giezen, M. (2013). Complexity and Uncertainty: Problem or Asset in Decision Making of Mega Infrastructure Projects? *International Journal of Urban and Regional Research*, 37(6), pp. 1984–2000.

Schaeffer, P.V. and Loveridge, S. (2002). Toward an Understanding of Types of Public-Private Cooperation. *Public Performance and Management Review*, 26(2), pp. 169–189.

Stadtler, L. (2014). Designing Public-Private Partnerships For Development. *Business & Society*, 54(3), pp. 406–421.

Stoker, G. (1998). Public-Private Partnerships and Urban Governance. In: J. Pierre, ed. *Partnerships in Urban Governance: European and American Experience*. London: Macmillan, pp. 34–51.

Swyngedouw, E. (2005). Governance Innovation and the Citizen: The Janus Face of Governance-Beyond-the-State. *Urban Studies*, 42(11), pp. 1991–2006.

Tasan-Kok, T. (2010). Entrepreneurial Governance: Challenges of Large-Scale Property-Led Urban Regeneration Projects. *Tijdschrift voore Economische en Sociale Geografie*, 101(2), pp. 126–149.

Velotti, L., Botti, A. and Vesci, M. (2012). Public-Private Partnerships and Network Governance. *Public Performance & Management Review*, 36(2), pp. 340–365.

Ysa, T. (2014). Governance Forms in Urban Public-Private Partnerships. *International Public Management Journal*, 10(7), pp. 35–57.

Zarco-Jasso, H. (2005). Public-Private Partnerships: A Multidimensional Model For Contracting. *International Journal of Public Policy*, 1(1–2), pp. 1–21.

7
THE URBAN POLITICS OF STRATEGIC COUPLING IN GLOBAL PRODUCTION NETWORKS

Rachel Bok and Neil M. Coe

Introduction: the intersections of urban politics and global production networks

> ...global production arrangements are concrete determinations of social and spatial divisions of labor. Places do not 'enter into' global supply chains; rather, places and global arrangements of production dynamically reproduce one another.
>
> *(Werner 2015: 184)*

This chapter develops the argument that a widely used conceptual apparatus from economic geography – the global production networks (GPN)[1] framework – can be used to reveal much about the nature, possibilities and limitations of urban politics in the contemporary era. Bringing these conceptual domains into dialogue is an appealing prospect given that both literatures attempt to offer a multi-scalar perspective, and hold the territorial and relational dimensions of their core phenomena in productive tension. While work on urban politics increasingly seeks to place territorial formations within wider relational webs of circulation and exchange, GPN research identifies networked global formations and evaluates the territorial developmental impacts on the places that are enrolled therein.

This conversation has great potential benefit for both relational economic geography and urban politics research. Over the past twenty years or so, global production networks – broadly defined as organizationally fragmented, spatially dispersed and globally coordinated production systems – have emerged as the "world economy's backbone and central nervous system" (Cattaneo *et al.* 2010: 7). Global production networks are thus organizational arrangements, coordinated by powerful transnational corporate actors known as lead firms, which produce goods and services across multiple geographical locations for global markets (Coe and Yeung 2015). The fact that some 60 per cent of global trade is now accounted for by exchanges of intermediate (i.e. unfinished) goods and services (UNCTAD 2013) provides a powerful indication of the prevalence of these globally networked production systems (also termed global commodity/value chains by some). From an urban politics perspective, therefore, understanding the economic development trajectories of cities – and struggles thereover – increasingly need to be seen from the context of how these cities are 'plugged in' to global production networks.

From a GPN perspective there is also much to gain. Many studies of global production networks focus primarily on the 'global', networked dimensions of production systems, at the expense of understanding the cities and territories that mutually constitute those global network formations; it is as though global production networks 'touch down' on the head of the pin. Foregrounding urban politics, however, allows us to reveal the concrete, sprawling, and multi-centric places that are enrolled into such networks, and the contestations and struggles over the economic value generated by these intersections (Levy 2008). As we will see in this chapter, these distributional struggles have both social and spatial dimensions; that is, between actors directly connected to global production networks and those who are excluded, and between territories that are variously 'plugged in' to a greater or lesser extent. In simple terms, the consideration of urban politics can help us move from profiling development 'in' a given locality to the thornier issue of development 'of' that locality.

Arguably, then (and as the opening epigram of this chapter alludes), global production networks and urban politics should be seen as mutually constitutive, or at least overlapping and intersecting, fields. Global production networks are not external to urban politics, but rather reflect the real-world production structures, processes and social relations through which economic development in urban territories is shaped. Similarly, urban politics are reflective of the contested processes of economic development surrounding globally networked production systems. In this chapter, we will explore these intersections in three stages. First, we will consider in greater detail how GPN theory is well suited to providing a framing for these discussions. Here the concept of *strategic coupling* will be introduced as a key analytic. Second, the strategic coupling concept will be unpacked to reveal the multiple modes and types of coupling that exist in the contemporary global economy, demonstrating the various ways whereby urban territories may be incorporated into wider regional and global production networks. Third, the urban politics of strategic coupling will be evaluated as the domain through which the inherent tensions and frictions of urban economic development and governance are negotiated and contested.

Using the GPN approach to frame the economic development of city-regions

GPN theory clearly does not offer the only starting point for conceptualizing urban economic development processes in the context of the global economy. Current debates are, arguably, heavily inflected by the World-City Network (WCN) approach (e.g. Taylor 2004; Taylor et al. 2010), which is mostly focused on mapping a narrow range of sectors of corporate activity (specifically, advanced producer/business services). This results in a rather partial understanding of the place of cities in global corporate networks, and how urban spaces are governed by a range of powerful economic agents operating at multiple spatial scales (Coe et al. 2010). Our contention is that GPN analysis provides a more flexible and nuanced way to investigate the dynamics of powerful economic actors (e.g. transnational corporations) in globalizing cities and can thus more deeply unpack urban politics in a global setting.

GPN theory has three important attributes that make it amenable to framing the notion of urban politics in relation to economic development in a globalizing economy. First, it is a multi-actor approach. Such networks are understood as being constituted through complex meshes of intra-, inter- and extra-firm networks – the latter of which brings non-firm actors such as the state, international organizations, labour groups, consumers and civil society groups into view as crucial elements of global production networks who may, in turn, shape firm activities within the particular places absorbed into the networks. In so doing, this approach also brings to the fore the institutional context of global production networks, as actors with different territorial responsibilities interact and intersect. Second, global production networks are systems of power relations in which ongoing negotiations and struggles occur over processes of value creation, enhancement (i.e. value adding) and capture (i.e. those actors and/or places where value is ultimately retained). Global production networks, therefore, are inherently and unavoidably political-economic formations (Levy 2008). From a GPN perspective, power is understood as a relational attribute; it is transaction-specific, and co-exists with relations of mutual interest and dependency.

Third, global production networks are conceptualized as simultaneously networked and territorial formations, combining both a 'vertical' dimension in which economic activities are organized at scales ranging from the local to the global, and a 'horizontal' territorial interface through which production is embedded in on-the-ground territorial formations such as clusters, cities and city-regions (Coe and Yeung 2015). In this regard, research on the territorial and developmental aspects of global production networks tends to revolve around the 'region' as a spatial setting for the interactions between transnational corporate networks and localized assets, even as these exchanges are not restricted to exclusively regional spaces alone. To this end, such a viewpoint perhaps underplays how global production networks are underpinned by urban dynamics, and the ways whereby they necessarily intersect with specific urban territories (Coe et al. 2010; Vind and Fold 2010).

Thinking of global production networks in terms of their 'vertical' and 'horizontal' dimensions pushes us to move beyond viewing the city as a point in Cartesian space, and to identify it as a territorial entity crucial to the rhythms of economic development. In particular, it leads us to reflect on how 'urban' politics can take on different spatial forms; as Scott (2001) describes, these can range across central metropolitan areas and their hinterlands, a set of overlapping urban areas (i.e. a conurbation) or a regional network of distinct cities. More broadly, it speaks to emerging debates within urban politics concerning the 'where' of urban conflicts, together with the distinctly urban dimensions of capitalism (Rodgers et al. 2014), which cannot be reduced to either the metropolis or the suburbs, as earlier research has tended to do (Young and Keil 2014). A wider perspective allows us to keep the sectoral diversity of the economy in view – as opposed to the advanced producer service/central business district lens perpetuated by WCN research – and highlight the full range of urban actors and spatialities that are relationally linked through global production networks. Seeing the world economy as an interconnected mosaic of porous urban-regional economies complicates the notion of 'the urban', and posits the functional city-region as the key analytical unit for understanding processes of economic development. As Scott et al. (2001: 11, emphasis added) describes, "it has become increasingly apparent that *the city in the narrow sense* is less an appropriate or viable unit of local social organization than city-regions or regional networks of cities".

More specifically, we argue that the notion of *strategic coupling* developed in the GPN literature (Coe et al. 2004; Yeung 2009) offers great analytical potential for comprehending the mutually constitutive intersections between urban politics and global production networks. In line with our aforementioned evoking of city-regions, GPN analysis posits that the key locus for understanding economic development is the subnational region (which here we equate with the city-region): firms are situated in particular places, not merely within national economies, and all regions have distinctive spatio-institutional conditions that shape development practices and processes. The city-region is thus the basic geographical building block through which patterns of economic growth and decline should be interpreted. Importantly, strategic coupling pays analytical attention to both endogenous growth factors within specific territories and the strategic needs of extra-local actors coordinating global production networks – notably lead firms. Development is thereby conceptualized as a dynamic outcome of the complex interactions between region-specific networks and global production networks within the context of evolving regional governance structures. Regional development is understood as being shaped by these patterns of interaction, rather than just inherent territorial advantages or deterministic top-down globalization processes.

In the strategic coupling framework, endogenous factors are seen as necessary but insufficient for generating growth in an era of increasingly global competition. For development to take place, a locality must be able to leverage the local human, technological and institutional resource base. In Figure 7.1, the term 'regional assets' (Storper 1997) is used to delineate these necessary preconditions for regional development. In general, these assets can produce two types of economies. On the one hand, economies of scale can be achieved in certain regions through highly localized, place-based concentrations of specific knowledge, skills and expertise. On the other hand, economies of scope can exist if these regions are able to reap the intangible benefits of learning and the co-operative atmosphere embedded in these agglomerations. GPN analysis suggests that these economies of scale and scope are only advantageous to regions – and are able to bring about regional development – insofar as they can complement the strategic needs of lead firms in global production networks. As shown in Figure 7.1, when such a complementary effect exists between regions and global production networks, a developmental process of *strategic coupling* will take place through which regional assets interact with the strategic needs of actors in these global production networks.

The notion of strategic coupling has several important characteristics (see also MacKinnon 2012). First, it is strategic in that it needs intentional and active intervention on the part of both regional institutions and global production network actors. Second, it transcends territorial boundaries as actors at different spatial scales interact. In other words, many of the key strategic decisions that determine the nature of coupling within a particular locality are undertaken beyond its boundaries by actors associated with other spatial scales (e.g. national, global). Third, it is time-space contingent in being a temporary coalition between local and non-local actors that is subject to change. The strategic coupling of local actors (firms and extra-firm organizations) with lead firms in global production networks should not be construed in functionalist terms because the coupling process is not automatic, nor is it always successful or beneficial for the actors and localities involved. In many cases, coupling is not so much strategic as it is structural: MacKinnon (2012), for instance, argues that the term 'structural coupling' crucially emphasizes the continued relevance of themes such as uneven regional development and

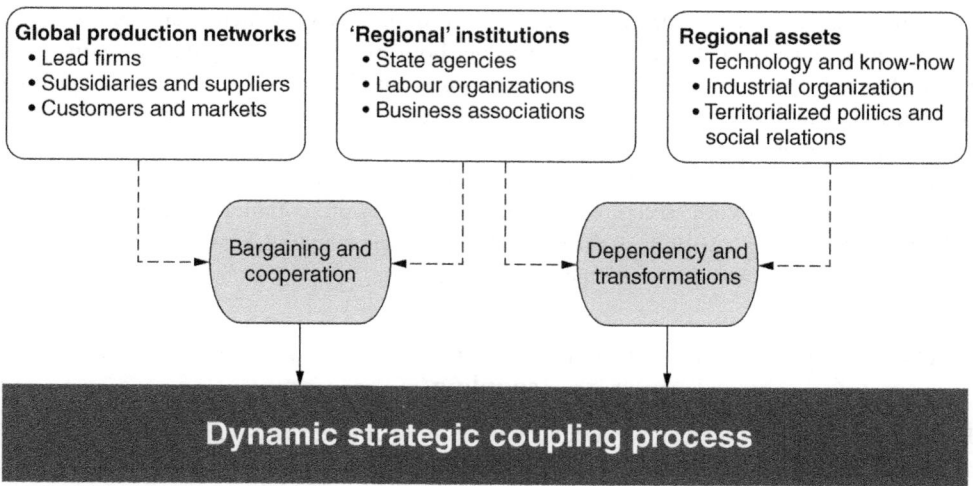

Figure 7.1 Urban politics and the strategic coupling process

corporate dominance, which require deeper interrogation of the exact nature of development in/of any given territory. The developmental process, then, needs to be critically evaluated because it changes across time and space in different geographical contexts. Hence, regional development can be fruitfully cast in evolutionary terms as being shaped by periods of strategic coupling in sequence with phases of decoupling and subsequent recoupling (Bair and Werner 2011).

In Figure 7.1, the coupling process is seen to work through the processes of value creation, enhancement and capture. Regional assets can become an advantage for development *only if* they fit the strategic needs of global production networks. This process requires the presence of appropriate institutional structures that simultaneously promote regional advantages and enhance the locality's articulation into global production networks. It is critical here that the notion of 'regional' institutions includes not only regionally specific institutions, but also local arms of national/supranational bodies (e.g. a trade union's local chapters), and extra-local institutions that affect activities within the region without necessarily having a physical presence (e.g. a national tax authority). These multi-scalar 'regional' institutions are pivotal because they can provide the 'glue' that ties down global production networks in particular localities. Generally, the more a region is integrated into global production networks, the more likely it is able to reap the benefits of economies of scale and scope in these networks, but the less likely it is able to control its own fate. However, when their region-specific assets are highly complementary to the strategic needs of global production network actors, regional institutions may be able to bargain with lead firms and negotiate the terms of entry in ways that mean power relations are not necessarily one-way, and in favour of the latter.

Even based on this short introductory account, the potential for connecting to debates on urban politics should be clear. The notion of multi-scalar institutions is directly linked to urban political actors and the ways whereby urban politics needs to be understood as shaped, in relational terms, by processes operating in other places and at multiple spatial scales. Strategic coupling captures how urban politics is simultaneously both an 'inward' (in terms of developing and managing localized assets) and 'outward' (in terms of negotiating with global production network actors) domain. Moreover, these two facets seem to align, respectively, with the spaces of dependence and engagement so powerfully delineated by Cox (1998). Therefore, as Figure 7.1 intimates, urban-regional politics are implicated in, rather than external to, the dynamics of strategic coupling. At the core – at least in the area of economic development – are concerns over the creation, enhancement and capture of value, both by the urban-regional economy as a whole, but also among the local and extra-local actors enrolled therein. And these interactions of bargaining and cooperation, and dependency and transformation, are two-way in nature; in some contexts, global production networks will drive key aspects of urban politics and institutional agendas, but equally in others, institutional strategies will lead to reconfigurations of global production network structures. In sum, for matters of economic development, urban politics constitutes the core domain through which strategic coupling between city-regions and global production networks is contested and negotiated.

Integrating city-regions into global production networks through strategic coupling

If the idea of strategic coupling powerfully explains how different city-regions can be variously articulated into the global economy (as networked sites of production and consumption), in this section we use this concept to illustrate these diverse modes of incorporation, together with the organizational dynamics that underlie such processes. As we have detailed, key to the strategic coupling concepts are the different ways whereby actors in city-regions may purposefully interact with those in global production networks to pursue particular developmental outcomes through mobilizing localized assets. As Yeung (2009) suggests, there is a certain element of *directionality* to such engagements. They can be broadly viewed from the standpoint of lead firms seeking to invest in particular localities and in the process incorporating regional actors into the wider production system, or regional actors who take the initiative to 'reach out' and forge transactional relationships with global production network actors. The former takes on an 'outside-in' nature, where the interaction is initiated by external actors, while the latter can be framed more as an 'inside-out' engagement, where coupling is set in motion by regional actors. From this we may begin to parse different forms of contemporary economic globalization to better comprehend the ways in which urban territories are entwined with the global economy, and with what implications.

In an effort to delineate this multi-faceted concept, Coe and Yeung (2015: 184) provide a nuanced account of the *modes* of strategic coupling that may exist between actors and territories, which can be categorized as:

Indigenous couplings: These are initiated by urban-regional actors, who reach beyond their domestic territories with the explicit aim of forging relationships with global production network actors; this direction of coupling allows local actors to hold a presumably greater degree of autonomy in dealing with external actors, which has implications for the nature and outcomes of bargaining between actors.

Functional couplings: These can be initiated by either urban-regional actors or global production network actors, where city-regions and their localized assets are envisioned as capable of fulfilling the functional requirements of global production networks; given the ambiguity in direction, the question of autonomy is an open one.

Structural couplings: These occur when global production network actors initiate interactions with urban-regional actors to enrol their territories into wider corporate networks; this direction of coupling suggests that the city-region is constituted primarily through a state of dependency and reliance on external actors and factors.

These three modes of incorporation are further illustrated in Table 7.1[2] through various *types* and real-world examples of strategic coupling between city-regions and global production networks.

In what follows, we draw on empirical examples from the wider literature to show the multiple ways diverse city-regions can be articulated into global production networks, and more generally how urban economic development is organized, influenced and shaped in a globalizing setting by various corporate and institutional interests. Where **innovation hubs** are concerned, these are positioned as city-regions which constitute a 'home-base' from which lead firms can construct and steer global production networks. There is a discernible notion of leadership – together with a relational interplay of power – as actors in these localities presumably possess more organizational power than others in shaping the network, and deciding the terms of entry by which other city-regions are to be enrolled. For example, the growth of the LCD (Liquid Crystal Display) industry in the Seoul Metropolitan Area reflects the multi-scalar entanglements of state, institutional and business interests within and beyond Seoul itself to develop suitable home-base advantages for lead firms such as Samsung Electronics and LG Display, enabling them to create and drive global production networks in the sector through different clusters (Lee *et al.* 2014). As Lee *et al.* (2014) further elaborate, the efforts of state governments at multiple scales overseeing these clusters are concentrated in the realms of financial and institutional support, with the aims of retaining Samsung and LG's investments in the clusters specific to their jurisdictions, and to ensure more generally a state of national

Table 7.1 Key types of strategic coupling between city-regions and global production networks

Type	Mode of coupling	Brief description	Examples of city-regions
Innovation hubs	Indigenous	Domestic territories of lead firms that drive global production networks	Silicon Valley, Seoul, Stockholm
Global cities	Indigenous	Spaces of convergence for the headquarters of lead firms and specialized service providers	London, New York, Tokyo
International partnerships	Functional	Spaces where partnerships between local and extra-local actors are negotiated and cultivated to meet the needs of global lead firms	Singapore, Taipei-Hsinchu, Rotterdam
Logistics hubs	Functional	Transportation infrastructure that coordinates global production networks through various forms of service provision	Singapore, Shanghai, Rotterdam
Market regions	Functional	Key locality to coordinate various global production networks through sales and distribution	Various city-regions in Pacific Asia
Assembly platforms	Structural	Standardized assembly of offshored goods and services for export	Pearl River Delta, Yangtze River Delta, Greater Bangkok

Source: derived from Coe and Yeung (2015, Table 5.3).

competitiveness in the LCD sector. For instance, not only has the national state provided significant amounts of funding to support Samsung's locational decisions in the Asan-Tangjung cluster, it has also established supportive regional institutions and technology centres to provide technological assistance to firms and stimulate mutually beneficial inter-firm and extra-firm exchanges. More generally, state governments at multiple scales have ensured beneficial planning outcomes and the construction of regional innovation systems, which collectively constitute favourable territorial assets for coupling. As a consequence of cluster formation in this context, display-related foreign suppliers, particularly those from Japan, have built operations in Gyeonggi Province to supply parts and materials to LG and Samsung. The creation and development of LCD clusters in Seoul both contributes to, and is reflective of, strategic coupling between South Korean transnational corporations (TNCs), regional actors and state institutions, and foreign suppliers and TNCs based in these clusters.

Similar to the sort of leadership position innovation hubs hold in global production networks, **global cities** would seem to provide a privileged space for the convergence of lead firm headquarters and specialized service providers. As Parnreiter (2010: 36) argues, global cities are "the places where producer services, in other words the means to *deal with and to control* complex cross-border networks of production and distribution, are provided, [allowing] each global city [to] constitute a significant juncture in numberless production networks" (emphasis added). Global cities, then, are viewed as the 'decision cities' from which the global economy is governed (Rossi *et al.* 2007). These complex processes are cast as mutually reinforcing linkages by Brown *et al.* (2010), who assert that the very function of global cities in production networks is to establish control over the generation of value, in so doing ensuring that their position in these transnational corporate networks is sustained through the further accumulation of value. As they go on to show in a case study of the coffee production network, major roasting companies constitute the lead firms in this context; these TNCs manage their complex global operations through strategically coupling with producer service intermediaries – particularly banking and financial institutions and advertising agencies – in New York and London. Through the forging of these transactional linkages, the coffee production network is not only 'disseminated', but also *centralized* through, global-relational urban networks in very specific parts of cities (Brown et al. 2010: 24).

In terms of **international partnerships**, the potential of urban-regional actors to integrate their territories into global production networks hinges on their ability to negotiate and cultivate transactional, yet beneficial linkages with lead firms. In order to achieve these mutually productive relationships, place-based actors must be capable of manipulating distinctive territorial assets to fit the strategic and functional requirements of global lead firms. It is perhaps unsurprising that territories which best fit this mode of coupling are those of high-growth city-regions (Yeung 2015). As an international business hub, Singapore is able to mobilize strategic resources such as good physical infrastructure, beneficial policies, tax incentives and easy access to knowledge institutes. In this context, Savage and Pow (2001) have commented that Singapore is able to offer a 'total package' to TNCs consisting of both economic and political security benefits. This ability, however, is predicated on Singapore's spatial scale as a global city state, wherein the political and economic priorities of the city and the nation-state converge, allowing the state government to transcend a great deal of the inter-locality competition and political gridlocking that impedes most other municipalities (Olds and Yeung 2004; Yeung 2010). State and institutional actors have almost absolute control over matters such as urban planning and labour flows, ensuring both a strategic capacity and flexibility to continually mould territorial resources in ways that fit the requirements of lead firms and their production networks.

This functional sense of strategic coupling would also seem to pervade **logistics hubs**, where cities and their wider regional hinterlands operate as sites for the construction and development of transport nodes such as seaports and airports. City-regions which are well-equipped with the requisite functional transportation and logistics infrastructure become promoted by state and institutional actors as favoured regulatory nodes through which global production networks run and are governed (Hesse 2008; Raimbault et al. 2016). As Hesse (2010: 85–86) argues, "cities have to rearrange site and situation to become part of the chain(s) … cities have to behave like organisational units and try to catch a certain position in the chain", laying emphasis on how the logistical capacities of city-regions are in essence strategic assets for articulating themselves in wider transnational corporate networks, and controlling the generation of value. To this end, a city-region's logistical capabilities and its economic developmental fortunes appear to be mutually entwined possibilities. For instance, Jacobs and Lagendijk (2014) detail how port expansions in Rotterdam are triggered, influenced and stabilized by broader shifts in the global shipping industry. Municipal port actors – primarily the city's port authority and trucking companies – have capitalized on these opportunities for coupling with external actors by upgrading the seaport infrastructure. Consequently, this is meant to improve Rotterdam's potential to receive the largest vessels in the business, strengthening its competitive position vis-à-vis neighbouring port cities such as Hamburg and Antwerp. What further consolidates the bargaining capacity of state and institutional actors is Rotterdam's 'Mainport' policy, which has historically served to garner support for port planning and expansion, and also serves to reflect how regional and national institutions rely on Rotterdam's position in global shipping and logistical networks to enhance their economic basis (Jacobs and Lagendijk 2014). Collectively, these material and institutional assets are localized and distinctive enough to distinguish Rotterdam as a prime logistics hub, securing its position in the global economy as a key logistics node, and enhancing its bargaining power and ability to shape the terms of coupling with global production network actors.

Another case of functional coupling exists in **market regions**, which exist as key localities for sales and distribution for a range of global production networks. These city-regions are integrated into global production networks for reasons related to their possession of a significant corporate and/or final customer base, and typically hold regional assets that are fairly generic, in which case they are more reliant on the demands of lead firms and their production networks. In the case of automotive industries, rapidly growing domestic markets – coupled with investment liberalization – in Asian city-regions have provided a strong customer base, and global lead firms in the automotive industry have sought to capitalize on such conditions. In Chinese city-regions, such as Chongqing, Guangzhou and Tianjin, strategic coupling between Chinese local governments and automobile firms and global lead firms in the industry (e.g. GM and Audi) has been facilitated through the low costs of factors of production like land and labour (Liu and Dicken 2006). Similarly, in the Rayong and Samutprakarn provinces of Thailand, the Thai national government has enabled strategic coupling through the development of subnational government organizations such as the Thailand Automotive Institute, the purpose of which is to enhance the skills of the Thai workforce and assist local companies in order to meet the needs of lead firms such as BMW (Coe et al. 2004).

If the earlier modes of strategic coupling can be characterized as cases where city-regions *potentially* hold a greater degree of autonomy – due to the relatively distinctive territorial assets they are able to control and leverage as a source of bargaining power – the case of **assembly platforms** demonstrates a structural form of coupling where city-regions are more dependent on the global production networks into which they are integrated. These city-regions typically feature the standardized assembly of offshored goods for export, and offer low costs of labour and land, as well as tax incentives, as key assets for coupling with lead firms. Personal computer (PC) investments in the Yangtze River Delta, for Yang (2009), results from municipal governments in the proximate city-regions promoting notebook clusters. City governments in Suzhou and Kunshan have, for instance, developed key policies such as land and tax incentives to attract leading Taiwanese PC firms. These approaches have been supplemented with different forms of institutional support, such as the construction of various industrial parks and development zones, as well as ensuring easy access to high-ranking government officials. Yang (2009) terms partnerships between municipal governments and Taiwanese lead firms as examples of 'explicit coupling' that indicate actions by local actors to place their localities into global production networks. However, given that the regional assets in this case are largely predicated on the low costs of land and labour – and are hence fairly generic – the risk of decoupling is high, as the Taiwanese lead firms continue to retain a high degree of bargaining power, which translates into an ability to play different city-regions against each other to secure the most favourable terms of entry (Yang 2009).

This range of examples provides a fairly comprehensive view of the diversities of urban economic development in a globalizing world. In this regard we believe that the strategic coupling concept has several advantages for urban scholars who wish to better grasp the functional and economic dimensions of global-relational urban linkages. First, it illustrates how specific urban-regional spaces may be variously integrated into global production networks, and how this is framed by notions of power and autonomy for the firm and extra-firm actors and territories involved. Second, it questions the assumption that city-regions need to function as production bases in order to be articulated into wider corporate networks. Depending on the sector in question, the needs of global production networks might require localities to take on different operations, such as logistical nodes or bases for advanced producer services. Third, it pushes us to more vigorously examine questions of power in these transactional partnerships by reinforcing the possibility that the mere presence of strategic coupling in no way guarantees positive developmental outcomes. It is to this territorial question of politics and tensions in/of strategic coupling that we now turn.

The politics of integrating city-regions into global production networks

By framing urban regional development as a dynamic outcome of exchanges of bargaining and negotiation between territories and global production networks, the concept of strategic coupling foregrounds the differentiated capacities of city-regional actors to mobilize capital and public resources in a fashion competitive enough to meet the needs, and facilitate the embedding, of incoming lead firms and their production networks. In this vein, GPN research has frequently been accused of underplaying the tensions and disarticulations manifested in the inherently uneven powers of city-regions to ground production networks (Levy 2008), which highlight the politics of growth and unevenness that emerge within and between regions. These issues of power within global production networks recall Christopherson and Clark's (2007) observations on the politics of network participation: namely that powerful actors – in this case lead firms – would not remain in networks were they not deriving disproportionately large benefits of participation, and that belonging to networks allows members to enjoy various forms of exclusivity that are largely inaccessible to non-participants. Clearly, and contra 'flat' and apolitical conceptions of network participation, there is a hierarchy and structure of actors and priorities that serve to unevenly (re)produce these networked forms of social and economic organization.

In the reproducing of global production networks and processes of strategic coupling, there are aspects of unevenness and inequity to the bargaining process that have begun to receive attention. In an exploratory discussion of the 'dark side' of strategic coupling, Coe and Hess (2011) point out the asymmetries of this economic integration. They note that strategic coupling is capable of "produc[ing] significant economic gains on an aggregate level, [but] in many cases it also causes *intra-regional disarticulations*" which reflect the politics of growth and unevenness that emerge within regions. For the authors, this

> 'dark side' of strategic coupling not only affects firms and their growth potential, but also, and maybe more importantly, the opportunities and livelihoods of people and households, and hence raises serious questions about the value generated, enhanced, and captured within the region.
>
> *(Coe and Hess 2011: 134, emphasis added)*

Following this, the negative consequences of territorial integration into global production networks – or 'disarticulations' – across longer-term regional growth trajectories are conceived through the notions of 'frictions' and 'ruptures' (Coe and Hess 2011). While the former draws on Anna Tsing's (2005) idea of 'friction' to delineate the sort of everyday discontinuities and tensions that are inherent to economic development, the latter is embodied in the form of more destabilizing forces that amount to the disintegration of inter-firm and extra-firm partnerships (see also MacKinnon 2012; Yeung 2015).

In both scenarios, conflicts speak to the different *struggles* over the distribution of value and capital between different actors and localities: who/what is implicated in the generation of value, and whom does this process ultimately serve? At their core these questions are both political and spatial. In the case of cities, these are situated within "increasingly volatile, financialised circuits of capital accumulation"; they are also "arenas in which conflicts and contradictions associated with historically and geographically specific accumulation strategies are expressed and fought out" (Brenner et al. 2009: 176). Given these aspects of capital-related conflict, we believe the concept of strategic coupling helps us understand struggles over value accumulation and distribution, and the politics of production, in city-regions. By illuminating different registers of the political more broadly when it comes to the development of urban regional economies in a globalizing context, strategic coupling is able to provide a lens into the urban politics of economic development. The fundamental point here is to examine, as Coe and Yeung (2015: 190) argue, how these developmental processes are framed by "wider influences that will also shape the character of regional economic growth (or decline) that results from strategic coupling with global production networks". As we show in what follows, there are three dimensions to this.

First, strategic coupling draws attention to the range of actors and their interactions that contribute towards economic development. More precisely, in conceptualizing regional development as the ongoing outcome of engagements between (translocal) firm and extra-firm organizations across different sites and scales, GPN research places these corporate and institutional actors at the forefront of economic activity. This is not to say, however, that other extra-firm actors are neglected; work on labour organizations as active shapers of economic developmental processes, for example, has been growing in recent years (e.g. Cumbers et al. 2008; Phillips 2011). For instance, Rainnie et al. (2011) consider how workers' actions are implicated in the restructuring of production networks, and the exploitation of vulnerabilities in lead firms' distribution systems; in other words, labour, as an embodied extra-firm agent, is able to significantly influence the rhythms of value extraction. Where the territorial and institutional contexts of city-regions are concerned, the range of actors capable of shaping the terms and outcomes of strategic coupling – and the 'fixing' of global production networks in particular places, as it were – seems similarly diverse. If urban governance is conceived as a two-way endeavour that is inflected not just by formal, top-down strategies emanating from the exclusive realm of elites, but also bottom-up tactics from other urban actors, then various stakeholders involved in the decision-making of cities come into play. This includes: municipal and regional government and institutional organizations, business elites, non-governmental bodies such as civil society organizations, and the ordinary citizens that make up urban populations.

Second, there remains the need to interrogate the *distribution* of value across the various actors and localities involved. Strategic coupling, as aforementioned, works through processes of value creation, enhancement and capture. It is imperative, however, to reiterate the fact that the presence of value creation and enhancement is often skewed in favour of powerful lead firms, and frequently comes at the expense of the actors and city-regions in which these production networks 'touch down' (Christopherson and Clark 2007). In this respect there remains the need to critically examine the uneven distribution of these supposed benefits, and how "positive value capture dynamics at the firm and regional level may or may not lead to improved terms and conditions for those that are labouring to produce that value" (Coe and Yeung 2015: 192). The case of the Shunde District in China's Pearl River Delta, for instance, shows that the successful coupling of latecomer domestic firms with lead Taiwanese firms in the information technology industry belies the weak bargaining power the locality actually holds due to its reliance on low cost factors of production. Consequently, local firms continue to remain captive to global lead firms for the import of key components and prospects of technological upgrading, and they remain vulnerable to the domination of lead firms in terms of price adjustments and technological change (Liu and Yang 2013).

Evidently, the intra-regional politics of growth reflect the unequal distribution of the advantages of strategic coupling across global production network actors within the same city-region. More fundamentally, greater consideration needs to be given to how the benefits of strategic coupling are distributed within the region across actors who might not be directly plugged into these networks (Coe and Yeung 2015). In other words, are local actors and institutions able to retain the resources obtained from their integration into global production networks for the development *of* the region, rather than continuing to satisfy conditions of coupling for development *in* the region? Kelly (2009) suggests instead the term 'global *re*production networks' to foreground the actors and spaces of social reproduction that are in many cases integral to regional development, even as they may be less directly plugged into global production networks. This would, for example, shift our view to households as a unit of socio-economic analysis, together with intersecting dynamics of gender and race. In any case, this pushes us to take a more multifaceted approach towards regional development and value capture. In so doing, these distributional elements raise fundamental questions of who/what exactly captures the gains of strategic coupling between city-regions and global production networks, which in no small part constitute the urban politics of regional development. They also highlight the need to remain cognizant of how exactly processes of strategic coupling are always already situated within *pre-existing* social, institutional and regulatory conditions of city-regions, for these contribute towards an understanding of broader regional development impacts.

Third, we should remember that even as city-regions are increasingly positioned as producers and recipients of globalizing dynamics, they continue to be embedded in their national space economies, and subject to the motivations and policies of their national states. It follows that national states continue to influence how institutional strategies and policies at subnational levels are designed and implemented across space, as well as determining the national terms of entry for lead firms. These all have profound impacts on how processes of strategic coupling are conceived and wrought. More generally, the importance of the national state is more evident in some policy realms than others, for instance, when it comes to macro-regional trade and industrial policy, as well as constructions of spatial developmental zones that are intended to ground lead firms and their production networks. For example, Lee *et al.* (2014) call for more attention to forging connections between global production networks and the national state, showing how Korean national agendas of growing competitiveness and boosting industry growth have been the driving force behind the development of industry clusters and parks in the Seoul Metropolitan Area. These in turn have served as an attractive regional asset to strengthen domestic lead firms and their production networks in the LCD industry, as well as embedding foreign lead firms and their supplier networks.

Arguably, the significance of the national state is evidently more prevalent in contexts of centralized governance – wherein national governments continue to retain a high degree of power – as opposed to federations with more decentralized governance. In any case, it is imperative to consider the state "as a multi-scalar intermediary that profoundly affects regional-level developmental outcomes" (Coe and Yeung 2015: 194; see also Smith 2015). At a broader level, the politico-ideological orientation of the national government (e.g. neoliberal versus developmental) also regulates the nature and its degree of involvement in subnational matters of development. Collectively, these conditions have wider implications for how nation-states are able to influence the terms and outcomes of strategic coupling within their national territories.

Conclusion: the urban politics of global production networks

In this chapter we have sought to draw useful connections between two bodies of work on relational economic geography and urban politics. Using the global production networks framework, we have elaborated on its relevance for discussions of the politics of urban economic development in globalizing contexts. As we have argued, the GPN framework is a multi-actor and multi-scalar one that lays emphasis on the crucial involvement of non-firm actors for growth, and the politics of their participation in processes of economic development. Furthermore, it also leads us to examine the city-region as a territorial and institutional setting for the grounding of global production networks. In particular, we have focused on the concept of strategic coupling to unpack the integration of subnational city-regions into wider regional and global production networks through qualitatively different modes of coupling, and examined some of the implications of these transactional exchanges.

As we have shown, these relational linkages may be manifested in distributional struggles over growth; at a deeper level, they are inescapably bound to uneven outcomes of bargaining, dependency and power. These are issues which are inherent to contemporary processes forms of economic globalization and development, but also critically constitute the urban politics surrounding the insertion and governance of city-regions into regional and global production networks, together with their longer-term developmental trajectories. Where cities and their broader hinterlands are concerned, at least, the consideration of GPN theory should serve to strengthen our understanding of their urban politics and evolutionary processes of economic growth and decline.

Notes

1 In this chapter, we use the acronym 'GPN' to denote the theoretical framework, and 'global production networks' to refer to the empirical phenomenon.
2 The examples from Table 7.1 are extracted from a more extensive list of cases of strategic coupling that are not restricted to city-regions alone (see Coe and Yeung 2015, Table 5.3).

References

Bair, J. and Werner, M. (2011). Commodity Chains and the Uneven Geographies of Global Capitalism: A Disarticulations Perspective. *Environment and Planning A*, 43(5), pp. 988–997.

Brenner, N., Marcuse, P. and Mayer, M. (2009). Cities For People, Not For Profit: Introduction. *City*, 13(2/3), pp. 176–184.

Brown, E., Derudder, B., Parnreiter, C., Pelupessy, W., Taylor, P.J. and Witlox, F. (2010). World City Networks and Global Commodity Chains: Towards a World-Systems' Integration. *Global Networks*, 10(1), pp. 12–34.

Cattaneo, O., Gereffi, G. and Staritz, C. (2010). Global Value Chains in a Postcrisis World: Resilience, Consolidation, and Shifting End Markets. In: O. Cattaneo, G. Gereffi and C. Staritz, eds. *Global Value Chains in a Postcrisis World: A Development Perspective*. Washington, DC: World Bank, pp. 3–20.

Christopherson, S. and Clark, J. (2007). *Remaking Regional Economies: Power, Labor, and Firm Strategies in the Knowledge Economy*. London and New York: Routledge.

Coe, N.M. and Hess, M. (2011). Local and Regional Development: A Global Production Networks Perspective. In: A. Pike, A. Rodriguez-Pose and J. Tomaney, eds. *Handbook of Local and Regional Development*. London: Routledge, pp. 128–138.

Coe, N.M. and Yeung, H.W-C. (2015). *Global Production Networks: Theorizing Economic Development in an Interconnected World*. Oxford: Oxford University Press.

Coe, N.M., Hess, M., Yeung, H.W-C, Dicken, P. and Henderson, J. (2004). Globalizing Regional Development: A Global Production Networks Perspective. *Transactions of the Institute of British Geographers*, 29(4), pp. 468–484.

Coe, N.M., Dicken, P., Hess, M. and Yeung, H.W-C. (2010). Making Connections: Global Production Networks and World City Networks. *Global Networks*, 10(1), pp. 138–149.

Cox, K.R. (1998). Spaces of Dependence, Spaces of Engagements and the Politics of Scale, or: Looking For Local Politics. *Political Geography*, 17(1), pp. 1–23.

Cumbers, A., Nativel, C. and Routledge, P. (2008). Labour Agency and Union Positionalities in Global Production Networks. *Journal of Economic Geography*, 8(3), pp. 369–387.

Hesse, M. (2008). *The City as a Terminal: The Urban Context of Logistics and Freight Transport*. Burlington, VA: Ashgate Publishing Company.

Hesse, M. (2010). Cities, Material Flows, and the Geography of Spatial Interaction: Urban Places in the System of Chains. *Global Networks*, 10(1), pp. 75–91.

Jacobs, W. and Lagendijk, A. (2014). Strategic Coupling as *Capacity*: How Seaports Connect to Global Flows of Containerized Transport. *Global Networks*, 14(1), pp. 44–62.

Kelly, P.F. (2009). From Global Production Networks to Global Reproduction Networks: Households, Migration, and Regional Development in Cavite, the Philippines. *Regional Studies*, 43(3), pp. 449–461.

Lee, Y.S., Heo, I. and Kim, H. (2014). The Role of the State as an Inter-Scalar Mediator in Globalizing Liquid Crystal Display Industry Development in South Korea. *Review of International Political Economy*, 21(1), pp. 102–129.

Levy, D.L. (2008). Political Contestation in Global Production Networks. *Academy of Management Review*, 33(4), pp. 943–963.

Liu, W. and Dicken, P. (2006). Transnational Corporations and 'Obligated Embeddedness': Foreign Direct Investment in China's Automobile Industry. *Environment and Planning A*, 38(7), pp. 1229–1247.

Liu, Y. and Yang, C. (2013). Strategic Coupling of Local Firms in Global Production Networks: The Rise of the Home Appliance Industry in Shunde, China. *Eurasian Geography and Economics*, 54(4), pp. 444–463.

MacKinnon, D. (2012). Beyond Strategic Coupling: Reassessing the Firm-Region Nexus in Global Production Networks. *Journal of Economic Geography*, 12(1), pp. 227–245.

Olds, K. and Yeung, H.W.C. (2004). Pathways to Global City Formation: A View from the City-State of Singapore. *Review of International Political Economy*, 11(3), pp. 489–521.

Parnreiter, C. (2010). Global Cities in Global Commodity Chains: Exploring the Role of Mexico City in the Geography of Global Economic Governance. *Global Networks*, 10(1), pp. 35–53.

Phillips, N. (2011). Informality, Global Production Networks and the Dynamics of 'Adverse Incorporation'. *Global Networks*, 11(2), pp. 380–397.

Raimbault, N., Jacobs, W. and Van Dongen, F. (2016). Port Regionalisation from a Relational Perspective: The Rise of Venlo as Dutch International Logistics Hub. *Tijdschrift voor Economische en Sociale Geografie*, 107(1), pp. 16–32.

Rainnie, A., Herod, A. and McGarth-Champ, S. (2011). Review and Positions: Global Production Networks and Labour. *Competition and Change*, 15(2), pp. 155–169.

Rodgers, S., Barnett, C. and Cochrane, A. (2014). Where Is Urban Politics? *International Journal of Urban and Regional Research*, 38(5), pp. 1551–1560.

Rossi, E., Beaverstock, J. and Taylor, P.J. (2007). Transaction Links Through Cities: 'Decision Cities' and 'Service Cities' in Outsourcing by Leading Brazilian Firms. *Geoforum*, 38(4), pp. 628–642.

Savage, V.R. and Pow, C.P. (2001). 'Model Singapore': Crossing Urban Boundaries. In: R.J. Stimson, ed. *International Urban Planning Settings: Lessons of Success*, Vol. 12. Oxford: JAI, pp. 87–121.

Scott, A.J., ed. (2001). *Global City-Regions: Trends, Theory, Policy*. Oxford: Oxford University Press.

Scott, A.J., Agnew, J., Soja, E.W. and Storper, M. (2001). 'Global City-Regions'. In: A.J. Scott, ed. *Global City-Regions: Trends, Theory, Policy*. Oxford: Oxford University Press, pp. 11–32.

Smith, A. (2015). The State, Institutional Frameworks and the Dynamics of Capital in Global Production Networks. *Progress in Human Geography*, 39(3), pp. 290–315.

Storper, M. (1997). *The Regional World: Territorial Development in a Global Economy*. New York: Guilford Press.

Taylor, P.J. (2004). *World City Network: A Global Urban Analysis*. London: Routledge.

Taylor, P.J., Ni, P., Derudder, B., Hoyler, M., Huang, J. and Witlox, F. (2010). *Global Urban Analysis: A Survey of Cities in Globalization*. London: Earthscan.

Tsing, A.L. (2005). *Friction: An Ethnography of Global Connection*. Princeton, NJ and Oxford: Princeton University Press.

UNCTAD (2013). *World Investment Report 2013: Global Value Chains: Investment and Trade for Development*. New York: United Nations.

Vind, I. and Fold, N. (2010). City Networks and Commodity Chains: Identifying Global Flows and Local Connections in Ho Chi Minh City. *Global Networks*, 10(1), pp. 54–74.

Werner, M. (2015). *Global Displacements: The Making of Uneven Development in the Caribbean*. Chichester: Wiley Blackwell.

Yang, C. (2009). Strategic Coupling of Regional Development in Global Production Networks: Redistribution of Taiwanese Personal Computer Investment from the Pearl River Delta to the Yangtze River Delta, China. *Regional Studies*, 43(3), pp. 385–407.

Yeung, H.W-C. (2009). Regional Development and the Competitive Dynamics of Global Production Networks: An East Asian Perspective. *Regional Studies*, 43(3), pp. 325–351.

Yeung, H.W.C. (2010). Globalising Singapore: One Global City, Global Production Networks, and the Developmental State. In: T.H. Tan, ed. *Singapore Perspectives 2010*. Singapore: World Scientific, pp. 109–121.

Yeung, H.W.C. (2015). Regional Development in the Global Economy: A Dynamic Perspective of Strategic Coupling in Global Production Networks. *Regional Science Policy & Practice*, 7(1), pp. 1–23.

Young, D. and Keil, R. (2014). Locating the Urban In-Between: Tracking the Urban Politics of Infrastructure in Toronto. *International Journal of Urban and Regional Research*, 38(5), pp. 1589–1608.

8
THE SKY IS NOT THE LIMIT
Negotiating height and density in Toronto's condominium boom

Ute Lehrer and Peter Pantalone

Introduction

As we now have become accustomed to hear: the twenty-first century is the urban century where the majority of the world population lives in cities and will increasingly do so. And with this continuous migration to urbanized areas the question emerges of how to expand in already existing cities. In order to save the countryside from being overbuilt, many regions have introduced policies that steer growth away from certain places to those that are considered to be more appropriate. Under the term of sustainability, smart growth policies, which mandate the intensification of urbanized areas, have become the norm.

It is widely agreed that neoliberalism as the new political economy, or as it is specified for the context of Toronto as "'common-sense' neoliberalism" (Keil 2002: 578), is changing the practices not only on how to build cities but also on how to strategically use cities in the remaking of political-economic space (Brenner and Theodore 2002; Wilson 2004). With the disjuncture between the ideology and the reality of the roles of state and market, respectively, new regimes and practices are introduced to the construction process of built environments, where economic development is the main interest of the local state (Kipfer and Keil 2002). And while urban planning per se has been used as a motor for economic development within capitalism, it is even more pronounced in an era of neoliberalism. As Jamie Peck (1995: 16) had observed some twenty years ago, "businessmen are playing an increasingly important part in setting agendas for urban development". We are suggesting here to narrow Peck's observation to developers who increasingly play a fundamental role in defining city building processes and who deal with planning regulations as if these regulations are just mere suggestive guidelines and not clear rules. Everything seems to be negotiable in this climate, as long as the developer is willing to 'give back' to the municipality, either in direct payments or in-kind contributions and in exchange can go beyond of what the regulations allow. In that sense, urban development is a form of economic development.

Selling of air rights in exchange for more building height has been around for a while, but since the 1970s we have seen an increased number of new instruments emerging that all were motivated by linking planning with economic development and consequently travelled, in slightly adapted versions, to other places (Ward 2006). What in 1970 started as a strategic beautification strategy of a small district in Toronto, where businesses would pay an annual levy that consequently would be used for the improvement of the streetscape in order to compete against the attractiveness of the suburban malls (Yang 2010), the Business Improvement Area has now become a practice that is broadly used throughout North America, parts of Europe, South Africa, New Zealand and Japan. Known under a variety of names, this aestheticization strategy is mainly applied in the context of already well-established business communities that put their own interests forward (Rankin and Delaney 2011). On the other spectrum is Tax Increment Financing (TIF), which seeks new investments for a development area, built on the promise of future increased assessed values of the land. This tool has been widely utilized in Chicago, financing a variety of economic development projects (Weber 2010), as well as in many parts of the USA; it also resonates in Canada, UK and Australia.

The planning tool of density bonusing is increasingly exercised as an enticement, which creates a regime that organizes city building processes along economic incentives for the developer and the municipality. In exchange for density that is beyond site-specific zoning regulations and land controls, developers agree to return some profit that is gained through land uplift, back to the community, through an agreement that specifies monetary and/or non-monetary assets. The rather technical decision that often guides planning decisions within its institutional framework, is embedded in a political environment. The usual actors are: developers, who are interested in squeezing the highest exchange value out of pieces of land; politicians, who are motivated to stay in power; and inhabitants, whose everyday life is affected by new urban development and therefore are also interested in the use value of built environments.

We will discuss density bonusing in the context of Toronto's condominium boom, where this instrument has been around within the legislative framework since 1983 (Pantalone 2014: 20). Yet only over the past decade has this become a major tool within the politics of negotiations for community benefits from urban development processes. Density bonusing has its origins in betterment levies, which are direct taxes to recapture the wealth accruing to private landowners due to public investments in the built environment and other planning initiatives. The exact planning tools that enable density bonusing vary among different jurisdictions, and bonusing is also referred to as 'Land Value Recapture', 'Bonus Zoning' and 'Windfalls Capture' (Alterman 2012; Calavita and Wolfe 2014). All reflect a mechanism that is quite similar: the value of land is enhanced through planning policies and/or public investments to the benefit of the landowner, and courts have recognized that there is a firm legal basis for governments to take back some value from this 'uplift'. We argue that while the specificities of density bonusing in Toronto are particular in its legal and practical context, our analysis here is also relevant for other places.

Although density bonusing is presented as ostensibly benefiting the community, it is our opinion that first, typically density bonusing agreements are negotiated in backroom deals between politicians and developers; second, these decisions are not necessarily guided by the principles of 'good planning' but by individual interests; and third, since the public is usually not included in these negotiations, there is a clear disjuncture between what politicians versus community define as beneficial to a particular neighbourhood (Lehrer 2008).

In this contribution we start with a general discussion of the condominium boom in Toronto which provides crucial context to our exposition. Next, we discuss density bonusing historically and geographically, followed by an explanation of how density bonusing has become part of the development process in Toronto. Our conclusion draws attention to density bonusing as a practice and the complexity of negotiation processes, and we end with an overall critique on density bonusing as another form of 'let's make a deal' planning.

Condo boom in Toronto and its policy framework

After we have witnessed a long period of expansion into the countryside we now see a revived interest of developers to invest into inner cities. As Sassen (2015) describes, this has become a worldwide phenomenon where large-scale projects are changing urban fabrics: "In the post-2008 period, much buying of buildings involves destroying and replacing them with far taller, more corporate and luxurious buildings – basically, luxury office and luxury apartments." The vertical multiplication of homeownership has become a global phenomenon. While it typically concentrates in city centres, it is increasingly found in suburban areas. In the case of the Toronto region, this is supported, as we argue, by a policy framework, which asks specifically for the protection of green space while redirecting growth to particular areas.

Since the late 1990s Toronto has experienced a massive building boom that has involved especially condominium towers. This happened in the wake of deregulation of zoning and land controls in the downtown core as well as a real estate market with low interest rates and minimum down payments, where the small condo units became an alternative to the tight rental market (Lehrer and Wieditz 2009a, 2009b; Lehrer et al. 2010; Rosen and Walks 2013, 2015). From 'The Condo Generation' (Toronto Life 2006) to 'Boomtown: The Incredible, Unstoppable Growth of Downtown' (Toronto Life 2010), the condo boom has regularly made the front pages of newspapers and magazines over the last decade despite warnings of a possible overheating of the real estate market.

In a strict sense, condominium does not refer to a building form but to a legal institutional framework that regulates ownership of individual and common areas within larger complexes (Harris 2011). This property regime allows a hybrid arrangement between public and private spaces that creates "real life behind the gates" (Kirby 2008: 92). While the condominium structure is widely present in gated communities in the USA (Lasner 2012), it now proliferates in the form of high-rise towers in cities, as well as in suburban regions throughout the world. In Toronto it has become normalized to live in condos for a population whose idea of ownership used to be the single family home, if not in the city then in the suburban region. And while there is a continuation of constructing single family homes, the condominium tower has also caught the interest of developers. Increasingly, condo towers are being erected and with it, the morphology of the city and entire region is changing. Warnings of overheating have been around ever since the financial crisis hit the real estate market on a global scale; but with little impact on building activities, developers in Toronto have continued to follow this formula. Headlines such as 'Canada's Condo Boom Spikes to Heights Not Seen Since 1971' (Argitis 2015) are a clear expression of not only its historical significance but also its physical magnitude and currency. It is both the sheer volume of condominium units that increase from year to year, and the height of the developments themselves. With almost every new proposal, developers push the limits of what is possible. While ten years ago the norm was between twenty and forty storeys, now several proposals ask for towers in the eighties and even nineties range. Currently the tallest residential tower is the 'Aura', a condominium building with seventy-eight floors, but this might soon be toppled from first place by other projects, which propose to be ninety and more storeys high when finished (including one by star architect Frank Gehry). To be sure, there are egotistical aspects for this building up, but the dominant motivation for developers is the financial return they can gain from it. And often, they use the existing policy framework as a legitimization strategy for their demands for increased heights.

Since 2005, the policy directives on the provincial level have come from two laws: the Greenbelt Act and the Places to Grow Act. The Greenbelt Act is the basis for the Greenbelt Plan, which has the mandate to "protect about 1.8 million acres of environmentally sensitive and agricultural land in the Golden Horseshoe region from urban development and sprawl" (Greenbelt Act 2005). Working in tandem with this legal binding provincial document is the Places to Grow Act, which seeks to "plan for growth in a coordinated and strategic way". It identifies specific areas where development should be directed and "makes sure that growth plans reflect the needs, strengths and opportunities of the communities involved, and promotes growth that balances the needs of the economy with the environment" (Places to Grow Act 2015). These two legal frameworks guide development in the so-called Golden Horseshoe, an area comprised of several cities around Toronto of about 33,500 km^2 large, of which approximately a fifth is protected green space. As the most southern area within Canada, continued urban growth challenges the need to preserve fertile agricultural land. With these two laws, the conflict between different interests – the speculative thinking of developers and some landowners whose dream of good pensions are based on agricultural land rezoned, and the protectionist thinking of environmentalists and landowners who desire to continue working their land – was redirected to new areas in the inner city, the underused industrial sites and empty lots. While the impetus for developing guiding principles for growth came from the political level and the implementation was done via the legislative level, it is interesting to analyse how these new policies that first were fought by the development industry, now are more and more used as a legitimization strategy to ask for higher density and more sellable units of housing. As we can chronicle below, Section 37 of the provincial Planning Act provides developers with an instrument that encourages them to ask for more in exchange of so-called community benefits. But first let us look at the Provincial guiding principles which define good planning.

All locally generated master plans (or official plans as they are called in the Canadian context) need to be consistent with the Provincial Policy Statement, which

> provides policy direction on matters of provincial interest related to land use planning and development. As a key part of Ontario's policy-led planning system, the Provincial Policy Statement sets the policy foundation for regulating the development and use of land. It also supports the provincial goal to enhance the quality of life for the citizens of Ontario.
>
> *(Provincial Policy Statement 2014: 1)*

In other words, the Provincial Policy Statement is the overarching policy framework to which all other municipal plans need to be consistent with. This point is important, because most of the developers are asking for an adjustment of their site-specific plans by proposing developments that are well beyond the usual building envelope. This has led to an overall height increase and what has been considered tall ten years ago is now normal in the most intensified parts of Toronto.

In hearings on the local level (Committee of Adjustment) as well as provincial level (Ontario Municipal Board) these developers more often than not use the provincial policy framework for growth centres as a legitimization strategy for their proposals that are regularly well beyond the allowed building envelope. Any discussion about the politics of density bonusing in Toronto needs to also include the provincial Ontario Municipal Board (OMB). This quasi-judicial institution has as its mandate to hear disputes over land use issues and to decide over proper amendments to official plans and zoning by-laws, after balancing all the particular interests against the legal framework and what constitutes 'good planning'. While it does not shape actors' interests, it most likely "influences their behaviour given its potential to redistribute resources in the politics of urban development" (Moore 2013a: 35).

Betterment and density bonusing: a brief history

Density bonusing has its origins in betterment capture policies which date back at least as far as Britain's 1909 Town Planning Act. Betterment refers to "the rise in land values directly caused by a planning or public works decision" (Alterman 2012: 760) and goes hand-in-hand with its inverse counterpart, 'worsement'. Britain's 1909 legislation imposed a 50 per cent levy on betterment resulting from public planning initiatives, although from the outset the levy was beset with implementation challenges, in particular difficulties with collecting the tax from landowners. Over the subsequent decades, waves of Labor and Tory governments vacillated between different iterations of betterment levies, the most ambitious of which was an 80 per cent betterment levy introduced in 1976 called the Development Land Tax, only to be scrapped by the Thatcher government nine years later. With that, betterment disappeared off the British public policy radar for some time until the Labor government introduced the Community Infrastructure Levy in 2010 (Alterman 2012). The CIL, which is an optional planning tool local councils may elect to adopt, allows councils to impose a flat fee on eligible new development to fund new infrastructure in the area. Before a council can use the CIL it must formulate levy charges setting out rates for different development classes and exemptions (Carpenter 2016).

Betterment levies can and do take different forms in different localities. Additional schemes that fall under the betterment levy umbrella include density bonusing, windfall recapture, uplift capture, land value capture, bonus zoning, transfer of development rights, and many others (Alterman 2012; Calvita and Wolfe 2014; Walters 2012). The essence of these programmes is the same; a landowner or developer's land value windfall is recaptured, to varying degrees, by the state (local or regional) either in the form of 'community benefits' or a monetary contribution. In their white paper on public benefits zoning in the San Francisco Bay Area, Calvita and Wolfe (2014) summarize the two implementation approaches to a community benefits programme. A *plan-based* bonusing regime ties specific increases in development density to specific community benefits that are then given to the local jurisdiction in return. This arrangement affords certainty for both the landowner/developer as well as the locality and community actors. Alternatively, a *negotiation-based* bonusing regime, rather than being grounded in a plan or formula, entails a site-specific negotiation between a landowner/developer and the local jurisdiction, in order to determine the bonus density conferred and the community benefits provided in exchange.

Vancouver, British Columbia uses a planning tool called the 'Community Amenity Contribution' (CACs) to collect in-kind or cash contributions from developers when the city council authorizes re-zoning applications. CACs are employed to secure community benefits such as parks, libraries, childcare facilities, community centres, transportation services, cultural facilities and heritage preservation (City of Vancouver 2016). CACs are levied in two different, and sometimes overlapping, ways. In the city's downtown core and other high-density areas, CACs are negotiated on a case-by-case basis – the greater the windfall profit to the developer due to the bonus density or height conferred, the larger the in-kind or cash contribution the developer must pay to the city. Outside the downtown area and key nodes, all other city areas have a flat fee for re-zoning applications that determines the value of the CAC to be provided in exchange for development approval. One typical example of a larger CAC is a development approved in 2014 including two twenty-eight and thirty-storey towers with 620 residential condominium units, which was granted by council in exchange for a C$19.6 million CAC to facilitate a new dragon boat racing facility and acquisition of land for a new social housing facility (Lam 2014). In Vancouver, CACs are negotiated between developers and city planning staff – not local councillors – a practice which former chief planner Larry Beasley observes "[takes] negotiations out of the hands of politicians and maintains a focus on corporate policy and management of equity among all kinds of public goods, while depersonalizing the negotiation and cutting down on [political] abuse" (Beasley 2006). This approach is in stark contrast to Toronto, where local politicians often take the lead in negotiations regarding density bonuses and community benefits.

Some municipalities apply bonus zoning to generate one specific public/community benefit. Chicago introduced its downtown affordable housing zoning bonus on 1 November 2014, which offers additional square footage for residential development projects within a delineated geographical area, in exchange for the provision of on-site affordable housing or a cash-in-lieu payment to the city's Affordable Housing Opportunity Fund (Chicago Department of Planning and Development 2016). When authorizing such a bonus, Chicago's Department of Planning and Development requires a developer to either make a financial agreement or enter into an affordable housing agreement *before* a building permit is issued, and also requires a letter of credit or other securities to ensure the affordable units are delivered as agreed.

New York City has an extensive tool kit for bonus zoning, with the most well-known of these being Transfer of Development Rights (TDR). TDR enables a landowner to sell any rights to future development on their parcel of land to another nearby proprietor to use for developing their separate site (New York City Department of City Planning 2015). A TDR renders the donor site undevelopable while allowing the benefiting property bonus development rights in the form of additional height or density. A classic example is Grand Central Station, which was designated as a landmark in 1967 in an effort to forestall a redevelopment proposal. The landmark designation, which essentially 'down-zoned' the Grand Central Station property, was declared by the courts a constitutional taking (i.e. *expropriation*) of land because the unused development rights (i.e. air) over the terminal could be sold to nearby developers (Bagli 2015). A stark implication of TDR is that we have reached a new frontier where the commodification of air is not only possible but practiced.

Density bonusing and development in Toronto

The legislative basis for density bonusing in Toronto is found within Section 37 of the Province of Ontario's Planning Act (1990), which states that a municipality may grant additional height or density to a developer, beyond prevailing zoning restrictions, in exchange for the provision by the developer of community benefits or cash-in-lieu of benefits. Density bonusing was officially legislated into the Planning Act in 1983, although ad hoc bonusing in Ontario had taken place for decades prior. All municipalities in Ontario are permitted to use density bonusing but only if they include policies governing the scope and protocol for bonusing within their local 'official plans' (City of Toronto 2007).

As already mentioned, the condominium boom in Toronto has vastly redefined the so-called 'highest and best use' for urban land throughout the region, creating a situation where residential development is significantly more profitable than any other land uses. Provincial policy directing growth inward to already built-up areas, combined with unabated population growth in the Greater Toronto Area, has encouraged developers to seek permission to build higher and denser than envisioned by the zoning regime and other comprehensive planning policies. In fact, it is more often the case that developers introduce proposals that *exceed* zoning parameters, rather than *conform* to them. Consequently, the use of density bonusing has skyrocketed in Toronto and several outlying municipalities eager to generate community benefits from the development frenzy.

Density bonusing agreements in Toronto are more commonly referred to as 'Section 37 agreements' – reflecting that section of the Planning Act through which they are enabled – and are in theory required to reflect a rational nexus between the bonus density granted to a developer and the community benefits provided in return (see Morgan 2013). The most widely understood expression of this rational nexus is a reasonable geographic and planning relationship between the density bonus and the community benefits. However, Section 37 agreements have been extensively misapplied in Toronto over the duration of the condominium boom, which we argue has been to the benefit of developers and politicians and to the detriment of good planning principles and community actors.

The development approvals process in Toronto has become a precarious balancing act between the developer, the municipality as the local approval authority, and the Ontario Municipal Board (OMB) as the provincial appellate tribunal that retains final jurisdiction over planning decisions across the province. Traditional local political economy theory (i.e. Logan and Molotch 1987), which focuses on the behaviour of developers, planners, politicians and community actors at the local level, cannot account for the urban development politics of Toronto, due to the introduction of the OMB – an appellate body – as the final arbiter of land use decisions (Moore 2013). When the local authority rejects a development proposal because it does not comply with the city's official plan or zoning by-laws, a developer can appeal to the OMB and obtain permission to proceed with their proposal in spite of the municipality's refusal. This creates an incentive for local politicians to negotiate deals at the local level with developers over contentious development proposals, in order to avoid an OMB appeal that could take away any bargaining leverage from the municipality altogether.

In Toronto, the precariousness of the development approvals process has given rise to what has been described as 'let's make a deal planning' (Devine 2007: 13; 2013) between developers and local politicians. Toronto's city council is comprised of forty-four wards, each with an elected ward-specific councillor who retains much oversight over neighbourhood urban development. In theory, the city's official plan requires planning rules to be applied evenly and consistently across the city, but some observers have noted there is a tendency for councillors to treat their individual wards like personal fiefdoms (Moore 2013a). Incoming land development applications are reviewed by city planning staff, who assess the planning merits of the proposal against the official plan and any other relevant planning policies that apply. But ultimately it is city council that makes a decision whether to allow or refuse a development application. Notwithstanding the planning staff recommendation on a particular development application, the local ward councillor will often insert themselves into negotiations with the developer over community benefits to be provided to the city in exchange for approval of the application. Ever cognizant of developers' threats to appeal to the OMB if the city does not support their proposal, politicians have embraced deal-making as a way to position themselves as being willing to compromise with developers on the one hand (to grant a density bonus) and responsive to the demands and needs of the community and electorate on the other (to secure sorely needed community benefits in return). The concept of 'good planning' is rendered highly elastic indeed when development decisions are negotiated in a manner akin to legalized bribery.

As identified above, Alterman (2012) has highlighted the distributive-justice dilemma inherent to recapturing land uplift. In Toronto, the vast majority of hundreds of millions of dollars in Section 37 revenues that have accumulated since 2000 were generated through development approvals in only five of the city's forty-four wards, due to the uneven geography of growth (Moore 2013b; Pantalone 2014). Because of the requirement for community benefits to have a reasonable geographic relationship to their originating development project, these five wards have such enormous reserves of community benefits and in-lieu cash payments that the city does not even have the staff capacity to budget for the funds. Conversely, those wards where development activity is not taking place, although also very much in need of new community facilities and amenities, benefit very little from the condominium boom that is so graphically reshaping the urban landscape.

Pantalone (2014) has argued that 'community benefits' are a somewhat dubious descriptor for benefits secured through Section 37 agreements, given that most community actors are systematically excluded from the negotiating table altogether. Interviews with councillors of development-rich wards in Toronto revealed that it is often a select handful of local community elites – leaders of residents' associations and 'business improvement areas', for example – who have insider access to the councillor's office and the ability to shape Section 37 negotiations. And by virtue of their prominence in the community, these elites are afforded an elevated bargaining chip to shape the outcome of local urban development, in comparison to other, more marginalized members of the community. One promising counterpoint to this trend was a 2014 participatory budgeting pilot project by Toronto Councillor Shelley Carroll, who invited her constituents to cast a vote to determine how a financial contribution from a Section 37 agreement with a local developer should be spent (Dale 2014). The ballot listed several options for community benefits and constituents could vote for whichever project they believed most desirable. However, once density bonusing revenues are put to a community vote, the nexus requirement goes out the window, and there is certainly no legal sanction for using participatory budgeting to determine the eligibility of particular community benefits.

In the case of Toronto, we identify several reforms that could remediate the most egregious misapplications of Section 37 agreements. First, the city's implementation criteria and negotiating protocol for Section 37 agreements – which are merely *guidelines* – should be elevated to full policy status so that they cannot be ignored by ward councillors who prefer to undertake their own particular brand of deal-making. Second, a quantum approach, or formula, should be used to determine the value of the community benefits given by a developer in exchange for their density bonus. This would mean that a specific percentage of the developer's windfall would be consistently recaptured by the city when the re-zoning is allowed. Third, the city should adopt contingencies that allow unspent Section 37 funds to be reallocated if they are not budgeted and disbursed for their intended community benefit in a reasonably timely manner. Fourth, a determination needs to be made regarding whether it is permissible for local councillors to support development that has been opposed by city planning staff, in order to obtain community benefits from development that might not reflect the principles of good planning. Lastly, we call for a reconceptualization of the 'nexus requirement' in density bonusing, which presently requires any Section 37 benefits to bear a clear and demonstrable planning relationship including, at minimum, a close geographical proximity between the community benefit and its originating development. There is an undeniable correlation between new urban development and the changing socio-economic composition of the city, which can be called 'new-build gentrification' (Lehrer and Wieditz 2009b). As we suggest, the nexus requirement must be broadened conceptually to account for more complex and insidious relationships between new urban development and geographies of displacement, and to acknowledge the obligation of developers to compensate those who are displaced in various ways through new development projects (Pantalone 2014: 96) as well as addressing other negative externalities. Whereas the vast majority of Section 37 benefits in Toronto have been earmarked for cosmetic enhancements and "desirable visual amenities" (Moore 2013b), there is a desperate need for funding of existing and new social services and facilities.

Conclusion

Toronto's ongoing condominium boom illuminates how a well-intentioned planning tool, density bonusing, can become highly unwieldy when development decisions depart from principles of 'good planning' and convert into systematic deal-making. We believe that the underlying rationale for density bonusing is the same internationally, and therefore some important lessons from Toronto's experience with density bonusing can be extrapolated to other localities grappling with similar issues and dilemmas.

Torontonians have become accustomed to construction cranes in the sky, the disruptions that go hand-in-hand with urban development and the Manhattanization of the city's downtown core and suburban growth centres. In the neoliberal city, the condominium boom has become an economic development strategy, and countless subsidiary industries have sprung forth to piggyback on the relentless condofication of the city. Lobbyist organizations like the Building Industry and Land Development Association and the Ontario Homebuilders Association have boisterously inserted themselves into public policy debates about the merits of urban development and the contributions of the construction sector and its spinoff industries to the economy at large (Altus Group Economic Consulting 2013; Ontario Home Builders' Association 2014). Development charges and other taxes on new homes and condominiums are kept artificially depressed to ensure the growth machine rumbles on and land development remains a profitable enterprise (Blais 2010). But the trite refrain that 'growth pays for growth' is only a half-truth in Toronto, where new development has far outstripped existing community infrastructure and public amenities. Section 37 agreements, a planning tool to encourage density bonuses in exchange for community benefits, has not facilitated a proportional growth in the community facilities necessary to accommodate the impacts and externalities of the new urban development it has engendered. 'Let's make a deal planning' is yet another expression of the incremental deregulation of planning controls and the growth of a more flexible, neoliberal model of city-building. Like 'Transfer of Development Rights', density bonusing has proven that it is possible to commodify air and sky in the persistent pursuit of extracting ever-increasing levels of profit from urban land.

The merit and effectiveness of a bonusing regime in exchange of increased density and/or height is debateable. Producing public benefits through private development feeds into costs of intensification but also into the diminishing of public funds. Because of its inconsistent application it also has a tendency towards poor planning decisions. Therefore, two questions remain: what are the limitations of this planning tool that has become more and more a strategy in tight budgetary times for the re-election of local politicians? And second, is this mechanism an adequate source of income in order to offset the planning impacts of private development?

References

Alterman, R. (2012). Land Use Regulations and Property Values: The 'Windfalls Capture' Idea Revisited. In: N. Brook, K. Donaghy and G.-J. Knapp, eds. *The Oxford Handbook of Urban Economics and Planning*. Oxford: Oxford University, pp. 755–786.

Altus Group Economic Consulting. (2013). Government Charges and Fees on New Homes in the Greater Toronto Area. Prepared for the Building Industry and Land Development Association, 23 July. Available at: www.bildgta.ca/BILD/uploadedFiles/Media/Releases_2013/Gov_chargesJune6.pdf [accessed 10 January 2014].

Argitis, T. (2015). Canada's Condo Boom Spikes to Heights Not Seen Since 1971 – and That's Making Economists Nervous. *Financial Post*, 8 December. Available at: http://business.financialpost.com/personal-finance/mortgages-real-estate/canadas-condo-boom-spikes-to-heights-not-seen-since-1971-and-thats-making-economists-nervous [accessed 12 January 2016].

Bagli, C. (2015). Owner of Grand Central Station Sues Developer and City for $1.1 Billion Over Air Rights. *New York Times*, 28 September. Available at: www.nytimes.com/2015/09/29/nyregion/owner-of-grand-central-sues-developer-and-city-for-1-1-billion-over-air-rights.html?_r=0 [accessed 6 January 2016].

Beasley, L. (2006). Amenities for Density: Section 37 of the Planning Act. All About Planning Symposium, 6 December. Munk Centre for International Studies. Toronto.

Blais, P. (2010). *Perverse Cities: Hidden Subsidies, Wonky Policy, and Urban Sprawl*. Vancouver: University of British Columbia Press.

Brenner, N. and Theodore, N. (2002). Cities and the Geographies of 'Actually Existing Neoliberalism'. *Antipode*, 33(3), pp. 349–379.

Calvita, N. and Wolfe, M. (2014). White Paper on the Theory, Economics and Practice of Public Benefits Zoning (San Francisco Bay Area). Available at: www.thecyberhood.net/documents/projects/whitepaper14.pdf [accessed 6 January 2016].

Carpenter, J. (2016). CIL Watch: Who's Charging What? *Planning Resource*. Available at: www.planningresource.co.uk/article/1121218/cil-watch-whos-charging-what [accessed 23 February 2016].

Chicago Department of Planning and Development. (2016). City of Chicago's Affordable Housing Zoning Bonus – Administrative Regulations and Procedures. Available at: www.cityofchicago.org/content/dam/city/depts/dcd/general/housing/AdmRule.pdf [accessed 4 December 2017].

City of Toronto. (2007). Implementation Guidelines for Section 37 of the Planning Act and Protocol for Negotiating Section 37 Community Benefits. Available at: www1.toronto.ca/city_of_toronto/city_planning/sipa/files/pdf/s37_consolidation_080117.pdf [accessed 10 January 2014].

City of Vancouver. (2016). Community Amenity Contributions. Department of Urban Planning, Sustainable Zoning and Development. Available at: http://vancouver.ca/home-property-development/community-amenity-contributions.aspx [accessed 4 December 2017].

Dale, D. (2014). How Would You Spend $500,000? Big Money at Stake in Participatory Budgeting Exercise. *The Toronto Star*, 20 June. Available at: www.thestar.com/news/city_hall/toronto2014election/2014/06/20/how_would_you_spend_500000_big_money_at_stake_in_participatory_budgeting_experiment.html [accessed 30 July 2014].

Devine, P. (2007). An Update on 'Let's Make a Deal' Planning. Fraser Milner Casgrain LLP.

Devine, P. (2013). Section 37 – A Further Update and a Question: 'Is It a Tax?'. Dentons LLP.

Greenbelt Act. (2005). Available at: www.ontario.ca/laws/statute/05g01 [accessed 10 January 2018].

Harris, C.D. (2011). Condominium and the City: The Rise of Property in Vancouver. *Law and Social Inquiry*, 36(2), pp. 694–726.

Keil, R. (2002). 'Common-Sense' Neoliberalism: Progressive Conservative Urbanism in Toronto, Canada. *Antipode*, 34(2), pp. 578–601.

Kipfer, S. and Keil, R. (2002). Toronto Inc? Planning the Competitive City in the New Toronto. *Antipode*, 34(2), pp. 3–17.

Kirby, A. (2008). The Production of Private Space and Its Implications For Urban Social Relations. *Political Geography*, 27(1), pp. 74–95.

Lam, R. (2014). Vancouver Council Approves Two Residential Tower North of Cambie Bridge. *The Vancouver Sun*, 12 June. Available at: www.nationalpost.com/m/vancouver+council+approves+residential+towers+north+cambie/9932685/story.html [accessed 30 July 2014].

Lasner, M.G. (2012). *High Life: Condo Living in the Suburban Century*. New Haven, CT and London: Yale University Press.

Lehrer, U. (2008). Urban Renaissance and Resistance in Toronto. In: L. Porter and K. Shaw, eds. *Whose Urban Renaissance? An International Comparison of Policy Drivers and Responses to Urban Regeneration Strategies*. London: Routledge, pp. 147–156.

Lehrer, U. and Wieditz, T. (2009a). Condominium Development and Gentrification: The Relationship Between Policies, Building Activities and Socio-Economic Development in Toronto. *Canadian Journal of Urban Research*, 18(1), pp. 140–161.

Lehrer, U. and Wieditz, T. (2009b). Gentrification and the Loss of Employment Lands: Toronto's Studio District. *Critical Planning*, 16(1), pp. 138–160.

Lehrer, U., Keil, R. and Kipfer, S. (2010). Reurbanization in Toronto: Condominium Boom and Social Housing Revitalization. *The Planning Review*, 46(1), pp. 81–90.

Logan, J.R. and Molotch, H.L. (1987). *Urban Fortunes: The Political Economy of Place*. Berkeley: University of California Press.

Moore, A. (2013a). *Planning Politics in Toronto: The Ontario Municipal Board and Urban Development*. Toronto: University of Toronto Press.

Moore, A. (2013b). Trading Density For Benefits: Toronto and Vancouver Compared. Institute on Municipal Finance and Governance, Paper No. 13. Munk School of Global Affairs, Toronto.

Morgan, E. (2013). The Sword in the Zone: Fantasies of Land Use Planning Law. *University of Toronto Law Journal*, 62(2), pp. 163–199.

New York City Department of City Planning. (2015, 26 February). A Survey of Transferable Development Rights Mechanisms in New York City; Department of City Planning. Available at: www1.nyc.gov/assets/planning/download/pdf/plans-studies/transferable-development-rights/research.pdf [accessed 16 January 2018].

Ontario Home Builders' Association. (2014). Land Use Planning and Appeal System Response to the Consultation Document. Ontario Minister of Municipal Affairs and Housing. Available at: www.bildgta.ca/BILD/uploadedFiles/Government_Relations/2012/OHBA%20Land%20Use%20Planning%20and%20Appeals%20Submission.pdf [accessed 10 January 2014].

Pantalone, P. (2014). Density Bonusing and Development in Toronto. Faculty of Environmental Studies, York University, Toronto. Available at: http://yorkspace.library.yorku.ca/xmlui/bitstream/handle/10315/30269/MESMP02347.pdf?sequence=1 [accessed 4 December 2017].

Peck, J. (1995). Moving and Shaking: Business Elites, State Localism and Urban Privatism. *Progress in Human Geography*, 19(1), pp. 16–46.

Places to Grow Act. (2005). Available at: www.placestogrow.ca/index.php?option=com_content&task=view&id=4&Itemid=9 [accessed 12 January 2016].

Provincial Policy Statement. (2014). Available at: www.mah.gov.on.ca/AssetFactory.aspx?did=10463 [accessed 10 January 2018].

Rankin, K.N. and Delaney, J. (2011). Community BIAs as Practices of Assemblage: Contingent Politics in the Neoliberal City. *Environment and Planning A*, 43(6), pp. 1363–1380.

Rosen, G. and Walks, A. (2013). Rising Cities: Condominium Development and the Private Transformation of the Metropolis. *Geoforum*, 49, pp. 160–172.

Rosen, G. and Walks, A. (2015). Castles in Toronto's Sky: Condoism as Urban Transformation. *Journal of Urban Affairs*, 85(1), pp. 39–66.

Sassen, S. (2015). Who Owns Our Cities – And Why This Urban Takeover Should Concern Us All. *Guardian*, 24 November. Available at: www.theguardian.com/cities/2015/nov/24/who-owns-our-cities-and-why-this-urban-takeover-should-concern-us-all [accessed 13 February 2016].

Toronto Life. (2006). The Condo Generation: Cool Owners, Hot Buildings, Great Designs. *Toronto Life*, February.

Toronto Life. (2010). Boomtown: The Incredible, Unstoppable Growth of Downtown. *Toronto Life*, February.

Walters, L.C. (2012). Land Value Capture in Policy & Practice. Proceedings of the World Bank Conference on Land and Poverty, Washington, DC. Available at: www.landandpoverty.com/agenda/pdfs/paper/walters_full_paper.pdf [accessed 10 January 2014].

Ward, K. (2006). Policies in Motion, Urban Management and State Restructuring: The Trans-Local Expansion of Business Improvement Districts. *International Journal of Urban and Regional Research*, 30(1), pp. 54–75.

Weber, R. (2010). Selling Urban Futures: The Financialization of Urban Redevelopment Policy. *Economic Geography*, 86(3), pp. 251–274.

Wilson, D. (2004). Toward a Contingent Urban Neoliberalism. *Urban Geography*, 25(8), pp. 771–783.

Yang, J. (2010). The Birthplace of BIAs Celebrates 40 Years. *Toronto Star*, 18 April. Available at: www.thestar.com/news/gta/2010/04/18/the_birthplace_of_bias_celebrates_40_years.html [accessed 16 July 2016].

9
DIGITAL TECHNOLOGIES AND RECONFIGURATION OF URBAN SPACE

Barney Warf

Without doubt, the microelectronics revolution, digital technologies and the various manifestations of the Internet have had, and continue to have, profound impacts on the structure of urban space. Computer code has become so woven into the fabric of urban space as to be indispensable, profoundly shaping the contours of work, shopping, education, travel and everyday life (Dodge and Kitchin 2005). So central have webpages, email, social media and the like become that simple dichotomies such as online/offline fail to do justice to the ways in which the real and the virtual are interpenetrated. As Graham and Marvin (1996) note, if cities arose to overcome time by using space, information and communications technologies (ICTs) overcome space with time. Hudson-Smith et al. (2009: 271) argue that "computers in cities exist in abundance, of course, but it is cities inside of computers that now define the digital frontier". Rather than claim that ICTs restructure urban space – a statement that reeks of simplistic technological determinism – it is more accurate to note that ICTs and urban social relations co-evolve, shaping one another in a mutually determinative relationship. As the real and virtual worlds have become shot through with each other, it is more accurate to say that cities and telecommunications are co-evolving.

The impacts of ICTs are felt both within urban areas and across the urban hierarchy. Although large cities typically have much better developed telecommunications infrastructures than do small ones, the technology has rapidly diffused through most national urban hierarchies across the world. In the future, therefore, the competitive advantages based on telecommunications will diminish, forcing competition among localities to occur on the cost and quality of labour, taxes and local regulations. Regions with an advantage in telecommunications generally succeed because they have attracted successful firms for other reasons.

This chapter has an ambitious goal: to provide a broad overview of several major ways in which ICTs are reshaping urban areas. It focuses on four major topics. The first part notes the role of ICTs in the emergence of smart or intelligent cities, telework and new transportation systems, giving rise to new divisions of labour and modes of work and travel. Second, it examines urban e-government, or the use of the Internet to provide public services, which has helped to streamline interactions with the state, improve efficiency and enhance transparency. The third section turns to the changing nature of urban planning in light of the growth of Web 2.0; rather than simply constituting the domain of privileged experts, interactive websites allow far greater levels of public input, helping to democratize the planning process. The fourth part addresses the digital divide, or uneven urban access to the Internet and to digital technologies more broadly. Although the vast majority of urbanites in the developed world regularly use digital media, significant populations remain of the poor, elderly and uneducated, who enjoy few of the benefits of digitization. Social and spatial inequalities, which have mushroomed under neoliberalism, are thus firmly reinscribed in cyberspace, particularly in terms of access to the Internet.

Before proceeding, because ICTs are frequently portrayed in ahistorical terms, it is worth noting that they have long played a role in the growth and commodification of urban space. In the nineteenth century, first the telegraph, then the telephone, ushered in dramatically new ways of communicating, forming an integral part of the time-space compression of that era. The telegraph – the Internet of the nineteenth century (Standage 1998) – unleashed history's greatest communications revolution since the invention of the printing press. "In a historical sense, the computer is no more than an instantaneous telegraph with a prodigious memory, and all the communications inventions in between have simply been elaborations on the telegraph's original work" (Marvin 1988: 3). Telegraphy became vital to stock markets, warehousing, shipping and wholesaling. Pred's (1977) analysis of the nineteenth century American city-system emphasized the role of telegraphy to the differential growth of urban areas, which he theorized in terms of circular and cumulative feedback loops. It was vital to the development of the multi-establishment corporation and the spatial differentiation of the office from the factory. In world cities, the telegraph was central to the transmission of vast quantities of data that coordinated the increasingly complex world economy. Similarly, the telephone, following its introduction in 1876, made all places equidistant from one another, or at least almost so. It also had enormous social effects. "The introduction of the telephone did more than enable people to communicate over long distances: it threatened existing class relations by extending the boundary of who may speak with whom; it also altered modes of courtship and possibilities of romance" (Poster 1990: 5). The telephone was also central to the "delocalization debate", in which one view held that, like the automobile, the telephone accelerated the individualism, social withdrawal and disengagement from public life so widespread in American society, while the opposing view maintained that telephones allowed additional convenient contacts, especially for the isolated and lonely (e.g. in rural areas), furthering the formation of "communities without propinquity", that is, telephones complemented, not substituted for, face-to-face interaction. In short, digital ICTs, which began to appear *en masse* in the late twentieth century, arose within the context of a long tradition of pre-digital urban informatics.

Intelligent cities, telework and transportation systems

As ICTs have become woven into the fabric of urban space, many urban areas have sought to transform themselves into 'smart cities' or 'intelligent cities' (Allwinkle and Cruickshank 2011; Caragliu et al. 2011). Definitions vary, but essentially smart cities are those that utilize integrated ICT networks in their infrastructures to minimize urban problems and improve quality of life on a large scale basis (Holland 2008). The term connotes high degrees of interconnectedness, including the Internet, networked sensors, cameras, kiosks, cell or mobile phones, and even implanted medical devices (e.g. heart monitors). The emergence of smart or intelligent cities is often motivated by the desire to minimize energy use, address global climate change, reduce oil dependence and encourage a focus on renewable energy and sustainable development. These issues reflect the origins of smart cities within urban and regional planning, which has become increasingly preoccupied with urban competitiveness, and they facilitate the growth of gentrified inner cities with populations employed in business services. These comments serve to illustrate one means by which information technology and neoliberal urbanism are deeply intertwined.

Smart cities offer abundant opportunities for generating savings in energy, time and capital. For example, smart grids are used to manage electricity efficiently, including smart meters to inform households of their energy use and that charge higher prices during periods of peak demand (Calvillo et al. 2016). Smart grids manage energy packets in much the way that the Internet manages data packets, routing them across optimal network paths to minimize time and cost. Health care in smart cities is improved through the use of integrated sensors that provide publicly available data about temperature, humidity, allergens and air pollution (Solanas et al. 2014). Traffic management in smart cities can reduce congestion, notably at rush hour, by using smart traffic lights and by optimizing usage of underutilized roads (Kostakos et al. 2013), a process complemented by recent internet-based services such as Uber and Lyft. Smart parking strategies allow users to reserve parking spots in advance. Other benefits include: smart water meters that identify water mains that are likely to break ahead of time; minimizing the number of building inspectors as smart buildings report their maintenance status; and dispatching road crews only to those smart bridges that report icy conditions. Many airports now routinely utilize fingerprint, facial recognition and retinal scans to identify passengers. Most such systems draw upon sophisticated artificial intelligence algorithms that deploy neural networking, Markov chains and entropy maximizing strategies. Moreover, by fusing the virtual and digital worlds, they exemplify how the Internet of Things (IoT) – a world of interconnected devices – has become a powerful force in shaping urban life.

The most advanced examples in this regard are found in the European Union. Santander, Spain, has 20,000 sensors embedded in buildings and the infrastructure that monitor pollution, noise, traffic and security (e.g. identity management). Barcelona has implemented smart irrigation systems for parks that adjust water use based on temperatures and rainfall. When Amsterdam initiated its TrafficLink web-based traffic management system in 2014, hours lost due to congestion dropped by 10 per cent within a year. Residents there can use the app Mobypark to rent out parking spaces they own. Stockholm's Green IT strategy includes smart buildings that conserve energy, e-services for publicizing information and open access GPS that allows residents to plot shortest paths to their destinations.

Another important effect of ICTs is 'telework' or 'telecommuting', in which workers substitute some or all of their working day at a remote location (almost always home) for time usually spend at the office (Golden 2009; Haddon and Brynin 2005). Telework is most appropriate for jobs involving mobile activities or routine information handling such as data entry or directory assistance. The self-employed do not count as teleworkers because they do not substitute it for commuting. Proponents of telework claim that it enhances productivity and morale, reduces employee turnover and office space, and leads to reductions in traffic congestion and accidents, air pollution and energy use. Critics contend that it is simply a digital version of the neoliberal labour market in which the boundaries between home and work disappear (Greenhill and Wilson 2006). The extent to which telecommuting has become popular is unclear, but some estimates hold that there were roughly 2.9 million full-time teleworkers in the USA, or 2 per cent of the labour force, in 2010 (Lister and Harnish 2011). Many others, however, may telecommute part-time. Future growth of telecommuting will likely encourage more decentralization of economic activity in suburban areas.

However, the growth of teleworking may ironically increase the demand for urban transportation services rather than decrease it, as was widely expected. While telecommuters travel to their workplace less frequently than do most workers, many have longer *weekly* commutes overall, leading to a rise in aggregate distances travelled. In addition, the time released from commuting may be utilized travelling for non-work related purposes. Thus, while telecommuters may have different reasons for travelling, the frequency or volume of trips may not change. Moreover, even if telecommuting did reduce some trips, the associated reductions in congestion could simply induce other travellers onto the roads. In short, whether or not an actual trade-off between telecommunications and commuting occurs (their substitutability rather than complementarity) is not clear.

Similarly, another impact of ICTs concerns transportation informatics, or intelligent transportation systems, including smart metering, electronic peak load road pricing, automated toll payments, navigation and travel advisory systems, and remote traffic monitoring and displays (Hounsell *et al.* 2009). In contrast to traditional, centralized traffic management systems, which are difficult to implement as the volume of traffic increases, ICT-based decentralized ones such as agent-based models are more flexible, adaptable and efficient (Chen and Cheng 2010). For example, intelligent traffic management systems can use RFID tags in cars to inform managers about congested roads, stolen vehicles and offer dynamic navigation to identify least congested paths (Wen 2010). Seamless flows of information between intelligent vehicles and the road infrastructure dramatically improve transport speeds and reduce accidents (Bishop 2005). Similar applications include bus and truck fleet and freight management, ramp meters, intelligent traffic lights, improved emergency response systems and mobile confirmation messages that provide drivers with information about road conditions in real time (Giannopolous 2004). Such systems demand enormous degrees of interoperability and compatibility among various computer systems. The Intelligent Vehicle Highway System currently under way in southern California represents the next generation of this technology. Finally, the rapid growth of wireless technologies, particularly cellular phones, allows commuters stuck in traffic to use their time more productively.

ICTs and urban e-government

The deployment of ICTs is a central part of e-government. The Internet can be used to deliver government information and services, improve administrative procedures, and raise citizen input and participation. Examples include the digital collection of taxes; electronic voting; payment of utility bills, fines and dues; applications for public assistance, permits and licences; online registration of companies and automobiles; and access to census and other public data. Interactive municipal sites give residents access to information about schools, libraries, bus schedules and hospitals, and to download forms, maps and data. As Jaeger and Thompson (2004: 94) note, "e-government is quickly becoming 'simply the way things are done' in technologically advanced nations". Not surprisingly, the topic has drawn considerable scholarly attention (for a review, see Rocheleau 2007). Often e-government is divided into government-to-business (G2B), government-to-government (G2G) and government-to-citizens (G2C) forms (Fountain 2001). G2B e-government includes digital submissions of bids, contracts and bills. Finally, G2G e-government enhances communication and interaction among different government agencies.

E-government is frequently alleged to increase citizen accessibility to government services, improve efficiency and transparency, create synergies and generate economies of scale in the delivery of government services. Online access facilitates acquisition of information and reduces trips to and waiting times in, government offices. E-government may also encourage a democratization of public bureaucracies by moving them from classic hierarchical forms of control to more horizontal, collaborative models (Ndou 2004).

Far from comprising some homogeneous whole, the set of ideas and practices that lie at the heart of e-government are actually quite diverse, varying over time, space and institutional context. Layne and Lee (2001) offered a well-received conception of developmental stages of e-government, ranging from simple online presence (i.e. a webpage); interfaces that allow citizen access to data and services; to vertical integration in which citizens can actively participate (e.g. for licence applications); and finally to horizontal integration, in which one or a few centralized websites offer a broad range of government functions and purposes. Empirical assessments of e-government initiatives typically focus on the quality of websites, including criteria such as missing links, readability, the publications and data displayed, contact information for public officials, number of languages that provide access, sound and video clips, ability to use credit cards and digital signatures, security and privacy policies, and opportunities for citizen feedback. The growth of cellular or mobile phones and social media outlets such as Facebook and Twitter likewise has paved the way for mobile governance, or what Linders (2012) calls the shift from e-government to "we-government". To the extent that e-government improves the abilities to deliver services, it may increase satisfaction with existing political regimes; conversely, in societies in which public sector jobs are often allocated through patronage networks, the increased efficiency resulting from e-government may minimize the growth of public employment. Similarly, digital hotlines for submission of citizen complaints give voice to those who are typically voiceless in the circles of governance. Concerns over e-government include the potential invasions of privacy that it invites, possible problems concerning local and national security (i.e. hacking of government files) and the inequality of access generated by digital divides (about which more later) (Belanger and Carter 2008). The role of e-government varies widely among and within countries; its impacts tend to be felt most heavily in economically advanced countries in which internet use is high, although many countries in the developing world are also adopting it (Warf 2014a, 2014b).

Web 2.0, neogeography and urban planning

One arena in which ICTs have had significant impacts in cities is in urban planning, largely due to the growth of Web 2.0, a diverse set of software applications that have revolutionized usage of the Internet. Key components of this technology are Asynchronous Javascript and XML (AJAX) and Application Programming Interfaces (API), which facilitate the creation of websites that allow instantaneous user interactions. The functionality offered by Web 2.0 has precipitated significant changes from traditional approaches to internet usage, making the web markedly more user-centric. In this sense, it has fostered an unprecedented democratization of knowledge. Goodchild (2007: 27), focusing on "citizen sensors", maintains that whereas

> the early Web was primarily one-directional, allowing a large number of users to view the contents of a comparatively small number of sites, Web 2.0 is a bi-directional collaboration in which users are able to interact with and provide information to central sites, and to see that information collated and made available to others.

The interactive websites characteristic of Web 2.0 allow users to upload information about locations into online content and apply their data in diverse ways, including, for example, simple displays of locations (e.g. favoured routes for a proposed bike trail) or lists of attributes of a place near a user equipped with a GPS. This approach lies at the heart of mapping websites such as Google Maps, Yahoo!Maps, OpenStreetMap and Bing Live Maps.

The term neogeography, which has been used in several ways with varying meanings, points to the process by which people use online geospatial tools to describe and document aspects of their lives and environment in terms that are meaningful to them (Hudson-Smith *et al.* 2009).

As Web 2.0 has enabled growing legions of people to interact with one another, and with government agencies, neogeography has enjoyed explosive growth. As a consequence, geographic knowledge has increasingly escaped the confines of academia or urban planning professionals and has been embraced by an enthusiastic and rapidly growing public of amateurs and hobbyists. Rather than rely on state or corporate-produced data, neogeography generates volunteered data/content, relocating the centre of knowledge production from a handful of self-appointed experts to large numbers of people with limited formal geographic training.

For example, Malczewski (2004) suggests that geographic information systems (GIS) have moved from a scarce, specialized commodity operating on mainframe computers to a common tool of planners, in the forms of simple desktop programs such as MapInfo GIS and ArcView, and in the process facilitated collaborative decision making in planning circles. Sui (2008) labels these changes the "wikification of GIS", after Wikipedia, the famously popular, user-generated, online encyclopedia. For example, user-generated maps may be those of endangered bird species sightings, publicly accessible restrooms, ideal camping locations, accident-prone roads and green buildings. Analogously, Sieber (2006) argues that Public Participation GIS has markedly expanded the circle of stakeholders involved in planning decisions. As a consequence, GIS has been embraced by an enthusiastic and rapidly growing public of non-professional users, with results that have yet to be well understood. Kolbitsch and Maurer (2006) hold that the participatory qualities of Web 2.0 allow ordinary users to forge a collective intelligence, so that planning outcomes are "emergent" properties of multiple actors. Adopted on a large scale, Web 2.0 and neogeography may encourage public bureaucracies to modernize their administrative practices, increase responsiveness and transparency, and empower citizens to shape local government actions (Noveck 2009). In short, neogeography facilitates collective, bottom-up place-making rather than top-down, state-mandated designs of locales (Beyea et al. 2009).

The idea of neogeographic-inspired municipal e-government is far from science fiction, but has begun to materialize in fact. While simple, one-way government websites have become almost universal among municipalities in the industrialized world, the shift towards transactional ones has been much slower and spatially uneven. Few cities offer e-government services beyond the most basic provision of digital forms and information (Reddick 2004). Moon (2002) suggests that larger cities are more likely to have interactive e-government portals and to offer a broader array of applications than do smaller ones. In Tampere, Finland, municipal Web 2.0 sites allow citizens to provide urban planners with their views and experiences (Jaeger 2003). In Greece, the nation's first 'digital city', Trikala, was launched in 2006, giving 70 per cent of its residents internet access and giving them an ability to participate in telework, on-line library and school programmes, emergency response systems, environmental and transportation information, and demographic data that can be utilized through publicly available geographical information systems. A few American municipalities (e.g. Bakersfield, CA) broadcast city council meetings on the web; some (e.g. Durham, NC; Scottsdale, AZ; Fort Lauderdale, FL) offer online geographical information systems that permit location-based searches and interactive mapping.

Urban digital divides

Despite early utopian expectations that the Internet would obliterate the importance of social and spatial differentials, there remain significant discrepancies in terms of access to the Internet. Even in the USA, where 81 per cent of the population had internet access in 2015, significant minorities have not logged on. Even in the most wired of cities, pockets of offline populations remain. Indeed social inequalities are inevitably reinscribed in cyberspace. Everywhere in the world, access to the Internet, either at home or at work, is highly correlated with age, income, educational level and employment in professional occupations. 'Access' and 'use' are admittedly vague terms, and embrace a range of meanings, including the ability to log on at home, school, cybercafé or work (DiMaggio et al. 2001). Rather than a simple access/non-access dichotomy, it is more useful to think of a gradation of levels of access, although data of this subtlety rarely exist. Thus, it is increasingly common to speak of 'digital differentiation' rather than a divide (Selwyn 2004).

Difficulties in procuring the skills, equipment and software necessary to get internet access threaten to exclude economically and socially marginalized groups from the benefits of cyberspace, a phenomenon that reflects the growing inequalities generated by labour market polarization. Thus, the microelectronics revolution has accentuated the digital divide between groups that are proficient with computer technology and those who neither understand nor trust it, reinforcing and deepening the inequalities in broader social relations. The digital divide is a serious obstacle to upwards social mobility, enhancing the vulnerability of long-disenfranchised populations (Korupp and Szydlik 2005; Warf 2001). As Howard et al. (2010: 111) point out, "The causes and consequences of the digital divide have become a contested area of research. Understanding the digital divide is crucial to understanding the role of the Internet in contemporary social development."

The digital divide is also a deeply geographical phenomenon. Everywhere, large urban centres tend to exhibit higher rates of connectivity than do rural areas (Mills and Whitacre 2003). Because they are often not well served by telecommunications firms, rural areas often have slow or substandard connections. Rural areas are often shunned by internet service providers because their populations are distributed over broad areas, making it difficult to generate economies of scale, and are frequently poorer, older and less educated than those in urban areas.

As the uses and applications of the Internet have multiplied, the costs sustained by those denied access rise accordingly. At precisely the historical moment that contemporary capitalism has come to rely upon digital technologies to an unprecedented extent, large pools of the economically disenfranchised are shut off from cyberspace. As the Internet erodes the monopolistic roles once played by the telephone and television, and as the upgrading of required skill levels steadily render ICT skills necessary even for low wage service jobs, lack of access to cyberspace becomes increasingly detrimental to social mobility. The digital divide is also a major obstacle to the successful adoption and implementation of e-government (Yigitcanlar and Baum 2006). After all, digital provision of government forms and services is useless to those without access to the Internet; thus e-government and the digital divide are deeply intertwined phenomena (Helbig et al. 2009).

Public schools are a critical domain within which the digital divide is manifested, reproduced and sometimes overcome (Monroe 2004). Variations in public school funding are reproduced in terms of the quality of internet access within their classrooms (Becker 2000): while 99 per cent of American schools offer children access to networked PCs, the speed and reliability of the ability to log onto the Internet vary significantly, with important effects: "students with Internet-connected computers in the classroom, as opposed to a central location like a lab or library, show greater improvement in basic skills" (Kaiser Foundation 2004). Unsurprisingly, the digital divide in public schools is racialized: white students are much more likely than are minorities to use the Internet in the classroom or school library (US Department of Education 2006). However, measuring internet access via simple availability to PCs at a school fails to do justice to the extent of the digital divide: poor students are considerably less likely to have them at home or to possess the skills necessary to install and maintain them, or to navigate the accompanying software.

Another change that has amplified the complexity of urban digital divides is the rapid growth in wireless and mobile broadband services. In 2014, approximately 90 per cent of Americans owned a mobile phone, of which 64 per cent were smart phones connected to the Internet (Pew Research Center 2015). One-third of American internet users used their phones as their preferred means of internet access. Predictably, these rates varied by income and educational level. As internet access shifts increasingly to mobile devices, digital divides may gradually close, and thus reduce the social inequalities that they reflect and in turn sustain.

Concluding thoughts

There can be no doubt that ICTs are part of an ongoing, massive reconfiguration of urban space, an integral part of the globalized, neoliberal city that has become the norm across most of the world. The digital world has extended deep tentacles into urban labour markets, municipal governance, traffic planning, retailing and everyday life, altering the ways in which urbanites work, live and move people, goods and information. Used correctly, ICTs help to make urban life safer, cleaner, more efficient and convenient, and easier to manoeuvre. In an age in which information is by far the central input and output of most economic activity, urban infrastructures have responded by becoming intelligent and networked, allowing for better energy use and traffic management. The diffusion of Web 2.0 technologies has facilitated citizen input into urban governance to an unprecedented degree. The gradual growth of smart cities holds great promise for reducing energy use and encouraging sustainable development.

Although often heralded as such, ICTs are not a panacea for urban ills. While most people in the developed world have access to the Internet, pockets of the most economically and politically vulnerable remain deeply entrenched, and are likely to be ever more excluded from the benefits of widespread digitization. The situation in the developing world remains even more troubling: less than half of the world uses the Internet, and the deployment of urban ICTs in many countries remains a distant dream. In this light, it is imperative to consider digitization not simply as a technical process, but as a profoundly political and social one.

References

Allwinkle, S. and Cruickshank, P. (2011). Creating Smart-er Cities: An Overview. *Journal of Urban Technology*, 18(2), pp. 1–16.
Becker, H. (2000). Who's Wired and Who's Not: Children's Access to and Use of Computer Technology. *The Future of Children*, 10(2), pp. 44–75.
Belanger, F. and Carter, L. (2008). Trust and Risk in E-Government Adoption. *Journal of Strategic Information Systems*, 17(2), pp. 165–176.
Beyea, W., Geith, C. and McKeoen, C. (2009). Place Making Through Participatory Planning. In M. Foth, ed. *Urban Informatics: The Practice and Promise of the Real-Time City*. Hershey, PA: IGI Global, pp. 55–67.
Bishop, B. (2005). *Intelligent Vehicle Technology and Trends*. Norwood, MA: Artech House.
Calvillo, C., Sanchez-Mirales, A. and Villar, J. (2016). Energy Management and Planning in Smart Cities. *Renewable and Sustainable Energy Reviews*, 55(1), pp. 273–287.
Caragliu, A., Del Bo, C. and Nijkamp, P. (2011). Smart Cities in Europe. *Journal of Urban Technology*, 18(2), pp. 85–82.
Chen, B. and Cheng, H. (2010). A Review of the Applications of Agent Technology in Traffic and Transportation Systems. *IEEE Transactions on Intelligent Transportation Systems*, 11(2), pp. 485–497.
DiMaggio, P., Hargitai, E., Newman, W. and Robinson, J. (2001). Social Implications of the Internet. *Annual Review of Sociology*, 27(1), pp. 307–336.
Dodge, M. and Kitchen, R. (2005). Code and the Transduction of Space. *Annals of the Association of American Geographers*, 95(3), pp. 162–180.

Fountain, J. (2001). The Virtual State: Transforming American Government? *National Civic Review*, 90(3), pp. 241–251.

Giannopolous, G. (2004). The Application of Information and Communication Technologies in Transport. *European Journal of Operational Research*, 152(2), pp. 302–320.

Golden, T. (2009). Applying Technology to Work: Toward a Better Understanding of Telework. *Organization Management Journal*, 6(4), pp. 241–250.

Goodchild M. (2007). Citizens as Voluntary Sensors: Spatial Data Infrastructure in the World of Web 2.0. *International Journal of Spatial Data Infrastructures Research (IJSDIR)*, 2, pp. 24–32.

Graham, S. and Marvin, S. (1996). *Telecommunications and the City: Electronic Spaces, Urban Places*. London: Routledge.

Greenhill, A. and Wilson, M. (2006). Haven or Hell? Telework, Flexibility and Family in the E-Society: A Marxist Analysis. *European Journal of Information Systems*, 15, pp. 379–388.

Haddon, L. and Brynin, M. (2005). The Character of Telework and the Characteristics of Teleworkers. *New Technology, Work, and Employment*, 20(1), pp. 34–46.

Helbig, N., Gil-Garcia, J. and Feerro, E. (2009). Understanding the Complexity of Electronic Government: Implications From the Digital Divide Literature. *Government Information Quarterly*, 26, pp. 89–97.

Holland, R. (2008). Will the Real Smart City Stand Up? *City*, 12(3), pp. 303–320.

Hounsell, N., Shrestha, B, Piao, J. and McDonald, M. (2009). Review of Urban Traffic Management and the Impacts of New Vehicle Technologies. *Institution of Engineering and Technology, Intelligent Transport Systems*, 3(4), pp. 419–428.

Howard, P., Busch, L. and Sheets, P. (2010). Comparing Digital Divides: Internet Access and Social Inequality in Canada and the United States. *Canadian Journal of Communication*, 35, pp. 109–128.

Hudson-Smith, A., Milton, R., Dearden, J. and Batty, M. (2009). The Neogeography of Virtual Cities: Digital Mirrors into a Recursive World. In M. Foth, ed. *Urban Informatics: The Practice and Promise of the Real-time City*. Hershey, PA: IGI Global, pp. 270–291.

Jaeger, P. (2003). The Endless Wire: E-Government as Global Phenomenon. *Government Information Quarterly*, 20(4), pp. 323–331.

Jaeger, P. and Thompson, K. (2004). Social Information Behavior and the Democratic Process: Information Poverty, Normative Behavior, and Electronic Government in the United States. *Library & Information Science Research*, 26(1), pp. 94–107.

Kaiser Foundation. (2004). Children, the Digital Divide, and Federal Policy. Available at: www.kff.org/other/issue-brief/children-the-digital-divide-and-federal-policy/ [accessed 12 January 2018].

Kolbitsh, J. and Maurer, H. (2006). The Transformation of the Web: How Emerging Communities Shape the Information We Consume. *Journal of Universal Computer Science*, 12(2), pp. 187–213.

Korupp, S. and Syzdlik, M. (2005). Causes and Trends of the Digital Divide. *European Sociological Review*, 21, pp. 409–422.

Kostakos, V., Ojala, T. and Juntunen, T. (2013). Traffic in the Smart City: Exploring City-Wide Sensing for Traffic Control Center Augmentation. *Internet Computing*, 17(6), pp. 22–29.

Layne, K. and Lee, J. (2001). Developing Fully Functional E-Government: A Four Stage Model. *Government Information Quarterly*, 18, pp. 122–136.

Linders, D. (2012). From E-Government to We-Government: Defining a Typology for Citizen Coproduction in the Age of Social Media. *Government Information Quarterly*, 29, pp. 446–454.

Lister, K. and Harnish, T. (2011). The State of Telecommuting in the U.S.: How Individuals, Business and Government Benefit. Telework Research Network. Available from: www.workshifting.com/downloads/downloads/Telework-Trends-US.pdf [accessed 2 January 2016].

Malczewski, J. (2004). GIS-based Land-Use Suitability Analysis: A Critical Overview. *Progress in Planning*, 62(1), pp. 3–65.

Marvin, C. (1988). *When Old Technologies Were New: Thinking about Electric Communication in the Late Nineteenth Century*. Oxford: Oxford University Press.

Mills, B. and Whitacre, M. (2003). Understanding the Non-Metropolitan–Metropolitan Digital Divide. *Growth and Change*, 34(2), pp. 219–243.

Monroe, B. (2004). *Crossing the Digital Divide: Race, Writing, and Technology in the Classroom*. New York: Teachers' College.

Moon, M. (2002). The Evolution of E-Government Among Municipalities: Rhetoric or Reality? *Public Administration Review*, 62(4), pp. 424–433.

Ndou, V. (2004). E-Government For Developing Countries: Opportunities and Challenges. *Electronic Journal on Information Systems in Developing Countries*, 18(1), pp. 1–24.

Noveck, B. (2009). *Wiki Government: How Technology Can Make Government Better, Democracy Stronger, and Citizens More Powerful*. Washington, DC: Brookings Institution.

Pew Research Center. (2015). Mobile Technology Fact Sheet. www.pewinternet.org/fact-sheets/mobile-technology-fact-sheet/ [accessed 3 January 2016].

Poster, M. (1990). *The Mode of Information*. Chicago: University of Chicago.

Pred, A. (1977). *City-Systems in Advanced Economies*. London: Hutchinson.

Reddick, C. (2004). A Two-Stage Model of E-Government Growth: Theories and Empirical Evidence for U.S. Cities. *Government Information Quarterly*, 21(1), pp. 51–64.

Rocheleau, B. (2007). Whither E-Government? *Public Administration Review*, 67(3), pp. 584–588.

Selwyn, N. (2004). Reconsidering Political and Popular Understandings of the Digital Divide. *New Media & Society*, 6(3), pp. 341–362.

Sieber, R. (2006). Public Participation Geographic Information Systems: A Literature Review and Framework. *Annals of the Association of American Geographers*, 96(3), pp. 491–507.

Solanas, A., Patsakis, C., Conti, M., Vlachos, M., Ramos, V., Falone, F., Postolache, O., Oereez-Martinez, P., Pietro, R., Perrea, D. and Martinez-Balleste, A. (2014). Art Health: A Context-Aware Health Paradigm Within Smart Cities. *IEEE Communications Magazine*, August, pp. 74–81. Available at: www.researchgate.net/profile/Victoria_Ramos/publication/265091720_Smart_Health_A_context-aware_health_paradigm_within_smart_cities/links/542aa4af0cf277d58e874ecd.pdf [accessed 3 January 2016].

Standage, T. (1998). *The Victorian Internet*. New York: Walker and Company.

Sui, D. (2008). The Wikification of GIS and its Consequences: Or Angelina Jolie's New Tattoo and the Future of GIS. *Computers, Environment and Urban Systems*, 32, pp. 1–5.

US Department of Education. (2006). Computer and Internet Use by Students in 2003. Available at: http://nces.ed.gov/pubs2006/2006065.pdf?loc=interstitialskip [accessed 4 December 2017].

Warf, B. (2001). Segways into Cyberspace: Multiple Geographies of the Digital Divide. *Environment and Planning B: Planning and Design*, 28(3), pp. 3–19.

Warf, B. (2014a). Geographies of E-Governance in Latin America and the Caribbean. *Journal of Latin American Geography*, 13(1), pp. 169–185.

Warf, B. (2014b). Asian Geographies of E-Government. *Eurasian Geography and Economics*, 55(1), pp. 94–110.

Wen, W. (2010). An Intelligent Traffic Management Expert System with RFID Technology. *Expert Systems with Applications*, 37(4), pp. 3024–3035.

Yigitcanlar, T. and Baum, S. (2006). E-Government and the Digital Divide. In: M. Khosrow-Pour, ed. *Encyclopedia of E-Commerce, E-Government, and Mobile Commerce*. Hershey, PA: IGI Global.

PART III

Spaces of the environment and nature

Byron Miller and Andrew E.G. Jonas

A couple of decades ago, the notion of a dedicated section of an urban handbook dealing with the environment and nature would have been met with a measure of incredulity in many circles. Traditionally, the notion of the city has been posed in opposition to 'nature'. Nature, once seen as pristine and distinct from the world of humans, has gradually given way to a view of nature as not 'natural', but constructed. While this view is not new – it can be traced back to Marx and still others before him – contemporary understandings of the 'production of nature' have their direct antecedent in Neil Smith's (1984) classic book, *Uneven Development: Nature, Capital, and the Production of Space*. Smith argued that capitalism survives through its capacity to produce space and nature. Smith's argument was not taken up immediately. When William Cronon (1995) asserted that it was time to "rethink wilderness", he was met with a furious backlash from mainstream environmentalists. Nature as pristine and distinct had become the dialectical other to an anthropocentric perspective of nature as a realm under the dominion of 'man'. Failure to uphold the ontological boundaries of pristine nature deprived those concerned with the degradation and destruction of the planet's ecosystems of a stable foundation from which human-caused environmental crises could be critiqued, or so many thought. Many still believe so today. But the notion of nature as something apart from humans has weakened in light of the adoption of relational and process-focused ontologies that call into question static and categorical understandings of the world, including what we have traditionally called 'nature'. In this vein David Harvey provocatively has asserted that there is nothing unnatural about New York City. Harvey's point, and the point of many scholars working in the rapidly expanding field of political ecology, is that the conceptual boundaries of human and natural systems are of our own construction, that humans are part of nature and vice versa (although the precise ontological qualities of this relationship is a matter of considerable debate), and that human influence on ecosystems must be understood as the exercise of differentiated social and political-economic power.

As our conceptualization and understanding of human-environment relations have evolved, so has our conceptualization and understanding of cities. The notion of cities as distinct self-contained entities that could be studied in isolation has also given way to relational, process-oriented understandings. Henri Lefebvre (1970) was the first to call for the study of 'urban society' as distinct from cities and his prediction of the development of 'planetary urbanism' is now reality, according to Neil Brenner (2014) and others. While Brenner's structuralist approach has its critics, the notion of largely unbounded urbanization has been embraced widely. Tellingly, the cover of Brenner's 2014 *magnum opus*, *Implosions/Explosions* is not a picture of a sprawling mega-city, but rather a massive open-pit strip mine. The message of intertwined environmental and urbanization processes, breaching the boundaries of cities, is abundantly clear.

The idea of cities as distinct from nature is no longer tenable. Contemporary understandings focus on the city as a nexus of socio-ecological processes, processes that extend well beyond the city's boundaries. Increasingly attention to the 'metabolism' of cities is coupled with urban political ecology as scholars examine the ways in which capitalist urbanization creates urban landscapes and urban nature (Heynen et al. 2006). But advances in epistemology often starkly contrast with the politics and policies of politicians and administrators. The promise of an emancipatory urban 'sustainability' agenda – simultaneously addressing cities' intertwined social, economic and environmental processes – often gives way to the techno-managerialism of ecological modernization. Not infrequently politicians, developers, administrators and even citizens divide the indivisible into distinct spheres, addressing the three dimensions of sustainability individually and sometimes not at all. Perhaps of even greater concern, ecological modernization presumes environmental crises can be addressed without serious political debate and hard political questions that call into question societal structures and power relations. This 'postpolitics' entails "a consensus … built around the inevitability of neoliberal capitalism as an economic system [and] parliamentary democracy as the political ideal" (Swyngedouw 2007: 24). More radically democratic and ecologically grounded alternatives are foreclosed.

The six chapters in this section explore and elucidate many of the central questions surrounding the politics and power relationships of urban socio-ecological processes. In 'Climate science and the city: consensus, calculation, and security in Seattle, Washington', Jennifer Rice provides a nuanced analysis of the ongoing relationship between science and the state, focusing on how selective framings of climate science have been employed by the state to support policy objectives concerning greenhouse gas emissions reduction and climate change risk mitigation. In portraying climate change policy as a narrowly scientific and technocratic issue, expert knowledge has been privileged over that of citizens. The result is that particular people and ideas are disenfranchised and questions of how to equitably distribute the burdens of climate change mitigation and adaptation have gone unanswered.

In 'Democratizing the production of urban environments: working in, against, and beyond the state, from Durban to Berlin', James Angel and Alex Loftus continue the examination of the role of the state in managing socio-ecological relations. They stress that "the social relations that comprise the state are not rigid and stable but, rather, fluid and contested". Instead of focusing on the role of scientific expertise in state policy making, they emphasize "everyday environmentalism", those "everyday practices and the common sense understandings involved in the production of nature". Drawing on examples from Durban and Berlin, they show how the insights and demands that arise from everyday environmentalism can provide the basis for advancing claims to the right to the city, and for the creation of a new urban commons.

Marit Rosol addresses a particular form of everyday environmentalism in 'Politics of urban gardening'. Rosol points out the multiple, complex and sometimes contradictory roles urban gardening may play in urban politics. While urban gardening has its roots in the counter-culture of the 1970s, today it may play a number of contradictory roles including the burnishing of city-image, in turn contributing to gentrification; the voluntary provision of food and community services, buffering the service shortfalls of the neoliberal state; promoting neo-communitarianism in line with the "neoliberal ethos of self-responsibilization"; creating a non-commodified form of urban commoning; and providing places for activist networking and mobilization. The complexities and contradictions of urban gardening are in many ways a microcosm of urban socio-ecological processes and politics, generally.

In 'Just green spaces of urban politics: a pragmatist approach', Ryan Holifield examines protests in Istanbul's Gezi Park to illustrate the spatialities of democratic politics in urban green spaces. Holifield argues that green spaces are not only objects of political struggle, but also places that are crucial to the production of cities' "ecologies of communicative space". His melding of French and American pragmatism with urban political ecology provides the theoretical foundation for his analysis of the role of green spaces in the formation of publics, spatial strategies and strategic shifts in political argumentation.

In 'From sustainability to resilience: the hidden costs of recent socio-environmental change in cities of the Global North', David Saurí offers a convincing analysis of the decline of sustainability policy and planning – particularly its social dimension – and the rise of resilience as an alternative ecological planning framework. Saurí argues that, in practice, sustainability planning rarely addressed social relations adequately and its concern with social conditions was never compatible with the logic and "chronic instability" of capitalism. Instead, resilience returns us, ideologically speaking, to the naturalization of environmental problems, a focus on technical fixes, and survival of the fittest thinking. The disproportionate suffering of the less well-off is hidden behind naturalistic metaphors, although counter-hegemonic resistance to the neoliberal socio-ecological order also occurs.

Finally, Eliot Tretter and Elizabeth Mueller examine the social, environmental and economic dimensions of neighbourhood transformation in 'Transforming Rainey Street: the decoupling of equity from environment in Austin's smart growth agenda'. Austin's smart growth agenda was, in its early days, aligned with mainstream sustainability thinking but over time lost its social dimension. Tretter and Mueller's detailed narrative of a multi-decade struggle over neighbourhood change shows how a struggle that originally addressed affordable housing for low-income residents in the end became a "paradigmatic example of ecological gentrification". The Austin example provides lessons in the uncritical acceptance of planning agendas that fail to take social needs and social justice seriously. These six chapters, taken collectively, provide us with a wealth of lessons for a politics of just socio-ecological change.

References

Brenner, N., ed. (2014). *Implosions/Explosions: Towards a Study of Planetary Urbanization*. Berlin: Jovis.
Cronon, W. (1995). *Uncommon Ground: Rethinking the Human Place in Nature*. New York: W.W. Norton & Company.
Heynen, N., Kaika, M. and Swyngedouw, E., eds. (2006). *In the Nature of Cities: Urban Political Ecology and the Politics of Urban Metabolism*. London and New York: Routledge.
Lefebvre, H. (1970 [2003]). *The Urban Revolution*. Minneapolis: University of Minnesota Press.
Smith, N. (1984 [2010]). *Uneven Development: Nature, Capital, and the Production of Space*. Athens and London: University of Georgia Press.
Swyngedouw, E. (2007). Impossible 'Sustainability' and the Postpolitical Condition. In: R. Krueger and D. Gibbs, eds. *The Sustainable Development Paradox: Urban Political Economy in the United States and Europe*. New York and London: The Guildford Press, pp. 13–40.

10
CLIMATE SCIENCE AND THE CITY
Consensus, calculation and security in Seattle, Washington

Jennifer L. Rice

Introduction

More than a decade ago, then mayor of Seattle, Greg Nickels, proclaimed during an interview with *Grist* magazine:

> [W]e have become aware ... as to the fact that global warming is happening. It was even on the front page of *USA Today* that the science really is not in dispute. And we are going to have to take action in our individual lives to change that.
>
> *(Little 2005)*

This was part of a series of public statements made by Nickels in support of his newly created US Mayors Climate Protection Agreement, which helped propel cities to the forefront of the climate change policy debate (Rice 2010). It is important to remember that Nickels' statement was two years *before* the Fourth Assessment Report of Intergovernmental Panel on Climate Change declared that "warming of the climate system is unequivocal" (Intergovernmental Panel on Climate Change (IPCC) 2007), and it was also several years into the administration of President George W. Bush, who systematically denied that there was any agreement among scientists as to whether climate change was even happening. So, what benefit, if any, would this stance in the climate science wars provide for Seattle's urban governance? What effect, if any, did this have on the practice of climate science? The analysis provided here attempts to begin to unravel the relationship between science and urban governance, using a case study of Seattle's climate planning from the late 1990s until 2013, a period of aggressive climate action by a series of three mayors (Paul Schell, Greg Nickels, Mike McGinn).

Scholarship on the role of cities in climate change governance is now a robust field of social science research (Bulkeley 2010; Jonas *et al.* 2011; Rice 2014; Rutland and Aylett 2008; Sharp *et al.* 2011; While and Whitehead 2013). Yet, very little has been said about the use of climate change *science* in the creation and implementation of urban climate policies and programmes. The importance of standardized methods, quantifiable goals and measureable outcomes based in scientific ways of knowing has been widely noted as a central part of the technocratic nature of urban governance (Kaika and Swyngedouw 2012). City officials utilize, challenge and embrace science on a variety of issues with regularity, and climate governance is no exception. The effect this practice has on the production and circulation of scientific knowledge, as a constitutive feature of the state-science relationship in urban governance, requires more careful attention by urban scholars.

To address this need, this chapter analyses Seattle's framings of climate science used in debates about urban governance. By using the term 'scientific framings', I am referring to the selective use and representation of scientific and technical knowledge for the purpose of supporting a particular policy position or governing technology. The focus here is on better understanding how scientific claims shape the governing technologies that Seattle uses, as well as the influence of urban governance on the practice of science itself. It will be shown that urban governance is a key site in the production, circulation and utilization (Goldman and Turner 2011) of scientific ways of knowing, which directs researchers' attention more directly and explicitly to the urban arena for a more careful consideration of the ways that urban programmes and policies coproduce scientific ways of knowing.

Furthermore, after a brief review of technocratic governing in the context of green governmentality studies, this chapter makes three central arguments about the state-science relationship in urban climate governance. First, Seattle's use of science in political debates about climate change illustrates the state's reliance on formal expert knowledge in policy formation, which legitimates certain forms of scientific and technical inquiry (e.g. greenhouse gas (GHG) inventorying protocols) over other forms of knowledge and experience of the urban environment (e.g. experiential, non-science). Second, this state-science relationship is fundamentally based on the "*science effect*" (Whitehead 2009) where the practices of science and the state are constructed as separate and distinct to serve technocratic interests, when they are, in fact, co-produced through the practices of climate governance in Seattle. Finally, Seattle's reliance on formal knowledge and expert authority, while providing the justification for action, also works to *depoliticize* contentious ethical and social debate about how and why climate policy should be enacted in the city (Swyngedouw 2010). It is important to note, however, that these findings are not meant to delegitimize science or dismiss experts and expertise – science and scientific ways of knowing are critical aspects of good governance. Instead, this analysis shows how and why forms of governing that privilege expert and technical knowledge above all else can disenfranchise particular ideas and people, while also diminishing the potential for conflict and disagreement that might produce more robust democratic processes (see also Rice 2016).

Urban technocracy: power and expertise in urban climate politics

The technocratic nature of governance has been well noted, and in the context of critical social science, heavily critiqued. Jane Jacob's 1961 book *The Death and Life of Great American Cities* was highly critical of 'rationalist' urban planners, while more recent scholarship by James Ferguson (1990), James Scott (1998) and Timothy Mitchell (2002) has pioneered new ways of thinking about the deeply unjust outcomes that result from expert-oriented modes of governing. Since then, scholars have gone on to note the technocratic nature of governing across a range of policy arenas in, for example, in the provisioning of public (though increasingly privatized) services (Larner and Laurie 2010), the application of social programmes such as welfare/workfare policies (Peck 2002), and water resources management (Budds 2009). As Jessica Budds writes, "Technocracy is an important component of neoliberal thought, based on the idea that policy should be directed by technical expertise instead of political partisanship", which utilizes the "technical and scientific facts necessary for the state to make administrative decisions" (2009: 421). It is no surprise, therefore, that urban methods for governing climate change have fully embraced technocratic ways of thinking. Davidson and Gleeson (2015: 34) note about the C40 Cities Climate Change Leadership Group, for example, that the framework "promotes technocratic, economically positioned programs as the conceptual framework to combat climate change", while Bee *et al.* (2015: 341) argue that "technocratic regimes of climate governance emphasize expert (i.e. scientific and technical) understandings of climate change". Erik Swyngedouw (2010, 2013) has even gone on to examine this as a form of 'post-politics', where democratic debate and disagreement are replaced by technocratic consensus on neoliberal ways of governing.

These technocratic practices have not gone without criticism. Budds (2009: 427, emphasis added) identifies the key problem of technocracy as a "neoliberal system [that] privileges technical expertise" and "offers *no scope for non-specialist contributions* in decision-making processes". This demonstrates that highly technical and scientifically based modes of governing often marginalize non-scientists' role (and legitimacy) in the decision-making process, likely reducing the effectiveness of urban governance towards achieving transformative goals. In terms of urban spaces, Rob Kitchin's (2014: 9) detailed analysis of the new 'smart city' puts a very fine point on this problem for urban governance:

> [T]echnocratic forms of governance are highly narrow in scope and reductionist and functionalist in approach, based on a limited set of particular kinds of data and failing to take account of the wider effects of culture, politics, policy, governance and capital that shape city life and how it unfolds. Technological solutions on their own are not going to solve the deep rooted structural problems in cities as they do not address their root causes. Rather they only enable the more efficient management of the manifestations of those problems.

Applying this to the context of urban governance, the complex socio-environmental problem of climate change is, also, framed in terms of scientific facts and the corresponding communities of experts that provide those facts, rather than larger social, political and economic issues, such as who is responsible and what to do about carbon-intensive capitalist growth. This limits possible solutions to those that are technical in nature, such that they can address the "build-up of carbon in the atmosphere, *not* the economic processes which produce it, as the source of the problem" (Schlemback *et al.* 2012: 814).

Green governmentality and the 'science effect' in urban climate governance

Where technocracy has been noted as a key (and problematic) feature of urban governance, other scholars have examined the role of science in (re)producing power relations more broadly. Understanding the practice of science as "interwoven into the fabric of rule of authority" (Lovbrand *et al.* 2009: 8) has been a central concept in green governmentality studies. Stephanie Rutherford (2007: 295) describes the importance of green governmentality as:

> The ways in which the environment is constructed as in crisis, how knowledge about it is formed, and who then is authorized to save it become important for understanding the ways that the truth about the environment is made, and how that truth is governed.

Drawing on this framework and a historical case study of air pollution policy in Britain, Mark Whitehead (2009: 15, emphasis added) notes the rise of "government *with* science" to show that defining, quantifying and monitoring atmospheric pollution requires multiple experts – those from science and government – work to collectively define relevant knowledge about air quality, which determines what technologies of government are employed and scientific practices legitimated. Three key features make up government with science, including intensified relations between the state and science, the primary significance of scientific expertise in policy-making, and governmental direction and influence of scientific inquiry. At the same time, he notes that there is an active *separation* of the two institutions (science and the state), such that there is the "existence of two already demarcated zones of political and scientific existence, untied only temporarily in the necessary pursuit of a common good" (Whitehead 2009: 126). Whitehead (2009: 220) identifies this as the "science effect" in modern environmental governance, stating: "it is also critical to recognize (after Mitchell 2006) the *effects* that the ideological construction of and homogeneous entities labeled 'the State' and 'science' have had in ordering and directing atmospheric government in Britain".

Though 'the state' and 'science' are intrinsically enmeshed through practices of inquiry, technology and governing, science and the state must *appear* as autonomous spheres of practice. In doing so, governments can select and represent science in the ways most appropriate for its desired outcomes, while the scientific community can maintain an 'objective' research agenda that both supports or refutes current political practice, as a fully technocratic state–science apparatus.

In the case of urban climate governance in Seattle, the science effect results in a clear demarcation of what types of questions can be asked about the problem of climate change. Central policy questions are: Where do scientists agree on the causes of climate change? How can we better understand the impacts of climate change on Seattle and the region? What are the best metrics for understanding risk? Discussion related to more complex and non-technical ethical and social issues of climate change are rarely incorporated into political debate. Questions left off the table under the current state–science arrangement are: Who (or what economic activities) are the biggest emissions offenders in our city? How can we equitably change urban development to less carbon-intensive forms? Who should bear the burden of mitigation and adaptation efforts in our city and who should not?

Another key feature of the science effect is that it obscures the ways in which governance practices, local officials and city employees, themselves, are actually influencing the production of scientific knowledge. Hughes and Romero-Lankao (2014: 1025) note that "[d]eveloping a local climate-change policy can be highly data intensive, and requires a city to undertake new ways of using, generating, and synthesizing scientific information". Corburn (2009: 414) has also shown that "global climate science … is being 'localised' and legitimized by municipal planners in collaboration with global scientists and how the credibility of science policy is crafted in the 'localisation' process". This is a critical arena for urban scholars to examine – how urban governance is (re)shaping and legitimating scientific and technical expertise.

The remainder of this chapter explores how this relationship between science and the state provides a particular way of knowing the climate change problem and particular ways of governing climate change (through scientifically sound metrics, standards and protocols) in Seattle, Washington. Showing how the evolution of Seattle's climate policy is rooted in three specific framings of scientific knowledge – what I identify as the science of consensus, calculation and security – the case study demonstrates how science is used to justify political action in a way that can work to narrow policy options to include only those that are technical and scientific in nature, producing important effects on the urban landscape.[1] At the same time, Seattle's policies influence and legitimate the production of climate science, including the need for more localized climate impacts predictions, the continued use and refinement of GHG measurement and inventorying technologies, and the role of risk assessment in adaptation planning.

Information for this analysis was collected via thirty-three semi-structured interviews with city and county workers, public officials, university research scientists and community activists in Seattle, along with extended and ongoing observation of city operations related to climate change. Extensive archival materials related to the development of Seattle's climate-related programmes and policies were also collected, and the text of all ordinances and resolutions related to climate change was gathered from archival and online databases.

Climate science and urban governance: the case of Seattle, Washington

Table 10.1 provides an overview of the relationship between science and technologies of urban governance underlying Seattle's climate-related policies. Between the late 1990s and 2013, three framings of science have been used by city officials in the crafting of local climate policy. Each of these framings, in turn, evoked particular scientific and technological practices that have become increasingly urban in focus.

Three distinct framings of climate change science used in Seattle's urban climate governance are presented in Table 10.1, but there are elements that do overlap between characterizations. Progression from one discourse to the next does generally occur through time, setting the stage for climate policies in Seattle to move from mitigation to adaptation, but all three discourses can (and do) exist simultaneously. Table 10.1 highlights the multi-directional relationship between science and urban governance: science is influencing what policy responses are created, while urban governance priorities also influence the types of science being produced.

Consensus science: climate change is happening here and now

In contrast to the discourse of uncertainty in science being propagated at the national level in the USA, Seattle public officials drew on claims about scientific *consensus* on the causes and consequences of climate change in the making of urban climate policy. Ordinance 122574[2] (passed in 2007), which requires that GHG emissions are quantified for all public and private development proposals as part of the State Environmental Policy Act (SEPA), states in its preamble: "The International Union of Concerned Scientists, the Intergovernmental Panel on Climate Change and numerous other international organizations have reached consensus that climate change is being negatively affected by human behavior and governmental policy."

Table 10.1 Climate science and governance in Seattle, WA

Discursive framing of climate science	Focus of policy mechanisms	Implications for climate science and technology
Consensus: International and local scientists, alike, agree that climate change is happening	Quantitative GHG reduction targets set for the entire city	Use of localized climate change impacts analysis to show that climate change is occurring in the region
Calculation: Measurement of GHG emissions should be accurate and standardized	Adoption of GHG inventorying mechanisms	Use of emissions accounting science and technologies to measure and monitor progress towards policy goals
Security: The threats of climate change on social and ecological systems are wide-ranging and significant	Risk assessment, tolerance and management guidelines	Use of specific scientific risk assessments of climate impacts on city operations and populations

Ordinance 122876 (passed in 2008), which established a subfund for Seattle Climate Action Now (Seattle 'CAN') public educational programme, also states that "there is now unprecedented consensus among international scientists that the earth's average annual temperature is rising and that human activities are increasing the release of greenhouse gases into the atmosphere that contribute to global warming". City workers echoed the importance of establishing scientific consensus on climate change in the early days of Seattle's efforts to implement climate policy during interviews. As a city worker stated in an interview: "One nice thing in Seattle is that we've totally stopped having the conversation about whether climate change is real. We don't need to talk about that anymore."

Consensus science allowed the local government to establish climate change as an environmental problem of significance, but then *more* science was needed to show how it takes place in the local community. Former Seattle Mayor Greg Nickels, for example, frequently referred to a lack of snow pack in the winter of 2004–2005, and the associated potential threats to the region's water supply that can result from lower than normal snowfall, as the impetus for him to create the US Mayors Climate Protection Agreement (Little 2005). Scientists at the Climate Impacts Group (CIG) at the University of Washington had been engaged in the production of regionally relevant climate science for use by local and regional policy-makers since the 1990s, but this provided a new opportunity for them to create this science *with* and *for* the local government. Using a combination of General Circulation Models, hydrologic models and water supply models, climate scientists and water managers in Seattle determined that rising temperatures would affect the amount and timing of snowmelt in the Pacific Northwest, which could negatively impact municipal water supplies. Local newspapers picked up on this new state-science collaboration in the following excerpt from a 2002 *Seattle Times* article:

> But that's the warning of climatologists, who forecast that within 20 years even a slight warming could dramatically – and with surprising speed – shrink the snows that blanket Northwest mountains. And since that snowpack plays a crucial role in dispensing precious water in dry summer months, salmon, farms and people [in and around Seattle] could compete even more for something the region often takes for granted.

With widespread political agreement about the urgency and severity of climate change in the Seattle community (based on this new, city-focused science), city workers could go about making policy. In line with the technocratic approach to policy, Seattle's climate mitigation goals focused on the establishment of measureable and quantifiable mitigation goals. In 2009, those mitigation reduction goals were set at: 7 per cent below 1990 levels by 2012 (the USA's Kyoto Protocol commitment); 30 per cent below 1990 levels by 2024; 80 per cent below 1990 levels by 2050 (City of Seattle 2011). More recently, Seattle has addressed the issue through its desire to have net zero GHG emissions by 2050 (i.e. be 'carbon neutral') by adopting Resolution 31312 in 2011, which, again, directly references scientific consensus by stating:

> [T]here is growing consensus among the scientific community that the quickest path to stay below the United Nations Copenhagen Accord-identified 2 degrees centigrade maximum increase (equivalent to 80% below 1990 emissions levels by 2050) may be insufficient to address the risks of climate change because its assumption that global emissions will peak in 2011.

This approach to climate mitigation emphasizes *quantitative* targets for acceptable levels of warming and GHG concentrations, but does *not* take into account variability of GHG emissions based upon socio-economic status, economic sector or geographic location, which would differentiate who should be held responsible for emissions reductions. Seattle's use of climate science, centred on the notion of consensus, focused debate on whether or not climate change is occurring, how it is experienced in the region and what emissions reduction targets should be, rather than the social and ethical questions of who is most responsible. One-size-fits-all mitigation targets are set for the entire city, with far less specification of how those goals will be equitably met by the members of the community.

Calculation science: measurement as the pathway to policy

Seattle officials and government employees also produce a framing of climate change deeply embedded in the science of calculation. This is centred on the idea that *accurate* measurement of GHG emissions is the primary way that climate change should be understood and addressed. The use of emissions accounting science is fundamental in the city's efforts to measure and monitor progress towards its stated reduction goals, which requires, first, that emissions are calculated in the most accurate way possible, and second, that those protocols are standardized through time and across space. This notion of scientific calculation argues that the City's GHG inventory "follows the best available science and methodologies in the published literature" (City of Seattle 2005), and that the establishment of standardized inventorying protocols is a scientific and technical priority. A city worker provided the following statement about the quantification of GHG emissions in the State Environmental Policy Act (SEPA) review process:

> [E]very project that undergoes SEPA review has to at least quantify what the life cycle greenhouse gas impacts will be ... So we have a worksheet where they fill in some simple numbers and it calculates the operational impacts of heating air conditioning etc., as well as the transportation, the indirect impacts of the project, and the embodied energy like construction and materials used, etc.

As policy-makers in Seattle embraced scientifically based methods of GHG calculation, GHG inventorying technologies are co-produced with local policies. For example, input was sought from a range of local experts as part of City's Green Ribbon Commission on Climate Change in the early 2000s. They were asked questions about which GHGs should be included, how the boundaries of the city's inventories are established, and specific units by which GHGs are measured and reported (City of Seattle 2005). A city worker stated about GHG inventorying process during an interview in 2008:

> [We had a] series of meetings, different expertise, it's fair to say, well staffed ... For example the carbon footprint we would bring graphs of it to them [the Green Ribbon commission] and answer questions and ask their advice on things like should the airport be in or out, should we do it by car emissions ... we identified the key issues that are complex at the local level. All these issues that are now going to be, we hope, sort of decided upon in a consistent way.

In each case, Seattle relies on experts to help define emissions types and boundaries, further entrenching the city's reliance on formal technical expertise. At the same time, emissions inventorying technologies are adapted to the local context by Seattle's Office of Sustainability and Environment as they update the City's community inventory every few years (since the first inventory was calculated in 2002).

Despite this technical specification, however, GHG inventories are notoriously complex and incomplete – abstracted GHG inventories rarely tell the entire story of who is producing emissions, where and why. For example, including the embodied emissions in the things people buy is currently not standard practice in Seattle's GHG inventory, but research suggests this inclusion would greatly elevate the average per capita emissions of Seattle residents (Erikson et al. 2012). New technology industries located in Seattle, such as Amazon and Facebook, furthermore, drive emissions far beyond the boundaries of Seattle and are not captured in typical urban GHG measuring protocols. This points to the fact that there are complex scalar problems of GHG emissions and climate governance that require much more careful attention and debate. Yet, Seattle's policies place the policy emphasis on 'getting the numbers right', rather than larger political economic questions about how and why GHG emissions are produced as part of a carbon-intensive economy or the spatiality of carbon emissions that vary from neighbourhood to neighbourhood, house to house and person to person.

Security science: reduction of risk through assessment

More recently, as Seattle has moved into climate adaptation planning, a scientific framing of security has also become prevalent. Understood as 'threats' (City of Seattle 2010), climate change impacts of sea level rise, more summer days over 90 degrees, and changes in how the region receives precipitation, all have the potential to disrupt the social and ecological fabric of the Seattle region. In a 2008 Senior City of Seattle Staff Briefing Memo obtained as part of the archival documents collected for this research, it is stated that:

> Preparing for climate change's impacts is necessary because scientists believe that the impacts expected over the next century are generally irreversible … While cities can ward off catastrophic climate change by reducing climate pollution, cities will also need to adapt to the climate changes of the next century by developing adaptation policies that ensure cities are resilient to the projected climate impacts.

As part of an internal 'Climate Change Impacts Questionnaire' survey distributed to eleven different city departments during the summer of 2008, information was collected about the potential vulnerabilities city government agencies expect may arise due to climate change. To determine the effect that climate change might have on their day-to-day activities, the survey included a scientific overview of predicted climate impacts for temperature, precipitation and sea level rise before asking respondents to think about the effect of climate change on their department's responsibilities. Departments and agencies that work with vulnerable populations (e.g. the elderly), for example, will need to have excessive heat day procedures to ensure that those without air conditioners can survive increasing summer temperatures. Similarly, coastal roads and buildings may face increased erosion and loss of seafront. Figure 10.1, taken from an Adaptation Planning Briefing Memo used by a city worker in the Office of Sustainability and Environment, shows the entire list of municipal vulnerabilities compiled as a result of the internal impacts survey. Of central importance is a concern for the security of human life, ecological health and infrastructural resilience.

This effort has resulted in the creation and assessment of *new* city-wide planning metrics of risk assessment. For example, Seattle's Office of Sustainability and Environment is working to create new 'risk tolerance' guidelines for sea level rise when planning and assessing city developments near the shoreline (see Figure 10.2). Seattle Public Utilities has also produced its own sea level rise map to determine areas most likely to flood in the year 2050 based upon data provided by the

> **What are the potential impacts on the city of Seattle?**
>
> This summer, OSE surveyed City departments to identify the anticipated impacts on City planning, operations, facilities, and residents.[1] In general, City staff identified impacts in the seven area:
>
> - **Building energy**: Increased energy demands and associated costs for cooking.
> - **Emergency response**: Challenges in reaching and sheltering vulnerable populations during extreme weather events.
> - **Stormwater management**: Increases flooding events and water quality impacts from run-off and sewer overflows.
> - **Shoreline management**: Loss of shoreline properties and parks from erosion, storm surges, and sea-level rise.
> - **Land use and building codes**: Development and building codes that exacerbate impacts on other properties.
> - **Urban forest and landscaping**: Increased demand for irrigation and more frequent incidents of insect infestation.
> - **Public health**: Increased incidents of health problems due to heat events and poor air quality.

Figure 10.1 Information about climate impacts from departmental briefing memo

Risk tolerance	SLR by 2050	SLR by 2100	Description of risk tolerance factors
High	3"	6"	Facility has a relatively short life span (10–20 years) and/or can be easily/cost-effectively modified to accommodate higher SLR. Little risk to facility from storm surges.
Medium	6"	13"	Facility has medium life span (30–50 years) and/or could be modified with a moderate investment. Facility may be affected by significant storm surges.
Low	22"	50"	Facility has very long life span (>50 years) and/or could only be modified with significant investment. Facility is likely to be damaged with storm surge or high tides.

Figure 10.2 Reproduction of 'risk tolerance' guidelines provided by the city of Seattle to department heads

University of Washington Climate Impacts Group. Estimates of sea level rise in the region range from 3 to 22 inches, which will be exacerbated by storms and high tides (Mapes 2013).

This framing of climate-related science as assessments of impacts on city operations and populations to promote local socio-environmental security has integrated a risk management approach into government operations more broadly. As a CIG researcher said about her experience describing climate impacts to government institutions:

> Putting it [climate adaptation] in the context of risk management, it's something that, some governments are already dealing with … you're not asking for money to study this or to retrofit that on the basis of a green issue but rather as a risk issue.

As security has become the central justification for climate adaptation planning in Seattle, the creation of standardized planning metrics (like those in Figure 10.2) are a primary policy mechanisms utilized in current adaptation efforts. These risk assessments require an understanding of the likelihood that a climate-related event or change will occur (e.g. sea level rise) and an estimation of the associated losses or consequences (e.g. damage to infrastructure and cost to repair or replace), to help prioritize adaptation projects to the areas most vulnerable. The use of scientifically based assessments, furthermore, legitimates science as the primary form of expertise in understanding the nature of climate risk, rather than the production of uneven vulnerability as part of the urban political economy. With the development of broad risk management guidelines, the underlying production of vulnerability is de-emphasized in favour of a more widely applicable metric of adaptation planning. It should be noted that Seattle's Office of Sustainability and Environment has proposed that 'Race and Social Justice' be included as one of its Climate Adaptation Impact Areas by stating the goal that "resiliency benefits everyone" (City of Seattle 2010). This has primarily taken the form of employee training and budget considerations, but if it were to be included in formal policy, it would represent a significant shift towards better understanding uneven production of vulnerabilities, even though it does not directly address the fundamental political economic processes that produce it.

Conclusion: urban governance and the depoliticization of climate politics

The theoretical and empirical analysis provided here illustrates both the state–science relationship that underlies urban climate governance and the co-production of climate science and policy in Seattle. Scientific discourse and expert knowledge serve as the basis for justification and implementation of climate policy in Seattle because experts agree climate change is happening, measurement of GHGs should be accurate and standardized, and threats of climate change are wide-ranging for both social and ecological systems in the region. As policy-makers begin to assert that climate change is an issue of central and urgent concern, the practice of scientific inquiry is legitimated and advanced: more regionally specific climate change impacts science is needed, newer and more 'accurate' GHG emission data are required, the creation of risk assessment metrics becomes imperative.

Even though the practices of city workers legitimate and reinforce the importance of climate science, scientists cannot act as advocates to those involved in urban climate governance in Seattle – a key feature of the science effect (Whitehead 2009). Science and the state must appear separate, even though they are highly intertwined through urban governance. As one regional climate scientist said about providing recommendations to policy-makers, "You [as a scientist] know better than that. You're not a politician." Another scientist said, "It's really important for your credibility [as a scientist] to keep that [separation] … and that's just the way you have to be." This presentation of science and policy as autonomous spheres of practice requires and proliferates certain types of governance technologies (e.g. metrics and standards) that adhere to the standards of technocratic governing. This has the effect of overshadowing other framings of climate change from community members, concerned citizens and environmental groups. For example, forms of knowledge not addressed by Seattle's techno-scientific approach to climate governance include the lived experiences about the uneven causes and consequences of climate change, alternative measures of how best to understand and measure climate mitigation and impacts, and critical discourse about acceptable levels of risk tolerance. In each of the three framings of science described as part of Seattle's climate policy, technical and scientific aspects of policy are emphasized at the expense of debates about *who is responsible* for climate change, the *highly differentiated GHG emissions* by people and economic sectors at a variety of scales, and the production of *uneven climate vulnerability* that accompanies uneven urban development. This shows urban governance is becoming further entrenched in its reliance on specific forms of scientific and technical expertise that *depoliticize* complex socio-environmental problems, even places like Seattle – a city thought to be at the forefront of progressive climate governance.

It is beyond the scope of this analysis to speculate on what alternative ways of knowing and governing climate change should be. Yet, the inclusion of diverse forms of knowledge and experience in climate policy is an important area of inquiry that should be further explored by urban scholars. Efforts to open up what knowledge 'counts' in urban policy debate and to break down the divide between science and the state should not be seen as an attack on either, but rather, as a pathway to more inclusive and collective efforts in addressing pressing environmental concerns. At the same time, this may provide room for more democratic negotiations of environmental policy, where many forms of expertise, from community experiences to scientific inquiry, are incorporated into the making of urban climate policy.

It is also essential to note, in closing, that research on the politics of knowledge and expertise may be entering into a new time of transition. Where critical studies of expert-oriented governing, like the one provided here, are meant to call for more political and democratic forms of governing, it appears they are increasingly being used in new ways by conservative political interests, including the far right in the United States, to delegitimize the practice of science and the idea of an 'expert'. While we must remain focused on opening up technocratic urban governance to become more inclusive of non-scientific ways of knowing, we stay vigilant of contortions of our assessments against the practice of science altogether.

Notes

1 It should be noted that while narrowing of policy options towards technical solutions is often the case in the interaction between science and policy, there have been notable exceptions to this. For example, the Campaign for Nuclear Disarmament drew on the scientific ideas of Linus Pauling and other scientists to call for a complete elimination of nuclear weapons as scientifically sound policy.
2 The text of all Ordinances and Resolutions was retrieved from: http://clerk.seattle.gov/~public/CBOR1.htm.

References

Bee, B.A., Rice, J. and Trauger, A. (2015). A Feminist Approach to Climate Change Governance: Everyday and Intimate Politics. *Geography Compass*, 9(6), pp. 339–350.
Budds, J. (2009). Contested H_2O: Science, Policy and Politics in Water Resources Management in Chile. *Geoforum*, 40(3), pp. 418–430.
Bulkeley, H. (2010). Cities and the Governing of Climate Change. *Annual Review of Environment and Resources*, 35, pp. 229–253.
City of Seattle. (2005). *2005 Inventory of Seattle Greenhouse Gas Emissions: Community and Corporate*. Available at: www.seattle.gov/environment/documents/2005%20Seattle%20Inventory%20Full%20Report.pdf [accessed 1 December 2016].
City of Seattle. (2010). *Memorandum: Climate Adaptation SLI Response*. Available at: www.seattle.gov/environment/documents/SLI_Response_31-1-A-1.pdf [accessed 1 December 2016].
City of Seattle. (2011). *Seattle Climate Action Plan*. Available at: www.seattle.gov/archive/climate/ [accessed 1 December 2016].
Corburn, J. (2009). Cities, Climate Change and Urban Heat Island Mitigation: Localising Global Environmental Science. *Urban Studies*, 46(2), pp. 413–427.
Davidson, K. and Gleeson, B. (2015). Interrogating Urban Climate Leadership: Toward a Political Ecology of the C40 Network. *Global Environmental Politics*, 15(4), pp. 21–38.
Erikson, P., Chandler, C. and Lazarus, M. (2012). *Reducing Greenhouse Gas Emissions Associated with Consumption: A Methodology for Scenario Analysis*. Stockholm Environmental Institute Working Paper No. 2012–05.
Ferguson, J. (1990). *The Anti-Politics Machine: 'Development', Depoliticization, and Bureaucratic Power in Lesotho*. Cambridge: Cambridge University Press.

Goldman, M. and Turner, M. (2011). 'Introduction'. In: M. Goldman, P. Nadasdy and M. Turner, eds. *Knowing Nature: Controversies at the Intersection of Political Ecology and Science Studies*. Chicago: University of Chicago Press.

Hughes, S. and Romero-Lankao, P. (2014). Science and Institution Building in Urban Climate-Change Policymaking. *Environmental Politics*, 23(6), pp. 1023–1042.

Intergovernmental Panel on Climate Change (IPCC). (2007). Summary for Policy Makers. In: S. Solomon, D. Qin, M. Manning, Z. Chen, M. Marquis, K.B. Averyt, M. Tignor and H.L. Miller, eds. *Climate Change 2007: The Physical Science Basis: Contribution of Working Group I to the Fourth Assessment Report of the Intergovernmental Panel on Climate Change*. Cambridge: Cambridge University Press.

Jacobs, J. (1961). *The Death and Life of Great American Cities*. New York City: Random House.

Jonas, A.E., Gibbs, D. and While, A. (2011). The New Urban Politics as a Politics of Carbon Control. *Urban Studies*, 48(12), pp. 2537–2554.

Kaika, M. and Swyngedouw, E. (2012). Cities, Natures and the Political Imaginary. *Architectural Design*, 82(4), pp. 22–27.

Kitchin, R. (2014). The Real-time City? Big Data and Smart Urbanism. *GeoJournal*, 79(1), pp. 1–14.

Larner, W. and Laurie, N. (2010). Travelling Technocrats, Embodied Knowledges: Globalising Privatisation in Telecoms and Water. *Geoforum*, 41(2), pp. 218–226.

Little, A. (2005). An Interview with Seattle Mayor Greg Nickels on his pro-Kyoto Cities Initiative. *Grist*, 16 June. Available at: www.grist.org/article/little-nickels/ [accessed 1 December 2016].

Lovbrand, E., Stripple, J. and Wiman, B. (2009). Earth System Governmentality: Reflections on Science in the Anthropocene. *Global Environmental Change*, 19, pp. 7–13.

Mapes, L.V. (2013). Seattle Calculates How Climate Change Will Redraw Its Shores. *Seattle Times*. Available at: http://seattletimes.com/html/localnews/2020120911_climatechange13m.html [accessed 1 December 2016].

Mitchell, T. (2002). *Rule of Experts: Egypt, Techno-Politics, Modernity*. Berkeley: University of California Press.

Peck, J. (2002). Political Economies of Scale: Fast Policy, Interscalar Relations, and Neoliberal Workfare. *Economic Geography*, 78(3), pp. 331–360.

Rice, J.L. (2010). Climate, Carbon, Territory: Greenhouse Gas Mitigation in Seattle, Washington. *Annals of the Association of American Geographers*, 110(4), pp. 929–937.

Rice, J.L. (2014). Public Targets, Private Choice: Urban Climate Governance in the Pacific Northwest. *The Professional Geographer*, 66(2), pp. 333–344.

Rice, J.L. (2016). 'The Everyday Choices We Make Matter': Urban Climate Politics and the Postpolitics of Responsibility and Action. In: H. Bulkeley, M. Paterson and J. Stripple, eds. *Towards a Cultural Politics of Climate Change: Devices, Desires and Dissent*. Cambridge: Cambridge University Press, pp. 110–126.

Rutherford, S. (2007). Green Governmentality: Insights and Opportunities in the Study of Nature's Rule. *Progress in Human Geography*, 31(3), pp. 291–307.

Rutland, T. and Aylett, A. (2008). The Work of Policy: Actor Networks, Governmentality, and Local Action on Climate Change in Portland, Oregon. *Environment and Planning D: Society and Space*, 26(4), pp. 627–646.

Schlembach, R., Lear, B. and Bowman, A. (2012). Science and Ethics in the Post-Political Era: Strategies Within the Camp for Climate Action. *Environmental Politics*, 21(5), pp. 811–828.

Scott, J.C. (1998). *Seeing Like a State: How Certain Schemes to Improve the Human Condition Have Failed*. New Haven. CT: Yale University Press.

Sharp, E.B., Daley, D.M. and Lynch, M.S. (2011). Understanding Local Adoption and Implementation of Climate Change Mitigation Policy. *Urban Affairs Review*, 47(3), pp. 433–457.

Swyngedouw, E. (2010). Apocalypse Forever? Post-Political Populism and the Spectre of Climate Change. *Theory, Culture & Society*, 27(2–3), pp. 213–232.

Swyngedouw, E. (2013). The Non-Political Politics of Climate Change. *ACME: An International E-Journal For Critical Geographies*, 12(1), pp. 1–8.

While, A. and Whitehead, M. (2013). Cities, Urbanisation and Climate Change. *Urban Studies*, 50(7), pp. 1325–1331.

Whitehead, M. (2009). *State, Science and the Skies: Governmentalities of the British Atmosphere*. Malden, MA: Blackwell Publishing.

11
DEMOCRATIZING THE PRODUCTION OF URBAN ENVIRONMENTS

Working in, against and beyond the state, from Durban to Berlin

James Angel and Alex Loftus

Introduction

The city is neither an enemy of nature, nor should it be seen as its 'other'. Instead – just as a bee crafts its own structure through labouring with its fellow insects – the city can be viewed as a produced nature. One of the key normative questions that emerges from such a claim revolves around how that production could or should be organized in ways that benefit rather than harm human and non-human lives. For many political theorists this question can only be addressed through reference to state institutions, which are seen – in often-contradictory ways – as help or hindrance to an environmental politics, or as the basis for more or less inclusive environmental decision-making. Strangely, though, political ecologists are often relatively silent on the question of the state (although for a notable exception see Whitehead *et al.* 2007). Yet as the potentialities, limitations and contradictions of the twenty-first century capitalist state are made newly visible through the advent of left electoral projects across Europe, there is a need for a more in-depth political ecological account. In what follows we want to take the question of the state more seriously than it has been in relation to urban political ecology, while simultaneously arguing that the state is not the 'thing' that many take it to be. Instead it is a concrete form that emerges out of a broader set of political ecological relations. Or, put differently, the material apparatus of the state – the parliament, the court, the police and so on – is a product of complex and interconnected social, economic and ecological processes.

In this chapter we focus on two sets of struggles targeted at ensuring fairer, more equitable and more ecologically just urban environments in which the state appears as a somewhat ambiguous 'actor'. We focus first on the provision of energy within Berlin and second on the practices involved in accessing water in Durban. In both examples 'the state' – understood in both its local and its national manifestation – can be seen as both an enabler and as an obstacle to a more just socio-ecological future. It appears to open avenues for broader involvement in environmental decision-making while granting rights to its citizens, and yet it simultaneously narrows the horizon for change and frustrates any broader interpretation of the very rights it appears to grant. Thus, we look at how struggles for energy democracy and struggles for the right to water develop in, against and beyond the state, as well as in, against and beyond the city.

Berlin's struggle for energy democracy

In November 2013, the people of Berlin voted for a radical reversal of neoliberal privatization policies, through the participatory re-municipalization of the city's electricity sector. While 600,000 people – a resounding 83 per cent majority – voted for re-municipalization, the result was not binding, with a narrow failure to reach the voter quorum required by a mere 21,000 votes. Since the mid-2000s, Germany has seen a rising tide of re-municipalization, reversing the privatization and liberalization policies of the 1990s on the back of frustration at rising prices, the expiry of distribution network concessions and the German state's planned transition towards low-carbon energy (Becker *et al.* 2015). Sixty new non-profit municipally owned utility companies (*Stadtwerke*) were established between 2007 and 2012, with over 190 distribution network concessions returning to municipal hands (Hall *et al.* 2013). It was in this context, alongside the expiry of Berlin's electricity grid concession with Swedish state-owned company Vattenfall, that the Berliner Energietisch campaign was formed in 2011.

The Energietisch is a broad civil society coalition of fifty-six groups, ranging from radical left organizations to environmental NGOs and trade unions. Making use of the citizens' initiative referendum – which allows citizens to call referenda on proposed laws, backed by 200,000 signatories – the Energietisch seized the opportunity to attempt an ambitious transformation of Berlin's energy landscape. The draft law proposed by the Energietisch (see Taschner n.d.) stipulated, first, the establishment of a new 'Stadtwerke': a publicly owned, non-profit utility company that would invest in 100 per cent renewable energy to sell on to consumers, operating with an explicit commitment to tackling fuel poverty and rising bills. This, though, was not to be a traditional Stadtwerke, but rather a democratized company controlled, as much as possible, by the city's inhabitants. It would operate with total transparency on its internal operations. It would be governed by a board of directors comprised of one-third local politicians, one-third Stadtwerke workers and one-third democratically elected ordinary citizens. Annual neighbourhood assemblies would offer all Berlin inhabitants the opportunity to raise questions and concerns with representatives of the Stadtwerke. The draft law also stipulated that the local state must table a bid for the ownership of the electricity distribution network. Should the state be successful, the governance of the network would then be subject to the same democratic criteria as the Stadtwerke.

The radical changes demanded by the Energietisch proved unacceptable to the Berlin Senate, ruled by a coalition of the centre-left Social Democrat Party (SPD) and the conservative Christian Democratic Union (CDU). With four months to the referendum remaining, the Senate decided to move the date of the vote away from the general election, in a move designed to undermine the chances of reaching voter quorum. With the Senate's opposition proving decisive, the referendum was lost. Now, two years on, the Energietisch continues to campaign around the issue of energy re-municipalization, contesting the concessions for Berlin's electricity and gas supply networks.

Durban's struggles for the right to water

The struggle for the right to water in Durban is, ostensibly, a very different example of an urban socio-ecological politics. The immediate post-apartheid period in South Africa represented a moment of experimentation with forms of municipal service delivery and, in a more limited manner, with governance arrangements. After its victory in the first democratic elections of 1994 the ANC launched some of the experimental initiatives promised within its Reconstruction and Development Programme (RDP) while beginning to disappoint many in its paradoxical conservatism when it came to financing municipal service provision. By 1996 an apparent shift seemed to take place as the RDP was quietly supplanted by a Growth Employment and Redistribution strategy (GEAR) that, for many, represented a neoliberal turn (Bond 2000). At the same time, through the uniting of struggles across union groups, environmental justice activists and social movements, municipal service provision came to be seen as one of the most pressing of post-apartheid *environmental* concerns (McDonald 1998). Profound paradoxes began to open up between the claims of the ANC to represent the many, the directions it appeared to be taking when it came to adopting principles of full cost-recovery, and the demands of activists, many of who had been central in the struggle against apartheid.

The right to water came to be enrolled within this paradoxical situation. Thus, in one of the first examples of the concretization of the right to water at a national level, the 1996 constitution guaranteed all citizens the right to a safe secure supply of drinking water. A 1998 Water Act appeared to support this right; nevertheless the contradictory tug of GEAR meant that principles of full cost-recovery increasingly guided the financing of water provision. A spate of disconnections for non-payment was then followed by the development of pre-payment technologies in which those unable to pay for their right to water had to forgo access to potable supplies. A cholera outbreak in KwaZulu Natal in 2000 appeared to demonstrate that the consequences of a relentless pursuit of payment were grave (Cottle and Deedat 2002). In Durban, meanwhile, the municipal government pioneered a low-cost option for water delivery that paved the way for an experiment in providing lifeline supplies of 200 litres of water for every household in the municipality every day. Nevertheless, in yet another twist, those who weren't able to pay for the amount above and beyond their lifeline supply faced disconnection. The municipality was thereby hailed internationally as a pioneer of the right to water, while the rising rate of disconnections appeared to defy such claims. Court cases, wildcat protests and an emerging band of 'struggle plumbers' reconnecting the disconnected, were all symptoms of this paradoxical situation.

In trying to make sense of both the Berlin and Durban cases, we will review the contribution of urban political ecology to re-interpreting the nature of the city before seeking to develop a dialogue with literature on the right to the city, the struggle for the commons and the struggle in, against and beyond the state. We turn first to urban political ecology.

Urban political ecology

For urban political ecologists, as in post-apartheid South Africa, valorizing a pristine nature as the antithesis of the corrupting urban form is problematic; instead it is imperative to understand those environments lived in by the majority of the world's population. In emphasizing the production of *urban* natures or, as Swyngedouw and Kaika (2000) refer to it, the "urbanization of nature", urban political ecology focuses attention on the historical specificity of urban ecologies – always understood as socio-natural *achievements* rather than natural givens. This attention to historical specificity can, in part, be traced to Marx's critique of political economy, which positions humans within the historical and geographical environments of which they are a co-evolutionary agent. Neil Smith (1984) has done much to excavate such an approach, emphasizing the close connections between the development of the urban form and a historically specific produced nature. Capitalist societies thereby produce a nature that embodies and expresses the increased emphasis placed on the exchange abstraction.

If the urban form in capitalist societies is one example of such an abstract nature it is still no less natural than the national parks or wilderness areas that are often the concern of wildlife organizations or environmental NGOs. Although often a dirty, degraded and even dangerous urban environment, there is, Harvey (1996: 94) writes "nothing particularly unnatural about New York City". For just like a national park, the city is produced and reproduced via complex and co-evolving interconnected human and non-human processes. Building on such a claim, Swyngedouw and Kaika (2000: 569) argue that: "In the city, society and nature … are inseparable, integral to each other, infinitely bound up, yet simultaneously, this hybrid socionatural 'thing' called city is full of contradictions, tensions, and conflicts." For many urban political ecologists – often building, again, on Smith (1984) – there are profound political implications to the notion of the city as one historically and geographically specific production of nature. If the city is one example of an abstract nature, this form is a historical achievement and the outcome of a specific moment within a struggle between groups to shape that nature.

For Loftus (2012) enhancing the democratic content of socio-environmental change requires a philosophy of praxis that builds on the everyday practices and the common sense understandings involved in the production of nature. This is what he refers to as "everyday environmentalism". A simple example would be the type of critique that sometimes emerges from the act of ensuring that a household has access to safe, sufficient supplies of potable water. Within a South African context, that act needs to be situated within a gendered division of labour and within an increasingly commercialized water service. The legacies of apartheid, of the struggle against apartheid and of social movements in the post-apartheid era have all shaped access to water in different ways. Nevertheless, within the act of negotiating these different relations lie conditions of possibility for an immanent critique – not only of the politics of water provision but a more general critique of the injustices of the post-apartheid urban environment.

Of course the anger that one might feel at being denied access to water need not necessarily translate into a critique of urban ecological injustice. It might instead translate into shame, fear or xenophobia. If there is a target for one's anger, it might be 'the rich', 'the neighbour' or those of a different 'race'. Or, anger over unjust political ecological relations and outcomes may be targeted at the state, in which case it can be expressed as a demand on the state, a demand for change within the state, or a simple request for the state to treat its citizens with greater dignity. Although urban political ecologists have become increasingly interested in the conditions of possibility for democratizing urban ecological processes, the role played by the state as a conduit for such expressions of anger is generally ignored. In the following section we therefore turn to the way in which the state figures within the call for the right to participate in democratizing the urban commons.

The right to the city, the commons and the state

In recent years, Henri Lefebvre's notion of the *right to the city* has captured the imagination of urban social movements and critical urban scholars alike. Yet the connection between the right to the city and the perspective of urban political ecology is not so straightforward. Lefebvre (1968) counterposed the right to the city with the "pseudo-right" to nature (158), suggesting that transformative political action to re-make the city is "displace[d]" in favour of attempts to escape the "noise, fatigue, the concentrationary universe of cities" for the evening or weekend in the country (158). It is little wonder, then, that Smith's (2003) foreword to the *Urban Revolution* critiques Lefebvre's treatment of nature as "radically closed as a venue for political change" (xv). The problem for Lefebvre is a failure to afford 'nature' the kind of sophisticated process-based understanding that he devotes to the production of space. Lefebvre, in short, evokes a pristine, once untouched nature "ravage[d]" by urbanization (158); a notion that we argue is deeply problematic.

When we extend our process-based lens beyond urban space towards nature, we arrive at an interesting re-reading of the right to the city as a right to partake in the production of nature. Lefebvre (1968) was not interested in a right to visit or access the city in its current or past form but, rather, the right to a transformed future city re-claimed for human pleasure and use, rather than alienated commodity exchange. Purcell (2002) distinguishes two aspects to Lefebvre's vision: (1) *appropriation*: the right to use and occupy urban space in ways that satisfy our needs and wants; and (2) *participation*: the right to partake in decisions regarding the production of urban space. Putting this understanding of Lefebvre into conversation with our dialectical reading of nature, the right to the city becomes the right to participate centrally in the production of urban natures that are conducive to human use and desire.

Understood as such, enacting the right to the city requires the development of new collective organizational forms that allow for participatory democratic control over the production of urban nature. Harvey (2012) suggests that notions of the *commons* can prove illuminating in this regard. Federici (2012a), furthermore, documents how, in recent years, the commons has become a new unifying lexicon for the "cooperative society that the radical Left is striving to create" (405). Commons are created when things we need – from land, water and air to languages, the digital world and electricity – are de-commodified and subjected to democratic control for popular use. The project of the commons, for Federici, must be about creating new shared, collective forms of social reproduction: the making and re-making of ourselves as human subjects through the vital everyday labour, historically undertaken by women, of feeding and clothing ourselves, creating and maintaining our homes, caring for one another.

For Federici – as with other open and autonomous marxists such as Hardt and Negri (2009) and Holloway (2010) – this means the creation of collective and non-commodified relations outside of capital and, furthermore, outside of the state. The politics of the commons, argues Federici, emerges from "the demise of the statist model of revolution that for decades has sapped the efforts of radical movements to build an alternative to capitalism" (2012a: 404). Indeed, many of the social movements associated with the new politics of the commons – from the anti-globalization movement of the 1990s and early 2000s to the Occupy movement of 2011 – are highly sceptical of engagement with the state or formal politics. For these movements, the goal is not to take or, in some cases, even influence state power. Rather, to deploy Holloway's (2010) metaphor, these movements have sought to create emancipatory "cracks" in capitalist social relations through the creation of temporal moments and spaces outside of capital and state, in which self-determined and autonomous human activity can flourish, for instance self-managed squats and urban gardens, workers' cooperatives and solidarity economies.

Offering an alternative take, Cumbers (2012, 2015) and Harvey (2012) reject the eschewal of the state from the politics of the commons. Harvey points to a pertinent social need for forms of coordination and redistribution that transcend the micro-scale of the squat, community garden or the cooperative. Fetishizing the local, suggests Harvey, plays into the hands of the neoliberal project of undermining the redistributive functions of the state, won through decades of working class struggle. Instead, Harvey proposes a dual strategy for the commons: push to extend state provision for social reproduction, while simultaneously endeavouring to appropriate state provision such that it embodies the participatory, non-commodified logic of the commons articulated by Federici, Holloway and others.

What, then, of the prospects for the democratization of the production of nature through the state? Critical scholars quite rightly point to the state's role in the perpetuation of dominant relations of exploitation and oppression: from the coercive enforcement of private property rights to facilitate what Harvey terms accumulation by dispossession, through to the imposition of bureaucratic procedures and categories in order to "make legible" populations to facilitate colonial expansion (Scott 1998) and the repressive and violent policing of bodies and borders. Yet simplistic fetishizations of the state as a mere instrument of domination – "the executive committee of the bourgeoisie", as a particularly crude marxist understanding might have it – must be avoided. After all, Bob Jessop (1982, 2007) makes the obvious point that actually existing states under shared relations of domination are very different: the German state and the US state are both capitalist states, yet have very different forms and functions. Rather than reifying the state as an independent subject or as a thing-like instrument to be wielded by whichever particular social group claims power, Jessop – drawing on Gramsci, Foucault and Poulantzas – argues that we must understand the state as, like capital, as a process, a social relation. The material institutions of the state – the police, the military, parliament and so on – are enmeshed within broader networks of social relations such as relations of capital, patriarchy and white supremacy. However, the social relations that comprise the state are not rigid and stable but, rather, fluid and contested. As such, for Jessop (2007), states are *strategically selective*: state institutions are structurally biased towards the reproduction of the dominant relations that comprise them, yet the possibility remains for oppositional struggle inside or outside of the state's apparatus to shift social relations within state institutions in ways that challenge the status quo. An obvious example might be the advent of the post-war Welfare State which, its many imperfections notwithstanding, still issued in substantial gains for labour against capital.

If, then, the state must be understood as one aspect of a broader ensemble of social (or, more precisely, *socio-ecological*, relations, then it makes little sense to isolate state institutions as the supreme locus of struggle. At the same time it makes little sense to insist upon an eschewal of the state from our political strategies. Enacting the right to the city demands an approach to the state that accounts for its contradictions and complexities. Our suggestion here, drawing on our previous work (Angel 2017), is that the orientation towards struggle *in, against and beyond* the state might prove fruitful in this regard.

For the London Edinburgh Weekend Return Group (LEWRG) – a collective of scholar-activists working within British state institutions in the 1970s – the state embodies a major contradiction: "Resources we need involve us in relations we don't" (1980, Chapter 2). For while the state plays a vital role in social reproduction via the provision of basic goods like health care and housing, our experience of the state as workers and 'service-users' is often demeaning and dissatisfying. On the basis of the distinction between the state *qua* apparatus – the institutional cluster of the police, parliament and so on – and the state *qua* social relation (as set out by Jessop), the LEWRG propose the following political task: "The problem of working in-and-against the state is precisely the problem of turning our routine contact with the state apparatus against the form of social relations which the apparatus is trying to impose upon our actions" (1980, Chapter 3).

More recently, scholars such as John Holloway (2002; Holloway and Wainwright 2011) and Andrew Cumbers (2015) have expanded the in-and-against imaginary to within-against-and-beyond, suggesting a form of politics that not only opposes the state form but, moreover, transcends this. The struggles in both Berlin and Durban considered here can be read as working in-against-and-beyond the state, albeit in historically and geographically specific ways. In what follows, we attempt to develop this reading with a view to the implications for the right to the city and the democratization of urban natures.

In-against-and-beyond the state in Berlin

Becker *et al.* (2015) situate the Berliner Energietisch campaign within debates around the right to the city and the urban commons: while the Energietisch did foreground questions of energy affordability and cut-offs, their political aspirations went beyond the right to *access* energy, portending a more radical desire to lay claims towards the "material and social forms of the urban", to render the city as both a "site" and an "object" of struggle (84). The authors suggest that when we consider the commoning expressed by the Energietisch as at once "in-against-and-beyond … the neo-liberal urban paradigm", we "adjust the expectations associated with the 'commons' to the messy realities of urban struggles" (87). Likewise, the Energietisch can productively be understood as working in-against-and-beyond *the state*, in order to account for the messy reality of this particular aspect of the urban political landscape.

First, the Energietisch very clearly worked in the state, using a mechanism of formal politics – the citizens' initiative referendum – to demand the extension of state provision in the city's energy system. While Energietisch activists argued that state actors' oppositional responses to the referendum campaign was, in part, a result of extensive lobbying by incumbent utility company Vattenfall, they also spoke directly of their disagreement with a radical politics that construes the state as the enemy, or the instrument of corporate power or capital. While there was broad acknowledgement within the campaign that long-term success would require ongoing struggle to guard against state strategies of dilution and co-optation, activists believed that using formal politics could advance their goals. The state, in sum, was seen as a fruitful terrain of struggle. In the words of one activist: "the state in fact is a field of controversy … which always shifts power to the capitalist factions inside, but where also marginalised working class demands or movements can win things". For a number of Energietisch activists, this understanding of the state – and the imperative to struggle within state institutions – emerged out of the direct influence of Gramscian and Poulantzian theory: "Poulantzas and Gramsci, they are very good guidelines. Because they understood what is the bourgeois state and how it works and that it's not only a repressive apparatus, it's a question of the relationship of forces and permanent fight."

The Energietisch's desire to struggle in the state has an interesting scalar dimension. There was a broad consensus within the coalition that emancipatory energy transition requires action at a variety of scales, with a role for local co-operative ownership, the local and central state and supranational governance. Yet there was a decision to focus energies at the level of the local state, emerging out of a range of interconnected factors. Activists saw the expiry of the local state's contract with Vattenfall for the management of the city's electricity distribution grid as presenting a novel political opportunity for a progressive energy intervention (Miller 1994). But, further, the local state was seen as a fruitful scalar terrain for the vision of *energy democracy* that the Energietisch sought to enact. Operating within a smaller spatial constellation than the central state and with far fewer citizens to include, novel forms of participatory democracy were seen as more easily incorporated at the local than national level. And while a separate campaign named BürgerEnergie Berlin sought to craft urban energy democracy through the incorporation of partial control of the electricity distribution grid by a citizen-owned cooperative, the cooperative model was rejected by the Energietisch on political grounds. While the cooperative's primary responsibility is the participation and benefit of their membership, the Energietisch favoured a *universal* vision of energy democracy run in the interests of all urban inhabitants, who each have an equal chance to participate (Becker *et al.* 2015).

If, then, the Energietisch can be read as working *in* the local state, the ways in which it struggled *against* the state should also be attended to. Returning to the contradiction posed by the LEWRG, while the local state facilitated access to a resource deemed necessary by the

Energietisch – the possibility to win universalistic energy democracy – the state was simultaneously implicated in 'relations we don't need'. This was made plain, most obviously, in the Senate's strategic selectivity towards the reproduction of prevailing relations, seen through repeated attempts to undermine the campaign, most obviously the referendum date change. Moreover, in the desire for the democratization of state institutions, the Energietisch in a sense worked against the state: they betrayed an understanding of state institutions' role within the reproduction of dominant relations and an aspiration to disrupt this. For one activist: "the goal of fight inside the state is to weaken the state structure and to destroy the original idea of the state". By struggling to incorporate the interests of "normal people or oppressed people ... who wouldn't have influence in the normal state", the Energietisch believed that internal relations within state institutions could be shifted, with transformative implications for the state's form and function.

Finally, in many ways, the Energietisch recognized the necessity of struggle *beyond* the state's material apparatus. Activists understood that the state could not offer the Energietisch everything they wanted: there were legislative restrictions, for instance, on voting on 'financial matters' such as energy prices, while EU competition directives prevented the possibility of a direct vote on re-municipalizing the grid. Thus, there was a broad acknowledgement that political action beyond the state was necessary. Yet, for Holloway (2002; Holloway and Wainwright 2011), movements can easily become "sucked in" to the state's way of doing things, a tendency that, perhaps, was evident in Berlin through the de-prioritization of struggle beyond the state. Many Energietisch activists expressed frustration at the post-referendum standing of the campaign, suggesting that much of the momentum and political ground gained had been lost. For some, this resulted from an over-privileging of formal politics over the building of grassroots power. For one activist, the campaign lost sight of the notion that "when you win the referendum it doesn't mean that they do what you want". Consequently, "the people who voted for us were so quiet" after the referendum: popular anger at the decisive date change was limited because no popular movement existed.

Relatedly, the Energietisch encountered difficulties on account of the limited attention devoted to questions of social reproduction. Feminist scholars such as Federici (Federici 2012b) have called attention to the devalued yet integral role of unpaid reproductive labour of caring and domestic endeavours – historically gendered as 'women's work' – in maintaining social movements seeking to build emancipatory alternatives. Energietisch activists depicted participation in the campaign as "really hard", "hard work" and a "tough time", sadly even forcing people to "put their family aside". Arduous winter days collecting signatures and weeks of late nights spent toiling over the draft law were described, and many activists spoke of exhaustion and burn-out, forcing them to drop out of the campaign after the referendum. Perhaps greater attention to the quality of activists' relations and lives within and outside of the campaign might have better equipped the Energietisch for sustained struggle. Here, again, a limited concern for struggle beyond the state was evident.

In sum, the Energietisch can be understood as attempting to enact the right to the city and create urban commons through the process of struggle in-against-and-beyond the local state. The experience of the Energietisch demonstrates the promise of this political orientation, in that it offers one approach to the thorny task of navigating the complex realities of the state without fetishizing it as a blunt instrument of ruling class power or else the only fruitful site of struggle. Yet the problems encountered by the Energietisch on account of under-attentiveness to matters beyond the state's material apparatus, should be kept in mind for future endeavours towards the democratization of urban natures.

In-against-and-beyond a rights-based politics in Durban

It would be wrong to over-extend a comparison between the Durban and Berlin struggles for a democratization of energy and water production; nevertheless, reading the cases alongside one another enables one to better understand and, indeed, to rethink some of the shibboleths of urban political ecology. In particular, it enables a more considered approach to the role of the state, as well as a reconsideration of the political possibilities within struggles to remake the urban commons in more democratic ways.

On the one hand Durban's struggles over water can be read as an effort to make claims on the state. Given that the apparent denial of a right to water has been one of the key motivating factors for quotidian struggles over water, the struggles of Durban residents can be understood, in line with other readings of the right to water, as a first step in pressuring states to fulfil their obligations to citizens. Interpreting the right to water in this light, the UN Special Rapporteur on the right to water has framed the UN's role as the developer of a legislative framework, without which "the State cannot be held accountable by the individuals, or 'rights-holders', who live within its jurisdiction" (de Albuquerque 2014: 21). Within such a framework, the right to water appears meaningless if it is not accepted by and acted upon by states. Nevertheless, it is precisely on these grounds that the right to water has also been criticized. Thus, as Bustamante et al. (2012: 227) write:

> If we are to consider how rights can be recognized and employed by the state, we are recognizing and justifying the state as responsible for ensuring compliance. A rights-based approach means that other institutional and organizational forms are not recognized, even though they may occupy spaces for interaction and rights that don't necessarily originate from the state.

In one of the more high-profile examples of the struggle to ensure that Durban municipality fulfilled its obligations to citizens of the municipality over the right to water, Christina Manquele, supported by the Concerned Citizens Group, embarked on a court battle, claiming that the municipality had acted unlawfully in relation to the 1998 Water Services Act through disconnecting her water supply for non-payment. Challenging such a claim, an advocate representing the municipality went on to argue that the state was only obliged to fufil such rights if it was within its budgetary capabilities to do so. Thus, in not paying, Manquele had undermined her claim to the right to water. Although winning the first round of the case and having her water supply reconnected, when the Durban High Court released its final judgment a year later it fully supported the legality of the municipality's right to disconnect. On the one hand the Manquele case represented a partial victory in the struggle to ensure that the state fulfil the right to water; however, on the other, it frustrated activists through chanelling energies into a legal process that seemed stacked against them. Bond (2012) contextualizes the legal challenges to pressure the South African state to fulfil its obligations over the right to water, demonstrating that the state has continually sought to read the right to water in its narrowest sense. For him it becomes necessary to move beyond a narrow juridical reading of the right to water and, instead, to mobilize a participatory understanding (see also Clark 2012). Bond finds hope for such a radical and participatory re-reading of the right to water within Lefebvre's understanding of the right to the city.

Notwithstanding the somewhat dualistic reading of nature and the city within Lefebvre's writings that we noted above, we would concur with such a position. Nevertheless, it remains something of an abstract claim if divorced from quotidian understandings of the struggle to access water, which, again, is often directed towards the state, sometimes directed against the state, and is also sometimes suggestive of a struggle beyond the state. In order to make sense of what might be referred to as the lived contradictions of everyday life, we find it helpful to return to a gramscian understanding of the fractured and fragmented understandings emerging within 'common sense'. An understanding of the right to water might thereby emerge from people's everyday practices of producing and reproducing the needs of an individual household.

Thus, in an example that one of us has developed elsewhere (Loftus 2007) it is possible on a micro-scale to see how a gendered division of labour might influence the situated knowledges through which people make sense of a sudden interruption of water supply. In the informal settlement of Amaoti, these situated knowledges shaped a protest, which began with a set of demands addressed towards the local municipality for clean drinking water, before quickly escalating into a more radical protest that went beyond simple demands, calling instead for free water. The protest enabled an embryonic movement that suggested a move beyond the state form as protestors called for control and ownership over the provision of water. Throughout such efforts to achieve fairer access to water, understandings of the ANC continue to overdetermine how people view themselves in relation to the state. Having struggled to bring the ANC to power, common sense understandings are often contradictory. Does the ANC represent the state? To what extent to lived realities suggest a betrayal of immediate post-apartheid hopes? As Loftus and Lumsden (2008) have argued, a fragile situation has emerged in which the continual disappointments with the ANC over service delivery have not translated into a clear critique. Struggles over water are typical of this contradictory situation and suggest a need to understand how everyday practice produces situated knowledges that are influenced by historically sedimented common sense.

Within such fragmented and contradictory conceptions of the world in Durban, there are the possibilities for a critique that might transform the social relations upon which both the state form and the narrow juridical reading of the right to water depend. Likewise, returning to Berlin, the Energietisch campaign gained traction because of its ability to tap into the frustrations of people's everyday interactions with an energy system broadly seen as unfair and exploitative. From rationing heating in the winter through to the shame and injustices of energy debt incurred to sustain the corporate utility Vattenfall's profit margins, the Energietisch's vision of energy democracy gained support precisely because of their connection to the rhythms of everyday life in the city and the common sense beliefs and aspirations that shape this.

Conclusion

The urban practices considered here – of energy provision in Berlin and water access in Durban – must be attended to in all their historical and geographical specificity. Attempts to draw parallels between the European and the post-colonial, post-apartheid context must be treated with caution. Yet, there are, we believe, productive comparisons to be made, which tell us something of interest about how we might navigate the apparently ossified forms of the city and the state in both theory and praxis. In both Berlin and Durban, urban struggle over basic aspects of social reproduction – electricity and water – resisted confinement within the narrow framing of access rights. Rather, in both instances, struggles for universal, decent quality municipal services portend desires for a transformed future city: an urban commons in which social reproduction within the city is collectivized – or, put differently, where the production of urban environments is fully democratized. Yet the radical imaginary of the urban commons or the right to the city seems hopelessly inadequate, without immediate material relevance. Just as the right to water or the right to clean energy must allude to the right to the city, the right to the city must speak to the basic necessities of survival.

It is the need to connect transformative politics to – in the words of the LEWRG – the "resources we need", that forces us to confront the state. Thus, in both Berlin and Durban, while political aspirations did not end at improved state provision, this was the starting point for urban resistance. Yet, in both cases, the state's tendency to preserve unjust and unsustainable socio-ecological relations – relations that, resoundingly, we do not need – was made plain, from the Senate's referendum date change in Berlin to the municipal government's role in water disconnections in Durban. Struggle was directed *within and against* the state on account of such frustrations – and on account of a radical attempt to transform the state through the incorporation of greater levels of direct control for the city's inhabitants. Further, the importance of struggle *beyond* the state's material appartus resounds from both cases. In Durban, it is through the everyday practices of providing one's family with clean water – a task undertaken principally by women – that the possibility of transformative critique emerges. In Berlin, the limited attention to the everyday practices entailed by the ebbs and flows of the Energietisch campaign, was one factor in leaving activists under-equipped for sustained mobilization. The city, then, can be remade through interventions in the processes of local statecraft; yet it must also be remade through the transformation of the everyday behaviours and relations that the historically specific urban form – shot through with processes of exploitation and domination – imposes upon our lives.

The *in-against-and-beyond the state* imaginary is helpful because it offers a way out of dichotomizing and fetishizing thinking. The state is not the be-all-and-end-all of emancipatory struggle, nor is it merely a dangerous instrument wielded by our enemies. The state, rather, is a set of socio-ecological processes, dynamic and in flux; always tending towards the reproduction of dominant relations, yet always the outcome of struggle, leaving open the possibility of rupture. Endeavours to democratize the production of urban environments cannot afford to ignore the state. Yet our political imaginations ought not be constrained by the restrictions of the state form – and political action cannot be located solely at the level of state institutions.

References

Angel, J. (2017). Towards an Energy Politics In-Against-and-Beyond the State: Berlin's Struggle for Energy Democracy. *Antipode*, (49)3, pp. 557–576.
Becker, S., Beveridge, R. and Naumann, M. (2015). Remunicipalization in German Cities: Contesting Neo-liberalism and Reimagining Urban Governance? *Space and Polity*, 19(1), pp. 76–90.
Bond, P. (2000). *Elite Transition: From Apartheid to Neoliberalism in South Africa*. London: Pluto Press.
Bond, P. (2012). The Right to the City and the Eco-Social Commoning of Water: Discursive and Political Lessons from South Africa. In: F. Sultana and A. Loftus, eds. *The Right to Water*. Abingdon: Routledge.
Bustamante, R., Crespo, C. and Walnycki, A. (2012). Seeing Through the Concept of Water as a Human Right in Bolivia. In: F. Sultana and A. Loftus, eds. *The Right to Water*. Abingdon: Routledge.
Clark, C. (2012). The Centrality of Community Participation to the Realization of the Right to Water: The Illustrative Case of South Africa. In: F. Sultana and A. Loftus, eds. *The Right to Water*. Abingdon: Routledge.
Cottle, E. and Deedat, H. (2002). *The Cholera Outbreak: A 2000–2002 Case Study of the Source of the Outbreak in the Madlebe Tribal Authority Areas, uThungulu Region, KwaZulu-Natal*. Durban: Health Systems Trust.
Cumbers, A. (2012). *Reclaiming Public Ownership: Making Space for Economic Democracy*. London: Zed Books.
Cumbers, A. (2015). Constructing a Global Commons In, Against and Beyond the State. *Space and Polity*, 19(1), pp. 62–75.
de Albuquerque, C. (2014). *Realising the Human Rights to Water and Sanitation: A Handbook*. Available at: http://unhabitat.org/series/realizing-the-human-rights-to-water-and-sanitation/ [accessed 19 December 2017].

Federici, S. (2012a). Feminism and the Politics of the Commons. In: S. Federici, ed. *Revolution at Point Zero: Housework, Reproduction, and Feminist Struggle*. Oakland, CA: PM Press.

Federici, S. (2012b). *Revolution at Point Zero: Housework, Reproduction, and Feminist Struggle*. Oakland, CA: PM Press.

Hall, D., Lovina, E. and Terhorst, P. (2013). Re-municipalisation in the Early Twenty-First Century: Water in France and Energy in Germany. *International Review of Applied Economics*, 27(2), pp. 193–214.

Hardt, M. and Negri, A. (2009). *Commonwealth*. Cambridge, MA: Harvard University Press.

Harvey, D. (1996). *Justice, Nature and the Geography of Difference*. Oxford: Blackwell.

Harvey, D. (2012). *Rebel Cities: From the Right to the City to the Urban Revolution*. London: Verso.

Holloway, J. (2002). *Change the World Without Taking Power*. London: Pluto.

Holloway, J. (2010). *Crack Capitalism*. London: Pluto.

Holloway, J. and Wainwright, H. (2011). Crack Capitalism or Reclaim the State? *Red Pepper*, 15 April. Available at: www.redpepper.org.uk/crack-capitalism-or-reclaim-the-state/ [accessed 5 December 2017].

Jessop, B. (1982). *The Capitalist State: Marxist Theories and Methods*. Oxford: Blackwell.

Jessop, B. (2007). *State Power*. Cambridge: Polity Press.

Lefebvre, H. (1968). The Right to the City. In *Writings on Cities*, translated by E. Kofman and E. Lebas. Cambridge, MA: Blackwell.

Lefebvre, H. (1970). *The Urban Revolution*. Translated by R. Bononno. Minneapolis: University of Minnesota Press.

Loftus, A. (2007). Working the Socio-Natural Relations of the Urban Waterscape in South Africa. *International Journal of Urban and Regional Research*, 31(1), pp. 41–59.

Loftus, A. (2012). *Everyday Environmentalism Creating an Urban Political Ecology*. Minneapolis: University of Minnesota Press.

Loftus, A. and Lumsden, F. (2008). Reworking Hegemony in the Urban Waterscape. *Transactions of the Institute of British Geographers*, 33(1), pp. 109–126.

London Edinburgh Weekend Return Group. (1980). *In and Against the State*. Available at: https://libcom.org/library/against-state-1979 [accessed 5 December 2017].

McDonald, D. (1998). Three Steps Forward, Two Steps Back: Ideology and Urban Ecology in South Africa. *Review of African Political Economy*, 25(75), pp. 73–88.

Miller, B. (1994). Political Empowerment, Local-Central State Relations, and Geographically Shifting Political Opportunity Structures: Strategies of the Cambridge, Massachusetts, Peace Movement. *Political Geography*, 13(4), pp. 394–406.

Purcell, M. (2002). Excavating Lefebvre: The Right to the City and its Urban Politics of the Inhabitant. *GeoJournal*, 58, pp. 99–108.

Scott, J.C. (1998). *Seeing Like a State: How Certain Schemes to Improve the Human Condition Have Failed*. New Haven, CT and London: Yale University Press.

Smith, N. (1984). *Uneven Development: Nature, Capital and the Production of Space*. Oxford: Blackwell.

Smith, N. (2003). Foreword to H. Lefebvre, *The Urban Revolution*. Minneapolis: University of Minnesota Press.

Swyngedouw, E. and Kaika, M. (2000). The Environment of the City … or the Urbanization of Nature. In: G. Bridge and S. Watson, eds. *A Companion to the City*. Oxford: Blackwell.

Taschner, S. (n.d.). *Taking Over Berlin's Energy Supply: How the Berliner Energietisch Is Campaigning to Democratize the Local Energy Sector*. Berlin: Rosa Luxemburg Foundation. Available at: http://rosalux.gr/sites/default/files/taschner_the_question_of_power_-_alternatives_for_the_energy_sector_in_greece_and_its_european_and_global_context.pdf [accessed 5 December 2017].

Whitehead, M., Jones, R. and Jones, M. (2007). *The Nature of the State: Excavating the Political Ecologies of the Modern State*. Oxford: Oxford University Press.

12
POLITICS OF URBAN GARDENING

Marit Rosol

Introduction

> Thus, I think all these projects in this so-called third sector at the moment move along a very thin line. A very thin line between social change on the one hand and, on the other hand, complete co-optation by the system.
>
> (Community Gardener Berlin, interview Rosol)

Community gardens are urban green spaces that are collectively operated. They differ from both uniform institutionalized public green spaces as well as from private gardens. In contrast to city parks, they are community-managed, i.e. collectively designed, built, and maintained by local residents. Unlike other forms of urban gardening, such as the well-known German allotment gardens (*Schrebergärten*) that are only open to their members and offer individual plots for lease, they are at least occasionally open to the general public. They provide a public service through community work and community organizing and therefore benefits beyond just the gardeners themselves. They differ in size (from small abandoned building lots to several hectares), user group (local residents, migrants, children) and appearance (landscaped park, organic vegetable garden, brown fields with spontaneous vegetation). Some are centred on food production; others serve mostly as a green and non-commercial socializing space. They heavily depend on voluntary work, but different to other forms of voluntary engagement as, for example, stewardship for existing green spaces or sporadic volunteers' days, the involved residents create new green areas according to their own ideas.

Central to much research on community gardens are the many social, ecological, economic and political benefits. In the quote cited above, the gardener from Berlin complicates this picture by referring to the internal contradictions of community gardening: on the one hand, it is characterized by its grassroots origins and progressive ambitions. On the other hand, the commitment and voluntary work also has the potential to serve as a resource for cushioning the outsourcing of former (local) state responsibilities in neoliberalizing cities. In that context, community gardens can be seen as serving the neoliberal idea of self-contained communities and the privatization of the service-sector.

This chapter discusses the politics of urban community gardening and its ambiguous positioning between the poles of neoliberal and austerity urbanism on the one hand and progressive urban politics on the other. It argues that urban community gardening can never be understood as either a sole expression of neoliberal urbanism nor of pure resistance. Instead, the different forms and examples of community gardening should be analysed carefully and by critically acknowledging its inevitable contradictions. These societal contradictions are time and place specific. Thus, one important task of critical scholarship on urban gardening is to contextualize it within the historically changing trends and processes of urban politics. In the post-industrial cities of the Global North this context consists of the cycles of urban disinvestment and gentrification, tied partly to a general movement from the Keynesian welfare state of the 1970s towards the neoliberal and austerity urbanism of today.

The chapter starts by starting by showing the roots of community gardens in urban counter-cultural movements of the 1970s. Next, urban gardening is discussed in the context and as part of neoliberal urban governance in the new millennium. Different forms of the relations and connections between urban gardening and neoliberal governance will be shown using examples from Germany and the USA. Finally, the progressive potential of urban community gardening is discussed in regards to the current debate around urban commoning. The chapter closes with a brief summary, conclusions and reflections on further research.

Roots of urban gardening in counter-cultural movements

One important root of community gardening is the counter-cultural and environmental urban social movements of the 1970s.[1] The New York City case is well known in that respect. Here, community activists in 1973 founded the first community garden in NYC in the Lower East Side as well as the support organization Green Guerillas. In reaction to urban decline and disinvestment, fiscal crisis and social unrest in the 1970s, they wanted to appropriate urban space and to form it according to a different logic. Their actions included throwing 'seed bombs' over fences on vacant lots as a symbol for the openly 'illegal' re-appropriation. Residents of the Lower East started to use (or occupy), clean and turn vacant lots, most of them in public ownership,[2] into gardens without legal permission nor governmental assistance (Smith and Kurtz 2003). By the mid-1980s, already 1,000 community gardens existed in NYC (Smith and Kurtz 2003). As long as these city properties were regarded as having low value, community gardening on public property was tolerated, legalized and since 1978 even supported by the City's 'Operation Green Thumb'. In the early 1990s, this changed when in the course of an 'urban renaissance', i.e. the return of capital and the middle classes (Porter and Shaw 2009), land once again became attractive for investors (Schmelzkopf 1995, 2002; Smith and Kurtz 2003; Smith and DeFilippis 1999; Staeheli et al. 2002). When Mayor Giuliani's administration put several hundred gardens located on city property on sale, the gardeners reacted with massive protest and lobbying and formed the NYC Community Gardens Coalition. As a result of those protests, about 500 gardens were saved, partially through private acquisition until 2002 (for a critique of this solution see Eizenberg 2012b).

Similarly, in West Berlin, Germany, community gardens arose out of counter-cultural and urban protest movements in the late 1970s and early 1980s. For example, a still existing urban farm for children in Berlin-Kreuzberg was started in 1981 by squatting on derelict land right beside the Berlin Wall (Rosol 2010). A registered association (Kinderbauernhof Mauerplatz e.V.) was founded in the same year. Organized mainly by single mothers, the aim was to create an educationally supervised green space for small children in the densely built-up inner-city borough. Simultaneously, the project was part of a broader social movement of squatters and other social activists against the predominant urban renewal policies of that time (*Kahlschlagsanierung*, the clearance of turn-of-century buildings to make way for new high rise buildings).

The project evolved from the resistance against a misguided urban development policy and in direct confrontation with urban planners and local politics. With the founding of the children's farm the group wanted to criticize these policies as well as demonstrate alternative ways of creating a city. Members of the group have been actively engaged not only in environmental and educational topics, but also in local politics in general. The project has been an integral part of the neighbourhood for over twenty-five years. The plot of about one hectare in size had remained squatted on for twenty years, until, in 2001, after a long battle with the borough's politicians and administration, the association obtained a five-year-contract from the borough of Friedrichshain-Kreuzberg. For more than twenty years, the *Kinderbauernhof* remained highly contested.

This shows how some community gardens are the result of long battles by local residents who are eager to influence the shape and the functioning of city parks and who want 'other spaces', decentralized, non-bureaucratic solutions. In this sense, community gardens have been an expression of active and progressive appropriation of urban spaces by citizens and thus of 'grassroots urbanism' that oppose state-centric as well as neoliberal development. Residents are not only the decision-makers of how to use an empty lot, but also responsible for the creation and maintenance of the open green space. In this way, they exercise a civic control over urban space – that may, however, also take on exclusive forms, as will be shown below.

Urban gardening in the context of neoliberal and austerity urbanism

Some community gardens thus arose in opposition to top-down planning and neoliberal urbanism. However, they can also be co-opted by the neoliberal city. In this section, I identify three forms and relations of urban community gardening with the (local) state and the neoliberalization of urban governance.

Gentrification and image policies

Community gardens are affected by and part of the changing historical-geographical trends of urban development. The New York City case is instructive in that it shows that community gardens that once "flourished through a kind of benign neglect by capital accumulation" (Pudup 2008: 1232) in the 1970s and 1980s turned into contested space as soon as land exchange values rose again. There has been substantial research on conflicts over land use that typically arise when cities prioritize profitable economic development and building projects over social uses. The privatization of public lots in New York was strongly connected to the understanding of cities as 'enterprises'. This involves the re-orientation of the local state away from redistribution and towards competition in order to attract investors (Harvey 1989). In the current climate of 'austerity urbanism', cuts in public spending as well as privatization of public services, land and infrastructure have intensified again. Triggered or reinforced by the economic crisis since 2008, government downsizing and privatization of public resources, including community gardens on public properties, are justified as "fiscal necessities" (Peck 2012: 626).

At the same time, community gardens are being discovered in many cities by developers as an instrument for urban development. Community gardening and urban agriculture are used as marketing strategies for upgrading the image of neighbourhoods in order to increase profits (Dooling 2009; Quastel 2009). This puts gardening activists in a difficult position: they are not responsible nor are they profiting from development (that remains the privilege of property owners and developers), on the contrary they are often the ones that suffer rising housing costs (e.g. Rosol 2006: 259). But if strategies to improve living conditions in a neighbourhood are not combined with mechanisms that prevent displacement of residents and keep housing affordable, even the most well-meaning projects can become engines of gentrification. Thus, the political task of today is not only to struggle for the acceptance and support of urban gardening, but also against displacement and thus form alliances with other urban political struggles.

Voluntary work

Peck and Tickell introduced a helpful analytical differentiation of neoliberalization processes, distinguishing the "roll-back" of the Keynesian welfare state in the 1980s (based on privatization, de-regulation and financial cutbacks) from the "roll-out" of neoliberal institutions in the 1990s, which they see as a response to neoliberalism's immanent contradictions (Peck and Tickell 2002). This roll-out neoliberalism involves "new state forms, new modes of regulation, new regimes of governance, with the aim of consolidating and managing both marketization and its consequences" (Peck and Tickell 2007: 33, see also summary p. 34). Crucial for the discussion of community gardens is the rising significance of "governance-beyond-the-state" (Swyngedouw 2005), i.e. the increasing participation of non-state actors in (local) state decision-making and the transformation of roles, responsibilities and institutional configurations of the (local) state and citizens in urban spatial politics.

In many cases, inclusion of non-state actors is geared less at citizens' participatory rights, but rather at the outsourcing of traditional state functions to civil society organizations (e.g. Fyfe 2005). This is especially obvious in the shift of responsibilities for service provision towards the profit-making and the non-profit sector and to volunteering citizens (Bondi and Laurie 2005; Fyfe and Milligan 2003; Milligan and Conradson 2006). The role of the voluntary sector then, is to meet the shortfalls in services and benefits that result from service reductions and withdrawal of funding (Wolch 1989, 1990). This again leads to an increasing professionalization of the voluntary sector, the adoption of a managerial rationality and increasing competition between voluntary organizations at the expense of co-operation (Wolch 2006).

The use of the gardeners' unpaid labour can be characterized as a neoliberal roll-out strategy because it seeks to outsource and offload responsibility for green infrastructures through calls for more civic engagement and participation in maintaining urban green spaces, including through gardening projects (Ghose and Pettygrove 2014; Rosol 2012; regarding urban agriculture programmes McClintock 2014). Often, roll-back and roll-out neoliberalism strategies go hand in hand. In Berlin for example, 'roll-back' neoliberalism prepared the ground for these new forms of creating and maintaining public green spaces. Many of the gardening projects came into existence due to the withdrawal of the local state from adequate service provision and due to the severe fiscal crisis of the City of Berlin for two reasons. First, lots became available for interim uses because important public amenities were not built because of a lack of resources (specifically: a public playground, a police station, a school garden). Second, the engagement of the residents is a response to lack in quality and maintenance of existing public parks resulting from their inadequate funding (Rosol 2012). Pudup also points out that the early analysis of Thomas Bassett (Bassett 1979) showed that community gardening had been serving as "a buffering mechanism" by helping to accommodate crisis, emergency and change (Pudup 2008: 1229). In effect, these gardens are a response to roll-back neoliberalism, but they have been encouraged by roll-out neoliberalism. In other words: community volunteering bridges roll-back neoliberalism – filling in for former state's responsibilities – and roll-out neoliberalism – doing so under neoliberal principles.

Beyond just the care for urban green spaces, we also see a relocation of responsibility for social welfare onto community organizations such as gardening groups. In her work on community gardens in Buffalo, Knigge (2009) observed that community gardening organizations provide different types of social services and care beyond those typically associated with community gardening, e.g. refugee services, food pantries, after school programmes, tutoring, emergency and other placement services, as well as food and clothing giveaways. She ascribes this expanded role of community gardening organizations to the relocation of responsibility for social welfare and care to individuals and the community, which especially affects women, ethnic minorities and immigrants. She analyses community gardening in the general context of state restructuring and welfare devolution and as self-help in order to compensate for the devastating effects of neoliberal restructuring. Pudup makes the same connection and criticizes that "voluntary and third sector initiatives organized around principles of self-improvement and moral responsibility stand in for state sponsored social policies and programs premised on collective responses to social risk" (Pudup 2008: 1229).

Governing through community and technologies of the self

The continuous emphasis on self-help in regard to urban green spaces becomes even more problematic when civic engagement is exclusively focused on certain groups and values. This may promote exclusionary or even "revanchist" (Smith 1996) spaces. Allowing or even encouraging community gardens in some parts of the city and not in others and under conditions like a temporary use only, can be seen as a state strategy to maintain control over space as well as the population (Ghose and Pettygrove 2014; for the case of Berlin see Rosol 2010). When the local state, in contrast to Fordist times, 'allows' intensified participation in the shaping of urban green spaces in order to rid itself of responsibilities, certain groups, usually with a middle class background, are in a better position to appropriate these projects and spaces as they are able to better articulate their demands and needs. Of a similar concern, although from a different origin, are observable tendencies of the local administration spending the little money it still disposes of on those green spaces and parks where civic engagement already takes place. This is a valuable form of approval of these activities, but as a result, money is withdrawn from other areas with less engagement. Those areas are often the very areas with the greatest needs, lacking in public and voluntary infrastructure and with a population under great economic pressure and time constraints so that providing voluntarily for neighbourhood greening is far from possible.

Thus, the problem of community-organized collective consumption does not lie in a potential outsourcing of communal tasks only, but in the more fundamental shifts in the logics of government. Open public spaces created and maintained by communities bring about a new form of social control and are part of a new technology of governing, that is, "governing through community" (Rose 1996). This means that, for example, the possibilities for marginalized urban residents to become visible and to participate in political decision-making processes "are increasingly defined in terms of their own abilities to govern themselves as a community" (Maskovsky 2006: 77). This strategy of governance combines the political rationalities of a neoliberal ethos of self-responsibilization of the individual and the neo-communitarian ideal of active citizenship and the promotion of community sense (Rose 1996). As Dean emphasizes, the neoliberal focus on the self-governing individual is not antithetical to that of community. If the main rational of liberal thought is 'freedom' of 'individuals', in a world, in which "there is no such thing as society" (Thatcher 1987), the exercise of this freedom for responsible and autonomous subjects is through voluntary associations and voluntary work, i.e. in and for the 'community' (Dean 1999: 152). This way, 'community' is not only compensating for neoliberal failures as discussed above (see also Jessop 2007; Mayer 2003). It is the *necessary* connection of governing the self and governing a population and thus a cornerstone of neoliberal governmentality (Rosol 2013).

Regarding the topic of urban gardening, Pudup, in her work on organized garden projects in the San Francisco Bay area – often in an institutional setting like a school, clinic or jail – observes a "rise of gardens ... as spaces in which gardening puts individuals in charge of their own adjustment(s) to economic restructuring and social dislocation through self-help technologies" (Pudup 2008: 1228). Thus, they condition participants to pursue change through individual endeavour instead of collective resistance and mobilization. She points out that these projects rest upon the promise that "direct contact with nature, through gardening, will transform people who are otherwise poor and socially and culturally marginalized" (Pudup 2008: 1230). She interprets this as a form of neoliberal governmentality with its emphasis on technologies of the self. Also Eaton discusses local food projects in Ontario as a form of neocommunitarism and neoliberal governance (Eaton 2008).

Urban gardening as green urban commoning

Commons suggest alternative, non-commodified means to fulfil social needs, e.g. to obtain social wealth and to organize social production. Commons are necessarily created and sustained by 'communities', i.e. by social networks of mutual aid, solidarity and practices of human exchange that are not reduced to the market form (De Angelis 2003: 1).

At present, the progressive potentials of community gardening are often discussed as 'urban commoning'. The reawakened interest in the commons can be understood as a response to neoliberalism and the devastating effects tied to its main tenets of liberalization, privatization and retrenchment of the welfare state. In light of neoliberal transformations actively fostered by national states dismantling the regulatory framework of capital accumulation and facilitating and promoting privatization, there is a search for ways to provide for social wealth and organize social reproduction outside the private (= commodity and profit based) versus public (= state) dichotomy. In this context, the commons have much appeal. The concept of the commons expresses the claim for collective rights and interests not limited to those that have to be paid for in the market, tied to individualized private property rights, or controlled by states, although calls for more state regulation in order to protect them might be part of it (McCarthy 2009: 509). It is a search for alternative social relations and values. Importantly, commons consist not only in material resources, but rely on communities and democratically negotiated rules and regulations. Thus, there are no commons without 'commoning' and the

> common is not to be construed, therefore, as a particular kind of thing, asset or even social process, but as an unstable and malleable social relation between a particular self-defined social group and those aspects of its actually existing or yet-to-be-created social and/or physical environment deemed crucial to its life and livelihood.
> *(Harvey 2012: 73)*[3]

In a radical sense and going beyond just the management of common-pool resources, the 'urban commons'[4] can be defined as places and processes of collective (not necessarily open-access), cooperative, non-commodified creation, maintenance, protection and transformation of urban spaces (Chatterton 2010; for a different definition see Colding *et al.* 2013). They seek to influence urban politics and potentially urban societies that confront the ongoing enclosure and privatization – or new rounds of "accumulation by dispossession" (Harvey 2003) – processes within the neoliberal city.

In her work on community gardens in New York City, Eizenberg points out that "manifestations of an alternative political project" (Eizenberg 2012a: 764), i.e. 'actually existing commons', are "never complete and perfect" but nevertheless do exist (Eizenberg 2012a: 765). They have to be produced and negotiated in the everyday – based on collaboration, cooperation and communication (Eizenberg 2012a: 766). This requires concern with complex questions such as: Should urban space be used for dwelling, recreation, food production, as social space? By whom? By whom not, i.e. who or what use may have to be excluded in order to maintain the common successfully? How? According to what rules – and how can they be modified? How can sanctions against users who violate the rules be enforced? And even: How is this related to other scales? The commons have to be constantly maintained and protected too since capitalism perpetually seeks to expand the market logic and thus to banalize, degrade, destroy and 'enclose'[5] these spaces and efforts (Harvey 2011). To protect and maintain the commons thus also includes the search for legal solutions to protect this kind of neither private nor public spaces – within a legal system that negates such spaces and forms of social relations.

Eizenberg was one of the first to discuss in detail community gardens (in NYC) as an example of reproducing the urban commons. To ground her claim, first she refers to Karl Linn (Linn 1999), who regards the collective management of community gardens and that they enable some degree of self-sufficiency as indicator of a common. However, she seeks to go beyond Linn's mostly just material understanding and develops further the theoretical framework of the commons. For this purpose, she presents a detailed analysis of NYC community gardens and the actual practices, experiences, subjectivities, knowledge production and social relations involved, including the transition from gardening to activism in the 1990s. She concludes that the actually existing urban commons can create and promote alternatives to capitalist social relations by providing a mechanism for redistribution, by challenging the hegemonic social order of organization and planning of urban space, as well as by challenging the market and profit logic that dominates it. Commons challenge the exchange-value and (abstract)[6] private property logic of how urban space is produced and used and can transform and de-commodify former private and commodified spaces. And they can create alternative spaces in a very concrete and in the far-reaching Lefebvrian sense of a "right to the city" (Eizenberg 2012a: 779).[7]

But not only in New York are community gardens debated as new urban commons. For example, in Berlin, the 'Allmende-Kontor'-Community Garden project ('*Allmende*' being actually the old-fashioned German word for the not yet enclosed village common) seeks to contribute to the debate on property regimes and the re-production of the urban commons (Meyer-Renschhausen 2015). It was founded in 2011 on the former airfield of Berlin-Tempelhof and is currently being run by about 500 community gardeners.[8] It serves as a main node in the community gardening network in Berlin and Germany in general and sees itself as a learning-ground for the collective self-governing of an urban common. It aims to create spaces for alternatives to the hegemonic "consumption, growth and throwaway society" (Martens *et al.* 2014: 49).

In Cologne, in 2012 a community garden ('*NeuLand*') was created as a political project on a highly contested site. Its initial trigger was not the wish to garden but to affect the use of a particular derelict site and of urban space in general. A garden was chosen as a medium for claiming a greener and more just city. Its aim was specifically to experiment with new forms of urban commons and for claiming a "right to the city" (Follmann and Viehoff 2015). Follmann and Viehoff discuss the claims but also the challenges this involves (e.g. communicating *NeuLand* as an open and inclusive urban common, neoliberal co-optation and land use conflicts, *NeuLand* as enclosure) and how the *NeuLand* gardeners engage with these challenges. They interpret NeuLand as "an example of an actually existing urban common in-the-making" (Follmann and Viehoff 2015: 1164) that uses practices of gardening and brings together a broad range of activists and users with very different backgrounds aiming at creating political awareness and establishing a new claim to the city. Thus, it not only vocally protests against the neoliberalization of urban policy and spaces, but also challenges the neoliberal logic that drives the current redevelopment of the site and the area in practice. Follmann and Viehoff conclude that even these imperfect and ever-evolving and in-the-making urban commons can support a transition towards a more democratic governance of the city as a whole (2015: 1169).

Also in Germany, in 2012, a manifesto called 'The city is our garden' was initiated as a political action and intervention in urban politics by activists from the above two gardens and others. They seek to (re-)position the growing community gardening movement in Germany politically. They criticize urban administrations on the one hand for embracing community volunteering and urban gardening and on the other placing the monetary considerations above all else. This thinking results in a precarious legal status and lack of a long-term perspective for many community gardens. The activists also contest the increasing use of urban gardening locations by big corporations for their PR and advertisement activities and generally for commercial interest (for a similiar development in Vancouver, see Quastel 2009). Against these tendencies they proclaim urban gardening as commoning and as 'right to the city'. In line with this, the manifesto defines community gardens as "Common goods, opposing the increasing privatisation and commercialisation of public space".[9] As of November 2016, about 150 gardening groups and some other organizations in Germany have signed it.

All these examples show the progressive and transformative potential of community gardening, which can counter the neoliberal tendencies of current urban politics – but also the difficulties in such endeavours.

Conclusion

In this chapter, I discussed the contradictory politics of urban community gardening. On the one hand, I identified three ways of how community gardening may reinforce neoliberal urbanism: land use struggles and gentrification, the role of volunteering within neoliberal urban governance, and governing through community and technologies of the self.

On the other hand, community gardening has roots in counter-cultural and oppositional urban movements and at present, urban commoning has gained significant attention within the debate around community gardening and amongst activists themselves. This provides new perspectives and enthusiasm about urban gardening because as urban commons, their potential goes much beyond the actual gardening itself. Nevertheless, the return to the commons also involves certain traps and limits: As McCarthy points out, the rejection of the state and the privileging of communities as most appropriate for the organization of social reproduction within the commons-movement bears an uncanny resemblance of neoliberal ideologies that they claim to reject (McCarthy 2005; 2009: 511–512; see also Blackmar 2006). There might be other solutions to rampant neoliberalism, for example by radically rethinking and democratizing public ownership (for example, see Cumbers 2012).

What these examples of urban gardening as urban commoning show us nevertheless, is that these projects exist and serve as alternatives in the here and now, often in only small-scale projects but committed to bigger claims of regaining collective control over urban space and urban politics. They challenge the notion of private property and provide space for non-commodified leisure activities, and for hands-on social learning of solidarity, mutuality and trust. They encourage encounters across social and ethno-religious divides (for the examples of Belfast and Dublin see Corcoran and Kettle 2015) and can become nodes of solidarity and increased political awareness not least because the practice of gardening allows 'hands-on' forms of learning and non-verbal forms of interaction between social groups. This way they can provide (learning grounds for) alternatives to neoliberalizing urbanism as well (Rosol 2010).

Because of these contradictions, I argue that a universalizing question, if urban gardening is an expression of urban neoliberalism or of commoning, poses a misguided binary. In any case, this question can never be answered in the abstract but requires thorough analysis of these projects and their context on several scales. The interesting aspect about community gardens is precisely that they are not either/or, but always have both potentials. As Ghose and Pettygrove write, grassroots community gardens can "simultaneously contest and reinforce local neoliberal policies" (2014: 1092). Also McClintock writes that urban agriculture (and as I would add, urban gardening) can be and necessarily is "radical and neoliberal at once", because "contradictory processes of capitalism both *create opportunities for* urban agriculture and *impose obstacles* to its expansion" (McClintock 2014: 157, emphasis in the original).

Thus, I agree with McClintock who demands we move beyond the above-mentioned dualism. Instead he suggests focusing on the how and where of specific projects to better understand its functions, forms, tensions and contradictions – that more often operate on different scales. Similar to him, I argue for detailed empirical analysis of both specific projects and the general context of their activities. This means, for example, to ask about their aims, practices, their reflections on their own acting and the context they work in, as well as the effects of their doing (e.g. that these projects may improve lives for some residents but cannot tackle broader structural conditions). This cannot be done without reflecting the historically changing relation to the (local) state and market-based activities. Thus, it has to be asked, for example, in how far these gardens and the engagement of gardeners serves purposes of increased state control or economic interests of corporations.

Unlike McClintock who cautions us against the potentially counterproductive effects of critiques of urban agriculture's entanglement with neoliberalism, I believe these critiques are important since it is impossible to escape the historically and geographically specific social, economic and institutional context. It has to be acknowledged, that in some instances, such as in the case of Greece, the critique of voluntarism in a neoliberal context cannot ignore "the level of bare necessity that needs to be covered on the spot [by grassroots-initiatives based on voluntary work] and not in some opportune future moment" (Vaiou and Kalandides 2015). But this important observation again strengthens my argument for careful and geographically specific analysis of the projects as well as the context they act in. A careful multi-scalar analysis of the economic and political context in which these gardens act is, in my view, one of the main tasks of critical and at the same time solidary scholarship. Knowing and being aware of the social, economic and institutional context should help to address these contradictions in the daily practices and ultimately transform the current (neoliberal) logic.

Notes

1 Other roots are war and victory gardens in the USA, initiated and subsidized by the state with the aim of food provision for the (urban) population in times of war and economic recessions (Lawson 2005).
2 As a result of abandonment and non-payment of taxes, many inner-city properties reverted to city ownership. Some buildings were destroyed by arson; others were levelled by the city, resulting in numerous vacant lots (Schmelzkopf 2002).
3 See also Harvey's excellent critique of Garrett Hardin's classic article 'The Tragedy of the Commons' (2011) that reminds us that the problems identified by Hardin (degradation of the land due to overuse) are the result of private property and the coercive laws of maximizing profitability within unregulated capitalist accumulation regimes. In his view the real tragedy lies in the loss of the commons.
4 The urban is of course not the only scale where the commons are reclaimed and debated (McCarthy 2009), but of special interest to this chapter.
5 Referring to the enclosure, i.e. privatization, of the British commons in the eighteenth century (McCarthy 2009: 499–502).

6 Abstract in the sense that the owner does not use it.
7 There are further works on community gardens and the 'right to the city', that for example, conceive of urban food production as a right to the city (Purcell and Tyman 2014).
8 www.allmende-kontor.de/index.php/gemeinschaftsgarten.html [accessed 10 October 2016].
9 An English version here: http://urbangardeningmanifest.de/pulsepro/data/files/Urban-Gardening-Manifest_english.doc, more on the background here: http://urbangardeningmanifest.de/hintergrund [accessed 27 November 2015].

References

Bassett, T. (1979). Vacant Lot Cultivation: Community Gardening in America, 1893–1978. Unpublished Master's Thesis, University of California, Berkeley.
Blackmar, E. (2006). Appropriating 'the Commons': The Tragedy of Property Right Discourse. In: S. Low and N. Smith, eds. *The Politics of Public Space*. New York: Routledge, pp. 49–80.
Bondi, L. and Laurie, N. (2005). Working the Spaces of Neoliberalism: Activism, Professionalisation and Incorporation. Introduction. *Antipode*, 37(3), pp. 394–401.
Chatterton, P. (2010). Seeking the Urban Common: Furthering the Debate on Spatial Justice. *City*, 14(6), pp. 625–628.
Colding, J., Barthel, S., Bendt, P., Snep, R., van der Knaap, W. and Ernstson, H. (2013). Urban Green Commons: Insights on Urban Common Property Systems. *Global Environmental Change*, 23(5), pp. 1039–1051.
Corcoran, M. and Kettle, P. (2015). Urban Agriculture, Civil Interfaces and Moving Beyond Difference: The Experiences of Plot Holders in Dublin and Belfast. *Local Environment*, 20(10), pp. 1215–1230.
Cumbers, A. (2012). *Reclaiming Public Ownership: Making Space for Economic Democracy*. London: Zed Books.
De Angelis, M. (2003). Reflections on Alternatives, Commons and Communities or Building a New World from the Bottom Up. *The Commoner*, 6, pp. 1–14.
Dean, M. (1999). *Governmentality*. London: Sage.
Dooling, S. (2009). Ecological Gentrification: A Research Agenda Exploring Justice in the City. *International Journal of Urban and Regional Research*, 33(3), pp. 621–639.
Eaton, E. (2008). From Feeding the Locals to Selling the Locale: Adapting Local Sustainable Food Projects in Niagara to Neocommunitarianism and Neoliberalism. *Geoforum*, 39(2), pp. 994–1006.
Eizenberg, E. (2012a). Actually Existing Commons: Three Moments of Space of Community Gardens in New York City. *Antipode*, 44(3), pp. 764–782.
Eizenberg, E. (2012b). The Changing Meaning of Community Space: Two Models of NGO Management of Community Gardens in New York City. *International Journal of Urban and Regional Research*, 36(1), pp. 106–120.
Follmann, A. and Viehoff, V. (2015). A Green Garden on Red Clay: Creating a New Urban Common as a Form of Political Gardening in Cologne, Germany. *Local Environment*, 20(10), pp. 1148–1174.
Fyfe, N. (2005). Making Space For 'Neo-Communitarianism'? The Third Sector, State and Civil Society in the UK. *Antipode*, 37(3), pp. 536–557.
Fyfe, N. and Milligan, C. (2003). Out of the Shadows: Exploring Contemporary Geographies of Voluntarism. *Progress in Human Geography*, 27(4), pp. 397–413.
Ghose, R. and Pettygrove, M. (2014). Urban Community Gardens as Spaces of Citizenship. *Antipode*, 46(4), pp. 1092–1112.
Harvey, D. (1989). From Managerialism to Entrepreneurialism: The Transformation in Urban Governance in Late Capitalism. *Geografiska Annaler: Series B*, 71(1), pp. 3–17.
Harvey, D. (2003). *The New Imperialism*. Oxford: Oxford University Press.
Harvey, D. (2011). The Future of the Commons. *Radical History Review*, 109, pp. 101–107.
Harvey, D. (2012). *Rebel Cities*. London: Verso.
Jessop, B. (2007). From Micro-Powers to Governmentality: Foucault's Work on Statehood, State Formation, Statecraft and State Power. *Political Geography*, 26, pp. 24–40.
Knigge, L. (2009). Intersections Between Public and Private: Community Gardens, Community Service and Geographies of Care in the US City of Buffalo, NY. *Geographica Helvetica*, 64(1), pp. 45–52.
Lawson, L. (2005). *City Bountiful: A Century of Community Gardening in America*. Berkeley: University of California Press.
Linn, K. (1999). Reclaiming the Sacred Commons. *New Village*, 1(1), pp. 42–49.

Martens, D., Zacharias, M. and Hehl, F. (2014). Gemeinschaftsgärten? Ja, Bitte – Aber Wie? In: S. Halder, D. Martens, G. Münnich, A. Lassalle, T. Aenis and E. Schäfer, eds. *Wissen wuchern lassen. Ein Handbuch zum Lernen in urbanen Gärten*. Neu-Ulm: AK SPAK, pp. 48–93.

Maskovsky, J. (2006). Governing the 'New Hometowns': Race, Power, and Neighbourhood Participation in the New Inner City. *Identities*, 13(1), pp. 73–99.

Mayer, M. (2003). The Onward Sweep of Social Capital: Causes and Consequences For Understanding Cities, Communities and Urban Movements. *International Journal of Urban and Regional Research*, 27(1), pp. 110–132.

McCarthy, J. (2005). Commons as Counterhegemonic Projects. *Capitalism Nature Socialism*, 16(1), pp. 9–24.

McCarthy, J. (2009). Commons. In: N. Castree, D. Demeritt, D. Livermann and B. Rhoads, eds. *A Companion to Environmental Geography*. Oxford: Wiley-Blackwell, pp. 498–514.

McClintock, N. (2014). Radical, Reformist, and Garden-Variety Neoliberal: Coming to Terms with Urban Agriculture's Contradictions. *Local Environment*, 19(2), pp. 147–171.

Meyer-Renschhausen, E. (2015). *Die Hauptstadtgärtner – Anleitung zum Urban Gardening – Tipps vom Allmende-Kontor auf dem Tempelhofer Feld*. Berlin: Jaron.

Milligan, C. and Conradson, D., eds. (2006). *Landscapes of Voluntarism: New Spaces of Health, Welfare and Governance*. Bristol: Policy Press.

Peck, J. (2012). Austerity Urbanism. *City*, 16(6), pp. 626–655.

Peck, J. and Tickell, A. (2002). Neoliberalizing Space. *Antipode*, 34(3), pp. 380–404.

Peck, J. and Tickell, A. (2007). Conceptualizing Neoliberalism, Thinking Thatcherism. In: H. Leitner, J. Peck and E. Sheppard, eds. *Contesting Neoliberalism: Urban Frontiers*. New York: Guilford Press, pp. 26–49.

Porter, L. and Shaw, K. eds. (2009). *Whose Urban Renaissance? An International Comparison of Urban Regeneration Strategies*. London: Routledge.

Pudup, M.B. (2008). It Takes a Garden: Cultivating Citizen-Subjects in Organized Garden Projects. *Geoforum*, 39(3), pp. 1228–1240.

Purcell, M. and Tyman, S. (2014). Cultivating Food as a Right to the City. *Local Environment*, 20(10), pp. 1132–1147.

Quastel, N. (2009). Political Ecologies of Gentrification. *Urban Geography*, 30(7), pp. 694–725.

Rose, N. (1996). The Death of the Social? Refiguring the Territory of Government. *Economy and Society*, 25(3), pp. 327–356.

Rosol, M. (2006). *Gemeinschaftsgärten in Berlin. Eine qualitative Untersuchung zu Potenzialen und Risiken bürgerschaftlichen Engagements im Grünflächenbereich vor dem Hintergrund des Wandels von Staat und Planung*. Berlin: Mensch-und-Buch-Verlag.

Rosol, M. (2010). Public Participation in Post-Fordist Urban Green Space Governance: The Case of Community Gardens in Berlin. *International Journal of Urban and Regional Research*, 34(3), pp. 548–563.

Rosol, M. (2012). Community Volunteering as Neoliberal Strategy? Green Space Production in Berlin. *Antipode*, 44(1), pp. 239–257.

Rosol, M. (2013). Regieren (in) der neoliberalen stadt. Foucault's analyse des neoliberalismus als beitrag zur stadtforschung. *Geographische Zeitschrift*, 101(3–4), pp. 132–147.

Schmelzkopf, K. (1995). Urban Community Gardens as Contested Space. *Geographical Review*, 85(3), pp. 364–381.

Schmelzkopf, K. (2002). Incommensurability, Land Use, and the Right to Space: Community Gardens in New York City. *Urban Geography*, 23(4), pp. 323–343.

Smith, C. and Kurtz, H. (2003). Community Gardens and Politics of Scale in New York City. *Geographical Review*, 93(2), pp. 193–212.

Smith, N. (1996). *New Urban Frontier: Gentrification and the Revanchist City*. New York: Routledge and Kegan Paul.

Smith, N. and DeFilippis, J. (1999). The Reassertion of Economics: 1990s Gentrification in the Lower East Side. *International Journal of Urban and Regional Research*, 23(4), pp. 638–653.

Staeheli, L., Mitchell, D. and Gibson, K. (2002). Conflicting Rights to the City in New York's Community Gardens. *GeoJournal*, 58(2–3), pp. 197–205.

Swyngedouw, E. (2005). Governance, Innovation and the Citizen: The Janus Face of Governance-Beyond-the-State. *Urban Studies*, 42(11), pp. 1991–2006.

Thatcher, M. (1987). Interview by Douglas Keay for *Woman's Own*, 23 September. Published as 'Aids, Education and the Year 2000!' in *Woman's Own*, 31 October, pp. 8–10.

Vaiou, D. and Kalandides, A. (2015). Practices of Collective Action and Solidarity: Reconfigurations of the Public Space in Crisis-Ridden Athens, Greece. *Journal of Housing and the Built Environment*, 31, pp. 457–470.

Wolch, J. (1989). The Shadow State: Transformations in the Voluntary Sector. In: J. Wolch and M. Dear, eds. *The Power of Geography*, Boston: Unwin Hyman, pp. 197–221.

Wolch, J. (1990). *The Shadow State: Government and Voluntary Sector in Transition*. New York: The Foundation Centre.

Wolch, J. (2006). Foreword: Beyond the Shadow State? In: C. Milligan and D. Conradson, eds. *Landscapes of Voluntarism*. Bristol: Policy Press, pp. xii–xv.

13
JUST GREEN SPACES OF URBAN POLITICS
A pragmatist approach

Ryan Holifield

Introduction

Public parks and green spaces have long been important spaces of urban politics. Not only are they fought for and fought over, but they also often serve as settings for speeches, demonstrations and conflicts typically regarded as fundamental to democracy. Unsurprisingly, then, they have attracted attention in the growing research field of urban political ecology (see, e.g. Brownlow 2006; Byrne and Wolch 2009; Gandy 2012; Heynen *et al.* 2006). Urban political ecology, although shaped profoundly by its origins in radical political economy and science and technology studies, has become a diverse field with multiple varieties (see Heynen 2014, 2015). Consequently, it now brings a wide range of theoretical tools and concepts to the analysis of urban environmental politics, building on Marxist, Foucauldian, feminist, queer and other critical theories. In combination, this increasingly rich body of research has shown how urban environments are products of uneven power relations and political contestation at multiple scales, as well as how struggles over the urban environment connect with and contribute to broader projects of social and political change. However, while shedding light on the contested production, development and uses of parks and green space, this literature has said less about the ways that green spaces themselves function as spaces of urban democratic politics.

This chapter sketches the outline of an emerging *pragmatist urban political ecology* approach to analysing the politics that take place in and over urban green space. It builds on two distinct pragmatist traditions: first, American pragmatism, and in particular the revived engagement in geography and related social sciences with John Dewey; and second, the various strands of sociology that together compose 'French pragmatism'. I join others in arguing that these two traditions are complementary, and that in combination they provide useful and innovative concepts for analysing the politics of urban green space (e.g. Holden *et al.* 2013). Specifically, I contend that recent conceptualizations of the spatiality of democratic politics, grounded in American pragmatism, can deepen an emerging urban political ecology based on French pragmatic sociology. In turn, the latter provides a well-established, specific analytical framework that can help develop richer empirical accounts of this spatiality. In order to illustrate the potential for this combination, I turn to one of the most famous recent examples of political struggle in and over urban green space: the 2013 protests over and within Gezi Park and Taksim Square, in Istanbul, Turkey. These protests made international headlines when localized demonstrations against efforts to redevelop the park as a shopping mall escalated into protests against the national government, extending throughout the country and beyond.

To be clear, my argument is *not* that pragmatism should replace existing approaches to urban political ecology in the analysis of the politics of urban green space. On the contrary, this is an argument for pluralism, suggesting that pragmatist approaches can both open new lines of inquiry and help strengthen currently dominant approaches through mutual engagement and dialogue. In turn, I propose that a pragmatist urban political ecology should adopt the normative orientation to which other urban political ecologies have long been committed: favouring the rectification of injustices and the flourishing of democracy in the shaping of urban environmental futures.

A pragmatist urban political ecology

American pragmatism and urban political ecology

After a period in which it fell out of fashion in the academy, American pragmatism – in particular, the work of John Dewey – has recently undergone a revival. Although the early American pragmatists – including Charles Sanders Peirce, William James, George Herbert Mead, Dewey and others – were by no means a unified front, Bernstein (1992) identifies a set of five major characteristics that they shared in common: anti-foundationalism, or the notion that truth is not inherent or transcendent, but a function of what works in practice; the social character of knowledge; the central role of chance and contingency in the emergence and uptake of ideas; a commitment to fallibilism and experimentation; and an embrace of epistemological pluralism (see also Barnes 2008; Hepple 2008). Pragmatism profoundly influenced social and urban research in the early twentieth century, including the Chicago School of urban ecology, but by the 1950s it had become marginalized by the rise of analytical philosophy and the embrace of positivism in the social sciences. The radical social science that emerged in the 1970s also critiqued pragmatism for its evident inattention to 'macro' structures and forces, such as capital accumulation (see Bridge 2005; Barnes 2008). However, the 'neo-pragmatism' of philosopher Richard Rorty reignited interest in the tradition, and since the 1980s its influence on many fields has been growing.

Although engagement with pragmatism remained sparse within human geography until recently (see Cutchin 2008; Hepple 2008), during the past decade the outlines of distinctive pragmatist geographies, including urban geographies, have begun to emerge. Building on concepts from Dewey, Bridge (e.g. 2005, 2008) has argued that the city, and in particular the decentralized postmodern city, presents fertile conditions for the emergence and flourishing of a distinctive *transactional rationality*. This is a rationality of communicative engagement, oriented towards not instrumental ends, but enquiry and understanding. Moreover, Bridge argues that it is not confined to homogeneous communities or lifeworlds, and it is thus capable of addressing the heterogeneity that characterizes urban space. This rationality goes beyond the representational and discursive dimensions of communication, encompassing performativity, as well as relations with non-humans and non-discursive objects. For Cutchin (2008: 1565), the Deweyan notion of transaction also lends itself to a distinctive conception of *place*: first, "one in which the human and natural are continuously in transactive processes together", and second, one that "contains both the local natural/societal situation and the effect of extralocal situations to which it is connected". In turn, Bridge (2008: 1581) suggests that the concept of transaction points to "a different type of politics of place", in which the city becomes "an ecology of communicative spaces (both agonistic and accommodative)".

Meanwhile, American pragmatism has also profoundly influenced recent political thought within the interdisciplinary field of science and technology studies. Of particular importance is the work of Marres (2007), who identifies a basis for deepening this field's conceptualization of politics in Dewey's engagement with what Walter Lippmann famously called the "phantom public". Marres argues that for both Dewey and Lippmann, when problems can no longer be solved within established institutions, or by technocrats and experts alone, they require the democratic participation of a public as a practical necessity. For the purposes of analysis, this insight "directs attention to precise moments in which issues are opened up for outside involvement, and attempts are made to move processes of issue formation beyond institutional settings", to the everyday world (Marres 2007: 771–772). The public, however, does not exist prior to problems, or what Marres calls issues. Instead, the public emerges when transactions create problems, becoming "all those who are affected by the indirect consequences of transactions to such an extent that it is deemed necessary to have those consequences systematically cared for" (Dewey and Rogers 2012: 48). But problems or issues do not automatically bring publics into existence; instead, publics are constituted as claims are made about issues and their effects, in a process Marres calls *issue articulation*. As Barnett and Bridge (2013) elaborate, this process of claims-making is characterized by a distinctive transactional rationality: oriented towards enquiry and collective coordination, but also often shaped by competition among conflicting interests. Marres' idea of politics in terms of the 'trajectories' of issues has also influenced Latour (e.g. 2007, 2013), who has conceptualized politics as a distinctive movement of issues through time, in which the contested formation of publics constitutes a central moment (see also Harman 2014).

Pragmatist conceptions of transactional rationality and the issue-oriented constitution of democratic publics are central to recent reconceptualizations of radical democracy (e.g. Barnett and Bridge 2013; Bridge 2014; Marres 2012). For example, Barnett and Bridge (2013) argue that pragmatism offers a corrective to the single-minded focus on extra-institutional conflict that characterizes dominant conceptions of radical democracy. They contend that Deweyan pragmatism supports a broader conception of "radical democracy as a process of debate, discussion, and persuasion in public and oriented to concerted, collective action" (2013: 1024). However, this does not imply a return to consensus-oriented versions of deliberative democracy. Deliberation, so often cast only as a "vehicle for consensus formation", is in fact itself a "space of agonistic encounter" among "competing rationalities" (2013: 1024). In addition, pragmatism also allows theories of radical democracy to attend "to experimental practices through which alternative institutional designs are developed" (2013: 1024). While Barnett and Bridge build on Dewey to relocate agonistic encounters *within* institutions and deliberative processes, Marres (2012: 15) brings Dewey together with the radical democratic theory of Chantal Mouffe to identify both how "political conflict and strife unfold on the plane of objects" and how this object-oriented agonistic politics can itself "be productive for democracy".

Barnett and Bridge further propose that Deweyan pragmatism suggests distinctive conceptualizations of the *spatiality* of democratic politics. One dimension of this is the spatiality of processes of public formation, as described above. When a public comes into being, through claims-making about who or what is affected by a particular issue, its spatiality is frequently an essential characteristic. Sometimes, as Cutchin (2008: 1566) contends, the process of public formation focuses on problems in particular places: "place processes and problems often implicate a place-based public, their agents, and political action". On the other hand, the formation of a public might involve advancing claims about how a seemingly localized issue affects a much wider, more spatially dispersed range of actors. Allen (2008) cites the example of anti-sweatshop activism, in which activists constituted new, widely dispersed oppositional publics by calling attention to the entanglement of local consumption in global chains of production.

The spatiality of public formation also encompasses the transactional spaces in which claims circulate, ranging from discussions and meetings in private homes or businesses, to demonstrations or performances in public spaces, to the more dispersed spaces produced by networks of media and communication. Barnett and Bridge (2013: 1032) argue that these "different transactional spaces of public action" are characterized by different "intensities" and can be analysed in terms of different combinations of "concentration, dispersal, and distribution".

As "spaces of heightened transactional intensity", urban spaces play a distinctive role not only in the formation of publics, but also in a second aspect of politics emphasized by Barnett and Bridge (2013: 1031): democratic experimentation and innovation. The focus here is on efforts to translate publics into institutions and collectivities capable of bringing about change. Again, what distinguishes the significance of urban space in this argument is not its scale, but the unique ways it concentrates, clusters and juxtaposes heterogeneous consequences and claims. Because urban space provides distinctive opportunities for "learning to live together with difference through ordinary exposure to alterity" and thus for developing "capacities to acknowledge the claims of others", it is fertile ground for innovation in democratic institutions, practices and processes (Barnett and Bridge 2013: 1034). Of course, cities are not the only spaces important for public formation or democratic innovation. Nonetheless, Barnett and Bridge (2013: 1036) maintain that urban space plays a distinctive role in democratic politics, by virtue of the "diverse qualities of publicness" it gathers together.

Although urban political ecology and related research on the spaces of urban environmental politics have begun to show the influence of readings of American pragmatism developed within urban geography and science studies, at this point the literature remains small and thus ripe for further development. Few existing studies aim explicitly to bring this pragmatist tradition into urban political ecology; one recent example is the effort of Holifield and Schuelke (2015) to trace the constitution of publics in conflict over the future of an urban river, drawing on the object-oriented political thought of Marres and Latour. But recent scholarship that develops the idea of politics as a trajectory of issues, such as the research of Donaldson *et al.* (2013) on flood risk management, also provides analytical frameworks that could inform a pragmatist urban political ecology. Others have built on Deweyan pragmatism to rethink communicative engagement in the planning of sustainable cities (e.g. Holden 2008). In addition to these, there exists a wider body of research on the urban environment that clearly resonates with the experimentalism and object-oriented politics of pragmatism, such as the influential "cosmopolitical experiment" of Hinchliffe *et al.* (2005) to sketch an emergent politics of urban wildlife. Another important source of theory about the parts non-humans play in the formation of publics for environmental politics is Marres's (2012) conception of "material participation", which emphasizes the changing roles of technological devices. I will return to the prospects for an engagement between Deweyan pragmatism and urban political ecology in the conclusion, but first I turn to the development of another pragmatist tradition that has begun to influence thinking on the spaces of urban environmental politics: French pragmatic sociology.

French pragmatist sociology and urban political ecology

French pragmatism, shorthand for the pragmatic sociology of critique, the sociology of critical capacities or the sociology of engagement, originated with a set of departures from the critical sociology associated with Pierre Bourdieu. Rather than seeking to unveil relations of domination through a critique of its own, pragmatic sociology analyses how ordinary actors challenge perceived injustices by critiquing others and justifying their own claims, in situated processes of argumentation and negotiation (Boltanski and Thévenot 2006). In order to give legitimacy to these critiques and justifications, actors draw on a plurality of co-existing, historically specific and sometimes incommensurable *orders of worth*, or logics of the common good. Drawing on extensive empirical research, Boltanski and Thévenot (2006) identified and elaborated six major orders of worth characterizing contemporary Western society: *civic, market, industrial, domestic, inspiration* and *fame*; they later added *projective* and *green* (Thévenot *et al.* 2000). In situations of conflict, actors draw not on a shared *habitus* – a key concept in Bourdieu's critical sociology – but on the distinctive principles of valuation, personal qualities and actions, and tests of legitimacy associated with these orders of worth.

There is growing interest in French pragmatic sociology as an approach to conceptualizing and analysing the spaces of urban politics (e.g. Fuller 2013, Holden and Scerri 2015). Fuller (2013), for example, argues that French pragmatism can address deficiencies and blindspots in both 'assemblage' conceptions of urban politics and in recent accounts of neoliberalization and post-political urban governance. Specifically, pragmatist sociology offers concepts for analysing "everyday negotiation between actors and how social coordination is temporarily created" (Fuller 2013: 639). By studying both the arguments deployed by urban political actors and the forms of justification to which they appeal, we can better understand "how certain actors acquire hegemonic status and the mechanics of consensus building, subordination and compromise" (Fuller 2013: 652). In addition, Fuller contends that French pragmatism allows us to connect the "micro-level practices" at the heart of much analysis of assemblages to "broader causal factors", including the capital accumulation process (2013: 642, 653). The latter claim requires further scrutiny and development. It is not clear, for instance, how analysing the values and orders of worth prioritized within capital accumulation processes, and then embedded in economic actors' arguments and critiques, would provide an adequate basis for *causal* explanations of urban politics. Nonetheless, Fuller contends persuasively that French pragmatism offers new ways of linking localized debates that produce configurations of urban space in particular sites with processes shared in common by diverse cities and trajectories of urbanization.

An urban political ecology grounded in French pragmatism is beginning to take shape. Blok and Meilvang (2015), for example, apply French pragmatism in their analysis of political activism in response to a large project for sustainable urban development in Copenhagen. In the hands of planners and government officials, the project has been dominated by compromise among a set of orders of worth: primarily green, market and civic, but also fame (the city's image) and industrial. Although the project has not generated significant public protests, media coverage or partisan conflict, it has nonetheless sparked a diverse array of small-scale challenges, each deploying images to support the claim that the plans would threaten recreational activities, the aesthetics of the cultural landscape or local biodiversity. What links these heterogeneous engagements together, the authors argue, is that they constitute efforts to make the personal public: that is, to translate familiar attachments into critiques and justifications oriented towards legitimating an alternative vision of the common good. Centemeri (2015) argues that valuations of the environment based on familiar attachments resist translation into arguments for legitimacy, in part because they are grounded in unsubstitutable human-environment relations. However, Blok and Meilvang (2015: 27) argue that these engagements, "[e]ven when ostensibly 'failing', … succeed in opening up new public spaces and bringing forth conflicting valuations as to what should count as green and 'sustainable'". Such engagements constitute experiments in a "new urban political ecology" in which "justice in the green city hinges on finding ways of building commonality, and modes of differing, which will be more hospitable to a wider set of personal attachments to more-than-human urban ecologies" (2015: 34).

Although the origins and emphases of American and French pragmatist approaches differ, they complement each other in ways that could inform a possible "transatlantic" urban political ecology. In their argument for a "trans-Atlantic alliance" of pragmatisms, Holden *et al.* (2013: 3) find "in the interstices of the French and American approaches a strong program that founds a social critique of injustice in the ways that actors contribute to 'qualifying' the 'reality' that encompasses public political debates". Both traditions, they note, embrace democracy, pluralism and an optimism about the capacity of ordinary people to contribute to the shaping of urban futures. Thévenot (2011: 61) has argued that despite the differences among the traditions (and between them and critical sociology), there are many good reasons for them to "join forces" (see also Blok 2013).

Holden *et al.* (2013) emphasize several advantages of French pragmatism, all of which could also benefit the analysis of spaces of urban environmental politics. First, it provides a conceptual framework for analysing the specific "grammars of public disagreement" deployed in argumentation over problems, focusing on the models of justification to which participants appeal (Holden *et al.* 2013: 6). Second, it attends to the assemblages and arrangements that enable, constrain and condition this argumentation, including the ways some modes of justification can exercise "tyranny" over others (Thévenot 2011). French pragmatism also provides the helpful concept of the *test*, which refers to the means through which actors determine whether their existing knowledge can effectively inform action in the real world. But Holden *et al.* (2013) note that the pragmatism of Dewey lends itself to a normative approach to analysis, which French pragmatism has historically avoided. In other words, American pragmatism insists that analysis cannot detach itself completely from a position on what the "common good" ought to entail. This does not mean arguing for a specific set of institutions or arrangements, but rather for maintaining the ideal of a public committed to democratic engagement and experimentation in the interest of finding a shared common good.

A brief illustrative example: Gezi Park as a space of urban politics

How might such a synthesis address the *spatial* dimensions of urban politics – and in particular the politics of green space? Here, I return to the twofold concept of transactional space introduced by Barnett and Bridge (2013), using the case of the Gezi Park/Taksim Square demonstrations of 2013 as an example to explore potential paths for an analysis drawing on both American and French traditions of pragmatism. The first dimension to consider is the transactional space of public formation. One empirical question would be how a heterogeneous set of actors transformed Gezi Park – in the first instance, a specific, localized urban green space and the site of protests over redevelopment – into a space for the formation of a much larger, broader and diverse public, united in opposition to a diverse set of changes at multiple geographical scales (Özen 2015). Certainly, a violent crackdown by police in response to the initial demonstrations played the most immediate role in sparking the transformation of the protest from the defence of Gezi Park into demonstrations against the national government (Kuymulu 2013). Kuymulu (2013) also contends that the capacity of activists to "jump scales" and form this more spatially diffuse public was facilitated because the Gezi Park project was part of a citywide, and indeed nationwide, hegemonic project of capital accumulation through urban redevelopment – in fact gentrification. But of course the formation of this spatially extensive public required spreading the news, primarily via the technology of social media networks, and it drew energy from the accumulation of numerous other "simmering grievances" against the increasingly authoritarian national government (Inceoglu 2014: 25; Özen 2015). Ultimately, the public that threats to Gezi Park helped bring into being occupied a wide, dispersed network of sites in multiple cities throughout and beyond the country.

The second spatial dimension of democratic politics that Barnett and Bridge highlight is the spatiality of institutional experimentation and democratic innovation. Again, the argument is that cities are primary sites for such experimentation and innovation, and the "urban" is significant not because it represents a smaller or "more local" scale, but because it is a "figure for practices of learning to live together with difference through ordinary exposure to alterity" (Barnett and Bridge 2013: 1034). Numerous accounts identify the latter as one of the central characteristics of the Gezi Park/Taksim Square protests, including this rich description by Inceoglu (2014: 26):

> The two-week long occupation of Gezi Park and Taksim Square converted this recreational area into a radical democratic public sphere. Testimony from the time records astonishment among some protesters at moments when they encountered the 'other' and were not disturbed by this. Powerful examples of such moments from the occupation include Kurdish groups holding Kurdish flags with Öcalan's photo in collaboration with Turkish nationalists holding Turkish flags with Ataturk's photo, or football fans reputed to be homophobic and sexist standing on the barricades next to LGBTIs.

As this heterogeneous public concentrated in the spaces of Gezi Park/Taksim Square and other public green spaces, the latter subsequently became spaces for democratic experimentation and innovation. Inceoglu (2014: 26), for example, emphasizes the neighbourhood park forums that took place following the initial protests: events in parks "where individuals took turns to speak on an open microphone and shared their reflections". Although the park forums declined, as place-based experiments they inspired subsequent efforts to create more inclusive, less hierarchical forms of democratic politics in Turkey, including "participatory forms of electoral monitoring and … the emergence of #occupyCHP, a movement to bring the spirit of horizontal activism into one of the mainstream Turkish political parties" (Inceoglu 2014: 24). What I am suggesting here is that a pragmatist urban political ecology needs to attend to contested green spaces not simply as spaces that actors 'fight over', but also as spaces that through their distinctive properties can foster both public formation and democratic experimentation.

What, then, would concepts from the French pragmatic sociology of critique add to such an analysis? For a start, they would suggest additional questions: How were efforts to develop the park, and to protect the park, justified in appeals to a common good? What kinds of arguments and critiques were made as the park's redevelopment became an issue, and how did some orders of worth prevail over others? One intriguing aspect of the trajectory of Gezi Park is the changing role of distinctively *environmental* politics, or the 'politics of trees', at different stages of the controversy. Using the language of French pragmatism, the initial protests critiqued market values and emphasized green and civic orders of worth, focusing on threats to the space and ecology of the park. However, appeals to green justifications with respect to a single localized space were clearly insufficient on their own to mobilize sufficient numbers of protesters to this vision of the common good. Ete and Taştan (2014: 22) include the following intriguing quote from a protester:

> I did not participate in the protests in the first four days. During these first four days, the protests were about trees. Ok, I have respect for trees, in the environment, but since Friday there has been a grassroots movement here and I have been here since that day. … Ok, the fight broke out because of a tree but this has nothing to do with the tree itself. Well, it is, in a sense, related to the environment but the people protesting here generally want the overthrow of this government.

The protester's words reflect the ambiguity of the park as a space of urban environmental politics. On the one hand, the generation of a broader, more heterogeneous and spatially dispersed public required moving away from the green world of justification, and to a lesser degree even from the critique of market values, towards a critique of the authoritarian government in the widest civic terms. As Kuymulu (2013: 276) observes: "Slogans like 'Capital be gone, Gezi Park is ours' were replaced by 'Erdogan resign, government resign' and 'shoulder to shoulder against fascism'."

On the other hand, as summer ended and the protests died down, the park's constitution as a green, ecological space – or its 'trees' – would ultimately remain pivotal to the issue's trajectory. Analyses of the shifting politics of the protests suggest that the national government was able to reassert control over the situation through an approach that involved both changes in the world of justification and an explicitly spatial strategy. As Kuymulu (2013: 276) suggests, the government ultimately had to recognize that its effort to justify the value of its project in market terms had failed: "whatever arguments its proponents put forth to justify the construction of another shopping mall in the historic city centre, they could not legitimate building the 94th shopping mall in Istanbul". In addition, as Özen (2015) points out, part of the AKP government's approach to deflating the oppositional movement was to shift attention back to the space and ecology of the park itself, accepting the legitimacy of the initial 'environmentalist' opposition to the park's redevelopment – and thus the green world of justification – but rejecting as illegitimate the critiques of the government that arose from the heterogeneous public sparked by the violent actions of the police. So even if the park's threatened trees initially faded from view and discourse as the protests grew, they returned as the government determined that the only legitimate oppositional critiques were the original appeals for ecological protection of the park as green space.

Although this brief illustration can only hint at the possibilities for deeper analysis using a pragmatist urban political ecology, it suggests not only that attending to strategic shifts in argumentation can help make sense of spatial strategies in urban ecological politics, but also that attending to spatial strategies can help us make sense of strategic shifts in argumentation. In this case, a shift from green justifications to civic justifications helped the protesters transform a strategy focused on the occupation of a particular green space to the creation of a widely dispersed extra-local space of opposition that, in turn, enabled numerous additional public spaces to become sites of democratic experimentation. For the government, however, the abandonment of its market justification and the acceptance of the protesters' original green justifications helped it refocus attention on the space of the park itself and thereby delegitimize the new spaces of democratic dissent that the protests engendered. Investigating the links between modes of justification and such 'scale-jumping' or space-claiming strategies would be at the heart of a pragmatist urban political ecology approach.

Concluding remarks

The aim of this chapter has been to lay out some basic principles for a pragmatist urban political ecology approach to analysing the politics of urban green space. This is a pluralist approach that draws on both Deweyan pragmatism and French pragmatism, acknowledging their differences but accepting the proposition that they can be brought together in productive ways. Building on the former tradition, the chapter proposes following Marres and Latour in analysing the politics of urban green spaces in terms of the trajectories of issues, paying particular attention to the processes through which publics are formed, and to the distinctive roles that non-humans of various kinds play in these processes. It also adopts the premise of Barnett and Bridge that public formation and democratic experimentation are inherently spatial processes, and that this spatiality should be a focus for empirical analysis. In turn, the chapter suggests that we can gain a richer understanding of this spatiality by drawing on French pragmatic sociologies and attending to the specific critiques, justifications, forms of engagement and orders of worth deployed in political controversies and compromises. It is also an approach that emphasizes not only how political processes shape urban green spaces, but also the converse.

Again, the argument is for the inclusion of pragmatism in a pluralist urban political ecology, rather than a call for pragmatism to replace existing approaches, or a call to synthesize other approaches with pragmatism. However, I also promote a productive tension within this pluralism. On the one hand, pragmatist approaches can probe the limits of analyses that explain the politics of urban green space in terms of class hegemony, capital accumulation or structural categories of various kinds. Similarly, it would be productive to situate the concept of metabolism, which emphasizes the transformation of nature into use values (see Heynen 2014), with respect to the broader concept of transaction, which addresses an arguably wider range of human-environment relationships. In addition, the French pragmatist approach invites urban political ecology to compare and calibrate its own critiques with the critiques and justifications of actors on the ground.

On the other hand, other forms of urban political ecology issue important challenges to the proposed pragmatist approach. A pragmatist urban political ecology needs to demonstrate that it can account for the inequalities of power, capacity and access that political ecology has long emphasized; recent rapprochements between the sociology of critique and critical sociology provide a possible pathway to such accounts (e.g. Thévenot 2011). If, for example, constituting publics requires not just the emergence of problems but also the making of claims, as Marres suggests, then how can and should we address individuals or communities indirectly affected by transactions, yet unable to articulate or mobilize a public? A pragmatist urban political ecology should therefore attend both to processes of public formation and to conditions that constrain or prevent these processes from taking place. Given its premise that urban space plays a distinctive role in democratic politics, the pragmatist approach would also need to engage with the critique that urban political ecology has emphasized the analysis of cities at the expense of urbanization processes (Angelo and Wachsmuth 2015). Furthermore, it needs to probe more deeply into the means and ends of different kinds of experimentation and innovation in urban green space, since sometimes urban 'experiments' simply perpetuate the status quo (Karvonen *et al.* 2013). Finally, urban political ecology accounts typically also embrace a normative commitment to overcoming such inequalities. I advocate for a pragmatist urban political ecology that shares this commitment, but a key frontier and challenge – not just for pragmatism, but for all varieties of urban political ecology – will be finding ways to translate this commitment into meaningful interventions that can make a difference beyond the academy.

References

Allen, J. (2008). Pragmatism and Power, or the Power to Make a Difference in a Radically Contingent World. *Geoforum*, 39(4), pp. 1613–1624.

Angelo, H. and Wachsmuth, D. (2015). Urbanizing Urban Political Ecology: A Critique of Methodological Cityism. *International Journal of Urban and Regional Research*, 39(1), pp. 16–27.

Barnes, T.J. (2008). American Pragmatism: Towards a Geographical Introduction. *Geoforum*, 39(4), pp. 1542–1554.

Barnett, C. and Bridge, G. (2013). Geographies of Radical Democracy: Agonistic Pragmatism and the Formation of Affected Interests. *Annals of the Association of American Geographers*, 103(4), pp. 1022–1040.

Bernstein, R.J. (1992). The Resurgence of Pragmatism. *Social Research*, 59(4), pp. 813–840.

Blok, A. (2013). Pragmatic Sociology as Political Ecology: On the Many Worths of Nature(s). *European Journal of Social Theory*, 16(4), pp. 492–510.

Blok, A. and Meilvang, M.L. (2015). Picturing Urban Green Attachments: Civic Activists Moving Between Familiar and Public Engagements in the City. *Sociology*, 49(1), pp. 19–37.

Boltanski, L. and Thévenot, L. (2006 [1987]). *On Justification: Economies of Worth*. Translated by Catherine Porter. Princeton, NJ: Princeton University Press.

Bridge, G. (2005). *Reason in the City of Difference: Pragmatism, Communicative Action and Contemporary Urbanism*. London and New York: Routledge.

Bridge, G. (2008). City Senses: On the Radical Possibilities of Pragmatism in Geography. *Geoforum*, 39(4), pp. 1570–1584.

Bridge, G. (2014). On Marxism, Pragmatism and Critical Urban Studies. *International Journal of Urban and Regional Research*, 38(5), pp. 1644–1659.

Brownlow, A. (2006). An Archaeology of Fear and Environmental Change in Philadelphia. *Geoforum*, 37(2), pp. 227–245.

Byrne, J. and Wolch, J. (2009). Nature, Race, and Parks: Past Research and Future Directions for Geographic Research. *Progress in Human Geography*, 33(6), pp. 743–765.

Centemeri, L. (2015). Reframing Problems of Incommensurability in Environmental Conflicts Through Pragmatic Sociology: From Value Pluralism to the Plurality of Modes of Engagement with the Environment. *Environmental Values*, 24(3), pp. 299–320.

Cutchin, M.P. (2008). John Dewey's Metaphysical Ground-map and its Implications for Geographical Inquiry. *Geoforum*, 39(4), pp. 1555–1569.

Dewey, J. and Rogers, M.L. (2012). *The Public and its Problems: An Essay in Political Inquiry*. University Park, PA: Penn State Press.

Donaldson, A., Lane, S., Ward, N. and Whatmore, S. (2013). Overflowing with Issues: Following the Political Trajectories of Flooding. *Environment and Planning C: Government and Policy*, 31(4), pp. 603–618.

Ete, H. and Taştan, C. (2014). *The Gezi Park Protests: A Political, Sociological, and Discursive Analysis*. Istanbul: SETA.

Fuller, C. (2013). Urban Politics and the Social Practices of Critique and Justification: Conceptual Insights from French Pragmatism. *Progress in Human Geography*, 37(5), pp. 639–657.

Gandy, M. (2012). Queer Ecology: Nature, Sexuality, and Heterotopic Alliances. *Environment and Planning D*, 30(4), pp. 727–747.

Harman, G. (2014). *Bruno Latour: Reassembling the Political*. London: Pluto Press.

Hepple, L.W. (2008). Geography and the Pragmatic Tradition: The Threefold Engagement. *Geoforum*, 39(4), pp. 1530–1541.

Heynen, N. (2014). Urban Political Ecology I: The Urban Century. *Progress in Human Geography*, 38(4), pp. 598–604.

Heynen, N. (2015). Urban Political Ecology II: The Abolitionist Century. *Progress in Human Geography*, 40(6), pp. 839–845.

Heynen, N., Perkins, H.A. and Roy, P. (2006). The Political Ecology of Uneven Urban Green Space: The Impact of Political Economy on Race and Ethnicity in Producing Environmental Inequality in Milwaukee. *Urban Affairs Review*, 42(1), pp. 3–25.

Hinchliffe, S., Kearnes, M.B., Degen, M. and Whatmore, S. (2005). Urban Wild Things: A Cosmopolitical Experiment. *Environment and Planning D*, 23(5), pp. 643–658.

Holden, M. (2008). The Tough Minded and the Tender Minded: A Pragmatic Turn for Sustainable Development Planning and Policy. *Planning Theory and Practice*, 9(4), pp. 475–496.

Holden, M. and Scerri, A. (2015). Justification, Compromise and Test: Developing a Pragmatic Sociology of Critique to Understand the Outcomes of Urban Redevelopment. *Planning Theory*, 14(4), pp. 360–383.

Holden, M., Scerri, A. and Owens, C. (2013). More Publics, More Problems: The Productive Interface Between the Pragmatic Sociology of Critique and Deweyan Pragmatism. *Contemporary Pragmatism*, 10(2), pp. 1–24.

Holifield, R. and Schuelke, N. (2015). The Place and Time of the Political in Urban Political Ecology: Contested Imaginations of a River's Future. *Annals of the Association of American Geographers*, 105(2), pp. 294–303.

Inceoglu, I. (2014). The Gezi Resistance and its Aftermath: A Radical Democratic Opportunity? *Soundings*, 57, pp. 23–34.

Karvonen, A., Evans, J. and van Heur, B. (2013). The Politics of Urban Experiments: Radical Change or Business as Usual? In: M. Hodson and S. Marvin, eds. *After Sustainable Cities*. London: Routledge, pp. 104–115.

Kuymulu, M.B. (2013). Reclaiming the Right to the City: Reflections on the Urban Uprisings in Turkey. *City*, 17(3), pp. 274–278.

Latour, B. (2007). Turning Around Politics: A Note on Gerard de Vries' Paper. *Social Studies of Science*, 37(5), pp. 811–820.

Latour, B. (2013). *An Inquiry into Modes of Existence*. Cambridge, MA: Harvard University Press.

Marres, N. (2007). The Issues Deserve More Credit: Pragmatist Contributions to the Study of Public Involvement in Controversy. *Social Studies of Science*, 37(5), pp. 759–780.

Marres, N. (2012). *Material Participation: Technology, the Environment and Everyday Publics*. London: Palgrave Macmillan.

Özen, H. (2015). An Unfinished Grassroots Populism: The Gezi Park Protests in Turkey and Their Aftermath. *South European Society and Politics*, 20(4), pp. 533–552.

Thévenot, L. (2011). Power and Oppression from the Perspective of the Sociology of Engagements: A Comparison with Bourdieu's and Dewey's Critical Approaches to Practical Activities. *Irish Journal of Sociology*, 19(1), pp. 35–67.

Thévenot, L., Moody, M. and Lafaye, C. (2000). Forms of Valuing Nature: Arguments and Modes of Justification in French and American Environmental Disputes. In: M. Lamont and L. Thévenot, eds. *Rethinking Comparative Cultural Sociology: Repertoires of Evaluation in France and the United States*. Cambridge: Cambridge University Press.

14
FROM SUSTAINABILITY TO RESILIENCE
The hidden costs of recent socio-environmental change in cities of the Global North

David Saurí

Introduction

'Sustainability' has encapsulated the response of capitalism to current socio-environmental challenges at all scales, during the last four decades. With over 300 definitions, sustainability remains perhaps one of the more vexing but at the same time attractive theoretical constructs ever formalized to describe what, for many, is an oxymoron (Krueger and Gibbs 2007: Redclift 2005): the idea that (economic) growth can be perfectly compatible with environmental well-being. This principle finds an operational arm in the postulates of ecological modernization, the use of specific technical and managerial skills to overcome environmental conditions that could threaten future growth. To gain further legitimacy, the construct has been painted green and, as Naredo and Gómez-Baggethun (2015) argue, the initial idea of growth versus the environment has been changed to a notion of growth for the environment. Technology and the market have been the two pillars upon which this change of emphasis has been made possible.

One of the common shortcomings attributed to sustainability is its lack of attention to social issues and particularly to the relationships between poverty and environmental degradation. In part, this may have been induced by a certain lack of interest in social and distributive matters by proponents of ecological modernization, probably due to the origins of ecological modernization in the affluent and relatively egalitarian societies of Northern and Western Europe. Hence the harmonization of economic growth and environmental protection may have been achieved under certain circumstances that allow for the achievement of social welfare and social cohesion. Beyond these societies, however, the social leg of sustainability seems to have been left limping.

One major argument of this chapter is that sustainable development has not fared well in the face of the economic and environmental turbulence of the last decades and, as a result, its former hegemonic role in mainstream environmental planning and management may be losing ground to more powerful concepts such as resilience. Sustainability, with its implicit promise of a better world for all generations (current and future) appears to be ill equipped to deal with the calamities of all sorts affecting the world economy and environment. Still more significantly, sustainability cannot confidently deliver solutions for the socio-environmental uncertainties of the future, with climate change leading the pack. Resilience, in contrast, abandons the promise of better presents and futures for all (and along these, the hope in collective projects) and puts the capacities of individuals at the forefront of actions to manage socio-environmental risks and uncertainties. Long-sought human and social aspirations, such as security and solidarity, are thus sacrificed and we appear to be back to the Darwinian survival of the fittest.

Another major point of this chapter is that, in many cities of the Global North, the change from sustainability to resilience occurs in a context of a citizenry already disciplined in socio-environmental terms through the application of various policies devoted to improve the urban landscape and the efficiency of resource use. Therefore, and despite assuming the costs of ecological modernization, citizens and especially the less well-off are not only deprived of the benefits of urban environmental improvement but, worse still, are told that current and future risks and uncertainties preclude any improvements in their situation. As an example of this double punishment, first to bear the costs with the promise of a better future and, second, to abandon all expectations about this future, this chapter briefly discusses the issues of water and energy poverty in the European Union. I seek to show how policies of ecological modernization, and especially those based on economic instruments, not only may have increased urban inequalities but may have outrun their usefulness for reformist urban governance as well.

My main argument will be that since urban populations, and among them the urban poor, can hardly be accused of being the leading edge of unsustainable practices and, given the persistence (real, perceived or plainly manipulated) of environmental threats, arguments based on sustainability that include a promise of reward if certain rules are followed (i.e. to reduce resource consumption), are being abandoned. Simultaneously, arguments based on resilience are gaining more and more strength, especially concerning climate change adaptation and related fields (see for example Folke 2006; Folke et al. 2003; Walker et al. 2004).

This chapter is organized as follows. I begin with a brief discussion of the effects of sustainability and ecological modernization on urban environments, focusing on basic resource flows such as water and energy. I argue that of the three traditional legs attached to sustainability, the economic and environmental legs have generally performed reasonably well in many respects, especially when comparing the current state of certain parameters with their recent past. Thus the economic engine fuelling the expansion of cities has not been challenged while environmental quality has improved in certain respects. Sustainability promised somehow a brighter social future as well. Social sustainability, however, has been left behind and does not generally show similar successful performances. This failure can be traced to the crises and turbulences of capitalism which increasingly are associated with the crises and turbulences of extreme 'natural' events. Sustainability has ceased to be an effective guiding logic and is now being replaced with the more convenient concept of resilience. Much of this chapter is devoted to a discussion of this now hegemonic concept which not only fails to give adequate credit to the real promoters of urban sustainability for their achievements, but also punishes them by denying any hope for achieving just socio-ecological change.

Sustainability as ecological modernization

In many cities of the Global North, ecological modernization (see Buttel 2000 and York and Rosa 2003) has left a visible imprint on urban landscapes. On the one hand, this imprint has been left at the level of the urban landscape itself, especially through programmes (usually with generous public funding) of urban renewal and revitalization. Thus previously degraded or derelict land in low-income neighbourhoods has been transformed into amenable urban space often under the banner of sustainable urban development. In turn, improved urban areas have whetted the appetite of real estate interests that have been quite successful in capturing the (positive) externalities of large renovation projects, thus creating patterns of 'green gentrification' (Anguelovski 2014) in which sustainability and all accompanying paraphernalia (from 'ecological shops' to other amenities) have become items for the enjoyment of the more well-off, usually to the detriment of the poorer former residents.

On the other hand, ecological modernization has left a deep impact on urban flows. The observation of recent trends in the consumption of resources such as water and energy appears to indicate a certain slowdown in the prodigal metabolic processes of urban areas of the Global North. Although not easily attributable to single causal explanatory factors, declining resource consumption is seen in many cases as primarily a product of sustainability policies that promote objectives of conservation, savings and increased efficiency, thereby easing pressures on the environment and fulfilling the promises of a gentler relationship with nature now and, especially, in the future.

Disciplining consumers through prices and taxes has been the favoured course of action, in perfect harmony with one of the basic tenets of ecological modernization: the application of economic instruments to solve environmental problems. Technological improvements, with costs also mostly passed on to consumers, have constituted another stepping stone of the sustainability quest. Since there is a general mistrust of the public's willingness to change in the desired direction, disciplining through the forced imposition of costs has been seen as the more convenient and effective path towards sustainability.

Environmental restructuring through ecological modernization, however, may be producing other unintended consequences for the providers and consumers of resource flows. In the water cycle, for example, municipalities and private companies are becoming increasingly stressed by falling revenues caused by declining consumption. To compensate for this, municipalities and companies may obtain from regulators additional increases in price which in turn may push consumption further downward, especially among the urban poor. During periods of economic crisis, low or no income households may cease to pay their water bills and run the risk of supply cut-off. The same argument could be developed for energy, although in this case there may be important differences with water; there are limits to the restriction of water service (water has been declared a 'human right', energy has not). Much more extensive than water poverty, energy poverty stems from price increases leading to declining consumption leading to declining revenues leading to further price increases. This vicious circle makes energy, in the form of fuel or electricity, ever more unaffordable to the poor. In both cases policies of ecological modernisation, targeting basic urban flows but blind to distributional impacts, increase social inequalities in access to basic resources.

Competing notions of resilience

There may be continuities between sustainability and resilience (see for example the reviews by Brown 2014; Brown and Westaway 2011; MacKinnon and Driscoll Derickson 2012), but in many respects it can be argued that the notion of resilience and its hegemonic interpretation within environmental science and policy implies a fundamental shift away from common understandings of sustainability. Resilience introduces us to an uncertain and dangerous world more akin to the survival of the fittest than to the cornucopian view of happiness and prosperity for all (in due time) that can be implicitly read from sustainability. While sustainability emphasizes steady social and environmental progress, resilience is more inclined to serve as guide for chaotic and turbulent futures. Hence, resilience could be understood as 'The Art of Living Dangerously' as Evans and Reid (2014) subtitle their book on the subject. But depending on different intellectual and disciplinary traditions, resilience is also open to multiple interpretations.

Resilience comes from at least three different intellectual and disciplinary traditions (for recent reviews see, for instance, Andres and Round 2015; Brown 2014; Olsson et al. 2015).

First, there is the engineering view which centres on single, inanimate objects; in this view resilience refers to the property of certain materials to revert to their original form after some sort of external shock. A second view emerges from the physical and life sciences and particularly from the systems-theory work of ecologist C.S. Holling. According to Holling (1973: 14), resilience (in opposition to stability) "measures the persistence of systems and of their ability to absorb change and disturbance and still maintain the same relationships between populations or state variables". In ecology the concept of resilience represents a fundamental theoretical shift from traditional notions of ecosystem equilibrium, succession and climax to notions of chaos, non-linear behaviour and abrupt ecological change (see, for example, Holling 1978 and Lomolino 2000). Finally, a third view of resilience, that of psychology, favours a more human and social approach. For psychology, resilience revolves around two key components: adversity and positive adaptation (Fletcher and Sarkar 2013). While adversity refers mostly to external stimuli, positive adaptation is said to depend on the social, political and cultural contexts of the individual. Furthermore, research on resilience in children, for example, emphasizes that resilience may not be just a purely individual characteristic, rather it exists and develops within the social and political environment of the child (Ungar 2008).

Of the three views of resilience summarized above, the ecological view stands undoubtedly as hegemonic in current socio-environmental discourse (Adger 2000). There are several reasons that may explain this dominance. First of all, the ecological view treats resilience as a natural condition and thus immune to human and social conditions. Second, and related to the above, the naturalization of the concept, one of the many concepts imported to social sciences from physical and life sciences (Andres and Round 2015), provides a thick armour against any criticism and especially against criticism of deregulated environmental governance. Walker and Cooper (2011) develop a convincing argument to explain the widespread uptake of the resilience concept in the environmental community. The chronic instability of modern capitalism, periodically affected by downturns and crises, fits poorly with explanatory frameworks based on equilibrium and (predictable) steady progress. Nature, coincidentally, is also intrinsically chaotic and unpredictable. Resilience (in the ecological interpretation of the concept), therefore, has emerged as a plausible guiding principle for these new times. Contrary to sustainability and its emphasis in the well-being of current and future generations, resilience abandons long-term expectations, including those based on equity and fairness. Resilience implies a politics of despair. Additionally, Walker and Cooper conclude that a future plagued with perturbations impossible to anticipate makes planning appear not only useless but, worse still, a dead weight on the ingenuity of (certain) human minds to find ways out of dangerous situations.

The relentless expansion of resilience

The rapid adoption of the concept of resilience during the last decade can be compared to the heyday of sustainability two or three decades earlier. In part this is related to the increasing attention given to climate change adaptation (already ahead of mitigation in many agendas), particularly in the management of large natural disasters that traditional engineering measures no longer adequately address. Resilience is increasingly occupying the position that sustainability used to have as a referential expression in urban environmental research and policy and, similarly to sustainability, is becoming an all-encompassing concept for the natural and social sciences alike (Adger 2000; Walker et al. 2004).

The causes behind the progressive decline of sustainability and the parallel uptake of resilience are many, ranging from yet another manifestation of the crisis and instability of capitalism since 2007, to the worsening of global socio-environmental conditions – both contradicting the supposed steady progress (environmental and else) implicitly associated with sustainability. In fact, the crises of the last three decades of neoliberal rule have broken the promise that democracy and capitalism would bring prosperity and much improved environmental conditions for all, as argued by the proponents of ecological modernization. The inability of capitalism to fulfil these expectations is now disguised through the turn to resilience theory, naturalizing socio-ecological conditions in the form of chaos, non-equilibrium patterns, looming socio-environmental disasters and the like – conditions that are intrinsically socially produced.

What is somehow intriguing is the speed with which resilience has been able to infiltrate and gain hegemonic status in the arena of international organizations, particularly those focusing on environmental and climate change issues. A few examples give us an idea of this very fast advance in ascendancy and hegemony.

One of the most influential documents on resilience and urban resilience in particular was published by the World Resources Institute (WRI) in 2008 with the cooperation of the United Nations and the World Bank; this report asks the private sector to lead the way towards the creation of economic, social and environmental resilience in response to the challenges of poverty and climate change (WRI 2008). In another influential report in 2010, ICLEI and other international groups working at local scales launched the 'Resilient Cities' Programme (renamed in 2012 as the Global Forum on Urban Resilience and Adaptation), with the objective of connecting local government, business interests and scientific experts in order to advance resilience objectives in urban policies. Likewise, the '100 Resilient Cities' Programme, funded among others by the Rockefeller Foundation and the World Bank's Resilient Cities Programme, aims to help cities become more resilient, while insisting on the multi-dimensionality of the approach (similar to the framing of sustainability a few decades ago). In yet another example, the 2010–2011 Worldwide Disaster Reduction Campaign, inspired by the United Nations International Strategy for Disaster Risk Reduction (UNISDR), explicitly asks local governments to develop 'resilient' initiatives to confront extreme natural events.

A very significant 'alliance' is also being forged between 'resilient' and 'smart' cities (Symantec 2013). The latter is yet another example of an ascendant term being used in conjunction with 'sustainability', 'adaptation' and 'resilience' – this one referring to the use of technological fixes (mostly ICTs) to solve a number of challenges (March and Ribera-Fumaz 2016). In the words of a top executive in Suez Environment (one of the largest private utilities in the world), "in order to guarantee service continuity and integrity, the ICT systems that oversee and control a 'smart city' need to be designed, from inception, with cyber security, robustness, reliability, privacy, information integrity, and crucially, resilience, in mind" (Hervet, Suez Environment Group 2013: 1).

The shift to resilience has accentuated the need to provide for oneself because collective planning and collective action may be totally ineffective in the face of uncertainty and potentially catastrophic tipping points. Therefore it is not at all surprising that resilience, and more so if accompanied by the term 'smart', fits perfectly within the neoliberal ideology that individuals, more so than collectives, must have the capacity to face and survive challenges of all sorts. After all, the distributional issues implied in natural metaphors are largely irrelevant because the scale of analysis is always at the level of the undifferentiated species. One example of this in the mainstream environmental literature would be the frequent use of 'we' or 'us' when facing supposedly external threats. But this 'we' or 'us' represents of course a simple aggregation of individual actions, never a collective and emancipatory project. Hence the insistence that the best (and most resilient) way out of uncertain socio-environmental futures is not some sort of communitarian approach to social life, but individual entrepreneurship. Collective security is no longer possible nor desirable; everybody must provide for him/herself and become an entrepreneur, no matter how small the activity might be (see Watts 2011 for the African case). Resilience, then, may come as an opportunity for some but as a cost for many more, especially in contexts where significant segments of the urban population are impoverished by economic crisis and then doubly victimized by successful ecological modernization policies that discipline the resource consumption of the very same people.

As with many other concepts, geographers and planners have attempted to develop a spatial interpretation of resilience (Christopherson et al. 2010; Davoudi 2012; MacKinnon and Driscoll Derickson 2012). One standard argument in this respect is that spaces of resilience are needed by capital to provide protection against increasingly frequent and devastating economic crises. Resilience therefore would fit well within the uneven capitalist development theory of Neil Smith (Smith 1984) in which cities and regions either triumph over, or succumb to, socio-environmental turmoil. Following from the ecologically inspired view of resilience, issues of power and justice are overlooked because they are seen as irrelevant, given the magnitude of external threats. Resilient citizens can no longer count on redistributive measures to assist in adaptation to natural threats, but are forced to become individually responsible for addressing threats. One of the most symptomatic examples of this approach can be found in the EU Green Paper on the Insurance of Natural and Manmade Disasters (European Commission 2013). In the text, citizens are admonished for relying on public relief when attempting to recover from a natural disaster. For the Commission,

> Governments could continue to absorb a large share of the costs of mitigation and public relief by continuing to generously compensate victims. But this is likely to exacerbate governments' budget difficulties and encourage undesired development in risk-prone areas. Alternatively, public authorities could withdraw resources from this area, control development in risk-prone areas and rely more heavily on market forces to encourage individual responsibility for reducing losses and insuring against them.
>
> *(European Commission 2013: 11)*

This paragraph nicely encapsulates the neoliberal resilience approach to 'external threats'. Public security in the face of socio-environmental calamities is no longer possible or desirable, not in the least because of the human propensity to succumb to 'moral hazards' (i.e. supposedly selfish behaviour). Instead, resilience can only be guaranteed if everyone carries their corresponding costs, an approach that perfectly suits the current times of economic and social austerity and the pervasiveness of neoliberal life. Again, this is an example of the naturalization of phenomena having important and fundamental human and social components. Nature is conveniently used to disguise social processes and implicitly place blame on the victims, who should pay for protection. Rather than resistance, resilience thinking calls for adjustment to perturbations, all conveniently stripped of their social origins (Lampis 2015). In sum, and as MacKinnon and Driscoll Derickson (2012) argue, in these socio-ecological times the 'social' has been devoured by the 'ecological'.

Another relevant shortcoming of ecological resilience is that by focusing on single extraordinary threats (i.e. an extreme natural or social event) it ignores the shocks and stresses of everyday life. This point was already raised by Kenneth Hewitt in his critique of hazard research in geography in the 1970s (Hewitt 1983); he essentially argued that the pressures of daily living (i.e. not having a job or having a poorly paid one) may be more challenging than a flood or a sudden downfall in the stock market (Andres and Round 2015). As it currently stands in the socio-ecological literature, resilience is theoretically ill prepared to deal with these situations, although this may not be the case if the social psychological view is adopted, particularly in what concerns community resilience (Olsson et al. 2015).

Water and energy consumption in cities: successes and casualties in sustainability and resilience

Overall, water and energy consumption is actually decreasing and not increasing in most cities of the Global North. Looking at several examples in Europe and North America, decreases in both absolute and relative (i.e. per capita) water consumption is apparent. In some cases reductions are the consequence of purely technological fixes in grossly inefficient distribution networks (e.g. as in many capitals of Eastern Europe after the collapse of the former socialist regimes); in other cases, the spread of efficient fixtures (in taps, toilets, etc.) and of 'green' domestic technologies (washing machines, dishwashers, etc.) may have also contributed to declining consumption (Saurí 2013). Awareness campaigns promoting conservation attitudes and especially behaviours among users have also become quite common, and not just in traditionally water-stressed areas. But as economists are keen to point out, there is nothing like price signals to curb consumption. The argument that water is too cheap and must have a price reflecting true costs has become a mantra in many public and private circles and water prices have risen accordingly, especially in cities and countries where the costs of ecological modernization in the water sector (supply, purification, treatments, etc.) have been passed on to consumers. Facing declining revenues because of diminishing consumption in countries such as Spain, companies and regulators have pushed prices and taxes higher, even in the context of a devastating economic crisis which has contributed to a growing number of households defaulting on payments, leading to the interruption of service (nearly 300,000 households according to the Spanish Association of Water Supply and Sanitation Companies). Although in some cities 'social tariffs' have been created to help with payments, these measures cannot compensate for recent annual water price increases – more than 12 per cent on average in 2014, according to the Spanish Statistical Institute (El Diario 2014).

In the case of energy, the picture is more complex given the heterogeneity of energy sources and markets. In the EU, for example, during the period 2000–2012 energy consumption in households decreased annually by 1.5 per cent. As with water, the combination of efficiency and pricing, coupled with the recent economic recession, appear as the main causes of this decrease. As a result of the liberalization of the energy sector in the EU, energy prices increased 64 per cent between 2004 and 2012 (Dromaque 2013). At the same time, average household incomes grew less than 12 per cent. In Southern European countries especially, reductions have been more intense since 2008, with Spanish households reducing their annual energy consumption nearly 2 per cent per year, Portuguese households some 4 per cent annually and Greek households nearly 6 per cent (Lapillone et al. 2015). Again, efficiency gains, in the form of better insulation and the proliferation of low consumption lightning and appliances explain an important part of this trend.

Still, a considerable number of European households have become 'energy poor'. According to EU Statistics on Income and Living Conditions (EUROSTAT 2015), more than 50 million European citizens (10.2 per cent of the EU population) could not keep their home adequately warm in 2014, with similar figures in arrears in the payment of utility bills. Southern, Eastern and Mediterranean countries were least likely to keep homes adequately warm (40.5 per cent of households in Bulgaria, 32.9 per cent of households in Greece and 28.3 per cent of households in Portugal), while figures approached 10 per cent in richer countries such as the United Kingdom. The recent economic crisis combined with higher prices has significantly increased these percentages. In Spain, for instance, households with late payments of utility bills went from 4.5 per cent of the total in 2005 to 9.2 per cent in 2014 (EUROSTAT 2015).

Thus, at least part of the urban population contributing to a better future (in terms of diminishing consumption of resources) is not only not rewarded for their improved behaviour, but rather continues to be disciplined through higher prices and taxes. Sometimes this may lead to paradoxical situations in which different strands of neoliberalism may stand in opposition. Witness, for example, the debate on renewable energy in California with right-wing groups rather opportunistically denouncing progressive liberals for imposing higher costs, associated with renewable energy sources, on the urban poor (Bryce 2015).

As Bouzarovski *et al.* (2012) argue, energy (and water) poverty can no longer be considered a problem of the Global South. In Europe, for instance, energy poverty and particularly fuel poverty, is affecting growing segments of the urban population (the elderly, the unemployed, poorly paid workers, the disabled, single parent families, etc.). Depriving society's most vulnerable groups of essential services strips bare the foundations of urban sustainability, especially when ecological modernization measures are largely paid for by consumers through prices and taxes that are not income sensitive. These measures may help to reduce tensions in urban environmental well-being, but they also create large pockets of socio-ecological injustice which the shift to resilience may very well aggravate.

From resilience to resistance: concluding remarks

This chapter has attempted to trace the rise of resilience as an overarching socio-ecological concept and its implication for cities and citizens of the Global North. My narrative began with the relative success of sustainability policy in improving the environmental quality of urban settings and in increasing the efficiency of resource consumption, mainly through economic and financial disciplining techniques associated with ecological modernization. This success, however, has not been accompanied by similar improvements in the social aspects of sustainability; low- and even middle-income segments of the citizenry may have financial difficulties living in increasingly green-gentrified neighbourhoods, or may have difficulties in affording basic resources such as water and, especially, energy. But I have also argued that sustainability with its implicit faith in a bright future for all may not be useful anymore in front of the challenges being faced by cities, especially after the most recent global crisis of capitalism. As several scholars emphasize, the shock and subsequent turbulence affecting the world economic and financial system shook the roots of a just and equitable future and put risk and uncertainty at the top of urban agendas. It is in this context that resilience has come to occupy a central position in academic and policy debates particularly, but not only, regarding major threats such as climate change. In the policy arena perhaps more than in academia, it appears that sustainability has run its course and that resilience is a much more appropriate concept for the current age of austerity.

Resilience, however, has different meanings according to different scientific disciplines. The definitions of certain disciplines, such as ecology, have been preferred over others, such as psychology. In this context, the change in emphasis from sustainability to resilience means increased acceptance of the burden of socio-environmental shocks and stresses on the less well-off. Worse still, this conception of resilience does not provide guidance towards a more socially just future since resilience, understood in ecological terms, is a natural condition immune to social change. Individuals are encouraged to build their own resilience to risky and uncertain futures based on their aptitudes, for instance by becoming 'entrepreneurs' or by assuming previously socialized costs (for example, the European Commission's enthusiasm for private insurance). Resilience frames naturalize water and energy poverty and ignore the tremendous pain inflicted on low-income and increasingly middle-income groups, groups that are punished with higher prices and taxes for basic services, despite being leaders in energy and water conservation.

Nonetheless, resistance to the hegemony of (neoliberal) resilience does occur, despite its tendency to disguise power, injustice and inequality through naturalistic metaphors and the language of systems, external forces and the like. Although one can be quite sceptical about the power of resilience thinking to alleviate poverty and improve the livelihoods of citizens (Olsson *et al.* 2015), there are attempts to explore the emancipatory potential of the concept by examining "transformative" or "persistent" rather than "adaptive" notions of resilience (Andres and Round 2015). In the same vein, some analysts ask the obvious question of "resilience for whom" (also asked regarding sustainability some decades ago) and attempt to rescue the concept by adding layers of social concern (Andres and Round 2015; Cretney 2014; Vale 2014). In the end, however, resilience may be running the same course as sustainability. No matter how progressive readings of society and the environment may be, resilience has been conveniently reduced to an ecological logic which suits the needs of a neoliberal order well and remains incapable of coherently incorporating issues of social relations and, above all, social progress. (Ecological) resilience, in sum, is the embodiment of despair, negating visions of socially just and sustainable futures. If resilience has still a place in the progressive agenda it will probably have to be sought in the psychological literature, largely forgotten in current socio-environmental debates (Lee *et al.* 2008). Although also meriting critical scrutiny, psychological resilience is at least more prepared to deal with what is perhaps the most fundamental shortcoming of the ecological approach: the negation of hope for true socio-environmental progress.

References

Adger, W.N. (2000). Social and Ecological Resilience: Are They Related? *Progress in Human Geography*, 24, pp. 347–364.
Andres, L. and Round, J. (2015). The Role of 'Persistent Resilience' Within Everyday Life and Polity: Households Coping with Marginality Within the 'Big Society'. *Environment and Planning A*, 47, pp. 676–690.
Anguelovski, I. (2014). *Neighborhood as Refuge: Community Reconstruction, Place Remaking, and Environmental Justice in the City*. Cambridge, MA: MIT Press.
Bouzarovski, S., Petrova, S. and Sarlamanov, R. (2012). Energy Poverty Policies in the EU: A Critical Perspective. *Energy Policy*, 49, pp. 76–82.
Brown, K. (2014). Global Environmental Change I: A Social Turn for Resilience? *Progress in Human Geography*, 38(1), pp. 107–117.
Brown, K. and Westaway, E. (2011). Agency, Capacity, and Resilience to Environmental Change: Lessons from Human Development, Well-being, and Disasters. *Annual Review of Environment and Resources*, 36(1), pp. 321–342.
Bryce, R. (2015). California's Energy Policies: The Poor Are Hit Hardest. *National Review*, August. Available at: www.nationalreview.com [accessed 21 December 2015].
Buttel, F.H. (2000). Ecological Modernization as Social Theory. *Geoforum*, 31, pp. 57–65.
Christopherson, S., Michie, J. and Tyler, P. (2010). Regional Resilience: Theoretical and Empirical Perspectives. *Cambridge Journal of Regions, Economy and Society*, 3, pp. 3–10.
Cretney, R. (2014). Resilience for Whom? Emerging Critical Geographies of Socio-Ecological Resilience. *Geography Compass*, 8(9), pp. 627–640.
Davoudi, S. (2012). Resilience: A Bridging Concept or a Dead End? *Planning Theory and Practice*, 13, pp. 299–333.
Dromaque, C. (with the assistance of A. Bogacka) (2013). *European Residential Energy Price Report 2013*. Helsinki: VaasaETT.
El Diario. (2014). El agua provoca los desahucios más silenciosos, 31 December. Available at www.eldiario.es/sociedad/agua-provoca-desahucios-silenciosos_0_340866018.html [accessed 20 November 2015].
European Commission. (2013). *Green Paper on the Insurance of Natural and Man-made Disasters*. Strasbourg.
EUROSTAT. (2015). *Statistics in Income and Living Conditions*. Available at: http://appsso.eurostat.ec.europa.eu/nui/show.do [accessed 4 January 2016].
Evans, D. and Reid, J. (2014). *Resilient Life: The Art of Living Dangerously*. Cambridge: Polity Press.
Fletcher, D. and Sarkar, M. (2013). A Review of Psychological Resilience. *European Psychologist*, 18(1), pp. 1–23.

Folke, C. (2006). Resilience: The Emergence of a Perspective for Social-Ecological Systems Analyses. *Global Environmental Change*, 163, pp. 253–267.

Folke, C., Colding, J. and Berkes, F. (2003). Synthesis: Building Resilience and Adaptive Capacity in Social-Ecological Systems. In: F. Berkes, J. Colding and C. Folke, eds. *Navigating Social-Ecological Systems: Building Resilience for Complexity and Change*. Cambridge: Cambridge University Press, pp. 352–387.

Hervet, B., Suez Environment Group. (2013). *Building the Smart and Resilient City*. Available at: www.newcitiesfoundation.org/building-the-smart-and-resilient-city/ [accessed 7 January 2016].

Hewitt, K. (1983). The Idea of Calamity in a Technocratic Age. In: K. Hewitt, ed. *Interpretations of Calamity*. Boston: Allen and Unwin.

Holling, C.S. (1973). Resilience and Stability of Ecological Systems. *Annual Review of Ecology and Systematics*, 4(1), pp. 1–23.

Holling, C.S. (1978). *Adaptive Environmental Assessment and Management*. Chichester: John Wiley and Sons.

Krueger, R. and Gibbs, D., eds. (2007). *The Sustainable Development Paradox: Urban Political Economy in the U.S. and Europe*. New York: Guilford Press.

Lampis, A. (2015). Resilience and Cities: Critical Thoughts on an Emerging Paradigm. Available at: https://ugecviewpoints.wordpress.com/2015/10/08/ [accessed 15 July 2016].

Lapillonne, B., Pollier, K. and Samci, N. (2015). *Energy Efficiency Trends for Households in the EU*. Available at: www.odyssee-mure.eu/publications/efficiency-by-sector/household/household-eu.pdf [accessed 30 November 2015].

Lee, E-K.O., Shen, C. and Tran, T.V. (2008). Coping with Hurricane Katrina: Psychological Distress and Resilience Among African-American Evacuees. *Journal of Black Psychology*, 35, pp. 5–23.

Lomolino, M. (2000). A Call for a New Paradigm of Island Biogeography. *Global Ecology and Biogeography*, 9, pp. 1–6.

MacKinnon, D. and Driscoll Derickson, K. (2012). From Resilience to Resourcefulness: A Critique of Resilience Policy and Activism. *Progress in Human Geography*, 37(2), pp. 253–270.

March, H. and Ribera-Fumaz, R. (2016). Smart Contradictions: The Politics of Making Barcelona a Self-sufficient City. *European Urban and Regional Studies*, 23(4), pp. 816–830.

Naredo, J.M. and Gómez-Baggethun, E. (2012). Río+20 en perspectiva. Economía verde: nueva reconciliación virtual entre ecología y economía. *Anexo a la edición española del informe de World Watch Institute. La Situación en el Mundo 2012*, Fuhem, pp. 347–370.

Olsson, L.A., Jerneck, H., Thoren, J., Persson, J. and O'Byrne, D. (2015). Why Resilience is Unappealing to Social Science: Theoretical and Empirical Investigations of the Scientific Use of Resilience. *Science Advances*, 22 May, 1(4). DOI: 10.1126/sciadv.1400217.

Redclift, M. (2005). Sustainable Development (1987–2005): An Oxymoron Comes of Age. *Sustainable Development*, 13, pp. 212–227.

Saurí, D. (2013). Water Conservation: Theory and Evidence in Urban Areas of the Developed World. *Annual Review of Environment and Resources*, 38, pp. 227–248.

Smith, N. (1984). *Uneven Development*. Oxford: Blackwell.

Symantec (2013). *Transformational 'Smart Cities': Cyber Security and Resilience*. Available at: https://eu smart cities.eu/sites/all/files/blog/files/Transformational%20Smart%20Cities%20-%20Symantec%20Executive%20Report.pdf [accessed 4 January 2016].

Ungar, M. (2008). Resilience Across Cultures. *British Journal of Social Work*, 38(2), pp. 218–235.

Vale, L.J. (2014). The Politics of Resilient Cities: Whose Resilience and Whose City? *Building Research and Information*, 42, pp. 191–201.

Walker, B.H., Holling, C.S. and Carpenter, S.R. (2004). Resilience, Adaptability and Transformability in Socialecological Systems. *Ecology and Society*, 9(2), p. 5.

Walker, J. and Cooper, M. (2011). Genealogies of Resilience from Systems Ecology to the Political Economy of Crisis Adaptation. *Security Dialogue*, 43, pp. 143–160.

Watts, M. (2011). On Confluences and Divergences. *Dialogues in Human Geography*, 1(1), pp. 84–89.

World Resources Institute (WRI) in collaboration with United Nations Development Programme, United Nations Environment Programme and World Bank (2008). *World Resources 2008: Roots of Resilience*. Washington, DC: WRI.

York, R. and Rosa, E. (2003). Key Challenges to Ecological Modernization Theory: Institutional Efficacy, Case Study Evidence, Units of Analysis, and the Pace of Eco-Efficiency. *Organization and Environment*, 16(3), pp. 273–288.

15
TRANSFORMING RAINEY STREET
The decoupling of equity from environment in Austin's smart growth agenda

Eliot Tretter and Elizabeth J. Mueller

Introduction

The 1987 Bruntland report, *Our Common Future*, is noted for galvanizing international attention on the environmental limits to growth through the concept of sustainability (Keil 2007). While the aspirations of the original concept – to foster development that meets present needs without compromising future generations – were potentially radical, over time the term has come to be embraced so widely as to be emptied of any meaning (Rosol 2013; Swyngedouw 2007). Instead, sustainability is invoked apolitically to justify 'techno-managerial' approaches to urban development, where aspirations have retreated from changing the nature of development to shifting the pattern of development (Miller 2015). 'Smart Growth' and/or 'New Urbanism' are primarily terms used to describe a range of 'sustainable' urban planning techniques, policies, initiatives and strategies that attempt to achieve sustainable growth by encouraging denser urban development, particularly in proximity to mass transit lines (American Planning Association 2000, 2002). While tensions between the three pillars of sustainability – economy, environment and equity – are acknowledged by some advocates of Smart Growth and New Urbanism, attention to equity concerns has been modest and far from the strong commitment to 'just sustainability' advocated by critics (Agyeman 2003; Campbell 1996; Krueger and Agyeman 2005; Oden 2016).

An increasing number of cities are being reshaped in the name of urban sustainability, with special concern for ecology, but a growing body of scholarship documents the ways that an urban sustainability agenda has simply monetized improved environmental conditions, rather than integrating environmental health with economic prosperity and social justice. In particular, environmental gentrification or "the implementation of environmental or sustainability initiatives that lead[s] to the exclusion, marginalization, and displacement of economically marginalized residents" (Pearsall and Anguelovski 2016: 2) has become increasingly common. Nearly all of the literature on environmental gentrification focuses on the recent push by cities across the USA and Canada to incorporate ecological concerns in the development process through the adoption of broad environmental sustainability goals, targets, agendas and strategies (Checker 2011; Gibbs and Krueger 2007; Krueger and Agyeman 2005). In addition, the literature points out how rapid neighbourhood revitalization, including economic and demographic transformation, is often preceded by the replacement of an environmental hazard with new environmental amenities (Anguelovski 2015; Bryson 2013; Curran and Hamilton 2012; Dooling 2009).

In other words, the tripartite sustainability agenda addressing the economy, environment and social equity is giving way to ecological gentrification in which issues of social equity fall by the wayside.

Environmental rationales figured into attempts to redevelop Austin's downtown Rainey neighbourhood long before the adoption of an urban sustainability agenda and the implementation of a series of Smart Growth and New Urbanist initiatives by the City in the late 1990s. Beginning in the 1960s, development plans for the neighbourhood consistently focused on improving the area's environmental quality, especially thorough the upgrading or creation of new environmental amenities, but with social equity as a policy consideration. Starting in the late 1970s, more intensive housing development began to be presented as a strategy for both neighbourhood revitalization and as a means to combat suburban sprawl. In the last decade, the push to develop denser, high-rise housing in the neighbourhood intensified as the neighbourhood's redevelopment became integral to the city government's sustainability agenda, tethered to its efforts to promote a more compact urban form. It is at this time that these two environmental rationales began to gradually converge into a single environmentally informed planning vision supporting the rapid upgrading and reuse of the CBD's environmental amenities as a means to attenuate residential suburban development. As support for higher-density luxury housing in the Rainey neighbourhood (and the city's adjacent CBD) has increased in recent years, it has become decoupled from an earlier sustainability agenda that focused on achieving more equitable urban redevelopment. In fact, concern for environmental priorities in the redevelopment of the Rainey neighbourhood have persistently undermined the demands of the neighbourhood's low-income residents who have long advocated for improvements to local environmental conditions that they could enjoy, while remaining in the neighbourhood.

The first part of this chapter outlines the historical context for Austin's recent urban-sustainability agenda, its efforts at Smart Growth and New Urbanism, and their impacts on the city's development. This is followed by a detailed historical account of the evolution of the Rainey neighbourhood, particularly how it was targeted by different local planning and redevelopments efforts since the late 1960s and how the area's fate became increasingly tied to broader discussions about housing density and residential growth in the CBD. The third section connects the first and second sections by focusing on how the recent changes in the Rainey neighbourhood are related to the city's current urban-sustainability agenda, particularly focusing on the connection between the neighbourhood's redevelopment and environmental sustainability. Finally, we discuss the implications of having a local sustainability agenda that is decoupled from a concern for equity. The construction of luxury high-rise housing is represented as serving the needs of all city residents, including low-income households; the reality does not quite match the rhetoric.

The new Austin model: smarter growth and newer urbanism

In the 1990s, the idea of Smart Growth took hold in Austin with force. A new political alliance had formed between mainstream environmentalists interested in diverting growth away from the city's karst aquifer, located under the hilly, affluent west side of town, and a coalition of business interests that wanted to transform Austin's CBD (Tretter 2016). The environmental and business communities had engaged in a nearly twenty-year battle over the appropriate scale of urbanization for Austin's western suburbs and the impact future development in this zone would have on the quality of the city's water supply and on the habitat of local wildlife (including endangered species) (Swearingen 2010). By 1997, this conflict had been partially resolved through a political arrangement offered by Mayor Kirk Watson. Building on earlier planning efforts, Watson had run on a platform that offered a mix of environmental and economic goals – most notably a promise to revitalize portions of Austin's existing urbanized area. Shortly after entering office, the new mayor convened a taskforce comprised of equal numbers of environmentalists and developers to develop a set of rules to govern future development (Moore 2007: 42). One outcome of this taskforce was the 'Smart Growth Matrix', which laid out criteria for incentivizing development in designated 'desired development zones', including the CBD and a large area to its east, historically home to the city's African American and Hispanic communities and colloquially known as 'East Austin'.

The Smart Growth platform had the benefit of being both anti-sprawl and pro-development. It favoured denser and more compact development within certain existing urbanized areas of the city as a trade-off for less development on greenfield suburban sites to the west. The business community liked its pro-growth stance and its emphasis on incentives, fee rebates and subsidized infrastructure (Barna 2002). Mainstream environmentalists liked the matrix because it promoted growth away from the less urbanized environment in Austin's western hills. It was a 'win-win' proposal.

Equity, in particular affordable housing, was presented as one of the three central planks of Austin's Smart Growth efforts, but it soon became clear that equity concerns would be subordinated to economic and environmental priorities (Portney 2013: 182). The city had established the Safe, Mixed-Income, Accessible, Reasonably Priced, Transit-Oriented (SMART) housing programme in 1999 to offer incentives for developers to produce affordable housing units, but the incentives offered, on their own, have never been enough to induce significant inclusion of affordable units in the CBD (City of Austin (COA) 2007; Community Action Network 1999). In particular, the way that the city's Smart Growth matrix treated residential development in the CBD is particularly illuminating. Projects earned points for maximizing density (with maximum points for at least 12–25 units/acre downtown), for the total number of units per project (maximum points for 200+ units) and for inclusion of residential units in mixed-use buildings. These features were linked to urban design features associated with high-rise housing such as parking structures and pedestrian-oriented streetscape features (COA 2001). Several high-end residential projects in the CBD received Smart Growth incentives during this period (Barna 2002: 24). Initially, points were offered for providing a mix of housing types, but no incentives were offered for inclusion of units affordable to low-income households. By 2001, points had been added for 'reasonably priced' housing, but still no affordable units were produced in the CBD.

In many ways Austin's Smart Growth efforts reproduced the equity problems found in the city's previous growth-management and planning efforts. Smart Growth priorities aligned with the basic patterns of growth suggested and codified in the city's 1979 comprehensive plan, *Austin Tomorrow*, and also with the mid-1990s recommendations of the Citizen's Planning Committee's (CPC). The CPC encouraged redevelopment of the CBD and other urbanized areas in the city, simplifying and reforming the land development code to promote infill and density (outside of existing single-family neighbourhoods), and encouraging more transit-oriented mixed-use projects. Yet like these past planning efforts, the vision and priorities embodied in the city's Smart Growth initiative reflected minimal input from and consultation with leaders of Austin's non-white communities (Busch 2015). It was their communities, especially in East Austin, that were targeted for substantial redevelopment (Moore 2007). Unsurprisingly, a number of studies have documented the impact of the city's Smart Growth efforts on selected neighbourhoods, specifically how they have encouraged the gentrification of lower-income and/or non-white neighbourhoods in Austin (Busch 2013; McCann 2007; Mueller and Dooling 2011; Tretter 2016).

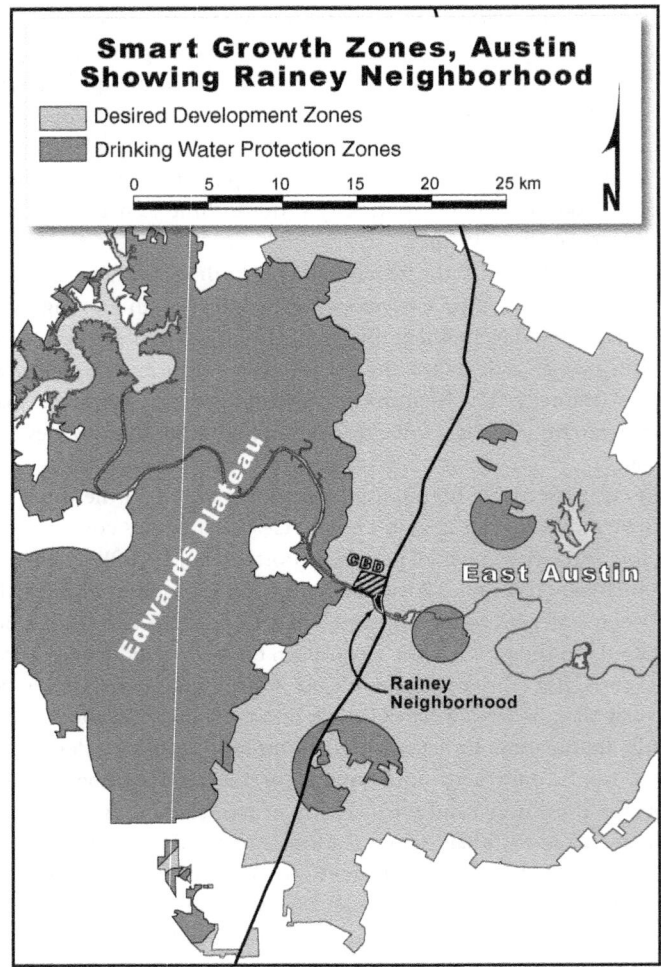

Figure 15.1 Smart growth zones, Austin

In 2000, official support for Smart Growth incentives began to wane but its sustainability goals remained embedded in future planning initiatives. The municipality has continued to strongly support mixed-use, transit-oriented development planning and higher-density projects for their ecological, fiscal and economic development benefits (Ellinor 2013). While local planning focuses on ways to encourage density on a project basis throughout the city, the CBD remains the primary focus of such efforts.

The Rainey neighbourhood 1970–1990: transitioning into the ambit of downtown

In 1967 Austin's Planning Department undertook an urban-renewal study of twenty areas in Austin that were "beginning to show some signs of structural and environment blight" (COA 1968: II-25). One of the areas studied was the Rainey neighbourhood, an area that had "experienced very little physical change or growth until 1964 when the influence of large-scale private and government development began to change the character of the neighbourhood". In particular, the report noted, "new high-density, high-rise structures" were leading to significant land-use changes in the area. Two new high-rise motels and one public-housing tower for the elderly had been built along the neighbourhood's perimeter. The report suggested that new development would continue to focus on the edges of the neighbourhood because of problems with the interior sections such as "lack of access, unpaved streets, small parcels, and existing structural blight. [But] before this area can develop to its fullest potential", the report went on, "these blighted influences must be removed" (COA 1968: 19-E). The planning department, therefore, recommended the relocation of "110 families [of the 166 families living in the area] and 33 individuals with 40 percent of the families having an annual income of $3,000 or less and one-third ... with five members of more" (COA 1968: 19-F). After the removal of these people and their houses, the city's planners projected that there would be an additional 400 housing units in high-density dwellings and 1,000 new people in the renewed area (a net increase of 216 units and 260 people).

An Urban Renewal programme was never undertaken in the Rainey neighbourhood, but in the late 1970s, Austin's municipal government became increasingly concerned about the quality of development in the city's central neighbourhoods – particularly those areas in and around the CBD. A 1977 report observed that "since 1940 various parts of the core of the City of Austin have been losing population" because of rising utility costs and, more importantly, the "threat of busing" i.e. school integration (COA 1977: 2–5). However, the CBD still retained its strategic regional advantage because "living in the core is more convenient in terms of getting to and from work ... and most types of live entertainment" (COA 1977: 3).

The movement of whites to Austin's suburbs – driven in part by racist fears – reinforced racial division within what was already a highly segregated city. By 1928 a pattern of racial and income segregation had been established through private (deed restrictions) and public zoning efforts (Tretter and Sounny 2013). Subsequent public planning overlaid industrial zones and highways over these existing spatial patterns and protected whiter and more affluent, mostly west side, neighbourhoods at the expense of lower-income and mostly non-white districts. For instance, in the 1950s, the development of a highway along the Rainey neighbourhood's eastern edge led to the encroachment of industrial and commercial uses.

Environmental and social factors also interacted to undermine the value and quality of residential property in the Rainey neighbourhood, as in other neighbourhoods on Austin's east side (Tretter and Adams 2012). Intense and frequent flooding repeatedly destroyed and damaged many houses in the Rainey neighbourhood and, beginning in the 1940s, settlement by ever-greater numbers of Hispanics occurred (Dase and Ward 2000; Feit 2012). White supremacy limited housing choices and vocational opportunities for Hispanics, contributing to the neighbourhood's overcrowded and poor housing conditions (McDonald 2012). Yet despite these pressures towards segregation, in 1967, a neighbourhood survey found the Rainey neighbourhood to be almost evenly split between 'Anglos' and 'Latin Americans'; it had no 'Negro' residents (COA 1968: 19-A).

Although in close proximity to downtown, the Rainey neighbourhood was not considered part of the CBD until the 1970s, when a series of city-commissioned reports explored the redevelopment potential of the CBD and increasingly "incorporated [the Rainey neighbourhood] into its ambit" (COA 1980: 1). The report titled *Strategies For the Economic Revitalization of Central Austin* (1978) was the most significant. This report offered specific recommendations for the redevelopment of the Rainey neighbourhood – or what it called the Town Lake Redevelopment District. Echoing and going beyond earlier studies the report emphasized the neighbourhood's potential for mixed-use, high-density, high-quality residential and commercial development. This potential, the report stressed, was driven by both "the excellent environmental amenities provided by Town Lake, Town Lake Park, Waller Creek and associated improvements and an abundance of large trees", and the amount of undeveloped or "underutilized" land, in addition to five acres of city-owned land (COA 1978: 85–86). The report went on to recommend that these "environmental amenities should be bolstered by public investments" (COA 1978: 91).

The Department of Planning's 1978 report, and the subsequent urban-renewal plan proposed for the CBD by James Rouse's American Cities Corporation, was strongly opposed by the Rainey Street Association, which produced a 1978 bilingual comic book that claimed the proposed revitalization plans were a means to undermine Chicano power by displacing the area's large Hispanic population. Moreover, the comic book charged the city government with using "taxpayers money to make downtown attractive to developers" and wanting to build "townhouses, condominiums, and apartments in the Rainey Street area" (Lowry 1978: 5). These efforts were successful and the City Council ultimately rejected the wholesale adoption of the recommendations in the *Strategies* report and the American Cities Corporation's plan (Kreps 1979; Real 1979).

However, efforts to redevelop the Rainey neighbourhood did not stop. In 1980, Austin's Planning Department released a study titled *Development Alternatives for the Rainey Area*, which described three competing options for the neighbourhood's redevelopment. The first would provide the "greatest protection possible for current lower income residents". Called the "Barrio Plan", this scenario recommended

> the development of low-density housing on vacant land ... that is affordable by low and moderate income households [and] rehabilitation and homeownership assistance to current and future lower income residents ... Extensive zoning rollbacks or land acquisition ... to reverse current market forces in the area. ... [And] the establishment of neighbourhood level services especially local retail and recreational facilities.
> *(COA 1980: 22)*

In contrast, the second scenario recommended that redevelopment maximize market-rate uses and minimize government support for preservation of the existing income mix of the neighbourhood. It called for "the single-family housing ... [to] be preserved and improved, vacant or under-utilized land surrounding this area [to] be developed to the maximum potential under current and anticipated zoning, including market rate multi-family housing" (1980: 23) Moreover, it proposed that municipally controlled land be reused "as a public facility, cultural centre, or market rate multi-family housing" (1980: 57). The third option explored a mixture of proposals from options one and two. It emphasized

> [the construction of] new housing [replacing existing houses and] more intensive uses, such as general retail ... along the boundaries of East First Street and Interstate 35. [But] [s]ome zoning rollbacks to moderate density residential uses would be required for vacant land adjacent to the single-family.
> *(1980: 23)*

It also called for the city-owned land to be repurposed for "the construction of a mix of subsidized and market rate housing" (1980: 23). The third alternative also highlighted the potential environmental benefit of creating an "ideal environment for pedestrian and bicycle travel" (1980: 57). Importantly, all three scenarios envisioned the preservation of "the existing single-family housing along Rainey Street, and none propose[d] full-scale CBD zoning and

Figure 15.2 Revitalization and East Austin

Source: artist: Carlos Lowry.

development for the neighbourhood" (The Austin Chronicle 1999). Instead, the focus of this dispute was over the appropriate scale and type of the residential and commercial redevelopment and its potential impact on the existing residents (it was only later, as the focus shifted to more intensive high-rise high-end housing, that environmental factors became significant). In the end, the City Council failed to endorse any of the proposed scenarios in 1981, but it did adopt a number of "development safeguards contained in Alternatives II and IV [an additional set of compromise proposals]" (Lillie 1981).

Nevertheless, in 1981 ground was broken in the Rainey neighbourhood on a new high-rise condominium complex called the Towers of Town Lake. Significantly the developer (Pence) noted that while the high-rise "may be impacting the area, this is what the city is trying to encourage as an alternative to urban sprawl" (Douthat 1981). Aware that increasing the amount of market-rate multi-family housing, particularly in formerly commercial-zoned areas, would place pressure on the area's low-rise residential character by increasing property values, Pence added a new environmental rationale for the redevelopment of the Rainey neighbourhood, suggesting that such development was beneficial precisely because it aligned with the city's growth-management efforts. In contrast, project opponents focused on the potential negative social fallout on the neighbourhood's local environment, arguing that the condominium project "compromised the integrity of the neighbourhood" (Hernandez 1981: 6). Despite concerns about the development's impact, when the project was completed in 1984 a mere thirty-seven units were sold and the building's owners went bankrupt (Breyer 1998).

The Pence development brought to the fore possible options for the neighbourhood's existing single-family bungalows (The Community Economic Development Policy Research Project 1981). The Urban Renewal study had found "[Of] the 158 residential structures 67 percent [are] needing major repairs or dilapidated" (COA 1968: 19-B), but the *Strategies* report claimed, "[Some] residences in the Rainey neighbourhood have deteriorated but seventy-five percent of the substandard structures are economically suitable for rehabilitation" (COA 1978: 107). Even in the mid-1980s, as some single-family homes were demolished and replaced by multi-family dwellings or vacant land, a large stock of occupied single-family dwellings remained that could be rehabilitated for existing residents (COA 1985: 7). The prospects for preserving these bungalows received a significant boost in 1985 when thirty-two homes along a strip of Rainey Street were included in a Historic District by the National Register of Historic Places, a designation that was "primarily based on the qualities of [the area's] residential architecture" (Dase and Ward 2000: 6). Barrio Plan advocates believed the preservation of single-family homes was the easiest way to preserve the neighbourhood's Hispanic and working-class character.

Moreover, there was also the contentious issue of whether subsidized or non-market rate housing would be built in the area. Neighbourhood activists were demanding the reuse of municipally controlled land for affordable housing. The 'Barrio Plan' called for the City Council to give the land to the East Austin Chicano Economic Development Corporation (EACEDC) so that it could build houses that would be "available for purchase by low and moderate income people such as current tenants in the Rainey neighbourhood" (Martinez 1980: 5). Some residents, owners and members of the City Council resisted this proposal, but in January 1982 the City Council voted to "commit" the city's land to "a low and moderate income housing project" (Wilson 1982). However, this was contingent on the city's ability to raise $1 million to move its existing facilities to an alternative location – and funds for such a move had been rejected by the electorate in a ballot referendum in August 1981. By 1985 the City Council remained committed to the construction of a "small lot subdivision for low and moderate income households", though the central partners were now members of the Austin Home Builders Association rather than the EACEDC (COA 1985: 15). Yet nothing was built.

In the face of increasing uncertainty, in 1985 Mayor Frank Cooksey proposed locating a new civic centre on the city's land. The idea of building a civic centre had been floated since at least 1969, but in 1984 the idea gained new life when several proposals for a "publically funded convention centre" were released (Cannon 1989: 2). The mayor wanted to locate the civic centre in the Rainey area because it only required minimal amounts of additional land and there were already some landowners eager to sell. However, resistance among some residents and the city's business community doomed this proposal (Slusher 1986). Nevertheless, the proposal succeeded in undermining the affordable housing plan by offering a viable alternative use for the site.

By 1989 no houses had been built and instead the city's parcel was made into parkland (COA 1989). The Town Lake Comprehensive Plan, a plan initiated by the City's Parks and Recreation Department in 1986, suggested action in its recommendations:

> The Rainey Street neighbourhood has been the focus of considerable attention in the past decade, most notably for efforts to preserve its historic character and its Mexican-American legacy. Despite these considerable efforts, this once-thriving neighbourhood has experienced physical deterioration. Only 36 houses, fewer than half of which are owner-occupied, remain in the neighbourhood today. South of River Street, the land is virtually vacant ... Significantly, the Rainey Street Neighbourhood Association has recently abandoned its preservation stance in favour of redevelopment at a higher intensity. [This report] endorses this position and recommends careful redevelopment into mixed uses featuring small-scale shops and offices combined with residential living. Commercial activity should animate the edge of Waller Creek, and the mature trees that shade the neighbourhood should be protected so that they continue to dominate the character of the area. All the land south of River Street is proposed to be acquired and dedicated as parkland to protect its pecan forest and enhance the neighbourhood. Similarly, the City's Street and Bridge Yard will be better used as parkland than for its present function.
>
> *(COA 1987: 70–71)*

The dedication of this land for a park came at about the same time a study funded by the City Council had determined the parcel would be the best location for a Mexican American Cultural Centre (MACC) (Garza 1998). Certainly the creation of such a cultural centre had been a demand of local activists for some time, but their proposals for the site had always included housing (Marban 1996). When the MACC finally opened in 2007, the site contained no housing.

The Rainey neighbourhood since 1998: the new urban housing transition

By the late 1990s, proposed redevelopment plans increasingly focused on the environmental benefits of the Rainey neighbourhood's redevelopment, especially its potential impact on arresting suburban sprawl and promoting walkability, while also increasingly emphasizing how property owners would take maximum benefit from the private-market-led wholesale remaking of the area. As development pressures continued to rise in the 1990s, two local developers, Perry Lorenz and Robert Knight, assembled land in or near the neighbourhood and met with a number of landowners interested in redevelopment; by 1999, fifty-one of the about sixty-nine property owners had signed letters declaring their intent to be party to a large-scale redevelopment plan (Rivera 1999). Initially, these developers helped property owners up-zone their properties, but by 2000 they had become advocates for a New Urbanist-inspired plan that called for the demolition of all single-family homes and the creation of a mixed-use neighbourhood with thirty-five detached row houses (Austin American-Statesman 2000; Breyer 1996). But before this plan could be officially considered, a different developer purchased and rezoned a key site for a condominium development and he refused to include it in this redevelopment effort (Clark-Madison 1999). However, in 2001 both this housing project and the neighbourhood's redevelopment plan collapsed (Novak 2001).

The area's architectural heritage initially presented a barrier to plans to rezone the properties, but under the neighbourhood redevelopment plan proposed in 2000 it was reimagined as a tool for economic development. As mentioned before, in 1985, historic preservationists had succeeded in placing thirty-two houses along a strip of Rainey Street on the National Register of Historic Places. This designation posed some impediments to wholesale rezoning since any demolition permit would trigger review by the local Historic Landmark Commission. Moreover, if this commission recommended historic zoning, then six of seven City Council members would have to vote in favour of a demolition for it to proceed (Osborne 2003). In the 2000 redevelopment plan the new housing was organized around a central square made up of these historic houses, but they would be "converted to uses compatible with a central pedestrian area. Coffee shops, small restaurants, delicatessens and offices have been suggested" (Rainey Street Neighorhood Association 2000: 28).

In 2003, the Downtown Commission, an organization seeking to revitalize downtown, put forward a formal plan that incorporated a reimagining of the use of single-family historic housing. This plan, strongly influenced by New Urbanist planning principles and echoing the plans from the 1980s, called for intensive housing development and claimed that "The development of additional housing in the Rainey Street area will be key to both the future of the area and to Downtown as a whole [because a] sizable residential population would help support ground-level retail" (Downtown Commission 2003: 3). In 2005 the City Council largely embraced the Downtown Commission's redevelopment plan and, marking the end of more than two decades of negotiating future land-use patterns in the neighbourhood, "finalized Rainey Street's rezoning as a [mixed use] business district" (Carney 2011). In turn, property owners sold their parcels to developers to build luxury apartments or condominiums and, if their homes had historic value, they would be transformed into restaurants, bars or coffee shops. The existing low-income residents would be priced out of the neighbourhood and the single-family quality of the area, while remaining constant in appearance, would be altered in its use and meaning. While a handful of these bungalows would still be affordable homes, they would not remain in the Rainey neighbourhood. In 2013, the municipal government funded the relocation of six historical bungalows from Rainey to the Guadalupe neighbourhood, where they became part of a community land trust for low-income homeowners run by the Guadalupe Neighborhood Development Corporation (Guadalupe Neighborhood Development Corporation 2013; Semuels 2015).

The final plans for redevelopment of the Rainey neighbourhood incorporated New Urbanist principles that encouraged better environmental stewardship by promoting walkability, denser living and recreational amenities but without concern for equity, particularly affordable housing and a commitment to the existing lower-income community. A density bonus programme adopted for the Rainey neighbourhood in 2005 required that 5 per cent of the 'bonus' units that a developer built be affordable to households with incomes at or below 80 per cent of the regional median income – well above the income levels of renters in the neighbourhood when neighbours first organized in the 1980s (COA 2005). More importantly, however, due to an oversight, the programme did not require that the units remain affordable after first occupancy (COA 2001). In practice, this meant that affordable units could become market-rate units as soon as the units were re-rented. Thus the programme would provide no ongoing inclusion of low-income residents in the transformed neighbourhood and, in effect, it did nothing to ensure that existing low-income residents remained. By 2014, the effects of this strategy were striking: the population in the Rainey neighbourhood had almost doubled, going from 702 in 2000 to 1,357 in 2010 (US Census Bureau 2000, 2010). By 2014, half of neighbourhood households had incomes above $125,000 per annum – almost double the regional median income and of the 212 households with incomes below $25,000 in the neighbourhood, 164 lived in the Lakeside public housing development (American Community Survey 2014; COA 2016). In short, the social and environmental transformation of the area has been dramatic.

Concluding thoughts

The evolution of proposals for the redevelopment of Austin's Rainey neighbourhood highlights the vulnerability of low-income neighbourhoods near the CBD. In many respects, the transformation of the Rainey neighbourhood is a paradigmatic example of ecological gentrification (Lees *et al.* 2016). On the one hand, its redevelopment was fuelled by a persistent 'rent-gap', a situation where sizeable profits could be earned based on the difference between the actual and potential price of a neighbourhood's improved land. At the same time, the influx of younger people desiring a more urban lifestyle has supported the neighbourhood's redevelopment. Furthermore, the neighbourhood's change was stimulated by a convergence of local government actions (and inactions) and private-market actors. While environmental priorities were part of initial community proposals for redevelopment, they were ultimately transformed and used to undermine political support for the housing vision that would have ensured the continued presence of low-income residents in the area. Since the early 1980s, discussions regarding the neighbourhood's redevelopment have become increasingly intertwined with the city's urban sustainability strategy: the neighbourhood's redevelopment has been promoted as an antidote to urban sprawl, where new ecological amenities and improvements are used to attract new higher income residents for high-density urban living.

But the Rainey neighbourhood case study also points to a thornier decoupling of environmental and equity goals from the city's Smart Growth (and to a lesser extent, New Urbanist) agenda. While the need for affordable housing was initially identified in Austin's sustainably agenda, over time this has been weakened and is not directed towards particular communities. In recent years, the Rainey neighbourhood's transformation into a series of large-scale luxury high-rise residential complexes is presented as benefiting all city residents – even more modestly resourced households. Some proponents claim that adding these high-end housing units frees up other housing units further down the housing-income ladder. Of course, verifying this claim requires empirical research. What's clear in the Rainey neighbourhood case is that low-income residents were never going to be able to occupy the new, high-rise housing or the preserved and renovated bungalows.

A concern for and commitment to equity at a neighbourhood scale has remained elusive. Increasingly, the Rainey neighbourhood's historic role as home to a community of lower-income residents is seen as unrelated to the city's pressing need for affordable housing. City Council has not valued Rainey's historic homes for their affordability to low-income residents, but rather for their potential to be redeveloped into commercial space to serve nearby luxury high-density residential development. Moreover, in the early 1980s a majority of City Council members agreed that the city-owned land in the Rainey neighbourhood should include some non-market housing, but this housing was never built and it was not included in any of the bonds floated for the future redevelopment of the CBD. Even in the 2000s, efforts to produce affordable housing for the neighbourhood failed when the city's density bonus programme neither targeted existing residents nor attempted to ensure ongoing affordability. Instead, the Rainey neighbourhood is seen as integral to the economic development of the CBD, enhanced by the improvement of its local environmental amenities. Social sustainability has, in essence, been dropped from Austin's sustainability agenda for the Rainey neighbourhood.

References

Agyeman, J. (2003). *Just Sustainabilities*. Boston: MIT.
American Community Survey. (2014). *Austin, Texas 5-Year Estimates, Household Income, T56A and B*. Social Explorer. Accessed 15 November 2016.
American Planning Association. (2000). *Policy Guide on Planning for Sustainability*. New York: APA.
American Planning Association. (2002). *Policy Guide on Smart Growth*. New York: APA.
Anguelovski, I. (2015). From Toxic Sites to Parks as (Green) LULUs. *Journal of Planning Literature*, 31(1), pp. 23–36.
Austin American-Statesman. (2000). A Center-City Test. *Austin American-Statesman*, 10 January.
Barna, J. (2002). The Rise and Fall of Smart Growth in Austin. *Cite*, Rice Design Alliance, 53, pp. 22–25.
Breyer, R.M. (1996). Urban Housing with a Lake View. *Austin American-Statesman*, 28 November.
Breyer, R.M. (1998). Pre-Sales Keep Condo Developers in Line with Market. *Austin American-Statesman*, 18 September.
Bryson, J. (2013). The Nature of Gentrification. *Geography Compass*, 7(8), pp. 578–587.
Busch, A. (2013). Building a City of Upper-Middle Class Citizens. *Journal of Urban History*, 39(5), pp. 975–996.
Busch, A. (2015). The Perils of Participatory Planning Space. *Journal of Planning History*, 15(2), pp. 87–107.
Campbell, S. (1996). Green Cities, Growing Cities, Just Cities? *Journal of American Planning Association*, 62(3), pp. 296–312.
Cannon, S. (1989). *The Austin Convention Center Site Selection and Feasibility Analysis*. Austin, AHC.
Carney, H. (2011). Priced Out in the Shadow of Downtown Austin. *Texas Observer*, November.
Checker, M. (2011). Wiped Out by the Greenwave. *City and Society*, 23(2), pp. 210–229.
City of Austin. (1968). *Area Analyses: Austin Community Renewal Program*. Department of Planning. Austin, AHC.
City of Austin. (1977). *Downtown Austin: Its Decline and Potential For Revitalization*. Huffman & Company. Austin, AHC.
City of Austin. (1978). *Strategies For the Economic Revitalization of Central Austin*. Department of Planning. Austin, AHC.
City of Austin. (1980). *Development Alternatives For the Rainey Area*. Department of Planning. Austin, AHC.
City of Austin. (1985). *Rainey Street Area Update*. Office of Land Development and Services. Austin, AHC.
City of Austin. (1987). *Town Lake Park: Comprehensive Plan*. Parks and Recreation Department. Austin, AHC.
City of Austin. (1989). *City Council Ordinance No. 890126: An Ordinance Adopting the Town Lake Park Comprehensive Plan*. Austin, COA.
City of Austin. (2001). *Smart Growth Criteria Matrix*. Planning and Design Department Transportation. Austin, COA.
City of Austin. (2005). *City Council Ordinance No. 20050407–063: An Ordinance Amending Section 25-2-739 of the City Code Relating to the Rainey Street Subdistrict of the Waterfront Overlay Combining District*. Austin, COA.
City of Austin. (2007). *The Affordable Housing Incentives Taskforce Final Report*. Austin, COA.
City of Austin. (2016). *Lakeside Apartments*. Available at: www.hacanet.org/communities/lakeside.php [accessed 15 November 2016].
Clark-Madison, M. (1999). A Change of Scenery. *The Austin Chronicle*, 16 July.
Community Action Network. (1999). *Through the Roof*. Community Action Network. Austin.
Curran, W. and Hamilton, T. (2012). Just Green Enough. *Local Environment*, 17(9), pp. 1027–1042.

Dase, A. and Ward, R. (2000). *The Peculiar Genius of Rainey Street*. Austin, AHC.
Dooling, S. (2009). Ecological Gentrification. *International Journal of Urban and Regional Research*, 33(3), pp. 621–639.
Douthat, B. (1981). Condominium Tower Proposed. *Austin American-Statesman*, 16 May.
Downtown Commission. (2003). *Rainey Street: Recommendations for Action*. Austin, AHC.
Ellinor, B. (2013). *The Costs of a Growing City*. Master's thesis, University of Texas, Austin.
Feit, R. (2012). The Ghosts of Developers Past. *Austin Chronicle*, 25 May.
Garza, M. (1998). Tracking the MACC. *The Austin Chronicle*, 12 June.
Gibbs, D. and Krueger, R. (2007). Containing the Contradictions of Rapid Redevelopment. In: R. Krueger and D. Gibbs, eds. *The Sustainable Development Paradox*. New York: Guilford Press, pp. 95–122.
Guadalupe Neighborhood Development Corporation (2013). *What is a Community Land Trust*. Austin: GNDC.
Hernandez, P. (1981). *An Experiment in Anti-economic Development in the City Core of Downtown Austin*. East Austin Chicano Economic Development Corporation, AHC.
Keil, R. (2007). Sustaining Modernity, Modernizing Nature. In: R. Krueger and D. Gibbs, eds. *The Sustainable Development Paradox*. New York: Guilford Press, pp. 41–65.
Kreps, M. (1979). Downtown Revitalization Plans Changed. *Daily Texan*, 14 November.
Krueger, R. and Agyeman, J. (2005). Sustainability Schizophrenia or Actually Existing Sustainabilities? *Geoforum*, 36(4), pp. 410–417.
Lees, L., Shin, H.B. and López-Morales, E. (2016). *Planetary Gentrification*. London: Wiley.
Lillie, R. (1981). *Rainey Street*. Councilman Deuser's request for information on the City Council's 1980 action related to the Rainey Street area. Austin, AHC.
Lowry, C. (1978). *What Are You Going to Do When They Run You Out?* Rainey Street Association, AHC.
Marban, A. (1996). The Great Brown Hope. *The Austin Chronicle*, 28 June.
Martinez, P. (1980). *Rainey Barrio Preservation Plan*. Rainey Area Neighborhood Association. East Austin Chicano Economic Development Corporation, AHC.
McCann, E. (2007). Inequality and Politics in the Creative City-Region. *International Journal of Urban and Regional Research*, 31(1), pp. 188–196.
McDonald, J. (2012). *Racial Dynamics in Early Twentieth-Century Austin, Texas*. Lanham, MD: Lexington.
Miller, B. (2015). Sustainability For Whom? Sustainability How? In: D. Wilson, ed. *The Politics of the Urban Sustainability Concept*. Champaign, IL: Common Ground, pp. 107–116.
Moore, S.A. (2007). *Alternative Routes to the Sustainable City*. Lanham, MD: Lexington.
Mueller, E.J. and Dooling, S. (2011). Sustainability and Vulnerability. *Journal of Urbanism: International Research on Placemaking and Urban Sustainability*, 4(3), pp. 201–222.
Novak, S. (2001). Foreclosure Ends Town Lake Condo Project. *Austin American Statesman*, 11 September.
Oden, M. (2016). Equity: The Awkward E in Sustainable Development. In: S. Moore, ed. *Pragmatic Sustainability*. New York: Routledge, pp. 31–49.
Osborne, J. (2003). City Urged to Rezone Historic Rainey Street. *Austin American-Statesman*, 16 October.
Pearsall, H. and Anguelovski, I. (2016). Contesting and Resisting Environmental Gentrification. *Sociological Research Online*, 21(3).
Portney, K. (2013). *Taking Sustainable Cities Seriously*. Cambridge, MA: MIT Press.
Rainey Street Neighborhood Association. (2000). *Rainey Street Neighborhood Redevelopment Strategy*. Austin, AHC.
Real, D. (1979). Groups Voice Reservations About Austin's Revitalization. *The Daily Texan*, 26 April.
Rivera, D. (1999). A Neighborhood Puts Itself Up For Sale. *Austin American-Statesman*, 1 July.
Rosol, M. (2013). Vancouver's 'Ecodensity' Planning Initiative. *Urban Studies*, 50(11), pp. 2238–2255.
Semuels, A. (2015). Affordable Housing Always. *The Atlantic*, 5 July.
Slusher, D. (1986). Cookey Covets Rainey Street. *The Daily Herald*, 5 June.
Swearingen, W. (2010). *Environmental City*. Austin: University of Texas.
Swyngedouw, E. (2007). Impossible Sustainability and the Postpolitical Condition. In: R. Krueger and D. Gibbs, eds. *The Sustainable Development Paradox*. New York: Guilford Press, pp. 13–40.
The Austin Chronicle. (1999). Past Planning Efforts. *The Austin Chronicle*, 16 July.
The Community Economic Development Policy Research Project (1981). *Community Economic Development*. Community and Regional Planning Program. LBJ School, University of Texas Austin.

Tretter, E. (2016). *Shadows of a Sunbelt City*. Athens: University of Georgia Press.
Tretter, E. and Adams, M. (2012). The Privilege to Stay Dry. In: S. Dooling and G. Simon, eds. *Cities, Nature, and Development*. New York: Ashgate, pp. 187–206.
Tretter, E. and Sounny, M.A. (2013). *Austin Restricted*. The Institute for Urban Policy Research and Analysis. Austin: University of Texas.
US Census Bureau (2000). *Travis County, Total Population, Tract 11, Block Group 7*. Social Explorer. Accessed 15 November 2016.
US Census Bureau (2010). *Travis County, Total Population, Tract 11, Block Group 2*. Social Explorer. Accessed 15 November 2016.
Wilson, J. (1982). City Commits Land to Housing Project. *Austin American-Statesman*, 22 January.

PART IV

Spaces of governing and planning

David Wilson and Byron Miller

The topic of this section, spaces of governing and planning, is a prominent presence in current cities as a spatiality that cuts through and across vast constellations of districts: residential zones, commercial enclaves, industrial districts, downtowns, gentrified blocks and infrastructural corridors. Borrowing from Henri Lefebvre, this presence delivers a welter of spaces that come to litter cities, e.g. 'exclusionary upgrade spaces', 'liminal poverty spaces', 'bourgeois cultural districts', 'human struggle ghettos spaces', 'racialized no-go working class enclaves'. For governing and planning sear the entire ground of cities, encompassing land and property as regulatory and planning strategies are enacted. Planning strategies, the brain-trust of the operation, codify a vision of how a city morphology and its accompaniment of social relations are to be and evolve. Planning, a volatile domain for formal and informal politics, has a necessary transparency that renders it vulnerable to interrogation and contestation. Governing involves the processural unfolding of formal and informal rules and regulations that are served up to ensure that planning goals are met. Governing, it follows, is the on-the-ground cudgel of the city management dynamic.

The chapters in this section collectively show the pervasiveness and power of these spaces as they come to dot the physical and social fabric of cities. All of the essays reveal a politics of formal and informal regulating that at times can be remarkably elusive and fleeting but also poignantly graphic and visible. Each essay chooses its own conceptual orientation to address and refine widely accepted notions of how spaces of governance emerge and become sustained. Keen insights materialize from the essays as city power formations and institutional amalgams are shown to navigate the turbulent waters of neoliberal, transnational and austerity days. Cities, caught in the grip of historically specific structural frames, toil to convert complicated and shifting terrains of politics and evolving places of meaning into stable, planned economic apparatuses. Governing and planning, pierced by concerns about spatial realities and spatial outcomes, become active builders of arrays of spaces.

Martin Murray begins by examining the contagion of "fast-track urbanism" in African cities. Martin demonstrates a new urbanism that defies the standard model of incremental growth and transformation that rejects the process of pieces being retrofitted onto existing physical and social infrastructures. Here spaces of governance and planning are rife: these instant cities are frequently dramatic interventions onto city form. Simultaneously making and destroying city morphology, they embed flagrant and concealed socio-cultural symbols that strive seek to distance the dramatically new from the past and to forecast a progressive future. A prominent message abounds: these new 'instant cities', built from scratch, promise a future free from the eviscerated infrastructure, chaotic and congested streets, and ugly, out-of-control sprawl that scar these "on the margins of modernity" cities. A political opening is seized: the once deep belief in a modernist teleology – of a gradual and inevitable march to progress – has vanished. In its place, a brute, dressed-up neoliberalism is sold to publics. Murray ultimately informs us of a new kind of spatialized governance and planning, one that regulates through the production of deeply symbolic cities.

Jason Hackworth critically examines a new regime of political governance and planning, the "right-sizing" of cities. After years of pursuing counter-productive, growth-oriented planning strategies, Hackworth suggests, many cities across the Global North have begun to deliberately downsize their infrastructure and housing stock to meet projected demographic and fiscal realities. Posing the questions – is right-sizing a post-growth process, and if so in what sense? – Hackworth narrates this as a trope which meets immediate political and economic sensibilities. Planning and governance latch onto a scheme that decisively plays to concerns and perceptions but never address the structures and processes that spawned the new growth paradigm. This paradigm, to Hackworth, is as much a dodge as a clear intervention, a vision deliberately filled with absences and silences that lends it a credibility to publics too often seduced by easy solutions and technocratic fixes to city problems. Most poignantly erased, to Hackworth, is the routinized sanctioning of capital mobility and planning rituals within a capitalist order.

Phil Ashton studies the fluid and shifting spaces of local governance of housing markets and neighbourhood stabilization in the wake of the US foreclosure crisis. Ashton chronicles local constructions of contingent legal space that re-works property rights to displace responsibility for tax foreclosed properties to 'guilty' banks and lenders. Expanded rules involving property upkeep, revamped applications of eminent domain and land acquisition powers, and new notions of legal eminent domain and proper land acquisition are at the heart of this transformed legal reality and the new political spaces they constitute. From this, active space makers, i.e. public-private partnerships, builders, developers, work through ritualized issues of property ownership and property registration to shift the burden of responsibility of disinvestment and tax foreclosure realities to banks. A politics of legal interpretation, joined to a process of state-making as localities construct themselves as legal agents, introduces a new legal-political space that creates new residential, commercial and industrial content and zones.

David Giband chronicles the ironic reality of an active and expanding state apparatus in France's current neoliberal days (where retrenched government is promised). A recent fixation on provision of housing, manifested in a specific space, the *banlieues*, involve a renewed effort to manage the urban poor. The goal, to Giband, is to "middle size" the urban poor, i.e. have these populations conform to social, economic and moral middle-class norms while they are to be physically removed from the creation of re-entrepreneurialized cities. A therapeutic ethos is served up to the French public that centres the need for the poor to be cultural and ethically upgraded. At the same time, this population is to be spatially banished to the margins with the identifying of them as problematic bodies and presences in zones of economic and social upgrade. A popular discourse of social mix – where the poor in the *banlieues* will be transformed by benevolent contact with the middle income – carries these themes for public consumption. Neoliberal governances and the new political spaces they produce, to Giband, often bear little resemblance to the content of their (neoliberal) pronouncements.

Each chapter in this section grows our appreciation for the complexity and diverse qualities of current governance and planning space making. This section's essays form a unity in their incisive examining of distinctive kinds and elements of this manufactured space that is so central in the contemporary city. Current drives to govern and manage cities, we learn, are as relentless as ever as politically infused spaces are made to populate the urban. Current cities, in this unearthing, are seared by politics in their spatial unfolding as the 'force of governance' structures the content of residential spaces social spaces, economic spaces and the city in its entirety. Institutions in cities today negotiate a welter of powerful forces – globalization, speeded-up transnational realities, deepening neoliberalism, hegemonic structures of austerity – ever sensitive to how they can be mobilized and managed in the ongoing project of city building. It is in this spirit of desiring to more deeply understand this process that we offer this focus on governance and planning spaces of the city.

16
CITIES ON A GRAND SCALE
Instant urbanism at the start of the twenty-first century

Martin J. Murray

The unprecedented scale and scope of fast-track urbanism in the Persian Gulf and the Asia Pacific Rim at the start of the twenty-first century raises serious challenges to the traditional image of the modern metropolis – and even the very "idea of the city" (Marshall 2003: 191). These 'instant cities' defy the conventional understanding of urban growth and development as a slow process of layering and retrofitting, where creative destruction of the built environment, incremental re-building, and constant erasure and re-inscription produce an ever-evolving urban form that combines vastly different building typologies and architectural styles where the old and the new are cobbled together in uneven and hybrid configurations (Cuff 2002; Harvey 2006; Page 1999). The rapid transfers of capital, information technologies and expert advice (the so-called 'travelling ideas') have created the possibilities for 'compressed' or 'telescoped' urban development, enabling city builders to effectively start over by bypassing existing derelict environments and leapfrogging over the unwanted detritus of past experiments with modernity (Tomlinson 2007).

From the outset, what should be clear that these 'instant cities' have not followed established routes or pathways laid down elsewhere and at earlier times (Gubitosi 2008; Lerner 2008). The experimental city-building processes that have produced 'instant cities' and fast-track urbanism are, in short, not at all reducible to antecedents in North America and Europe – or anywhere else for that matter (Bunnell *et al.* 2012: 2786). What defines this 'instant urbanism' – or what some have labelled 'test bed urbanism' (Halpern *et al.* 2013) or ready-made 'cities-in-a-box' (Boudreau 2010) – is not linear temporal evolution or incremental gradualism, but spatial transformation that conforms to the dreamscape of starting afresh with a clean slate (Chen and De'Medici 2010; Wright 2008). Unlike the sometimes painstakingly slow processes of piecemeal evolution which shaped urban transformation in earlier centuries, 'instant cities' are the product of a "super-fast urbanism", where *tabula rasa* experiments with holistic master-planning have become the new prototype for city building (Bagaeen 2007: 174).

Planning theorists have no ready-made models or standardized formulas which are appropriate for dealing with this new type of fast-track urbanism. In seeking to locate themselves at the cutting-edge of new approaches to city building, these 'instant cities' both embrace and reject those modernist principles that informed architectural innovations and planning protocols over the past century. By fostering the separation of urban landscapes into precincts for work, dwelling, recreation and transportation, the design specialists who build 'instant cities' implicitly endorse the rigid modernist principles of land use zoning and functional specialization. Yet with no clear hierarchy between dominant centre and dependent peripheries, these 'instant cities' confound the modernist expectations of spatial differentiation and heterogeneous urban form (Irving 1993; Shatkin 2008). These new 'instant cities' are produced by industrial 'machine-age' processes so cherished by modernist city-builders, but they are built with such speed and are too dynamic to be controlled by the rigid modernist principles of order, functional specialization and spatial ordering (Barnett 2016; Kim 2013).

Large-scale, master-planned and holistically designed 'satellite cities' have become an increasingly common urban development concept that originated in the Persian Gulf and the Asia Pacific Rim, and quickly spread to India, Africa, the Middle East and elsewhere (Murray 2015a). For the most part, real estate developers – operating oftentimes in rough synchronization with municipal authorities and sometimes working almost completely on their own – have followed a similar spatial pattern of building entirely from scratch, starting *de novo* instead of trying to retrofit the existing built environment or inserting their mega-projects into existing urban landscapes already in place. What distinguishes these new master-planned satellite cities – or what Gavin Shatkin (2008) has called "bypass-implant urbanism" – from the conventional understanding of city building is that they are constructed entirely free from the spatial, institutional and logistical constraints associated with the existing urban fabric. Constructing these large-scale urban redevelopment projects does not conform with the modernist principles of erasure and re-inscription whereby city builders incrementally rebuild the urban social fabric by tearing down outdated parts of the existing built environment to make way for something new, but instead are master-planned and holistically designed enclaves – what can be called "urban integrated mega-projects" (Shatkin 2011) – where a deliberately conceived assemblage of distinct parts fit into a functionally integrated whole. Building at such a large scale and at such a quick pace has produced a kind of 'fast-track' urbanism where all sorts of 'instant cities' seem to emerge out of nowhere like alien spaceships unexpectedly coming to ground without much forethought or preparation. As a general rule, these new satellite cities are strategically located outside the historic central core of existing metropolises, often with easy access to sparkling new airports and tourist-entertainment centres designed to attract corporate business travellers and the cosmopolitan consumerati. In a departure from past efforts at state-driven master planning and 'new town' development, these new large-scale, self-contained cities are constructed as profit-making undertakings, often the brain-child of a single prophetic real estate developer or a consortium of property investors, sometimes in partnership or alliance with municipal or state authorities. While many of these imagined future-cities have not advanced beyond the initial planning stages (Murray 2015a; Watson 2014), a number of exemplary prototypes for this kind of city building – including Masdar City (Abu Dhabi), Songdo IBD (outside Seoul), Astana (Kazakhstan), Dongtan Eco-City (outside Shanghai), Lavasa (outside Mumbai), Lusail City (Qatar), Dholera Smart City (Gujarat) and Waterfall City (Johannesburg) – have come into existence (Chang and Sheppard 2013; Cugurullo 2013; Datta 2012; Herbert and Murray 2015; Jazeel 2015; Joss 2011; Kim 2010; Murray 2015b; Shwayri 2013).

The main objective behind fast-track urban experiments is not simply to rationalize urban space by bringing it into conformity with up-to-date 'international best practices', but to create opportunities for maximizing profitability. 'Instant cities' coincide with a model of urban development that entails the synchronized assemblage of spatial components: hard and soft infrastructure, the provision of requisite services and pre-packaged building typologies are rolled out simultaneously as a fitted ensemble of integrated parts that combine to produce a pre-conceived whole. What propels this kind of 'fast-track' urbanism is the wish-image of contemporary global capitalism: the wholesale transformation of urban space into a profit-making machine (Herbert and Murray 2015; Murray 2015b). Whether the requisite capital investment originates from local sources or is mobilized from far afield, city builders have become fixated with using the sights and sounds of *faux*-urban life – the spectacle, enchantment and excitement of what Richard Lloyd and Terry Nichols Clark (2001) have called the "city as entertainment machine" – as a source of capital accumulation in and of itself (Koch 2012).

Early prototypes: historical antecedents

Holistically designed experiments in planned urban living are certainly not new in the history of city building. The origins of these holistic re-urbanizing efforts at the start of the twenty-first century can be traced to a variety of earlier prototypes. The utopian desire to create entirely new cities set free from persistent disorder and instability of modern urban life has its roots in the past. One particularly important early prototype for master-planned urban enclaves can be found in the construction of philanthropic 'company towns' in Europe and North America.[1] Margaret Crawford (1995: 1–2), for example, has explored the historical transformation of American company towns as a distinct urban form spanning a 150-year time-frame. The planning behind 'company towns' evolved from a vernacular building activity to a professional design task, undertaken by architects, design specialists and town planners. All in all, company towns were (and have remained) the spatial manifestation of a social ideology of paternalism embedded in the crass economic rationale of profit-maximization. Similarly, the Garden City movement that blossomed in the early twentieth century sought to offer an idyllic, community-based alternative to the teeming, over-crowded, polluted, chaotic and miserable modern cities of industrial capitalism. Inspired by Sir Ebenezer Howard, the Garden City movement conjoined two very different types of late nineteenth-century experimental communities, creating a deeply entrenched tension that has never been fully resolved. The first type of community – which was at once utopian, idealistic and radical in nature – challenged conventional laissez-faire values by endorsing forms of small-scale, networked social life that rested on cooperation, fraternity and shared economic progress. In contrast, the second type of community found expression in the model industrial towns of 'enlightened' capitalists – an approach that reinforced the values of rule-oriented hierarchies epitomized by top-down paternalism (Batchelor 1969; Buder 1969, 1990; Burgess 1993). As the Garden City model was exported to colonial Africa, it brought together a racially inscribed paternalism with rigid controls over social life (Bigon 2013). Finally, the New Towns initiative in post-war Great Britain (like Milton Keynes outside London) that flourished from the 1940s to the 1960s was one of the largest public housing building programmes of its kind (Degen *et al.* 2008). The New Towns approach to city building represented yet another iteration of large-scale planned communities that offered an alternative to disorderly industrial cities (Alexander 2009; Clapson 1998; Simson 2001).

The utopian goal of reducing – or even eliminating – disorder "reached an unprecedented level at the high-noon of modernism in the [early] twentieth century" (Schindler 2015: 18). Le Corbusier's all-encompassing plans to reshape cities in ways which conformed to the principles of efficiency and rational use of space appealed to municipal authorities "eager to display their modernist credentials" (Schindler 2015: 18). To this end, new cities built from scratch, such as Brasília and Chandigarh, were over-burdened with an excess of symbolic importance meant to indicate the kind of forward-looking progress associated with the modernist age. In the latter half of the twentieth century, urban planners infused with the spirit of high modernism sought to "decongest" historic city centres by uprooting the working poor and relocating industry to the urban periphery (Schindler 2015: 18–19).[2] This idea of planned settlements as self-contained enclaves has a contemporary counterpart in company towns attached to corporate enterprise and extractive industries like oil, minerals and agricultural commodities (Appel 2012; Ferguson 2005).

These early utopian city-building efforts and the new master-planned satellite cities that blossomed at the start of the twenty-first century have much in common: both are large-scale, holistically designed, integrated and purpose-built enclaves attached to the edges of existing metropolises as stand-alone (self-contained) entities. Both are very much the product of their time, each in their own way offering a forward-looking antidote to the perceived distressed urbanism that surrounded them. But ultimately both the new master-planned satellite cities at the beginning of the new millennium and earlier efforts at building 'off-grid' communities amounted to open-ended and quasi-utopian experiments. The new corporate-led city-building projects that originated in the Persian Gulf and the Asia Pacific Rim and quickly spread everywhere have come into existence as aspirational epicentres for privileged stakeholders striving for commercial success in the highly competitive circumstances of global capitalism. But with so many of these new master-planned mega-projects in the pipeline, it is difficult to predict the outcome of so much feverish city building at a time of intense competition for a place in the rank-order of aspiring world-class cities. A vast global urban experiment is currently under way, with not nearly enough forethought or analysis as to how these satellite cities will eventually affect the economies, physical environments and everyday lives of people who will live both in and outside of them (Lumumba 2013; Watson 2009).

Examining this new prototype for city building at the start of the twenty-first century requires a counter-intuitive excursion into a retro-futuristic global urbanism that brings to mind surrealist painting or science fiction novels. These new 'instant cities' are eerily reminiscent of such widely divergent urban experiments as walled medieval cities and city-states, the confederation of merchant towns assembled under the banner of Hanseatic League that dominated the Baltic maritime trade from the fourteenth to the seventeenth centuries, the colonial 'free ports' of the eighteenth and nineteenth centuries, the master-planned Garden Cities popularized by Ebenezer Howard at the start of the twentieth century, the high-modernist cities built out of whole cloth and that served as symbols of national destiny (Brasília, Abuja and Chandigarh), the fast-track urbanism of the Persian Gulf (and its spin-offs loosely termed 'Dubaization') and the towering 'skyscraper cities' built virtually overnight in the Pearl River delta of southern China. Yet what makes these 'instant cities' strikingly different from earlier experiments with master-planned, holistically designed city-building efforts is that they consist of a seemingly novel mixture of enclave capitalism, unregulated neoliberalism and generic modular city designs. These cities built from scratch bring to mind the phrase "Nothing dates faster than people's fantasies about the future" – an aphorism coined by Robert Hughes (1980) to describe the fabrication of Brasília. Without continuously reinventing promises of the future-yet-to-come, the 'instant cities' – as exemplary expressions of global urbanism in the neoliberal age – become passé and outdated (see Ferguson 2006: 17). These efforts at large-scale re-urbanism gesture towards genuine city-ness, but they seem more like self-contained platforms for projecting a roseate image of urbanity that can only exist in microcosm (Ong 2011; Sidaway 2007).

Reading fast-track urbanism

Reading these fast-track satellite cities as narratives, or more precisely, as spatial stories embedded as they are in utopian visions about the future, enables us to decipher the deeply rooted reservoir of cultural conventions associated with conceptions of novelty and progress. As Walter Benjamin (1999: 546) put it, "every epoch appears to itself [as] inescapably modern". Seen from this angle, these instant cities contain hidden socio-cultural messages that seek to distance themselves from the past and to foretell the radiant future. These self-contained, cocooned enclaves envision an 'integrated lifestyle' approach to city living, where material scarcity is non-existent and where strict regulatory regimes ensure predictability, stability and rational use of space. What is perhaps not surprising is that past 'utopias of social space' (City Beautiful, the Garden City, the Radiant City and/or the City of Tomorrow) have shown themselves to be remarkably resilient as models of city building that can be bent, folded and re-packaged to serve contemporary purposes. The acknowledgement that holistic master-planning lies at the heart of these new instant cities suggests that the modernist-inspired impulse towards 'starting over' – the relentless production of the new by dispensing with what came before – has remained a powerful force in imagining the Future City. Yet the history of city building is cluttered with the abandoned ruins of once-vibrant dreamscapes, discarded relics of bygone eras that no longer resonate and that have lost their *raison d'etre* (Buck-Morss 1995; Edensor 2005).

What is striking about the spatial development models, architectural designs and artist imagined renditions that appear in glossy promotional materials and visual presentation is the singular message about how these new 'instant cities' built from scratch promise a future free from the broken-down infrastructure, overcrowded streetscapes and unplanned sprawl that mark the present situation for struggling cities at the margins of modernity. A once-enduring faith in the modernist teleology of gradual-yet-inevitable march of progress has evaporated. Nostalgia for the future – to borrow out of context an apt phrase from Charles Piot (2010) – reflects deep-seated anxieties about the troubled urban present. With the visible ruins of past failures blocking the way forward, the language of 'building from scratch' and starting afresh provides an available (and uplifting) discourse for imaging a future unencumbered by the past. These calls to wipe the slate clean, to leap over the past and to achieve the coveted status of world-class city reflect a longing for a frictionless utopia, or a worldly paradise where cities buzz and hum with excitement, offering ceaseless opportunities for prosperity and advancement (see Glaeser 2009, 2011).

Experimental urbanism: building cities from scratch

While modernist approaches to a comprehensively planned urbanism as a strategy and method of state-led intervention has come under severe attack, the modernizing impulses that gesture towards the imagined ideal of a future urban utopia have not disappeared (Swyngedouw 2005: 127–128). At a time when state-sponsored modernist visions of city building have lost traction around the world, large-scale corporate enterprise has stepped into the breach, promising to start afresh and build new satellite cities that are efficient, sustainable and well-managed. At the start of the twenty-first century, large-scale corporate enterprises have unveiled a plethora of audacious schemes that include ambitious visions for master-planned, holistically designed 'private cities' cocooned under the umbrella of autonomous zones, privately sponsored infrastructure projects on a grand scale, and the unprecedented empowerment of corporate actors in urban governance (Acuto 2010; Angotti 2013; Datta 2012; Hogan and Houston 2002; Murray 2015b).

Constructing entirely new cities from scratch represents a powerful new trend in city building at the start of the twenty-first century. While tinkering with the incremental rebuilding of existing urban landscapes remains the dominant mode of urban transformation in cities around the world, the turn towards building master-planned, holistically designed satellite cities has significantly reshaped thinking about urbanism and urban life. Despite the celebratory praise showered on such acclaimed 'instant cities' as Dubai, Doha, Abu Dhabi and Shenzhen, this experimental approach to city building remains largely in its infancy (Bagaeen 2007; Dempsey and Cameron 2014; Elsheshtawy 2008). Stories of success are paralleled by sobering tales of failure, where over-confident city builders have been unable to make good on initial promises. What remains an open question is the extent to which construction of completely new cities *de novo* foreshadows city building to come over the foreseeable future (Watson 2014).

Time, chronology and history

Instant urbanism conflates history and chronology. Unlike the fits and starts and reversals that characterize the temporality of history, chronological time conveys the idea of sequential movement along a linear pathway. While it may appear as natural and value-neutral, linear time usually conveys a judgement – "new is better than the old; recent is superior to the past; and the future will be better" (Tanaka 2015: 162). Instant urbanism conceives of city building as a continuous process unfolding in linear or chronological time along a preordained route from inception to completion.

'Instant cities' and 'fast-track urbanism' project images of sudden, almost instantaneous transformation where something is created (as if by magic) out of nothing. These depictions of building on a blank slate provide grounds for dismissal as exemplary expressions of a 'false modernity': in this reading, 'instant cities' are not 'real', but instead they are artificial creations, or an otherworldly kind of hallucinogenic phantasmagoria. As Natalie Koch (2012: 2447) has suggested, the "Disney stigma casts the city as a theme park", and not an actual place. 'Instant cities' sorely lack the layered accretions of history and collective memory that shape and characterize conventional urban settlement patterns (Elsheshtawy 2008: 258; Hubbard and Lilley 2004).

In certain respects, the master-planned, holistically designed 'instant cities' that have blossomed at the start of the twenty-first century bear a striking resemblance to their modernist and high-modernist forbearers (Schindler 2015: 19). The dreamscape of building brand new cities is strongly linked to the persistent myth that the cure for the illness of urban disorder can be found in simply abandoning existing built environments, walling out the unwanted and starting anew.

Following Zack Lee (2015), it can be argued that 'instant cities' can be fruitfully understood as diverse assemblages of *worlding* practices, where city builders cobble together diverse ideas, features and programmes to fashion their peculiar version of world-class urbanity linked into the global economy. At a material level, city builders often use architecture "as a surface for projections to another real space, more perfect and better arranged, that should become actualized in reality" (Mattsson 2013: 124). In thinking about up-to-date design, they often tap into popular 'travelling ideas' and 'international best practices' as sources of inspiration (McCann 2011; Peck and Theodore 2010).

The libertarian option: charter cities and the dreamscape of blank slate urbanism

The thinking behind 'instant cities' cannot trace its origins to a single fountainhead of ideas, but instead has borrowed promiscuously from a wide variety of different sources. Extreme libertarian versions of pure public choice theory periodically appear in idealistic proposals to create new 'start-up' cities that are not – in theory at least – encumbered by the clumsiness and inefficiencies of interest-group politics that function through bargaining and compromise. These libertarian thinkers wish "to wipe the messy slate clean" (Storper 2014: 117). This enduring fantasy of starting over with self-governing enclaves has its roots in utopian thinking that curiously has animated both free-market libertarians and communitarian anarchists.

A prominent US economist-turned-policy-entrepreneur, Paul Romer, has proposed the concept of Charter Cities as a radical solution to the problems associated with rapid and unregulated urbanization on a global scale. As a framework for addressing dysfunctional urbanism, the Charter Cities initiative focuses on the potential for so-called 'startup cities' to fast-track the kinds of institutional reforms conductive to private enterprise and entrepreneurship. Charter Cities function in accordance with strict adherence to institutional frameworks which establish new market-based 'rules of the game' governing market-based transactions and structuring social interaction. The ultimate goal of the Charter Cities movement is to establish privately owned and managed urban enclaves which are geographically, socially, fiscally, administratively and, ultimately, politically separate from the host nations within which they are located (Freiman 2013).

The key to the success of the Charter Cities model is to establish clear institutional rules beforehand that apply uniformly to all who decide to participate and to effectively enforce those rules so as to reinforce long-term commitments (Fuller and Romer 2010). The effective enforcement of these rules – primarily geared towards fostering competition and choice – creates the solid bedrock of trust that private investors and property-holding individuals require in order to make long-term commitments to engaging in marketplace exchange. In the model developed by Romer (and grounded in post-neoclassical endogenous growth theory), Charter Cities bear a striking resemblance to Special Economic Zones, that is, cocooned enclaves that adopt strict rules for managing specific economic goals in ways that differ from those governing their 'host' countries. Yet unlike special economic zones, Charter Cities move beyond the focus on the narrow management of specific economic goals by expanding the scope of institutional rules that allow for the support of a broader range of market transactions and exchanges in a modern capitalist economy and to structure social interactions appropriate for a well-run city (Romer 2009). In short, these Charter Cities function as special administrative zones (or special reform zones) with sound rules and legal institutions, pro-growth and investment-friendly policies, and environmentally sustainable practices. Outfitted with such ring-fencing operations, privately managed Charter Cities belong to that distinct subcategory of self-governing enclaves that exist in the extra-territorial realm of pure administrative autonomy.

This attachment to laissez-faire ideas underpinned an ambitious scheme (initiated in 2011–2012 but eventually abandoned amidst a great deal of public outcry and protest) to launch several model 'free market cities' in Honduras (Rodas and Pineda 2013: 7–10, 36–38, 50).[3] Sometimes referred to as Free Cities, Free Towns, Charter Cities, Future Cities, Enterprise Zones and Special Development Regions, these utopian laissez-faire city-building projects were envisioned as "special development regions" located on unoccupied land roughly the size of Hong Kong and Singapore and administered "by an extra-governmental committee that would not be subject to any existing [regulatory and] institutional constraints" (Rodas and Pineda 2013: 50; Storper 2014: 117).

While the construction of 'free cities' may have stalled in Honduras, real estate developers have already begun construction of a private city on the outskirts of Guatemala City. Hailed as a safe haven from the crime-ridden capital, this new self-enclosed city, called Paseo Cayala, consists of a full range of amenities, including apartments, boutiques, retail shops, nightclubs, restaurants and storefronts designed in a Spanish Colonial architectural style contained inside a 14-hectare compound encircled by high white stucco walls. The cheapest home prices are seventy times the average yearly wage in Guatemala. Developers have already unveiled plans to expand the size of the privately owned and privately managed city into a fully operational 352-hectare alternative to the dangerous and congested streets of Guatemala City. The single gated access point leads to an underground parking garage, from which residents and visitors proceed through gazebo-covered escalators decorated like art-nouveau Parisian Metro stations onto uncluttered streets patrolled by armed private security guards (Ruiz-Goiriena 2013).[4]

Varieties of master-planned cities built from scratch

There is no single formula, or shared template, that captures the variety of possible iterations for this *tabula rasa* approach to city building. In some cases, these 'instant cities' are expressions of simply starting over, building an entirely new metropolis from the ground upward. The feverish construction of Astana, the capital city of Kazakhstan, reflects this pattern (Koch 2010, 2013; Köppen, 2013; Schatz 2004). In other cases, these fast-track satellite cities offer the promise of smoothly functioning alternatives to the messy realities of everyday urbanism that surrounds them. In still others, they represent the creation of well-functioning economic hubs, technologically advanced 'full service' cities that can compete for enterprise and talent in the global marketplace. Master-planned, holistically designed 'smart cities' (or eco-cities, 'intelligent cities', 'ubiquitous cities' and whatever other trendy name seems to fit) are projected to be the new urban utopias of the twenty-first century (Bunnell 2015; Datta 2015b; Rapoport 2014). By integrating comprehensive spatial planning with up-to-date digital technologies and information systems, 'smart cities' (and their similarly conceived brethren) have become the new urban laboratories for technical innovation, surging to the forefront of worldwide marketing efforts as solutions to the challenges of rapid urbanization and sustainable development (Datta 2012). In China, South East Asia and India (with over a hundred 'smart cities' under construction or in the advanced stage of planning), city builders have turned to some variant of 'smart cities' as launching pads for spurring economic growth (Bunnell and Das 2010; Datta 2015a, 2015b; Hollands 2008).

'Smart cities', eco-cities and other variants of the current fixation with sustainable urbanism are part of a longer genealogy of utopian urban planning ideas that have emerged and taken root as a response to the difficult challenges of development and modernity in the contemporary era. Built in the shadow of Dubai and other successful 'instant cities' that came of age during the contemporary era of neoliberal globalization, these master-planned, holistically designed 'smart cities' reflect a turn towards 'entrepreneurial urbanism' – a wholesale shift in urban governance that signals the triumph of the new urban imaginary of cities as business enterprises rather than experimental models for promoting social justice and social inclusion (Acuto 2010; Datta 2015a, 2015b; Murray 2015b).

The corporate cities paradigm

The steady growth of privately managed 'corporate cities' is a disturbing harbinger of twenty-first-century urbanism. As part of a global phenomenon, real estate developers and corporate property-owners assume an expanded role in determining what form the built environment takes and in creating regulatory (management) regimes that serve their narrow preoccupations over and above any consideration of the public interest (Easterling 2008: 30–45). In the typical case, their spatial layout consists of assortments of districts, zones or thematic enclaves, with names that evoke familiar images of rewarding work, domiciliary tranquillity and leisurely playfulness. Real estate developers market these self-contained corporate enclaves as mythologized 'fairy-tale urbanism', where daily life is carefree and where residents and visitors alike are spoiled for an abundance of equally enviable choices.

Corporate cities, entrepôts, bubble urbanism, sequestered enclaves – the names are interchangeable but the formats are strikingly similar. These corporate cities occupy a discrepant territory between two different species of urbanity: open and cosmopolitan, on the one side, and secretive and opaque, on the other (Easterling 2007: 75). These corporate precincts have succeeded in combining the 'connected expansiveness' of the networked global marketplace and transnational business with the sequestered enclave format that enables those inside to maintain both physical and psychological distance from what happens outside their perimeter. These corporate enclaves operate within the realm of extra-territorial sovereignty in the historical tradition of free trading ports, Maroon communities, offshore banking, pirate hideouts and frontier colonies (Easterling 2007, 2008). As an urban paradigm, these large-scale urban enclaves function as pre-programmed, self-regulating 'city-states' with inexhaustible possibilities for business and trade, precisely because of the lack of public authority and municipal oversight. Such 'smart eco-cities' like Songdo IBD, Dholera Smart City (Gujarat) and Masdar are marketed as problem-free enclaves (Carvalho 2015). Corporate cities have manufactured their own regulatory powers outside conventional norms of public administration that enable them to bypass procedural rules of democratic decision-making (Herbert and Murray 2015; Hogan and Houston 2002; Murray 2015b).

Corporate cities cannot operate without duplicity, secrecy and subterfuge. Theories of globalization – with their fanciful fables about the waning nation-state and the triumph of a borderless world – are the perfect camouflage for a globally ascendant corporate business culture "that clearly prefers to manipulate *both* state and non-state sovereignty, alternatively releasing and laundering their power and identity to create the most advantageous political or economic climate" (Easterling 2007: 76). Corporate cities operate between state and non-state jurisdictions, seeking the extra-jurisdictional spaces (like export processing zones, free trade zones or special economic zones) that provide a relaxed atmosphere of legal immunity (Bach 2011; Barry 2006; Easterling 2007, 2008). They aspire to lawlessness, not in the sense of anarchy or the non-recognition of authority, but in the legal tradition of a state of exception – a circumstance where regulatory regimes operate with rules that apply inside but not outside their own borders (Easterling 2007, 2008).

The expansion of privately managed corporate cities depends upon the adoption of new modes of urban governance premised on the delegation to private business interests of decision-making powers once reserved for public authorities. The implementation of what Ananya Roy, Gavin Shatkin and others have described as 'new planning regimes' have transferred the provision of large-scale infrastructure projects and urban management services into the hands of private real estate developers (Roy 2009; Shatkin 2011). This shift of urban planning functions into corporate hands creates a new geometry of spatial production which privileges privatized spaces for elite and corporate consumption, large-scale mega-projects intended to attract corporate investments, and deregulated localities such as free trade zones, enterprise zones and other extra-territorial areas. Whatever the differences between these places, the suspension of conventional regulatory powers enables these 'zones of exception' to provide special benefits for private corporations (Bach 2011; Easterling 2007, 2008).

Building instant cities from scratch: trends and possibilities

Constructing new cities *de novo* represents a new type of city building that marks a significant break from the conventional stress on retrofitting existing urban landscapes or regenerating neglected or under-utilized areas. Discourses linked with 'fast-track urbanism' are deeply rooted in seductive and normative visions of a bright future where building from scratch seems to offer a way around urban dysfunctionalities of all past experiments with city building. 'Instant cities' represent an extreme and complex expression of global urbanism at the start of the twenty-first century. 'Instant cities' like Dubai, Doha and Abu Dhabi represent an emerging prototype for market-led city building in the twenty-first century: 'prosthetic and nomadic oases' that assume the form of vast urban enclaves, either standing alone in defiant isolation or inserted incongruously into existing urban landscapes. Lacking in any historical identity and with few gestures to the past, these hyper-active cities thrive on newness, ephemerality, bigness and 'unprecedented optimism': spectacle and enchantment replace the everyday urbanism of compact neighbourhoods with their integration of mixed land uses, slow pace and quaint streetscapes. Built to resemble large-scale theme parks or gigantic stage-sets, these 'instant cities' consist of clusters of 'destination-precincts' held together by virtue of proximity and defined by their functional specialties: work and business, home and privacy, recreation and distraction, leisure and entertainment. Under circumstances where fast-tracked urbanism is the operative principle, the serial reproduction of standardized modules that are cobbled together to resemble a functional city has increasingly become the dominant mode of city building (Ponzini 2013).

The boosterist branding that accompanies these 'instant cities' boasts that these places are at the forefront of sustainable urbanism with up-to-date infrastructure and cutting-edge information and communications technologies. But these claims about ecological sustainability are largely justifications or legitimizations that often conceal as much as they reveal. At the end of the day, these 'instant cities' are primarily corporate business platforms that promote entrepreneurialism and profit-making activities. City builders use the rhetoric of sustainable urbanism, including energy-saving technologies and re-usable resources, as a way to create an aura of authenticity about themselves (Kim 2013).

Notes

1 Examples of master-planned 'company towns' include the proto-industrial 'Werksiedlungsbau' in the German Ruhr area (like housing estates for workers built by Krupp), the planned settlements constructed by Pullman (Pullman Town in Chicago, 1880), Cadbury (Bourneville near Birmingham, 1880) and Lever (Port Sunlight near Liverpool, 1887).
2 For New York City, see Robert Caro (1974), which chronicles the grandiose ambitions of Robert Moses, perhaps the quintessential modernist city builder of the twentieth century.
3 See 'Honduras Shrugged', *The Economist*, 10 December 2011; and 'Charter Cities: Unchartered Territory', *The Economist*, 6 October 2012.
4 See also 'Guatemalan Capital's Wealthy Offered Haven in Gated City', *Guardian* (UK), 9 January 2013.

References

Acuto, M. (2010). High-rise Dubai Urban Entrepreneurialism and the Technology of Symbolic Power. *Cities*, 27(4), pp. 272–284.
Alexander, A. (2009). *Britain's New Towns: From Garden Cities to Sustainable Communities*. New York: Routledge.
Angotti, T. (2013). *The New Century of the Metropolis: Urban Enclaves and Orientalism*. New York: Routledge.
Appel, H. (2012). Offshore Work: Oil, Modularity, and the How of Capitalism in Equatorial Guinea. *American Ethnologist*, 39(4), pp. 692–709.
Bach, J. (2011). Modernity and Urban Imagination in Economic Zones. *Theory, Culture, & Society*, 28(5), pp. 98–122.
Bagaeen, S. (2007). Brand Dubai: The Instant City; or the Instantly Recognizable City. *International Planning Studies*, 12(2), pp. 173–197.
Barnett, J. (2016). *City Design: Modernist, Traditional, Green and Systems Perspectives*. 2nd ed., New York: Routledge.
Barry, A. (2006). Technological Zones. *European Journal of Social Theory*, 9(2), pp. 239–253.

Batchelor, P. (1969). The Origin of the Garden City Concept of Urban Form. *Journal of the Society of Architectural Historians*, 28(3), pp. 184–200.

Benjamin, W. (1999). *The Arcades Project* [Howard Eiland and Kevin McLaughlin, translators]. Cambridge, MA: The Belknap Press, Harvard University.

Bigon, L. (2013). Garden Cities in Colonial Africa: A Note on Historiography. *Planning Perspectives*, 28(3), pp. 477–485.

Boudreau, J. (2010). Cisco Systems Helps Build Prototype For Instant 'City in a Box'. *Washington Post*, 9 June.

Buck-Morss, S. (1995). The City as Dreamworld and Catastrophe. *October*, 73, pp. 3–26.

Buder, S. (1969). Ebenezer Howard: The Genesis of a Town Planning Movement. *Journal of the American Institute of Planners*, 35(6), pp. 390–398.

Buder, S. (1990). *Visionaries and Planners: The Garden City Movement and the Modern Community*. Oxford: Oxford University.

Bunnell, T. (2015). Smart City Returns. *Dialogues in Human Geography*, 5(1), pp. 45–48.

Bunnell, T. and Das, D. (2010). Urban Pulse – A Geography of Serial Seduction: Urban Policy Transfer from Kuala Lumpur to Hyderabad. *Urban Geography*, 31(3), pp. 277–284.

Bunnell, T. Goh, D., Lai, C-K. and Pow, C-P. (2012). Introduction: Global Urban Frontiers? Asian Cities in Theory, Practice and Imagination. *Urban Studies*, 49(13), pp. 2785–2793.

Burgess, P. (1993). City Planning and the Planning of Cities: The Recent Historiography. *Journal of Planning Literature*, 7(4), pp. 314–327.

Caro, R. (1974). *The Power Broker: Robert Moses and the Fall of New York*. New York: Alfred Knopf.

Carvalho, L. (2015). Smart Cities From Scratch? A Socio-Technical Perspective. *Cambridge Journal of Regions, Economy and Society*, 8(1), pp. 43–60.

Chang, C.I-C. and Sheppard, E. (2013). China's Eco-Cities as Variegated Urban Sustainability: Dongtan Eco-City and Chongming Eco-Island. *Journal of Urban Technology*, 20(1), pp. 57–75.

Chen, X. and De'Medici, T. (2010). Research Note – The 'Instant City' Coming of Age: Production of Spaces in China's Shenzhen Special Economic Zone. *Urban Geography*, 31(8), pp. 1141–1147.

Clapson, M. (1998). *Invincible Green Suburbs, Brave New Towns: Social Change and Urban Dispersal in Post-War England*. Manchester: Manchester University Press.

Crawford, M. (1995). *Building the Workingman's Paradise: The Design of American Company Towns*. New York and London: Verso.

Cuff, D. (2002). *Provisional City: Los Angeles Stories of Architecture and Urbanism*. Cambridge, MA: MIT Press.

Cugurullo, F. (2013). How to Build a Sandcastle: An Analysis of the Genesis and Development of Masdar City. *Journal of Urban Technology*, 20(1), pp. 23–37.

Datta, A. (2012). India's Ecocity? Environment, Urbanisation, and Mobility in the Making of Lavasa. *Environment and Planning C*, 30(6), pp. 982–996.

Datta, A. (2015a). A 100 Smart Cities, a 100 Utopias. *Dialogues in Human Geography*, 5(1), pp. 49–53.

Datta, A. (2015b). New Urban Utopias of Postcolonial India: 'Entrepreneurial Urbanization' in Dholera Smart City, Gujarat. *Dialogues in Human Geography*, 5(1), pp. 3–22.

Degen, M., DeSilvey, C. and Rose, G. (2008). Experiencing Visualities in Designed Urban Environments: Learning from Milton Keynes. *Environment and Planning A*, 40(8), pp. 1901–1920.

Dempsey, M. and Cameron, M. (2014). *Castles in the Sand: An Urban Planner in Abu Dhabi*. Jefferson, NC: MacFarland & Company.

Easterling, K. (2007). The Corporate City is the Zone. In: C. de Baan, J. Declerck, V. Patteeuw, eds. *Visionary Power: Producing the Contemporary City*. Rotterdam: NAi Publishers, pp. 75–85.

Easterling, K. (2008). Zone. In: I. Ruby and A. Ruby, eds. *Urban Transformation*. Berlin: Ruby Press, pp. 30–45.

Edensor, T. (2005). *Industrial Ruins: Space, Aesthetics, and Materiality*. London: Berg.

Elsheshtawy, Y. (2008). Cities of Sand and Fog: Abu Dhabi's Global Ambitions. In: Y. Elsheshtawy, ed. *The Evolving Arab City: Tradition, Modernity and Urban Development*. New York: Routledge, pp. 258–304.

Ferguson, J. (2005). Seeing Like an Oil Company: Space, Security, and Global Capital in Neoliberal Africa. *American Anthropologist*, 107(3), pp. 377–382.

Ferguson, J. (2006). *Global Shadows: Africa in the Neoliberal World Order*. Durham, NC: Duke University Press.

Freiman, C. (2013). Cosmopolitanism Within Borders: On Behalf of Charter Cities. *Journal of Applied Philosophy*, 30(1), pp. 40–52.

Fuller, B. and Romer, P. (2010). Cities from Scratch. *City Journal*, 20(4). Available at: http://city-journal.org/2010/20_4_charter-cities.html [accessed 7 December 2017].

Glaeser, E. (2009). *Triumph of the City: How Our Greatest Invention Makes Us Richer, Smarter, Greener, Healthier, and Happier*. New York: Penguin Books.

Glaeser, E. (2011). Engines of Innovation. *Scientific American* (September), pp. 50–51, 54–55.

Gubitosi, A. (2008). Fast Track Cities – Dubai's Time Machine. In: A. Moustafa, J. Al-Qawasmi and K. Mitchell, eds. *Instant Cities: Emergent Trends in Architecture and Urbanism in the Arab World: The Third International Conference of the Center for the Study of Architecture in the Arab Region*. Sharaj, UAE: CSAAR Press, pp. 89–102.

Halpern, O., LeCavalier, J. and Calvillo, N. (2013). Test Bed Urbanism. *Public Culture*, 25(2), pp. 272–306.

Harvey, D. (2006). Neo-Liberalism as Creative Destruction. *Geografiska Annaler: Series B, Human Geography*, 88(2), pp. 145–158.

Herbert, C. and Murray, M. (2015). Building New Cities from Scratch: Privatized Urbanism and the Spatial Restructuring of Johannesburg After Apartheid. *International Journal of Urban and Regional Research*, 39(3), pp. 471–494.

Hogan, T. and Houston, C. (2002). Corporate Cities: Urban Gateways of Gated Communities Against the City: The Case of Lippo, Jakarta. In: T. Bunnell, L. Drummond and K.C. Ho, eds. *Critical Reflections on Cities in Southeast Asia*. Singapore: Times Academic Press, pp. 243–264.

Hollands, R.I.G. (2008). Will the Real Smart City Please Stand Up? *City*, 12(3), pp. 303–320.

Hubbard, P. and Lilly, K. (2004). Pacemaking the Modern City: The Urban Politics of Speed and Slowness. *Environment and Planning D*, 22(2), pp. 273–294.

Hughes, R. (1980). *Trouble in Utopia, Episode 4: The Shock of the New*. BBC and Time-Life Films. New York: Ambrose Video Publishing, 2001, DVD.

Irving, A. (1993). The Modern/Postmodern Divide in Urban Planning. *University of Toronto Quarterly*, 62(4), pp. 474–488.

Jazeel, T. (2015). Utopian Urbanism and Representational City-ness: On the Dholera Before Dholera Smart City. *Dialogues in Human Geography*, 5(1), pp. 27–30.

Joss, S. (2011). Eco-Cities: The Mainstreaming of Urban Sustainability; Key Characteristics and Driving Factors. *International Journal of Sustainable Development and Planning*, 6(3), pp. 268–285.

Kim, C. (2010). Place Promotion and Symbolic Characterization of New Songdo City, South Korea. *Cities*, 27(1), pp. 13–19.

Kim, J.I. (2013). Making Cities Global: The New City Development of Songdo, Yujiapu and Lingang. *Planning Perspectives*, 29(3), pp. 329–356.

Koch, N. (2010). The Monumental and the Miniature: Imagining 'Modernity' in Astana. *Social & Cultural Geography*, 11, pp. 769–787.

Koch, N. (2012). Urban 'Utopias': The Disney Stigma and Discourses of 'False Modernity'. *Environment and Planning A*, 44(10), pp. 2445–2462.

Koch, N. (2013). Why Not a World City? Astana, Ankara, and Geopolitical Scripts in Urban Networks. *Urban Geography*, 34(1), pp. 109–130.

Köppen, B. (2013). The Production of a New Eurasian Capital on the Kazakh Steppe: Architecture, Urban Design, and Identity in Astana. *Nationalities Papers: The Journal of Nationalism and Ethnicity*, 41(4), pp. 590–605.

Lee, Z. (2015). Eco-Cities as an Assemblage of *Worlding* Practices. *International Journal of Built Environment and Sustainability*, 2(3), pp. 183–191.

Lerner, I. (2008). Instant City: City as Recreation/Re-Creation. In A. Moustafa, J.A. Qawasmi and K. Mitchell, eds. *Instant Cities: Emergent Trends in Architecture and Urbanism in the Arab World: The Third International Conference of the Center for the Study of Architecture in the Arab Region*. Sharaj, UAE: CSAAR Press, pp. 33–44.

Lloyd, R. and Nichols Clark, T. (2001). The City as Entertainment Machine. In: K.F. Gotham, ed. *Critical Perspectives on Urban Redevelopment: Volume 6*. New York and Amsterdam: Elsevier, pp. 357–378.

Lumumba, J. (2013). Why Africa Should be Wary of its 'New Cities', Informal City Dialogues (Rockefeller Center), 5 February. Available at: http://nextcity.org/informalcity/entry/why-africa-should-be-wary-of-its-new-cities [accessed 7 December 2017].

Marshall, R. (2003). *Emerging Urbanity: Global Urban Projects in the Asia Pacific Rim*. London and New York: Spon.

Mattsson, H. (2013). Staging a Milieu. In: J. Nilsson and S.O. Wallenstein, eds. *Foucault, Biopolitics, and Governmentality*. Stockholm: Södertörn Philosophical Studies, pp. 123–132.

McCann, E. (2011). Urban Policy Mobilities and Global Circuits of Knowledge: Toward a Research Agenda. *Annals of the Association of American Geographers*, 101(1), pp. 107–130.

Murray, M. (2015a). City Doubles: Re-Urbanism in Africa. In: F. Miraftab, D. Wilson and K. Salo, eds. *Cities and Inequalities in a Global and Neoliberal World*. New York: Routledge, pp. 92–109.

Murray, M. (2015b). Waterfall City (Johannesburg): Privatized Urbanism *in Extremis*. *Environment & Planning A*, 47(3), pp. 503–520.

Ong, A. (2011). Hyperbuilding: Spectacle, Speculation, and the Hyperspace of Sovereignty. In: A. Roy and A. Ong, eds. *Worlding Cities: Asian Experiments and the Art of Being Global*. Malden, MA: Wiley-Blackwell, pp. 205–226.

Page, M. (1999). *The Creative Destruction of Manhattan, 1900–1940*. Chicago: University of Chicago Press.

Peck, J. and Theodore, N. (2010). Mobilizing Policy: Models, Methods, and Mutations. *Geoforum*, 41(2), pp. 169–174.

Piot, C. (2010). *Nostalgia for the Future: West Africa After the Cold War*. Chicago: University of Chicago Press.

Ponzini, D. (2013). Branded Megaprojects and Fading Urban Structure in Contemporary Cities. In: G.D.C. Santamaría, ed. *Urban Megaprojects: A Worldwide View [Research in Urban Sociology, Volume 13]*. Bingley, UK: Emerald Group Publishing Limited, pp. 107–129.

Rapoport, E. (2014). Utopian Visions and Real Estate Dreams: The Eco-City Past, Present and Future. *Geography Compass*, 8(2), pp. 137–149.

Rodas, L. and Pineda, G. (2013). *The Making of a Free City: The Foundation of Laissez-faire Capitalist Free Cities in Honduras in the Juncture of Globalization*. Department of Society and Globalization, Roskilde Universitet [Denmark].

Romer, P. (2009). Why the World Needs Charter Cities, TED Talks. Available at: www.ted.com/talks/paul_romer.html [accessed 7 December 2017].

Roy, A. (2009). Why India Cannot Plan its Cities: Informality, Insurgence, and the Idiom of Urbanization. *Planning Theory*, 8(1), pp. 76–87.

Ruiz-Goiriena, R. (2013). Paseo Cayala: Guatemala Builds Private 'Cayala City' for Rich to Escape Crime. *The Huffington Post*, 8 January. Available at: www.huffingtonpost.com/2013/01/08/paseo-cayala-guatemala-private-city_n_2434644.html [accessed 10 February 2016].

Schatz, E. (2004). What Capital Cities Say About State and Nation Building. *Nationalism and Ethnic Politics*, 9(4), pp. 111–140.

Schindler, S. (2015). Governing the Twenty-First Century Metropolis and Transforming Territory. *Territory, Politics, Governance*, 3(1), pp. 7–26.

Shatkin, G. (2008). The City and the Bottom Line: Urban Megaprojects and the Privatization of Planning in Southeast Asia. *Environment and Planning A*, 40(2), pp. 383–401.

Shatkin, G. (2011). Planning Privatopolis: Representation and Contestation in the Development of Urban Integrated Mega-Projects. In: A. Roy and A. Ong, eds. *Worlding Cities: Asian Experiments and the Art of Being Global*. Malden, MA: Blackwell, pp. 77–97.

Shwayri, S. (2013). A Model Korean Ubiquitous Eco-City? The Politics of Making Songdo. *Journal of Urban Technology*, 20(1), pp. 39–55.

Sidaway, J. (2007). Enclave Space: A New Metageography of Development? *Area*, 39(3), pp. 331–339.

Simson, A. (2001). The Post-Romantic Landscape of Telford New Town. *Landscape and Urban Planning*, 52(2–3), pp. 189–197.

Storper, M. (2014)."Governing the Large Metropolis. *Territory, Politics, Governance*, 2(2), pp. 115–134.

Swyngedouw, E. (2005). Exit 'Post' – the Making of 'Glocal' Urban Modernities. In: S. Read, J. Rosemann and J.V. Eldijk, eds. *Future City*. London and New York: Spon, pp. 125–144.

Tanaka, S. (2015). History Without Chronology. *Public Culture*, 28(1), pp. 161–186.

Tomlinson, J. (2007). *The Culture of Speed: The Coming of Immediacy*. London: Sage.

Watson, V. (2009). 'The Planned City Sweeps the Poor Away…': Urban Planning and 21st Century Urbanisation. *Progress in Planning*, 7(3), pp. 151–193.

Watson, V. (2014). African Urban Fantasies: Dreams or Nightmares? *Environment and Urbanization*, 26(1), pp. 215–231.

Wright, H. (2008). *Instant Cities*. London: Black Dog.

17
URBANIZATION, PLANNING AND THE POSSIBILITY OF BEING POST-GROWTH

Jason Hackworth

Introduction

It has been more than ten years since the publication of the *Youngstown 2010 Citywide Plan* (City of Youngstown 2005), but it is still mentioned as an important transition point by urban political economists (Bernt 2009; Rhodes and Russo 2013; Schatz 2013). The document was the first in the United States to publicly acknowledge, and begin to plan around, the reality of population shrinkage. By publicly acknowledging that "Youngstown is a smaller city", officials were able to embark on "A strategic program ... required to rationalize and consolidate the urban infrastructure in a socially responsible and financially sustainable manner" (City of Youngstown 2005: 18). Since 2005, at least four other American cities and several European cities have initiated 'right-sizing' plans similar to Youngstown (Bernt 2009; Hackworth 2015). These plans broach what was previously seen as politically toxic – openly acknowledging that growth might not be happening or be likely to occur in the future. The erstwhile growth derangement syndrome has largely been responsible for counterproductive and expensive measures like massive infrastructure projects, and unmoored suburban growth, so plans like Youngstown 2010, and *Detroit Future City* (Detroit Works 2012), are understandably celebrated by progressive scholars as a welcome circumvention of the power and structure of growth. To Schindler (2016: 820) this marks a shift to "degrowth machine politics whose objective is to improve the quality of life in the city rather than simply augment the value of land and spur economic growth". More broadly, some have begun to suggest that the notion of urban politics being completely dominated by a growth machine is outdated (Purcell 2000). The ideology and structure of growth has been disrupted, and as a consequence urban politics is different now than it was when Molotch (1976) introduced urban politics students to his growth machine concept.

Such statements are provocative and suggest that capitalist urbanization no longer responds to the supremacy of growth at all costs. But are such assessments justified? Is right-sizing a circumvention of growth and growth machine politics, and if so in what way? Addressing this question is a useful window into wider matters of local political autonomy, capital mobility and planning within the capitalist system. This chapter parses and critiques the project of right-sizing along these lines. It does so by building from the insights of the urban political economy literature. The thesis is that while planning for non-growth can be seen as a local political accomplishment, suggesting this is sufficient to over-ride the structures undergirding the growth paradigm in American cities is misleading. Right-sizing varies in its application but in no place has it meaningfully challenged the growth paradigm. In fact, in some cases it is simply a cynical packaging for an even more ruthless form of growth.

What is urban growth?

Defining urban growth is not always a simple straightforward matter. We can begin by dividing growth into three forms: population, area and economic. Much analysis in urban geography conflates these into one uncomplicated amalgam and concludes that acknowledging the lack of one (usually population) suggests the circumvention of all other forms. It is worth briefly parsing these forms of growth and their use in the literature to illustrate this point.

Population data is the most commonly used metric within the shrinking cities literature. Not only has it been used recently to document the extent of decline worldwide (Beauregard 2009; Oswalt 2005), but it has classically been used by political economists as a proxy for other forms of growth (see for example Molotch 1976). At a broad level this is completely justifiable. Not only is population data easily comparable across jurisdictions, population growth is generally a proxy for economic growth. All else equal, more people bring greater tax revenue, political influence and wealth. But using population data to carefully document the extent of decline is a much trickier matter for several reasons. First, household sizes have fallen by about a third in American cities since the mid-twentieth century. Theoretically then, a city would have to have a population loss of greater than a third for there to be pressure to abandon housing. Thus in landlocked cities, populations could have conceivably – and in fact have in many cases – dropped without any net outmigration or major land abandonment. It is precisely when population losses reach that point – where there are far fewer households than houses – that problems begin to occur for urban managers. Second, neighbourhood-level population gains or losses may not always be a proxy for economic growth. Temporary micro-scaled population losses are not necessarily a sign of decline. Similarly, temporary upticks are not necessarily a sign of growth. The final corollary point to this is that growth and decline are very uneven. While it may be easy to detect decline in a place like Detroit, some neighbourhoods in Indianapolis and Louisville are equally abandoned and impoverished, even though their aggregate populations have not fallen. In short, while population loss is quite popular in the shrinking cities literature, it has clear limitations. Nevertheless, much of our understanding of economic decline is influenced by population shrinkage numbers. Moreover city managers pursue population growth almost reflexively and avoid population decline with desperation.

Historically, cities have attempted to grow their way out of population loss by expanding their boundaries (Jackson 1985). This is especially the case in the USA where suburbanization and jurisdictional fragmentation is so much more acute than other parts of the world. The thought is that if a city expands its boundaries it can capture suburbanized populations and their tax revenues. This putative 'best practice' for cities in the American West – to annex as much land as possible pre-emptively – can presumably control the building practices and tax revenue of (generally more affluent) populations on the fringe (Rusk 2006). But the ability of cities in the American Midwest and Northeast to adopt this policy is severely circumscribed given that most are completely surrounded by administratively separate jurisdictions (Gordon 2008; Teaford 1979). Inner cities are not able to capture the property tax revenue of their suburban satellites and state legislatures (where annexation policy is set) have generally supported the ability of such satellites to remain separate in every possible sense – tax revenue, schools, infrastructure in particular. Globally, the United States is different than most nations in this respect. Elsewhere in the world, suburban fragmentation is more often seen as untenably expensive, and inefficient; it isolates inner core urban problems from the funding sources that could ameliorate them. Even in Canada – a country that is more similar than different to the USA globally speaking – provincial governments have not only allowed annexation; they have insisted on it to save resources (Hackworth 2016).

Cities, especially in the USA, have long tried to circumvent annexation restrictions by attempting to capture value created elsewhere – in a sense, a 'virtual' annexation. The simplest of these measures involves trying to attract firms whose profits are derived from a wide field, but invested in particular places. FIRE – finance, insurance and real estate firms – remain the most popular version of this strategy. Particularly in a globalized economy, such firms capture profits from production elsewhere in the region or world by lending and investing in ventures. Some of those revenues get invested in actual places – usually the headquarters location of a firm – so there is intense competition between places to attract such firms. A second method for widening the virtual boundaries of a city involves direct investments by city managers. By leveraging the tax revenue that they do have, often cities, or parts of city finances such as pension funds, can multiply their revenue by capturing value produced elsewhere, again via investments in ventures outside of their borders. In the most exotic of such efforts, cities engage in 'arbitrage' – essentially leveraged investment – as a way to increase their 'profits'. But as the much discussed Orange County debacle of the 1990s showed, such potential rewards come with enormous risks that can bankrupt a municipality (Baldassare 1998).

Ultimately such measures are marginal to city management efforts. Cities in the United States possess the greatest relative influence (in the toolkit of urban governance) over *land use* so much of their 'growth' or 'smart decline' policies remain centred on this sphere. Molotch (1976) was among the first to argue that governance is dominated by a rentier class whose main interest centres on increasing the value of their fixed investment – urban land. City officials have responded to these pressures through a variety of means. At the most extreme level they have expropriated land from private individuals and transferred it to more organized capital to increase their tax revenue (Jacobs and Bassett 2010; Niedt 2013). At a more marginal level, they have engaged in all manner of property-led regeneration to increase tax revenue. At the more sinister level, they have engaged in land use policies which seek (or at least function) to lower the number of poor people while increasing the number of wealthier people (Pedroni 2011). Though Molotch and many others have used population change as a proxy for economic development, this need not be the case. And in declining cities it is often *not* the case. Cities like Detroit are engaging in open campaigns to get more middle-class people to live there, while simultaneously penalizing the precariat with water service billing and tax collection crack-downs (Moskowitz 2015).

In short, while the language of 'growth culture' or 'growth machine' is common in discussing cities, the conceptual definition of growth is not straightforward. The way one defines growth is centrally important for assessing the veracity of claims that growth has been circumvented. Population data is used frequently but needs to be carefully considered. Population growth is not automatically the same as economic growth. Acknowledging population decline does not mean a city gives up on economic growth. Conflation of these two leads more easily to the conclusion that the logic of growth has been punctured when such declarations are made. Finally, all forms of growth are highly uneven between and within cities. Acknowledging the need to downsize one neighbourhood does not mean that there are not simultaneous efforts to build in another. Often, for example, downsizing efforts occur in concert with central business district revitalization.

Right-sizing in context

However one measures it, it is hard to escape the conclusion that places like Detroit, Cleveland and Flint are declining. They have all lost more than 50 per cent of their peak twentieth-century populations, within more-or-less the same municipal boundaries, and have experienced a collapse of employment and tax bases. They were substantial industrial cities in the early to mid-twentieth century, but have now shrunk into shells of their former selves. Infrastructure and service networks designed to meet the needs of larger populations are now spread across vast areas of low density, making them highly inefficient and expensive for already financially crippled cities to maintain. Given this set of conditions, these cities and others have embarked on right-sizing – an effort to manage the literal and figurative effluent of their former industrial landscapes. These efforts are varied in practice but they generally consist of several principles (Schilling and Logan 2008). First, there is an effort to repurpose vacated land in the form of green infrastructure. That is, rather than encouraging growth in every part of the city, right-sizing planners suggest converting some vacated lots into parks, gardens and spaces for storm run-off. Not only will this repurpose long-abandoned spaces, argue its proponents, it could help re-green cities. Second, there is an effort to concentrate existing populations in order to provide better access to jobs and infrastructure and lower cost of service delivery. Youngstown, Ohio was the first US city to formally invoke the language of right-sizing in a plan, but now at least five cities have done so (Hackworth 2015). These efforts share the acknowledgement that growth may not occur soon or at all, and some have concluded, based on these efforts, that the underlying impulse of growth has been punctured. To approach that point, it is useful to divide right-sizing into three ontological levels: ideational, statutory and actualized.

Ideational right-sizing

At its core, the *idea* of right-sizing is rooted in a good faith effort by planners, academics and journalists to reorient urban landscapes that have suffered from massive population loss. This strategy is communicated and debated at multiple levels, including popular books (Gallagher 2010), academic articles (Schilling and Logan 2008), documentaries (e.g. Detropia) and of course planning documents and meetings (Hackworth 2015). At this abstract level, the idea is uncontroversial, even hopeful. To many, these landscapes have already been abandoned by (middle-class) people, (organized) capital and the state (except for policing). Suggesting that greener, more accessible, town-like structures should replace isolated semi-rural poverty provides hope in an area that is largely ignored in other corners of the policy universe. Yet when the idea is suggested at the community or legislative level, it is far more controversial and becomes subject to many other ideational influences and material forces.

Some research suggests that shrinking cities often take ten or more years simply to accept that they are losing population and may not have the negotiating strength that they once did (Mayer and Greenberg 2001). On the pathway to accepting this reality, cities often enact counterproductive policies like building sports or entertainment complexes where little demand exists. Such cities are left with a huge bill in addition to their longer-term shrinking pattern. Others have suggested that the idea of growth is just so foundational at a cultural level in the United States that it is difficult for anyone to even broach the possibility of 'smart shrinkage'. When, for example, Frank and Deborah Popper of Rutgers University went on a book tour in the 1980s to suggest that the already-shrinking Great Plains should be partially renaturalized, their ideas were met with great hostility – even death threats (Matthews 1992). In this case and in others, the authors were simply suggesting that communities acknowledge their current realities and begin to plan around it.

Complicating these issues at the urban level is race and the experience of past urban renewal schemes. In a number of cities (New York City in the 1970s, Detroit in the early 1990s), public officials proposed the 'planned shrinkage' of neighbourhood services, and even wholesale dismantling of certain low-density neighbourhoods to save on infrastructure costs (Dewar and Weber 2012). These efforts follow similar ones during the mid-twentieth century to 'modernize' the urban landscape in the name of progress. In all of these historical antecedents, people of colour have borne the brunt of these programmes' mistakes and conceptual flaws. Black neighbourhoods were and remain the targets of 'downsizing'. When plans were actually implemented, as they were in the urban renewal period, money for demolition was always more plentiful than money for community rebuilding (Thomas 2013). Urban renewal and planned shrinkage did far more to remove communities of colour than to build new-and-improved neighbourhoods as promised (Fullilove 2005; Talen 2014). Many of the same neighbourhoods, sometimes with the same residents, are the target of right-sizing. There is understandable scepticism in believing that the aims of right-sizing will benefit more than the land owners in the central business district.

Whatever the benefits of ideational right-sizing, actualized programmes are subject to pressures and histories that are different to and sometimes more powerful than the idea itself (see also Bernt et al. 2014). Getting communities like Youngstown to acknowledge decline is indeed an accomplishment for all of the reasons listed above. But such efforts are always partial, often temporary and rarely progress beyond the idea phase. In many cases the legacies of distrust between city officials and poor communities are so polluted that right-sizing never gets debated or proposed.

Statutory right-sizing

In some locales, right-sizing dialogue has progressed to the legal form so it is perhaps apt to speak of a statutory right-sizing as distinct from its ideational form. By statutory right-sizing, I simply mean the ways in which the idea has been codified into a legally enforceable plan or policy. Local land use plans are the most straightforward example. In addition to the aforementioned Youngstown exercise, four other cities in the American Rust Belt – Detroit, Flint, Rochester and Saginaw – have formally invoked the language of right-sizing in either an official plan or a highly influential unofficial plan (Hackworth 2015). Though these documents acknowledge decline and gesture to the type of idealized city outlined in right-sizing ideational conversations, they have a notable asymmetry between demolition and building. The ideas of culling existing housing through intensified demolition are far more specific (and funded) than the lofty goals of green infrastructure, clustered affordable housing or environmental remediation. In this sense the first wave of planning documents reflects the values of austerity as much as idealized right-sizing. And in this way, it very much resembles the role that planning played in the much-repudiated 'planned shrinkage' proposals in 1970s New York City (Hollander and Nemeth 2011).

But while existing right-sizing documents fail to live up to the complete menu of ideas outlined by its theoretical proponents, and have only appeared formally in a few American cities, there are other ways that right-sizing has been transformed into a legal form. First, there a suite of local by-laws – property registration systems, crack-downs on slumlord housing, and intensified demolition – that seek to improve neighbourhood and commercial conditions particularly around 'salvageable' nodes. Though few of these laws directly invoke the language of right-sizing, they are unmistakably similar in their emphasis on improving conditions amidst decline by facilitating clustered stability. A second legal form has taken place largely at the county and state level in the United States. A number of important legal changes have made the creation of county-level 'land banks' – whose purpose is largely to bank and plan for land in a way that is more rational and community oriented than the market (Hackworth and Nowakowski 2015). These measures have been combined with tax reversion reforms that allow for cities and counties to acquire land before it falls into speculative hands, and to capture salvable housing before it has been exposed to speculators, scavengers and criminals. Like local by-laws, the land bank movement often does not formally invoke 'right-sizing' in a direct way, but its *raison d'etre* is very similar, and for the right-sizing movement to work, these measures will be integral.

In short, ideational right-sizing has begun to take direct and indirect legal form. These legal forms bear resemblance to the ideational form, but the codified version emphasizes penality, demolition and pragmatism more than the most abstracted versions of right-sizing. Moreover, like all forms of urban policy, however, there are meaningful limits to what can be done in practice. For that reason it is useful to speak of actualized right-sizing as distinct from its ideational and statutory form.

Actualized right-sizing

In practice, right-sizing has materialized as a mix of fairly conventional economic development – property-led CBD development in particular – and accelerated demolition of 'blight'. The more progressive goals of environmental remediation, affordable housing around economic nodes, and public open space creation have been backgrounded or rejected. Why has the recognition of decline morphed into this? In some ways the culprit is the nature of capitalist urbanization within a more or less liberal (and multi-scalar) democracy. *Ideas* of right-sizing spoken at meetings are not the same as what gets written into law. Legal plans or policies typically vary from what gets converted into practice. The structural and institutional filters governing this process bend towards the values of economic growth. These forces are organized and operate at multiple scales.

Within cities, it has been suggested that by broaching the topic of decline the forces of growth have been meaningfully superseded. But, the rentier class that formed Molotch's growth machine forty years ago still operates today. A handful of large land owners have, and continue to assert, disproportionate influence on city politics. In Detroit, Mike Ilitch and Dan Gilbert own much of the real estate in Downtown (Eisinger 2015). They are active in city politics, funding city politicians and influencing policy, including but not limited to right-sizing proposals. In other cities, the figures are perhaps less famous but no less influential in relative terms. Though their specific interests vary of course, they are unified by their desire to increase the exchange value of their considerable real estate holdings. Much effort is focused on the one zone perceived to still have considerable economic potential: the central business district. In this zone, growth interests have centred on commercial redevelopment often with state assistance. Within neighbourhoods, by contrast, the emphasis is on demolition – on removing blight. Though blight eradication efforts use the same language as CBD redevelopment – i.e. to increase property values – neighbourhoods themselves are increasingly left to their own devices.

Forces at the state and federal level have only amplified these tendencies. Much of what counts as 'urban policy' in the USA is in fact significantly influenced if not dictated by state legislatures, which are disproportionately rural and suburban in representation. They script laws that are shaped by their understanding of urban issues in that state. In many cases, that understanding is hostile to the interests of large formerly industrial cities (Greenblatt 2014). State legislators in Ohio and Michigan, for example, regularly get elected by running against the foil of Cleveland and Detroit respectively. The prevailing assumption in such legislative bodies is that places like Detroit and Cleveland are in despair because they are corrupt, and spend too much money. The notion that they are structural victims of austerity and racism is often summarily and angrily dismissed. These tropes are amplified by an active network of right-leaning think-tanks that influence and even write legislation at the state level (Akers 2013; Hackworth 2014). Thus, any state legal change, whether it has to do with transfer payments or the structuring of land banks, must confront, and often times reflects, an anti-urban hostility or a market fundamentalism. Cities and land banks set up under the right-sizing paradigm are, for example, forced by state legislatures to 'show their worth' and 'pay for themselves' rather than being a line item paid for by government. It is not difficult to recognize the facile reasoning behind such demands. In the mind of many Midwestern state legislators, the problem with cities is that they spend too much on wasteful items. Thus new state programmes for cities are increasingly market-oriented. Land banks must fund their own activities in exchange for the legal authority to exist (Hackworth 2014). Cities must reduce their staffing and spending levels in exchange for state funding (Peck 2014). In this context, the claimed intent to build new parts and greenways using unidentified state monies seems particularly far-fetched. Cities, in short, may control right-sizing plans but they often do not control the resources necessary to do it the way they see fit. Within this atmosphere of austerity, the more socially progressive aspects of right-sizing have been discarded.

And these are just the visible, legislative forces affecting actualized right-sizing. There are less visible but no less influential forces at play too. Bond rating agencies, for example, reinforce not just the ethos of growth in cities but the practical reality of it (Hackworth 2002). Cities that dare use their general fund to increase the size of the social economy, face the ire of very powerful rating agencies that can and will lower their rating and dramatically increase their borrowing costs. Suburban municipalities, moreover, continue to fight (sometimes subtly, sometimes not) the idea of regionalization which might make the loftier goals of right-sizing more fiscally attainable. To outspoken figures like Oakland County's (wealthy suburban) manager L. Brooks Patterson, the only thing to which Detroit is entitled is further austerity (Williams 2014). States have supported metropolitan fragmentation for more than a hundred years (Gordon 2008; Teaford 1979); it is unlikely that continued inner core despair is going to motivate them to force a regionalization in Detroit or anywhere else.

Within this multi-scaled quagmire sits urban governance. Right-sizing was indeed introduced and some of its goals, at the ideational level, are laudable. But the institutions and structures reinforcing growth assure that, short of some major political reorganization in the United States, the only goals of right-sizing that will be pursued are those that promote future growth or reinforce existing growth. In practice, actualized right-sizing is an awkward mix of demolishing houses in poor neighbourhoods and investing in CBD redevelopment.

(Declining) city limits

When Molotch (1976) published his famous essay, 'City as a Growth Machine', he described an urban political sphere dominated by a small cadre of land owning elites. They disproportionately influenced city politics by controlling land, funding political campaigns and promoting the notion that growth equals success, while decline was failure. Their activities and interests were supported by multiple institutions including state legislatures, bond rating agencies and federally supported development agencies. It resonated with the concept that urban space is not just a product of local decisions. Cities lie at the centre of a web of political and economic forces – many of which are outside of their direct control. Almost forty years later, some have begun to suggest that the power of growth machines is diminished, and that right-sizing plans represent a philosophical and actualized departure from the now counterproductive growth machine. Much of this conversation takes place within the narrow confines of urban planning, but in many ways, it is not really a planning question. It is a question about local autonomy and the ability of cities and citizens to dictate or influence a pathway that contradicts the multi-scaled machinery of economic growth. On this note, it is dubious that right-sizing marks anything other than a short-lived acknowledgement of population decline. Some cities have temporarily begun to include language of 'smart decline' and 'right-sizing' in their planning documents, but none of the codified versions meaningfully challenge the structures that Molotch pointed to four decades ago.

Urban growth is not just a mentality or discourse. It is an institutionally reinforced process. Even within the realm of discourse, finding evidence of 'post-growth' language does not automatically mean it is true or the only discourse present. Language of hope and post-growth ideas is severely limited by the dictates of multi-scaled growth machines, mentalities and institutions. City politics are, by definition, nested in a hierarchy of higher political and economic scales and forces. Some of these structures are more visible than others but most are more powerful the political potential of merely acknowledging that a place is losing population.

References

Akers, J. (2013). Making Markets: Think Tank Legislation and Private Property in Detroit. *Urban Geography*, 34, pp. 1070–1095.
Baldassare, M. (1998). *When Government Fails*. Berkeley: University of California Press.
Beauregard, R. (2009). Urban Population Loss in Historical Perspective: United States, 1820–2000. *Environment and Planning A*, 41, pp. 514–528.
Bernt, M. (2009). Partnerships for Demolition: The Governance of Urban Renewal in East Germany's Shrinking Cities. *International Journal of Urban and Regional Research*, 33(3), pp. 754–769.
Bernt, M., Haase, A., Grobmann, K., Cocks, M., Couch, C., Cortese, C. and Krzysztofik, R. (2014). How Does(n't) Urban Shrinkage Get onto the Agenda? Experiences from Leipzig, Liverpool, Genoa, and Bytom. *International Journal of Urban and Regional Research*, 38(5), pp. 1749–1766.
City of Youngstown. (2005). *Youngstown 2010 Citywide Plan*. www.cityofyoungstownoh.com/about_youngstown/youngstown_2010/plan/plan.aspx [accessed 12 August 2015].
Detroit Works (2012). *Detroit Future City: Detroit Strategic Framework Plan*. http://detroitworksproject.com/the-framework/ [accessed 13 August 2015].
Dewar, M. and Weber, M. (2012). City Abandonment. In R. Crane and R. Weber, eds. *The Oxford Handbook of Urban Planning*. Oxford: Oxford University Press.
Eisinger, P. (2015). Detroit Futures: Can the City Be Reimagined? *City and Community*, 14(2), pp. 106–117.
Fullilove, M. (2005). *Root Shock: How Tearing Up City Neighborhoods Hurts America and What We Can Do About it*. New York: Random House.
Gallagher, J. (2010). *Reimagining Detroit: Opportunities for Redefining an American City*. Detroit: Wayne State University Press.
Gordon, C. (2008). *Mapping Decline: St. Louis and the Fate of the American City*. Philadelphia: University of Pennsylvania Press.
Greenblatt, A. (2014). Rural Areas Lose People but not Power. *Governing*, April. Available at: www.governing.com/topics/politics/gov-rural-areas-lose-people-not-power.html [accessed 13 August 2015].
Hackworth, J. (2002). Local Autonomy, Bond-Rating Agencies and Neoliberal Urbanism in the US. *International Journal of Urban and Regional Research*, 26(4), pp. 707–725.
Hackworth, J. (2014). The Limits to Market-based Strategies for Addressing Land Abandonment in Shrinking American Cities. *Progress in Planning*, 90, pp. 1–37.
Hackworth, J. (2015). Right-sizing as Spatial Austerity in the American Rust Belt. *Environment and Planning A*, 47(4), pp. 766–782.
Hackworth, J. (2016). Why There Is No Detroit in Canada. *Urban Geography*, 37(2), pp. 272–295.
Hackworth, J. and Nowakowski, K. (2015). Using Market-based Policies to Address Market Collapse in the American Rust Belt: The Case of Land Abandonment in Toledo, Ohio. *Urban Geography*, 36(4), pp. 528–549.
Hollander, J.B. and Nemeth, J. (2011). The Bounds of Smart Decline: A Foundational Theory for Planning Shrinking Cities. *Housing Policy Debate*, 21(3), pp. 349–367.

Jackson, K. (1985). *Crabgrass Frontier: The Suburbanization of the United States*. Oxford: Oxford University Press.
Jacobs, H. and Bassett, E. (2010). After 'Kelo': Political Rhetoric and Policy Responses. *Land Lines*, 22(2), pp. 14–20.
Matthews, A. (1992). *Where the Buffalo Roam: The Storm over the Revolutionary Plan to Restore America's Great Plains*. New York: Grove Press.
Mayer, H. and Greenberg, M. (2001). Coming Back from Economic Despair: Case Studies of Small- and Medium-sized American Cities. *Economic Development Quarterly*, 15(3), pp. 203–216.
Molotch, H. (1976). The City as a Growth Machine: Toward a Political Economy of Place. *The American Journal of Sociology*, 82(2), pp. 309–332.
Moskowitz, P. (2015). The Two Detroits: A City Both Collapsing and Gentrifying at the Same Time. *Guardian*, 5 February. Available at: www.theguardian.com/cities/2015/feb/05/detroit-city-collapsing-gentrifying [accessed 12 August 2015].
Niedt, C. (2013). The Politics of Eminent Domain: From False Choices to Community Benefits. *Urban Geography*, 43, pp. 1047–1069.
Oswalt, P., ed. (2005). *Shrinking Cities, Vol. 1, International Research*. Ostfildern-Ruit, Germany: Hatje Cantz.
Peck, J. (2014). Pushing Austerity: State Failure, Municipal Bankruptcy and the Crises of Fiscal Federalism in the USA. *Cambridge Journal of Regions, Economy and Society*, 7, pp. 17–44.
Pedroni, T. (2011). Urban Shrinkage as a Performance of Whiteness: Neoliberal Urban Restructuring, Education, and Racial Containment in the Post-industrial, Global Niche City. *Discourse: Studies in the Cultural Politics of Education*, 32, pp. 203–215.
Purcell, M. (2000). The Decline of the Political Consensus for Urban Growth: Evidence from Los Angeles. *Journal of Urban Affairs*, 22(1), pp. 85–100.
Rhodes, J. and Russo, J. (2013). Shrinking 'Smart'?: Urban Redevelopment and Shrinkage in Youngstown, Ohio. *Urban Geography*, 34(3), pp. 305–326.
Rusk, D. (2006). Annexation and the Fiscal Fate of Cities. *Brookings Institute Online Report*. Available at: www.brookings.edu/research/reports/2006/08/metropolitanpolicy-rusk [accessed 12 August 2015].
Schatz, L. (2013). Decline-oriented Urban Governance in Youngstown, Ohio. In M. Dewar and J.M. Thomas, eds. *The City After Abandonment*. Philadelphia: University of Pennsylvania Press, pp. 87–103.
Schilling, J. and Logan, J. (2008). Greening the Rust Belt: A Green Infrastructure Model for Right-sizing America's Shrinking Cities. *Journal of the American Planning Association*, 74(4), pp. 451–466.
Schindler, S. (2016). Detroit After Bankruptcy: A Case of Degrowth Machine Politics. *Urban Studies*, 53(4), pp. 818–836.
Talen, E. (2014). Housing Demolition During Urban Renewal. *City and Community*, 13(3), pp. 233–253.
Teaford, J. (1979). *City and Suburb: Political Fragmentation of Metropolitan America, 1850–1970*. Baltimore: Johns Hopkins University Press.
Thomas, J.M. (2013). *Redevelopment and Race: Planning a Finer City in Postwar Detroit*. Detroit: Wayne State University Press.
Williams, P. (2014). Drop Dead Detroit! The Suburban Kingpin Who Is Thriving off the City's Decline. *The New Yorker*, 27 January. Available at: www.newyorker.com/magazine/2014/01/27/drop-dead-detroit [accessed 12 August 2015].

18
TROUBLED BUILDINGS, DISTRESSED MARKETS
The urban governance of the US foreclosure crisis

Philip Ashton

Introduction

The post-2006 foreclosure crisis has been a signal moment in US urban history, and in many cities the long-term effects are only now becoming apparent. For instance, the city of Chicago experienced 128,596 foreclosure filings between 2008 and 2015, of which 66,868 went to foreclosure auction and 61,726 ended up transferred to the lender as a 'real-estate owned' (REO) property (Woodstock Institute 2016). Concentrated foreclosures are not new in many urban neighbourhoods, and some cities have extensive experience managing abandoned properties through the tax foreclosure process. However, during the housing boom from 2002 through 2007 distressed owners could avoid losing their homes by selling or refinancing their troubled mortgages; this kept vacant, bank-owned properties to a minimum. Since the onset of the foreclosure crisis, the proliferation of REO properties, and the corresponding eviction of occupants, has brought with it new forms of blight in the form of vacant, boarded homes. Collapsing housing values trap other property owners who face few short-run options beyond digging in or walking away. Economic costs to residents are magnified by strains on localities, including increased service costs, a declining property tax base, as well as new demands for social welfare services for those losing their homes to banks (Apgar *et al.* 2005).

Whereas there is now a robust academic focus on the geographies of foreclosure activity (Immergluck 2010) and their relation to racial economies of subprime mortgage lending (Wyly *et al.* 2009) and transformations in the financial sector (Newman 2009), there has been a dearth of attention paid to the urban governance of the foreclosure crisis – which I understand broadly as a process of mobilizing social forces and capacities to manage the effects of foreclosures on households, neighbourhoods and the urban economy. This contrasts with an extensive literature on governance of the crisis as one of systemic circulation, which has focused on emergency interventions around 'distressed assets' (as foreclosures appear within financial circuits) (Ashton 2012). The irony here is that, in spite of obvious scalar differences, both local and systemic perspectives deal with a similar object: residential mortgages and their breakdown. Critical scholars have long noted the disjuncture between the 'global-ness' of the mortgage as a financial claim (represented by the promissory note, pooled into mortgage-backed securities and circulated within global bond and derivatives markets) and the inherent locality of the collateral (residential property) that secures that claim (Gotham 2006). And yet, even as the housing boom relied on some fundamental transformations in claims against local residential property, and even as the legal regulation of property-as-collateral is one of the few substantive policy areas over which municipal governments have broad rein under US federalism, localities have been marginalized in studies of the governance of the foreclosure crisis. That is, despite the locality being a crucial venue for the unfolding of the foreclosure crisis, it has been absent as a focus of analysis.

This lacuna may be due to prevailing accounts of the local state, which emphasize the dependent position of US local government (Frug 1980). In spite of the fundamentally urban dimensions of the crisis, these 'city bound' arguments portray localities as both resource constrained and lacking the legal authority to intervene in what have become matters of financial stability and banking regulation (Frug and Barron 2008). These accounts are joined by a parallel analysis of the dependent position of the local state from within urban political economy, wherein localities are only granted passive police powers – such as zoning/building codes, and nuisance law – necessary to ensuring orderly private markets in property (Fogelsong 1986). This situates the local state in a weak bargaining position, producing a strategic bias in favour of property capital or setting up unwinnable battles against entrenched financial sector interests over local initiatives that might disrupt lenders' control over property as collateral (Christophers and Niedt 2015). Correspondingly, within these arguments local governance of the foreclosure crisis can only aspire to a limited set of strategic choices available within the constraints imposed by state and federal programmes (Immergluck 2013).

Both 'city bound' and political economy accounts are thus prone to a determinate account of local police powers as they relate to the governance of property. In this chapter, I argue that such accounts are complicated when we open up property as a broad field of contestation within some of the very legal doctrines that are meant to protect it. I choose a narrow focus on nuisance law, partly due to space constraints, but also because within this legal sphere a *local* distributional politics of the foreclosure crisis comes into focus, one that seeks to redistribute costs and responsibilities back to lenders rather than being borne by homeowners and localities.

Shifting away from 'city bound' or political economy accounts requires a more specific understanding of both public and private law than is often evident in studies of local governance. I tentatively develop such an approach by examining law as an under-determined field of local statecraft (Jessop 1990), drawing long-standing critical concerns over urban governance into conversation with legal geography, socio-legal studies and cultural economy. Even as consideration of, for instance, single cases by municipalities against banks might conclude that local legal strategies are desperate efforts or prone to failure (cf. Rosenman and Walker 2016), the broader perspective I propose focuses on two elements of statecraft evident within nuisance law and its application to REO properties.

The first element of statecraft examines how localities are reworking elements of private property to forge a plural legal space – drawing together doctrines of nuisance, civil rights and sovereign power – within which vacant, bank-owned properties are rendered governable through law. This confronts a conventional understanding of the housing boom and subsequent bust, wherein *only* private agents engaged in legal play with property. For instance, in the hands of agents such as the Mortgage Electronic Recording System (MERS) – a consortium that centralizes mortgage deeds to facilitate their transfer in secondary markets – lenders were seen as exercising novel approaches to the organization of property claims within proprietary digital systems (Peterson 2010). In the wake of the crisis, however, developments such as widespread litigation against MERS by county property recorders, employing doctrines such as fraud and unjust enrichment, have shown how 'relational' legal doctrines (Valverde 2011) can be brought to bear on the governance of property in new and creative ways (Zacks 2013). Following insights from American institutionalism (Commons 1924) and legal geography (Blomley 2005;

Delaney 2010), this requires that we treat real property law as built on certain interpretive ambiguities that have been put 'in play' within attempts to wield vacant, abandoned properties into objects of governance.

Second, however, this politics of legal interpretation is necessarily joined to a process of state-making, as localities construct themselves as legal agents, manage adjudicative processes and take on new roles in governing property through legal instruments. This approach engages Sbragia's (1996) counter to 'city bound' arguments, seeing private law as a sphere of manoeuvre for localities within structural constraints on public powers – but one that sets in motion new, unexpected institutional trajectories (cf. Baptista 2013). This casts law not as a fixed set of "tools" that are wielded by public agents over private ones, but rather as one of "a set of practices and strategies, governmental projects and modes of calculation, that operate on something called the state" (Jessop 2007: 37). Closer attention to these practices – in mundane areas such as receivership, property registration and mass litigation – bring into focus a constellation of legal venues, third parties, instruments and techniques, and calculative capacities through which local state powers are constructed over foreclosed properties. One aspect of the *legal complexes* (Rose and Valverde 1998) that become evident within this analysis is the fusing of public enforcement powers to private legal mechanisms, highlighting how these emerging forms of local governance often work by re-wielding the very legal instruments – such as the mortgage contract itself – that sustained the long mortgage boom.

Wielding vacant property into a governable object: the jurisdictional politics of foreclosure governance

At the heart of the urban governance of the foreclosure crisis has been the extension of land use and property regulatory powers to REO properties. These powers, developed through the long evolution of federal-state and state-local relations (Frug 1980; Valverde 2011), have produced an "inherent authority" on the part of US municipal corporations "to regulate private parties in order to protect public health, safety, and welfare within their jurisdiction" (Martin 2010: 27–28). Rather than representing that jurisdiction as a static property of public law (Ashton 2014), however, this section examines the local governance of the foreclosure crisis through attempts to expand local powers to wrest control – a "jurisdictional practice of sovereignty [creating] legal relations" around vacant and abandoned properties (Dorsett and McVeigh 2012: 32). Looking broadly at examples across a number of cities – including Cleveland, Baltimore and Chicago – I argue that this jurisdictional practice has been evident in three different areas: first, in attempts to produce a *force of law* sufficient to displace costs and burdens back onto lenders; second, in attempts to *know property* through the complexities of the foreclosure process; and, third, within the *complexes* of legal and administrative means used to extend foreclosure governance. Together, I argue, these evidence a jurisdictional politics of foreclosure as localities attempt to wield vacant, abandoned properties into objects of governance.

Troubled buildings: grounding the force of nuisance law

First, the law of public nuisance has been a primary strategic field to articulate foreclosure governance. Under standard building code ordinances, localities can intervene where a property, "in relation to its existing use, constitutes a hazard to the public health, welfare, or safety by reason of inadequate maintenance, dilapidation, obsolescence, or abandonment" (Johnson 2008: 1191). This makes nuisance a "capacious and rather fuzzy category" (Valverde 2011: 292) given the relational manner that defines this. The process of governing property through nuisance law could hardly be more mundane. Conventionally, enforcement occurs through the civil nuisance abatement lawsuit (often argued before a special housing court), where local government agencies or "interested parties" (such as a mortgage noteholder with an enforceable claim against a property) seek an injunction against an owner to force them to properly secure the property and bring it up to code. Without owner compliance, courts can levy civil fines. However, these sanctions can be insufficient to incentivize lenders or servicers – many with no local presence – to respond to civil summons over their REO properties, let alone invest in upkeep and maintenance. Not surprisingly, many absentee owners simply fail to show up in court to respond to nuisance complaints (Johnson 2008).

A more potent enforcement tool for unresponsive owners is the judicial process of receivership. Here judges appoint and empower a local government agency or a delegated third-party to take temporary control of the property to make necessary repairs. In cities such as Chicago and Cleveland, receivership has become a significant field of institutional expansion, as city agencies have worked through housing courts and in partnership with third-party code enforcement contractors, utility providers, police and community organizations to intervene around problem buildings (Frater *et al.* 2009). Chicago's Troubled Buildings Initiative (TBI) draws on Illinois statutes that allow a court-appointed receiver to temporarily abrogate an owner's rights, including "[using] the rents and issues of [a nuisance] property toward maintenance, repair and rehabilitation of the property", suspending any claims on rent or the proceeds from sale of a property held by either owners or mortgage noteholders (Simpkins *et al.* n.d.). Further, Chicago has developed financial partnerships with specialized lenders to fund receivership work, working through provisions in Illinois law that authorize receivers to recover costs by issuing and selling interest-bearing receiver's notes that are secured by a priority first lien on the property in question.

Even receivership is a weak option where vacant properties offer no rent or have little market value against which to recoup a receivers' lien. In this sense, legal strategies are spatially selective within a broader neighbourhood geography of concentrated foreclosures, helping to shape "territorialities of rule" through law's variegated force (Baptista 2013: 41). As a response, civil proceedings around nuisance and receivership have increasingly had to 'jump' doctrines to produce a force to law sufficient to displace costs and responsibilities onto the shoulders of absentee owners. In cities such as Buffalo and Cleveland, housing court judges have turned to issuing criminal summons to banks, with the discretion to proceed with a trial *in abstentia* should defendants not respond. By escalating nuisance claims into criminal proceedings, owners can receive more substantial fines and penalties for non-compliance – as high as $40–50,000 per property (Johnson 2008: 1196). This has proven especially effective to extend the reach of nuisance enforcement to chronically unresponsive lenders or servicers.

At stake within these legal proceedings, then, is not criminal prosecution of banks but rather the search for a legal grounding to the nuisance concept that can produce a *force of law* capable of incentivizing compliance with local building codes – whether it is through appropriating and re-wielding the legal powers of private ownership directly through a receivership process, or through sanctions that compel owners to exercise those powers themselves.

Knowing property: addressing 'legal limbos'

These legal strategies around nuisance have come up against some important technical and scalar limits. One particular technical challenge is to locate an owner or legal agent responsible for the property – that is, the fundamental problem of *knowing property* as a condition for governance (Newman 2010; Pritchard *et al.* 2015). Both civil and criminal nuisance proceedings depend on liberal notions of ownership as involving a single, identifiable owner (or agent) and a clear and uncontested chain of title (Rose 1994). These are notions that were upended by the integration of real property into financial modes of property (cf. Maurer 1999), as ownership claims were fragmented among investors in mortgage-backed securities, and opened to continual resale and assignment to new owners, even after foreclosure proceedings have begun. They are further fragmented by the commodity chain of mortgage-backed securities, which reassign certain aspects of the debt relationship – such as mortgage servicing and local REO upkeep – to third-party entities (Newman 2009). The rapid expansion of the Mortgage Electronic Registration System (MERS) in the 2000s meant that assignment of mortgage claims within secondary markets occurred digitally – through a private, proprietary database (Peterson 2010) – cutting off one of the primary tools of property governance available to localities: its ability to use control over the public land title system to structure property-level interventions (Newman 2010).

Property registration has proven to help match concepts of nuisance with the legal complexities of vacant, abandoned properties. Within the long process of foreclosure, there is often a temporal mismatch between when a bank assumes ownership and the incidence of abandonment. The serving of a notice of foreclosure often coincides with an owner vacating their home, and yet default does not confer outright legal ownership on a lender or servicer, creating a legal 'limbo' prior to the foreclosure sale. As court processing times have stretched into months and years, this period often stretches into one where "the home sits empty, the borrowers are gone, and the lender won't take responsibility for it … [and] a great deal of damage and deterioration can occur" (Leeper 2008: 96). These problems are further exacerbated by the problem of "toxic title" – situations where the lender completes foreclosure proceedings but does not record the transfer of ownership, often precisely to avoid *de jure* ownership and responsibility for upkeep and property taxes (Johnson 2008: 1186).

Whereas registration ordinances start from the pragmatic need to assemble real-time contact information for titleholders or their delegated agents, they also often extend anti-nuisance enforcement – that is, they add enforcement powers against negligent owners to "halt and reverse the negative impact of vacant, abandoned, and problem properties" (Center for Community Progress 2015). These powers can include 'affirmative duties' for owners of vacant buildings – such as board-up and security requirements – as well as new penalties and 'strategic' enforcement mechanisms that can be wielded by local building departments (Martin 2010). By creating additional standards that can apply to foreclosed properties whether they are in compliance with extant building codes or not, registration ordinances further expand the jurisdiction for nuisance enforcement.

Campaigns for better knowledge of property have also taken a litigious turn. With critical real-time information about deed holders hidden within MERS' proprietary databases, county recorders have initiated civil lawsuits against MERS Corp (Zacks 2013). Whereas the grounds for these suits often focuses on lost recording fees (to the tune of $8 billion in one illustrative class action case involving all urban counties in Kansas), they have also once again 'jumped' doctrines, mobilizing long-standing legal concepts of unjust enrichment, conspiracy and *prima facie* tort to discredit MERS' business model and restore the primacy of the public land title recording system.

However, courts have proved unwilling to invalidate MERS' practices, and forcing the issue into the courts has resulted in the affirmation of the legal principles underlying MERS' business model – that being the separation between the mortgage note and the mortgage deed and MERS' ability to circulate the latter without repeat public recording (Zacks 2013). This has not ended the legal proceedings but transformed them into contractual negotiations to share MERS' proprietary information systems to better track ownership and servicing rights in real time – partnering with the very entities producing the fragmentation of property in the first place in the search for greater knowledge about vacancy and abandonment (Schilling 2009).

Legal complexes: governing foreclosure through other means

The local governing of the foreclosure crisis can be seen not only in the formal application of legal norms through public law, but also in the institutional mechanisms and "particular practices of authority" that "[hold] the law together" (Dorsett and McVeigh 2012: 34) within jurisdictional practices around nuisance. The discussion thus far has already raised a number of examples of how the force of law is produced through "legal complexes" – "hybrid [compositions] of elements with diverse histories and logics" (Rose and Valverde 1998: 543) that fuse legal authorizations with other non-legal forms of social power or capacity. In addition to code enforcement partnerships that organize temporary ownership of problem properties through receivership, an emerging set of partnerships with MERS facilitates data analysis to identify REO 'targets' at the property or neighbourhood market level (Schilling 2009). Rosenman and Walker (2016) further introduce the concept of the "demolition coalition" to describe the constellation of public and private actors and interests shaping particular approaches to vacant property management – an approach which might be usefully engaged with the jurisdictional practices of foreclosure governance broached here.

At the same time, the construction of legal complexes extend past these examples of "governance-beyond-the-state" (Swyngedouw 2005) to encompass the mobilization of diverse resources *within* the law itself – notably those embedded in private contracts or mortgage deeds. In some property registration ordinances (such as that enacted by Chula Vista, south of San Diego), localities have sought to bring REO properties "into the law" by expanding the statutory definition of "abandonment" to encompass "a property that is vacant and is under a current notice of default or notice of trustee's sale and is not currently being offered for sale, rent or lease" (Chula Vista Municipal Code, Sec 15.60.020, City of Chula Vista n.d.). In New York civil and criminal proceedings, prosecutors and judges have similarly argued that "by sending letters threatening eviction or foreclosure against defaulting homeowners, the lenders have asserted control over the property, triggering a responsibility to maintain the home after the homeowner vacates" (Johnson 2008: 1195).

In practical terms, this has forced lenders or servicers to exercise powers embedded within clauses in the mortgage deed, which enable lenders to take extraordinary measures to secure their collateral in situations of property "abandonment and waste". These measures include "the right to enter and maintain a property if it is vacant in order to secure [it] against vandalism or decay" (Martin 2010: 20). Instead of acting as a private, background, collateral claim, then, this allows mortgage deeds to anchor notions of public nuisance around bank-owned properties. In another example, recent vacant property ordinances have tied registration requirements to the legal process of foreclosure – that is, the filing or serving a notice to foreclose to an owner. As lenders exercise their (private) contractual clauses within county courts, then, they create a stream of information about properties likely to be REO (Martin 2010). In both of these examples, public enforcement works by engaging the legal instruments implicated in foreclosure in the first place, managing the excesses of REO governance by filling the 'black holes' of ownership limbo with various forms of economy.

As noted with receivership, legal strategies that harness mortgage contract provisions or other mechanisms of private ownership are likely to be of varying effectiveness depending on localized market calculations of profit-loss on the part of owners. Further, this wielding of private legal instruments is not effective with all property owners. With the nationalization of the government-sponsored enterprises Fannie Mae and Freddie Mac in September 2008, REO properties owned by these quasi-public agencies became federal property and thus outside the reach of local nuisance abatement law or registration requirements – a legal topology confirmed by litigation by the Federal Housing Finance Agency (FHFA, the federal conservator). In other words, jurisdictional practices around foreclosure need to be appreciated not only for the novel ways they bring vacant, abandoned properties "into the law", but also for the "territorialities of rule" (Baptista 2013) and market- and state-derived patterns of spatial selectivity (Jones 1997) they set in motion.

Territorializing foreclosed property: mass nuisance and *parens patriae*

If these jurisdictional practices seek to render vacant, abandoned properties into objects of governance, another set of legal strategies seek to *territorialize* property, deploying law to draw together fragmented parcels under a unified, legally responsible agent. In the form of mass nuisance litigation against lenders or anti-discrimination class action cases, these strategies respond to a set of technical and scalar problems that hamper the prosecution of concepts such as nuisance.

On the technical side, the process of redefining owners' obligations through nuisance has triggered due process requirements, including statutory waiting periods. These create a temporal mismatch between the problems of foreclosure and the deployment of governance mechanisms – an average of eight months in one estimate from Baltimore (Johnson 2008: 1195). Just as vexing is the scalar mismatch between the focus on parcels of property and the broader extent of the foreclosure crisis. For instance, receivership actions are pursued through individualized proceedings against legally delineated parcels of property – a labour-intensive approach that "[reduces] the number of properties that cities and nonprofit organizations are able to rehabilitate" (Johnson 2008: 1195). Registration offers a broader means to anticipate future nuisances, but enforcement still devolves to parcel-by-parcel interventions, in contrast to the mass approach necessary to address the combined and uneven effects of concentrated foreclosures.

This has opened up other innovative legal strategies for consideration, namely mass litigation and class action lawsuits against large banks, servicers or their counterparties. This litigation aims to compel those institutions to maintain their abandoned properties, or, to compensate the locality for lost tax revenue and increased service and monitoring costs. A handful of cities have pioneered this approach – Cleveland and Baltimore being two examples – although there is a growing list of cities and counties pursuing such litigation along a variety of doctrinal lines including nuisance and disparate impact (discrimination) (Balser and Slovensky 2014).

Subprime lending as a mass nuisance

Mass nuisance litigation *territorializes* property by causally connecting lenders, the harm of their subprime mortgage products, and the broader effects of concentrated foreclosures and REO properties. Cleveland first pursued this tactic by filing a lawsuit in 2008 against nearly two dozen large lenders, arguing that their lending produced a public nuisance in the form of "proliferating toxic sub-prime mortgages, [which] under the circumstances ... made the resulting spike in foreclosures a foreseeable and inevitable result" (City of Cleveland v. Deutsche Bank Trust Co. 2008: 26). The City's complaint referenced over 16,000 foreclosures filed by the defendants from 2002 to 2006. A parallel lawsuit filed by a housing advocacy group challenged the business practices of "dumping" dilapidated homes at low prices, arguing they created a "foreclosure death spiral" in neighbourhood housing values (CHRP v. Wells Fargo 2010).

The City of Cleveland did not prevail in its case, as the court held that the City had not adequately established lender liability for mass foreclosure. Effective legal territorialization requires linking both "the wrongdoer's actions to the increase in foreclosures and abandonment", and "the foreclosures and abandonment to injuries claimed by the city" (Johnson 2008: 1198). The City of Cleveland tripped over the technical challenges involved with making this link. Publicly available data may be insufficient to prove that, at the moment it was issued, each mortgage was *automatically* a harmful product by virtue of negligent underwriting or predatory loan terms. This evidentiary burden is necessary for "absolute" claims of nuisance – "intentional or unlawful conduct by the defendant that is so inherently dangerous that it cannot be conducted without damaging someone else's property rights or causing harm, no matter the care utilized" (Johnson 2008: 1217). Instead, Cleveland's case relied on a "collective or market-share theory of liability" that simply assumes damages are equally distributed among participants in an industry according to their market share of subprime mortgages or mortgage-backed securities. Whereas courts have begun to accept these mass nuisance claims in cases against gun manufacturers (Johnson 2008: 1216), they have yet to do so with foreclosures and REO properties.

Parens patriae *and the city as quasi-sovereign*

To characterize this case as a failure, however, misses how jurisdictional practices provide broader openings to wield together loans, properties and harm to borrowers into a legally coherent field of action. Indeed, mass litigation has demonstrated the "capaciousness" (Valverde 2011) of nuisance by showing how it can be switched among a variety of doctrinal bases. In the wake of Cleveland's 2008 case, it and other cities have changed the grounding for their mass litigation from public nuisance law to federal civil rights legislation and the doctrine of disparate impact (or indirect racial discrimination). The City of Baltimore, in an innovative move, filed a class action suit against Wells Fargo in 2008 alleging that the bank's racially discriminatory lending practices produced concentrated foreclosures in minority neighbourhoods. Since then, similar suits have been filed in Atlanta, Chicago (Cook County), Cleveland, Los Angeles, Memphis (Shelby County), Miami and Providence (Balser and Slovensky 2014).

Baltimore's primary innovation was its legal strategy of pursuing quasi-sovereign standing, or *parens patriae* (literally 'parent of the fatherland'). Since the 1970s, courts have recognized states' ability to engage in broad-based class action suits against corporations on behalf of aggrieved citizens, and parens patriae suits have been used in policy areas such as antitrust regulation, public health and consumer protection (Ratliff 2000). Whereas the question of whether cities can constitute themselves as quasi-sovereigns to recover damages on behalf of residents is still legally murky (Engel 2014), cities have found ways to adopt the *form* of parens patriae lawsuits by finding statutes that give them legal standing to pursue claims as "aggrieved persons" whose interests have been harmed by a discriminatory housing practice. In earlier litigation over blockbusting or racial steering, courts accepted as persuasive that cities had standing in instances where "racial steering by real estate sales agents destabilized the community, increased the burden on the city in the form of increased crime, and eroded the city's tax base" (Johnson 2008: 1200).This potentially represents "a viable alternative theory of recovery to cities that have seen neighborhoods comprised of minorities ravaged by predatory subprime loans" (Johnson 2008: 1198).

As with other doctrinal paths to governing foreclosures, these forms of mass litigation hinge on some critical doctrinal or technical issues. A primary challenge is the technical need to prove causation. As with mass nuisance claims, cities pursuing disparate impact cases have to "demonstrate that the subprime loans at issue were predatory and doomed to fail, and, therefore, would naturally lead to foreclosures and abandonment and subsequent harm" (Johnson 2008: p. 1198). This involves detailed and highly technical models of mortgage price discrimination, such that "the defendants' lending practices and loan terms [could be shown to be] 'unfair' and 'predatory'" or had "'a disparate impact on the basis of race'" (Johnson 2008: 1201) relative to other products prevailing in the market when the loans in question were made.

In addition to the technical capacities and data necessary to produce these models, we need to appreciate the selectivities they set in motion. The City of Baltimore's initial case was rejected based on failings in its technical model, which the court argued drew together too many disparate kinds of loans to demonstrate a plausible causal connection between the widespread damages sought by the City and Wells Fargo's lending practices. The case was allowed to proceed in 2011 only after its terms were narrowed to focus on a specific set of borrowers who qualified for prime mortgages but instead were issued subprime loans – transforming the case from (in the words of the presiding judge) "a 'macro' one, focusing upon systemic issues, [to] a 'micro' one, focusing upon specific borrowers and properties" (Mayor and City Council of Baltimore v. Wells Fargo Bank 2011). Here, the force of law is produced through a set of technical capacities to model financial behaviours, with the model itself becoming a key aspect of the legal complex while simultaneously writing certain territories out of its calculations of harm.

Conclusion: Juridification and Urban Governance

This chapter has examined the local governance of the foreclosure crisis through a set of jurisdictional practices aiming to render vacant, abandoned properties into objects of governance through nuisance doctrine, and aiming to empower law to move costs and burdens back to lenders. In contrast to 'city bound' accounts of municipal powers, this has opened up for consideration how legal and administrative tools extend foreclosure governance through plural legal doctrines and channels, noting that those often work through a fusion of public law with the same private legal instruments (such as the mortgage note/deed or the foreclosure filing) implicated in the production of the crisis. This reverses some of the conventional understandings of property circulating around the housing boom and subsequent bust, whereby *only* private agents engaged in this kind of legal play, as well as the "methodological tendency to regard legal and governance inventions (such as the notion of land use) as tools chosen to implement a fixed political project" (Valverde 2011: 280).

This is by no means the last word in understanding the local governance of the foreclosure crisis. Indeed, nuisance law represents a rather mundane position from which to assess the crisis and its shifting politics. A fuller account should also consider emerging legal strategies and political battles over community-led eminent domain and land-banking (Christophers and Niedt 2015), as well as broader forms of "governance-beyond-the-state" arrangements emerging around demolition (Rosenman and Walker 2016), eviction (Schneider 2015) and neighbourhood stabilization (Immergluck 2010). My narrow focus on nuisance has served to emphasize particular property practices and techniques, both as interesting mechanisms and as entries to a more sophisticated understanding of law and its relation to US urban governance. In this vein, Valverde (2011: 292–294) offers nuisance as one example of how to "see like a city" and appreciate how "relational categories" – ones with "fuzzy boundaries" capable of drawing together different social forces – have become "a necessary component of contemporary urban governance". The stakes in such an approach to law are "less the concrete success or failure of a given set of practices … and more what an analysis of its deployment reveals about the underlying dilemmas and conditions of contemporary urban governance" (Baptista 2013: 11).

At the same time, such an approach offers significant insights into tensions and contradictions in that "drawing together", ones that I have only been able to hint at in this chapter. These include tensions in the transforming of municipal roles as legal complexes emerge and change in concrete local circumstances, including how differences in coordination, capacity and context produce highly variegated 'forces' to law. They also include contradictions surfaced by the very legal instruments, techniques and forms of calculation put into play within attempts to wrest control over vacant, abandoned properties. In other words, jurisdictional practices around foreclosure need to be appreciated not only for the novel ways they bring vacant, abandoned "into the law", but also for the "territorialities of rule" (Baptista 2013) and market- and state-derived patterns of spatial selectivity that they enable.

References

Apgar, W., Duda, M. and Gorey, R. (2005). *The Municipal Cost of Foreclosures: A Chicago Case Study*, report, Homeownership Preservation Foundation, Minneapolis.

Ashton, P. (2012). Troubled Assets: The Financial Emergency and Racialized Risk, *International Journal of Urban and Regional Research*, 36(4), pp. 773–790.

Ashton, P. (2014). The Evolving Juridical Space of Harm/Value: Remedial Powers in the Subprime Mortgage Crisis. *Journal of Economic Issues*, 48(4), pp. 959–979.

Balser, D. and Slovensky, L. (2014). *Predatory Lending Claims by Cities and Counties Against Financial Institutions Escalate in 2014*. Available from King & Spaulding LLC, 25 January 2016.

Baptista, I. (2013). Practices of Exception in Urban Governance: Reconfiguring Power Inside the State. *Urban Studies*, 50(1), pp. 39–54.

Blomley, N. (2005). The Borrowed View: Privacy, Propriety, and the Entanglements of Property. *Law & Social Inquiry*, 30(4), pp. 617–661.

Center for Community Progress. (2015). *Strategic Code Enforcement*, report, CCP, Washington.

Christophers, B. and Niedt, C. (2015). Resisting Devaluation: Foreclosure, Eminent Domain Law, and the Geographical Political Economy of Risk. *Environment and Planning A*, 48(3), pp. 485–503.

Chula Vista Municipal Code, Sec. 15.60.020, City of Chula Vista. (n.d.). Available at: www.codepublishing.com/CA/ChulaVista/#!/ChulaVista15/ChulaVista1560.html [accessed 27 December 2015].

City of Cleveland v. Deutsche Bank Trust Co., 571 F. Supp. 2d 807 (N.D. Ohio 2008).

Cleveland Housing Renewal Project, Inc. v. Wells Fargo Bank, NA, 188 Ohio App. 3d 36, 2010 Ohio 2351 (Ct. App. 2010).

Commons, J.R. (1924). *Legal Foundations of Capitalism*. New York: Macmillan.

Delaney, D. (2010). *The Spatial, the Legal and the Pragmatics of World-Making: Nomospheric Investigations*. New York: Routledge.

Dorsett, S. and McVeigh, S. (2012). *Jurisdiction*. New York: Routledge.

Engel, K. (2014). The State of Play in City Claims Against Financial Firms. *Fordham Urban Law Journal City Square*, 40(1), pp. 82–94.

Foglesong, R. (1986). *Planning the Capitalist City: The Colonial Era to the 1920s*. Princeton, NJ: Princeton University Press.

Frater, M., Gilson, C. and O'Leary, R. (2009). *The City of Cleveland Code Enforcement Partnership*, report, Center for Community Progress, Cleveland, OH.

Frug, G. (1980). The City as a Legal Concept. *Harvard Law Review*, 93(6), pp. 1057–1154.

Frug, G. and Barron, D. (2008). *City Bound: How States Stifle Urban Innovation*. Ithaca, NY: Cornell University Press.

Gotham, K. (2006). The Secondary Circuit of Capital Reconsidered: Globalization and the US Real Estate Sector. *American Journal of Sociology*, 112(1), pp. 231–275.

Immergluck, D. (2010). Neighborhoods in the Wake of the Debacle: Intra-Metropolitan Patterns of Foreclosed Properties. *Urban Affairs Review*, 46(1), pp. 3–36.

Immergluck, D. (2013). Too Little, Too Late, and Too Timid: The Federal Response to the Foreclosure Crisis at the Five-Year Mark. *Housing Policy Debate*, 23(1), pp. 199–232.

Jessop, B. (1990). *State Theory: Putting the Capitalist State in its Place*. State College: Penn State University Press.

Jessop, B. (2007). From Micro-Powers to Governmentality: Foucault's Work on Statehood, State Formation, Statecraft and State Power. *Political Geography*, 26(1), pp. 34–40.

Johnson, C. (2008). Fight Blight: Cities Sue to Hold Lenders Responsible for the Rise in Foreclosures and Abandoned Properties. *Utah Law Review*, 2008(3), pp. 1169–1253.

Jones, M. (1997). Spatial Selectivity of the State? The Regulationist Enigma and Local Struggles Over Economic Governance. *Environment & Planning A*, 29(5), pp. 831–864.

Leeper, D. (2008). Testimony Before the House Subcommittee on Domestic Policy, House of Representatives, 110th Congress, 21 May. Available at: www.gpo.gov/fdsys/pkg/CHRG-110hhrg49971/html/CHRG-110hhrg49971.htm [accessed 3 March 2016].

Martin, B. (2010). Vacant Property Registration Ordinances. *Real Estate Law Journal*, 39(6), pp. 6–43.

Maurer, B. (1999). Forget Locke? From Proprietor to Risk-Bearer in New Logics of Finance. *Public Culture*, 11(2), pp. 365–385.

Mayor and City Council of Baltimore v. Wells Fargo Bank, Civil No. JFM-08–62 (D. Md. 28 September 2011).

Newman, K. (2009). Post-Industrial Widgets: Capital Flows and the Production of the Urban. *International Journal of Urban and Regional Research*, 33(2), pp. 314–331.

Newman, K. (2010). Go Public! Using Publicly Available Data to Understand the Foreclosure Crisis. *Journal of the American Planning Association*, 76(2), pp. 160–171.

Peterson, C. (2010). Foreclosure, Subprime Mortgage Lending, and the Mortgage Electronic Registration System. *University of Cincinnati Law Review*, 78(4), pp. 1359–1407.

Pritchard, S., Wolf, S. and Wolford, W. (2015). Knowledge and the Politics of Land, *Environment and Planning A*, 48(4), pp. 616–625.

Ratliff, J. (2000). Parens Patriae: An Overview. *Tulane Law Review*, 74(2), pp. 1847–1858.

Rose, C. (1994). *Property and Persuasion: Essays on the History, Theory, and Rhetoric of Property*. Boulder, CO: Westview.

Rose, N. and Valverde, M. (1998). Governed by Law? *Social & Legal Studies*, 7(4), pp. 541–551.

Rosenman, E. and Walker, S. (2016). Tearing Down the City to Save It? 'Back-Door Regionalism' and the Demolition Coalition in Cleveland, Ohio. *Environment and Planning A*, 48(2), pp. 273–291.

Sbragia, A. (1996). *Debt Wish: Entrepreneurial Cities, US Federalism, and Economic Development*. Pittsburgh, PA: University of Pittsburgh Press.

Schilling, J. (2009). Code Enforcement and Community Stabilization: The Forgotten First Responders to Vacant and Foreclosed Homes. *Albany Government Law Review*, 2(1), pp. 101–162.

Schneider, V. (2015). Property Rebels: Reclaiming Abandoned Bank-Owned Homes for Community Uses. *American University Law Review*, 65(2), pp. 399–433.

Simpkins, A., Esenberg, B., Frydland, J., McKenzie, S. and Ladores, J. (n.d.). *Receivership: Addressing Vacant and Abandoned Properties*, report, City of Chicago, Department of Planning and Development.

Swyngedouw, E. (2005). Governance Innovation and the Citizen: The Janus Face of Governance-Beyond-the-State. *Urban Studies*, 42(11), pp. 1991–2006.

Valverde, M. (2011). Seeing Like a City: The Dialectic of Modern and Premodern Ways of Seeing in Urban Governance. *Law & Society Review*, 45(2), pp. 277–312.

Woodstock Institute. (2016). Foreclosure Data Portal. Available at: www.woodstockinst.org/content/foreclosure [accessed 25 January 2016].

Wyly, E., Moos, M., Hammel, D. and Kabahizi, E. (2009). Cartographies of Race and Class: Mapping the Class-Monopoly Rents of American Subprime Mortgage Capital. *International Journal of Urban and Regional Research*, 33(2), pp. 332–354.

Zacks, D. (2013). Revenge of the Clerks: MERS Confronts County Clerk and *Qui Tam* Lawsuits. *Banking & Financial Services Policy Report*, 32(1), pp. 17–24.

19
HOUSING THE BANLIEUE IN GLOBAL TIMES

French public housing and spaces between neoliberalization and hybridization

David Giband

In Europe, recent decades have witnessed the dismantlement of the welfare state and the restructuration of states that plays a crucial role in spreading and implementing neoliberal urban policies (Pinson 2010). Yet, far from disappearing or weakening, the state currently acts as an entrepreneur in favour of particular interests and asks cities to transform into spaces of economic attractiveness. The "remaking of cities", according to neoliberal perspectives, seems to be a global injunction nurturing local urban growth agendas (Harvey 1989). These changes are part of the overall increase of neoliberal strategies to entrepreneurize cities. Here there is a kind of build-on from a regime of Keynesian accumulation long replaced by a regime of flexible accumulation (Amin 1994). In this global neoliberalization of urban policies, public housing policies now occupy a central place. They seem to be one of the last pieces of a larger movement of urban neoliberalization and reordering that is fundamentally changing European cities.

In France, this housing centring now manifests in a specific urban space: the banlieue.[1] Here large and disadvantaged peripheral public housing complexes house the country's poorest and most downtrodden. Today, the banlieue problem cannot be understood without referring to the process of neoliberalization of French urban and housing policies. In 2003, the French state published a reformist agenda consisting of an urban renovation programme (PNRU[2]) focusing on social and spatial restructuring for disadvantaged peripheral public housing complexes. Taking into account systemic changes associated with globalization, restructurings in urban economics and rescaling of state power, the agenda seeks to enhance attractiveness of urban regions in opening suburban public housing spaces to urban entrepreneurialism. A vast reform has been launched to "radically treat these ghettos".[3] The drive is led by a national agency (ANRU) introducing an entrepreneurial mode of urban governance.

In this chapter I chronicle that this reform consists of a policy of settlement aiming to more deeply manage this poor population. Proposed changes are rooted in a "mechanism of neoliberal localization" inducing new productions of spaces. A "middle sizing process" is expected: have these populations conform to social, economic and moral middle-class norms while they are to be physically removed from the creation of re-entrepreneurialized cities. A rescaling of power occurs, characterized by coercion and a disciplinary urban planning, lowering contractual governance inherited from the welfare state and redrawing the shape of urban power. This process finally produces hybrid policies between a global neoliberal restructuring and the resistances of the 'local'.

Housing the banlieue in global times: neoliberalization and politics of settlement

Since the 1980s, banlieues in France have been increasingly controversial. These vast peripheral public housing complexes[4] concentrate a population stigmatized by unemployment, poverty, immigration and criminality, and growing numbers of people have questioned their value and utility. Criticism has been direct about their function in French society and the benefits that cities, regions and the country derive from their existence.

The urban renewal programme: a neoliberal reform

The banlieue issue cannot be understood without referencing the process of neoliberalization of traditional French urban and housing policies (Brenner and Theodore 2002). From the 1980s to 2003, public authorities first faced the banlieue issue following a national policy locally declined: *la politique de la ville*.[5] The purpose was to foster social development of these peripheral poor neighbourhoods in articulating locally different national public policies in education, public housing, culture and employment according to the French welfare state model (Chaline 2010). Despite more than twenty years of massive public investments in social urban redevelopment programmes, the situation worsened. The national observatory for the banlieue[6] (a public institution dedicated to the understanding of banlieues' issues[7]) denounced the failure of the national programmes and suggested a worsening of segregation, poverty and isolation among banlieue inhabitants. National media, pointing out the failure of modern urbanism and the concentration of poor and immigrant households, echoed such criticisms. The banlieue is thus depicted as the realm of failing public housing policies that have produced dependency on welfare programmes. Those criticisms nurtured political discourses of the right and of the National Front amalgamating public housing, immigration and advanced marginality (Wacquant 2006).

In 2003, under the Sarkozy administration, the French state published the *City Orientation and Programing Act* (2003), a vast reform of this national policy (Epstein 2015). A first deregulation/privatization agenda was set consisting of an urban renovation programme (PNRU) focusing social and spatial restructuring for 490 neighbourhoods in the French banlieues. This programme identifies the need for public housing complexes to be demolished and reconstructed. A total of 137,000 housing units in 450 neighbourhoods were targeted for demolition, 130,000 for housing rehabilitation and 341,000 private housing units were to be built as home owned structures.[8] Taking into account systemic changes associated with globalization, restructurings in urban economics and rescaling of state power, it seeks to enhance attractiveness of urban regions in opening peripheral public housing spaces to urban entrepreneurialism by facilitating 'constructive' social and cultural change. The 2005 riots accelerated the desire for these changes.[9] The Sarkozy administration accelerated these reforms to "radically treat" these so-called ghettos. This administration, above all else, wanted to now control these "turbulent spaces and populations" targeting immigrant families and second, following a general injunction of "remaking the city", to have these public housing poor neighbourhoods conform to neoliberal urban policies. Located in the first ring suburbs of French metropolitan areas, the banlieues occupy large parts of metropolitan areas where there is heavy pressure on real estate markets. Reform would facilitate the liberalization of the public housing system and land tenure in the banlieue; while the urban renewal programme would restructure and deregulate the national public housing policy in order to diversify the housing stock (new private constructions) and to promote homeownership. The impulse of such a policy is led by a new national agency for urban renewal (ANRU[10]) introducing a specific mode of urban governance. In centralizing all public subsidies, this agency controls national and local public housing policies, imposes entrepreneurial norms of governance to local authorities, defines for each locality strict criteria for the demolition of the public housing stock and the rebuilding of private homes.

PNRU: an urban, social and moral reshaping of the banlieue

Behind PNRU's numerical targets, three objectives were assigned to the urban renewal programme: residential diversification through social mix, mutability of land tenure and rehousing poor households outside the banlieue.

At the core of this reform, social mix is central in local urban restructurings. Social mix is expected to help diversify the banlieue housing stock. Settling of middle classes is seen as a vector of social stability and diversity. In particular, homeownership is supposed to change behaviours following a moral restructuring based on middle-class norms and values. A "middle sizing process" is expected (Epstein 2005), namely adapt and conform these turbulent spaces and populations to middle-class norms and values, give financial and real estate add-values to theses spaces, and a kind of magical social and moral transformation would follow.

Following international trends (Lelévrier 2006), social mix is defined as a process of shared values and consolidated trust made possible by spatial and residential interactions between inhabitants (public housing inhabitants) and newcomers (in private housing). Social mix is rooted in the manufacturing of social heterogeneity across classes and ethnicities with two frequently unstated goals. First, it would create attractive places for further investments for private developers in stressed housing markets. Second, diversifying the banlieue housing stock is a means to fight local concentration of poverty, unemployment and unproductive spaces and add values to land and real estate capital.

Land tenure is another key element here. ANRU subsidies provide the financing of a third of each demolition/reconstruction project asking for municipalities to self-finance and to multiply public-private partnerships. Many public housing authorities were asked to sell land at low prices to private developers. In return, the private sector must include a 25 per cent share of social housing units in each of its real estate developments. But at the same time, the upper limits for a household to access public housing have been changed, authorizing public housing authorities to increase their rental fees. This rule modification allows in the very stressed real estate markets (Paris, Lyon, Marseille) middle-class households to access public housing in the banlieue.

The new urban form produced follows a typical suburban middle-class housing pattern: single family homes, self-contained residential units with limited access via enclosure, commercial malls. Some scattered public housing units made of individual houses have been planned while a few public housing complexes still remain in the less valuable neighbourhoods. At the same time, PNRU requests that half of the households living in demolished public housing be rehoused and dispersed. Community leaders and groups of inhabitants denounced rehousing programmes as an ethnic removal focusing families with an immigration background, mostly Muslims.

Social mix seems to be both the horizon of housing policies and one of the moral structuring myth of contemporary neoliberal cities, hiding policies of settlement that favour 'middle sizing' of former disadvantaged neighbourhoods, dispersing poor households in the name of the social mix.

A neoliberal politics of settlement

More than just another reform of a sectorial policy dedicated to affordable housing, PNRU has to be understood as a policy of settlement: a strategic issue in a context of neoliberalization of urban policies and banalization of urban entrepreneurialism. The willingness to control and act on social occupation of the banlieue and to influence residential mobility is not, of course, new. But today strategies – implemented in the banlieue in order to "remake the city" – symbolize a "neoliberal urbanism" (Peck et al. 2009) and what Wilson calls a "contingent urban neoliberalism" (Wilson 2004). The neoliberalization process has led to an important reshaping of urban policies, especially for public housing policies in the way they conform to goals of settlement much more than housing. The intention in these former public housing realms is no longer to house decently and massively poor households but to set the conditions of a spatial relocation for poor and undesirable households elsewhere in the city and to make place for more desirable citizens (Desage et al. 2015). In this neoliberal project, the aim of policies of settlement is not to help individuals to be housed decently on a universal basis, but merely to ensure the transformation of public housing neighbourhoods now seen as physically, socially, economically obsolete spaces. This change is part of the overall increase of strategies of attractiveness for entrepreneurial cities with a neoliberal growth agenda focusing on attractiveness of capital and specific social classes (Harvey 1989). Reorganization of urban capital led to the spread of urban entrepreneurialism as an answer to inter-urban competition. As such, urban renovation is an effort to adapt the banlieues to the needs of competitiveness and flexibility of today's capitalism. It is characterized by mechanisms of public-private partnerships and by a speculative making of urban space rather than the improvement of urban living conditions (McLeod 2002). Images and representations of such deprived spaces have to be improved in order to facilitate their inscription in the real estate market, making them desirable for investors and new clients.

Public housing policies play a central role in implementing entrepreneurial strategies and in adapting former public housing neighbourhoods to a speculative construction of place. Urban renovation programmes can thus be understood as the bridgehead of urban policies neoliberalization. Political history of housing policies in the banlieue (*politique de la ville*) and its recent restructurings illustrate the political rescaling in the public housing sector, the neoliberalization process and the emergence of new power relationships between the urban (local) and the central power (national).

Redefining public housing for neoliberal needs

Obviously, public housing restructurings are central in the neoliberal project of remaking cities in conforming poor neighbourhoods to neoliberal norms and values.

From mass social housing to residualization – homeownership policy

If post Second World War was considered the golden age for public housing policies in Europe (Desjardins 2008), since then economic crisis and the neoliberal turn in Europe have changed a "mass social housing" policy into a "residualization" policy (Tellier 2007). The "mass social housing" model was characterized by a massive production of public housing in large complexes. In France, slogans such as "public housing for everyone", "housing the blue collar and the engineer", were used in the 1950s and 1960s to promote public housing complexes in the banlieue. Designed – according to modern architecture and urban functionalist theorists – as a modernist city, public housing complexes answer objectives of universal housing for a wide range of the population. Alternatively, since the 1980s and first neoliberal attacks on public housing, the residualization model prioritizes small public housing scattered sites only dedicated to the most disadvantaged. Homeownership programmes replaced construction and maintenance programmes of public housings. "Residualization is generally defined as a situation in which the public rental sector detaches from overall trends in society, being mainly destined to the more impoverished, destitute and difficult populations" (Valença 2015: 108). In the urban neoliberal project, public housing has not disappeared. Its role has changed while its downsizing conforms to a neoliberal moral perspective. Indeed, in the PNRU, following what has happened in the rest of Europe, downsizing is justified by the necessity for public housing authorities to focus on the most disadvantaged. In that perspective, multiplication of housing voucher programmes seeks to answer to different levels of social exclusion. The PNRU has generated new types of housing vouchers according to a specific zoning focusing different types of beneficiaries to a smaller part of the population (disabled persons, single-head families).

A social neoliberalism? Public housing new norms and values

Today amid a period of state decline, a new form of public action in the social sectors has emerged (housing, education, health). Following Foucault (2004), one can understand these changes as a neoliberal turn in public social policies. In neoliberal times, public action does not target interventions in the productive sphere but intends to set an appropriate framework for free concurrency in diverse sectors, including public housing. Once restricted to public housing authorities, the banlieues are invited to open to free real estate market and concurrency. Public housing authorities are authorized to merge with private investors while a new legislation deregulates access to public domain in the banlieues. In this framework, state action has not weakened; it takes the shape of a 'governmentality' in which the role of the economy and politics recompose (Foucault 2004). Objectives dedicated to public housing policies have changed. They focus on homeownership even for the least disadvantaged while the purpose of public housing itself has switched from universal housing to a social exclusion mandate. This redefinition of public housing policies takes place inside a wider restructuring of urban governance (Hackworth 2009). It results in a mutation in the role played by traditional actors of this sector (housing authorities, elected officials, inhabitants) in the organization, production and management of public housing and public subsidies. It asks people to conform to economically rational principles and to an entrepreneurial state of mind and to act according to neoliberal rules.

In France, state subsidies for public housing have not been downsized. From 1975 to 2015, the level of subsidies stabilized but the reallocation operates in favour of individual subsidies (*aides à la personne* – vouchers to people), i.e. housing vouchers can be used for homeownership, and minor subsidies for public housing complexes involve new construction or rehabilitation (*aides à la pierre*: vouchers to building public housing complexes). This shift in housing subsidies and vouchers is symptomatic of the emergence of a 'social neoliberalism' (Desjardins 2008) succeeding the Keynesian compromise. This 'social neoliberalism' relies on four principles: social mix, participatory technics and rhetoric, inhabitants' empowerment and 'middle sizing'.[11]

'Governance from afar'? Power, norms and (administrative) domination

Beyond these changes, a rescaling of power lowers the usual forms of contractual governance inherited from the Keynesian housing policies and the introduction of a 'governance from afar' conforms to urban entrepreneurial rules dictated by the ANRU (Epstein 2005). The ANRU mode of governance introduces not only a new scale of regulation but also a distance between national and local actors to respond to global trends. This neoliberalization of banlieue governances depends on the involvement and inclusion of local authorities and private partners. It also has to articulate with inherited local institutional actors and politics and locally produces hybrid-housing policies between neoliberal restructuring and resistance/resilience of local welfare state and residents.

Governing from afar

Following Epstein (2005), three periods characterize French public urban housing policies. First, a centralized government of the local, from 1950 to the mid 1970s, fits in a centralized managing of the local by the national state via a national planning of public housing development and uniform public action. It is mainly characterized by the building of large peripheral public housing complexes according to standardized architectural norms. Power is concentrated in the hands of centralized public administrations diffusing homogeneous technics of construction, trivializing banlieue monotonous landscape, in the name of universality and national equity. The economic crisis of the mid 1970s and the decentralization process of the French state inaugurated a second period in public housing policies: *la politique de la ville*. This corresponds to a negotiated government of the banlieue spaces relying on two articulated principles: project and contract. In a time of rapid urban and social deterioration of the banlieue complexes, and facing critics about a rigid and bureaucratic centralized power from 1981 to 1995,[12] the French state set a mode of negotiated government between the national and the local. It was based on the contractualization of public action. Local projects and contractualization between the French state and local authorities allow the emergence of sectorial answers (in the fields of social development, housing, education, culture) seen as efficient ways to deal with 'the new social question' (urban decay and social problems).

In the early 2000s, many observers pointed out the failure of this reformist policy; while a common discourse tended to associate the banlieue with 'French ghettos', asking for 'a radical treatment of those ghettos'. The French minister of housing and urban development, Jean-Louis Borloo, urged in 2003 to "break the ghettos in order to save a Republic in danger". Dramatization of urban and social challenges facilitates a shift towards a mode of neoliberal governance, opening a third period: a governance from afar (Epstein 2005). It relied on the state establishing a powerful agency (ANRU) that centralized public subsidies and came to house urban renewal programmes in banlieues. Its role not only consists of regulating public subsidies; it also includes setting norms, procedures, technics, goals and instruments for public action. ANRU puts cities in competition for financing local urban renewal programmes (PNRU) according to calls for project procedures, and allocates subsidies on a discretionary manner.

> Everything goes on as if the French state doesn't need to be present at the local level to act. On the contrary, it organizes its withdrawal from local territories by governing from afar. This distancing allows central power to protect itself from the local authorities' influences that limit its latitude.
>
> (Epstein 2005: 5)

The French state carefully implements a governmentality that strongly acts on territories and people not directly but from afar, forcing local actors to adapt to this neoliberal governance.

Coercion, symbolic violence and administrative domination

In this governance from afar, power and domination do not rely on exercising direct power or violence but rather use coercion, consent and hegemony. Following Davies (2014b), this mode of governance is structured by administrative domination. Indeed, ANRU strongly relies on local implementation and spread of entrepreneurial rationalities in the administrative control of local urban renewal programmes. Through a mix of inclusion strategies and administrative coercion, it depends on a spatialized set of technocratic/entrepreneurial norms and instruments such as: strong control of urban renewal projects submitted for subsidies, intense performance management for each housing demolition/rebuilding projects, micro-management of rehousing programmes, strict norms of participatory practices between local authorities and inhabitants. Administrative domination is assured by a myriad of collaborative institutions (between national and local authorities) mainly in the urbanism, housing (municipal offices of ANRU), and social development fields (social cohesion). It takes the shape of a dense network of locally based agencies and institutions, conforming to norms, values and rules nationally edicted by ANRU. In this process, the local state is the agent of coercive power. Coercion is not only a matter of national norms for local authorities in order to fit with requirements for ANRU subsidies. Coercion is integrated into daily administrative routines, in participatory procedures, in the rehousing of inhabitants, in social programmes implemented, etc. The power relationship has to be understood – as Hackworth noted (2006) – as an institutionally regulated disciplinary of urban localities to fit with global and national incentives and goals. Local consent – and the way it conforms in its upstream preparation to such coercion – helps us to understand governance from afar as a way for the State to "enroll participants to the neoliberal goals of central and local governments agency" (Davies 2014a: 3217). The local implementation of PNRU is an opportunity for ANRU to build arenas for political control in which the rules are still dictated by the state in order to spread and cultivate a neoliberal 'governmentality' of the banlieue neighbourhoods through a national agency (ANRU). The end result: a kind of silent violence that can be perceived in the rehousing procedures of tenants forced to move and to leave their neighbourhood and solidarity networks, and in the participatory techniques that eviscerate local democracy and debate (Deboulet and Lelévrier 2014).

Public housing authorities as urban entrepreneurs: re-scaling and merging

In this urban regulatory scheme, PNRU induces new strategies for public housing authorities. It doesn't lead to their privatization, like what has happened in some European countries (Van Kempen and Priemus 2002). It follows another neoliberal path, accelerating the links with private developers and also redefining their role on the urban scene as competitors in the banlieue. In this context, most local public housing authorities have merged into larger groups, the perimeter of which has moved from the municipal to the regional and sometimes national markets. Large public housing groups have been established – like 'I3F', managing 160,000 public housing units nationwide – and have engaged a competition with each other in order to attract a clientele to specific housing niches (first-time-buyer households; affordable homeownership for low-income households). In this market re-scaling, permeability has been set between public housing authorities and private developers. Public housing authorities run private sections dedicated to real estate activities in the banlieue and private developers join public housing authorities in real estate developments. In the scheme initiated by the 'governance from afar', public housing groups redefine themselves as general housing promoters involved in real estate activities (both in rental and home ownership programmes; Driant and Li 2012); showing a strong ability to answer the 'remake the city' injunction in the banlieue, and to shift it from unproductive spaces to productive/attractive spaces for capital.

Much more than a housing reform, PNRU can be understood as a political and moral reform of the banlieue disadvantaged neighbourhoods, in which local authorities and inhabitants are expected to act as entrepreneurs of their own destiny.

Coercion and domination is not policing the banlieue but depoliticization is!

Undoubtedly, urban renewal is an 'anti-ghetto' policy[13] rooted in class inequalities and modulated by ethnicity (for a population with a post-colonial background). However, objectives assigned to PNRU did not fully match with policing goals. Unlike previous urban renewal policies implemented in the industrial capitalism of late nineteenth century and in the 1950–1970s period – analysed by Lefebvre (1968) as counter-revolutionary – today neoliberalism doesn't look for systematic policing. Above all, it seeks to open a former public real estate market to free concurrency and thus spatially determines capitalistic conditions of its (free) urban production. In this context, private and public developers intend to secure space and investments. Whereas at the beginning of urban renewal projects, urban planners and local authorities referred to technics of crime prevention through environmental design and hired security consultants, aiming to fight delinquency through urban planning, many scholars noticed that there was no tacit repressive dimension (Epstein 2013). Crime fighting and space policing were not a clear priority. National and local authorities rely more on housing regulation and social mix as an operational leverage in order to implement a policy of settlement. Moreover, PNRU is engaged in 497 *zones urbaines sensibles* (urban sensible zones[14]), in which social mix is a strategic tool for spatial dispersion of poor households with an immigration background and their replacement by white middle-class families. Social mix and rehousing of former inhabitants in scattered public housing complexes produces a social violence in which victims are noiseless and local contest is considered as illegitimate in front of consensual and politically correct norms of social mix. The objective of PNRU is less to police public housing complexes than to "break the ghettos" with residential relocations of poor households and "visible minorities" (Mucchielli *et al.* 2010). Residential dispersal is part of a hidden agenda, set after 2005's riots, considering that policing and repressive approaches are counterproductive, nurturing further riots. This point was clear when, after 2005's riots, the French government, under public opinion pressure, added a social dimension to PNRU: CUCS (Contrat Urbain de Cohésion Sociale[15]), reinforcing social mix objectives and introducing social cohesion as an alibi for private housing developments.

PNRU benefits from a large and unanimous media coverage celebrating its virtues and its social progress and does not face any political or partisan opposition (even from the left parties) on the national and local political arenas. Many scholars pointed out the depoliticizing effect of PNRU (Carrel and Rosenberg 2011; Epstein 2015). The weakness of the political debate can be first explained by the injunction of social mix acting as a very effective break for further debates on the legitimacy of the urban renewal programmes. The social mix alibi and the emergency to deal definitively with disadvantaged neighbourhoods in the banlieue also produced a political neutralizing consensus among political leaders and city officials. The use of competitive procedures for cities to access ANRU subsidies dissuades city officials to contest ANRU goals and instruments. It leads them to conform to the politics initiated by ANRU and to adopt a 'no risk at all'. After 2005's riot, ANRU announced that it financed 450 local urban renewal programmes among the 750 eligible neighbourhoods. The shortage in subsidies decided local officials to generalize participatory procedures in order to neutralize public and citizens contests. Indeed, participatory procedures are strongly regulated by an ANRU administrative circular issuing norms, instruments and objectives to public participation in urban renewal programmes. This circular addressed by ANRU to *préfets*[16] and city officials carefully specifies goals and forms for participatory procedures. They are mainly limited to participatory forums in which citizen power is boiled down to a single observation. Municipal councils have adopted urban renewal projects, most of the time without any public discussion. Obviously, the depoliticization of urban renewal programmes reinforces the power of control and domination wielded by promoters of PNRU in two ways. On the one hand, the depoliticization legitimizes the shift from a contractual governance – inherited from the welfare state – to a neoliberal 'governance from afar' presented as the most efficient way to locally achieve the goals nationally set. On the other hand, it tends to eliminate and outmode any citizen opposition considered as sectarian, communitarian and old-fashioned. This argument is particularly relevant for the French public opinion in a country that defined itself as an egalitarian republic in which the national community has no colour or religion, and tends to conceal the socio-ethnic dimension of the banlieue. More than a denial to the right to the city for the inhabitants of this disadvantaged neighbourhood, it illustrates the denial of the ethnic and cultural identity of those urban spaces. "Urban renewal did not deprive public housing complexes inhabitants of their right to the city, for which they were already banned from. However, it has denied them a right on the city, urban renewal programs were defined without their consultation" (Epstein 2015: 104).

Neutralizing rhetoric and euphemisms as instruments of power

Domination and power are also visible in the planning rhetoric used. The leading edges are the use of neutralizing expressions, meaning colonized words like rehousing, social mix, public consultation (Lelévrier 2013). Many words euphemize social reality, minoring social violence of housing destruction and removal of banlieue inhabitants. As P. Marcuse recently noticed: "Such euphemisms are well-known results of public relations efforts intended by users to avoid criticism and preempt discussion" (Marcuse 2015a: 152). One widely used expression in the urban renewal rhetoric is symptomatic of this euphemization process: social mix. In national and city official discourses, social mix references multiple things. It refers to residential mixing through development of homeownership programmes offering a "wide range of residential opportunities" for inhabitants. It references here an uncontestable instrument to achieve social mix through concrete actions in the local. It also assumes that demolition and rebuilding of banlieue poor neighbourhoods are good for society. Implicitly restricting inhabitants to their homeowner status, it legitimates discourses and practices of poor families removals and settlement of newcomers. And second, these ecumenical objectives allow the idea that social mix can only be reached by urban growth strategies and city remaking projects. This one-dimensional usage eludes social mix broader cultural and socio-spatial dimensions, neutralizing racial dimension of the banlieue in a generic perspective of social mix. The purpose clearly excludes inhabitants with an immigration background and namely those with a post-colonial background: North Africans and Africans families considered as the source of the problem (poor, Muslims, unemployed, single-headed households, home renters). This one-dimensional rhetoric closes off examination of critical questions in today urban policies such as: racial dimension of urban settlement, the urban citizenship issue and finally what is really going on in the banlieue. An anonymization of actors occurs: 'large families' is a term commonly used to describe families with more than three children with an immigration background. Formula such as 'a deep restructuring of neighbourhoods' depicts strategies of racial restructuring focusing the on dispersal of large families (mainly Muslims) outside the banlieue, in which social mix rhetoric averts the threats of immigration and communitarization in the banlieue (Kirszbaum 2013). "The concern with language used in urban policy and power discussion is not so much with the repression of alternative content, but with the suppression of the very germination" (Marcuse 2015b: 1265).

To conclude: 'local' (little) resistances and neoliberal reflexiveness

These housing policies, characterized by a powerful administrative control and coercion and embedded in political consensus, damage political resistance. Nevertheless, in a few cities (Montpellier, Nanterre, Perpignan), (little) resistances, depicted as 'low noise', have arisen (Deboulet and Lelévrier 2014). In Montpellier, activists successfully opposed the demolition of a large housing complex and articulated the political claims that inhabitants were being denied their citizenship.[17] Yet this resistance was dimmed by procedures of participation and lack of interest by local and national media. But finally, in Perpignan, after the riots of 2005, a local commission for urban memory and heritage effectively channelled the anger of poor inhabitants into action. A workshop for urban memory and heritage joined Gipsy and North African residents with local urban planners to collect people's memory and valorize ethnic heritage and legacy. This workshop was rapidly denounced as a ruse used by ANRU promoters to divert people's attention from demolition issues. It has also been seen by urban activists as a way to pervert the social and ethnic heritage into an urban folklore, now part of a city storytelling reinventing social/cultural mix for attractiveness purposes.

But far away from ANRU expectations, most of what PNRU implemented did not fully reach its goals. PNRU failed to change the face of the banlieue (Epstein 2015). About half of the planned demolitions and rebuilding has been carried out and the social mix programmes – based on the settlement of newcomers and the diversification of the housing stock – did not really occur.[18] Except in a few towns, most of the housing relocations were done in the same neighbourhoods. This semi-failure can be explained by many factors: the reluctance of many cities to house these households, the shortage of the affordable housing stock in metropolitan areas and the households' preference for staying in their neighbourhood. Private investors were also reluctant to invest in risky markets where conditions for a complete removal of this turbulent population were not ensured.

However, as Wilson and Miraftab (2015: 12) recently pointed out, neoliberal governances are reactive and show "a place-specific reflexiveness in response to situated, unpredictable conditions and circumstances". Facing private sector and middle-class households' reluctance to invest in the banlieue real estate market, promoters of PNRU shifted their objectives and adapted to this new context. A hybridization of housing and urban politics occurred. One current strategy implemented as part of PNRU II (second phase, 2013–2018) involves re-zoning social housing. On the one hand, ANRU seeks to reinforce public-private partnerships in some very dynamic peripheral housing markets where homeownership programmes work, allowing the settlement of the white middle class. On the other hand, PNRU now focuses on downtown's most disadvantaged neighbourhoods according to two leitmotivs: sustainability and the fight against slums. Both legitimate public abandonment of the most disadvantaged neighbourhoods in the banlieues and open the way to another wave of downtown gentrification using public incentives. ANRU programmes are redeployed from the banlieues to downtowns to 'officially' fight fuel poverty (the sustainable argument) and to revitalize deprived central urban areas through social mix (the social argument). For many urban activists in the banlieues, the redeployment of ANRU objectives and subsidies in downtown signifies the end of public policies in the banlieue and the normalization of urban segregation in contemporary France.

Notes

1. The term banlieue itself is deceptive. On the one hand, it merely designates an urbanized area on the outskirts of a large city. On the other hand, the term refers to urban deprivation, illiteracy, segregation, poverty, drugs and crimes, large immigrants families concentrated in public housing complexes.

(Body-Gendrot 2010: 657)

2. Programme national de rénovation urbaine.
3. In: Loi n° 2003–710 du 1er août 2003 d'orientation et de programmation pour la ville et la rénovation urbaine (Act N°2003–710, *City Orientation and Programming Act for Urban Renovation*), introduction, p. 3.
4. Hosting 4.5 million inhabitants (7 per cent of the French population).
5. Namely: city politics. It refers to a national policy in favour of the neighbourhoods in the banlieue (disadvantaged neighbourhoods consisting mainly of public housing complexes) implemented in 1981 and covering a wide range of housing and social programmes.
6. See: www.ville.gouv.fr/?observatoire-national-des-zus.
7. This observatory is part of the Ministry of Housing.
8. For further details: www.anru.fr/index.php/fre/ANRU/Objectifs-et-fondamentaux-du-PNRU.
9. In October 2005 violent riots gripped the banlieue spaces; see: http://riotsfrance.ssrc.org/.
10. Agence Nationale pour la Rénovation Urbaine – National Agency for Urban Renewal.
11. Namely: encouraging the settlement of middle classes.
12. During that period, the French Government was led by socialist administrations; in 1995, a new coalition was elected (liberal and conservative parties).
13. This 'anti-ghetto' policy is specifically mentioned as a main objective in the 2003 Act.
14. This zoning includes some criteria (such as spatial concentration of poverty, unemployment, delinquency) and allows firms to be eligible for tax exemptions.
15. Urban contract for social cohesion.
16. In France *préfet* is the local representative of the French state and government. The préfet's numerous responsibilities include: power of police, *politique de la ville*, control of urban and metropolitan plan, etc.
17. http://petit-bard.over-blog.com/.
18. See: ONZUS Report: www.anru.fr/index.php/fre/Actualites/Evenements/Rapport-de-l-ONZUS-dix-ans-de-PNRU.

References

Amin, A. (1994). *Postfordism: A Reader*. Oxford: Blackwell.
Body-Gendrot, S. (2010). Police Marginality, Racial Logics and Discrimination in the Banlieue of France. *Ethnic and Racial Studies*, 33(4), pp. 656–674.
Brenner, N. and Theodore, N. (2002). Cities and the Geographies of 'Actually Existing Neoliberalism'. *Antipode*, 34(3), pp. 349–379.
Carrel, M. and Rosenberg, S. (2011). Injonction de mixité sociale et écueils de l'action collective des délogés. Comparaison entre les années 1970 et 2000. *Géographie, économie, société*, 2(13), pp. 119–134.
Chaline, C. (2010). *Les politiques de la ville*. Paris: PUF.
Davies, J. (2014a). Rethinking Urban Power and the Local State: Hegemony, Domination and Resistance in Neoliberal Cities. *Urban Studies*, 51(15), pp. 3215–3232.
Davies, J. (2014b). Coercitive Cities: Reflexions on the Dark Side of Urban Power in the 21st Century. *Journal of Urban Affairs*, 36(2), pp. 590–599.
Deboulet, A. and Lelévrier, C. (2014). *Rénovations urbaines en Europe*. Rennes: PUR.
Desage, F., Morel-Journel, C. and Sala-Pala, V., eds. (2015). *Le peuplement comme politiques*. Rennes: PUR.
Desjardins, X. (2008). Le logement social au temps du néo-libéralisme. *Métropoles*, 4, pp. 26–45.
Driant, J.C. and Li, M. (2012). The Ongoing Transformation of Social Housing Financing in France: Towards a Self-Financing System? *International Journal of Housing Policy*, 12(1), pp. 91–103.
Epstein, R. (2005). Acte II, scène première: la fin de la politique de la ville? Au crible de la loi Borloo. *Informations sociales*, 1(121), pp. 88–97.
Epstein, R. (2006). Gouverner à distance. Quand l'état se retire des territoires. *Esprit*, 1(1), pp. 96–111.
Epstein, R. (2013). *La rénovation urbaine; démolition-reconstruction de l'état*. Paris: Presses de Sciences-Po.
Epstein, R. (2015). La gouvernance territoriale: une affaire d'État. La dimension verticale de la construction de l'action collective dans les territoires. *L'Année sociologique*, 2(65), pp. 457–482.

Foucault, M. (2004). *Naissance de la bio-politique. Cours au Collège de France, 1789–1979*. Paris: Gallimard.
Hackworth, J. (2006). *The Neoliberal City. Governance, Ideology, and Development in American Urbanism*. Ithaca, NY: Cornell University Press.
Hackworth, J. (2009). The Neoliberal City after the Death of Neoliberalism, APSA 2009 Toronto Meeting Paper. Available at: SSRN: https://ssrn.com/abstract=1450983.
Harvey, D. (1989). From Managerialism to Entrepreneurialism: The Transformation in Urban Governance in Late Capitalism. *Geografiska Annaler: Series B, Human Geography*, 71(1), pp. 3–17.
Harvey, D. (2005). *A Brief History of Neoliberalism*. New York: Oxford University Press.
Kirszbaum, T. (2013). La rénovation urbaine comme politique de peuplement. *Métropoles*, 13. Available at: http://metropoles.revues.org/4769 [accessed 8 December 2017].
Lefebvre, H. (1968). *Le droit à la ville*. Paris: Anthropos.
Lelévrier, C. (2006). La mixité dans la renovation urbaine: dispersion ou reconcentration? *Espaces et Sociétés*, 140–141, pp. 59–74.
Lelévrier, C. (2013). Social Mix Neighbourhood Policies and Social Interaction: The Experience of New Renewal Developments in France. *Cities*, 35(1), pp. 409–416.
Marcuse, P. (2015a). Depoliticizing Urban Discourse: How 'We' Write. *Cities*, 44(2), pp. 151–156.
Marcuse, P. (2015b). Gentrification, Social Justice and Personal Ethics. *International Journal of Urban and Regional Research*, 38(6), pp. 1263–1269.
McLeod, G. (2002). From Urban Entrepreneurialism to a 'Revanchist City'? On the Spatial Injustices of Glasgow's Renaissance. *Antipode*, 34(3), pp. 602–624.
Mucchielli, L., Olive, J.L. and Giband, D. (2010). *État d'émeutes, états d'exception, retour sur la question centrale des périphéries*. Perpignon: Presses Universitaires de Perpignan, pp. 257–272.
Peck, J., Theodore, N. and Brenner, N. (2009). Neoliberal Urbanism: Models, Moments, Mutations. *SAIS Review*, XXIX(1), pp. 49–66.
Pinson, G. (2010). La gouvernance des villes françaises. Du schéma centre-périphérie aux régimes urbains. *Pôle Sud*, 1(32), pp. 73–92.
Tellier, T. (2007). *Le temps des HLM, 1945–1975*. Paris: Autrement.
Valença, M. (2015). Social Rental Housing in HK and in the UK: Neoliberal Policy Divergence or the Market in the Making? *Habitat International*, 49(1), pp. 107–114.
Van Kempen, R. and Priemus, H. (2002). Revolution in Social Housing in the Netherlands: Possible Effects of New Housing Policies. *Urban Studies*, 39(2), pp. 237–255.
Wacquant, L. (2006). *Parias urbains. Ghettos – banlieue – etat*. Paris: La découverte.
Wacquant, L. (2008). *Urban Outcasts: A Comparative Sociology of Advanced Marginality*. Cambridge: Polity.
Wilson, D. (2004). Toward a Contingent Urban Neoliberalism. *Urban Geography*, 25(8), pp. 771–783.
Wilson, D. and Miraftab, F. (2015). Introduction. *Cities and Inequalities in a Global and Neoliberal World*. London: Routledge, pp. 1–13.

PART V

Spaces of labour

Kevin Ward and Andrew E.G. Jonas

According to UN Habitat (2016), 75 per cent of the world's cities have higher levels of income inequalities than two decades ago. On the one hand, many forms of paid employment in cities in the Global North have become more informal and less regulated. They are, to current terminology, 'precarious'. On the other hand, in cities in the Global South where the growth rates are highest, labour markets remain largely informal, unregulated and thus workers often labour under poor, unhealthy and unsafe terms and conditions. Urbanization and economic growth in many of these cities has not been accompanied by drop in income inequalities. Indeed wage inequalities continue to characterize cities around the world.

Of course, not all the work done in cities is paid. Much of it is unpaid, and much of this work is done by women. This is despite women doing an increasing amount of paid work in cities in both the Global North and South. Forty years ago, Castells (1977: 177–178) wrote about the critical importance of women's unpaid labour to the functioning of capitalist cities:

> In the end if the system still 'works' it is because women guarantee unpaid transportation ... because they repair their homes, because they make meals when there are no canteens, because they spend more time shopping around, because they look after others' children when there are no nurseries, and because they offer 'free entertainment' to the producers when there is a social vacuum and the absence of cultural creativity. If these women who 'do nothing' ever stopped to do 'only that', the whole urban structure as we know it would become completely incapable of maintaining its function.

In this fifth section of the book the chapters explores the types of work done in cities around the world, the conditions under which it takes places, the spaces in which it occurs and the politics around its representation. It starts with a chapter by Luis Aguiar and Yanick Noiseux who highlight developments in the ways in which workers are organized in the context of the longer post-industrialization of cities of the Global North and the more recent introduction of national and urban programmes of austerity. They emphasize some recent developments in worker organization that involves labour unions working with those outside of the workplace to challenge particular low-read models of economic development. The chapter builds upon the work of Meyer and Kimeldorf (2015) and their focus on the scales of the city and street-level organizing to argue that these matter most to the daily lives of workers and residents. It is here that the impact of change is understood, felt most deeply, and for longer periods of time.

Michelle Buckley and Emily Reid-Musson, in the second chapter in the section, demonstrate how the urbanization of the Greater Toronto Area (GTA) has occurred through the production of a labouring workforce differentiated on the basis of ethno-nationality and citizenship. For them this case illuminates the need to reanimate and refocus questions about the role that surplus populations play in the urbanization processes. In particular, it reveals the role of different branches of the state in mediating access to the different forms of employment generated through urbanization and how this access is highly differentiated along a range of ethnic and racial lines.

Continuing with the North American focus, Marc Doussard, in his chapter, discusses the challenges facing labour organizing activists in St Louis and Indianapolis. With an emphasis on community-labour models, similar in design to those discussed by Luis Aguiar and Yanick Noiseux, Doussard turns his attention away from those cities in which these models have been pioneered. Rather, his focus is on what limits their introduction elsewhere.

He argues that their limits to introduction in many cities stems from distinctive labour histories, political power and social services support. As a result, Doussard argues that expanding the geography of labour activism will require not just the exporting and reproduction of past organizing campaigns, but also the deliberate adaptation of organizing strategies to address fundamentally different pre-conditions.

In the fourth chapter in this section David Jordhus-Lier and Debbie Prabawati turn the empirical focus to cities in the Global South. They explore domestic workers and their employers in Kupang and Jakarta in Indonesia, arguing that cities are the basing points of multiple mobilities involving not just domestic workers but also those who might employ them. Each city may play multiple roles simultaneously in the demand, supply and mediation of mobile labour. In the cases of Kupang and Jakarta, this variegates how different domestic workers' rights are realized – as workers and citizens.

The fifth chapter in the section is by Ilda Lindell and the empirical focus remains on cities in the Global South. Using the example of Maputo in Mozambique, her chapter focuses on the urban politics of street work. Conflicts and negotiations around street work are arguably a central feature in the politics of most cities of the Global South, where the bulk of contemporary urbanization is occurring. However, Ilda argues this form of work is also of increasing importance in many cities in the Global North. In particular, the chapter attempts to uncover the nature of this politics, highlighting its *multifaceted, multilayered* and *multi-scalar* character in a particular urban setting. More generally, the example of Maputo suggests that the actual political practices and relations at work are not easily captured by available approaches to urban politics. She makes the case for an approach that acknowledges the wide range of relations, forms of work and political dynamics that generate and structure street work in cities of the Global South.

The final chapter in this section turns to activism around day labour in the USA. This is still a relatively poorly understood segment of the US labour market, despite both its growth in recent years and its wider disciplining effects. Nik Theodore argues that in promoting the regularization of informal hiring sites through transparency and collective decision-making, worker centres are challenging not only the economic imperatives that have been used to justify the systematic erosion of wages and workplace standards across the low-wage labour market (as well as the downloading of regulatory risks onto workers in the informal economy that has occurred through the state's highly selective enforcement of worker protections), but also conventional forms of labour politics that have seen the whittling away of employment security provisions in the wake of deregulatory initiatives administered in the name of labour-market flexibility. Day labour worker centres confront these political-economic dynamics, demonstrating that – even at a time when the politics of immigration reform has become polarized and vengeful, and when low-wage workers increasingly must rely on precarious jobs – marginalized workers can organize in defiance of low-road practices and routine violations of labour standards. Nik picks up on a question that is asked in a number of chapters in this section, most noticeably in the chapters by Luis Aguiar and Yanick Noiseux and by Marc Doussard. That is, what are the best ways for those in paid work and rendered most marginal by the neoliberal system to secure material gains in their terms and conditions? This relates to a wider question that is an issue for each of the chapters in this section. That is, whether inside or outside the workplace, how can the organization of paid workers be the platform for a more emancipatory and inclusive urban politics?

References

Castells, M. (1977). *The Urban Question*. London: Edward Arnold.

Meyer, R. and Kimeldorf, H. (2015). Eventful Subjectivity: The Experiential Sources of Solidarity. *Journal of Historical Sociology*, 28(4), pp. 429–457.

UN Habitat. (2016). *World Cities Report, Urbanization and Development: Emerging Futures*. Available at: http://wcr.unhabitat.org/ [accessed 8 December 2017].

20
ROLL-AGAINST NEOLIBERALISM AND LABOUR ORGANIZING IN THE POST-2008 CRISIS

Luis L.M. Aguiar and Yanick Noiseux

Introduction

But the fissures within the system are also all too evident.

(Harvey 2012: 14)

Rising social and economic inequalities across cities due to the high cost of housing is a serious problem. So is the high cost of education, food and house bills, transportation and debt (Auyero 2015; Pelosi 2015). This situation is exacerbated by frozen or poverty-level wages, labour market insecurities and precariousness (Standing 2011), and the expulsion of (black) workers from the formal economy (Wacquant 2007). Government austerity measures, the restructuring of social programmes disqualifying a growing number of people from accessing benefits or making eligibility to benefits more restrictive make life especially difficult for the working class and other vulnerable groups. In contrast, elites and the upper middle class increase their wealth and burrow themselves in exclusive enclaves[1] for cocooned relationships and interactions (Harvey 2012: 15) protected by high tech surveillance systems and private security firms (Caldeira 2001; Low 2003). All of this might lead one to say that these are fertile conditions for courageous initiatives on mobilizing the "collateral casualties" of the neoliberalizing economy (Bauman 2011). But a fertile socio-economic context and a sympathetic social climate, though helpful, even necessary, are rarely sufficient to mobilize people into action for meaningful social change. Creativity, vision, connections and a bit of abandonment may be the key elements in kick-starting the pursuit of unprecedented gains (Castells 2012). But "[c]apitalism will never fall on its own. It will have to be pushed. … The capitalist class will never willingly surrender its power. It will have to be disposed" (Harvey 2010: 260). This chapter focuses on describing a few instances where the legitimacy of neoliberalism is questioned and new organizing is an affront to the neoliberal order and its 'values' of our time. The case studies below capture a roll-against neoliberalism in a socio-politico climate where "roll out neoliberalism as political rhetoric has run it [*sic*] course in many jurisdictions" even if in practice it may yet have some sustenance (Keil 2009: 234).

According to Keil (2009: 240) resisting neoliberalism matters in terms of choosing between a roll-with-it neoliberalism of an authoritarian social formation and a roll-with-it neoliberalism of benevolence and reformism. This is because we have experienced, endured, imbued and been personally, subjectively and socially transformed by the routinization of neoliberalism over the last thirty to forty years (Dardot and Laval 2009). Roll-with-it neoliberalism is the latest phase of neoliberalism, and a phase which can only be overtaken by its own internal dynamics rather than by some pre-or-post neoliberal development. Mitchell and Sparke (2016), on the other hand, see the dialectic of neoliberalism at work through an emerging humanitarianism as the post-2008 crisis mode of regulation. A neoliberalism that is humane, caring, reinventing itself in light of the general outrage against its greed, self-interests and corruption. But Mitchell and Sparke's argument does not make clear what "economic base" this neoliberal humanitarianism stands on. If the regime of accumulation under crisis neoliberalism was a flexible labour process with just-in-time workers, weak unions and deregulatory mechanisms dismissing opposition and facilitating profitability through the liberalization of markets and the obsolessencialization[2] of social protection systems, what then does humanitarianism as a mode of regulation solidify? Neoliberalism is not only resilient but apt at reinventing itself (Peck 2010). Still, we propose the concept of roll-against neoliberalism to capture the ongoing mobilizations, organizing and resistance to neoliberalism's reinvention within the globalization and flexible production regime of accumulation. While the crisis has impacted neoliberalism as ideology and practice, we hesitate to endorse the view that something new and different has emerged and that it is already identifiable as a fixed socio-politico system. Whereas neoliberalism was able to roll through opposition in the past, this time around there is a general mood[3] that neoliberalism has failed and is no longer a regime to be trusted and endorsed (Birch and Mykhnenko 2010; Watkins 2010), even if writers (i.e. Calhoun 2013) are not quite ready to pronounce themselves on what is actually emerging. In the meantime, vigilance is needed to roll against its attempted reinvention and reassertion to dominate our lives. And if indeed the new grows within the old, today's citizens may be ambiguous and uncertain on what social formation they want, but they seem clear that neoliberalism is part of the problem not the solution[4] (Castells 2012; Hill 2016; Iglesias 2014; Taylor 2016; Wright 2010).

In the past few years there has been an impressive display of criticisms, protests, organizing and mobilizations against neoliberalism (Castells 2012; Taylor 2016). The cases described below are further examples of this development. It is true that the labour movement (on both sides of the border) remains in crisis (see Aronowitz 2014; Crevier et al. 2015; McAlevey 2015; Thomas and Tufts 2016), and we would be foolish to argue otherwise. But labour is in a better position today to assert its influence in the workplace and society than it has been at any other time during the roll-back and roll-out phases of neoliberalism. And, that the city remains key to roll-against neoliberalism campaigns in North American and elsewhere. This is the case in spite of the fact that union density in the USA continues to drop and has only recently stabilized in Canada (McAlevey 2015; Thomas and Tufts 2016; Vela 2013).[5] And while it is not the point of this chapter to explore this influence totally, we will describe a few examples where labour's influence is rising and comprise what we call the roll-against neoliberalism moment of today. We do this by observing Meyer and Kimeldorf's (2015: 432) "eventful subjectivity" argument which states that small-scale events are much more immediate to our everyday lives and they matter for subjectivity and social change purposes. In other words, engagement matters in the now and in the future. While other researchers fix on the big picture scale of organized labour, we prefer to follow Meyer and Kimeldorf's advice and focus on grassroots scale of organizing events and its long-lasting impact on activists and participants.

This chapter begins with a historical background to organized labour and the city. It ends with discussions of empirical cases of organizing where organized labour and spaces of the city intertwine.

Fordism, women and unions, and the city

At the tail end of Fordism signs began to emerge that the labour movement and the male industrial working class in Canada and the United States were in trouble (Cowie 2010; Davis 1986; Palmer 1992). The decline and re-location of manufacturing to the Global South and right to work states in the United States, capital flight and runaways shops leading to deindustrialization, a lower tax base, and the loss of union membership in manufacturing for hundreds of thousands of workers suddenly out of work and in a mid-life crisis about work, their career and the wage earning self, were some of the most obvious signs of this trouble. Also at that time the dismantling of the welfare state began and so did the turn-over of many social programmes to private companies and interests in providing 'protection' and 'security' to (white) citizens in the post-industrial city (CBC 1997; McBride and Shields 1997). In deindustrializing cities, slum lords, for instance, became prevalent aided by policies deregulating rental controls thereby putting many poor and marginalized peoples on the street (Wright 1997). A campaign of fear also emerged criminalizing the marginalized and vulnerable claiming unsafe city streets due to their presence and visibility. This campaign proved to be part of a scheme to sanitize city streets for yuppies moving back to the city into the gentrified spaces they help create with their dollars, tastes and lifestyles (Goffman 2014; Kern 2010; Mitchell 2003). Not to seem uncaring in a cosmopolitan milieu, yuppies were willing to tolerate a sprinkling of marginalized people to give their (yuppies) space colour, favour and authenticity (see Florida's (2002) ideas of new city spaces) (Barraclough 2011). But scapegoat populations – immigrants and racialized bodies as black – were increasingly harassed suffering more aggressive campaigns of intimidation, racism and even armed fascist attacks in cities across the USA and Western Europe (Browne 2015; Elliott-Cooper *et al.* 2014). And, labour became increasingly atomized, competitive and many were suspicious of fellow precarized workers (Standing 2011). Unions begin to haemorrhage members, but remained hesitant in embracing new immigrants and visible minority women for organizing purposes and improving people's standard of living (Hunt and Rayside 2000). Undocumented workers in the USA, for instance, were largely ignored by unions afraid of the political fall-out of organizing 'illegal' immigrants in America (Delgado 1994). This was at a time when the sense of what a city is was changing and the right to the city under declining Fordism and rising neoliberalism

> became redefined, in many instances, as the right of the consumer to privatized urban space and differential commodities on the marketplace rather than the right of the urban inhabitants to the possibilities the urbanized societies have to offer and to the historical achievements previous social struggles have yielded.
>
> *(Keil 2009: 237, paraphrasing Mitchell 2003)*

Changes were also happening in the realm of employment and especially in the public sector (the civil service for example). The shape of the city changed due to new and improved modes of collective consumption, and the large influx of (white) women into the nascent neoliberal labour market. In Canada women no longer content to remain servants to their husbands and the domestic order, left the monotony of the domestic sphere to take up jobs in the formal labour market (Armstrong and Armstrong 1978). As a result 'pink' labour ghettos quickly formed as women's skillsets were undervalued because of the naturalization of so-called feminine competences (England 1993). Women working in the public sector (which in Canada were allowed to organize in 1967), joined teachers' unions, civil service unions, public service unions, health care unions, and a few others as organizing targeted the large workplaces of the welfare state and the thousands of workers occupying jobs within various departments of the capitalist state (Briskin and McDermott 1993; Foley and Baker 2009; McElligott 2001). Unionizing occurred at a much slower pace in the private sector of the economy where many women laboured especially in the semi-skilled and deskilled jobs in the service delivery parts of workplaces (Shalla and Clement 2007). Here, organizing was especially difficult due to the small size of the workplaces, labour force turnover and the omnipresence of the employer (Leidner 1993). As sole wage earner or as part of a family pulling income together or in partnership with a lover, women in the economy led to changes in the physical structures of the city. For instance, apartment buildings and condos were erected to accommodate new residents with new lifestyles and social relations. So services emerged and industries arose to cater to women's needs as urban and industrial citizens (Kern 2010). But union organizing of the public sector soon stalled and union growth stagnated and then declined with the neoliberal dismantling of the welfare state and the restructuring of welfare services under the 'flexible firm' model (Harvey 1989) where, in Quebec for example, underfinanced social economy enterprises now provided services previously offered by the public healthcare sector. A massive number of workers – often women and visible minorities – in support staff positions working alongside professional workers (even within the public sector) – cleaners, caterers, domestic workers, teaching aids, fast food workers, collateral health care workers, etc. – remained largely unsolicited by organized labour as the movement declined rapidly. Faced with this development and the urgency of the maelstrom of the declining Fordism and its unionized worker, trade unions panicked and sought almost exclusively to secure and retain those workers already organized, as well as prioritizing male union members to boot. This strategy was pursued through concessions, mergers and raiding of less powerful unions. Unions were seen as opportunistic and greedy organizations who only protected their own and stood in the way of young workers' access to jobs through seniority and bumping rules. When young workers managed to secure a job, they often did so with a two-tier wage structure which punished new employees but protected – even rewarded – long-time employees who were also union members. This approach did little to push back capital's assault on labour but succeeded in diluting solidarity within its rank. Business unionism has come to define this politics of bread and butter issues to be resolved through legislative fiat rather than social and industrial militancy inside and outside places of employment.

The modern city had its own version of business unionism evident in rigid architectural forms of functionality, hierarchy and "planning by numbers" (Relph 1987: 140). Style was trumped as was aesthetics, creativity, innovation and vision reducing the likelihood of "happy accidents" forming part of city experience (Relph 1987: 143). The street grid system design allowed for more rapid movement and transportation of people and products but conformed to a rigidity and simplicity not unfamiliarity to the straight-jacket model of business unionism. Some argued that the 'soul' of the city was exorcized through these practices (Jacobs 1992). The 'alliance' of these models of organized labour and city design was further cemented by the intensification and diversification of the division of labour as surplus capital was re-invested in the city's infrastructure, including modes of collective consumption (Castells 1983). The growth in the middle class occupying positions in the private and public sectors relating to infrastructure development, education and government bureaucracy, segregated people especially in terms of housing and position in the class structure.

Meanwhile, having gained consensus status (Harvey 2005) neoliberalism reorganized the city and women and their families lived with decreasing wages, rising unemployment and aggressive union busting. The intensification of women's domestic labour hardly made a bent in the family economy as uncertainty and precarity increased with neoliberal labour market restructuring (Bezuidenhout and Fakier 2006). As a result women's double day of labour intensified and their labour market presence grew even more suspect to lay-offs, especially with declining workplace (and societal) union influence. Moreover, gentrification dispossessed and displaced fordist workers from their houses and neighbourhoods as the new petite bourgeoisie moved in putting its own stamp on place while earning high salaries in the cultural industries and FIRE[6] sectors of the economy (Sassen 1991). At the other end of the labour market, working class women and visible minorities endured low pay, few benefits and mistreatment in the industries of private security, domestic 'help', catering, cleaning and healthcare aides (Waldinger and Lichter 2003). New spaces in the city were re-valorized and the new economy of cultural industries (e.g. film, social media and entertainment) and high tech software development established their niche market presence in the gritty neighbourhoods where the working class formerly lived (Butler and Rustin 1996; Hutton 2010; Pelosi 2015; Scott 2000). By the end of the twentieth century, Park and Burgess' (1925) concentric circles were inverted, i.e. those in the outer rings were now occupying the inner rings and those in the inner rings had been expulsed to the outer edges of the expanding city no longer finding jobs, services and affordable housing in the centre of the city.

The 'feeling' of the city also changed as boundaries between neighbourhood and people were re-established not only through physical structures, but via culture as the new petite bourgeoisie quickly established its presence, tastes and markings in the city (Florida 2002; Sassen 1991; Schwanhausser 2016). The inner world of the new petite bourgeoisie emerged and this class ventured out to the few remaining working class spaces only when it sought 'culture' and 'entertainment' while still fearing the 'other' (Zukin 2010). This class faction were never part of the 'old' social movements, or left it due to its unresponsiveness to their interests, demands and identity issues. Others such as LGBTQ communities established their own new social movements defined around identity, choice and lifestyles. Reeling from the economy's changes, the restructuring of labour markets, the rehousing of workers elsewhere and their disconnection from work and neighbourhood, compounded with new immigrant workers with new and different status from those who anchored the old labour movement of yesteryear, the labour movement had few clues and innovative ideas on how to respond and reinvent itself in a restructuring labour market and changing social landscape in the city (but see Milkman (2006) on the SEIU (Service Employees International Union) in Los Angeles).

Global cities like Toronto, Vancouver and London imported foreign labour power to take up growth in low-skilled jobs abandoned by their local labour forces due to its low pay, limited benefits and uncertain future (Wills et al. 2010). In some cases workers have been restructured out of their local economy labour market altogether, as Wacquant (2007) explains in the case of Chicago, and Aguiar et al. (2011)[7] in the case of the Okanagan Valley of British Columbia. Workers are also recruited for high skilled jobs in the IT and finance sectors (Satzewich 2015) often living in secluded enclaves in the spatially divided cities of the Global North (Low 2003). This recent development is an indication of immigration regime change whereby immigration policy aligned with post-fordism, and only occasionally with human rights and humanitarian goals, notwithstanding the recent crisis of Syrian refugees. As a result an unprecedented number of migrants come to cities from the Global South with temporary status, defined contract limits and substandard accommodations (Aguiar et al. 2011; Austin 2013; Noiseux 2012). This movement creates a migrant division of labour with polarizing incomes, an ethnic division of labour and extremes between wages and benefits of the highly skilled workers and their low-skilled counterparts. The global city labour market has transformed from an egg-shaped configuration to one of an hour-glass (Wills et al. 2010). What is labour to do in this context? And where is it to do it?

Organizing campaigns

There is no alternative, coherent movement challenging neoliberalism. But there are examples of organizing against neoliberalism as the latter reshapes its mode of regulation to regain social legitimacy (Peck 2010). It is in this context that we propose the concept of roll-against neoliberalism to capture the myriad of activities and events problematizing the ideology of neoliberalism and its practices as an economic model. Over the last half dozen years, labour, community organizations (Tattersall 2010), organizations for the homeless (Wright 1997) and student anti-sweatshop mobilizations (Ross 2004), to name just a few, have denounced neoliberalism and argued for the need to move beyond it. And while in some circles neoliberalism has been repeatedly criticized, this is the first time since its implementation that arguments against neoliberalism are being heard and gaining social traction in mainstream society.[8] Roll-against neoliberalism refers to the many instances focused on undermining the hegemony of neoliberalism at a time of profound distrust of the model and the denouncement of elites who benefited from its corrupt set-up in the first place (Castells 2012; Iglesias 2014).

For the reasons expounded in the first section of this chapter, we argue that the reinvention of the North American labour movement must take place in the city. According to Turner (2007) labour renewal is evident in cities across the USA since it is in these locations that 'openings' and resources are available to support activism like living wage campaigns, the 'Fight for $15' and Jobs with Justice initiatives. In cities, unions find allies as they build social movement unionism on coalition partnerships.[9] Turner points out that it is the institutional structures of city governance that are key to understanding the urban as the new battleground for unions. For him, unions need to read the political landscape in order to ascertain the "institutional openings"[10] enabling to strike a blow to the existing power structure through organizing and mobilizing.

More radically, in his survey of insurrections and cities, Hobsbawm (1982) writes that resistance movements shape cities and in turn cities shape the form and actions of movements. For him (232) effective riots are "due not so much to the actual activities of the rioters, as to their political context". That is "riots have rarely aimed directly at any of the buildings of large corporations. Nor are they [corporate buildings] very vulnerable" (231). Instead, he offers that it is "'collectively' downtown [that] is vulnerable. The disruption of traffic, the closing of banks, the office staffs who cannot or will not turn up for work, the businessmen [sic] marooned in hotels with overloaded switchboards, or who cannot reach their destinations" (231). It is this strategy of disrupting everyday routines and clogging downtown arteries in protest of poor treatment or in the promotion of identity and dignity by groups, including labour, that have affect (Meyer and Kimeldorf 2015). This is surely borne out by the recent history of anti-globalization protests such as 'Teamsters and Turtles'[11] uniting in Seattle in 1999 (Starr 2005), and Occupy Movements across the USA and Canada (Castells 2012). A more recent example is the 'Maple Spring' student-led uprisings in Quebec where protests disrupted the downtown core for more than a hundred consecutive days in 2012. These protests are an integral part of social movement unionism and civil disobedience tactics of unions practising this type of unionism (Pineault 2012).

Cities have always been important to social movements organizing whether they organize for identity or to contest the status quo with respect to identity stereotypes. Thus, with Nicholls (2008), we argue that the city is where the interdependency of resources are found, located and then gathered. An interdependency of groups with strong or weak social links are knitted together to access resources of various types necessary for organized action. It is the dependence and reliance on the collection of different resources offered by different groups gathered in the city that is crucial to understand the role the city plays in emerging social movements. The post-industrial city is a construct of capital and its investment in rising industries, infrastructures and employment growth. But it is also the solicitation of foreign workers to do the low-skilled and specialized high-skilled labour in the post-industrial city of today that is too changing the metropolis. It is all happening in the city and organized labour can take advantage of this to effect change and reclaim political space lost with the emergence of neoliberalism.

New workers organizing in the city

Richard Florida (2002) lauds sidewalk shopping and café culture in the development of spaces for the creative types he sees inhabiting cities and energizing neighbourhoods, cities and the economy of the twenty-first century. In this development, chic and niche market shopping and entertainment corridors have emerged in some cases as independent business enterprises catering to a specific clientele with unique tastes. In other cases businesses are owned by corporations creating spaces for new residents in the city (Kern 2010). Sharon Zukin (1995) does not discuss outdoor developments of corporate spaces, but instead focuses on the development of front and back of the house workforces in elite or quirky restaurants and cafés made attractive to the tastes of the elite and new petite bourgeoisie in global cities. In these workplaces, Zukin says that the clientele desire front of the house employees possessing a cultural capital facilitating interaction between client and employee. Thus restaurant employers hire pretty faces with the ability to be at ease with middle and upper middle class customers and culture. This practice is an example of active discrimination against potential employees who aren't 'aesthetically pleasing' or possess the cultural capital to converse with elite clientele. It is a racialized and racist practice in another way too: non-white workers are hired to fulfil back of the house (e.g. kitchen) jobs. Workers in such workplaces are increasingly unsatisfied and many seek relief from agents outside their workplaces.

In the past coffee shop work was a temporary job for the transient young middle class students pursuing their career aspirations after university. Today this 'passing through' job experience is an increasingly tenuous path with tight labour markets and employers demanding service workers for their growing businesses (Klein 1999). The growth in service work and industry is the launching pad for the labour movement "making some inroads into typically low-wage, part-time, non-unionized workplaces" (Vela 2013). The "stigma of unionization" is "being lifted by young workers" as they "become frustrated with a job market that leaves them vulnerable or insecure with part-time work" (Vela 2013). Some of these themes were prominent in a campaign to organize coffee shop workers in Halifax, Nova Scotia. In a campaign to organize *Just Us!* coffee shops, workers pointed out the hypocrisy of their ethical employer who promoted 'people before profits' while mounting an anti-union stand against its workers and their wish to join Local 2 of the SEIU (Brickner and Dalton 2015). Reaching out and building up through their LBGTQ activist networks, organizers reminded the owner of the coffee shop that his/her motto of 'people before profit' had to stand for something concrete or otherwise be some catchy, frivolous marketing ploy devised only to draw politically sensitive millennial customers.

Another interesting organizing campaign in neoliberalized city spaces is the development of corporate street presence in the culture of convenient stores. The rise in sidewalk and street corner shopping defines much of city spaces as these provide intermediate locations between the large grocery stores and the household kitchen. 'Couche Tard'[12] stores are in-between spaces of impulse and convenient late-night shopping. They are often within easy access of new, single apartment dwellers since these stores are often on the first floor of apartment buildings or street corners where 'new' residents live. In downtown Montreal, Couche Tard cater to young city residents living in apartment buildings and to less mobile elderly shoppers also living in nearby apartment buildings. A campaign began in one downtown outlet and spread to the greater metropolitan area and to a less extent elsewhere in the province. It sought to organize these corner stores that often push aside mom-and-pop outlets and replace them with 'state of the art' convenience stores tied to large multinational corporations running late-night shopping habits for huge profits. These stores are difficult to organize because staff turnover is common, management practices exhibit features of "simple control" (Edwards 1979), and head office pursues hostile 'Walmart-like' anti-unionism. In addition, unions face the challenge of mounting campaigns to organize workers who will yield low dues as a result of their precarious position in corner store service work. But in spite of all of these challenges, the FC-CSN (Fédération du commerce de la Confederation des Syndicats nationaux) mounted an intense union drive to organize Couche Tard. With community and student and rank-and-file involvement, the union organized six Couche Tard franchises (Joly 2013), and coaxed the multinational to extend large portions of the benefits reached in the agreement to all company non-unionized Couche Tard employees in the province of Quebec.[13]

In some cases space and identity are interpreted as 'barriers' to mounting across the board support from organized labour and community organizations even though these obstacles can be overcome as new initiatives from workers' centres in association with traditional unions are slowly making inroads. One such example of community organizing is Montreal's Immigrant Worker Centre/Centre des travailleurs et travailleuses immigrants which has recently launched two initiatives – the Temporary Foreign Workers Association (TFWA) and the Temporary Agency Workers Association (TAWA). The Centre is bringing together temporary foreign workers and temp agencies workers with activists from the community and union professionals to assist them with immigration and labour problems and issues. The goal is to organize collectively those workers, within and beyond the scope of traditional union drives in various locations across Montreal where visible minority workers work under duress. These types of networking between community organizers and traditional union structures were also decisive in 'importing' the US-based 'Fight for $15' campaign to Canadian cities through the leadership of Toronto's Worker Action Centre and its '$15 and Fairness' campaign.[14]

Social movement unionism combines unions and community actors for the purpose of organizing a workplace in place. In some cases organizing remains conventional as campaigns move through legislative means to define specific workers and their labour rights including right to form a union). In the cases discussed above, a roll-against neoliberalism approach captures the resistance to conditions unfavourable to workers and their mistreatment by greedy employers pursuing profit over the well-being of people and communities. The pushback and new organizing campaigns being mounted across North American cities are initiatives of roll-against neoliberalism.

Organizing has to deliver more

Roll-against neoliberalism operates outside the labour movement too, though labour plays an important part. This is the case in the fight for $15 per hour and living wage campaigns in North America when unionized workers' wages are often barely above that which they already earned before unionization. Wage increases rise incrementally and over the length of a contract which these days can be up to five years. In this situation union wage increases make little difference in the budgets of workers, especially those in low-skilled work. Thus a push for more significant and meaningful wage increases is at the forefront of the fight for $15 across American and Canadian cities. Here issues about $10 burgers in order to offset wage increase have been pushed aside as have companies' claims of unions trying to hamstring their profitability and business operations. There is no attempt to socialize fast food enterprises or immediately organize workers into unions. The goal is more straightforward and immediate: to raise the wages of low-paid workers in the North American fast food industry so that they can live a life not so burdened by low pay. And while fast food workers may not be the group that garnishes the most sympathy by a buying public, the fact that there are so many so far from achieving a decent standard of living (Grusky and MacLean 2016) makes it a compelling campaign to join and support. This argument resonates with a public who by an overwhelming proportion think that a full-time worker should be able to earn a proper living income (Ehrenreich 2001; Tirado 2014). If Harvey (2005) argues that neoliberal consent was organized around freedom, then we can argue that in the new politics of labour in the USA, for example, it is the discourse of the American dream that is gathering steam and sympathy for fast food workers and their campaign.

The Fight for $15 campaign differs according to space and history. That is, in Ireland where the SIPTU (Services, Industrial, Professional and Technical Union) is thinking of developing a similar campaign but without a focus on the American dream and fast food workers. Instead the union is interested in highlighting a campaign on care workers as the Irish population ages and as commitments to work and nuclear family increasingly prevent people from having the time to care for their elderly parents. In Dublin, the SIPTU is trying to develop the idea that care work is about duty and responsibility to ageing or ill family members. The importance of care work for the elderly is a powerful discourse given the emphasis on the family in Irish culture and the influence of the Roman Catholic Church. So in this attempt, Irish characteristics are invoked to coalesce around the organizing campaign.[15] But as with the fight for $15 in the USA, they are not about bringing members into the union fold, though there is an aspiration that eventually these workers will join the SIPTU. Perhaps more importantly in the short term is that such campaigns are proof that unions are no longer only interested in their members only, even if they ultimately would like to grow their ranks with fast food workers members[16] and/or care workers.

Labour participates as a partner in living wage campaigns across North American communities (Tait 2005). In this instance organizing residents into unions is not the goal given that membership in these coalitions is various and not workplace-specific. Indeed, the reason people come together is to improve the standard of the city's working population rather than identifying one category of workers for specific improvement. There are many North American cities that have signed city ordinances stipulating that contracts with business must respect established living wages. These establish living standards beyond the minimum wage. And while this is important political work for unions, they are not always present or welcome in living wage campaigns. In one case in Revelstoke, BC the major unions in the city were not involved in the establishing of a living wage campaign. In this instance, business representatives were forceful in requesting that an academic (a sociologist) construct a model where a living wage would not be detrimental to business. The notion that institutionalizing a living wage will not affect profit margins is unrealistic, and the absence of labour unions from the discussion and negotiations is a false start.

Conclusion

We accept that the labour movements in Canada and the USA are in crisis, and that nationally specific austerity programmes will make even more demands on labour to conform to the 'needs' of the economy. What we have done in this chapter is highlight developments in organizing workers in the post-industrial and austerity-governed cities of today. We have done this through a moving scale analysis of micro events where labour organizing is evident and where the city with its infrastructures and governance approach plays an important role. By focusing on the scales of the city and street-level organizing, we follow Meyer and Kimeldorf (2015) and argue that these matter most to the daily lives of workers and residents. It is here that the impact of change is understood, felt most deeply, and for longer periods of time.

In this chapter we discussed the notion that the Fordist crisis-ridden labour movement hit hard male industrial workers as deindustrialization accelerated and the rise of service industries sought workers with other social and ethnic 'qualities' (Waldinger and Lichter 2003). White women entered the public sector labour markets on a scale not seen before, bringing with them a degree of independence even if it appears that the domestic division of labour remained unchanged for most. However, with the rise of neoliberalism, the downsizing of government and the restructuring of the welfare state, women workers and especially those racialized as black, their jobs and unions came under attack resulting in work intensification, loss of union memberships and benefits, in the process creating an uncertain economic future with ongoing implementation of austerity measures (Goffman 2014). Women in the private sector of the economy fared no better as their vulnerability to increased precariousness and precarity became the norm.

Today the focus for organizers is on identifying workers long excluded from the labour movement but increasingly important in the service industries in the city. Migrant and visible minority workers are organized by aggressively roll-against neoliberalism. This push is coming from within the communities these workers belong to and from the labour movement featuring organizing campaigns of non-traditional union members labouring in the service sector of the economy. Our concept of roll-against neoliberalism captures how marginalized workers in the post-industrial, service economy of the city are organizing in the midst of a neoliberalism seeking reinvention (Peck 2010). Perhaps the social base to overthrow neoliberalism lies in the experiences of migrant and visible minority workers, combining community expertise and engagement with workplace militancy (Therborn 2014).

Acknowledgements

The authors would like to thank Tina Marten and Kevin Ward for their comments and editorial assistance with this chapter. Some of Aguiar's and Noiseux's research contained in this chapter was supported by author-specific research grants from the Social Sciences and Humanities Research Council of Canada.

Notes

1 These spaces are racialized too. On this see the funny clip from comedian Chris Rock – www.youtube.com/watch?v=24PcF7LDQ7M [accessed 10 November 2016].
2 By this term we mean the process of making the social protection system outdated and incapable of protecting people engaged in new working arrangements (i.e. non-standard jobs).
3 Or at least in the political mainstream, see Sarkozy's comments in 2010 – www.cnn.com/2010/BUSINESS/01/27/sarkozy.davos.bank.regulation/ [accessed 1 November 2016].
4 For example the Occupy Movement and the Indignados.
5 In the US private sector unionization is at just 6.3 per cent (McAlevey 2015: 416). The decline in union density has not been as dramatic in Canada. But 31.5 per cent union density (16 per cent in private sector and 71 per cent in public sector) represents a stagnant position (www.cbc.ca/news/canada/unions-on-decline-in-private-sector-1.1150562 [accessed 21 December 2015]).
6 This stands for Finance, Insurance and Real Estate.
7 See also www.nytimes.com/2016/01/26/us/lawsuit-claims-disney-colluded-to-replace-us-workers-with-immigrants.html?smprod=nytcore-iphone&smid=nytcore-iphone-share [accessed 29 January 2016].

8 For instance see www.cnn.com/2010/BUSINESS/01/27/sarkozy.davos.reaction/, [accessed 2 November 2016].
9 In the formation of coalitions, Turner identifies bridge builders as key activists with varied experiences in different social movements as necessary to cohere a coalition for success.
10 By "institutional openings" he means "institutions are also power structures that can provide elements of an opportunity structure – opportunities that open up when institutions lose cohesiveness or legitimacy – that can be used to challenge institutionally embedded power" (Turner 2007: 4).
11 http://democracyuprising.com/2001/04/17/teamsters-and-turtles-unite-again/ [accessed 2 November 2016].
12 'Open late' is the translation of this label though these corner stores in the rest of Canada go by the name of Mac's convenience stores with a picture of an owl as the emblem.
13 On this topic, see L. Rivet-Préfontaine, forthcoming, www.gireps.org.
14 https://briarpatchmagazine.com/articles/view/everything-goes-up-but-pay [accessed 8 November 2016]. For a discussion of the Canadian SEIU Healthcare 'sweet $16' campaign in Ontario, see L. Hamel-Roy, forthcoming, www.gireps.org.
15 Aguiar interview with SIPTU official, August 2015.
16 On the fight for $15 see www.theatlantic.com/business/archive/2015/08/fifteen-dollars-minimum-wage/401540/ for the USA and http://15plus.org in Quebec [accessed 29 January 2016].

References

Aguiar, L.L.M., Tomic, P. and Trumper, R. (2011). *Mexican Migrant Agricultural Workers and Accommodations on Farms in the Okanagan Valley*. Working Paper Series, No. 1–04, Metropolis BC, Vancouver.
Armstrong, P. and Armstrong, H. (1978). *The Double Ghetto: Canadian Women and Their Segregated Work*. Toronto: McClelland and Stewart.
Aronowitz, S. (2014). *The Death and Life of American Labor: Toward a New Workers' Movement*. New York: Verso.
Austin, D. (2013). Guestworkers in the Fabrication and Shipbuilding Industry Along the Gulf of Mexico: An Anomaly or a New Source of Labor. In: D. Griffith, ed. *(Mis)managing Migration: Guestworkers' Experiences with North American Labor Markets*. Santa Fe: School of Advanced Research, University of New Mexico Press.
Auyero, J., ed. (2015). *Invisible in Austin: Life and Labor in an American City*. Austin: University of Texas.
Barraclough, L.R. (2011). *Making the San Fernando Valley: Rural Landscapes, Urban Development, and White Privilege*. Athens: The University of Georgia Press.
Bauman, Z. (2011). *Collateral Damage: Social Inequalities in a Global Age*. Malden, MA: Polity.
Bezuidenhout, A. and Fakier, K. (2006). Maria's Burden: Contract Cleaning and the Crisis of Social Reproduction in Post-apartheid South Africa. In: L.L.M. Aguiar and A. Herod, eds. *The Dirty Work of Neoliberalism*. Malden, MA: Blackwell.
Birch, K. and Mykhnenko, V., eds. (2010). *The Rise and Fall of Neoliberalism: The Collapse of an Economic Order?* New York: Zed Books.
Brickner, R. and Dalton, M. (2015). Organizing Baristas in Halifax Cafes: Class Identity in the Millennial Generation. Paper presented at the second annual meetings of the Canadian Association for Work and Labour Studies, 4–5 June.
Briskin, L. and McDermott, P., eds. (1993). *Women Challenging Unions: Feminism, Democracy, and Militancy*. Toronto: University of Toronto Press.
Browne, S. (2015). *Dark Matters: On the Surveillance of Blackness*. Durham, NC: Duke University Press.
Butler, T. and Rustin, M., eds. (1996). *Rising in the East: The Regeneration of East London*. London: Lawrence & Wishart.
Caldeira, T. (2001). *City of Walls: Crime, Segregation, and Citizenship in Sao Paulo*. Berkeley: University of California Press.
Calhoun, C. (2013). What Threatens Capitalism Now? In: I. Wallerstein, R. Collins, M. Mann, G. Derluguian and C. Calhoun, eds. *Does Capitalism Have a Future?* New York: Oxford University Press.
Canadian Broadcasting Corporation (CBC). (1997). *Insecurity: The New Age of Policing*. Director Robin Benger. CBC, Toronto.
Castells, M. (1983). *City and the Grassroots: A Cross-Cultural Theory of Urban Social Movements*. Berkeley: University of California Press.
Castells, M. (2012). *Networks of Outrage and Hope: Social Movements in the Internet Age*. Malden, MA: Polity.

Cowie, J. (2010). *Stayin' Alive: The 1970s and the Last Days of the Working Class*. New York: The New Press.
Crevier, P., Forcier, H. and Trepanier, S. (2015). *Renouveler le syndicalisme*. Montreal: Écosociété.
Dardot, P. and Laval, C. (2009). *La nouvelle raison du monde. Essai sur la société néolibérale*. Paris: La Decouverte.
Davis, M. (1986). *Prisoners of the American Dream*. Verso: New York.
Delgado, H.L. (1994). *New Immigrants, Old Unions: Organizing Undocumented Workers in Los Angeles*. Philadelphia, PA: Temple University Press.
Edwards, R. (1979). *Contested Terrain: The Transformation of the Workplace in the Twentieth Century*. New York: Basic Books.
Ehrenreich, B. (2001). *Nickel and Dimed: On (Not) Getting by in America*. New York: Picador.
Elliott-Cooper, A., Murrey, A., Kumar, A. and Younis, M. (2014). Labour and Resistance Across Global Spaces. *City*, 18(6), pp. 771–775.
England, K. (1993). On Suburban Pink Collar Ghettos: The Spatial Entrapment of Women? *Annals of the Association of American Geographers*, 83(2), pp. 225–242.
Florida, R. (2002). *The Rise of the Creative Class and How It's Transforming Work, Leisure, Community and Everyday Life*. New York: Basic Books.
Foley, J.R. and Baker, P.L., eds. (2009). *Unions, Equity and the Path to Renewal*. Vancouver: University of British Columbia Press.
Goffman, A. (2014). *On the Run: Fugitive Life in an American City*. Chicago: University of Chicago Press.
Grusky, D.B. and MacLean, A. (2016). The Social Fallout of a High-Inequality Regime. *Annals of the American Academy of Political and Social Science*, 663(1), pp. 33–52.
Harvey, D. (1989). *The Condition of Postmodernity: An Enquiry into the Origins of Cultural Change*. Malden, MA: Blackwell.
Harvey, D. (2005). *A Brief History of Neoliberalism*. New York: Oxford University Press.
Harvey, D. (2010). *The Enigma of Capital and the Crises of Capitalism*. New York: Oxford University Press.
Harvey, D. (2012). *Rebel Cities: From the Right to the City to the Urban Revolution*. New York: Verso.
Hill, M.L. (2016). *Nobody: Casualties of America's War on the Vulnerable, from Ferguson to Flint and Beyond*. New York: Atria Books.
Hobsbawm, E. (1982). Cities and Insurrections. In: E. Hobsbawm, *Revolutionaries: Contemporary Essays*. New York: Pantheon Books, pp. 220–233.
Hunt, G. and Rayside, D. (2000). Labor Union Response to Diversity in Canada and the United States. *Relations industrielles/Industrial Relations*, 39(3), pp. 401–444.
Hutton, T.A. (2010). *The New Economy of the Inner City: Restructuring, Regeneration and Dislocation in the Twenty-First-Century Metropolis*. New York: Routledge.
Iglesias, P. (2014). *Disputar la democracy: politica para tiempos de crisis*. Madrid: Akral.
Jacobs, J. (1992). *The Life and Death of Great American Cities*. New York: Vintage.
Joly, S. (2013). Couche-tard: une bataille inspirante. *Vie économique*, 4(4), pp. 1–10.
Keil, R. (2009). The Urban Politics of Roll-with-it Neoliberalization. *City*, 13(2–3), pp. 231–245.
Kern, L. (2010). *Sex and the Revitalized City: Gender, Condominium Development, and Urban Citizenship*. Vancouver: University of British Columbia Press.
Klein, N. (1999). *No Logo: Taking Aim at the Brand Bullies*. New York: Picador.
Leidner, R. (1993). *Fast Food, Fast Talk: Service Work and the Routinization of Everyday Life*. Berkeley: University of California Press.
Low, S. (2003). *Behind the Gates: Life, Security, and the Pursuit of Happiness in Fortress America*. New York: Routledge.
McAlevey, J. (2015). The Crisis of New Labor and Alinsky's Legacy: Revisiting the Role of the Organic Grassroots Leaders in Building Powerful Organizations and Movements. *Politics & Society*, 43(3), pp. 415–441.
McBride, S. and Shields, J. (1997). *Dismantling a Nation: The Transition to Corporate Rule in Canada*. Halifax: Fernwood Publishing.
McElligott, G. (2001). *Beyond Service: State Workers, Public Policy, and the Prospects for Democratic Administration*. Toronto: University of Toronto Press.
Meyer, R. and Kimeldorf, H. (2015). Eventful Subjectivity: The Experiential Sources of Solidarity. *Journal of Historical Sociology*, 28(4), pp. 429–457.
Milkman, R. (2006). *L.A. Story: Immigrant Workers and the Future of the US Labor Movement*. New York: Sage.

Mitchell, D. (2003). *The Right to the City: Social Justice and the Fight for Public Space.* New York: Guilford Press.

Mitchell, K. and Sparke, M. (2016). The New Washington Consensus: Millennial Philanthropy and the Making of Global Market Subjects. *Antipode,* 48(3), pp. 724–749.

Nicholls, W.J. (2008). The Urban Question Revisited: The Importance of Cities for Social Movements. *International Journal of Urban and Regional Research,* 32(4), pp. 841–859.

Noiseux, Y. (2012). Mondialisation, travail et précarisation: Le travail migrant temporaire au coeur de la dynamique de centrifugation de l'emploi vers les marchés périphériques du travail. *Revue recherches sociographiques,* 53(2): 389–414.

Palmer, B.D. (1992). *Working-Class Experience: Rethinking the History of Canadian Labour, 1800–1991.* Toronto: McClelland & Stewart.

Park, R. and Burgess, E., eds. (1925). *The City.* Chicago: University of Chicago Press.

Peck, J. (2010). Zombie Neoliberalism and the Ambidextrous State. *Theoretical Criminology,* 14(1), pp. 104–110.

Pelosi, A. (2015). *San Francisco 2.0.* HBO.

Pineault, E. (2012). Quebec's Red Spring: An Essay on Ideology and Social Conflict at the End of Neoliberalism. *Studies in Political Economy,* 90(1), pp. 29–56.

Relph, E. (1987). *The Modern Urban Landscape.* Baltimore: Johns Hopkins University Press.

Ross, A. (2004). *Low Pay, High Profile: The Global Push for Fair Labor.* New York: The New York Press.

Sassen, S. (1991). *The Global City: New York, London and Tokyo.* Princeton, NJ: Princeton University Press.

Satzewich, V. (2015). *Points of Entry: How Canada's Immigration Officers Decide Who Gets In.* Toronto: University of Toronto Press.

Schwanhausser, A., ed. (2016). *Sensing the City: A Companion to Urban Anthropology.* London: Birkhausser.

Scott, A.J. (2000). *The Cultural Economy of Cities: Essays on the Geography of Image-Producing Industries.* London: Sage.

Shalla, V. and Clement, W., eds. (2007). *Work in Tumultuous Times: Critical Perspectives.* Montreal: McGill-Queen's University Press.

Standing, G. (2011). *The Precariat: The New Dangerous Class.* New York: Bloomsbury.

Starr, A. (2005). *Global Revolt: A Guide to the Movements Against Globalization.* New York: Zed Books.

Tait, V. (2005). *Poor Workers' Unions: Rebuilding Labor from Below.* Cambridge: South End Press.

Tattersall, A. (2010). *Power in Coalition: Strategies for Strong Unions and Social Change.* Ithaca, NY: ILR Press.

Taylor, K-Y. (2016). *From #blacklivesmatter to Black Liberation.* Chicago: Haymarket.

Therborn, G. (2014). New Masses? Social Bases of Resistance. *New Left Review,* 85(Jan/Feb), pp. 7–16.

Thomas, M. and Tufts, S. (2016). Austerity, Right Populism, and the Crisis of Labour in Canada. *Antipode,* 48(1), pp. 212–230.

Tirado, L. (2014). *Hand to Mouth: Living in Bootstrap America.* New York: GP Putnam's Sons.

Turner, L. (2007). Introduction: An Urban Resurgence of Social Unionism. In: L. Turner and D. Cornfield, eds. *Labor in the New Urban Battlegrounds: Local Solidarity in a Global Economy.* Ithaca, NY: NLR Press.

Vela, T. (2013). Young Workers Listening to Union Pitch. *Globe and Mail,* 5 August, p. B3.

Wacquant, L. (2007). *Urban Outcasts: A Comparative Sociology of Advanced Marginality.* Malden, MA: Polity.

Waldinger, R. and Lichter, M.I. (2003). *How the Other Half Works: Immigration and the Social Organization of Labor.* Berkeley: University of California Press.

Watkins, S. (2010). Editorial: Shifting Sands. *New Left Review,* 61(1), pp. 5–27.

Wills, J., Datta, K., Evas, Y., Herbert, J., May, J. and McIlwaine, C. (2010). *Global Cities at Work: New Migrant Divisions of Labour.* London: Pluto.

Wright, E.O. (2010). *Envisioning Real Utopias.* New York: Verso.

Wright, T. (1997). *Out of Place: Homeless Mobilizations, Subcities, and Contested Landscapes.* Albany: State University of New York Press.

Zukin, S. (1995). *The Cultures of Cities.* Cambridge, MA: Blackwell.

Zukin, S. (2010). *Naked City: The Death and Life of Authentic Urban Places.* Oxford: Oxford University Press.

21
URBANIZATION AS A BORDERING PROCESS
Non-citizen labour and precarious construction work in the Greater Toronto Area

Michelle Buckley and Emily Reid-Musson

Introduction

We begin this chapter by considering three very different events that took place between 2012 and 2015. The first, in 2012, involved Canada's former Minister of Citizenship and Immigration appearing on a popular late-night Irish talk show to advertise the thousands of high-quality jobs available in Canada for hard-working Irish citizens interested in applying for a temporary work permit; dedicated Irish workers were, in his words, simply a 'natural choice' for Canadian employers looking for labour abroad. Two years later, in the late summer of 2014, officials from the Canada Border Services Agency conducted a series of coordinated raids on construction day labour spots across North Toronto. In less than forty-eight hours, border agents had detained and deported a total of twenty-two men, mainly undocumented migrants from Latin America. Finally, in 2015, a European engineering firm with the contract to oversee the expansion of one branch of Toronto's subway system was summarily terminated following major cost overruns and mismanagement. The subcontractor then fled the country, leaving many of its non-Canadian workforce suddenly unemployed and without legal permission to work and live in Canada.

While at first glance these three events may appear unrelated, in this chapter we draw connections between these seemingly disparate moments to demonstrate how they are each enrolled in shaping the political economy of urbanization across the Greater Toronto Area (GTA). The GTA is a dense chain of municipalities that wraps around the north-western edge of Lake Ontario, and which comprises the City of Toronto as well as a number of neighbouring cities such as Brampton, Newmarket, Mississauga and Markham. As one of the fastest growing metropolitan corridors across North America in recent years, the GTA's non-citizen construction workforce – comprising refugee claimants, foreign students, skilled foreign nationals on an array of temporary work visas, and non-citizens with no legal authorization to remain and/or work in Canada – have played an essential role in meeting recent demand across the region for new-build housing, residential renovations and dozens of state-funded infrastructure projects under way since the early 2000s. In particular, we argue that these three events each reveal how bordering processes have profoundly shaped the nature of construction work and employment for migrant workers in the GTA. Focusing on a partial but fundamental set of material practices that underpin processes of urbanization – namely, the production, demolition and renovation of the urban built environment – we highlight how important temporary, non-citizen construction labour has been to the GTA's urban growth in recent years, a foreign workforce which is itself highly stratified.

Over the past fifteen years, the Canadian government has dramatically expanded its temporary migration streams, which surpassed permanent immigration rates for the first time in 2008 (Valiani 2013). In view of this historically significant trend towards temporary migration in Canada, we use the term 'migrant' to refer to those whose legal status and rights to permanency in Canada are insecure. The term thus denotes a wide range of socio-legal positions upon which the GTA construction industry draws. Though some migrants to which we refer may well possess or gain more permanent forms of legal status, the catchall 'migrant' underscores current and lasting vulnerabilities associated with temporariness and non-citizenship that crosscut otherwise static legal classifications. Empirically, we draw on a recent case study on migrant work in the GTA comprising semi-structured interviews with twenty-five non-Canadian men employed in construction, as well as thirteen interviews with labour support agencies, political leaders, migrants' rights groups, labour lawyers and construction union representatives. In what follows, we chart a number of key shifts in the post-millennial labour market for precarious forms of non-citizen construction work in Canada. In particular, we undertake a more detailed discussion of the vignettes in the opening paragraph to highlight how the last decade has seen the rapid opening of temporary immigration channels for skilled workers in construction. Following this, we explore how this growth has also been accompanied by revanchist practices aimed at informalized and predominately low-waged construction day labourers in the region.

Analytically, our objective in this chapter is to intervene in conceptualizations of urbanization within critical urban theory to map some of the racialized and ethno-nationalized migration channels and local labour market practices which encode the urbanization process; these have tended to be elided by Marxian conceptualizations of urbanization which highlight class struggle but which evacuate, as Don Mitchell (1996) so forcefully demonstrated over twenty years ago, attention to the ways that all landscapes are fundamentally laboured. Following on from pathbreaking work illuminating the importance of state immigration policies in mediating urbanization processes (Chan 1994, 1996), and other research demonstrating how the livelihoods of migrants employed in construction are directly implicated in processes of urbanization as a spatial fix for unproductive surplus capital (e.g. Buckley 2012, 2014), we conceptualize urbanization in the Greater Toronto Area as a process that depends in part on the national *de jure* production and socio-spatial regulation of labour supply through immigration law, as well as on the variegated de facto production and negotiation of 'migrant' status by various state and private actors. In particular, we highlight two specific loci of regulation that directly mediate the urbanization process in the region: temporary labour migration schemes and border policing. We specifically highlight how the production and maintenance of forms of legal status, including temporary, insecure and illegal[1] forms of territorial residency, are immanent to material processes of urbanization.

We also argue here for more attention in the study of urban politics to the role(s) of diverse forms of urban labour, both unwaged and waged, in processes of urbanization. The privileged subject within much critical urban scholarship on the politics of urbanization – from predatory mortgage owners, to the perpetrators and victims of gentrification and urban redevelopment – has typically been some variant of what we shall call the *resident consumer*. In contrast, attention to the emergence, (re)production and policing of differentiated working subjects and their relevance to, for example, the creation of value or surplus value within urbanization processes is fundamentally important to conceptualizing the relations of power that undergird them. Yet despite an immense literature across the social sciences now foregrounding precarious and nonstandard employment, wageless life, global care chains or the 'uber-fication' of local economies, these perspectives have had little uptake within efforts to theorise urbanization processes and the politics, capitalist or otherwise, that transect them (though see Merrifield 2013). Beyond the analytical-empirical sites of citizenship or work in construction, grappling with the power relations of value creation, reciprocity or survival in which diverse (and differentiated) working subjects are enmeshed opens up exciting arenas to theorize and potentially (re)conceptualize urbanization.

Migration and urbanization: beyond the demographic-territorial relation

The literature documenting the relationships between human migration and the process of urbanization is vast. Since the 1970s, research has sought to understand how the rapid growth of cities was connected in particular to various forms of international, circular and internal in-migration. Much social sciences literature has highlighted the impacts of informal migration on urban landscapes and the conceptual challenges this posed for both urban theory and migration studies (e.g. Abu-Lughod 1975; Williamson 1986). Since then, this literature has widely noted the role played by rural to urban migration (Oberai 1987; Rogers and Williamson 1982) and intra- and international urban migration in growth in the late twentieth century (Satterthwaite 2005). More recently, the technocratic frameworks shaping global policy responses to a rapidly urbanizing planet – ones deeply entangled with 'Urban Age' discourse – have also focused heavily on rapid in-migration to urban centres (see for example UN-Habitat 2010; and UN-Habitat 1996 cf. Brenner 2013). Indeed, the now-clichéd celebration of entrepreneurial subjects migrating to metropolitan centres in search of a brighter future has become firmly entrenched not only in global policy discourse but also the popular imaginary through journalistic tomes about migration, urbanization and the informal ingenuity of urban newcomers (e.g. Neuwirth 2005; Saunders 2012).

Whether journalistic or not, much scholarship constructing this view of the urbanization-migration nexus, has – to borrow a valuable insight from a very different set of debates about urbanization currently under way – been overwhelmingly focused on framings of conventional definitions of urbanization as a phenomenon comprising population growth within a given metropolitan area (Brenner 2013). As a result, many of the assumptions within this literature on the relationships between migration and urban change have arguably been predominately oriented around the role that in-migrants play in urbanization as a process of *territorial-demographic* accretion and the subsequent expansion of municipal or metropolitan borders.

We propose a different understanding of capitalist urbanization as a nexus of a very different set of *bordering practices*. To do so, we take some preliminary steps to bring two groups of literature into conversation with each other to expand the socio-spatial terrain on which urbanization is seen to take place, and the attendant subjects on which it relies: this includes Marxian theories of urbanization as a process of capitalist accumulation, and scholarship on bordering practices, ethno-national difference and urban citizenship. In distinct contrast to the demographic-territorial framework governing migration studies and some strands of urban studies, some Marxian-inflected strands of critical urban theory have conceptualized urbanization not in demographic terms, but as a socio-material process of capital accumulation, crisis alleviation and class struggle. This literature has tended to be centred either explicitly or implicitly on the material object of the 'urban built environment' and the commodified relations of private property that encode it as a specific arena through which urbanization takes place. In a world economy struggling to find profitable destinations for large amounts of unproductive surplus capital, this scholarship has demonstrated how urban built environments have become crucially important circuits for capitalist accumulation, creative destruction and dispossession.

The 2008 global financial crisis was a particular flash-point for some of this scholarship. David Harvey's insistence of the distinctly urban dimensions of recent economic upheaval is in keeping with his wider efforts to theorize the ways that unproductive excess capital has historically tended to be directed into large, long-tenor investments in the built environment, a distinct realm of surplus accumulation in the capitalist space-economy which he terms the "secondary circuit of capital" (Harvey 1985; see also Aalbers 2008; Beauregard 1994; Gotham 2006, 2009). As many have noted, Harvey argues that metropolitan property bubbles often foreshadow impending crisis, as the large-scale urbanization of capital often serves as a "last-ditch hope" (Harvey 1978: 120) to counteract a crisis of overaccumulated capital developing in the economy. In the catastrophic wake of the collateralized mortgage market crash, he argues that the financial crisis was in this sense an *urban crisis* (Harvey 2010). These and other recent analyses of the global economic instability of the early twenty-first century have placed the social relations of urbanization at the very heart of recent capitalist crises, and more specifically and recently, crises engendered by the circulation of fictitious forms of capital through components of the built environment (see also Christophers 2011).

In doing so, this literature has crucially demonstrated that the regulatory arenas that shape urbanization require new objects of focus and methods of research beyond mapping population change. For example, the emergent literature seeking to theorize the 'real estate-finance' nexus demonstrates that the social relations of urbanization are currently playing out in part within secondary financial markets (Aalbers 2008; Buckley and Hanieh 2014). Scholarship on predatory mortgage lending, housing foreclosure and household debt (such as Crump *et al*. 2008; Walks 2013; Wyly and Ponder 2011), meanwhile, has carefully mapped the racialized, gendered and age-based forms of violence that have accompanied residential financialization and dispossession.[2] Additionally, research on the growing penetration of global capital into cities through the use of new municipal debt instruments for financing urban infrastructure development (for example, Ashton *et al*. 2012; Weber 2010) has shown how local states are not only active agents in the urbanization of capital, but also in financializing and privatizing the public infrastructures that undergird urban social reproduction.

Critical urban theory, however, has had comparatively little to say about the relationships between urbanization, migrant forms of work and employment, and the production of precarious legal status – with some key exceptions (Steil and Ridgley 2012; Varsanyi 2006; Wills *et al*. 2010). Beyond critical urban studies, meanwhile, research on the 'work-citizenship matrix' has shown how citizenship, immigration and migration policies (and the non-citizen categories such policies spur) generate precarious employment conditions. Unequal citizenship and immigration regimes often place migrant workers at a disadvantage relative to citizen-workers across a range of sectors in the Global North. Indeed, migrants are often situated in the lowest rungs of the labour market (Anderson 2010; De Genova 2010; Goldring and Landolt 2013; Strauss and McGrath 2016). In other veins of research, the notion that citizenship is not only a *de jure* form of membership in a national state, but a contradictory and dynamic relationship shaped in part through de facto practices by 'local' actors, has developed broad purchase within the social sciences over the past two decades (Holston and Appadurai 1999; Isin and Neilson 2008; Lowndes 1995; Staeheli 2008; Varsanyi 2006). More broadly, a growing body of critical urban border scholarship has intervened in recent debates specifically about urban citizenship for residents with precarious legal status, exploring how ethno-national forms of border policing are being undertaken by urban and suburban authorities, or how municipal actors are enacting local laws that can either enhance or preclude fuller forms of citizenship for non-citizen groups (Carpio *et al*. 2011; Steil and Ridgley 2012; Walker and Leitner 2014). While constraints on space preclude a more detailed overview of this compelling work (for a review of the scholarship, see Lebuhn 2013), the legal and practical acts criminalizing non-citizens in cities, coupled with counter-claims to belonging and entitlement by those with precarious legal status and their allies have, in Lebuhn's words, "[turned] the urban realm into a (conflictive and very place-specific) site of negotiating, shaping and interconnecting local practices of border control and urban citizenship; and in effect renders ... cities an uneven landscape of border spaces" (2013: 38).

While this literature has very saliently illuminated the changing forces that produce urban citizenship, and that subsequently shape foreign residents' rights to political participation, public space and urban governance, it has been less focused on the ways that these rescaled and re-engineered citizenship practices shape and are shaped by the social relations of capitalist urbanization itself. It is in this sense that Marxian geographical-materialist efforts to understand the political economy of contemporary urbanization on one hand, and Foucaultian, Agambenian or even Lefebvrian explorations of the urban realm as a contested arena of bordering practices that produce differentiated forms of citizenship on the other, arguably constitute two solitudes within critical urban research and theory.

Yet construction work undertaken by non-citizens has long been a key facet of urbanization processes. Non-citizen construction labour creates value and surplus value in the built environment through renovation work and brand new construction projects; helps realize state-funded collective consumption agendas in the upgrading and building of large infrastructure projects like public transit projects, social housing development or highway and bridge construction, and equally destroys and produces value within real estate assets through manual or mechanical demolition services, the conversion of factory buildings into loft condominiums or private rooming houses into boutique hotels. All of these processes require a set of manual and professional services, ranging from manual, 'biblical' trades like carpentry or masonry, to professional construction services such as project management and engineering services. Because the work of non-citizens in the construction sector is so crucial to the labour market that underpins urban development in the GTA, we contend that the production and maintenance of various forms of precarious legal status is immanent to the political economy of urbanization in the region. The production of 'probationary' (Goldring and Landolt 2013) and 'illegal' working subjects, as we aim to show in the following sections, is fundamentally regulated through bordering practices happening at a range of scales, and which are recalibrating the ways that some non-citizens are being incorporated into the regional metropolitan labour market.

Crises of urbanization and temporary (im)migration to the GTA

While construction markets in the westernmost and easternmost regions of Canada have been shaped significantly by the demand for industrial construction labour in the trades for oil extraction activities over the last decade, in Ontario, construction markets have been driven primarily by investment in urban and suburban real estate markets. A majority of this construction growth, moreover, has been taking place in the GTA; between 2000 and 2015, the total annual value of building permits issued in the City of Toronto alone rose consistently from just over C$2 billion to nearly C$8 billion, of which approximately half were for residential construction (City of Toronto 2016: 17). In 2013, there were 170 high-rise condominiums under construction in the GTA; this was the highest construction rate in North America at the time, and constituted the region's largest building boom in nearly two decades (City of Toronto 2014: 21).

At the same time, the migration of tradespeople out of Ontario to work for higher wages in the oil fields of western Canada, coupled with a rapidly ageing construction workforce meant that the GTA's booming residential construction sector suffered from a shortage of workers in recent years. In the early 2000s, new housing orders were backlogged for up to two years, with some employers turning to informal networks to find workers.[3] Conditions in the Toronto region, meanwhile, have been symptomatic of a larger national problem; with the average age of a skilled tradesperson in Canada being fifty-one, major labour shortfalls were forecasted for the industry in the next decades (BuildForce Canada 2014). These broader dynamics have incentivized the local government to re-regulate and expand a number of temporary immigration channels in recent years. Among these has been the International Experience Canada (IEC) visa, a working holiday programme that the Canadian government runs with European countries, such as Ireland, France and Italy, offering renewable and open work permits for people under thirty-five who wish to travel and work in Canada. Many IEC entrants have found construction jobs in Toronto and the GTA.

The expanded IEC visa has been heavily marketed to skilled tradespeople by the Canadian government, not only as a way of accessing high-paying jobs in the country, but as a route to permanent residency and citizenship for eligible and skilled applicants. In 2012, the former Minister of Citizenship and Immigration, Jason Kenney, appeared on a national Irish talk show to market Canadian jobs to Irish workers hoping to emigrate. Speaking to an audience of Irish citizens planning to emigrate in search of employment, when asked by the show's host what Canada was offering to those hoping to find jobs outside of Ireland, Kenny responded:

JK: Opportunity. Prosperity … the story of immigration from Ireland to Canada is longer than the story of immigration from Ireland to the United States. There are over 4 million Canadians of Irish descent. … There's a lot of opportunity…

PRESENTER: Is this why you've come to Ireland?

JK: It is … actually one of our biggest economic problems that we face in Canada is a labour shortage, it's a skill shortage … we have over fifty Canadian companies here for the 'Working Abroad' Expo that's here in Dublin this weekend, and they're looking to fill, I understand, over 3,000 positions…

PRESENTER: Why Ireland? Why Irish people?

JK: Well, the employers in Canada are increasingly identifying Ireland as a great source of talented, highly educated, hard-working folks who are culturally compatible – they can walk in and get to work the day they arrive. And … so I think it's just a natural choice for the employers.

The repurposing of temporary migration channels like the IEC – whose chief purpose historically was to foster 'cultural exchange' between countries primarily linked to Canada's colonial migration networks – has been one strategy that the government has used to capitalize on crises of urbanization occurring in other parts of the world. As a senior official with an Irish immigration services organization in Toronto explained:

…it [the IEC visa] used to be just go work abroad, have a year out, have a lark. But that completely changed and it's 100 per cent because of the economic downfall in Ireland. Ireland was as you know economically decimated and so they were coming out of college or the trades … And of course the first thing to go was building projects and it all just went down the tubes there.[4]

In the wake of crashing property markets and the halting or outright cancellation of construction projects across Ireland, thousands of tradespeople facing high unemployment in Ireland turned to the IEC as a route to both temporary and permanent employment in Canada (Hough

2012). Approximately half of the IEC migrants to Canada in 2012 were Irish, while annual quotas for Canadian IEC permits issued to Irish migrants to Canada have increased each year due to demand and the government's interests in addressing acute construction labour shortages domestically (Goodman 2014).

This opportunistic seizing on working holiday visas to address acute structural challenges in construction labour markets is illustrative of two further transformations within immigration law and policy that the Conservative federal government instituted over the last decade. In the next section, we highlight the move towards a 'two-track' immigration system since the 1980s in Canada which has institutionalized *de jure* differences between immigrants and temporary foreign workers in the GTA's construction sector.

Marco's story: employer-adjudicated legal status and the Federal Skilled Workers Programme

A second step the federal government has taken in recent years to re-engineer the border to attract temporary construction labour has been to create a number of temporary immigration channels specifically catering to workers in the trades. First, the Federal Skilled Workers Programme (FSWP) was a two-year visa which catered mainly to professionals in project management, architecture and engineering; tens of thousands of foreign workers were recruited into professional construction positions beginning in 2005 (Construction Sector Council 2011).

A second programme, the Federal Skilled *Trades* Program (FSTP) was introduced in 2013 because the FSWP was not well oriented to the needs of lower-skilled workers in the trades, as it placed significant emphasis on post-secondary education, and required high proficiency levels in Canada's official languages, English and French. Under the FSTP, the 100-point system was eliminated in favour of four basic criteria, which give much more weight to practical training, work experience, basic proficiency in English or French, and the level of demand for a specific trade. Both programmes are billed as a route to permanent settlement, offering emigrants a two-year work visa and the possibility to apply for permanent residency in the event of permanent employment. However, currently 150 Italian, Spanish and Portuguese construction workers who entered under the FSTP have launched a lawsuit against the federal government on the grounds that, compared to their Australian and Irish counterparts with the same skills, their applications for permanent residency were subject to ethno-linguistic forms of discrimination (Keung 2015).

Moreover, the story of one interviewee, 'Marco',[5] illustrates the unpredictable adjudication of these new employer-managed temporary immigration channels. Granted a visa under the FSWP, Marco had moved with his wife and daughter to Toronto from western Europe in 2012 after taking a job offered by a European subcontractor who specialized in tunnel digging; the company was a small firm hired by another foreign firm tasked with implementing a multi-billion-dollar transportation infrastructure project in Toronto. Working twelve hours a day, five or more days a week, Marco's monthly take home pay was high by Canadian standards. However, despite being a highly skilled and waged migrant who was legally (if temporarily) authorized to work in Canada, Marco and his co-workers had no formal contract with their European employer.

Additionally, despite working in an industry with the highest rate of traumatic and disease-related fatalities of any sector in the country, neither he or his co-workers were aware that they were not covered under the province's public health insurance, nor that it was their responsibility to pay privately for it while they were in Canada. This led to one worker at Marco's firm owing thousands of dollars in medical fees when his son fell sick and needed to be hospitalized.

He also spoke and read very little English when he arrived in Canada. When recruited by the company while still in Europe, Marco explained that:

> the company contacted me, asked for my name, if I was married, all of that. But I never … [got] papers, I never signed anything. And that's basically how we got here. I arrived in Toronto, they sent me to a hotel … I didn't sign anything. Up to today, I still haven't signed anything. That's why when I went to see the lawyer, everything he saw of mine, [he said] how can people work here two years [with no contract] … It was two years that he'd been paying me and he never gave me a pay stub or a cheque, nothing. … The … problem is that we got here and we didn't know anything. We didn't know the language. Our employer would pass us papers, and they would all be in English.

Marco's overwhelming dependence on his employer, coupled with his lack of understanding of Canadian employment standards and immigration processes ultimately led to the loss of his legal status. While his employer had told him they were submitting the paperwork to extend Marco's two-year skilled worker visa, according to Marco they deliberately allowed his visa to expire so that they could terminate his employment when the need arose. In his words:

> [a]t the end of August, I finished the work permit. The employer told me that I had three additional months [to work], while the paperwork got figured out. What happened next was that he told about twenty of us here with the work permit by November that our work permits hadn't yet arrived … [then] one day he just told us, "you can't work anymore, because your work permits didn't arrive and the government denied them" … I went to [a worker's advocacy group], and they referred me to a lawyer. He [the lawyer] requested my papers from the government, from immigration, in order to see what had happened with our permits that hadn't arrived. And he discovered that the employer had lied to us all.

As it turns out, a request to extend Marco's work visa had never been properly made; this deception was likely a result of the summary termination of this employer's contract by the City, allegedly for mismanagement of the contract and cost overruns. By early 2015, Marco, his wife and their child were living in Canada without legal status and were surviving on savings, as were many of his fellow co-workers.

Marco's story illustrates how the global market for temporary labour from which Canadian construction employers are increasingly drawing – including the Federal Skilled Workers Programme but also other temporary channels such as the IEC visa and the Temporary Foreign Workers Programme – operates through an immigration process in which permanent settlement under temporary immigration channels billed as 'routes to citizenship' are in fact probationary, in that the post-war regime of immigration in Canada which was predicated on permanent residency has been overtaken since the 1980s with temporary forms of immigration (Valiani 2013). More specifically, we point to the intersecting rise of employer-driven immigration policy with the proliferation of temporary forms of immigration status in Canada (Faraday 2012; Goldring and Landolt 2013; Valiani 2013).

Marco's story resonates with Goldring and Landolt's (2013) assertion that for many, immigration increasingly involves a probationary period; rather than a linear, unidirectional movement from temporary resident to permanent resident or citizen, they argue that "some will achieve permanent residence, others may move between temporary authorised categories or between these and unauthorised status, and thus continue to live with precarious status"

(Goldring and Landolt 2013: 8). Additionally, they note that the outcomes of this probationary status are now significantly determined not only by the state, but by employers and other local intermediaries. In highlighting the messy and multi-directional nature of immigration for temporary migrants under Canada's two-track regime, they argue that this model

> invites attention to the role of policies and institutional actors in precipitating movement along or across tracks ... Front-line workers, teachers, landlords, doctors, legal consultants, employers and other institutional actors may act as catalysts, moving people from one legal status category to another, and toward more or less secure status.
> (Goldring and Landolt 2013: 8)

While Valiani (2013) rightly notes that employer-driven immigration systems effectively privatize decision-making on migrants' legal status, from a geographical perspective, the growing importance of local, or in Marco's case, local transnational actors in shaping migrants' residency outcomes also arguably constitutes the partial rescaling of the immigration adjudication process. In this case, the choice of this transnational company to use workers' precarious legal status as a convenient tool for large-scale retrenchments had a profound impact on individual workers' immigration trajectories. Moreover, this case foregrounds how individual local actors responding to the conditions in local labour markets – and their concomitant decisions about whether they still 'need' foreign workers – are bound to have major impacts on some workers' ability to settle permanently. As employers become more integral actors adjudicating citizenship on the ground for construction migrants, so too do immigration bordering practices in these cases become increasingly inseparable from the capitalist logics of urbanization.

In the penultimate section, we turn to consider a third group of precarious foreign workers, one which arguably forms the largest component of the migrant labour market in the GTA: precarious and undocumented workers in the residential construction and renovations sectors.

'They don't need us anymore': deportability, illegality and the circulation of residential value

As a major gateway region for newcomers to Canada, construction employers in the GTA tend to have access to a large pool of cheap, non-citizen workers, including professionals who arrive in Canada but cannot find work in their field, refugee claimants or individuals who were once legally working in the country but whose legal status, for one reason or another, has lapsed. It has been widely recognized by both migrants' rights groups and the construction industry itself that informally resident, and informally employed workers, particularly those employed in the residential construction and renovations sector in the GTA, have been indispensable to the industry in recent years.

Since the mid-2000s, a number of efforts have been undertaken to address the growing employment of undocumented workers in the construction sector. Some of these have been fuelled by union and contractor interests in 'levelling the playing field' for residential construction, while others have been brought forward by labour rights activists and migrants' rights activists. For example, in response to political pressure by registered members of the Ontario Homebuilders Association, various mechanisms to regulate the 'underground economy' in construction have been under way over the last six years. For example, the 'Dean Report' – a 2010 Ontario health and safety regulatory review – advocated for information sharing between municipal building permit issuers and the Ministry of Labour (MOL) to find projects that have not submitted a notice to the MOL; these kinds of operations are very often ones in which workers without legal status might be employed.[6] Another initiative comprising formal legislative amendments tabled in 2004–2005 – an initiative led in part by Toronto-based undocumented workers with trade union membership – sought to create avenues to legalize the status of large numbers of undocumented workers in the construction industry. A plan to regularize construction workers' status based on their importance to the industry, however, was rejected by the federal government in 2006.

While information sharing between provincial ministries was touted by some as a means to better protect the industry's most vulnerable workers, for illegalized migrants working in the residential construction sector, information-sharing initiatives between different arms of the state – also undertaken through the rhetoric of 'safety' – have had the opposite effect. In the late summer of 2014, combined inter-jurisdictional efforts led to provincial transportation inspectors targeting contractors' vans under the thin pretext of 'commercial vehicle safety inspections'. The raids across Toronto involved cooperation by at least four arms of the state spanning the federal Canada Border Services Agency (CBSA), the Provincial Ministry of Transportation, the Ontario Provincial Police and police from the municipality of Halton-Peel. The CBSA raids predominantly targeted day labourers in the construction trades – occurring in and around the parking lots of strip malls in the Jane-Wilson, Keele-Wilson and Weston-401 areas. Vans were followed by Ministry of Transportation officials out of the lots, stopped on the highway for a purported vehicle safety infraction, at which point the passengers were asked for their identification, and detained by the Ontario Provincial Police. This detainment involved the deliberate removal of workers outside the borders of Toronto and into the hands of police from the regional municipality of Halton-Peel in order to proceed with immigration detainment and deportation; this strategy carefully side-stepped the City of Toronto's Sanctuary City policies, which preclude Toronto police and all City of Toronto staff from sharing information about an individual's legal status with Canadian Border Services officials or other government agencies.

As a number of scholars have argued, deportation and detention practices are not solely a product of securitizing borders that have accompanied a post-9/11 geopolitics; they are also very much a mechanism of labour control (De Genova 2010). Inseparable from the securitization of migration globally, De Genova (2010) argues that local forms of deportability of irregularized, informalized and/or criminalized migrant workers must be understood in part as a facet of the political economy of local labour markets. The very fact of unauthorized or illegalized migration is, in De Genova's (2010: 93) view, inseparably tied to processes of capitalist accumulation. In his words:

> … all 'illegal' or 'irregular' migrations, however constituted historically, must at least potentially be ultimately apprehensible to be 'regularizable' (or normalizable or routinizable). This is the case, however, only insofar as they may finally be (re)composed – as labour. Undocumented migrations are therefore best understood as distinct transnational manifestations of a global social relation of labour and capital, which is mediated by the regulatory authority and coercive force of territorially delimited 'national' states.

In other words, as migrants without legal status must secure their ability to reside in a locale in many cases *through their labour*, De Genova argues that the deportability of this labour is a part of "the broader political dynamics of labour subordination"; "[i]ndeed, it is their distinctive legal vulnerability, their putative 'illegality' above all else, which facilitates the subordination of undocumented or 'irregular' migrants as a highly exploitable workforce" (2010: 94).

As such, we assert that these coordinated forms of bordering by multiple government agencies levelled at the most precarious segments of the GTA's construction workforce – forms of bordering which specifically 'work' the territorial borders of the GTA – serve to secure and reproduce that same workforce on a hyper-precarious and deportable basis, one integral to the reproduction and maintenance of the GTA's built environment.

Conclusions: bordering practices, labouring subjects and urbanization

In exploring the reconfiguration of national immigration channels and the policing of migrant construction workers in the GTA, we have sought to conceptualize – in very preliminary terms – just some of the diverse practices of bordering that both inflect and are shaped by the process of urbanization in the GTA. Temporary citizenship regimes for highly-paid and -skilled workers, coupled with a highly deportable illegalized workforce at the bottom of the labour market, together shape a highly striated transnational workforce to serve an industry which, while currently forecasting enormous labour scarcity, is also an industry characterized by extreme spatio-temporal flexibility as it operates by assembling an ever-changing team of workers based on project by project needs which are then disbanded after project completion. It is also an industry that relies heavily on migrants as 'reserve' or 'surplus' labour supply, not only in times of construction market growth, but also in times of contraction, as an industry that not only goes through large cyclical and seasonal swings in activity, but which is prone to crisis, often needing to shed large numbers of workers quickly.

We suggest that bordering shapes the supply, the cost and the expendability of urban construction labour in a variety of crucial ways. The first is that broader agendas aiming to multiply the channels for short-term, impermanent migration into the trades while primarily reserving permanent settlement for highly skilled immigrants serves to further entrench certain colonial and racialized forms of labour market segmentation within the GTA's construction labour market. Former Immigration Minister Kenney's not-so-veiled references to white, Anglo-European-ness in his articulation of the appeal of 'culturally compatible' Irish workers for Canada's labour market feeds into a pernicious realignment of immigration policy with high-skilled, white settlers on temporary tracks billed as routes to permanent residency. Meanwhile, undocumented workers, providing a vital source of cheaper, flexibilized urbanization labour, have been the targets of migrant policing and urban retrenchments.

At the same time, tidy and familiar narratives of labour market segmentation on the basis of ethnicity, class and citizenship are complicated by the proliferation of probationary, employer-mediated immigration channels that affect even highly waged migrants. Marco's story highlights that in the move towards a more flexible, employer-driven model of immigration in Canada such as the Federal Skilled Workers Programme, the power to determine and shape individuals' immigration trajectories is moving further away from juridical arms of the federal state and towards employers. In this sense, some of the most important bordering practices are not ones operating at some federal scale, but through construction employers who, beyond threatening illegalized migrants with deportation, in many cases now hold sway over the immigration trajectories of migrants employed on a variety of insecure work visas.

In this chapter we have sought to show, in a partial and embedded manner, how processes of urbanization across the GTA have been unfolding in part through the production of a labouring workforce differentiated on the basis of ethno-nationality and citizenship. For us, this case illuminates the need to reanimate and refocus questions about the role that surplus populations play in urbanization processes, and attention to the racialized, linguistic and ethno-nationalized hierarchies of access to a more permanent place in these urban labour markets. Finally, it necessitates a deeper inquiry – theoretical and empirical – into the role of state actors in mediating local urbanization processes through immigration law and practice.

Notes

1 We use 'illegal' and variants thereof to de-naturalize this category and to underscore the ways illegality is produced, especially in relation to the coercive capacities of nation-states, but also shaped, negotiated and breached through a multiplicity of actors, relationships, institutions and processes. We follow De Genova (2010: 93) when he argues that "there is no such thing as ... migrant 'illegality' ... 'in general', and these analytic categories plainly do not constitute a generic, singular ... object of study" (see also Goldring and Landolt 2013: 10).
2 The constraints of space preclude a longer discussion of the relationships identified between contemporary forms of *rural* dispossession and urbanization, but see for example Arboleda (2015), Shin (2016) and Williamson (1986), as well as scholarship within peasant studies such as Chuang (2015).
3 Interview, former Member of Parliament and former Minister of Human Resources and Social Development Canada, Toronto, 10 July 2014; Interview, independent consultant, Toronto, 4 July 2014.
4 Interview, Executive Director of the Irish Canadian Immigration Centre, Toronto, 2 October 2014.
5 Interview subjects' names have been anonymized.
6 Interview, senior regional supervisor, Ministry of Labour, Toronto, 11 September 2014.

References

Aalbers, A. (2008). The Financialization of Home and the Mortgage Market Crisis. *Competition & Change*, 12(2), pp. 148–166.
Abu Lughod, J. (1975). The End of the Age of Innocence in Migration Theory. In: B.M. DuToit and H.I. Safa, eds. *Migration and Urbanization: Models and Adaptive Strategies*. Paris: Mouton Publishers.
Anderson, B. (2010). Migration, Immigration Controls and the Fashioning of Precarious Workers. *Work, Employment & Society*, 24(2), pp. 300–317.
Arboleda, M. (2015). Financialization, Totality and Planetary Urbanization in the Chilean Andes. *Geoforum*, 67(December), pp. 4–13.
Ashton, P., Doussard, M. and Weber, R. (2012). The Financial Engineering of Infrastructure Privatization: What Are Public Assets Worth to Private Investors? *Journal of the American Planning Association*, 78(3), pp. 300–312.
Beauregard, R. (1994). Capital Switching and the Built Environment: United States, 1970–89. *Environment and Planning A*, 26(5), pp. 715–732.
Brenner, N. (2013). Theses on Urbanization. *Public Culture*, 25(1), pp. 85–114.
Buckley, M. (2012). From Kerala to Dubai and Back Again: Migrant Construction Workers and the Global Economic Crisis. *Geoforum*, 44(2), pp. 250–259.
Buckley, M. (2014). On the Work of Urbanization: Migration, Construction and the Commodity Moment. *Annals of the Association of American Geographers*, 104(2), pp. 338–347.
Buckley, M. and Hanieh, A. (2014). Diversification by Urbanization: Tracing the Property-Finance Nexus in Dubai and the Gulf. *International Journal of Urban and Regional Research*, 38(1), pp. 155–175.
BuildForce Canada. (2014). *Construction and Maintenance Looking Forward*. Ottawa: BuildForce Canada. Available at: www.buildforce.ca/en/products/2014-construction-and-maintenance-looking-forward-highlights-reports [accessed 13 March 2016].
Carpio, G., Irazàbal, C. and Pulido, L. (2011). Right to the Suburb? Rethinking Lefebvre and Immigrant Activism. *Journal of Urban Affairs*, 33(2), pp. 185–208.
Chan, K.W. (1994). *Cities with Invisible Walls: Reinterpreting Urbanization in Post-1949 Hong Kong and New York*. Oxford: Oxford University Press.
Chan, K.W. (1996). Post-Mao China: A Two-Class Urban Society in the Making. *International Journal of Urban and Regional Research*, 20(1), pp. 34–150.
Christophers, B. (2011). Revisiting the Urbanization of Capital. *Annals of the Association of American Geographers*, 10(16), pp. 1347–1364.
Chuang, J. (2015). Urbanization Through Dispossession: Survival and Stratification in China's New Townships. *Journal of Peasant Studies*, 42(2), pp. 275–294.
City of Toronto. (2014). *Economic Dashboard*. City of Toronto: Economic Development Committee. Available at: www.toronto.ca/legdocs/mmis/2014/ed/bgrd/backgroundfile-72689.pdf [accessed 7 July 2016].
City of Toronto. (2016). *Economic Dashboard – Annual Summary, 2015*. City of Toronto: Economic Development Committee. Available at: www1.toronto.ca/City%20Of%20Toronto/Economic%20

Development%20&%20Culture/Business%20Pages/Filming%20in%20Toronto/Info%20for%20residents/backgroundfile-90508.pdf [accessed 7 July 2016].

Construction Sector Council. (2011). *Construction Employer's Roadmap to Hiring and Retaining Internationally Trained Workers*. Ottawa: Construction Sector Council. Available at: www.buildforce.ca/en/products/construction-employers-roadmap-hiring-and-retaining-internationally-trained-workers [accessed 7 July 2016].

Crump, J., Newman, K., Belsky, E., Ashton, P., Kaplan, D.H., Hammel, D.J. and Wyly, E. (2008). Cities Destroyed Again for Cash. *Urban Geography*, 29(8), pp. 745–784.

De Genova, N. (2010). Alien Powers: Deportable Labour and the Spectacle of Security. In: V. Squire, ed. *The Contested Politics of Mobility: Borderzones and Irregularity*. New York: Routledge, pp. 91–116.

Faraday, F. (2012). *Made in Canada: How the Law Constructs Migrant Workers' Insecurity*. Toronto: The Metcalf Foundation. Available at: http://metcalffoundation.com/wp-content/uploads/2012/09/Made-in-Canada-Full-Report.pdf [accessed 1 January 2013].

Goldring, L. and Landolt, P. (2013). *Producing and Negotiating Non-Citizenship: Precarious Legal Status in Canada*. Toronto: University of Toronto Press.

Goodman, L.-A. (2014). More Foreign Workers Coming under International Experience Canada Program. CTV News. Available at: www.ctvnews.ca/politics/more-foreign-workers-coming-under-international-experience-canada-program-1.1798171 [accessed 4 September 2015].

Gotham, K.F. (2006). The Secondary Circuit of Capital Reconsidered: Globalization and the U.S. Real Estate Sector. *American Journal of Sociology*, 112(1), pp. 231–275.

Gotham, K.F. (2009). Creating Liquidity out of Spatial Fixity: Globalization and the U.S. Real Estate Sector. *International Journal of Urban and Regional Research*, 33(2), pp. 355–371.

Harvey, D. (1978). The Urban Process under Capitalism: A Framework for Analysis. *International Journal of Urban and Regional Research*, 2(2), pp. 101–131.

Harvey, D. (1985). *The Urbanization of Capital: Studies in the History and Theory of Capitalist Urbanization*. Baltimore: John Hopkins University Press.

Harvey, D. (2010). *The Enigma of Capital and the Crises of Capitalism*. London: Profile Books.

Holston, J. and Appadurai, A. (1999). Introduction: Cities and Citizenship. In: J. Holston, ed. *Cities and Citizenship*. Durham, NC: Duke University Press.

Hough, R. (2012). The Celtic Invasion: Why the Arrival of Hundreds of Irish Construction Workers Benefits Toronto's Building Boom. *Toronto Life Magazine* [online]. Available at: www.torontolife.com/informer/features/2012/11/05/the-celtic-invasion/?page=all#tlb_multipage_anchor_1 [accessed 6 April 2014].

Isin, E.F. and Nielsen, G.M. (2008). *Acts of Citizenship*. New York: Palgrave Macmillan.

Keung, N. (2015). Migrant Construction Workers Sue Ottawa for Discrimination. *Toronto Star* [online]. Available at: www.thestar.com/news/immigration/2015/02/07/migrant-construction-workers-sue-ottawa-for-discrimination.html [accessed 7 July 2016].

Lebuhn, H. (2013). Local Border Practices and Urban Citizenship in Europe: Exploring Urban Borderlands. *City*, 17(1), pp. 37–51.

Lowndes, V. (1995). Citizenship and Urban Politics. In: D. Judge, G. Stoker and H. Wolman, eds. *Theories of Urban Politics*. London: Sage.

Merrifield, A. (2013). The Planetary Urbanization of Non-work. *City*, 17(1), pp. 20–36.

Mitchell, D. (1996). *The Lie of the Land: Migrant Workers and the California Landscape*. Minneapolis: University of Minnesota Press.

Neuwirth, R. (2005). *Shadow Cities: A Billion Squatters, a New Urban World*. New York: Routledge.

Oberai, A.S. (1987). *Migration, Urbanisation, and Development*. Geneva: International Labour Office.

Rogers, A. and Williamson, J.G. (1982). Migration, Urbanization, and Third World Development: An Overview. *Economic Development & Cultural Change*, 30(3), pp. 463–482.

Satterthwaite, D. (2005). *The Scale of Urban Change Worldwide 1950–2000 and its Underpinnings*. International Institute for Environment and Development. Human Settlements Division, London.

Saunders, D. (2012). *Arrival City: The Final Migration and Our Next World*. Toronto: Vintage Canada.

Shin, H.B. (2016). Economic Transition and Speculative Urbanisation in China: Gentrification versus Dispossession. *Urban Studies*, 53(3), pp. 471–489.

Staeheli, L. (2008). Citizenship and the Problem of Community. *Political Geography*, 27(1), pp. 5–21.

Steil, J. and Ridgley, J. (2012). 'Small Town Defenders': The Production of Citizenship and Belonging in Hazelton, Pennsylvania. *Environment and Planning D: Society and Space*, 30(6), pp. 1028–1045.

Strauss, K. and McGrath, S. (2016). Temporary Migration, Precarious Employment and Unfree Labour Relations: Exploring the 'Continuum of Exploitation' in Canada's Temporary Foreign Worker Program. *Geoforum* (early online view).

UN-Habitat. (1996). *An Urbanizing World: Global Report on Human Settlements*. United Nations Centre for Human Settlements (Habitat). Oxford: Oxford University Press.

UN-Habitat. (2010). *State of the World's Cities: Bridging the Urban Divide*. Nairobi: United Nations Human Settlements Programme. Available at: https://sustainabledevelopment.un.org/content/documents/11143016_alt.pdf [accessed 7 July 2016].

Valiani, S. (2013). The Shifting Landscape of Contemporary Canadian Immigration Policy: The Rise of Temporary Migration and Employer-driven Migration. In: L. Goldring and P. Landolt, eds. *Producing and Negotiating Non-Citizenship: Precarious Legal Status in Canada*. Toronto: University of Toronto Press, pp. 50–75.

Varsanyi, M.W. (2006). Interrogating 'Urban Citizenship' Vis-a-vis Undocumented Migration. *Citizenship Studies*, 10(2), pp. 229–249.

Walks, A. (2013). From Financialization to Sociospatial Polarization of the City? Evidence from Canada. *Economic Geography*, 90(1), pp. 33–66.

Walker, K.E. and Leitner, H. (2014). The Variegated Landscape of Local Immigration Policies in the US. *Urban Geography*, 32(2), pp. 156–178.

Weber, R. (2010). Selling City Futures: The Financialization of Urban Redevelopment Policy. *Economic Geography*, 86(3), pp. 251–274.

Williamson, J.G. (1986). *Migration and Urbanization in the Third World*. Cambridge, MA: Harvard Institute of Economic Research, Harvard University.

Wills, J., Datta, K., Evans, Y., Herbert, J., May, J. and McIlwaine, C. (2010). *Global Cities at Work: New Migrant Divisions of Labour*. New York: Pluto Press.

Wyly, E. and Ponder, C.S. (2011). Gender, Age, and Race in Subprime America. *Housing Policy Debate*, 21(4), pp. 529–564.

22
ORGANIZING THE RUINS
The thin institutional geography of labour in the US Midwest

Marc Doussard

Faced with the engineered demise of collective bargaining and the rapid obsolescence of the employment compact anchored by the New Deal, the US labour movement has found new life in so-called 'alt-labour' organizing strategies. Encompassing innovations that range from worker centres to informal strikes to renewed community organizing, these new labour techniques provide workplace and political bargaining leverage increasingly unavailable to workers and unions who confine their advocacy to the procedural and legalistic rules of federally sanctioned collective bargaining (Dixon 2014; Doussard 2013; Garrick 2014; Lichtenstein 2014; Milkman 2013). Once studied as experimental novelties, alt-labour strategies now command scrutiny for their sustained success in propagating a string of legal reforms, including city-level minimum wage laws, wage theft legislation, and scheduling and earned sick-time laws. The emergent question is no longer whether these experiments can work – it is how, where, when and to what end they do.

Labour activists remain engaged in a deliberate process of expanding alt-labour strategies. This goal rests on the ambitious project of inventorying and evaluating the highly heterogeneous local organizations, economic conditions and political institutions shaping employment advocacy. Here, unions, workers, advocacy organizations and community organizers can draw on an impressive and growing body of literature documenting the mechanics, methods and outcomes of new-labour organizing campaigns focused on (for example) black-market taxis, law-breaking manufacturing subcontractors, garment sweatshops, janitorial contractors, day labour, wage theft laws, and more (Milkman 2006, 2014; Milkman *et al.* 2010; Nissen and Russo 2006; Reich *et al.* 2014; Sonn and Luce 2008). These accounts and others provide a staggering volume of raw material from which new strategies, practices and domains of organizing can prospectively be charted. But successfully extracting a set of plans and principles for future labour organizing from these materials requires the evaluation and systematic comparison of the means and ends of literally hundreds of labour organizing campaigns. More subtly, but also more important, this enormous conceptual task requires engagement with two conceptual challenges rarely voiced by advocates or analysts of alt-labour practice.

First, alt-labour organizing techniques have been developed and tested primarily in New York, Los Angeles, Chicago and San Francisco – large urban areas with high organized labour representation and well-worn channels of access through which unions and community organizations can influence public policy. The dissimilarity between these sites and the smaller,

Southern and politically conservative cities to which community and labour organizers seek to move alt-labour strategies, poses a series of distinct empirical and theoretical barriers to comparison. Abstracted from the conditions of possibility that make community-labour advocacy potent in Los Angeles and state house advocacy effective in New York, the mechanics of community-labour partnerships and creative policy-making may offer limited, or misleading information for (to take a representative set of cities currently being organized) Nashville, Albuquerque, Houston or Kansas City. This is a third-sector variant on the much-scrutinized problematique of policy mobility (McCann and Ward 2012; Peck and Theodore 2010), which suggests basic challenges to translating labour innovations from one city to others.

Second, accounts of alt-labour organizing operate within the basic conceptual limits of a New Deal collective bargaining system that privileges the workplace and the state as sites of labour regulation and action. Even as alt-labour organizers have won substantial victories since 2010, the political and legal basis of those victories is vanishing. Nationwide, union membership rates annually hit new lows; four states in the Midwest have recently passed anti-union right-to-work laws, with several others moving to follow; supreme court decisions are whittling away at unions' ability to collect the dues on which their political influence and organizing budgets depend (Gould 2015); and fiscal austerity at the federal, state and local scales has shrunk budgets and goals for the service agencies and community organizations vital to organizing. Even if alt-labour strategies can be adapted to new places, the grounds for adjusting them to the continued demise of New Deal bargaining frameworks remain unclear.

Examining the development of alt-labour organizing strategies in mid-size cities with small labour movements provides the means to address both of these challenges. Accordingly, this chapter presents case studies of community-labour organizing activities in Indianapolis and St Louis – cities lacking the institutions, union density, funding streams and state-level political access that have to date been vital to successful alt-labour organizing. Drawing on more than seventy interviews conducted with community activists in these cities, I demonstrate that these smaller, formerly industrial cities lack not just vigorous community-labour organizing coalitions, but also the institutional, financial and political pre-conditions that make sustained organizing possible. These conditions speak directly to present efforts to expand community-labour organizing to new locations, and to the future viability of alt-labour strategies amid the continued organizational and financial contraction of labour unions.

The promise and limits of alt-labour organizing

Urban-based activism in the USA contends with a difficult paradox. Local issues animate citizens and provide the energy on which effective campaigns rest. But City Hall is rarely the ideal target for reform (Doussard 2015). As income inequalities widened in the early 2000s, the implications were dispiriting: reform mayors didn't win elections, and even when they triumphed at the ballot box, they were powerless in the face of organized corporate interests and mobile capital (Logan and Molotch 1987).

The long history of these problems made the success of community-labour coalitions in the 2000s extremely surprising. These successes include the diffusion of Justice for Janitors campaigns to sites far outside the traditional geography of organized labour, such as Houston, Indianapolis and Miami (Lerner *et al.* 2008); the assembly of a national network of locally focused Restaurant Opportunities Centers engaged in employer- and state-level activism (Jayaraman 2013); the passage of hundreds of living wage laws with the support of the now defunded Association of Community Organizations for Reform Now (Luce 2004); and dozens of other place- and industry-specific policy solutions to the problems encountered daily by low-wage and marginal workers. After 2010, these organizations began to record significant municipal policy victories, including wage theft laws in San Francisco, Chicago and Newark; double-digit minimum wages in Seattle, Los Angeles, San Diego, San Francisco, Chicago and Washington, DC; and, increasingly, legislation regulating non-waged aspects of work, such as scheduling, paid time off and earned sick time.

Several strategic and organizational innovations contribute to these alt-labour successes. Most significant, formalizing alliances between unions and neighbourhood-level community organizations allows unions to supplement community organizations' social networks and relationships with financial resources, and vice versa (Clawson 2003; Reynolds 2004). These partnerships also help to expand the reach of labour organizing, whose roots in the industrial era, and ties to historically white manufacturing workers, long left unions ill-equipped to expand their reach into service industries and meet the needs of the predominantly minority and immigrant populations those industries employ (Doussard 2013; Fine 2006; Jayaraman and Ness 2005).

Other advantages to alt-labour approaches are political and procedural. Alliances between unions and community organizations amplify the political reach of workplace organizing campaigns that need to exert public pressure on low-wage employers and industries in order to persuade them to comply with weakly enforced workplace organizing laws (Eaton and Kriesky 2009). In part, they do this by replacing the formalism and often alienating proceduralism of rules-bound National Labor Relations Board elections with organizing and political actions – spontaneous walk-outs from abusive workplaces, or snap delegations to law-breaking employers – that generate high levels of enthusiasm for participating workers (Garrick 2014).

However, the constituent components of these relationships are distributed unevenly across US cities. New York, Los Angeles, Chicago and other sites of movement development stand as increasingly rare locations of high union density. More than half of US states now have right-to-work laws that effectively undermine the capacity of workers to sustain labour unions. The large size of the urban areas currently generating alt-labour strategies also makes them numerically dominant in state legislatures whose reliance on urban votes and supports builds in clear channels of political access for community and labour organizations (Doussard and Gamal 2015). Just as urbanization produces a reliable legislative attentiveness to working issues in these states, the smaller size of urbanized electorates in Wisconsin, Indiana and other states increasingly hostile to labour repositions as antagonists the state assembles that appear as allies to labour organizers in existing accounts of alt-labour strategies.

As the successes of community-labour coalitions mount, the number of ad hoc local organizations and unions seeking to copy their mechanics continues to expand. The practical, operational question of *how* to do so thus takes on greater meaning. This question cannot be addressed from summary reviews of extant practices. Instead, it is to be resolved by adequately conceptualizing the day-to-day work in which they engage. Here, the entrenched focus on the goal of collective bargaining significantly limits the focus of scholarship. Developed under the historically anomalous institutional arrangements that defined post-war work, labour scholarship retains an explicit focus on the workplace as the core unit of analysis. But unions, manufacturing and commitment to the social wage have been in direct decline since the early 1970s (Harrison and Bluestone 1990). Two conceptual problems result. The first is that Fordist institutions themselves cover a progressively smaller portion of the economy, with the effect of collective bargaining and workplace-focused organizing proving increasingly difficult goals in terms of research and logistics (Doussard 2013). Among the many reasons why the recent Fight for $15 organizing campaign stands out, is its ability to reroute the focus of advocacy away from individual workplaces, and towards firms, employment law and regulators more broadly (National Labor Relations Board 2014; Vinik 2015).

Less frequently noted, the second limitation of focusing on the workplace may be more significant. While workplaces remain the key sites of labour negotiation, they constitute a limited portion of the lives of workers themselves. While the post-war settlement in the USA centred organizing and advocacy on the workplace itself, the social pillars of Fordism – including partriarchal nuclear families, working-class urban neighbourhoods and state maintenance of the social wage and collective consumption – were themselves in flux (Katznelson 1981; Lake and Newman 2002; McDowell 1991). Thus, labour advocacy today must contend not only with changes in the workplace and on the job, but also with changes in how labour reproduces itself *away* from the worksite: single-parent and multi-family households, suburbanization and the loss of neighbourhoods as sites of solidarity and, crucially, the withdrawal of federal support for human and social services, public education, public transportation and other programmes that allow workers to either meet the gap between pay and subsistence, or to withhold their labour altogether from jobs that fail to provide for life's basic needs (Lake and Newman 2002).

Re-focusing scholarship on the reproduction of labour, rather than labour itself, provides an important corrective to these oversights (Hanson and Pratt 1995; McDowell 2011; Peck 1996). For the purposes of organizing, a more systematic treatment of the pre-conditions of workplace advocacy comes from Herod (2001), who offers *labour geographies* as an alternative to economic and feminist geographies that focus on capital and the household, respectively. As a corrective, a labour geographies approach suggests shifting focus from the resources, plants and legal forums vital to capital, to the social, economic and institutional worlds of workers themselves.

This suggests two distinct tasks to be included in the project of assessing the potential expansion of alt-labour organizing techniques to places thoroughly dissimilar to New York, Chicago and Los Angeles. First, such an investigation should account for dissimilarities in social and institutional support for labour organizing between the sites in which alt-labour strategies originate and the sites to which they have been relocated. Second, evaluations of organizing in mid-size and de-industrialized cities – the affective 'ruins' of de-industrialization – should consider alternative measures labour and community advocates undertake to account for the absence of labour-supporting infrastructure.

Labour's fuller context: two post-industrial ruins in the Midwest

With very few exceptions, scholars of cities and labour pay scant attention to St Louis and Indianapolis. While growing interest in shrinking cities and abandonment in the US Midwest has strengthened interest in St Louis as a case of extreme population loss (Hackworth 2015; Tighe and Ganning 2015), this scholarship positions St Louis as an object capable of illuminating a broader trend, and not a site to be scrutinized on its own terms. Thanks to a history of entrepreneurial annexation and enthusiastic privatization, Indianapolis has been investigated as a potentially revealing site in which new forms of governance and policy are generated (Walcott 1999; Wilson 1989, 2007). As sites through which to generate urban theory, both cities are affectively ordinary and dissimilar to labour's historical centres in terms of union density, social service provision, public transportation and neighbourhood political organization (Doussard 2016). Each of these factors imposes extra barriers to labour activists seeking to organize, either in the workplace or politically.

The differences with the major sites of alt-labour strategic development stand out most directly on the issue of unionization. In 2014, New York, California and Illinois stood among the five most heavily unionized states, with urban workers accounting for a disproportionate share of state-wide union densities exceeding 15 per cent. By contrast, just 9 per cent of workers belong to unions in Missouri and Indiana (Hirsch and Macpherson 2017). These low figures are effectively vestigial, as both states passed a Right to Work laws often 2012. As union contracts grandfathered in from the prior era lapse, the increased difficulty of renewal under right-to-work should soon lead to a fast decline in union membership. In both cities, low union density limits the membership, resources and political connections on which alt-labour strategies elsewhere have productively drawn.

In each city, low union coverage stands as the most visible component of a legal and institutional system more systematically unfriendly to labour advocacy. In terms of both total not-for-profit organizations and aggregate organizational budgets, Indianapolis and St Louis are significantly smaller than in the main sites of alt-labour strategy development. As an activist in St Louis explained to the author, community-labour partnerships represent an attempt, however limited, to join labour advocacy to the kinds of basic service delivery that neither unions nor the state ensure at this point:

> In a sense, I kind of think the labour and community partnership is a return to that concept. It's become much more difficult to organize, because where workers live is so spread out. ... But if the unions continue to define their role narrowly in terms of representation campaigns, they're going to continue to shrink.
>
> *(Quoted in Doussard, 2016)*

Perhaps most significant, anti-union legislation and severe austerity in public programmes result from severely unfavourable state political conditions. Both cities lie in states with entrenched conservative legislatures. This results in activists consistently devoting their already-scarce organizational resources to policy issues requiring little such attention elsewhere. For example, activists in both states worked for years to secure Medicaid expansion, under the Affordable Care Act. While expansion represented a net financial gain for each state, and would have directly benefited low-income workers, annual organizing and advocacy that begun in 2010 did not secure expansion in Indiana until January 2015 (Radnofsky and Campo-Flores 2015). Missouri has yet to expand Medicaid. As an immigration activist explained, "our state legislature is just so insane. ... It really limits what we can do."

These limitations on organizational capacity are growing in tandem with the needs activist organizations attempt to address. In Indianapolis and St Louis, as elsewhere, the period following the 2007–2009 financial crisis has been characterized by growing poverty rates and the loss of mid-wage jobs (National Employment Law Project 2011). As state legislatures cut human services budgets in the 2010s, the human needs to be addressed with those budgets were expanding. This shortfall severely restricted workers' capacity to participate in burgeoning alt-labour experiments in St Louis and Indianapolis. As one social services provider memorably framed the problem: "Just think of the hierarchy of needs – if I don't know where I'm going to eat and sleep, am I going to go to the organizing meeting?"

"Climatizing to organize": how labour activists build the base

This growing gap between basic human needs and the capacity to fill them significantly circumscribes labour organizing potential in St Louis and Indianapolis. Each city possesses a fledgling community-labour coalition, built out of a small core of local union staff, advocacy organization members, scholars and lawyers. While those coalitions undertake some workplace organizing and policy advocacy campaigns, they do so primarily when international unions and national advocacy organizations initiate local campaigns. Recent examples include Fight for $15 fast food organizing campaigns, hotel worker organizing campaigns, and nationally funded efforts to pass living-wage, municipal wage and state-level minimum wage legislation.

This externally driven approach to organizing arises directly from limited internal capacity. Multiple organizers in each city indicated that the availability of external resources constitutes the main determinant of which campaigns they pursue. But external sponsorship also addresses the more fundamental problem of these campaigns' effective inability to identify, organize and mobilize populations of low-wage workers. With comparatively few neighbourhood-based organizations and limited social services support on which to draw, the taken-for-granted pre-conditions of community-labour organizing elsewhere do not hold. As a result, community-labour organizers do not just orchestrate campaigns in pursuit of individual workplace and political goals, they use their resources to create the pre-conditions on which such campaigns rest.

A senior labour activist in Indianapolis characterizes the movement's work memorably, as "climatizing to organize". Working through schools, churches and worker centres, activists in both cities seek to establish networks of workers and activists, and to develop conventions allowing them to collaborate (Doussard 2015). This represents an especially difficult task. Even in large, unionized cities, community-labour coalitions needed more than a decade of sustained cooperation to develop effective working conventions. And significantly, those successes built on the institutionalization of activism, either in the form of standing, funded organizations (Tattersall 2013), or vigorous community and neighbourhood organization activism (Milkman et al. 2010), or deep and long-established political connections between movement participants and elected officials in positions of power. With each of these types of institutionalization unavailable, activists in St Louis and Indianapolis instead turn to the more speculative work of bending incumbent institutions towards the same ends.

Churches

Labour activists have long valued churches for their ability to ground materialist economic claims in morally resonant messages that shift the discursive frame of organizing campaigns away from the zero-sum game of economic trade-offs (Bobo 2008; Sziarto 2008). Both within geography and the broader social sciences, religious institutions are understood in terms of values and political ideology. This is especially true in the USA, where politically conservative evangelical congregations have played a crucial role in organizing and supporting market-fundamentalist policies (Hackworth 2012).

Religious values, whether authentically held or strategically deployed, play significant roles in organizing in Indianapolis's Community, Faith and Labor Alliance and the St Louis-area Metropolitan Congregations United. Yet these values are functionally secondary to the role religious organizations play as interstitial institutions with territorial footprints spanning politically diffuse regions. In the sprawling, institutionally thin cities of St Louis and Indianapolis, churches (and to a lesser extent mosques and synagogues) solve the fundamental spatial and institutional problem of providing an infrastructure with which to reach workers across spatially diffuse neighbourhoods and suburbs. Taking advantage of these unique characteristics, organizers use religious institutions not as means of legitimizing or popularizing movements, but rather as humble spatial infrastructure through which to identify, meet and link workers (Cloke et al. 2016).

The work often begins with religious institutions' efforts to fill the basic human and social services gaps facing each city. As a labour-friendly church volunteer in St Louis explains, churches fill basic needs the state does not: "I'm not always able to provide what people need, but I can help connect them. Either some other church or a different group is probably able to help ... Most of the time it's basic needs: rent and utilities."

Religious organizations are heterogeneous in ways that complicate these roles and render their potential contribution to worker organizing uneven in space. In both cities, older, main-line religious organizations have proven the most accessible to community and labour organizers. In North Indianapolis, a series of main-line protestant churches endowed by the late pharmaceutical tycoon Eli Lilly – churches known as 'the Lilly pads', in reference to both their benefactor and the predominantly white, moneyed worshippers they bring to black neighbourhoods – participates actively in labour and organizing causes [Interviews, Indianapolis Organizers]. In St Louis, labour activists and elected officials alike ground advocacy discourse in references to their 'faith-driven' city. They coordinate religious outreach and organizing efforts through Metropolitan Congregations United, a social advocacy network with hundreds of member churches, synagogues and mosques, the most influential of which are drawn from old congregations in the City's politically and economically influential West End.

These religious organizations' contributions to alt-labour organizing draw on the elevated social status of their members. Members of wealthier city churches and synagogues possess significant political influence, and they often use it to direct elite and policy-maker attention to issues such as English as a Second Language programme funding, Low-Income Heat and Energy Assistance Program availability, food security and job safety [Interviews, St Louis and Indianapolis Religious Organization Members]. However, the low-income and undocumented immigrants who lie at the centre of efforts to organize the new economy, live elsewhere, in poorer, peripheral and far-flung suburban neighbourhoods in which religious organizations are both less well networked and less well funded. As a result, efforts to use religious institutions to coordinate systematic outreach in space, and to scale worker organizing up to the municipal level, continue to stall. An outreach volunteer in a suburban St Louis Parish serving many undocumented immigrant workers explains the restraints on this essential work:

> The undocumented population has a lot of needs. And they can't be served with public funds, which is our primary source of income. And let me tell you, they are vigilant about how you use those dollars. So we need to do local fundraising to make sure we have income that we can link to that population.

Despite these limitations, organizers seeking to work with undocumented and immigrant populations continue to prize religious institutions for the superior organization structure they provide in comparison to the alternatives. As a St Louis organizer explains, "When I started I tried to find other people to do this work with, i.e. I was just knocking on doors. It wasn't really happening."

Schools

Like religious organizations, schools provide a means to organize outreach and organizing, both through their dispersed spatial structure, and their organization into nested and place-specific political units. In principle, their spatial organization mimics that of the neighbourhood-based community organizations on which community-labour coalitions elsewhere depend (Doussard 2013; Milkman 2006, 2014; Shragge 2013). In each of these cities, that organizational infrastructure is lacking. Indianapolis has just ten neighbourhood community development organizations, most of which draw funding from the City and from elite foundations (especially the Eli Lilly Foundation) that do not prioritize social equity work. As a result, this small cohort of community organizations is less active in labour organizing causes than the numbers would suggest, as organizations avoid union and labour activism for fear of jeopardizing their funding

[Interviews, Indianapolis Community Organization Members]. In St Louis, community development organizations find their activities constrained by Aldermanic control of Community Development Block Grant funds, and the attendant devotion of those funds towards basic infrastructure and beautification, at the expense of service provision and organizing (Swanstrom and Guenther 2011).

In each city, community and union organizers have attempted to build relationships with schools as an alternate means of reaching and educating populations of low-income workers whose children were in attendance. The numerous community meetings schools hold on budget, safety and other issues provided the main vehicle for this outreach. Although speculative, these outreach efforts have proven effective elsewhere. In nearby Chicago, where the public school district had proposed and executed a plan to close dozens of neighbourhood schools, predominantly in low-income and African American neighbourhoods, community forums and hearings brought community and labour organizers into direct contact with neighbourhood residents, and catalysed organizing around interrelated issues of public spending, work, housing and policing (Gutstein and Lipman 2013).

However, the impact of school-based organizing in St Louis and Indianapolis remains limited, due to a number of factors. First, the lack of financial resources that limits community and union organizing capacity writ large, severely restricts the resources available for the intensive task of participating in multiple meetings across thousands of metropolitan schools [Interviews, Indianapolis and St Louis Community Organizers]. Second, the comparative thinness of neighbourhood advocacy organizations means that school meetings proceed without the clearly defined political issues and stakes taken for granted in larger cities [Interviews, Indianapolis and St Louis Community Organizers]. As a result, schools-based organizing in both cities was limited to a handful of organizers attending school meetings when time allowed.

Organizing conventions

A defining feature of community-labour coalitions in larger cities is the presence of standing institutions charged with brokering community and labour organizing efforts. Backed with permanent staff and dozens of labour and community members, these organizations – Make the Road New York, the Los Angeles Alliance for a New Economy, Chicago's Grassroots Collaborative – provide a stable basis for developing and evaluating organizing strategies and practices. They played central roles in recent organizing and policy victories in Chicago, New York and Los Angeles.

St Louis and Indianapolis do not just lack organizations of this type – they also lack the financial resources and potential membership base with which to establish them. The organizers themselves remain keenly aware of these limitations, and have set out to create social and organizational practices that deepen the close working relationships on which sustained alt-labour organizing depends. As is the case with schools, these practices remain speculative and experimental.

In Indianapolis, members of the community-labour coalitions in 2014 founded the Indianapolis Worker Justice Center as a means of building the basic relationships that will in time support workplace and policy campaigns (Quigley 2015). Interviews and the author's direct observation indicate two basic ways activists use the Worker Center to build these connections. First, the siting of the Worker Center on Indianapolis's Near West Side represents a deliberate attempt to territorialize advocacy in the neighbourhoods that are home to the city's growing population of immigrant workers from Mexico. In their strategies, flyering, door-knocking and organizing actions, Worker Center staff attempt to build a *neighbourhood-level* base of power through collaboration with migrant, hometown and legal services organizations catering to this population of vulnerable workers. Second, activists use the Worker Center to host soup kitchens, workers' rights trainings, film screening, anti-war meetings and legislative advocacy forums. Previously unavailable in a central location, these basic networking activities represent deliberate efforts to build a neighbourhood base and a city-wide network of organizations through which to undertake sustained advocacy.

In St Louis, many of these same functions are fulfilled by the city's Jobs with Justice affiliate. A national labour advocacy organization with thirty-eight local affiliates, Jobs with Justice provides institutional support for local chapters to coordinate labour advocacy. These local chapters are uneven in funding, constitution and scope: while the Central Indiana affiliate operating in Indianapolis lacks a full-time staffer, St Louis's Missouri Jobs with Justice has multiple staff members – one of whom is a minister donated by his church – weekly meetings and distinctive local conventions, such as slowing business at fast food franchises targeted in Fight for $15 strikes by paying for meals in pennies, one penny at a time. This affiliate organizes and carries out worker travel to the distant state capital to testify on public policy, and, in the words of an organizer deployed from elsewhere to organize fast food workers, it provides "the only table in town" at which to bring together community and labour organizations (quoted in Doussard, 2015).

In St Louis and Indianapolis, fledgling conventions and local institutions of this type represent the most fully realized efforts to create the routines, strategies and institutional capacity taken for granted in efforts to excavate effective community-labour organizing strategies in large, economically diverse and coastal cities.

Conclusion: the ruins as the future?

The basic, elemental difficulties labour organizing activists face in St Louis and Indianapolis represent substantial barriers to the expansion of community-labour models developed in large cities with distinctive labour histories, social services support and political power. Low union density, lightly funded social services and hostile state legislatures represent the norm for US cities, rather than the exception, but scholarship to date has remained focused on the fundamentally exceptional cities in which new organizing techniques have been pioneered. As a result, expanding the geography of labour activism will require not just the exporting and reproduction of past organizing campaigns, but also the deliberate adaptation of organizing strategies to address fundamentally different pre-conditions. The barriers faced by activists in St Louis and Indianapolis suggest several important components of any such adaptation.

First, community-labour organizing campaigns need to take into account the pre-conditions that make alt-labour organizing effective on the coasts and in Chicago. In the absence of strong unions, high-functioning neighbourhood organizations to provide services and political connections to legislators at higher spatial scales, community-labour organizers in smaller cities lack the basic context that makes workplace and legal organizing campaigns successful elsewhere. Here, a labour geographies analysis is both conceptually responsible and practically useful. Because organizing is conducted by workers and their institutions, and because it increasingly focuses on firms that lack the threat of capital mobility, focusing on the uneven geographies of labour and social reproduction, rather than the more familiar geography of capital, aligns empirical inquiry with the actual challenges facing the dwindling US labour movement.

Second, the funding and political limitations community-labour organizers face in smaller cities result from the more fundamental and stubborn lack of support institutions and political power. This means that alt-labour organizing campaigns need to focus on institutional creation and maintenance, rather than the campaign strategies and tactics that require strong institutional support. In St Louis and Indianapolis, experimentation with alternative institutional forms has been necessary but also necessarily uneven: while the social service delivery mission of churches has rendered them capable supporters of some organizing strategies, other institutions, such as schools, provide forums for organizing, without a mandate to tend to basic human needs. This suggests that the capacity of community-labour activists in smaller cities to compensate for the lack of institutional support for their causes will itself be uneven, and contingent upon existing religious, educational and organizing institutions.

Third, these limited examples suggest a need to more systematically inventory the uneven geographies of labour across US cities and regions. Deep and exacting written histories of social, public and work life exist for many, if not most, US metropolitan areas. But these are not easily convertible into the kinds of stylized facts and ready statistical comparisons that guide both labour organizing and the broader project of urban theory-building. As immigration, pro- and anti-labour reforms and changes in foundation support chart increasingly uneven geographies for US labour organizing, the long-neglected projects of mapping workers' social and everyday worlds, and of expanding urban theory beyond the large cities from which knowledge is produced, take on growing importance.

References

Bobo, K. (2008). *Wage Theft in America: Why Millions of American Workers Are Not Getting Paid, and What We Can Do About It*. New York: The New Press.
Clawson, D. (2003). *The Next Upsurge: Labor and the New Social Movements*. Ithaca, NY: Cornell University Press.
Cloke, P., May, J. and Williams, A. (2016). The Geography of Food Banks in the Meantime. *Progress in Human Geography*, pp. 1–24.
Dixon, M. (2014). Union Organizing and Labor Outreach in the Contemporary United States. *Sociology Compass*, 8(10), pp. 1183–1190.
Doussard, M. (2013). *Degraded Work: The Struggle at the Bottom of the Labor Market*. Minneapolis: University of Minnesota Press.
Doussard, M. (2015). Equity Planning Outside City Hall: Rescaling Advocacy to Confront Complex Problems. *Journal of Planning Education and Research*, 35(3), pp. 296–306.
Doussard, M. (2016). Organising the Ordinary City: Community-Labor Reform Strategies Travel to the U.S. Heartland. *International Journal of Urban and Regional Research*, 40(5), pp. 918–935.
Doussard, M. and Gamal, A. (2015). The Rise of Wage Theft Laws: Can Community-Labor Coalitions Win Victories in State Houses? *Urban Affairs Review*, 52(5), pp. 780–807.
Eaton, A.E. and Kriesky, J. (2009). NLRB Elections versus Card Check Campaigns: Results of a Worker Survey. *Industrial and Labor Relations Review*, 62(2), pp. 157–172.
Fine, J. (2006). *Worker Centers: Organizing Communities at the Edge of the Dream*. Ithaca, NY: Cornell University Press.
Garrick, J. (2014). Repurposing American Labor Law: Immigrant Workers, Worker Centers, and the National Labor Relations Act. *Politics & Society*, 42(4), pp. 489–512.
Gould, W.B. (2015). Organized Labor, the Supreme Court, and Harris v Quinn: Déjà Vu All Over Again? *The Supreme Court Review*, (1), pp. 133–173.
Gutstein, E. and Lipman, P. (2013). The Rebirth of the Chicago Teachers Union and Possibilities for a Counter-hegemonic Education Movement. *Monthly Review*, 65(2), p. 1–10.
Hackworth, J. (2012). *Faith Based: Religious Neoliberalism and the Politics of Welfare in the United States*. Athens: University of Georgia Press.
Hackworth, J. (2015). Rightsizing as Spatial Austerity in the American Rustbelt. *Environment and Planning A*, 46(4), pp. 766–782.
Hanson, S. and Pratt, G. (1995). *Gender, Work, and Space*. New York: Guilford Press.
Harrison, B. and Bluestone, B. (1990). *The Great U-turn: Corporate Restructuring and the Polarizing of America*. New York: Basic Books.
Herod, A. (2001). *Labor Geographies: Workers and the Landscapes of Capitalism*. New York: Guilford Press.
Hirsch, B.T. and Macpherson, D.A. (2017). Union Data Compilations from the 2016 CPS. Machine-readable database. Available at: http://unionstats.gsu.edu/CPS%20Documentation.htm [accessed 19 December 2017].

Jayaraman, S. (2013). *Behind the Kitchen Door*. Ithaca, NY: Cornell University Press.
Jayaraman, S. and Ness, I. (2005). *The New Urban Immigrant Workforce: Innovative Models for Labor Organizing*. London: Routledge.
Katznelson, I. (1981). *City Trenches: Urban Politics and the Patterning of Class in the United States*. Chicago: University of Chicago Press.
Lake, R.W. and Newman, K. (2002). Differential Citizenship in the Shadow State. *GeoJournal*, 58(2–3), pp. 109–120.
Lerner, S., Hurst, J. and Adler, G. (2008). Fighting and Winning in the Outsourced Economy: Justice for Janitors at the University of Miami. In: A. Bernhart, H. Boushey, L. Dresser and C. Tilly, eds. *The Gloves-off Economy: Workplace Standards at the Bottom of America's Labor Market*. Ithaca, NY: Cornell University Press, pp. 243–267.
Lichtenstein, N. (2014). Two Roads Forward for Labor: The AFL-CIO's New Agenda. *Dissent*, 61(1), pp. 54–58.
Logan, J. and Molotch, H. (1987). *Urban Fortunes: The Political Economy of Place*. Berkeley: University of California Press.
Luce, S. (2004). *Fighting for a Living Wage*. Ithaca, NY: Cornell University Press.
McCann, E. and Ward, K. (2012). Policy Assemblages, Mobilities and Mutations: Toward a Multidisciplinary Conversation. *Political Studies Review*, 10(3), pp. 325–332.
McDowell, L. (1991). Life without Father and Ford: The New Gender Order of Post-Fordism. *Transactions of the Institute of British Geographers*, 16(4), pp. 400–419.
McDowell, L. (2011). *Capital Culture: Gender at Work in the City*. Chichester: John Wiley & Sons.
Milkman, R. (2006). *LA Story: Immigrant Workers and the Future of the US Labor Movement*. New York: Russell Sage Foundation Publications.
Milkman, R. (2013). Back to the Future? US Labour in the New Gilded Age. *British Journal of Industrial Relations*, 51(4), pp. 645–665.
Milkman, R. (2014). Toward a New Labor Movement? Organizing New York City's Precariat. In: R. Milkman and E. Ott, eds. *New Labor in New York: Precarious Workers and the Future of the Labor Movement*. Ithaca, NY: Cornell University Press, pp. 1–22.
Milkman, R., Bloom, J. and Narro, V. (2010). *Working for Justice: The LA Model of Organizing and Advocacy*. Ithaca, NY: ILR Press/Cornell University Press.
National Employment Law Project. (2011). The Good Jobs Deficit: A Closer Look at Recent Job Loss and Job Growth Trends Using Occupational Data. Available at: www.nelp.org/page/-/Final%20occupations%20report%207-25-11.pdf?nocdn=1 [accessed 8 December 2017].
National Labor Relations Board. (2014). NLRB Office of the General Counsel Authorizes Complaints Against McDonald's Franchisees and Determines McDonald's, USA, LLC is a Joint Employer. Available at: www.nlrb.gov/news-outreach/news-story/nlrb-office-general-counsel-authorizes-complaints-against-mcdonalds [accessed 5 September 5 2014].
Nissen, B. and Russo, M. (2006). Building a Movement: Revitalizing Labour in Miami. *Working USA*, 9(1), pp. 123–139.
Peck, J. (1996). *Work-Place: The Social Regulation of Labor Markets*. New York: Guilford Press.
Peck, J. and Theodore, N. (2010). Mobilizing Policy: Models, Methods, and Mutations. *Geoforum*, 41(2), pp. 169–174.
Quigley, F. (2015). *If We Can Win Here: The New Front Lines of the Labor Movement*. Ithaca, NY: Cornell University Press.
Radnofsky, L. and Campo-Flores, A. (2015). Indiana Governor to Expand Medicaid Coverage. *Wall Street Journal*, 27 January. Available at: www.wsj.com/articles/indiana-governor-to-expand-medicaid-coverage-1422371729 [accessed 8 December 2017].
Reich, M., Jacobs, K. and Dietz, M. (2014). *When Mandates Work: Raising Labor Standards at the Local Level*. Berkeley: University of California Press.
Reynolds, D.B. (2004). *Partnering for Change: Unions and Community Groups Build Coalitions for Economic Justice*. London: M.E. Sharpe.
Shragge, E. (2013). *Activism and Social Change: Lessons for Community Organizing*. Toronto: University of Toronto Press.
Sonn, P. and Luce, S. (2008). New Directions for the Living Wage Movement. In: A. Bernhardt, H. Boushey, L. Dresser and C. Tilly, eds. *The Gloves-off Economy: Workplace Standards at the Bottom of America's Labour Market*. Champaign, IL: Labor and Employment Relations Association, pp. 269–286.

Swanstrom, T. and Guenther, K. (2011). *Creating Whole Communities: Enhancing the Capacity of Community Development Non-Profits in the St. Louis Region.* Public Policy Research Center, University of Missouri-St Louis.

Sziarto, K.M. (2008). Placing Legitimacy: Organising Religious Support in a Hospital Workers' Contract Campaign. *Tijdschrift Voor Economische en Sociale Geografie*, 99(4), pp. 406–425.

Tattersall, A. (2013). *Power in Coalition: Strategies for Strong Unions and Social Change.* Ithaca, NY: Cornell University Press.

Tighe, J.R. and Ganning, J.P. (2015). The Divergent City: Unequal and Uneven Development in St. Louis. *Urban Geography*, 36(5), pp. 654–673.

Vinik, D. (2015). Critics Were Completely Wrong about the 'Fight for $15' Movement. *The New Republic*, 20 May. Available at: www.newrepublic.com/article/121855/los-angeles-raising-minimum-wage-15-shows-fight-15-worked [accessed 8 December 2017].

Walcott, S.M. (1999). Bustbelt to Boomtown: Regime Succession and the Transformation of Downtown Indianapolis. *Urban Geography*, 20(7), pp. 648–666.

Wilson, D. (1989). Local State Dynamics and Gentrification in Indianapolis, Indiana. *Urban Geography*, 10(1), pp. 19–40.

Wilson, D. (2007). City Transformation and the Global Trope: Indianapolis and Cleveland. *Globalizations*, 12(1), pp. 29–44.

23
MOBILITIES AND MORALITIES OF DOMESTIC WORK IN INDONESIAN CITIES

David Jordhus-Lier and Debbie Prabawati

Introduction

Domestic workers constitute a significant part of the global workforce, but are seldom acknowledged as such. The experience of being a domestic worker and the role these workers take up in the societies in which they work is spatially variegated. While domestic work is always embedded in the micro-geographies of the household, the global domestic labour force is highly mobile and many, often young women, cross continents to work for other families. Domestic workers tend to migrate to urban labour markets in search of employment, but often find themselves culturally marginalized in foreign cities. This is why mobility and urban citizenship are both key concepts in which to understand domestic workers as subjects.

Based on recent research on domestic workers in Indonesian cities, this chapter will attempt to highlight a set of dynamics that speaks to the mobilities of domestic work but in ways that have received little focus in the literature. Our aim is threefold. First, we want to focus on migrant workers who do not cross a national border. Despite the magnitude of this group, most probably more than 10 million workers in Indonesia alone, internal domestic workers have somehow fallen off the radar of the human geography discipline. Second, we want to add nuance to the understanding of the urban for domestic work. We do so by looking at cities as senders, transit ports and destinations for domestic workers and ask how this informs the politics of domestic work in the city. Third, we pay attention to other groups of subjects constitutive of domestic work, particularly employers and recruitment agents, and discuss how their variegated mobilities shape the ability for domestic workers to claim citizenship.

The chapter will start out with a brief review of relevant academic scholarship on domestic work, with a particular focus on how this literature has framed domestic worker mobilities, on the one hand, and the role of the urban, on the other. We will describe two cases drawn from a comparative research project on citizenship and domestic worker organization in urban Indonesia, and discuss what these findings tell us about the possibilities for the political organization of domestic workers in the country.

Global geographies of domestic work

There is a small but significant number of academic publications on Indonesian domestic workers working abroad (e.g. Silvey 2004; Yeoh and Huang 1998; Yeoh *et al.* 1999), which place themselves in a broader academic literature on the transnational politics of migrant domestic labour (Anderson 2000; Bapat 2014; Chin 2003; Pratt 1999). In addition, international organizations such as Amnesty International, International Labour Office (ILO) and Human Rights Watch are important actors in mapping the extent and nature of domestic work worldwide. Because of the informal nature of most domestic work, and the often irregular practices of migration, much of this information is based on estimates.

Domestic workers represent a sizable workforce globally, estimated at 52.6 million workers in 2010, of which 83 per cent are female and a significant proportion under the legal working age (ILO 2013). This workforce is also becoming increasingly mobile, to the point where the demand for domestic workers is named as "the main reason for the mass migration of women from the southern hemisphere to cities in the North" by the ILO (2010: 9). The South-to-North notion is too simplistic, however, and human geographers have been instrumental in mapping and understanding particular migrant routes and their social implications, including research on Indonesian workers travelling to the Middle East (Silvey 2006), the experiences of domestic workers in Asian megacities such as Hong Kong and Singapore (Yeoh and Huang 1998), and of Filipino workers taking part in the Live-In Caregiver Program of the Canadian government (England and Stiell 1997; Pratt 1999). The socio-geographical implications of domestic worker migration are vast. Many researchers have justifiably taken a problem-oriented approach, focusing on the forced or unforced isolation of migrant workers in private homes, the stretching of nuclear families who see their mother disappearing for years (Pratt 2012).

Others have chosen to emphasize narratives of agency and opportunity, such as Williams's (2007) study of women on ships travelling from Eastern Indonesia, many of which were poor domestic workers whose mobility enabled them to renegotiate their identities and class status. The role played by remittances also serve to nuance our understanding of domestic work migration, as money sent home to families and regional economies in sending states. Whereas India, Mexico and the Philippines might be the top recipients of remittances, the Eastern regions of Indonesia can arguably claim another record – having the highest dependency level on remittances for family income (IRIN News 2009).

Despite the relatively rich literature on domestic workers in the human geography discipline and beyond, we would argue that the geographies of domestic work still suffer from certain omissions in the literature to date. Thus far, the academic treatment of domestic worker mobility has mainly focused on those who cross national borders. This bias invokes a moral geography wherein poor countries send domestic workers to rich countries, where they are being treated in unfair ways mainly due to their lack of legal citizenship status. This moral geography is mirrored in *realpolitik*, for instance in the occasionally strained diplomatic relations between sending and receiving countries, such as between Indonesia and Malaysia (Business Insider Australia 2015). However, millions of domestic workers also become migrants within their own countries, and this form of migration, like international migration, is also shaped by state policies and discourses, and invokes yet other moral geographies which complicate the categories of sending and receiving states. Another side effect of this bias is that it tends to ascertain the role of visitor, or foreigner, to the domestic workers. They are the ones who traverse space in search of happiness, but often find themselves 'out of place' in their work destination – in contrast to the employers who hire them. As we will show, this is not always the case.

Social scientists across disciplines have shown interest in how lack of legal citizenship circumscribes migrants' opportunities and segments labour markets for workers in precarious employment. This is what Goldring and Landolt (2011: 337–338) have labelled the work-citizenship matrix, in which a migrants' "shift to a more secure legal status may not necessarily be accompanied by a reduction in job precarity". This allows substantial inequalities to persist and deepen, both internal to each national economy and at a global scale, leading to what Castles (2011: 317) has termed "new hierarchies of citizenship". Domestic workers with college degrees leaving their country of origin to serve middle-class or elite households in mega-cities across the world epitomize these emerging hierarchies. Relative deterioration of class status can be painful, as migrant domestic workers find it difficult to recover previous occupational identities in their country of residence (Pratt 1999). Moreover, the mere fact that low-skilled and poor labour migrants lack citizenship in the country of work, makes them far easier to exploit (Castles 2011).

If we are to become more sensitive to the intra-national migration of domestic workers, and how this also reconfigures geographies of citizenship, we need to adopt a broader notion of 'the citizenship' concept itself. For many internal work migrants, the journey from the countryside to the city has substantial implications on their social inclusion and access to rights, legally and in practice. While the civil and political rights they are granted as citizens of a nation-state like Indonesia might remain intact, the rights of migrant domestic workers are unlikely to be substantiated unless they are actively acknowledged in the households where they work, and unless this large group of workers is acknowledged in the political community of the city. In short, we have to shift our focus from a more legalistic notion of citizenship to one which is sensitive to "the process of political engagement between diverse groups and individuals" in urban societies, to cite Painter's (2005: 7) definition of urban citizenship. Suffering from low social status and lacking regulation, domestic workers are typically among the most low-paid workers in highly unequal cities. Hence, they are firmly placed at one end of the differentiated urban citizenship that arises as a result of the coming together of two groups in many Asian cities: "the executive transnational elite in their protected communities of up-market housing [and] migrant workers in dormitories or peripheral or inner city ghettos" (Forrest 2008: 296). For workers who venture outside the household in which they work, cities are also public spaces which they have to negotiate as gendered and racialized subjects (Yeoh and Huang 1998). But while they might experience stigma and discrimination from employers and agents (Chin 1997), urban spaces are also sites of play and leisure where domestic workers can meet fellow expatriates from their home region (Williams 2007).

Our research on organizing domestic workers in Indonesian cities also reveals a more complex role of the urban in these migration patterns. In general, geographers have mainly portrayed cities as representing destinations for migrating domestic workers. True, multicultural world metropoles such as Hong Kong, Singapore and Toronto do offer international migrant opportunities of work in segmented labour markets. Such cities are indeed key destinations for female work migrants, not only due to their size, but due to their proportionally high levels of working middle-class women, creating a demand for paid domestic work (England and Stiell 1997). The paradoxical geographies of migrating domestic workers is well-formulated by Williams:

> [D]omestic workers cross the marginality of the region, locating themselves physically at the centre of global cities. There in the centre, they become marginal in their roles as domestics, while at the same time they occupy a central position in the family income generating activities at home on the margin.
>
> *(Williams 2007: 176)*

But cities are not only attractive destination labour markets. They should also be acknowledged as key nodes in more complex mobility networks of domestic workers. They are transit ports and they play an important coordinating function in sending regions. Often they play several of these roles at once in an urban hierarchy constituting a global labour market ladder of domestic work. In what follows, we will show that these different roles lead to contradictory discourses in particular cities, accommodating multiple moral geographies of work and exploitation.

Domestic work in Indonesian cities

In contrast to the literature on domestic worker migrants discussed above, academic scholarship on domestic workers working *in* Indonesia is very sparse, and mostly outside the human geography discipline. In recent years, international multilateral organizations have become engaged in the issue through campaigns and advocacy work. These tend to focus on the abject working conditions, widespread use of child labour and forced labour, and the abuse many domestic workers suffer from (Sheppard 2009), as well as the lack of legal protection (ILO 2013). This work has highlighted the apparent hypocrisy of the Indonesian government, being a vocal critic of the maltreatment domestic workers receive overseas without matching this concern with "an equally strong commitment to protecting domestic workers at home" (ILO 2006: 39).

There are long, historical continuities in the use of domestic work in Indonesia. Weix (2000) portrays Indonesian domestic workers as straddling between, on the one hand, feudal and familial employment relations and, on the other, capitalist wage work in a globalized economy. Key actors in the shift from traditional to contemporary employment practices are the recruitment agents. While domestic workers used to be recruited through kinship networks, commercial actors are increasingly performing this role in modern Indonesia. Recruitment agents might be state-licensed or unauthorized, and they perform a number of functions, from recruiting and releasing workers from their own families, to facilitating legal and illegal migrant flows, to supplying employers with domestic workers. Rudnyckyj (2004) also shows how recruitment agents discipline prospective domestic workers into subjects of modern servitude through training and skills development. Given the mobility imperatives facing Indonesian domestic workers – both those who work within and outside the country – some of these agents have become powerful actors, vis-à-vis local authorities.

According to Jala PRT, the largest NGO network devoted to domestic worker rights in Indonesia, there are more than 10.7 million domestic workers in the country.[1] Middle-class and upper working-class families across this vast country employ domestic workers. This creates a complex geography of sending regions, rural hinterland in general and Eastern Indonesian provinces in particular, and receiving regions, urban centres in general and the islands of Java and Sumatra in particular.

Domestic workers are also subject to the micro-geographies of the household, but in ways which differ between national contexts. In Indonesia, where workers traditionally were recruited through kinship networks, Weix (2000) argues that domestic workers express 'split subjectivities' by juggling their insider/outsider status as part of the home, but outside the family for which they work. This might set domestic workers working in Indonesia apart from Indonesians working abroad, as well as domestic work in countries such as South African or US households (see, for instance, Ally 2010; Varghese 2006), where the cultural distance between employer and employee tend to be greater. As we will show, the cultural and physical distance between employer and employee, or between public authorities and exploited workers, shape the moral geographies in question.

We have undertaken qualitative research on domestic workers and their organizations in five different Indonesian cities, including focus group research with domestic workers, interviews with policy-makers, bureaucrats and local activists and academics. Our findings in two of these cities, a provincial capital in Eastern Indonesia (Kupang) and the national capital of Jakarta, are particularly illustrating for the argument we present here. Whereas Kupang is located in a sending region, and has no domestic worker organizations, Jakarta is the destination of hundreds of thousands of domestic workers, who come to work for a transnational elite, and which has recently seen an embryonic organization of workers.

Contrasting moral geographies of domestic work in Kupang, East Nusa Tenggara

Kupang is the capital of the Nusa Tenggara Timur province (NTT). This Eastern province is among the poorest regions of Indonesia, and struggles with social and political problems such as illiteracy and corruption. The majority of the population are Christians, both Protestant and Catholic. Many people migrate out of the province in search of work, either to Java and other islands or abroad. Malaysia is the primary overseas destination for labour migrants, followed by Singapore and Hong Kong. While we were interested in the city of Kupang as a *sending node* for domestic workers to other parts of Indonesia, informants were quick to point out that Kupang acted as a *transit port* for domestic workers recruited in the rural areas of West Timor. On this point, both NGO activists and local government officials seemed to agree. Their opinions differed, however, on the legality of this migration and whether or not local government officials were implicated in trafficking practices or not. Some activists described Kupang as "the main place to exchange identity", referring to how prospective domestic workers from NTT get fake identity papers with new names and an "appropriate age". Some activists claim that local government officials have been actively involved in providing these new identities. The fact that many of the fake IDs seemed to have an address in a district of Kupang city was used as corroboration of this suspicion.

There are also two government-licensed private training centres in Kupang, where domestic worker recruits are taught how to cater for middle-class households, including English and Mandarin language training, how to use various electrical appliances and how to care for babies. These training centres are run by agents who also facilitate job placement in Hong Kong, Singapore or other Indonesian cities. In addition to these local agents, a number of recruitment agents based in Java have branches in Kupang, and engage in the recruitment of people in collaboration with government bodies.

While some of the recruitment undoubtedly is done in an orderly manner, with agents offering young women protection and sufficient preparation for their working experience outside NTT, we heard many stories of girls being sent on hazardous journeys without necessary information, resources and documentation. We also heard many stories of abuse and maltreatment of workers. Most of these stories were referred to as "human trafficking". In fact, several NGOs, government bodies and development cooperation projects were specifically devoted to the fight against human trafficking in NTT.

While the out-migration of domestic workers is an important dimension of domestic work in NTT, it is not the only one. Kupang city itself, with its 350,000 inhabitants, is also employing a significant, albeit unknown, number of domestic workers. Informants describe domestic worker employment practices in Kupang as mainly based on kinship or tribal relations. Domestic workers in Kupang typically live with their employers, and might instead of being fully wage-compensated perform household tasks in exchange for schooling or simply a place to live. The average domestic worker wage is estimated as lower than other regions of the country, and less than half of the regional minimum wage.

A striking observation when speaking to various informants in Kupang, was the contrasting ways in which locally based and migrating domestic workers were described. Migrants were seen, on the one hand, as vulnerable individuals who are taking risks in pursuit of happiness, and, on the other hand, as remittance senders and therefore potentially important contributors to local economies and facilitators of social mobility for their own children. Still, when discussing the problems of human trafficking, informants seemed to express implicit moral judgements over those who choose to leave NTT:

> Domestic helpers should go back to Timor, who are from Timor.
>
> *(Local activist 1)*

> If you ask someone to work as domestic worker, if it is in Kupang, they will refuse. But if they are offered to work as domestic workers in Bali, Surabaya, or Batam, they will gladly take it. To them it means they will travel to faraway places, different places, and with an airplane.
>
> *(Local activist 2)*

> How should we empower them, domestic workers, so that they are not sent overseas or elsewhere here in Indonesia?
>
> *(Local activist 3)*

> I said if we can protect the domestic workers [through regulation] then we will not become slaves in other countries because we appreciate them.
>
> *(Local activist 4)*

In some of these quotes, a problem-oriented attitude to out-migration is explicitly connected to the low wages and lack of protection of domestic workers in Kupang. But theses glimpses of reflexivity were exceptions from the norm. In most cases, and particularly when speaking to government officials, domestic worker employment cases in Kupang were depicted as unproblematic by our informants:

> So far in NTT, especially in Kupang, there have never been any problems with domestic helpers. ... It is more a question of social control. It is the task of society to make sure workers are treated well.
>
> *(Provincial government official 1)*

> In principle, domestic workers in NTT are different than on Java in general, or more specifically in big cities, because usually domestic workers here are also NTT people. We rarely hire those from Java. ... Since we are the same, and usually from the same tribe, for example Sabu people hire Sabu people, that eventually makes us treat them as our own.
>
> *(Provincial government official 1)*

> In general here in NTT, the workforce is not seen as a politically contested issue, but rather as an economic issue.
>
> *(Local activist 5)*

As an indication of the latter quote, there were neither any domestic worker organizations in the province nor any NGOs devoted to the rights of domestic workers. A representative of a women's crisis centre we spoke to, said that a few of the victims of domestic violence that sought help at the centre were domestic workers, but that she suspected this group in particular was under-represented, since domestic worker matters were seen as private matters in Kupang (see above quote on 'social control').

Another issue that most of our informants seemed to agree on was that paying the minimum wage to domestic workers was both unrealistic and undesirable. On this question, it became evident that government officials and middle-class activists alike were inhabiting multiple roles of interests. While they were politically involved in issues of women's rights and worker protection, they were also employers of domestic workers themselves. Hence, several informants stated that an introduction of a mandatory minimum wage would effectively drive a historic privilege out of the hands of the middle class, by making paid domestic work unaffordable. Female civil servants and other middle-class individuals we interviewed spoke of how difficult it was to juggle their professional commitments and expectations placed on them of being managers of a household.

By way of summary, domestic work in Kupang is understood as contrasting moral geographies. Domestic workers who leave the region are spoken of as vulnerable victims of opportunistic agents and corrupt government officials. Domestic workers who work in Kupang, however, are portrayed as a natural part of Eastern Indonesian society, embedded in cultural practices and protected by mechanisms of social control. Perhaps this distinction is understandable (not excusable), all the time the 'wrongdoers' in the human trafficking narrative are agents, local government officials and employers elsewhere, whereas the 'wrongdoers' in the local domestic helper narrative are the ordinary citizens of Kupang.

Organizing domestic workers in Jakarta expat communities

Jakarta is a very different city from Kupang, not only in its capacity as the national capital of Indonesia but also as one of the most populous metropolitan areas in the world. As Jakarta accommodates the national and regional headquarters of domestic and multinational companies, it constitutes a dynamic urban labour market which not only attracts thousands of people each year working in these companies, but also a lot of people who perform supply and support functions for these companies, for their employees and for their families. Business expats in Jakarta often arrive in the city with demanding professional commitments but without their social networks, and therefore rely on using domestic workers to perform household tasks. To satisfy this demand, poor women take up work with expat families in Jakarta – many of them are themselves migrants from rural areas of Java or other parts of Indonesia.

Among the most popular places to live for expats are the apartment blocks in South Jakarta, close to international schools and a relatively short commute from the central business district. Here, many expats tend to cluster together in residential areas, often inhabiting the same apartment complexes. Expats of particular nationalities also make use of a host of clubs and organizations, and these ethnic networks are crucial in embedding people quickly into social life in Jakarta. Expat life can also enjoy various online communities entirely devoted to offer information, facilitate discussions and mediate sales between expats.

The same networks are also useful for recruiting household staff and coordinating their employment. Because many apartment dwellers do not require a full-time live-in domestic worker, there is a need for coordination between different households over 'sharing' the same part-time staff. Informants told us of caretakers and other key people which functioned as domestic labour agents in particular apartment blocks, monitoring and managing household staff for their own ethnic community. The preference for ethnic homogeneity among expats is also projected onto their employment practices: online forums actively encourage expats to hire household staff from the same ethnic community to "maintain harmonious relationships" in the household (Living in Indonesia n.d.).

Domestic workers we spoke to told of mixed experiences with working for expats. On the one hand, expat employers were seen as paying better wages and offering more household aids, thus relieving manual work. They are also described as demanding, and even though domestic workers in expat households were required to speak English, there were both cultural and linguistic barriers to an open communication with their employers. Workers we interviewed made it clear that they viewed the expats as foreigners in Jakarta, in contrast to Indonesian employers and recruiters, who were referred to as "our people". Some nationalities had a reputation for being good employers, mainly countries with extensive worker rights protection in their own labour markets. Still, domestic workers expressed concern that many expat employers suspended their own normative assumptions of what decent employment should look like when living in Jakarta. In one case, an employer had referred to the lack of domestic worker regulations in Indonesian law when denying his staff pension payments.

It is in these close-knit expat communities that embryonic worker organizations have started to emerge among domestic workers. Encouraged and supported by the national domestic worker network Jala PRT, organizations such as Sapulidi Domestic Workers' Union have gone from zero to 150 members in a few years time (Gastaldi 2015). Because working for expats is seen as a good job for a domestic worker in relative terms, few workers we spoke to had thought of becoming a union member before they found themselves in a difficult situation with their employer. It was only when problems arose and trust was broken that the idea of seeking support or getting organized seemed to appear. Among the many problems domestic workers in South Jakarta told us of were financial penalties, harassment/sexual abuse and unpaid wages. Moreover, these workers were systematically denied work contracts and social security by their employers. We even were told stories of domestic workers who turned down offers of a formal contract from their employer, because they were afraid of sanctions from the employer or the police.

Workers who were involved in the early organizing efforts, emphasized how the networks of the expats provided a blueprint for their own organization:

> [T]hey talk to each other. Always. So, these Koreans have a group. And that's why I told my friend, "those Koreans have a group, why can't we domestic workers have one? Let's make an organization!"
>
> *(Union organizer 1)*

In a manner similar to their employers, the workers with support from the NGO network Jala PRT started using the apartment buildings as a basis for their recruitment efforts. In each apartment building, they recruit a 'community leader' who becomes in charge of representing existing members and recruiting new. This organizing model is similar to how unions in the hospitality industry recruit in large hotel facilities (Gray 2004; Tufts 2007). Sapulidi organizers 'map the community' of each block, and find out how and when the workers exit the house. In many ways, their organization is built on already existing social networks in these apartment buildings. Starting out as basic exchange of information between workers working in the same workplace, the organization now facilitates skills development like English courses and web courses.

Thanks to its close relationship with Jala PRT, Sapulidi has been able to refer workers to activists and lawyers who has settled problems members have had with their employers. This reputation has undoubtedly assisted its recruitment efforts. Given that Sapulidi organizes in expat communities, the cultural distance between employers and employees arguably makes it easier for workers to hide their political activism from their employer. Sapulidi attends regular meetings, often without the employers knowing, although this is more challenging for workers living with their employers. When asked about their experiences of joining the union, members of Sapulidi were eager to express a feeling of empowerment:

> At first I was not active. But once I got a problem, it made me realize that we should be together to become stronger. That's when the organization comes in handy, to help me with my case.
>
> *(Domestic worker 1)*

> Thanks to this organization, my friends now are able to stand up for themselves. We didn't have the guts before, including me.
>
> *(Union organizer 1)*

> Before joining this organization it was difficult to speak; now that we have joined, everytime we meet people we speak until we realize that we have talked too much.
>
> *(Domestic worker 2)*

> I used to be a helper. But now I am a worker.
>
> *(Domestic worker 3)*

Their progress notwithstanding, Sapulidi and other embryonic worker organizations in Jakarta still only organize a tiny fraction of the city's workforce, and face huge difficulties in representing domestic workers effectively. An important reasons is that domestic workers are still not seen as legitimate workers in Jakartan society and, consequently, that their representative organizations are still not seen as legitimate actors in industrial relations. Hence, Sapulidi is very weakly articulated with other trade unions. In fact, trade union members in Jakarta often employ domestic workers themselves, which raises serious questions around their solidarity for Sapulidi's members. Moreover the organization is still heavily reliant on the support of the Jala PRT activist network in order to have any media visibility or political influence. In other words, substantial urban citizenship is still a long way ahead for the domestic workers of Jakarta.

Conclusion

The importance of mobile labour for the growth of paid domestic work worldwide has already been acknowledged. But as we have tried to argue in this chapter, this is often done by ascribing relatively static and uniform roles to employers and employees, and to the cities themselves. The global mobility of labour is embedded in the urban in complex ways. Domestic workers use cities as starting points, transit ports and work destinations. So may their employers. Each city may play multiple roles simultaneously in the demand, supply and mediation of mobile labour. In the cases of Kupang and Jakarta, this variegates how different domestic workers' rights are realized – as workers and citizens.

Domestic worker rights in Kupang seem trapped in a moral geography between those who use the city as a transit port and those who seek work there. The former group is stigmatized as risk-taking, profitable and vulnerable job-seekers, but also as citizens whose well-being and protection is the moral obligation of political authorities and civil society. The latter group is rendered invisible in political life, and their exploitation is legitimized by reference to cultural practices and social code. Jakarta is an important place of work for an international elite of business people. Interestingly, it is among this group of mobile workers, who themselves lack formal citizenship, that an emergent domestic workers' movement has found the most fertile ground for recruiting members. By permeating expats' own tight-knit cultural communities, domestic workers have managed to build a small trade union in a hostile climate for workers.

Being subject to the atomized intimacy of the home is recognized as a key challenge for domestic workers worldwide. The socio-cultural proximity between employers and domestic employees in Indonesia is likely to create even stronger disincentives for rights-based or class-based politics. While we should be careful not to make sweeping generalizations based on the two cities presented in this chapter, it appears as if the physical and perceived distance that emerges with highly mobile employers, in the case of expats in Jakarta, or employees, in the case of labour migrants from Kupang, might be opening up political opportunities for domestic workers in urban Indonesia.

Note

1 According to national chairperson Lita Anggraini, cited in Jakarta Post (2014).

References

Ally, S. (2010). *From Servants to Workers: South African Domestic Workers and the Democratic State*, Scottsville, SA: University of KwaZulu-Natal Press.
Anderson, B. (2000). *Doing the Dirty Work? The Global Politics of Domestic Labour*. London: Zed Books.
Bapat, S. (2014). *Part of the Family? Nannies, Housekeepers, Caregivers and the Battle for Domestic Workers' Rights*. New York: Ig Publishing.
Business Insider Australia. (2015). Indonesia Wants to Stop Maids from Working Overseas after a Nightmare Incident Was Exposed in Hong Kong. *Business Insider Australia*, 17 February. www.businessinsider.com.au/r-indonesia-wants-to-stop-women-going-abroad-as-maids-after-abuse-media-2015-2 [accessed 7 January 2016].
Castles, S. (2011). Migration, Crisis and the Global Labour Market. *Globalizations*, 8(3), pp. 311–324.
Chin, C.B.N. (1997). Walls of Silence and Late Twentieth Century Representations of the Foreign Female Domestic Worker: The Case of Filipina and Indonesian Female Servants in Malaysia. *International Migration Review*, 31(2), pp. 353–385.
Chin, C.B.N. (2003). Visible Bodies, Invisible Work: State Practices Toward Migrant Women Domestic Workers in Malaysia. *Asian and Pacific Migration Journal*, 12, 49–73.
England, K. and Stiell, B. (1997). "They Think You're as Stupid as Your English Is": Constructing Foreign Domestic Workers in Toronto. *Environment and Planning A*, 29(1), pp. 195–215.
Forrest, R. (2008). Managing the Chaotic City? New Forms of Urban Governance and Challenges for East Asia. In: K.H. Mok and R. Forrest, eds. *Changing Governance and Public Policy in East Asia*. Abingdon, Oxford: Routledge.
Gastaldi, M. (2015). *Domestic Workers Organisation as a Tool to Reduce Social Exclusion: The Case of Domestic Workers Organisation in Java*. Master en Science de la population et du développement, Universite Libre de Bruxelles.
Goldring, L. and Landolt, P. (2011). Caught in the Work-Citizenship Matrix: The Lasting Effects of Precarious Legal Status on Work for Toronto Immigrants. *Globalizations*, 8(3), pp. 325–341.
Gray, M. (2004). The Social Construction of the Service Sector: Institutional Structures and Labour Market Outcomes. *Geoforum*, 35(1), pp. 23–34.
International Labour Office. (2006). *The Regulation of Domestic Workers in Indonesia: Current Laws, International Standards and Best Practice*. Jakarta: International Labour Office.

International Labour Office. (2010). *Decent Work for Domestic Workers. Report IV.* Geneva: International Labour Office.

International Labour Office. (2013). *Domestic Workers Across the World: Global and Regional Statistics and the Extent of Legal Protection.* Geneva: International Labour Office.

Irin News. (2009). Indonesia: Tough Times for Returning Labour Migrants. *Irin News*, 14 May. www.irin-news.org/report/84379/indonesia-tough-times-for-returning-labour-migrants [accessed 11 December 2017].

Jakarta Post. (2014). Modern-day Slavery Flourishing in RI, Says Report. *Jakarta Post*, 19 November. www.thejakartapost.com/news/2014/11/19/modern-day-slavery-flourishing-ri-says-report.html [accessed 14 January 2016].

Living in Indonesia – A site for expatriates. (n.d.). Practical Information. www.expat.or.id/info/info.html#Staff [accessed 7 January 2016].

Painter, J. (2005). *Urban Citizenship and Rights to the City.* Background Paper for the Office of the Deputy Prime Minister. Durham, NC: International Centre for Regional Regeneration and Development Studies.

Pratt, G. (1999). From Registered Nurse to Registered Nanny: Discursive Geographies of Filipina Domestic Workers in Vancouver, BC. *Economic Geography*, 75(3), pp. 215–236.

Pratt, G. (2012). *Families Apart: Migrant Mothers and the Conflicts of Labor and Love.* Minneapolis: University of Minnesota Press.

Rudnyckyj, D. (2004). Technologies of Servitude: Governmentality and Indonesian Transnational Labor Migration. *Anthropological Quarterly*, 77(3), pp. 407–434.

Sheppard, B. (2009). *Workers in the Shadows: Abuse and Exploitation of Child Domestic Workers in Indonesia.* Human Rights Watch. Available at: www.hrw.org/report/2009/02/11/workers-shadows/abuse-and-exploitation-child-domestic-workers-indonesia [accessed 25 May 2017].

Silvey, R. (2004). Transnational Migration and the Gender Politics of Scale: Indonesian Domestic Workers in Saudi. *Singapore Journal of Tropical Geography*, 25(2), pp. 141–155.

Silvey, R. (2006). Consuming the Transnational Family: Indonesian Migrant Domestic Workers to Saudi Arabia. *Global Networks*, 6(1), pp. 23–40.

Tufts, S. (2007). Emerging Labour Strategies in Toronto's Hotel Sector: Toward a Spatial Circuit of Union Renewal. *Environment & Planning A*, 39(10), pp. 2383–2404.

Varghese, L. (2006). Constructing a Worker Identity Class, Experience, and Organizing in Workers' Awaaz. *Cultural Dynamics*, 18(2), pp. 189–211.

Weix, G.G. (2000). Inside the Home and Outside the Family: The Domestic Estrangement of Javanese Servants. In: K.M Adams and S. Dickey, eds. *Home and Hegemony: Domestic Service and Identity Politics in South and Southeast Asia.* Ann Arbor: University of Michigan Press.

Williams, C.P. (2007). *Maiden Voyages: Eastern Indonesian Women on the Move.* Singapore: Institute of Southeast Asian Studies (ISEAS).

Yeoh, B.S.A. and Huang, S. (1998). Negotiating Public Space: Strategies and Styles of Migrant Female Domestic Workers in Singapore. *Urban Studies*, 35(3), pp. 583–602.

Yeoh, B.S.A., Huang, S. and Gonzalez III, J. (1999). Migrant Female Domestic Workers: Debating the Economic, Social and Political Impacts in Singapore. *International Migration Review*, 33(1), pp. 114–136.

24
STREET WORK AS A KEY SITE OF URBAN POLITICS

Ilda Lindell

Introduction

Street work in varied manifestations has become a global phenomenon (Bhowmik 2010; Cross and Morales 2007; McFarlane and Waibel 2016). Once believed to be a feature of cities in developing nations, street work has become a common element in many post-industrial urban societies, generating growing anxieties about the possible reversal of a well-established urban modernity. In cities of the Global South, these types of work were earlier seen as a transitory phenomenon and as the expression of an incomplete modernization. Contrary to expectations, however, such activities have only expanded and have in many cities become a widespread form of work (Lindell 2010a; Staples 2007). For example in some African countries informal employment is estimated at 70–80 per cent of urban employment, with trade constituting the largest sector (Vanek *et al*. 2014). This can be seen as part of a more general deepening of informality of urban living that, propelled by varied and contradicting forces, has become a dominant mode of urbanization (Roy 2005).

Despite academic recognition of the growing importance of informal street work in widely different urban contexts, these forms of work have remained at the margins of mainstream theorizing on urban politics. And yet, there are clear indications that varied forms of street work and related appropriations of urban space are increasingly at the centre of urban conflict in wide-ranging contexts. Examples in the Global North currently multiply. A well-known case is New York, where street vendors were reportedly being imposed a $1,000 fine for minor trespasses, as one of several measures to restrict and control street vending in the city (Wilmot 2014). In Swedish cities, at the time of writing, there is a mounting public angst around the emergence of street begging as a form of livelihood by EU migrants, something that has mobilized great attention in both urban and national politics. However, in such urban settings, the study of the politics of work in the city has traditionally prioritized issues related to more conventional forms of 'labour' and concomitant forms of representation, i.e. trade unions, and given less attention to other kinds of politics and other organizational forms that might emerge from street work.

In the cities of the Global South, where street-based livelihoods have often had a perennial existence, these activities have attracted long-standing scholarly attention, and this substantial body of research can offer insights that may be of relevance for understanding emerging street work in other contexts. But in this diverse body of research, the political dimensions of this type of work have seldom been explicitly addressed and have often been insufficiently problematized.

More generally, the politics of informality has often been understood in antagonistic terms, often emphasizing a politics of exit and avoidance or a politics of opposition (for a discussion, see Lindell 2010a, 2010b). Informed by a long-standing conception of informal practices as autonomous from and circumventing the state, many have described how such practices undermine state control and have far-reaching effects (see for example Bayat 2004; Centeno and Portes 2006). Engagement with the state is seen as inevitably oppositional. Recurrent conflicts in many cities between street workers and state actors seem to confirm the relevance of these perspectives. As growing numbers occupy sidewalks and interstitial spaces in the city, governing powers intervene in unfavourable ways to control these populations (Crossa 2009; Swanson 2007). These collisions have been described as a manifestation of "conflicting rationalities" opposing the mindsets of planners and elites to the survivalist needs of disadvantaged urbanites (Watson 2009). But these conceptual oppositions hide from sight the interactions and negotiations that often take place between seemingly opposing sides, as well as conflicts and divisions among practitioners of informality (Lindell and Ampaire 2016).

Other perspectives have emphasized instead the vertical networks that link together political elites and regular citizens, through which the state exercises considerable influence in segments of economy and society seemingly beyond its control (see Meagher 2009 for a discussion and Gay 2006 for a critique). Indeed, street workers and their representatives may engage in considerable informal political networking with influential politicians in their search for protection and influence (Prag 2010), particularly in contexts where formal channels of political participation are inaccessible or bypassed. However, given the often marginalized condition of these workers, they are frequently required to make use of a range of political strategies. The forms of political interaction that they engage in cannot be captured by frames of understanding that focus either on oppositional practices or on clientelistic connections. Rather, they enact what I have called elsewhere an "untamed politics of informality" (Lindell and Ampaire 2016) that makes use of multiple political practices to create space for negotiation, advance their positions or merely keep a foothold in the city. In this respect, the flexible tactics employed by their organizations tend to differ from the conventional strategies of traditional labour organizations (Lindell 2011a).

One additional limitation in current theorizations is that the politics of urban informality is often understood as something 'local'. Although many recognize the great exposure of informal and street workers to global forces (Cross 2007; Roy 2005), addressing the challenges that they face are normally discussed exclusively in terms of 'good' urban planning and 'good' urban governance. The agency of informal workers is usually depicted as a version of the 'localist traps', bound to the local, focusing mainly on practices of encroachment and of territorial transgression in the city (see for example Bayat 2004). Such local agendas and struggles are of course very important means by which people negotiate their existence in the city, but one should be cautious about reproducing understandings of the city as self-contained and of urban politics as something exclusively local. While many street workers' organizations remain quite localized, a share of them participates in associational networks that stretch far beyond their cities and countries (Lindell 2009, 2011b). This globalizing agency may have a bearing on the politics of their city and shape these workers' subjectivities.

This chapter asserts informal street work be understood as an important constituent of urban politics. 'Street work' is here understood as income-generating activities that involve the use or occupation of street or open urban spaces. The chapter aims to examine the *nature* of the politics of street work, through an empirical analysis of vending and of the practices of a vendors' organization in Maputo, Mozambique. The chapter attempts to move beyond the limitations identified above and argues that the politics of informal work is more complex than often considered. First, this politics will be described as being *multifaceted*, in terms of the multiple modes of political (inter)action employed by the vendors' association, which encompass both oppositional and clientelistic practices. Second, it is a *multilayered* politics because, rather than expressing itself in a simple dichotomy between the 'governed' and the 'governing', this politics takes place in and through multiple loci of power, including forms of authority beyond government and within the urban informal economy. This multilayered politics involves exclusions and divisions between different groups of street workers – as will be uncovered in relation to market and pavement vendors, as well as men and women. Third, the politics of informal work can be seen as a *multiscalar* politics, illustrated in this case by participation in a transnational associational network. Adopting a conception of place and local politics as being constituted by both relations of proximity and distance (Amin 2004; Massey 2004), the chapter discusses the political possibilities that such participation uncovered as well as the unequal gender practices and the new tensions it entailed. The chapter is based on long-term research in Maputo and several periods of fieldwork during which a large number of vendors, market and association leaders as well as state actors were interviewed.

The chapter proceeds with a short description of how vending opportunities have been evolving in a city undergoing dramatic transformation, including the removal of vendors from major avenues in central Maputo. This is followed by a discussion of the convoluted relations between a vendors' association and the city authorities, including struggles over urban space for vending activities, contests over the collection of revenues and competition over the governance of market spaces. The chapter then turns to the political strategies of the association to strengthen its influence, particularly its considerable political networking with influential political figures. Lastly, the chapter examines the experiences and political implications of participation in a transnational network.

A multifaceted and multilayered politics

Vending on the streets and in unplanned markets has long existed in Maputo but gained momentum from the 1980s on, in tandem with a pronounced informalization of the urban economy. People of highly diverse backgrounds today make a living in the so-called informal economy – redundant wage workers, rural migrants, unemployed youths, de-mobilized soldiers after the end of a long civil war, among others. A considerable share of them pursues a livelihood through vending in the city. Some operate without a stall, on sidewalks, roadsides and at the edges of markets – whom I will call 'pavement vendors', to distinguish them from 'market vendors', i.e. vendors with a stall. Others occupy vacant city spaces more permanently, putting up stalls and erecting self-built markets. These markets may gather thousands of vendors and consumers and become key nodes of intense economic activity in the city. Official attitudes have oscillated between indifference and forced displacement, but have generally been guided by negative views of these activities. Self-built markets, in spite of their importance, have often been treated as illegal or temporary. Among the vending population, pavement vendors have been additionally exposed to these unfavourable attitudes as they tend to occupy particularly visible spaces in the city. Pressures on this group have further intensified in recent years, in the context of renewed desires to improve the city's image and of a boom in speculative foreign investment. Large-scale residential and infrastructure projects have induced the relocation of hundreds of vendors from lucrative locations to inviable spaces (Lindell 2017). In the central areas of the city, pavement vendors have been chased from major avenues, as their work is seen as inappropriate for these noble city spaces. Their access to valuable vending spaces in the city centre is thus becoming more restricted and controlled.

The long-standing unfriendly attitude of the city authorities towards vendors prompted the creation of a vendors' association in the late 1990s (with the help of a national trade union centre). A major area of concern and work by the association pertained to the official recognition of vending as a legitimate form of urban work and the protection of urban vending spaces occupied by vendors. Indeed, in the past, the association attained some successes and made itself visible in local politics. On several occasions it was able to mobilize large numbers of people from markets across the city to contest the relocation or eviction of marketers (Lindell 2008). On other occasions, however, the association was merely able to mitigate the worst effects of official interventions or to negotiate unfavourable alternatives – often exacerbating tensions around issues of representation within the association (Kamete and Lindell 2010; Lindell 2017). In addition, although association leaders claimed that the organization represented urban vendors in general, in practice it tended to favour the concerns of market vendors, compared to those of pavement vendors. Market vendors often blame the more mobile pavement vendors for posing unfair competition – although the line between these two categories is fluid as many market vendors abandon their stalls for the street when their profits wane. Unsurprisingly, interviewed association representatives in some markets were explicit about disallowing pavement vending around the edges of their markets. Importantly, association leaders also explained that they collaborated with the municipal authorities in the removal of pavement vendors from key avenues in central Maputo and in their relocation to less visible sites. Since closeness to a large number of pedestrians is critical for these vendors (Lindell and Ihalainen 2014), the practices of the association towards this group potentially exclude and aggravate the already precarious conditions of a large group of vendors.

Besides claiming to represent vendors in the city, the association plays a key role in the governing of market spaces (Lindell 2008). The association was structured as a sort of federation of market committees in a number of city markets, whose leaders met on a regular basis at the association's headquarters. An executive committee was responsible for coordinating activities at city (and national) level and for negotiating with government when issues arose – although its legitimacy would be gradually undermined by an overly centralized leadership style among other things. Market committees became the institutions that carry out the day-to-day management of their respective markets: they allocate vending space, coordinate cleaning in the markets, provide some infrastructure and security, establish norms of conduct, resolve labour conflicts and other disputes, etc. Governance practices varied between markets but generally, market committees commanded considerable authority over the vendors (including the authority to apply sanctions). Sometimes they assumed a quasi-feudal character, whereby market committees enjoy a virtually unchecked authority and create their own sources of revenue (based on regular fees charged to all vendors and in some cases unsanctioned payments for stall allocation). Vendors may appear to accept this state of affairs, particularly if they see their market leaders as their sole protectors. But there have also been instances where vendors have contested the authority of their leaders (for example with accusations of embezzlement and failure to defend their interests) (Lindell 2008). The key point is that these associational structures constitute the de facto authority governing market spaces and relations within them.

The associational management of the markets was for long exercised mainly in antagonism with the local government. The latter was virtually absent in terms of actual regulation and provision of services and infrastructure, but was nevertheless reluctant to accept this new authority in the markets. A major feud developed around the charging of fees to vendors. While the association instructed vendors to pay their fees to the municipal collectors, presumably in exchange for the right to use an occupied space, the council contested the legitimacy of the association to charge its own fees – it argued that it undermined the ability or willingness of the vendors to pay the municipal fees, obviously without taking into account the creative accounting practices of municipal collectors. From the vendors' perspective, while fees paid to the market committees often contributed to some improvements in the market, vendors often wondered why they should pay fees to the municipal council who did not invest back in the markets. Recurrent crises ensued, including cases of physical violence and imprisonment of market representatives in the past (Lindell 2008). This contestation over the right to tax came to be partly resolved through an agreement reached between the local government and the association: the association was entrusted with collecting a single market tax, on behalf of the local government who would then transfer 10 per cent of the collected amount back to the association. This was an important concession by the local authorities, even if in practice the agreement has not always worked as expected. The association leaders regarded the agreement as a milestone in their relations with the local authorities.

The association sought to increase its political clout through various means. Membership in a trade union federation – which I have discussed elsewhere (Lindell 2008) – is one such strategy. However, a supportive but weak labour movement appears to have had limited effectiveness. Personal links with influential politicians were apparently regarded as more important. Indeed, a closer look at the politics of the association revealed the importance of political networking and patronage as a means to achieving some modest protection for its constituencies and support for its operations. In a political context where opportunities for formal political participation by marginalized groups are scarce, influence may instead be sought through informal channels. Association leaders have, from the start, invested in close relationships with the political elite, particularly with influential individuals in the political party that has, since independence and to the time of writing, ruled at both national and local levels. On several occasions, the association called upon the general secretary of the party – including a public meeting with the participation of hundreds of vendors – to curb the unfavourable conduct of the city council (Lindell 2008). This was perceived as an appropriate strategy as, in spite of previous decentralization reforms, centralized practices of authority and accountability have persisted. Occasionally, association leaders also nurtured relations with individual candidates to the municipal president post. Prior to the local elections, promises of mutual support were exchanged and expectations grew about an eventual partnership in the future – although the outcome of the electoral process would turn out differently from expected.

This strategy of close connections with elements of the political elite had a clear expression at the level of the markets (Lindell 2008). Not only did association leaders campaign for the ruling party and specific candidates in the markets, membership in the party was common among leaders at various levels in the association, including among members of market committees. This close identification with the ruling party had double-edged implications. For the association, it was referred to as the strongest card to play when under attack and as a means for creating some space for negotiation during unfavourable interventions. But as association leaders are compelled to negotiate compromises for the sake of maintaining such relations, they also risk alienating those vendors who bear the consequences of such deals. Such close relations of patronage potentially facilitate for the ruling party to (re)gain access to the vast constituency of vendors – what is often referred to as 'vote bank politics'. Ultimately, these practices may work against political renewal and the emergence of other forms of work politics.

Transnational agency and local politics

The politics of the vendors' association is not circumscribed to the local arena. A few years after its birth, the association joined a transnational network of vendors' organizations, which gathers membership-based associations and unions across the regions of the Global South (Lindell 2009, 2011b). The network takes the form of a formalized structure with elected bodies, a constitution and its own policies on several areas. The conditions of the most vulnerable groups of vendors are a stated priority. The network organizes international workshops and congress meetings for member organizations and facilitates exchange visits between them. Representatives of the association have had the opportunity to participate in several of these international conferences, in Africa and in other continents, as well as in exchange visits. At international meetings, vendors' organizations from a wide range of geographical contexts share their experiences and learn how other organizations deal with challenges often similar to those they face in their own cities. Exchange visits facilitate mutual learning between organizations and provide an opportunity to gain a deeper insight into how another vendors' organization works and its environment.

The association participated in at least one such exchange visit, with a vendors' organization based in South Africa (Lindell 2009). The delegates that visited South Africa reported how they were amazed about the developmental efforts of the local government towards vendors and the vending facilities that were provided. They were also impressed with the spirit of dialogue and respect that existed between the local authorities and the visited vendors' association – something that the Maputo association was yet to experience. This exchange visit opened for them new visions of the imaginable and they returned home with new plans. Back in Maputo, the delegation requested a meeting with the local government officials to inform them about the positive approach of the urban authorities in South Africa and about what could be achieved through good dialogue with a vendors' organization. They had concrete proposals to present on how to make Maputo a more hospitable city for vendors. In fact, this international experience seems to have strengthened their assertiveness in relation to the local government.

This was perhaps not unique to this association. A leader of the international network reported that participation by member-organizations in international visits and activities boosts their self-confidence during interactions with their city governments (Lindell 2009). This leader explained that the latter often look upon vendors as illiterate and incompetent and become impressed when vendors show knowledge of developments in other countries and thereby begin to take them more seriously. In fact, the primary motivation for the association to join the network seems to have been, as for many of the other member organizations, to obtain international support for their local struggles. Not only does the network have an explicit policy of supporting its affiliates in their local battles, member organizations are also able to mobilize support from partners in the network in situations of crisis – through support letters for example. This international solidarity makes city governments aware that information about unfavourable interventions circulates abroad and that their targets are not an isolated group but part of a transnational network (McFarlane 2008; Mittulah 2010).

Another telling example of the potential importance of these transnational linkages for local politics is worth noting. When the South African delegation visited Maputo as part of the exchange visit, besides giving interviews to the local media, it also participated at a meeting with the local authorities: they were outspoken and, among other things, they 'lectured' municipal officials that they should use the sizeable revenues that were collected from vendors to provide facilities for them. The pressure exercised by the visitors reportedly made an impression on the city officials (Lindell 2009). The long-standing political effects of these international interactions in Maputo should not be exaggerated, however, given that the local government is exposed to other, often more powerful, pressures and interests. But intangible effects on participants' self-perceptions and subjectivities seem to be important, shaping their identities, projects, discourses and visions.

While participation in the transnational network seemed to open new political possibilities, there is more to this story. First, there were constraints limiting the international mobility of association members, many of which lack material resources, passports, etc. So, participation in international activities was restricted to a limited number of people – mainly those occupying leadership positions in the association (Lindell 2011b). This also means that participation was mediated, in the sense that the few who attended network meetings and handled information exchange in the network were in a position to interpret and filter (and even withhold) information and policies emanating from the network's leading bodies and meetings, before it possibly reached regular members.

Partaking in the international activities of the network also had a gender dimension to it. This reflected a skewed gender structure in the association: while women constituted the largest share of the constituency, they were poorly represented among the leadership bodies. Parallel structures existed within the association to deal with 'women's issues' – the women's desk (at central level) and the women's committees (at the market level) – but leading women in these structures expressed that they aspired to greater autonomy and influence. Unsurprisingly, women's participation in international activities seemed particularly low, at least in the first years. In addition, the (largely male) association leadership was selecting which women to send abroad, sidelining for example the then city-level coordinator of the women's department. These practices of gendered gatekeeping, as I have called them elsewhere (Lindell 2011b), created tensions within the association. These gender differences in international involvement reflected unequal gender structures within the association – as mentioned, leadership positions were male-dominated and the women's department for example lacked sufficient autonomy to carry out its own activities and projects and access to the association's resources. The transnational network, on the other hand, has an explicit gender policy – this included an overt intention to strengthen women's leadership capabilities and gender quotas for elected bodies and attendance to international meetings. As the network leadership became aware of the above-mentioned gender discrepancies in international participation, it began to insist on female representation from Maputo.

Gender tensions would further accentuate in the context of interactions with the South African vendors' association (Lindell 2011b). The latter was a women's association with a conscious aim to empower women. Its representatives remarked the pronounced gender hierarchies in the Mozambican association. They reported that they were unable to meet only with the women and that at meetings the female members left the speaking to the men. The South African delegation confronted the Maputo association leaders on gender issues, speaking about 'women's rights' and women's leadership. These different attitudes to gender and women's role in associational structures generated further strains within the association and between the two associations.

Furtively and outside the formal meetings between the two associations, some of the Mozambican female leaders would approach the South Africans to consult them on possibilities to form a women's-only organization. Interviewed later on, however, women leaders in Maputo would often regard the creation of an autonomous association as unfeasible or even counterproductive in their current cultural and political context. In their view, male and female vendors faced the same problems and should thus struggle together for their recognition as workers.

Conclusion

This chapter gives centre stage to the urban politics of street work. Conflicts and negotiations around street work are arguably a central feature in the politics of most cities of the Global South, where the bulk of contemporary urbanization is occurring, but they are also of increasing importance in many post-industrial cities where this kind of work is on the increase. In particular, the chapter attempts to uncover the nature of this politics, highlighting its *multifaceted*, *multilayered* and *multiscalar* character in a particular urban setting. The configurations of this politics are of course shaped by the histories and political cultures of particular places. But some of the findings may partly resonate with political dynamics of street work in other places, particularly in urban settings where informality of work is pervasive and where the political marginalization of such populations is deeply entrenched. More generally, the example of Maputo suggests that the actual political practices and relations at work are not easily captured by available approaches to urban politics and call for a broader canvas that can encompass a wider range of relations, forms of work and governance of relevance for political dynamics of labour in the vast cities of the Global South and beyond.

The *multifaceted* nature of the politics of street work can best be illustrated by the multiple modes of political interaction between the vendors' association and the bureaucratic and political elite, which defied categorizations into either antagonistic or convivial relations. Antagonistic elements were clearly present. These were manifested in struggles around displacement and access to urban space for vending, around recognition of vending as a legitimate form of urban work, and around the governance of markets and their populations, particularly concerning taxation issues. But this only accounts for one side of these relations. The association representatives also cultivated links with elements of the political elites. While deeply unequal, these relations involved a degree of mutual dependence. Politicians generally disapprove of informal vending, yet need the political support of these crowds and the association has provided a channel to achieve this. The political importance of these workers lies not only in the large 'vote bank' they constitute, but also in the awareness that they are able to mobilize in large numbers. The above mutual dependencies and vertical relations have sometimes opened some space for negotiation and for some concessions from the authorities – which, for example, helped resolve the previous violent conflicts concerning the taxing of vendors. The relations between the association and the political and bureaucratic elites can thus best be described as ambivalent.

The politics of street work described in this chapter is also a *multilayered* politics. In the virtual absence of state regulation, the association is the entity that de facto regulates vending spaces and their occupants. It thus constitutes in itself a regulating power, partly resembling the state and operating through a more or less top-down logic, albeit with episodes of internal dissidence and contestation. This regulating force generates its own dynamics of power and exclusion. As described above, it discriminated against pavement vendors and facilitated their removal from central avenues in Maputo. Gender inequalities were also visible in the association. These power relations and divisions at grassroots level are relevant for understanding the multiple forces shaping the conditions and political possibilities for different groups of street workers.

Participation by the association in transnational networks speaks to the *multiscalar* dimensions of street work politics and challenges usual assumptions about an urban politics of informal work as played out exclusively on the local level. Such participation potentially opens new political possibilities – as exemplified by the intervention of the visiting South African association with the Maputo authorities and by instances of transnational solidarity between associations. These international encounters shape participants' subjectivities and visions in ways that transcend local frames of reference. Being part of the network seems to be a source of confidence, knowledge, inspiration and new ideas, and may even promote the endurance of workers' associations in adverse political settings (Lindell 2009).

However, transnational participation and its eventual benefits cannot be taken for granted, particularly for urban groups who face multiple and heavy constraints (on mobility, on collective organizing, etc.). The varying spatial reach of these connections thus should be in itself an element to consider in analyses of urban politics of informal work. It raises the question of *who* is being empowered by these transnational links. In the studied association, involvement in transnational activities was limited to a small number of individuals as well as mediated – similar to what has been found in other global networks with participants from the Global South (Routledge 2008). It is improbable that the association's transnational engagement was of any assistance to pavement vendors when they were faced with displacement from strategic locations in central Maputo. Participation in transnational activities also had a gender dimension, manifested in practices of gendered gatekeeping that potentially keep women 'in place' (Lindell 2011b). The tensions that emerged from such practices are also part of this extroverted politics of urban informal work. Those tensions would subside as livelihood concerns gained precedence over gender equality. This was a trade-off that the women perceived as necessary but one that consolidated men's position at the helm of the organization. These gender relations and dynamics are arguably a central aspect of street work politics, given that women constitute in many places a large share of the urban vending populations, particularly among groups operating at the lower income levels.

While the association has on several occasions played a critical role in defending vendors' right to a livelihood in the city, it fell short on several respects. Besides exclusionary practices towards pavement vendors and leaving gender hierarchies unaddressed, the association glossed over employer–employee relations in vending operations. In spite of its transnational engagement, there was little internal reflection about the global forces shaping the conditions of its constituencies or attempt to intervene in the commodity chains within which many of its members operate. The nature of its engagement with political elites as described above may contribute to the perpetuation of a politics of patronage that require compromises that may ultimately disadvantage street workers. A progressive politics of urban informal work, as we understand it, would entail a proper consideration of these limitations and of the domains discussed in this chapter, among others.

Acknowledgements

This research is partly funded by the Swedish Research Council (grant 2015-03474).

References

Amin, A. (2004). Regions Unbound: Towards a New Politics of Place. *Geografiska Annaler: Series B, Human Geography*, 86(1), pp. 33–44.
Bayat, A. (2004). Globalization and the Politics of the Informals in the Global South. In: A. Roy and N. Alsayyad, eds. *Urban Informality: Transnational Perspectives from the Middle East, Latin America and South Asia*. Lanham, MD/Boulder, CO/New York/Toronto/Oxford: Lexington Books.
Bhowmik, S., ed. (2010). *Street Entrepreneurs in the Global Urban Economy*. London: Routledge.

Centeno, M. and Portes, A. (2006). The Informal Economy in the Shadow of the State. In: P. Fernández-Kelly and J. Shefner, eds. *Out of the Shadows: Political Action and the Informal Economy in Latin America*. Pennsylvania: Pennsylvania State University Press.

Cross, J. (2007). Pirates on the High Streets: The Street as a Site of Local Resistance to Globalization. In: J. Cross and A. Morales, eds. *Street Entrepreneurs: People, Place and Politics in Local and Global Perspective*. New York: Routledge.

Cross, J. and Morales, A., eds. (2007). *Street Entrepreneurs: People, Place and Politics in Local and Global Perspective*. New York: Routledge.

Crossa, V. (2009). Resisting the Entrepreneurial City: Street Vendors' Struggle in Mexico City's Historic Center. *International Journal of Urban and Regional Research*, 33(1), pp. 43–63.

Gay, R. (2006). The Even More Difficult Transition from Clientelism to Citizenship: Lessons from Brazil. In: P. Fernández-Kelly and J. Shefner, eds. *Out of the Shadows: Political Action and the Informal Economy in Latin America*. Pennsylvania: Pennsylvania State University Press.

Kamete, A. and Lindell, I. (2010). The Politics of 'Non-planning' Strategies in African Cities: Unravelling the International and Local Dimensions in Harare and Maputo. *Journal of Southern African Studies*, 36(4), pp. 890–912.

Lindell, I. (2008). The Multiple Sites of Urban Governance: Insights from an African City. *Urban Studies*, 45(10), pp. 1879–1901.

Lindell, I. (2009). 'Glocal' Movements: Place Struggles and Transnational Organizing by Informal Workers. *Geografiska Annaler*, 91(2), pp. 123–136.

Lindell, I., ed. (2010a). *Africa's Informal Workers: Collective Agency, Alliances and Transnational Organizing in Urban Africa*. London: Zed Books.

Lindell, I. (2010b). Between Exit and Voice: Informality and the Spaces of Popular Agency. *African Studies Quarterly*, 11(2/3), pp. 1–11.

Lindell, I. (2011a). Introduction to the Special Issue: Organizing Across the Formal-Informal Worker Constituencies. *Labour, Capital and Society*, 44(1), pp. 3–16.

Lindell, I. (2011b). The Contested Spatialities of Transnational Activism: Gendered Gatekeeping and Gender Struggles in an African Association of Informal Workers. *Global Networks*, 11(2), pp. 139–158.

Lindell, I. (2017). Disjunctive Infrastructures and Fractured Subjectivities in Maputo's Periphery. Paper presented at the N-AERUS conference, 17–19 November 2017, Gothenburg.

Lindell, I. and Ampaire, C. (2016). The Untamed Politics of Urban Informality: 'Gray Space' and Struggles for Recognition in an African City. *Theoretical Inquiries in Law*, 17(2), pp. 257–282.

Lindell, I. and Ihalainen, M. (2014). The Politics of Confinement and Mobility: Informality, Relocations and Urban Re-making from Above and Below. In: W. Willems and E. Obadare, eds. *Civic Agency in Africa: African Arts of Resistance in the 21st Century*. Suffolk: James Surrey.

McFarlane, C. (2008). Postcolonial Bombay: Decline of a Cosmopolitan City? *Environment and Planning D: Society and Space*, 26(3), pp. 480–499.

McFarlane, C. and Waibel, M., eds. (2016). *Urban Informalities: Reflections on the Formal and Informal*. Aldershot: Ashgate.

Massey, D. (2004). Geographies of Responsibility. *Geografiska Annaler: Series B, Human Geography*, 86(1), pp. 5–18.

Meagher, K. (2009). *Culture, Agency and Power: Theoretical Reflections on Informal Economic Networks and Political Process*. Working Paper 2009:27. Danish Institute for International Studies, Copenhagen.

Mitullah, W. (2010). Informal Workers in Kenya and Transnational Organizing: Networking and Leveraging Resources. In: I. Lindell, ed. *Africa's Informal Workers: Collective Agency, Alliances and Transnational Organizing in Urban Africa*. London: Zed Books.

Prag, E. (2010). Women Leaders and the Sense of Power: Clientelism and Citizenship at the Dantokpa Market in Cotonou, Benin. In: I. Lindell, ed. *Africa's Informal Workers: Collective Agency, Alliances and Transnational Organizing in Urban Africa*. London: Zed Books.

Routledge, P. (2008). Acting in the Network: ANT and the Politics of Generating Associations. *Environment and Planning D: Society and Space*, 26(2), pp. 199–217.

Roy, A. (2005). Urban Informality: Toward an Epistemology of Planning. *Journal of the American Planning Association*, 71(2), pp. 147–158.

Staples, J., ed. (2007). *Livelihoods at the Margins: Surviving the City*. Walnut Creek, CA: Left Coast Press Inc.

Swanson, K. (2007). Revanchist Urbanism Heads South: The Regulation of Indigenous Beggars and Street Vendors in Ecuador. *Antipode*, 39(4), pp. 708–728.

Vanek, J., Alter Chen, M., Carré, F., Heintz, J. and Hussmanns, R. (2014). *Statistics on the Informal Economy: Definitions, Regional Estimates and Challenges*. Women in Informal Employment Globalizing and Organizing (WIEGO) Working Paper (Statistics) Number 2. Available at: www.wiego.org/sites/default/files/publications/files/Vanek-Statistics-WIEGO-WP2.pdf [accessed 25 May 2017].

Watson, V. (2009). The Planned City Sweeps the Poor Away ... Urban Planning and 21st Century Urbanization. *Progress in Planning*, 72(3), pp. 151–193.

Wilmot, D. (2014). The Street Vendor Project. In: H. Moksnes and M. Melin, eds. *Claiming the City: Civil Society Mobilisation by the Urban Poor*. Uppsala Centre for Sustainable Development, Uppsala University.

25
URBAN INFORMALITY AND THE NEW POLITICS OF PRECARITY
Day labourer activism in the USA

Nik Theodore

Introduction: informality and the politics of precarity

Low-wage labour markets in major US cities have become key sites of political activism and worker organizing in the early twenty-first century. The Fight for $15, which has seen strikes by fast-food workers pushing for higher wages; struggles against Walmart and the retailer's low-road business practices; and a range of minimum-wage and living-wage policy fights across the country are just some of the campaigns that have signalled a rejuvenated urban politics of precarity. Many of these campaigns are being spearheaded by so-called alt-labour groups – independent worker organizations engaged in "economic action organising" (Fine 2006) to raise standards at the bottom of urban economies (Cordero-Guzmán 2015; Jaffe 2015; Jayaraman and Ness 2005; Milkman and Ott 2014) – as workers who have been long regarded as 'unorganizable' because of their precarious status in the job market take the lead in challenging spread of substandard work.

Arguably, some of the most vibrant worker organizing is occurring within segments of the informal economy, particularly domestic workers who enter into verbal contracts to provide cleaning and care services for private households, and day labourers who assemble at informal hiring sites in public spaces to search for manual-labour jobs as construction workers, landscapers, movers and cleaners. Both domestic workers and day labourers are employed in 'grey markets' of urban economies where they are hired directly by employers with few labour protections and little government oversight of workplace conditions. As a result, domestic worker and day labourer occupations are characterized by widespread violations of labour standards, including violations of wage and hour laws as well as health and safety regulations (Burnham and Theodore 2012; Valenzuela *et al.* 2006). In response, these workers have sought to collectively redress workers' rights violations by organizing to raise labour standards and founding organizations that serve as sites of collective action. In the case of day labourers, the organizations they have formed – known as worker centres – have established a needed regulatory presence in informal construction labour markets and have become a platform for workers to defend their rights in the workplace. In an era of deteriorating standards in the low-wage sectors of urban economies, such initiatives can provide important mechanisms for the social regulation of contingent labour markets, as well as a being a vibrant site of social struggle against socio-economic inequities arising from the breakdown of employment protections in the USA.

This chapter examines the restructuring of the US residential construction industry and the efforts of immigrant day labours to staunch the deterioration of labour standards in the industry through collective action. The next section considers the role of the state in the spread of labour-market informality. Research on the informal economy has implicated the state in the spread of casualized employment relations, in large part because of the rolling back of government enforcement activities. This is followed by a brief summary of the failure of labour-standards enforcement and a lengthier exploration of the political economy of restructuring. The concluding section highlights organizing efforts by day labourers to place a floor under wages and working conditions in the informal economy.

Informality and the state

Economic informality is generally understood as arising from an absence of effective government intervention in the economy owing to institutional deficits, principally in the state's capacities to regulate economic activities occurring within its territory. The spread of labour-standards violations in day labour markets seems to confirm such assessments. Day labourers congregate in public spaces to search for work, typically with non-union construction contractors, landscaping companies and private households. Though these hiring sites are located along major thoroughfares, in front of home-improvement stores and in other highly visible locations in US cities, their regulation by government enforcement agencies charged with safeguarding labour standards has been virtually non-existent and violations of workplace standards have endured (Theodore *et al.* 2006; Valenzuela *et al.* 2006).

It also is widely accepted that the hard-and-fast distinction between the 'formal' and 'informal' sectors of advanced economies misrepresents the integral role informal activities play in supporting corporatized segments of the economy and underplays the ways in which informality is produced by wider processes of urban-economic restructuring (Mörtenöock *et al.* 2015; Roy and AlSayyad 2004; Santos 1979). Furthermore, the state is seen as central to how processes of restructuring unfold. Centeno and Portes (2006: 27–28, original emphasis) have expressed the prevailing view through a series of core propositions:

1 "An informal economy will develop *when and where it can*" though "the 'degrees of freedom' for this development to take place is affected both by the regulatory capacity of state agents and the scope of regulation they are expected to enforce."
2 "*The relationship between the informal economy and the state is, by definition, one of inevitable conflict.*"
3 The "relationship between state strength and informality" is impacted by "(a) the regulatory intent of the state, and (b) the social structure and culture of the population subject to it."

This third point is especially important for understanding the politics of precarity. *Regulatory intent* – the objectives of government enforcement of labour standards, including those that are clearly articulated through legislation and those that exist within the silences that occur when state agents use their discretion and do not fully enforce policy mandates – speaks volumes about a state's de facto enforcement priorities. Countries, such as the USA, with strong and well-resourced state apparatuses may tolerate the spread of informality and substandard employment if reining in degraded work runs counter to other, more privileged policy objectives and if the populations that would benefit by the state's greater regulatory reach into zones of informality are not favoured.

Immigrant workers and members of racial minority groups disproportionately occupy informalizing sectors within the US economy, where formal mechanisms for worker organization (such as labour unions) are largely absent and employment arrangements are precarious. These workers have little leverage to improve labour standards, and unscrupulous employers are afforded enormous latitude in devising contingent employment relations. This partly explains growing inequality within the USA and the decoupling of the historic link between labour productivity and wages. From the end of the Second World War until the early 1970s, productivity and wage increases virtually went hand in hand. Between 1973 and 2013, on the other hand, though worker productivity rose by 74.4 per cent, hourly compensation increased by just 9.2 per cent (Pitts 2015). Growing inequality has accompanied informalization and the spread of low-wage work, and violations of labour standards have become an entrenched feature of low-wage industries (Bernhardt et al. 2013).

Following Castells and Portes' (1989: 12) call to "look behind the appearance of social conditions (poverty, destitution, blight) to focus on the social dynamics underlying the production of such conditions", the next two sections consider how labour-market informalization has taken hold in the US construction sector, focusing on the changing dynamics of work and labour-market governance within the residential segment of the sector. With a focus on the state's failure to enforce labour standards while also erecting barriers to unionization, these sections describe how precarious work has proliferated across the residential construction industry, and immigrant workers, many of whom are undocumented, have become the 'workers of choice' for construction contractors who compete based on low costs and labour sweating.

The failure of labour standards enforcement

Core employment and labour laws that place a floor under wages, safeguard worker safety and maintain decent working conditions are failing to protect low-wage workers in the USA. In addition to providing vital worker protections, these laws also are designed to level the playing field for businesses in competitive industries by removing incentives for devising routes to profitability based on the degradation of working conditions. Unfortunately, however, the erosion of labour protections has been spurred by three self-reinforcing developments. The first is what David Weil (2014) has termed the "fissured workplace": the shift towards extended chains of subcontracting, the growing use of temp agencies and other labour brokers, and the proliferation of small enterprises. Workplace fissuring has seen employment, and with it much of the legal liability for abiding by worker protections, "shifted to a range of secondary players that function in more competitive markets and are separated from the locus of value creation" (Weil 2014: 14). Enterprises in these more competitive markets depend on extracting labour-cost savings from their workforces, often through strategies of informalization that regard long-term employment contracts, pension and health benefits, insurance for workplace injuries, and pay increases that accompany career progression as costs employers no longer can afford to bear.

The second factor has been the deregulation of the segments of the labour market through a four-decade-long assault on labour unions, one that has seen the unionization rate among private sector workers fall to just 6.6 per cent (BLS 2015). Through a series of decisions issued by the National Labor Relations Board and laws passed by state legislatures, as well as actions undertaken at the federal level by the executive branch of government, workers' rights to freedom of association have been eroded, remedies for unfair labour practices have been weakened, the right to strike has been abridged and the right to collectively bargain with one's employer has been undermined.

The damage to the system of worker protections caused by the fissuring of employment relationships along with the weakening of the regulatory framework of unionization has been compounded by a third factor, the erosion of labour-standards enforcement at all levels of government. Between 1980 and 2007, despite a 52 per cent increase in the private-sector labour force, the number of federal wage and hour inspectors actually declined by 31 per cent and the number of enforcement actions fell by 61 per cent (Bernhardt et al. 2013). The figures are equally as dismal in the area of workplace health and safety. The American Federation of Labor – Congress of Industrial Organizations (2015: 18) reports,

> At its current staffing and inspection levels, it would take federal OSHA [US Occupational Safety and Health Administration], on average, 140 years to inspect each workplace under its jurisdiction just once; at the federal and state levels there is one inspector for every 71,695 workers, well below the benchmark of one labor inspector for every 10,000 workers recommended by the International Labour Organization for industrialized countries.

The combined effects of these three factors – the fissuring of employment relations, de-unionization and the erosion of government capacity to enforce labour protections – have radically altered the competitive landscape for businesses. Increasingly firms are seizing opportunities within the institutional voids that have been created by the dismantling of regulatory frameworks, which in turn has energized processes of informalization. Small businesses are connected to high-value enterprises through extended supply chains, along which profits and risks travel, typically in opposite directions. High-margin activities undertaken in markets protected by high barriers to entry are retained at or near the top of subcontracting chains, while low-margin activities are outsourced to firms in highly competitive markets where marginal costs can be squeezed. With government resource commitments for labour-standards enforcement failing to keep pace with changes in the structure and composition of the US economy, opportunities are created for small firms, operating below the radar of workplace standards enforcement, to secure market share by holding down wages through contingent-work arrangements. This sets in motion a self-reinforcing process of restructuring: state withdrawal leads to the spread of smaller firms located on the margins of industry; because of this growth, the practices of these firms become even more difficult to monitor; which in turn leads to further rounds of restructuring, including the proliferation of substandard employment arrangements. The heightened cost pressures that are produced through successive rounds of restructuring result in processes of informalization placing downward pressure on wages and working conditions.

Informalization in construction

The construction sector is an important contributor to the regional jobs base of every major metropolitan area. Processes of political-economic restructuring, though, have resulted in the remaking of employment relations, most acutely in the sector's residential segment. These include: (1) the enactment of legal and regulatory changes that have restricted the ability of building trades unions to organize workers in the industry; (2) the exclusion of large segments of the labour force from unionized employment and apprenticeship-training opportunities; (3) the increasing capacity of non-union contractors to carry out high-quality work; and (4) the determined efforts of end users to reduce the influence of unions at construction worksites (Abernathy et al. 2012; Rabourn 2008). As a result, over the course of the last four decades, union density in the construction sector has fallen from approximately 40 per cent to just 13.9

per cent (BLS 2015), and there has been "the virtual elimination of unionized building trades in some metropolitan areas" (Weil 2005: 448).

The post-Second World War housing boom led to dramatic growth in homebuilding, and by the mid-1950s numerous non-union construction contractors entered the industry. As housing demand slowed towards the end of the decade and into the next, high unemployment rates in the industry led many "Unemployed union members ... to work for nonunion homebuilders, taking lower wages and forgoing union representation to avoid unemployment" (Rabourn 2008: 13). In subsequent years, growth in the non-union construction workforce continued, driven by population growth in Sunbelt cities (where union presence historically has been very low) and with it a shift in the locus of homebuilding to so-called 'right-to-work' states where anti-union policies make it difficult for trade unions to organize workers. Union membership in the building trades declined from 1.6 million in 1973 to 1.2 million in 2002, while the non-union construction workforce more than doubled from 2.5 to 5.5 million, trends that continued into the 2000s (Grabelsky 2007). Union density in residential construction declined sharply, from approximately 50 per cent in the 1950s to less than one in seven workers in the early 2010s. And as building trades unions lost their grip on the homebuilding industry, falling union density led to falling wages.

With competition increasingly based on downward pressures on labour costs, non-union residential construction contractors resorted to various tactics designed to contain costs by remaking employment arrangements. They deepened their dependence on contingently employed workers through the adoption of various flexible workforce arrangements, including misclassifying independent contractors, employing workers through temporary staffing agencies, increasing their reliance on subcontractors, and hiring labourers who are employed 'off the books'. This heightened employment instability as contractors sought to render quasi-fixed labour costs variable by abandoning stable employment relationships, while also avoiding costs associated with unemployment insurance and workers' compensation insurance to cover injuries on the job. The race to the bottom that ensued found many non-union firms chasing short-run profitability by sweating labour, cutting back on workplace safety provisions and pursuing other low-road forms of competition. As a result, the rate of workplace violations in the residential construction industry is extremely high (Bernhardt *et al.* 2013).

In the context of declining union density, building trades unions faced enormous obstacles in monitoring an increasingly decentralized residential construction industry composed of small contractors, and opportunities for non-union firms to enter the market increased. From the standpoint of wider industry practices, when profitability comes to depend on violations of labour standards and the subcontracting of low value-added work, the balance of influence shifts in the direction of low-road firms, leading to the further downgrading of employment conditions across the industry. As Erlich and Grabelsky (2005: 429) succinctly explain, "Legitimate employers – union or non-union – that provide a living wage and benefits for their employees are constantly looking over their shoulders at the legions of subcontractors that play by a different set of rules", subcontractors that collectively act as price-setters in an informalizing industry. When processes of labour sweating go unchecked and employers are allowed to reap the benefits of low-cost, flexible labour, the competitive dynamics of entire industries can be transformed, and pressures on reducing labour costs intensify. This is precisely what happened in the residential construction industry, as informality and expanded managerial discretion replaced union pay scales, hiring halls and apprenticeship programmes (Doussard 2013; Rabourn 2008).

With cost pressures in residential construction markets mounting and increasingly centring on wages, construction firms began devising strategies for reducing labour costs wherever they could, further diminishing the influence of unions on wages and working conditions. In the context of high union density, the influence of unions extends well beyond those workers covered by collective bargaining agreements and the contractors that employ them. When union density is high, non-union contractors are compelled to raise wages and improve working conditions in an effort to attract qualified workers. Therefore, even those workers who are not covered by union contracts enjoy some of the benefits of a strong union presence in the industry. However, when density falls to very low levels, this 'union effect' on wages and working conditions across the local labour market all but disappears, setting in motion processes of labour-market adjustment. Unionized workers leave the jurisdiction for areas where wage rates are higher, resulting in labour shortages in low union-density markets.

> With wages too low to attract the industry's traditional demographic base of recruits, contractors turned to immigrants to fill the vacuum ... [Today,] undocumented immigrants are clearly the [construction] industry's backbone in many areas and even in subsectors like residential and light commercial in still strong union markets.
> *(Erlich and Grabelsky 2005: 426)*

Within residential construction labour markets, Erlich and Grabelsky (2005: 428) have argued,

> The growing use of immigrant workers meshes with the structural trend toward subcontracting. As general contractors/construction managers shift on-site labor responsibilities to an array of mobile subcontractors, intermediaries such as temporary staffing agencies or individual labor brokers emerge that seek to provide non-union firms some of the referral services offered to unionized contractors by a union hiring hall.

There has been a proliferation of such intermediaries in the construction industry, the presence of which has been associated with low wages and workers' heightened exposure to substandard conditions. For example, the use of temps in the construction industry has been shown to be seasonal, with the number of temps dispatched to construction worksites swelling during peak periods (Mehta and Theodore 2006). From the standpoint of standards setting in de-unionized industries, peak seasons are crucial in that the resulting scarcity of labour creates leverage for workers and opportunities to increase wage rates and secure overtime pay; temp agencies alleviate wage pressures during these crucial periods, allowing employers to maintain lower hourly wage rates than might otherwise be the case.

A key aspect of the restructuring of the construction sector has been the emergence of new forms of labour market segmentation brought on by the diminished role of general contractors in directly structuring employment arrangements (Weil 2005). As employment arrangements are increasingly left to the market (e.g. to suppliers of contingent workers like temp agencies and informal labour brokers), the relative importance of internal labour markets is reduced, reservation wages fall and the reliance on subcontracting grows. "The increased use of undocumented workers also complements the growing presence of the underground economy in construction. It is a small step for an unscrupulous employer to move from hiring undocumented workers to operating entirely off the books – or vice versa" (Erlich and Grabelsky 2005: 428).

Erlich and Grabelsky (2005) and Rabourn (2008) reveal a complex relationship between the rise of an immigrant workforce and patterns of industrial change within the residential construction industry. Rather than being the cause of declining labour standards, as is sometimes argued, immigrant workers, and undocumented immigrants in particular (who perhaps comprised 17 per cent of the industry's workforce prior to the Great Recession (Passel and Cohn 2009)), have been drawn into the industry following several decades of declining union density, falling wages, the weakening of training systems, and intensifying competition among small contractors. Immigrant workers have responded to deteriorating conditions in the industry by developing robust job-search networks and informal systems through which skills can be developed and transferred (Hagan *et al.* 2015; Iskander and Lowe 2010; Lowe and Iskander 2016). These networks and systems act as mechanisms through which employment instability can be cushioned, and they represent an attempt to raise wages from the bottom up by demonstrating work quality to employers and by sharing information among workers regarding pay scales and job opportunities. This is crucial because, in an industry like construction with volatile production schedules and unstable employment, skill is a primary source of competitive advantage and bargaining power. Iskander and Lowe's (2013) analysis of 'tacit' skills development among migrants from Mexico who work in the residential construction industry shows that workers possess robust skill sets that span individual craft boundaries – even if these are not explicitly acknowledged by their employers, and thus poorly remunerated. Employers in regions with low union density are especially reliant on immigrant workers' tacit skillsets since, in the absence of union training programmes, new labour-market entrants have few avenues for acquiring trade-specific skills. The relative invisibility of these tacit skills, however, poses challenges for workers intent on using work quality to leverage wage gains. To overcome these challenges, skills

> are generally viewed on work teams as a shared resource ... immigrants [make] it difficult for employers to isolate any given worker in the team as the driver of innovation or quality. ... By pushing for collective pay increases, they safeguard the cohesion of their work group from employer attempts to create divisions by paying workers differently.
>
> *(Iskander and Lowe 2013: 798, 799)*

In short, contrary to the widely held perception that migrant workers passively accept degraded labour standards, as well as the erroneous argument that they are drivers of declining standards in the construction industry, Iskander and Lowe reveal a key strategy among workers to raise wages in the context of informalization and weakened institutional supports. Through modifications in labour processes and the strengthening of worksite social networks, they show how workers shore up pay standards in an industry where downward wage pressures are especially intense.

Important though they are, however, informal arrangements such as those described by Iskander and Lowe (2013) are not a substitute for institutionalized mechanisms of labour market regulation, especially in an industry where employers increasingly seek to suppress wages and circumvent employment and labour laws. The highly precarious and casualized segments of the construction industry continue to absorb the most intense cost pressures within the industry, the burdens of which fall squarely on the workforce. Immigrant workers have not created these segments nor do they dictate the low wage rates found there. Instead, they have been produced by sweeping processes of industrial restructuring that have unfolded over the course of several decades, and that have created opportunities for small firms to join an industry with low barriers to entry. These firms, in turn, have been aided by the emergence of a set of labour market intermediaries, such as temporary staffing agencies and other labour brokers, that help low-road employers access labour pools to fill vacancies on an as-needed basis. The spread of substandard employment that has occurred has been further enabled by underinvestment in government enforcement of labour standards. It is in this context that expansion of day labour, the most casualized form of employment in the residential construction industry, has occurred.

Worker centres: regulating informal labour markets from the bottom up

During the construction sector's 'long boom' of the 1990s and early 2000s, labour shortages created new openings for immigrant workers to enter the industry in large numbers. Many did so as day labourers, and the burgeoning just-in-time workforce grew. The lack of institutional supports for these workers – along with many workers' status as undocumented immigrants – meant that these jobs were precarious and poorly paid, and workers were unable to leverage labour-supply shortages to raise wages and improve working conditions. And then the real-estate bubble burst. During the Great Recession and the protracted jobless recovery that followed, reducing labour costs became an even greater priority for construction contractors. As the recession reverberated through residential construction markets, it contributed to the ongoing reworking of workforce systems in the industry, including the steady decline in union density that has been occurring over the last forty-plus years and the deepening reliance by many construction contractors on contingently employed workers.

In addition to being the targets for business expansion, however, informalized zones within the construction industry became sites of regulatory experimentation by day labourers and their allies. Faced with exploitative conditions in the residential construction industry, immigrant day labourers began organizing in the early 1990s to shun abusive employers, defend their rights to seek work in public spaces, and establish minimum wages at the informal hiring sites that had formed outside of home-improvement stores and along major thoroughfares (see Narro 2005; Nicholls 2016; Peck and Theodore 2012). This economic action organizing places direct pressure on employers in industries where labour-standards violations are widespread. Then, as now, economic action organizing to establish employment standards at informal day labour hiring sites relied on the cultivation of worker solidarity and self-policing measures to ensure that standards were established and maintained over time, especially during periods of slow employer demand. These efforts were significant because they marked some of the first attempts to set wage standards in casualized segments of the construction industry in the absence of high union density and robust government regulation of employment conditions. In one notable case in 2006, day labourers at an informal hiring site in Agoura Hills, on the outskirts of Los Angeles, established a $15 minimum wage – nearly three times that of the federal minimum wage at the time (Theodore 2015).

By the mid-1990s day labourers and advocates began establishing independent worker organizations – known as worker centres – to institutionalize labour-market rules. Day labour worker centres operate hiring halls that inject needed transparency into opaque employment relationships in the day labour market by setting and enforcing minimum wage rates, helping hold employers accountable for workplace injuries and assisting workers in redressing wage theft (Theodore et al. 2009). Worker centres are not employers and they do not retain portions of workers' earnings as profit. Day labourers and their employers enter into employment relationships directly, though the terms of employment are regulated by worker centre rules. Day labourers convene regular worker assemblies where the specific rules governing the centre are established, revisited and revised. Typical rules include the setting of minimum-wage rates, which increasingly is done by occupation in an effort to ensure that higher skilled work is properly compensated; systems of job allocation, which may rely on lotteries or lists; and general rules of conduct, say regarding sanctions for behaviours or speech that exhibit racism, sexism or homophobia. It is on this basis that worker centres provide a platform for day labourer organizing and for expanding the scope of worker leverage in the segment of urban economies that confers its benefits so unequally. Combining mechanisms for labour-market intervention (e.g. wage setting) with worker-organizing strategies rooted in Popular Education praxis (Theodore

2015) and programmes of social inclusion for migrant workers (Visser *et al.* 2017), day labour worker centres have established a regulatory presence in the informal economy, one that targets the very locus of substandard labour practices in the residential construction industry. So too have they modelled a highly democratic form of politics based on worker self-determination and voice.

Early efforts to create day labour worker centres began in Los Angeles, Portland, Seattle and the Washington, DC area before spreading to other US cities. One of the mechanisms for cultivating worker solidarity was the use of team sports to nurture cooperation, and friendly competition, among workers. In the mid-1990s, day labourers in Los Angeles formed a soccer league with more than twenty teams comprised of workers from various informal hiring sites (Alvarado 2014). Workers would use soccer matches as an opportunity to discuss employment conditions, as well as strategies for increasing their earnings. Through the game of soccer, workers who might be competitors for any given job assignment in the highly precarious day labour market were reminded of the importance of teamwork in achieving a goal. The metaphorical lessons learned were then translated into worker organizing, including the importance of collectively setting wage rates and denying labour to unscrupulous employers.

Many day labourers migrate between various cities to search for better employment opportunities. Through this migration, strategies for worker organization were transferred and developed, ultimately coalescing in the formation of the National Day Laborer Organizing Network (NDLON) in 2001, a national alliance of fifty-three workers' rights organizations that operate seventy day labour worker centres. In fact, the idea to establish a national network of day labourer organizations was first introduced following a soccer match between teams from several cities (Alvarado 2014). At its founding convention, NDLON's priorities included: (1) protecting workers' labour and civil rights; (2) creating day labour worker centres; (3) enhancing the education and organizing of day labourers; and (4) calling for an immigration naturalization programme to regularize the status of undocumented immigrants. This set of priorities continues to guide the organization as it nurtures the civic and political engagement of day labourers. The national network, in collaboration with affiliate organizations in US cities, has undertaken a range of initiatives to build worker power. These include the launching of the 'jornaler@ app', a digital application that enables day labourers and domestic workers to record on their mobile phones information that will be useful in contesting instances of wage theft (the non-payment of wages for work completed); the delivery of safety training programmes that assist workers in protecting themselves on the job; and developing strategic alliances with labour unions to press for immigration reform and stronger worker protections.

Conclusion

Day labour worker centres are an institutionalization of a locally scaled, worker-oriented intervention aimed at rebalancing asymmetrical power relations in zones of regulatory abandonment by the state. In promoting the regularization of informal hiring sites through transparency and collective decision-making, worker centres are challenging not only the economic imperatives that have been used to justify the systematic erosion of wages and workplace standards across the low-wage labour market (as well as the downloading of regulatory risks onto workers in the informal economy that has occurred through the state's highly selective enforcement of worker protections), but also conventional forms of labour politics that have seen the whittling away of employment security provisions in the wake of deregulatory initiatives administered in the name of labour-market flexibility. Day labour worker centres confront these political-economic dynamics, demonstrating that – even at a time when the politics of immigration reform has become polarized and vengeful, and when low-wage workers increasingly must rely on precarious jobs – marginalized workers can organize in defiance of low-road practices and routine violations of labour standards. Worker centres, therefore, are an important element of an emergent, alternative regulatory strategy, one that utilizes community-based labour-market intermediaries to perform crucial re-regulatory functions in the informal economy. And they are a key site of urban politics among disenfranchised immigrant workers who are prepared to contest substandard conditions, even those who are denied citizenship and who too often are regarded as lacking legitimate standing in the formal decision-making processes of the state. The deep democracy many worker centres practise, therefore, can be a model for new forms of urban politics in these divisive and uncertain times.

References

Abernathy, F., Baker, K., Colton, K. and Weil, D. (2012). *Bigger Isn't Necessarily Better: Lessons from the Harvard Home Builder Study*. Lanham, MD: Lexington Books.

Alvarado, P. (2014). Globalizing Struggle: From Civil War to Migration. In: C. Tzintzún, C. Pérez de Alejo and A. Manríquez, eds. *¡Presente! Latin@ Immigrant Voices in the Struggle for Racial Justice*. 1st ed., Oakland, CA: AK Press, pp. 39–49.

American Federation of Labor – Congress of Industrial Organizations. (2015). *Death on the Job*. Washington, DC: AFL-CIO.

Bernhardt, A., Spiller, M. and Theodore, N. (2013). Employers Gone Rogue: Explaining Industry Variation in Violations of Workplace Laws. *International Labor Relations Review*, 66(4), pp. 808–832.

BLS (US Bureau of Labor Statistics). 2015. *Union Members Summary*. Available at: www.bls.gov/news.release/union2.nr0.htm [accessed 30 August 2015].

Burnham, L. and Theodore, N. (2012). *Home Economics: The Invisible and Unregulated World of Domestic Work*. New York: National Domestic Workers Alliance.

Castells, M. and Portes, A. (1989). World Underneath: The Origins, Dynamics, and Effects of the Informal Economy. In: A. Portes, M. Castells and L.A. Benton, eds. *The Informal Economy: Studies in Advanced and Less Developed Countries*. 1st ed., Baltimore and London: Johns Hopkins University Press, pp. 11–37.

Centeno, M.A. and Portes, A. (2006). The Informal Economy in the Shadow of the State. In: P. Fernández-Kelly and Jon Shefner, eds. *Out of the Shadows: Political Action and the Informal Economy in Latin America*. 1st ed., University Park: Pennsylvania State University Press, pp. 23–48.

Cordero-Guzmán, H.R. (2015). Worker Centers, Worker Center Networks, and the Promise of Protections for Low-Wage Workers. *WorkingUSA*, 18(1), pp. 31–57.

Doussard, M. (2013). *Degraded Work: The Struggle at the Bottom of the Labor Market*. Minneapolis: University of Minnesota Press.

Erlich, M. and Grabelsky, J. (2005). Standing at a Crossroads: The Building Trades in the Twenty-first Century. *Labor History*, 46(4), pp. 421–445.

Fine, J. (2006). *Worker Centers: Organizing Communities at the Edge of the Dream*. Ithaca, NY: Cornell University Press.

Grabelsky, J. (2007). Construction or De-construction? The Road to Revival in the Building Trades. *New Labor Forum*, 16(2), pp. 47–58.

Hagan, J., Hernández-León, R. and Demonsant, J.-L. (2015). *Skills of the "Unskilled": Work and Mobility Among Mexican Migrants*. Oakland: University of California Press.

Iskander, N. and Lowe, N. (2010). Hidden Talent: Tacit Skill Formation and Labor Market Incorporation of Latino Immigrants in the United States. *Journal of Planning Education and Research*, 30(2), pp. 132–146.

Iskander, N. and Lowe, N. (2013). Building Job Quality from the Inside-out: Mexican Immigrants, Skills, and Jobs in the Construction Industry. *International Industrial Relations Review*, 66(4), pp. 785–807.

Jaffe, S. (2015). *Roll Back Low Wages: Nine Stories of New Labor Organizing in the United States*. New York: Rosa Luxemburg Stiftung.

Jayaraman, S. and Ness, I., eds. (2005). *The New Urban Immigrant Workforce: Innovative Models for Labor Organizing*. New York: M.E. Sharpe.

Lowe, N. and Iskander, N. (2016). Power Through Problem Solving: Latino Immigrants and the Inconsistencies of Economic Restructuring. *Population, Space and Place*, 23(7), pp. 1–13.

Mehta, C. and Theodore, N. (2006). Workplace Safety in Atlanta's Construction Industry: Institutional Failure in Temporary Staffing Arrangements. *WorkingUSA*, 9(1), pp. 59–77.

Milkman, R. and Ott, E., eds. (2014). *New Labor in New York: Precarious Workers and the Future of the Labor Movement*. Ithaca, NY: Cornell University Press.

Mörtenöock, P., Mooshammer, H., Cruz, T. and Forman, F., eds. (2015). *Informal Market Worlds: Reader*. Rotterdam: NAI010 Publishers.

Narro, V. (2005). Impacting Next Wave Organizing: Creative Campaign Strategies of the Los Angeles Worker Centers. *New York Law School Law Review*, 50, pp. 465–513.

Nicholls, W. (2016). Politicizing Undocumented Immigrants One Corner at a Time: How Day Laborers Became a Politically Contentious Group. *International Journal of Urban and Regional Research*, 40(2), pp. 299–320.

Passel, J.S. and Cohn, D. (2009). *A Portrait of Unauthorized Immigrants in the United States*. Washington, DC: Pew Hispanic Center.

Peck, J. and Theodore, N. (2012). Politicizing Contingent Work: Countering Neoliberal Labor Market Regulation ... From the Bottom Up? *South Atlantic Quarterly*, 111(4), pp. 741–761.

Pitts, S. (2015). Low-Wage Work in the Black Community in the Age of Inequality. In: S. Thomas-Breitfeld, L. Burnham, S. Pitts, M. Bayard and A. Austin, eds. *#BlackWorkersMatter*. 1st ed., Boston: Discount Foundation and Neighborhood Funders Group.

Rabourn, M. (2008). Organized Labor in Residential Construction. *Labor Studies Journal*, 33(1), pp. 9–26.

Roy, A. and AlSayyad, N., eds. (2004). *Urban Informality: Transnational Perspectives from the Middle East, Latin America, and South Asia*. Oxford: Lexington Books.

Santos, M. (1979). *The Shared Space: The Two Circuits of the Urban Economy in Underdeveloped Countries*. London and New York: Methuen.

Theodore, N. (2015). Generative Work: Day Labourers' Freirean Praxis. *Urban Studies*, 52(11), pp. 2035–2050.

Theodore, N., Valenzuela, A. and Meléndez, E. (2006). *La Esquina* (the Corner): Day Laborers on the Margins of New York's Formal Economy. *Working USA*, 9(4), pp. 407–423.

Theodore, N., Valenzuela, A. and Meléndez, E. (2009). Worker Centers: Defending Labor Standards for Migrant Workers in the Informal Economy. *International Journal of Manpower*, 30(5), pp. 422–436.

Valenzuela, A., Theodore, N., Melendez, E. and Gonzalez, A.L. (2006). *On the Corner: Day Labor in the United States*. Los Angeles: UCLA Center for the Study of Urban Poverty.

Visser, M.A., Theodore, N., Meléndez, E.J. and Valenzuela Jr, A. (2017). From Economic Integration to Socioeconomic Inclusion: Worker Centers and the Social Inclusion of Day Laborers. *Urban Geography*, 38(2), pp. 243–265.

Weil, D. (2005). The Contemporary Industrial Relations System in Construction: Analysis, Observations and Speculations. *Labor History*, 46(4), pp. 447–471.

Weil, D. (2014). *The Fissured Workplace*. Cambridge, MA and London: Harvard University Press.

PART VI

Spaces of living

David Wilson and Kevin Ward

The city has always been many things to many people. To a growing number around the world it is home. It is where they dwell, spending time on the everyday activities of social reproduction. Coming and going over the week, doing hours of paid and unpaid work, making their bit of the city work for them and those for whom they care and are responsible. Doing this in the context of the wider transformations in the international, national and urban space economies in which their cities are located and to which the cities contribute which structures their livelihoods is for many becoming hardier not easier.

According to Harvey (1978: 11), "Labour, in seeking to protect and enhance its standard of living, engages in the living place, in a series of running battles over a variety of issues which relate to the creation, management and use of the built environment." You do not have to be a Marxist to recognize the tensions which characterize living in many contemporary cities. Increasingly expensive housing, the pursuit of high-end consumption and retail redevelopment strategies, creaking public services and persistently low real wages are just some of the ways in which cities are being refashioned. How to carve out a living in an environmentally and socially sustainable manner for current and future generations? How to secure access to affordable and secure public services, such as energy, transport and water?

The first chapter by Joshua Evans and Geoff DeVerteuil takes aim at Skid Row in Los Angeles. It focuses on how urban poverty management is achieved through a mixture of caring, curing and controlling institutions operating at the local level. These include those by urban elites, voluntary sector organizations and the state and serve to 'regulate' urban poverty and, more generally, maintain civil order in the city. This is an example of how social welfare provision – understood as education, public health and social health – remains important and political in the contemporary city, even under conditions of ever deeper and persuasive neoliberalization.

The second and fifth chapters turn to greenery and the city. Rina Ghose and Margaret Pettygrove's chapter is about the emergence and role of community gardens in Milwaukee, Wisconsin. They explore the changing economic and social conditions in the city that generate the context out of which has emerged a focus on urban agriculture, and the different types of gardens that emerged and the work they have done in nurturing and sustaining active communities, particularly those experiencing multiple disadvantages. Nathan McClintock, Christiana Miewald and Eugene McCann in their chapter explore food production and, specifically, urban agriculture (UA) as a set of fundamentally political practices, both in terms of their role in neoliberal governance and 'sustainability' policy-making, and also as objects of contestation. They use case studies from Portland, Oregon and Vancouver, British Columbia to highlight the contentious nature of UA in cities that explicitly frame their policy-making in terms of sustainability, resilience, and 'greenness'.

The third chapter is that by Helen Jarvis. She takes aim at the notion of liveability. The chapter attends to the intangible 'soft' architectures of everyday life to recognize the importance of 'bottom up' visions of development and alternative forms of building and managing local community assets. In turn the chapter challenges the 'metrics' underpinning conventional notions of liveability and against which cities position one and other. It uses a couple of examples, one from the UK and the other from the USA.

The fourth chapter is by Charlotte Lemanski and uses the example of Cape Town in South Africa to consider forms of infrastructural citizenship that are integral to everyday spaces of urban life. Based on fieldwork undertaken over a long period (2004–ongoing) in a single state-subsidized housing settlement in Cape Town, South Africa, Charlotte uses a handful of residents' perceptions to indicate the potential utility of infrastructural citizenship as a lens through which to understand these experiences. What is clear in this state-subsidized housing settlement is that housing beneficiaries themselves demonstrate a strong connection between citizenship and infrastructure. In particular, residents frequently expressed a vivid association between the home that they now own and 'the end of the struggle'. In this sense, housing beneficiaries demonstrate explicit attachment to the physicality of the house and its associated infrastructure in terms of their identity as citizens of a post-apartheid South Africa and of Cape Town, a country and a city where they have potentially always lived, but have only recently received citizenship.

The sixth and final chapter of this section focuses on another aspect of the spaces of living. Kacper Pobłocki starts by challenging the assumption that all forms of property-anchored politics are reactionary. This assumption, he argues, stems from much of the field of urban studies taking the USA as its reference point. For example, the first Anglo-Saxon book on the right to the city (Mitchell 2003) deems the homeless as the true revolutionary subject of urban politics. However, outside of Anglo-Saxon countries the word property is *not* synonymous with real estate and the relationship between ownership, class and space follows a different logic. In his chapter on housing in Poland Kasper describes different lineages and dynamics of the relationship between political contention, class and space, showing yet another case of the material and daily struggles for the right to the city.

While there is much that divides these chapters, all are bound together by a common concern to render visible the various ways in which the city is made as a place in which people come together to live and the politics over this making. This is a politics rooted in the material relations of place, in which place is constituted through its global-local relations (Massey 1993). The ongoing urbanization of the planet is only likely to further sharpen this need to appreciate this dialectical understanding of place and the politics in its making.

References

Harvey, D. (1978). The Urban Process under Capitalism: A Framework for Analysis. *International Journal of Urban and Regional Research*, 2(1–3), pp. 101–131.

Massey, D. (1993). Power-Geometry and a Progressive Sense of Place. In: J. Bird, B. Curtis, T. Putman, G. Robertson and L. Tickner, eds. *Mapping the Future: Local Cultures, Global Change*. New York: Routledge, pp. 59–69.

Mitchell, D. (2003). *The Right to the City: Social Justice and the Fight for Public Space*. New York: Guilford Press.

26
THE POLITICAL SPACES OF URBAN POVERTY MANAGEMENT

Joshua Evans and Geoff DeVerteuil

Introduction

At first glance, the fifty square blocks of Skid Row to the east of Downtown Los Angeles are a welter of despair, castoffs and disarray, the last resort for the most deprived of populations, dumped unceremoniously on America's most misery-encrusted square mile. Pavements are impassable from encampments and lean-tos; the denizens themselves are tired, frustrated and desperate, under the gun from merciless structural forces and the controlling gaze of the LAPD; and every second storefront promises help for the destitute, the abandoned, the addicted and the lost. A quarter of all shelter beds in the entire county (10 million people) are in Skid Row (DeVerteuil 2006), along with a bewildering array of detoxes, flophouses, addiction treatment centres, mental health drop-in centres, and the like, some wedded to no-strings care and others to a 'recovery zone' model that aggressively seeks to shield clients from the temptations of the street (Stuart 2014).

As an exemplar space for how extreme poverty is managed in the Global North, Skid Row is a politically ambiguous setting marked by tension between control and containment on the one hand and the provision of stopgap care on the other. As DeVerteuil notes (2015: 153), "Skid Row treads a fine line between service saturation and warehousing on the one hand, and genuine support and recovery on the other." While easy to dismiss as a dumping ground for surplus populations that is highly convenient for the rest of the city yet highly stigmatizing for those who find themselves on the street, Skid Row is actually a complex landscape of care, life-giving sustenance and coercive abeyance (DeVerteuil and Wilton 2009; von Mahs 2013). Yet all too frequently, the last two impulses have been falsely subsumed within a neoliberal grammar, the sinister overtones of which obscure the more compassionate responses at street level.

This obscuring fits into the influential perspective informing current thinking on contemporary cities, that there is a 'new' type of urban politics reshaping urban landscapes, a politics in which 'urban entrepreneurialism' has displaced 'urban managerialism' as the primary logic driving urban development (Harvey 1989; MacLeod and Jones 2011). While scholarship conducted under the umbrella of this 'New Urban Politics' (NUP) has made numerous contributions to understandings of contemporary urban landscapes, particularly in relation to processes of neoliberalization, it has tended to privilege a politics of interurban competition over a politics of social reproduction (MacLeod and Jones 2011; McGuirk and Dowling 2011). As a result the

Figure 26.1 Image of Skid Row
Source: photo: Geoff DeVerteuil.

coalitions, regimes and partnerships sustaining forms of collective social welfare provision such as social housing, education and public health services have been sidelined. This risks perpetuating the false assumption that the purported shift away from managerial concerns over social welfare provision towards entrepreneurial concerns with place-promotion and economic growth reflects a zero-sum game in which shifting emphasis on entrepreneurialism is accompanied by a declining emphasis on redistribution and social welfare provision. It also obscures important residuals of collective consumption that still help to shape the politics of help within urban spaces (DeVerteuil 2015).

This chapter focuses on one such residual, a regime of practices we refer to as *urban poverty management*. Urban poverty management is achieved through a mixture of caring, curing and controlling institutions operated at the local level by urban elites, voluntary sector organizations and the state to 'regulate' urban poverty and, more generally, maintain civil order in the city (Wolch and DeVerteuil 2001). The political spaces of urban poverty management are multifaceted and their complexity defies attempts to reduce them to mere expressions of social control (Cloke *et al.* 2010; DeVerteuil *et al.* 2009). A different grammar is needed to make sense of these mixtures and their implications for urban (in)justice. We develop such a grammar in this chapter by conceptualizing poverty management as an urban apparatus traversing an assortment of discourses, practices and spaces that, in conjunction, produce strategic effects.

The urban poverty management perspective

Poverty, understood in terms of relative or material deprivation, is both an effect of capitalist social relations and constitutive of the capitalist system itself (Raphael 2009). In this regard, poverty has long been an enduring and complex governance problem for the capitalist state: namely, how to manage poverty in ways that quells social disorder, assimilates low-income communities into the existing regime of social relations, all while maintaining state legitimacy (Sloss et al. 2011). While on the surface, responses to poverty exist to temporarily relieve hardship and misery, they also function to encourage participation in low-wage work and maintain civil order, objectives that work to uphold the existing capitalist order more generally (Piven and Cloward 1993).

In liberal welfare regimes the governance of poverty is minimalist (Esping-Andersen 1990). Interventions occur only where necessary and they typically take the form of public benefits and services that exist to manage poverty by softening its hard edges and ameliorating its externalities. We refer to the urban expression of residual, liberal welfare regimes as *urban poverty management*. Here 'management' signifies the strategic imperative to govern social and economic inequality in a way that is congruent with social and economic relations in a liberal market economy. Urban poverty management is targeted towards what Tyner (2013) has called 'surplus populations': an extremely vulnerable class dispossessed through economic restructuring and abandonment within post-welfare state configurations, and which are seen as potentially disruptive. These 'surplus populations' include individuals dislocated from paid work (and by extension housing) due to fluctuations in the labour market along with those temporarily or permanently excluded due to disability, illness and addiction.

In Anglo-American cities the socio-spatial management of 'surplus populations' has taken a common form. Management is generally accomplished through an ensemble of income assistance programmes and (semi-)institutional sites and spaces co-located in low quality inner-city milieus, what Dear and Wolch (1987) described as 'service-dependent ghettos' in their landmark book *Landscapes of Despair* (see DeVerteuil and Evans 2009). In a majority of North American cities, these milieus are the primary outlet for last-resort social welfare services – of which Skid Row, Los Angeles is but one conspicuous example. Typically, they encompass a mixture of service spaces including: 'spaces of care', such as community-based emergency shelters, food banks and drop-in centres, offering sanctuary, moral support and essential subsistence needs; 'spaces of recovery', such as detox centres, addiction treatment facilities, residential group homes and hospitals, offering peer support, counselling and medical care; and 'spaces of control', such as municipal jails and state prisons, that function to remove and intern problematic populations (DeVerteuil 2006; DeVerteuil and Wilton 2009; DeVerteuil et al. 2009; Gowan 2010; Murphy 2009; Wolch and DeVerteuil 2001).

Operated at the local level by voluntary sector organizations and private for-profit corporations, in many cases contracted by the state, these milieus can be understood as one spatial configuration in a much longer historical sequence of poverty management, a sequence of responses that have similarly functioned to 'regulate' urban poverty and, more generally, maintain civil order in the city (Wolch and DeVerteuil 2001). In this regard, poverty management has involved sites such as the workhouses and asylums of the nineteenth century, the single-room occupancy (SRO) hotels and skid row districts of the mid-twentieth century, and, more recently, the 'service-dependent ghettos' of the late twentieth century. Over time many of these twentieth century management sites and milieus, such as the 'service-dependent ghetto', have been uprooted, disassembled and dispersed due to a variety of factors, most notably urban gentrification, producing new poverty management landscapes in the process (DeVerteuil 2011; Reese et al. 2010).

Urban poverty management and neoliberal governance

From a poverty management perspective, 'service-dependent gettos' are understood as spatial expressions of a powerful and pervasive logic of social control (DeVerteuil and Evans 2009). In this regard, they are often read as putative 'spatial fixes' insofar as they are seen to function as incremental, stopgap measures for supervising 'surplus populations' dispossessed through economic restructuring and abandoned within post-welfare state configurations (Fairbanks 2009; Gowan 2010). Here, poverty management regimes are potentially indicative of wider urban strategies associated with 'revanchism' (Smith 1996), the 'new urban politics' (Cox 1993) and 'roll-out' neoliberalism (Peck and Tickell 2002).

Yet geographers have critiqued this tendency to narrowly frame activities subsumed under 'poverty management' as expressions of social control (Cloke *et al.* 2010; DeVerteuil *et al.* 2009). One problem with this 'punitive and controlling orthodoxy' (Cloke *et al.* 2010) is that it clouds attempts to account for the presence and location of compassionate and caring responses that endure, and in some cases proliferate, in the supposedly 'mean and lean' neoliberal city (DeVerteuil 2006). For instance, the provision of emergency shelter and supportive services to homeless populations, most often by local, faith-based voluntary sector organizations, has in fact expanded and diversified in many cities, in most cases as a direct result of increasing (rather than decreasing) investment in shelters, transitional housing and homeless services (DeVerteuil 2006). Rather than completely disintegrating under the weight of welfare reform and punitive impulses, caring and compassionate responses have multiplied, diversified and become more complex (DeVerteuil *et al.* 2009). In this regard, these responses are, at first glance, somewhat out of place in relation to neoliberal urbanism and the logic of control embedded in post-welfarist configurations.

Attempts have been made to square the proliferation of compassionate and caring responses, originating largely within the voluntary sector, with the punitive leanings of neoliberal urbanism and post-welfare configurations more generally (Cloke *et al.* 2010; Sparks 2012). For example, Cloke *et al.* (2010) argue that poverty management is better seen as a 'messy middle ground' that epitomizes the re-working of social welfare by multiple rounds of neoliberalization. As a 'messy middle ground' spaces of poverty management are undoubtedly ambiguous and complex:

> romanticized, yet often in practice deeply unromantic; easily dismissed as merely upholding the status quo, yet powered by an urge to do something about the injustice of that status quo; a cog in the revanchist engine, yet engineered and operated by people for whom revenge is the last thing on their mind.
>
> *(Cloke et al. 2010: 11)*

In this regard, 'management' should, alternatively, be read as a field of articulation and contestation of neoliberal imperatives. Such spaces represent, in the words of Cloke *et al.* (2010: 11), "a potential nexus for resistance to such ideologies – either practicing alternative values from inside the system of governance, or fashioning spaces of resilient care in opposition to the joined-up orthodoxies of such governance". The challenge lies in how to conceptualize the 'incorporation' of charitable, voluntary activities and the 're-working' of social welfare by multiple rounds of neoliberalization while at the same time giving space to the deep-seated forces of care and their potential as a nexus for resistance to such ideologies.

This challenge is especially pronounced where one tries to reconcile the controlling and caring impulses of poverty management. Typically urbanists have attempted to resolve the two countervailing tendencies by placing them in relation to the structured coherence of urban neoliberalism (Fairbanks 2009; Gowan 2010). This unity, however, is dependent upon reducing care to control. This particular way of thinking views poverty management sites and spaces as fully determined by their relations in a whole – they are taken to be parts of a neoliberal system, or machine, organized to serve specific goals or ends (i.e. urban gentrification). This reductionist view is problematic because it cannot account for the way certain formations (such as urban neoliberalism) are held together despite the multiple differences, incoherencies and contradictions among their constituent spaces (spaces of care, for example) (McGuirk and Dowling 2011). Moreover, it cannot properly account for the agencies of particular individuals and sites without reducing their differences and casting them as externally determined, like 'cogs in a machine'. This view hides the complexity of urban neoliberalism as well as its emergent and contingent character (Cloke et al. 2010).

What is needed is a theoretical framework that is sensitive to how urban spatial orders form and endure through and in spite of the co-presence of different, irreducible parts without negating the agency and indeterminate potential of those parts. This resonates with previous calls for "a more complete grammar of contemporary urban politics" (McGuirk and Dowling 2011: 2611) and "a more balanced, *relational view* of urban in/justice that privileges a co-present and sometimes dependent set of practices" (DeVerteuil 2014: 875, emphasis added). DeVerteuil (2014: 875, original emphasis) argues that we need "to consider more expansive, as well as tempered, grammars that consider how injustice *co-exists with* and *depends upon* more supportive currents within urban space". In other words, to recognize that punitive approaches are always adulterated (at least in democracies) by more supportive currents, and that there can be genuine love and care in the enabling of survival and a vital stepping stone to a better life (Cloke et al. 2010; DeVerteuil 2014, 2015).

Rethinking urban poverty management with assemblage theory

Drawing on the critique of functionalist framings of poverty management presented above, we turn to an alternative conceptual framework: assemblage theory. This theoretical framework directs attention to the *processes of composition* by which urban poverty management landscapes are assembled and ordered, and how they endure despite internal differences, incoherence and contradictions (Anderson et al. 2012). Doing so helps avoid the trap of naturalizing poverty management landscapes, and the practices located within, in terms of an all-encompassing logic of control.

Assemblage thinking is now emerging in geography (Anderson et al. 2012) and in urban studies in particular (Farias and Bender 2010; McFarlane 2011) where it has been applied to urban policy (McCann and Ward 2011), urban planning (McGuirk and Dowling 2011) and gentrification (Rankin and Delaney 2011). The concept of assemblage concedes much more space for indeterminacy, emergence, becoming, processuality and the socio-materiality of phenomena (McFarlane 2011). An assemblage can be thought of as an arrangement held together through 'relations of exteriority' between heterogeneous entities whose individual capacities (or potential) cannot be known in advance but only through the emergent intra-action or co-functioning among entities that grant the composition a stability of form (see Anderson et al. 2012). Applied to cities, assemblage thinking posits urban space less as relational formations that express an a priori logic, such as neoliberalism, and more as self-generative, indeterminate 'gatherings' full of multiple projects, programmes and possibilities (Farias and Bender 2010).

The analytic value of assemblage theory is the way it allows one to rethink how a 'unity' or 'whole' acquires its coherence (DeLanda 2006). An assemblage is defined by the co-functioning of its individual elements, which can be stabilized or destabilized by the way they interact with other elements. Therefore it is the interactions between elements – interaction as mutually constitutive symbiosis – that generate a form of coherence. McFarlane (2011: 208) states "component parts may be detached and plugged into a different assemblage in which its interactions are different". In this sense, it is the 'relations of exteriority' between elements that are most important for understanding assemblages. This conceptual orientation emphasizes contingency and situatedness when it comes to how relations are formed. As DeLanda (2006: 11) explains, the relations of exteriority constituting an assemblage are *contingently obligatory* rather than *logically necessary*, as is the case in organic notions of 'wholes' or 'unities'.

In the remaining sections of the chapter we show how it is possible to rethink urban poverty management as an assemblage constituted by relations of exteriority. Moreover, we demonstrate how this particular approach moves urban studies forward in reconciling the issues introduced previously. One of the characteristics of an assemblage approach is that it "forces the researcher to conceptually move from structures to relationships, from temporal stability to uncertain periods of *emergence* and *heterogeneous* multiplicities, resisting the siren call of final or stable states, which are the foundations of classical social theory" (Marcus and Saka 2006: 106, original emphasis). To capture this temporal dimensionality we parse assemblage into three more specific concepts: problematization, apparatus (or dispositif) and assemblage.

First is the notion of problematization that featured prominently in Foucault's (2007, 2008) writings on governmentality. He described governmentality as a patterned style of political reasoning (encompassing procedures, calculations, reflections, critiques) within which problems of life "take on their intensity" and "assume the form of a challenge" (Foucault 2008: 317). Problematizations occur when a domain of experience loses its familiarity, becomes uncertain and registers as a practical difficulty or problem. His genealogies of liberal government were organized around moments in history when some aspect of living was rendered problematic, through power/knowledge, making life amenable to governance. Poverty, in the work of Foucault (2007, 2008) and in our chapter as well, can be thought of as one field of problematization that has contextualized regimes of modern government and the governance of cities in particular (Dean 1992).

Second, Foucault (1980) used the term 'dispositif' (or 'apparatus' in the English translation) to conceptualize these regimes of government. The apparatus is a strategic response to a historical problem (Rabinow 2003). The term figured prominently in Foucault's (1995, 2007) writings, particularly in his work around discipline and later political rule. Foucault (1980: 194–195) employed the concept apparatus to draw attention to a network of relations formed between discourses, institutions and practices, at a given historical moment, in response to an urgent need. In Foucault's (1980) descriptions, apparatuses are invested with strategy, meaning they are structured by explicit objectives, inscribed in specific webs of power relations and linked to specific types of knowledge. When it comes to analysing these apparatuses, Foucault (2008) asserted that analysts attend to these strategic logics by tracing the connections between the heterogeneous rather than try to homogenize the contradictory.

Finally, we reserve the term 'assemblage' to describe the process of relational composition that generate apparatuses which endure as more durable formations for governing. As Li (2007: 264) states, "assemblage flags agency, the hard work required to draw heterogeneous elements together, forge connections between them and sustain these connections in the face of tension". In this laborious process, problematization provides "particular schemes of thought, diagnoses of deficiency and promises of improvement" (Li 2007: 264) but do not necessarily determine processes of assemblage. In the following instructive passage, Rabinow (2003: 56) relates the matrix of problematizations-apparatuses-assemblage:

Assemblages are secondary matrices from within which apparatuses emerge and become stabilized or transformed. Assemblages stand in a dependent but contingent and unpredictable relationship to the grander problematisations. In terms of scale they fall between problematisations and apparatuses and function differently from either one. They are a distinctive type of experimental matrix of heterogeneous elements, techniques, and concepts. They are not yet an experimental system in which controlled variation can be produced, measured and observed. They are comparatively effervescent, disappearing in years or decades rather than centuries. Consequently, the temporality of assemblages is qualitatively different from that of either problematisations or apparatuses.

Here we get a better sense of the temporal distinctions between problematizations, which operate on a much grander scale, and apparatuses and assemblages which operate on the scale of years or decades. The relationship between processes of assembly and problematization is contingent and unpredictable. Finally, there is a close relationship between apparatuses and assemblage. Following Legg (2011: 131), we consider apparatuses as an ensemble more prone to territorialization and governing – the resultant formation – whereas assemblage is associated with the process of composition and assembly; apparatus and assemblage "thus emerge as one and part of each other, but in a continual dialectic".

Applying this matrix to rethink urban poverty management offers a way forward with regard to the aforementioned dilemma regarding the ambiguous, complex and 'messy' relationship between care and control in landscapes of urban poverty management. Central to this approach is attending to the strategic logic present in apparatuses and modified through processes of assemblage. By shifting the analytical approach to a *strategic* analysis of the co-incidence of 'care' and 'control' the focus is directed towards the 'relations of exteriority' among agents of care such as shelters and agents of control such as the police. A strategic analysis seeks to trace the linkages between heterogeneous elements while preserving their differences; in other words, connecting care and control without reducing one to the other.

Assembling urban poverty 'management': returning to Skid Row

Armed with this conceptual background, we can now revisit the landscape of Skid Row, Los Angeles. The initial description very much presented Skid Row as a messy middle ground, a field of articulation and contestation of neoliberal imperatives as expressed in the city of Los Angeles. A strategic analysis utilizing our conceptual matrix can assist in making better sense of the managerialism embedded in this complex living space. We begin first by briefly sketching the prevailing problematizations of poverty that serve as the backdrop to Skid Row's emergence and evolution. We then turn to consider Skid Row as an apparatus-assemblage.

The problematization of poverty

While forms of poverty have evolved over time in liberal welfare regimes, certain historical continuities have characterized responses (Sloss *et al.* 2011). These continuities reflect long-standing and deeply embedded problematizations of poverty. Gowan (2010) outlines three persistent and overarching constructions of poverty: moral, therapeutic and systemic. Moral constructions of poverty have dominated American politics since the colonial period, and locate the cause of poverty in the supposed character defects of the poor themselves, and thus deserving of punishment and exclusion rather than compassion or sympathy. These punitive responses are motivated by a strategy of deterrence. Second, therapeutic constructions of poverty have coloured American politics since the late nineteenth century when transient, vagrant populations, so-called 'tramps', came to be seen as suffering from a medical condition referred to as 'wanderlust'. In the time since, the psy-disciplines and medicine in particular have directed their gaze upon the urban poor, explaining the cause of poverty in terms of individual pathologies. From this point of view the poor are seen as deviant yet deserving of rehabilitation. These therapeutic responses are largely motivated by the strategy of the cure. Finally, emerging alongside therapeutic constructions of poverty were competing systemic constructions. Systemic constructions locate the cause of poverty in social structures beyond the control of individuals. Systemic constructions not only render the poor deserving of sympathy and public assistance, they are motivated by a strategy of social transformation.

As indicated above, each of these constructions of poverty – moral, therapeutic and systemic – are tied to particular strategies for managing the poor; namely, punishment, treatment and social transformation (Gowan 2010). Over time, these constructions have overlapped and intermingled, and in practice 'street-level' strategies reflect this hybridity. For example, contemporary state imperatives to regulate low-wage labour and quell civil disorder through means-tested relief (such as general assistance or social assistance) reflect both moral and systemic constructions (Piven and Cloward 1993). Importantly, these constructions of poverty, through their strategies, take spatial forms. As Wolch and DeVerteuil (2001: 152) explain:

> the imprints of poverty management upon the landscape are both ideological and physical: a specific strategy (e.g. rehabilitation, punishment, isolation) becomes geographically manifest in terms of its settings, including how these settings appear and how they are arranged in space (e.g. concentrated, dispersed, isolated, centralized).

Hence, responses to poverty have encompassed various spatial settings, some grounded in a charitable ethos (such as public shelter), some grounded in a therapeutic ethos (treatment facilities, and hospitals), and others grounded in a more punitive ethos (such as jails) (Katz 1996). Skid Row, Los Angeles, like many marginal urban spaces in other North American cities, has evolved amidst these well-historicized *problematizations* of poverty and as an urban landscape represents the "concretization of social relations at a particular moment ... a palimpsest of contested and restless processes and discourses" (Wolch and DeVerteuil 2001: 151). Next we describe this palimpsest using the concepts of apparatus and assemblage.

Skid Row as an historic and current apparatus-assemblage

The genesis of Skid Row can be traced to the development of commercial hotels in the early 1900s built to temporarily house a transient working population but later converted into long-term housing for low-income Los Angelinos (Reese et al. 2010). Skid Row's evolution into a poverty management apparatus occurred over several decades as a result of alignments between: (1) sympathetic and compassionate responses to homeless populations by charitable agencies providing basic necessities such as food, clothing and shelter; (2) entrapment of these agencies in Skid Row; and (3) quarantine of service-dependent populations through legal and police action.

First, beginning in the early 1980s the homeless population grew exponentially in Los Angeles and this dramatic growth initiated an expansion of homeless shelters and other poverty-related services in Skid Row (Wolch and Dear 1993). This expansion of services could not have occurred if it were not for the early success of advocates and activists in reframing homelessness as a structural problem as opposed to an individual failure (Gowan 2010). During the 1980s, the street homeless became seen by many as deserving of public support resulting in federal funding (e.g. the McKinney-Vento Homelessness Act passed in 1987) for local emergency shelters and, correspondingly, "the opening of thousands of emergency shelters across the country ... ranging from small volunteer-run operations in church basements to vast dormitories in hundreds of National Guard armories" (Gowan 2010: 46).

Second, despite successes in mobilizing sympathy for the homeless, moral problematizations of homelessness endured. Still heavily stigmatized, the homeless and the places that served them were seen as threats to downtown redevelopment. As homeless services expanded, Los Angeles County enacted an explicit policy of containment to appease real estate investors, intent on remaking the adjacent downtown core (Reese *et al.* 2010). During the 1980s and into the 1990s, social service facilities were purposefully relocated to Skid Row, in some cases subsidized by the City of Los Angeles, resulting in a number of 'mega-shelters' anchored in the heart of Skid Row (DeVerteuil 2006; Stuart 2011, 2014). This policy of containment was further reinforced by powerful NIMBY-sentiments in adjacent communities who systematically thwarted attempts to relocate social services outside of the downtown core (Law 2001).

Third, as the homeless crisis worsened and Skid Row's homeless population stretched the capacity of shelters and over-spilled the prescribed boundaries, a new breed of anti-homeless ordinances (Mitchell 2003) were created and selectively enforcement by police outside of Skid Row. This resulted in the expulsion of homeless people from surrounding neighbourhoods and a 'path of least resistance' to Skid Row where activities such as panhandling and street sleeping were tolerated (Reese *et al.* 2010). In Skid Row itself, policing largely took the form of 'rabble management' in which officers worked to quell disorder and prevent disruptive populations from 'spilling over' into neighbouring districts (Stuart 2014).

Given the heterogeneous mixture of institutional agents – the voluntary sector, the police, the City of Los Angeles, overlapping Business Improvement Districts, redevelopment agencies and housing activists (see Reese *et al.* 2010) – the response to need was far from unitary. Rather, it was a de facto response constituted by a diverse assortment of practices and spaces operating concurrently with little or no coordination (Wolch and DeVerteuil 2001). This inadvertent apparatus has proven durable. As Reese *et al.* (2010: 323) argue:

> Skid Row initially acted as a convenient pool for society's outcasts, and over time this socio-spatial arrangement became entrenched with the policy of containment, protecting and attracting Downtown investments in other areas. However, this very entrenchment created a series of contradictory crises, as it established a constituency of SRO residents, street people and homeless service providers that acted as a bulwark against overt incursions.

Having traced Skid Row's development into an enduring managerial apparatus, we can now shift forward in time, to the near present, to examine Skid Row's on-going mutation through processes of assemblage.

A consideration of Skid Row in terms of assemblage begins with an examination of how the managerial apparatus detailed previously, over the past decade, has been inflected by novel policy developments and renewed alignments among social service agencies, the surrounding community and the LAPD. This assemblage has involved the (1) reorientation of social services in relation to medical and rehabilitative goals; (2) further entrenchment of social service agencies in Skid Row; and (3) a shift in the LAPD's philosophy of policing Skid Row.

First, in response to federal reforms introduced in the early 1990s, reforms that incentivized the 'continuum of care' model, social service agencies re-vamped traditional charitable models of care by introducing a suite of therapeutic programmes designed to make individuals 'housing ready' by addressing individual deficiencies and pathologies (Lyon-Callo 2004). This programmatic shift in social service delivery reflects the growing influence of medicalized constructions of poverty that prescribe treatment and rehabilitation as solutions to homelessness (Gowan 2010). It also reflects the competitive realities faced by charitable/voluntary sector organizations, many of which struggle to remain financially viable. As Stuart (2014: 1915) explains,

> in order to remain competitive for both public and private funding, homeless service organizations across the country replaced the traditional model of 'three hots and a cot' with a more interventionist, rehabilitative and 'post-secular' (Cloke et al. 2010) model centered on intensive residential recovery programmes.

Second, the inertia of three decades of social service concentration in Skid Row has proven to be a force powerful enough to withstand growing gentrification pressures (DeVerteuil 2011; Reese et al. 2010). A number of local supportive initiatives involving a diverse array of elected officials, activists and business leaders have resisted the dispersal of services from Skid Row (DeVerteuil 2015). As Reese et al. (2010: 322) observe, "resistance and inertia has so far worked against a successful frontal assault; the vast majority of homeless services remain, with no plans to relocate any time soon". While efforts are underway to 'compress' Skid Row by 'cleaning up' its edges, its sheer size and entanglement with public funds has rendered it especially resilient (DeVerteuil 2015).

Third, a noticeable shift in the strategy of policing Skid Row has been observed involving a change from quarantining service-dependent populations to facilitating exit from Skid Row by coercing entry into recovery programmes (Reese et al. 2010; Stuart 2011, 2014). This shift was epitomized by the introduction of the 'Safer Cities Initiative' in the early 2000s, a policy collaboratively designed by the LAPD in partnership with some of Skid Row's largest shelter operators, which "redeployed 80 additional officers into the neighborhood's 50 square blocks" (Stuart 2014: 1910). As Stuart (2014: 1917) notes, these officers act primarily as "recovery managers ... This new role centres on the task of decreasing individuals' shelter resistance and increasing their willingness to enroll in rehabilitative services and programs". This new function of the patrol officer aligns with the rise of 'coercive care' among the largest shelters, what Johnsen and Fitzpatrick (2010) have characterized as "forcing people to seek care against their will, ostensibly for their own good" (DeVerteuil 2014: 881). This provides the impetus to recast Skid Row as a "recovery zone" (Stuart 2014), that is a place-based approach to addiction treatment that features a sacred 'inside', the facility in which treatment is administered, and a profane 'outside', where street temptations waylay the unwary.

Attending to the assembly of new alliances, practices and spaces in Skid Row provides a richer understanding of how this managerial apparatus has mutated in recent years. The politics of urban poverty management are located here amidst the strategic relations threading the caring, coercive and controlling elements of Skid Row. The containment of purportedly 'disruptive populations' in this fifty-block inner-city neighbourhood owes as much to the genuine feelings of compassion and care embodied by service workers and volunteers as it does to the antipathy encoded into anti-homeless laws that were once only applied selectively outside of Skid Row but now have found a place within. Moreover, this containment is as much a result of the entrapment of service organizations in Skid Row due to NIMBY forces as it is their entrenchment due to concerted resistance by advocates and activists against displacement pressures.

Re-examining Skid Row through our theoretical matrix yields a different view of urban poverty management that is sensitive to the 'messy middle ground' (Cloke et al. 2010). Urban poverty management can be understood as an economy of interests, forces and inertias, the balance of which produce a concentration-containment effect in the city. As far as management is concerned, Skid Row certainly functions to enclose a sizable portion of LA's homeless population; however, it does not acquire this functionality because each and every component part is constitutively encoded with an ethos of control. Containment is a composite effect that emerges as a result of co-functioning among a wide range of caring and controlling institutions. Hence, spaces of poverty management such as Skid Row may be better read as *a contingent expression of mutually constitutive symbiosis* rather than an all-encompassing logic of urban neoliberalism. This mutuality is exemplified by the fact that resistance to gentrification, reflected in the efforts of social service agencies to remain in Skid Row, *and* community opposition to relocation of social services elsewhere, have been integral to Skid Row's entrapment and entrenchment. Similarly, the trend towards coercive care in the social service sector placed alongside a shift in mentality among police towards 'recovery management' further modulates its containment effect.

Finally, Skid Row itself generates the conditions for symbiosis among its caring and controlling institutions in light of the fact that while every intervention achieves success, these are always accompanied by unintended effects and failures. The unintended or unplanned effects of one form of intervention, be it shelter, treatment or incarceration, provide a point of connection for other interventions often with asymmetrical, even conflicting, logics. As the unforeseen consequences of one set of practices (i.e. social services) become the domain of intervention of another (i.e. policing), an apparatus is stitched together through *concurrent operation*. Rather than logically necessary, the relations linking up these elements are better seen as contingently obligatory, especially in light of the complex histories and messy local circumstances of particular locales.

Conclusions

In this chapter we have endeavoured to theoretically address the ambiguous, complex and 'messy' relationship between care and control in landscapes of urban poverty management. To do so, we adopted a *strategic* analysis of the co-incidence of 'care' and 'control' by focusing on the 'relations of exteriority' among agents of care such as shelters and agents of control such as the police. Doing so recasts urban landscapes such Skid Row as a "palimpsest of contested and restless processes and discourses" (Wolch and DeVerteuil 2001: 151). By making use of the problematization-apparatus-assemblage matrix we provide a different vantage point for conceptualizing the 're-working' of social welfare by processes of neoliberalization while simultaneously acknowledging the deep-seated forces of care and their potential as a nexus for resistance to such ideologies. Our theoretical framework provides a new vocabulary for making sense of these mixtures and their implications for urban (in)justice. This involves conceptualizing urban poverty management as an apparatus traversing an assortment of discourses, practices and spaces that, in conjunction, produce strategic effects, some of which are aligned with urban neoliberalism, some of which are not.

References

Anderson, B., Kearnes, M., McFarlane, C. and Swanton, D. (2012). On Assemblages and Geography. *Dialogues in Human Geography*, 2(2), pp. 171–189.
Cloke, P., May, J. and Johnsen, S. (2010). *Swept Up Lives? Re-Envisioning the Homeless City*. London: Wiley RGS-IBG Series.
Cox, K. (1993). The Local and the Global in the New Urban Politics. *Environment and Planning D: Society and Space*, 11(4), pp. 443–448.

Dean, M. (1992). A Genealogy of the Government of Poverty. *Economy and Society*, 21(3), pp. 215–246.
Dear, M. and Wolch, J. (1987). *Landscapes of Despair*. Princeton, NJ: Princeton University Press.
DeLanda, M. (2006). *A New Philosophy of Society: Assemblage Theory and Social Complexity*. New York: Continuum.
DeVerteuil, G. (2006). The Local State and Homeless Shelters: Beyond Revanchism? *Cities*, 23(2), pp. 109–120.
DeVerteuil, G. (2011). Evidence of Gentrification-Induced Displacement Among Social Services in London and Los Angeles. *Urban Studies*, 48(8), pp. 1563–1580.
DeVerteuil, G. (2014). Does the Punitive Need the Supportive? A Sympathetic Critique of Current Grammars of Urban Injustice. *Antipode*, 46(4), pp. 874–893.
DeVerteuil, G. (2015). *Resilience in the Post-Welfare Inner City: Voluntary Sector Geographies in London, Los Angeles and Sydney*. Bristol: Policy Press.
DeVerteuil, G. and Evans, J. (2009). Landscapes of Despair. In: T. Brown, S. McLafferty and G. Moon, eds. *The Companion to Health and Medical Geography*. Oxford: Blackwell, pp. 278–300.
DeVerteuil, G. and Wilton, R. (2009). Spaces of Abeyance, Care and Survival: The Addiction Treatment System as a Site of 'Regulatory Richness'. *Political Geography*, 28(8), pp. 463–472.
DeVerteuil, G., May, J. and von Mahs, J. (2009). Complexity Not Collapse: Recasting the Geographies of Homelessness in a 'Punitive' Age. *Progress in Human Geography*, 33(5), pp. 646–666.
Esping-Andersen, G. (1990). *The Three Worlds of Welfare Capitalism*. Cambridge: Polity Press.
Fairbanks, R. (2009). *How it Works: Recovering Citizens in Post-Welfare Philadelphia*. Chicago: University of Chicago Press.
Farias, I. and Bender, T., eds. (2010). *Urban Assemblages: How Actor-Network Theory Changes Urban Studies*. New York: Routledge.
Foucault, M. (1980). *Power/Knowledge: Selected Interviews & Other Writings 1972–1977*. New York: Pantheon Books.
Foucault, M. (1995). *Discipline and Punish: The Birth of the Prison*. New York: Vintage Books.
Foucault, M. (2007). *Security, Territory, Population: Lectures at the College de France 1977–1978*. New York: Macmillan.
Foucault, M. (2008). *The Birth of Biopolitics: Lectures at the College de France 1978–1979*. New York: Macmillan.
Gowan, T. (2010). *Hobos, Hustlers and Back-Sliders: Homeless in San Francisco*. Minneapolis: University of Minnesota Press.
Harvey, D. (1989). From Managerialism to Entrepreneurialism: The Transformation in Urban Governance in Late Capitalism. *Geografiska Annaler: Series B, Human Geography*, 71(1), pp. 3–17.
Johnsen, S. and Fitzpatrick, S. (2010). Revanchist Sanitisation or Coercive Care? The Use of Enforcement to Combat Begging, Street Drinking and Rough Sleeping in England. *Urban Studies*, 47(8), pp. 1703–1823.
Katz, M. (1996). *In the Shadow of the Poorhouse: A Social History of Welfare in America*. New York: Basic Books.
Law, R. (2001). 'Not in my City': Local Government and Homelessness in Los Angeles Metropolitan Region. *Environment and Planning C: Government and Policy*, 19, pp. 791–815.
Legg, S. (2011). Assemblage/Apparatus: Using Deleuze and Foucault. *Area*, 43(2), pp. 128–133.
Li, T.M. (2007). Practices of Assemblage and Community Forest Management. *Economy and Society*, 36(2), pp. 263–293.
Lyon-Callo, V. (2004). *Inequality, Poverty, and Neoliberal Governance: Activist Ethnography in the Homeless Sheltering Industry*. Peterborough, Canada: Broadview Press.
MacLeod, D.G. and Jones, M. (2011). Renewing Urban Politics. *Urban Studies*, 48(12), 2443–2472.
Marcus, G.E. and Saka, E. (2006). Assemblage. *Theory, Culture and Society*, 23(2–3), pp. 101–109.
McCann, E. and Ward, K., eds. (2011). *Mobile Urbanism: Cities and Policymaking in the Global Age*. Minneapolis: University of Minneapolis Press.
McFarlane, C. (2011). Assemblage and Critical Urban Praxis: Part One. *City*, 15(2), pp. 204–223.
McGuirk, P. and Dowling, R. (2011). Governing Social Reproduction in Masterplanned Estates: Urban Politics and Everyday Life in Sydney. *Urban Studies*, 48(12), pp. 2611–2628.
Mitchell, D. (2003). *The Right to the City: Social Justice and the Fight for Public Space*. New York: Guilford Press.
Murphy, S. (2009). Compassionate Strategies of Managing Homelessness: Post-Revanchist Geographies in San Francisco. *Antipode*, 41(2), pp. 305–325.

Peck, J. and Tickell, A. (2002). Neoliberalizing Space. In: N. Brenner and N. Theodore, eds. *Spaces of Neoliberalism*. Oxford: Blackwell Press, pp. 33–57.

Piven, F.F. and Cloward, R. (1993). *Regulating the Poor: The Functions of Public Welfare*. New York: Vintage Books.

Rabinow, P. (2003). *Anthropos Today: Reflections on Modern Equipment*. Princeton, NJ: Princeton University Press.

Rankin, K.N. and Delaney, J. (2011). Community BIAs as Practices of Assemblage: Contingent Politics in the Neoliberal City. *Environment and Planning A*, 43(6), pp. 1363–1380.

Raphael, D. (2009). *Poverty in Canada: Implications for Health and Quality of Life*. Toronto: Canadian Scholars Press.

Reese, E., DeVerteuil, G. and Thach, L. (2010). 'Weak-Center' Gentrification and the Contradictions of Containment: Deconcentrating Poverty in Downtown Los Angeles. *International Journal of Urban and Regional Research*, 34(2), pp. 310–327.

Sloss, J., Fording, R. and Schram, S. (2011). *Disciplining the Poor: Neoliberal Paternalism and the Persistent Power of Race*. Chicago: University of Chicago Press.

Smith, N. (1996). *The New Urban Frontier: Gentrification and the Revanchist City*. London: Routledge.

Sparks, T. (2012). Governing the Homeless in an Age of Compassion: Homelessness, Citizenship, and the 10-year Plan to End Homelessness in King County, Washington. *Antipode*, 44(4), pp. 1510–1531.

Stuart, F. (2011). Race, Space, and the Regulation of Surplus Labor: Policing African Americans in Los Angeles's Skid Row. *Black Critiques of Capital: Marxism, Racism, and Exclusion*, 13(2), pp. 197–212.

Stuart, F. (2014). From 'Rabble Management' to 'Recovery Management': Policing Homelessness in Marginal Urban Space. *Urban Studies*, 51(9), pp. 1909–1925.

Tyner, J. (2013). Population Geography I: Surplus Populations. *Progress in Human Geography*, 37(5), pp. 701–711.

Von Mahs, J. (2013). *Down and Out in Los Angeles and Berlin: The Sociospatial Exclusion of Homeless People*. Philadelphia, PA: Temple University Press.

Wolch, J. and Dear, M. (1993). *Malign Neglect: Homelessness in an American City*. San Francisco: Jossey-Bass.

Wolch, J. and DeVerteuil, G. (2001). Landscapes of the New Poverty Management. In: J. May and N. Thrift, eds. *TimeSpace*. London: Routledge, pp. 149–168.

27
URBAN COMMUNITY GARDENS AS NEW SPACES OF LIVING

Rina Ghose and Margaret Pettygrove

Introduction

Manufacturing cities in the Global North have struggled with de-industrialization, disinvestment and poverty since the decline of Fordist-Keynesian economies in the late 1970s. In the USA, central city neighbourhoods occupied primarily by African Americans were particularly devastated. As 'inner-city' neighbourhoods, these are highly segregated spaces, characterized by high levels of poverty and unemployment, and are synonymous with decay, despair and 'urban blight'. The emergence of vacant lots caused by the razing of ill-maintained and abandoned buildings is particularly symbolic of such blight, as these are sites of garbage accumulation, crime and vice. Further, food insecurity in US cities tends to be most acute in central city neighbourhoods (Zenk *et al.* 2005). These neighbourhoods face significant disparities in access to nutritious food for urban residents, as corporations in charge of grocery stores and food industry resolutely avoid establishing stores in impoverished areas, on the basis of race and class divisions (Frank *et al.* 2006; Sharkey *et al.* 2009; Zenk *et al.* 2005). These 'food deserts' put residents at greater risk for food insecurity and diet-related disease (Larsen and Gilliland 2008; Larson and Moseley 2010).

Increasingly, community gardens are established on such vacant lots, transforming them from spaces of despair to new spaces of living for the residents of central city neighbourhoods. Urban gardens emerge as productive spaces that alleviate urban food insecurity, revitalize 'blight', increase citizen participation and build social capital. Yet the production of gardens is shaped by neoliberalization and politics of turf, and requires active engagement with actors-networks. Sustainability of urban gardens is a challenge, yet the increasing establishment of gardens in the vacant lots of 'inner city' indicates that they are seen as spaces of hope.

In analysing the politics of urban governance, Cox's notions of "spaces of dependence", "spaces of engagement" and "networks of association" are particularly useful. Spaces of dependence "are defined by those more-or-less localized social relations upon which we depend for the realization of essential interests" (Cox 1998: 2). These spaces contain local classes and interest groups, defined by social divisions of labour or other social geographies such as residence. Struggles for territorial control emerge to protect local interests. In response, agents then organize themselves and engage with other centres of power, creating spaces of engagement, defined as "the space in which the politics of securing a space of dependence unfolds" (2). These include those sets of relations that extend into spaces of dependence, but also beyond them to construct networks of association, exchange and politics at multiple scales, that help to protect local interests. These networks of association then form spaces of engagement. Networks of association are built and used by agents from different local interest groups to influence (directly or indirectly) territorially powerful local state agencies.

In this chapter, we use this framework to conceptualize community gardens as 'spaces of engagement' arising out of the 'spaces of dependence' created in the 'inner city', through which citizens can claim rights to the city. We aim to unpack the complex narrative of community gardening as new 'spaces of living'. Our findings are derived from seven years of longitudinal research conducted on this topic in Milwaukee, Wisconsin. We conducted document analysis of city government's policies regarding urban agriculture and community gardening. We conducted semi-structured interviews with the staff at government agencies and non-profits related to urban agriculture and with selected community gardening groups. Finally, we collected data from participant observations at public meetings and gardening events and volunteered in community gardens.

Community gardens in the neoliberal city

Shaped by the ideologies of free market, entrepreneurialism and privatization, neoliberalism has been a dominant policy influence at all levels of US government (Harvey 2007; Peck and Tickell 2002). At the urban scale, promotion of the collaborative governance model encouraging citizen participation and volunteerism has been a key neoliberal strategy in implementing a lean government (Ghose 2005; Jessop 2002; Lepofsky and Fraser 2003). Built upon the principle of public-private partnership, the collaborative governance model has significantly restructured the role of the traditional government through involvement of the private sector in governance, budgetary cutbacks on public welfare and devolution of state responsibilities to citizen groups. As state provision of basic welfare entitlements wanes, rights traditionally afforded by citizenship accrue only to individuals able and willing to voluntarily work for them (Perkins 2009). By linking volunteerism to discourses of place-making, empowerment and local autonomy, neoliberalism has legitimized conditional citizenship. These discourses are simultaneously individualistic, in that they promote self-help, and communitarian, because they draw on notions of participation in a community (Fyfe 2005; Van Houdt et al. 2011). Resource-poor minority communities are disproportionately burdened by state welfare retrenchment, which compels communities to compensate through voluntary or grassroots community development projects (Perkins 2010). However, the capacity to participate in voluntary organizing or in formal government processes varies contextually and opportunities for participation are not equally accessible to all (Fyfe 2005; Kearns 1995; Staeheli 1999). The rhetoric of collaborative governance simultaneously obscures and reproduces race and racism as organizing principles of society through discourses about individual responsibility and the supposed colour-blindness of market-based systems (Roberts and Mahtani 2012). Citizens practising localized community development can become complicit in the construction of neoliberal hegemony, acting as neoliberal citizen-subjects who alleviate the state from service.

Urban community gardens have proliferated as localized strategies to manage impacts of neoliberalization, including social service cutbacks, reductions in food stamps, environmental degradation, and disinvestment (Pudup 2008). Community gardens function as sites of place-based community development that provide access to affordable nutritious foods and safe green space and create new economic activities and employment opportunities related to urban agriculture (Armstrong 2000; Baker 2004; Irazábal and Punja 2009; Kurtz 2001; Quastel 2009;

Schmelzkopf 1995; Staeheli *et al.* 2002). Community gardening addresses key inner-city revitalization goals such as bringing new investment into disinvested neighbourhoods, increasing neighbourhood safety, creating new green space for recreation and improving declining housing stock. Such effects can lead to an increase in property values. Given that property values plummeted drastically due to disinvestment, community leaders and local state agencies perceive increased property values as a positive step. But this can also lead to gentrification and displacement of original residents, as transformed central city neighbourhoods have the amenities that attract the attention of buyers, real estate investors, and developers (Quastel 2009).

Simultaneously, neoliberalization constrains community garden development, as it emphasizes land use competition and creates pressure to generate revenue (Rosol 2012; Smith and Kurtz 2003). Community gardens are often established in vacant lots that are owned by local state agencies, which prioritizes commercial development in such lots, as the financial returns from commercial development are perceived to be high enough to counter the negative outcomes of disinvestment. Through restrictive lease agreements, local state agencies maintain control over these vacant lots so that if a lucrative development opportunity arises, a community garden can be evicted. Even where local state agencies promote green space as a revenue-generating use, consumption-oriented spaces tend to be prioritized and community gardens marginalized, through land tenure restrictions, landscape regulations or steep rents (Domene and Saurí 2007; Perkins 2009, 2010; Schmelzkopf 2002).

Further, the neoliberal emphasis of devolution of state responsibilities upon civil society within a climate of continuous fiscal reductions has also challenged community garden development by increasing competition amongst grassroots organizing (Elwood 2004; Ghose 2005; Rosol 2012). While the shift to the public-private partnership model ostensibly increases opportunities for civil society control, it also compels groups to compete for limited material resources and prioritize survival over political activism, leading to fragmentation of community organizing (Hackworth 2007; Newman and Lake 2006). Neighbourhood-based community garden groups, like many central city community-based organizations, tend to be small, vulnerable and resource poor, and struggle to operate in a climate of limited funds and material resources (Ghose 2005; Ghose and Pettygrove 2014a, 2014b). Community gardening across a city can be fragmented and limited to neighbourhoods where citizens already have access to material and social resources, resulting in challenges to establishing larger advocacy campaigns (Kurtz 2001; Smith and Kurtz 2003).

In this context, social network formation is a critical strategy for grassroots community organizing and a mechanism by which actors construct spaces to defend their interests (Cox 1998; Ghose 2007). The metaphor of network is used to describe how human actors interact with their social relations and refers to individual or group actors related through economic, social or political connections (Gilbert 1998; Smith and Kurtz 2003). Network formation allows impoverished organizations and individuals to gain political power and economic resources while it also provides opportunities for powerful actors to influence the community (Cox 1998; Gilbert 2008). Community garden groups have developed strategies to address the constraints imposed by neoliberalism primarily through engaging in supportive networks of relationships (Armstrong 2000; Smith and Kurtz 2003). These groups construct networks to negotiate land use conflicts, acquire material resources and bolster advocacy (Irazábal and Punja 2009; Schmelzkopf 2002; Staeheli *et al.* 2002). These networks involve citizens, established non-profit organizations, government agencies and private funders. They enable community garden groups to acquire necessary organizational resources and technical knowledge (Armstrong 2000; Pudup 2008; Schmelzkopf 1995). Garden groups use social networks to expand the scale of the conflict beyond the local and beyond traditional circuits of capital by creating new 'spaces of engagement' (Ghose and Pettygrove 2014a, 2014b; Smith and Kurtz 2003).

Community gardening in Milwaukee

De-industrialization and urban renewal efforts in the 1960s and 1970s contributed to the geography of urban agriculture in Milwaukee by disrupting predominantly black neighbourhoods and furthering uneven development. The legacies of these spatial processes are reflected in disproportionately high rates of poverty, food insecurity and land vacancy in Milwaukee's 'inner-city' black neighbourhoods. Community gardening emerged as a form of activism to combat entrenched poverty and decline in the central city (Ghose and Pettygrove 2014a; Roy 2010). The evolution of community gardens in Milwaukee's central city over time is explained in Table 27.1.

In the 1960s, the University of Wisconsin Extension (UWEX) through its Milwaukee County office first established its community garden rental programme to champion urban agriculture as a part of its educational and community outreach goals. The office provided plots of land, access to water and technical assistance for groups seeking to develop community gardens in Milwaukee City and in Milwaukee County. In the 1970s, this programme assumed ownership of a garden on a land parcel in the central city that later became the principal site of Alice's Garden.

Established in 1972, the now named Alice's Garden is the oldest, a two-acre farm with 127 rental garden plots. Its goal is to provide a community space where residents can gather to learn how to grow and cook organic food. It also offers yoga classes, cooking classes, reading circles and educational and wellness programmes such as 'Healthy Mom, Healthy Kids', 'Reclaiming and Nourishing Family Traditions' and 'Seedfolks Roots and Shoots Club'. Alice's Garden has blossomed due to its strong community engagement, and is now a role model for other organizations to emulate. In recognition of its contributions, the City Hall of Milwaukee has identified this garden as a strategic partner in its goal of advancing urban agriculture in the central city.

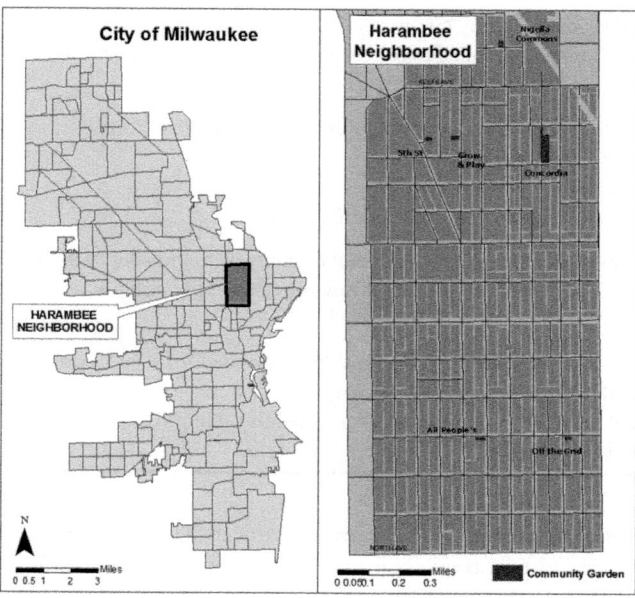

Figure 27.1 Harambee neighbourhood. The left-hand map locates Harambee in the City of Milwaukee, Wisconsin. The right-hand map displays the Harambee neighbourhood, with community garden sites labelled and symbolized in dark grey

Table 27.1 History of Milwaukee food projects

Date	Event
1963	University of Wisconsin Extension (UWEX) Milwaukee County community gardens programme begins
1969	Black Panther Party (BPP) free breakfast programme for children in Milwaukee (1969–1970)
1970	Center Street Market (predecessor to Fondy Farmers Market) begins UWEX-Milwaukee County takes over community garden that later became Alice's Garden
1990	Walnut Way Conservation Corps (WWCC) is established Growing Power is established
2000	Milwaukee Urban Gardens (MUG) forms Department of City Development (DCD) establishes permitting process for community gardens on vacant city-owned residential lots All People's Church establishes community garden in Harambee neighbourhood CORE/El Centro Garden and nutrition programme begins
2006	Alice's Garden established (under leadership of SeedFolks Youth Ministry)
2007	Milwaukee Food Council (MFC) begins Milwaukee Urban Agriculture Network (MUAN) begins Groundwork Milwaukee begins organizing community gardens
2008	Lindsay Heights Healthy Corner Store Initiative begins
2009	Milwaukee Childhood Obesity Prevention Project (MCOPP) begins Victory Garden Initiative forms Increasing volume of home foreclosures in Milwaukee (peaks 2010–2012)
2010	Latinos por la Salud Healthy Grocery Store Campaign Milwaukee Common Council passes ordinance allowing beekeeping within city limits (with permit and other restrictions)
2011	Milwaukee Common Council passes ordinance allowing hen keeping within city limits (with permit and other restrictions) CORE/El Centro rooftop garden opens EPA Region 5 conducts UA code audit (published 2012)
2013	City of Milwaukee produces ReFresh sustainability plan, formally establishes HOME GR/OWN as programme within Office of Environmental Sustainability Mayor launches Strong Neighborhoods Plan
2014	UA code changes passed within city, enabling expanded scope of UA practices on urban lots Groundwork Milwaukee and MUG merge HOME GR/OWN pilot projects implemented (including Gillespie Park) UWEX Urban Discovery Farm begins

Since the 1990s, numerous food and nutrition projects have emerged in Milwaukee, caused by the unequal access to nutritious food in its 'inner city' (Pettygrove 2016; Pettygrove and Ghose 2016). Urban agriculture in Milwaukee received a major boost in 1993 when Growing Power was established by noted African American basketball athlete Will Allen. Its headquarters are located on a three-acre farmland in the northern edge of Milwaukee, to "provide equal access to healthy, high quality, safe and affordable food for people in all communities" (www.

Urban community gardens

Figure 27.2 Urban agriculture in Milwaukee, Wisconsin

growingpower.org). Today it has urban farms in multiple locations in Milwaukee and Madison, Wisconsin as well as Chicago, Illinois. The farms grow fruits, micro greens and vegetables. Fresh water fish are also grown through aquaponics projects. Growing Power also produces its own nutrient-rich soil through vermicomposting and has its own beehives to produce honey. Growing Power has collaborated with many large and small community garden groups in Milwaukee and Chicago, providing technical expertise in urban agriculture, aquaponics and composting soil. Such efforts have helped countless community gardens to grow and flourish in central city neighbourhoods. It also has youth programmes that aim to provide urban agriculture education, leadership development and job training skills. Growing Power has received many national awards for advancing urban agriculture and Will Allen has been lauded by President Obama and First Lady Michelle Obama.

Walnut Way Conservation Corps (Walnut Way) formed in 1998 as a community organization in central city Milwaukee's Johnson's Park neighbourhood with a goal of undertaking "a comprehensive revitalisation strategy rooted in resident driven social, economic and environmental advancement" (www.walnutway.org). Walnut Way undertook community gardening in vacant lots as an explicit community development strategy to combat crime, poverty and neighbourhood decline (Roy 2010). Through strategic planning and partnerships with critical actors, Walnut Way has successfully pursued its revitalization goals. By actively participating in the city's revitalization programme, and by partnering in formal collaborations with educational institutes, for-profit organizations and philanthropic foundations, Walnut Way has emerged as a critical actor in revitalization projects through public-private partnerships. It has reshaped its neighbourhood by building a large campus where residents have created production gardens and orchards on vacant lots. It has hoophouses, rainwater storage, composting and apiaries. It also has a health and wellness centre that promotes healthy living through meditation, exercise and healthy eating. Produce is shared with neighbours, sold at local markets, restaurants and grocery stores. It has youth programmes that provide urban agriculture education, leadership development and job training. It also provides employment opportunity through its landscape service. The gardens are also important social spaces and its annual harvest festival with live music and fresh food is now a celebrated cultural event. Because of Walnut Way's significance in both community gardening and neighourhood revitalization, it is now a formal community partner in the City of Milwaukee's urban agriculture programme.

Several larger non-profit organizations have also played significant intermediary roles of support that helped establish and sustain neighbourhood gardens. Milwaukee Urban Gardens (MUG) was founded in 2000 "by a group of individuals who lost their gardens when the properties were used for development. They incorporated and began advocating for the long-term protection of community gardens and neighborhood green space in urban Milwaukee" (www. groundworkmke.org/milwaukee-urban-gardens/). Arising out of this activism, MUG was established as a non-profit land trust, to champion the creation of community gardens city-wide. Due to MUG's advocacy, the first municipal permitting process for community gardens was established through the City of Milwaukee, allowing citizen groups to build community gardens on vacant city owned lots with temporary leases of up to three years. The following years then saw the emergence of several community garden projects across the central city (Figure 27.1).

Another key organization, Groundwork Milwaukee, was formally established in 2007, as the local chapter of the Groundwork USA network of independent, not-for-profit, environmental businesses called Groundwork Trusts. Locally organized and controlled, it creates "private-public-community partnerships to provide cost effective project development services focused on improving their communities' environment, economy, and quality of life" (www.groundworkmke.org). MUG and Groundwork Milwaukee formally merged in 2013, and have collectively established over ninety-two community gardens in central city neighbourhoods. Its Green Team programme employs 14- to 17-year-olds to learn environmental leadership and gardening skills. The Green Team, along with volunteers, assist in the building and maintaining of community gardens (personal communication, Groundwork Milwaukee 2016).

These two organizations are part of the key supportive networks that have helped citizens in establishing community gardens by identifying suitable vacant lots for them, guiding gardening groups through city permitting procedures to obtain land and water, and collecting annual permitting fees for payment to the city planning department. They also purchase land for community gardens, provide insurance (up to $1 million), ongoing training in farming techniques, loan tools and provide labour for garden establishment. They also connect gardeners with private businesses and city agencies that provide lumber, woodchips, seeds and soil, so that citizens can create gardens in raised platforms to avoid soil contamination. Most garden groups are small, informal, dependent entirely on volunteer labour, and lack the financial resources and organizational infrastructure associated with formal non-profits. Thus, supportive networks that connect grassroots garden groups to city government agencies, philanthropic institutes and private businesses, have been crucial for garden development.

Most garden development in Milwaukee has been community-driven, but the ongoing activism for food justice and neighbourhood revitalization prompted the City to launch HOME GR/OWN (HG) in 2013, the first municipal programme devoted solely to urban agriculture. Initiated by Milwaukee's Mayor and operated by the Environmental Collaboration Office (ECO) within the city government agency, the programme was established (with city budget funding) as a pilot project for Milwaukee's 2013 sustainability plan (ReFresh Milwaukee). ECO aims to:

- transform targeted neighbourhoods by concentrating city and partner resources, catalysing new, healthy food access and greenspace developments;
- make it easier to access local food and re-purpose city-owned vacant lots … to streamline processes, permitting and ordinances, making it easier to grow and distribute healthy food, start new food-based businesses and improve vacant lots;
- work within Milwaukee's community food system to link local growers to local markets, increase urban food infrastructure (water, access, compost) and support new urban farms and healthy food retailers and wholesalers.

(city.milwaukee.gov)

HG projects are designed collaboratively between community groups and ECO. Construction and maintenance during the first year is provided through city-contracted labour, and gardening groups are expected to fund long-term garden maintenance costs. HG operates through an explicitly public-private partnership model, involving local philanthropic foundations, local businesses, larger urban agriculture non-profits, and coalition groups with the goal of developing vacant lots into food production green spaces that will come under the long-term care of community organizations. It is funded by a network of philanthropic foundations such as the Greater Milwaukee Foundation, Northwestern Mutual Foundation, the Fund for Lake Michigan and the Zilber Family Foundation. HG has installed 154 community gardens that include orchards and parks on city-owned vacant lots in this manner. HG has also implemented land use and building code changes that will enable a broader range of urban agriculture. The establishment of HG reflects a significant shift in the policy of the local state agency, from the previous tolerance of community gardens to the active promotion and financing of urban agriculture projects on city-owned land.

Community gardens as emergent spaces of living

Urban community gardens as new spaces of engagement in central city neighbourhoods perform multiple functions. Since food insecurity in US cities tends to be most acute in central city neighbourhoods, community gardens were created for contesting the prevalent food deserts of the central city and providing access to fresh organic produce, which otherwise is unaffordable to the impoverished residents. But community gardens also serve as a means to grassroots control that enable residents to shape their urban landscapes, strengthen collective identities and engage in governance (Armstrong 2000; Baker 2004; Kurtz 2001; Schmelzkopf 1995). Based on our research into selected community gardens, we share empirical findings to illustrate these functions. In this research, we examined the two most prominent community gardens (Alice's Garden, Walnut Way) that serve as role models. Additionally, we studied six community gardens located in the central city neighbourhood of Harambee.

We find that neighbourhood gardens serve multiple functions besides food production. Residents of Harambee neighbourhood built the Grow and Play garden because they wanted to create a safe play space to children.

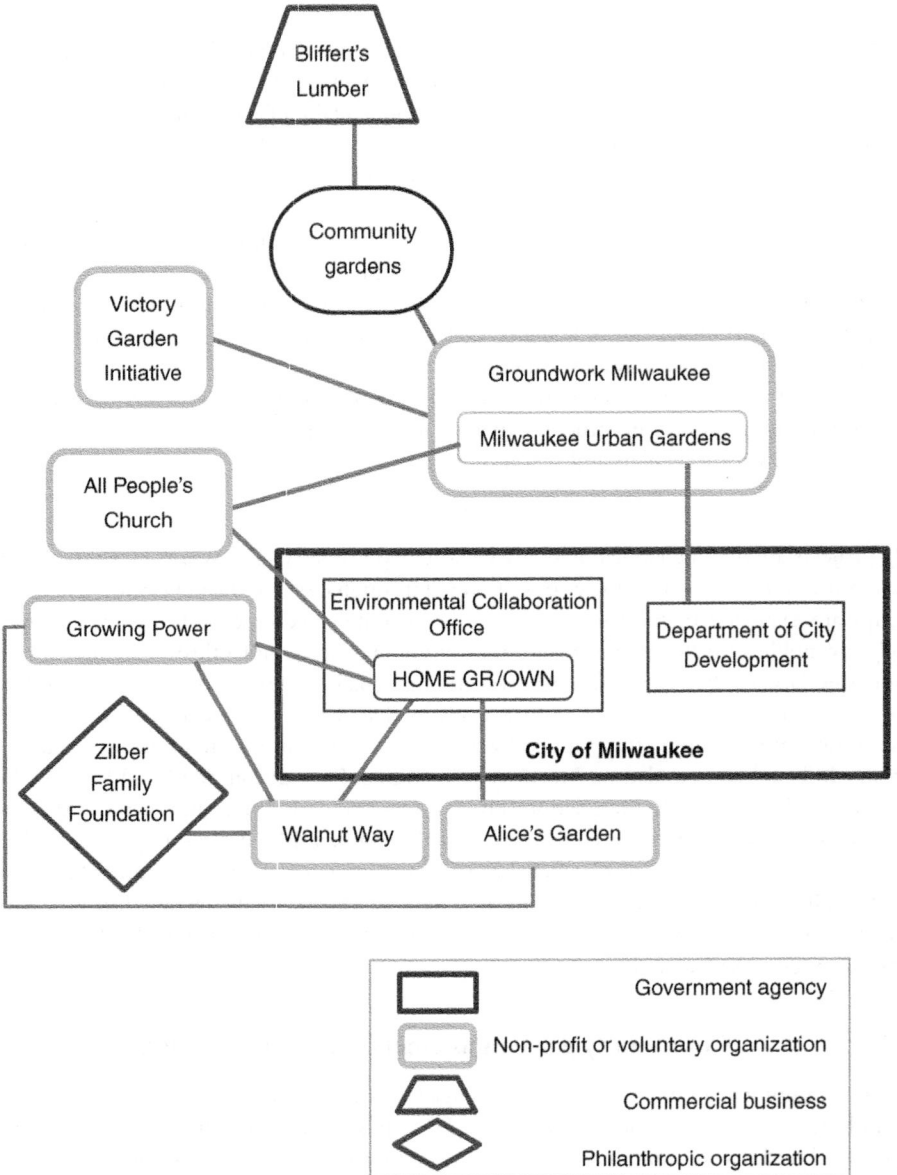

Figure 27.3 Networks of association for urban agriculture in Milwaukee, Wisconsin

Nigella Commons garden was built as a response to the residents' needs for green space. 5th Street garden developed from a desire to create a space where residents could engage in gardening activities. All People's garden grew as an educational space for youth and community space for residents, but also to contest food injustice. Concordia garden evolved as part of an environmental justice movement, emphasizing an off the grid approach. All People's garden is seen as "a place of justice" by its leader because it serves a neighbourhood dominated by renters who do not otherwise have ownership of farmable land (personal communications). Such a 'food justice' mission is echoed by other community gardens in Milwaukee that actively contribute fresh produce to local food pantries, which serve homeless people. Fresh produce is one of the most popular items that food pantries distribute, yet is the most expensive food the pantries purchase. Accordingly, receiving donations of organic produce greatly benefits the pantry and its beneficiaries (personal communications).

In shaping neighbourhood space, community groups assert control and enact local politics. While larger organizations such as Walnut Way and Alice's Garden participate formally in the City's many collaborative planning projects, smaller gardening groups too challenge hegemonic control. This is best exemplified by the case of All People's garden. Utilizing a vacant lot owned by the City, All People's Church had established the All People's garden in 1992 as a space for youth education, food production and community green space. In a neighbourhood consisting of mainly apartment buildings with limited green space, the garden became a place of beauty for its residents. Despite its popularity, the City evicted the thriving garden in 2005 in order to attract commercial development for that site. However, development opportunities fell through and the lot sat abandoned until 2008 when All People's Church reclaimed the space and rebuilt the garden. Its pastor conceives the rebuilding of the garden as "an act of civil disobedience" contesting the City's definition of the space (personal communication). Faced with unified community support for the garden, the City reissued All People's Church a gardening permit, and the garden is thriving once again. By reclaiming control over the space, the church and the residents spatially enacted their right to control the forms and functions of local space.

Community gardening is also tied to cultural production as residents suggest that the ability to grow food is an important cultural practice. As one neighbourhood resident responded, "that's African American culture: we grow our own foods" (personal communication). Another resident contends that growing food is important for older adults because "a lot of the folks, they've grown their own food before ... they're the part of the generation that they remember doing that with their parents" (personal communication). For these citizens, community gardening is meaningful because it provides a connection to family tradition. Alice's Garden explicitly demonstrates the connection between food, cultural traditions and history. Its 'Fieldhands and Foodways' programme particularly celebrates the historic food traditions of African Americans. It also honours its historical setting as the birthplace of the Underground Railroad in Wisconsin, a significant political movement to resist slavery.

Urban community gardens are also spaces for civic engagement and social interaction. Large gardening organizations such as Groundwork Milwaukee, MUG and Growing Power have provided extensive support to small gardens through technical instruction and education in horticulture, nutrition and environmental stewardship. Small gardens too share their knowledge with each other during formal and informal meetings. Further, many small garden participants begin with minimal gardening skills and learn how to garden, as they go along, from other participants. Because residents teach and mutually support each other in small community gardens, those with little or no gardening expertise can participate, and these community gardens become more socially inclusive. MUG has further facilitated such knowledge sharing in its ninety plus gardens, by creating its 'cluster' programme, a gardening network where one garden acts as the resource centre for a 'cluster' of proximal gardens.

Community gardens provide safe, green spaces of interaction, a rare commodity in the 'inner city'. Neighbourhood residents "want the neighborhood to be a good place ... where their kids are safe", and they see the garden as contributing positively to these ends (personal communication). As its name suggests, the Grow and Play garden is designed to allow children to play safely while their parents garden in the same space. Children "are always around" and have been eager to be involved in Nigella Commons garden because there is "so little green space" in the neighbourhood.

Many gardening organizations have created formal youth education programmes, through funding from philanthropic foundations. Organizations such as Growing Power, Walnut Way, MUG, and All People's Church have prioritized youth education and job training programmes for teenage children. This is a valuable asset for central city neighbourhoods to combat their high rate of unemployment.

Community gardens are also greatly valued by residents because they are perceived as contributing to the safety, stability and aesthetic quality of the neighbourhood. By involving the residents in improving the built landscape and perceived cultural qualities of the neighbourhood, the gardens have come to function as spaces of collective identity and cultural meaning. Alice's Garden, Walnut Way, and MUG gardens celebrate neighbourhood cultures and traditions and have strong support from the residents.

Community gardens are multigenerational, multicultural spaces in which youth and adults interact, collaborate and share knowledge. As one gardener explained, "my grandchild can play with the grandchild of the man who lives next door; that's so important to me" (personal communication). Community gardens also address the problem of social isolation. As the founder of Alice's Garden noted, "there's no question to me that part of what Alice's does is bring folks out of isolation. The garden introduces people who would otherwise never meet." Nigella Commons garden has brought together many formerly unacquainted residents who had lived for nearly thirty years without knowing each other.

Community gardens also function as significant multiracial spaces in which black and white residents work together. Given Milwaukee's long history of hyper-segregation and racial tensions, gardening provides an opportunity for cultural exchange and racial harmony. Our observations indicate that this is due to the interaction and collaboration based on shared interest from the initial planning stages of the gardens. This is best exemplified in the case of Nigella Commons garden, which was created by two white female residents who had recently moved to the neighbourhood and invited all interested residents to join them in building a community garden; the group became popular and is now predominantly black. Over time, opportunities have emerged for residents to explore cultural and racial understandings through exchange of knowledge about plants and food. This garden thus provides a shared space for white and black residents to discuss cultural differences while building connections between each other. Other gardens too have seen such positive social interactions.

Challenges to community gardening in the neoliberal city

Community gardening in the neoliberal city is a challenging proposition. In the context of austerity and budgetary cutbacks within the city government, it relies greatly on volunteerism. Volunteers decide how to assign responsibility for garden work, what to plant in garden beds and how to distribute garden produce.

Volunteers organize workdays, transport garden tools, raise funds and perform regular maintenance tasks, such as mowing grass in warmer months and shovelling snow in winter. The ability to mobilize neighbourhood residents to serve as volunteers as well as recruit volunteers from other neighbourhoods requires strong leadership. The success of Walnut Way and Alice's Garden is largely attributed to the leadership qualities of these organizations. Both are led by educated, African American women who reside in the neighbourhood, have strong visions of neighbourhood development and have the ability to translate their visions into action. Using community gardening as a neighbourhood building strategy, both leaders mobilized their residents, involving them in community redevelopment activities. They also successfully built strategic partnerships with businesses, city government, educational institutes and philanthropic foundations, which provided both discursive and material support. Similar leadership quality is found in the All People's garden, where the church and the garden are an established neighbourhood community site. Here residents interact frequently and take ownership of the garden. The pastor has built many networks of association to support the church's garden, youth activities and food pantry in the neighbourhood. Similar leadership is seen in Nigella Commons garden, led by two educated, white female residents, who in their spare time write grant applications to solicit funding, store tools at their homes, participate and lead in gardening activities and organize other volunteers into regular group workdays at the garden. But it is their ability to successfully engage their African American neighbours in the gardening activities that has created a vibrant and racially diverse garden in a hyper-segregated city.

Such leadership qualities are not always evident as community gardening can evolve as a grassroots venture, and the ability to sustain a community garden can thus be uneven. Reliance on volunteerism requires extracting material and labour resources from already resource-poor citizens, who struggle to fulfil basic survival needs. Many residents live below the poverty line, and must work at several jobs to make ends meet. Time and energy for volunteering in gardens can be quite limited for many residents. Small gardening groups must know how to acquire material resources and navigate specific procedures for obtaining land and water use permits, through partnerships with large non-profit organizations, funding agencies or informal social networks. Such awareness is created through personal social networks, and lack of such awareness can halt gardening activities. This suggests that while the opportunity for grassroots community garden development exists, the ability to take advantage of the opportunity depends on having knowledge acquired through specific channels. In recent years, the City, Growing Power, MUG and food justice organizations have made particular efforts to facilitate network formation and knowledge sharing, in order to alleviate some of these issues.

The capacity to procure material resources is equally crucial to successful community garden development. This includes knowing what companies are likely to provide in-kind donations and how to apply for monetary grants. Groups with relatively good access to material resources can more easily develop community gardens, while those with poor resource access may be unable to afford basic infrastructural needs, such as lumber for raised beds or clean soil. Many garden groups in Milwaukee have obtained donations or grant funding, but in varying quantities and from different sources. Because each group relies on its own knowledge, social connections and skills to secure material resources, there is no guarantee that all groups will be equally successful. All People's Church has a network of institutional connections and organizational experience that enables it to find and apply for grant funding, which has allowed it to hire a part-time garden coordinator. Nigella Commons, conversely, has obtained funding only to build raised beds. A Nigella Commons leader explains the importance of funding to the initial success of the garden: "Building the garden was a huge expense ... if we had not been in the ten neighborhoods [eligible for this particular grant], I'm still not even sure how we would have managed..."

Although Nigella Commons successfully secured grant funding, this was principally due to the knowledge, skills and commitment of two lead volunteers. The uncertainty and variability of funding underscores one of the principal ways in which grassroots community garden development is inherently uneven and context-dependent.

Dependency on volunteerism as a strategy for navigating the effects of urban disinvestment and political marginalization has its limitations. Issues of volunteer turnover, lack of skill among volunteers or unreliability of volunteers can limit organizational capacity and scope of feasible of action (Sheriff 2009). While the large, well-established gardens tend to have a steadier supply of volunteers, small gardens tend to suffer. In the Concordia, Grow and Play and Nigella Commons gardens, levels of volunteer participation have fluctuated over time, particularly during each garden's first year. Participants tend to be more involved in maintaining personal plots, to the neglect of essential communal tasks. Grow and Play leaders struggled for two years to convince participants to share responsibilities for tasks like mowing grass. Participation levels dropped in the Nigella Commons garden when leaders stopped organizing group workdays. Even highly committed volunteers often have limited time to dedicate to the garden and are restricted in what they can feasibly accomplish because they support themselves through regular employment. Without adequate volunteer participation groups lack the labour necessary to establish and maintain community gardens.

Additional problems occur when volunteers lack the skills or physical abilities necessary for garden work. In the Grow and Play and Nigella Commons gardens, the most actively involved participants are elderly residents with limited mobility or children too young and unskilled to work efficiently. Consequently, the burden of ensuring the survival of these gardens typically falls on a small number of skilled individuals. A Concordia garden organizer notes that with volunteers "tasks are not always going to happen exactly when or how you want them to happen" (personal communication). Although the director values the volunteer basis of the garden, she acknowledges that paid employees would be able to tend it more efficiently and effectively. By privileging volunteerism as the basis of community garden development, we are accepting potential reductions in capacity for food production and potentially excluding those who are unable to volunteer.

Community gardens are thus constrained by a range of factors, some caused by neoliberalism, some of which pre-date its emergence in Milwaukee. Critics contend that community gardens can thus cultivate neoliberal citizen-subjectivity by conditioning participants to behave as consumers and to pursue change through individual endeavour (Guthman 2008; Pudup 2008). But participation in community gardens can also enable individuals to negotiate alternative meanings of citizenship and cultivate alternative political imaginaries (Staeheli 2008). Regardless of how they are understood the growing popularity of community gardens in central city neighbourhoods suggest that these are spaces of engagement and living.

Conclusion

Citizen participation through gardening in the context of neoliberalization can simultaneously empower and challenge citizens (Elwood 2004; Kofman 1995; Trudeau 2008). Community gardens can be sites through which citizens can more actively reshape their urban landscape and develop alternative citizen subjectivities (Baker 2004; Staeheli 2008). Community gardens are grassroots spaces that exist because citizens have supported their development and volunteered to build them. Such community gardens provide numerous benefits to participants, including opportunities to grow fresh food, to interact with other residents and to control small pieces of urban space. To the extent that community gardens represent efforts to resuscitate degenerating urban space from forces of economic and political marginalization, they represent prospects for democratic citizenship practice. These community gardens create claims to space and resist local government policies that prioritize for-profit development, whether or not they concretely impact state policies.

Yet, there are many challenges to community gardening. While the rights to build and participate in community gardens exist for all, not all citizens are equally able to do so. Groups that lack access to material resources or organizational capacity face relatively greater barriers to participation in community garden development. In central city neighbourhoods, where many residents already face poverty and food insecurity, acquiring resources for grassroots organizing is a significant challenge. Further, grassroots activities based on volunteerism may inadvertently support the interests of the local state in its deployment of neoliberal policies.

Conceptualizing urban community gardens without reference to the broader political economic contexts and networks in which they are embedded obscures entrenched systems of power and difference (Barraclough 2009). Community gardens can have a more substantial systematic impact if the basic structural conditions underlying resource scarcity are addressed and the restrictions on community gardens are eliminated. It is crucial to target activism towards improving the conditions in which community organizations operate, such that competition for resources does not impinge on the capacity of these organizations to advocate for radical social and political reforms. For community garden development in Milwaukee, this means that the barriers to participation in the process need to be levelled, such that residents who lack financial resources and social connections can directly enjoy the benefits of community gardens. Scaling up or linking together various individual community garden projects may be beneficial, as a step towards ensuring that the impacts of grassroots community gardening extend beyond particular garden spaces.

References

Armstrong, D. (2000). A Survey of Community Gardens in Upstate New York: Implications for Health Promotion and Community Development. *Health and Place*, 6, pp. 319–327.
Baker, L.E. (2004). Tending Cultural Landscapes and Food Citizenship in Toronto's Community Gardens. *Geographical Review*, 94(3), pp. 305–325.
Barraclough, L.R. (2009). South Central Farmers and Shadow Hills Homeowners: Land Use Policy and Relational Racialization in Los Angeles. *The Professional Geographer*, 61(2), pp. 164–186.
Cox, K. (1998). Spaces of Dependence, Spaces of Engagement and the Politics of Scale, or: Looking for Local Politics. *Political Geography*, 17(1), pp. 1–23.
Domene, E. and Saurí, D. (2007). Urbanization and Class-Produced Natures: Vegetable Gardens in the Barcelona Metropolitan Region. *Geoforum*, 38, pp. 287–298.
Elwood, S. (2004). Partnerships and Participation: Reconfiguring Urban Governance in Different State Contexts. *Urban Geography*, 25(8), pp. 755–770.
Frank, L.D., Sallis, J.F., Conway, T.L., Chapman, J., Saelens, B.E. and Bachman, W. (2006). Many Pathways from Land Use to Health: Associations between Neighborhood Walkability and Active Transportation, Body Mass Index, and Air Quality. *Journal of the American Planning Association*, 72(1), pp. 75–87.
Fyfe, N.R. (2005). Making Space for 'Neo-communitarianism'? The Third Sector, State and Civil Society in the UK. *Antipode*, 37(3), pp. 536–556.
Ghose, R. (2005). The Complexities of Citizen Participation through Collaborative Governance. *Space and Polity*, 9(1), pp. 61–75.
Ghose, R. (2007). Politics of Scale and Networks of Association in Public Participation GIS. *Environment and Planning A*, 39, pp. 1961–1980.
Ghose, R. and Pettygrove, M. (2014a). Actors and Networks in Urban Community Garden Development. *Geoforum*, 53, pp. 93–103.
Ghose, R. and Pettygrove, M. (2014b). Urban Community Gardens as Spaces of Citizenship. *Antipode*, 46, pp. 1092–1112.
Gilbert, M.R. (1998). 'Race', Space, and Power: The Survival Strategies of Working Poor Women. *Annals of the Association of American Geographers*, 88(4), pp. 595–621.
Guthman, J. (2008). Neoliberalism and the Making of Food Politics in California. *Geoforum*, 39(3), pp. 1171–1183.
Hackworth, J. (2007). *The Neoliberal City: Governance, Ideology, and Development in American Urbanism*. Ithaca, NY: Cornell University Press.
Harvey, D. (2007). *A Brief History of Neoliberalism*. Oxford: Oxford University Press.
Irazábal, C. and Punja, A. (2009). Cultivating Just Planning and Legal Institutions: A Critical Assessment of the South Central Farm Struggle in Los Angeles. *Journal of Urban Affairs*, 31(1), pp. 1–23.

Jessop, B. (2002). Liberalism, Neoliberalism, and Urban Governance: A State-Theoretical Perspective. *Antipode*, 34(3), pp. 452–472.

Kearns, A. (1995). Active Citizenship and Local Governance: Political and Geographical Dimensions. *Political Geography*, 14(2), pp. 155–175.

Kofman, E. (1995). Citizenship for Some but Not for Others: Spaces of Citizenship in Contemporary Europe. *Political Geography*, 14(2), pp. 121–137.

Kurtz, H. (2001). Differentiating Multiple Meanings of Garden and Community. *Urban Geography*, 22(7), pp. 656–670.

Larsen, K. and Gilliland, J. (2008). Mapping the Evolution of 'Food Deserts' in a Canadian City: Supermarket Accessibility in London, Ontario, 1961–2005. *International Journal of Health Geographics*, 7(16), pp. 1–16.

Larson, J. and Moseley, W.G. (2010). Reaching the Limits: A Geographic Approach for Understanding Food Insecurity and Household Hunger Mitigation Strategies in Minneapolis-Saint Paul, USA. *GeoJournal*, 77(1), pp. 1–12.

Lepofsky, J. and Fraser, J. (2003). Building Community Citizens: Claiming the Right to Place-making in the city. *Urban Studies*, 40, pp. 127–142.

Newman, K. and Lake, R.W. (2006). Democracy, Bureaucracy and Difference in US Community Development Politics since 1968. *Progress in Human Geography*, 30(1), pp. 44–61.

Peck, G. and Tickell, A. (2002). Neoliberalizing Space. In: N. Brenner and N. Theodore, eds. *Spaces of Neoliberalism: Urban Restructuring in North America and Western Europe*. Malden, MA: Blackwell Publishing, pp. 33–57.

Perkins, H.A. (2009). Out from the (Green) Shadow? Neoliberal Hegemony Through the Market Logic of Shared Urban Environmental Governance. *Political Geography*, 28, pp. 395–405.

Perkins, H.A. (2010). Green Spaces of Self-Interest within Shared Urban Governance. *Geography Compass*, 4(3), pp. 255–268.

Pettygrove, M. (2016). *Food Inequities, Urban Agriculture and the Remaking of Milwaukee, Wisconsin*. Doctoral dissertation, University of Wisconsin-Milwaukee.

Pettygrove, M. and Ghose, R. (2016). Community Engaged GIS for Urban Food Justice Research. *International Journal of Applied Geospatial Research*, 7(1), pp. 16–29.

Pudup, M.B. (2008). It Takes a Garden: Cultivating Citizen-Subjects in Organized Garden Projects. *Geoforum*, 39(3), pp. 1228–1240.

Quastel, N. (2009). Political Ecologies of Gentrification. *Urban Geography*, 30(7), pp. 694–725.

Roberts, D.J. and Mahtani, M. (2010). Neoliberalizing Race, Racing Neoliberalism: Placing 'Race' in Neoliberal Discourses. *Antipode*, 42(2), pp. 248–257.

Rosol, M. (2012). Community Volunteering as Neoliberal Strategy? Green Space Production in Berlin. *Antipode*, 44, pp. 239–257.

Roy, P. (2010). Analyzing Empowerment: An Ongoing Process of Building State–Civil Society Relations – The Case of Walnut Way in Milwaukee. *Geoforum*, 41(2), pp. 337–348.

Schmelzkopf, K. (1995). Urban Community Gardens as a Contested Space. *Geographical Review*, 85(3), pp. 365–381.

Schmelzkopf, K. (2002). Incommensurability, Land Use, and the Right to Space: Community Gardens in New York City. *Urban Geography*, 23(4), pp. 323–343.

Sharkey, J.R., Horel, S., Han, D. and Huber Jr, J.C (2009). Association Between Neighborhood Need and Spatial Access to Food Stores and Fast Food Restaurants in Neighborhoods of Colonias. *International Journal of Health Geographics*, 8(9), pp. 1–17.

Sheriff, G. (2009). Towards Healthy Local Food: Issues in Achieving Just Sustainability. *Local Environment*, 14(1), pp. 73–92.

Smith, C. and Kurtz, H. (2003). Community Gardens and Politics of Scale in New York City. *The Geographical Review*, 93(2), pp. 193–212.

Staeheli, L.A. (1999). Globalization and the Scales of Citizenship. *Geography Research Forum*, 19, pp. 60–77.

Staeheli, L.A. (2008). Citizenship and the Problem of Community. *Political Geography*, 27, pp. 5–21.

Staeheli, L.A, Mitchell, D. and Gibson, K. (2002). Conflicting Rights to the City in New York's Community Gardens. *GeoJournal*, 58, pp. 197–205.

Trudeau, D. (2008). Towards a Relational View of the Shadow State. *Political Geography*, 27, pp. 669–690.

Van Houdt, F., Suvarierol, S. and Schinkel, W. (2011). Neoliberal Communitarian Citizenship: Current Trends Towards 'Earned Citizenship' in the United Kingdom, France and the Netherlands. *International Sociology*, 26, pp. 408–432.

Zenk, S.N., Schulz, A.J., Israel, B.A., James, S.A., Bao, S. and Wilson, M.L. (2005). Neighborhood Racial Composition, Neighborhood Poverty, and the Spatial Accessibility of Supermarkets in Metropolitan Detroit. *American Journal of Public Health*, 95(4), pp. 660–667.

28
ENVISIONING LIVEABILITY AND DO-IT-TOGETHER URBAN DEVELOPMENT

Helen Jarvis

Introduction

A common challenge for cities and urban politics today is to restore a culture of liveability. What this entails in practice is heavily contested. Liveability is typically defined by the extent to which a place can attract and retain its population (homes, jobs, transport and the like); one with a good quality of life that makes provision for children and older people, clean air, green spaces, social and spiritual belonging, heritage and cultural sites. In many respects the concept of liveability reiterates parallel efforts to model urban development around ideas of "sustainable" and "harmonious" cities, "gender mainstreaming", UNICEF's "child-friendly city" and the WHO's "age-friendly city" global initiative (Buffel *et al.* 2012: 598). Fundamental to each place-based policy is an incomplete shift from "top-down" physical environmental development to "bottom-up governance", through resident participation. This chapter extends this transition to suggest that "a city for all" implies the right to "make and remake our cities and ourselves" through opportunities to *appropriate* urban space (such as ordinary streets) and to *participate*, individually, collectively and collaboratively, in decision-making surrounding the production of urban space (Harvey 2009: 315; Purcell 2003: 577).

According to the Philips Liveable Cities Think Tank (2011), for a city to be liveable it should combine resilience, inclusiveness and authenticity. But the qualities that make an area attractive to invest and live in are not always readily aligned with a progressive politics of social justice, or the freedom for all people to access urban public space equally. For many commentators, liveability conjures up the kind of relaxed, vibrant, walkable character which, in 'successful' high-growth cities, such as San Francisco and London, coincide with hostile forms of gentrification. This is where affluent population groups with high disposable income buy into and upgrade an area, pushing up living standards and costs that in turn displace vulnerable people and small businesses.

Ambiguity over the purpose of liveability, whether it is oriented to the inclusive 'good city' or the commercial attributes of a 'good life' for some, is due in part to a tendency for it to be measured by institutional actors and 'competitive' urban regeneration (Coe and Bunnell 2003; Tait and Jensen 2007). Mainstream research also tends to study large-scale 'flag-ship' macro-economic development rather than the meso-scale of ordinary streets and neighbourhoods. This skewed attention neglects the informal grassroots social spaces that foster less tangible qualities of informal urbanism. Sophie Watson (2006: 5) captures a flavour of what is missed in her portrait of "magical urban encounters ... of buzzing intermingling ... [in] scruffy, unplanned and marginal public spaces"; describing from a visit to a local city farm in London "a space [that is] cut out of the railway sidings and abandoned land". This intermingling of different cultural groups and questions of connection and solidarity resonate with what Doreen Massey (2005) calls the 'throwntogetherness' of disenfranchised people living together in marginal spaces (see also Amin 2004 on the politics of propinquity).

Whereas formal planning and urban design tend to measure social and material influence and impact in terms of profit and loss; in reality, how people inhabit a particular place involves many intangible qualities that are rooted in social connections which may not be measurable. The implication is that when formal planning fails to recognize or value this informal 'glue', it may inadvertently remove or destroy it, even when contriving to 'engineer' the qualities of liveability that allude the formal realm. As Scott (2012: 15) observes:

> the more highly planned, regulated, and formal a social or economic order is, the more likely it is to be parasitic on informal processes that the formal scheme does not recognize and without which it could not continue to exist, informal processes that the formal order cannot alone create and maintain.

In the academic literature it is widely accepted that places are 'made' according to a vision that is variously imposed 'top-down' or that grows 'bottom-up' or indeed represents a hybrid of both formal planning and informal participation. More recently, academic debate highlights a trend of 'localism' that is evident on a number of intersecting geographic scales; as the currency of popular 'post-material' social movements seeking to reverse the decline of civic influence in local concerns; as a platform for issue-specific community-led development, such as locally defined housing; and as the substance and rhetoric of national planning and policy frameworks (Jarvis 2015a). This trend suggests broad political, cultural and practical alignment around the positive connotations of community participation. This is considered to summon forth new forms of citizenship that offer the potential for local residents to shape and influence neighbourhood development in socially progressive and sustainable ways. The expected outcome of citizen involvement is to validate the values of a shared identity, whether this is real or imagined.

Arguably, we need a new analytic approach that might think about local place-making influence more critically and, by the same token, to challenge what is deemed 'relevant' for planners and politicians to learn from. This is similar to the argument that Jennifer Robinson (2006) makes when she calls for an urban theory that accounts for a wider variety of 'ordinary' cities and, crucially, for analytical approaches that bridge between research on 'planning' in global or world cities and 'informality' in the small cities of 'less developed' countries (Bell and Jayne 2006: 5). This follows the many ways historically that towns, cities and urban places have come to be narrowly defined and 'measured' according to fixed categories, whether in relation to building density and scale, employment and wealth, transport and communication networks, social structures and governance, or tourism.

This chapter proposes a more nuanced urban analytic approach; one that pays greater attention to the intangible 'soft' architectures of everyday life; to recognize the importance of 'bottom-up' visions of development and alternative forms of building and managing local community assets (Jarvis 2015a). When citizens take over the management or ownership of public spaces or amenities in their neighbourhood, for instance, there is the potential to challenge mainstream development beyond the market and the state. A shift from 'top-down' to 'bottom-up' action can influence what counts (and what gets counted) and consequently the material changes that unfold and evolve in a given place. This in turn can challenge the 'metrics' underpinning conventional notions of liveability. The remainder of this chapter picks up this debate in the context of public space and civil society. This culminates in two case study examples of 'do-it-together' activism and community-led development: neighbours closing their street to allow their children to 'play out' in the UK (Box 28.1); and the Los Angeles Ecovillage project of grassroots activism (LAEV), USA (Box 28.2).

Public space and ordinary streets

Much has been said in the literature about the end of public space. In contemporary cities, the public sphere is increasingly being seen to comprise dead public spaces, privatized shopping malls and gated communities, eroding the essence of city life (Low and Smith 2013; Paddison and Sharp 2007: 87). But significant gaps remain in our understanding of the fine-grained engagement with, and understanding of, particular public spaces by increasingly divided, diverse publics. Just thinking in terms of the complex and contradictory needs and concerns of young people, paid adult workers and older people; it is clear that efforts to create liveable cities for all ages are problematic (van Vliet 2011). The gaps in critical understanding and intervention are particularly apparent in prominent public spaces. Urban parks are still widely maintained, despite the pressures on local authorities to sell off playing fields. Yet these tend to be physically and culturally constructed around specific leisure activities, membership-based social groups and taxpayer interests. There are very few spaces or opportunities for ordinary citizens to collectively envision the public space of their imagination (for the seminal study of People's Park, California, see Mitchell 1995).

Public space functions on a continuum of ownership and access. For example, Clare Cooper Marcus (2002: 32) identifies five categories of outdoor space: private spaces owned by individuals (which are accessible only to them and their guests), corporate public spaces such as shopping centres (which are privately owned but open to the public), public spaces such as neighbourhood streets and parks (which are publicly owned and open to the public), shared spaces which are enclosed or gated (a community space owned by a group of residents which is usually accessible only to that group or their invited guests) and shared spaces which are open or porous (a collectively owned and managed landscape which allows the general public to wander through while at the same time providing a 'place-setting' for community life). These types of outdoor spaces can be expressed not only in terms of their legal ownership and associated governance (contrasting elected and non-elected representative bodies with collective, participatory decision-making) but also to the quality of social relations and encounters cultivated within them.

Since the nineteenth century, if not before, 'the street' has been regarded as a lively and contested public domain, the site of popular protest and political struggle. Yet, as a public space, the street is not especially accessible to people for multiple uses. In Britain, under the Highways Act of 1980, streets are provided and maintained solely for the passage of motorized traffic and any other activity can technically be described as an obstruction (Barrell and Whitehouse 2004: 262). This highlights the tensions that exist between streets for motorized transport and space for pedestrian activity. Contradictions are apparent today both in the desire for liveable streets and the material development of government-sponsored pedestrian-friendly commercial streets. As observed by Bell and Jayne (2006: 160) "the street itself can be either pedestrianised or given over to traffic, but streets remain highly regulated spaces where legal or social conventions delineate good behaviour".

Recent years have witnessed repeated attempts by campaigning organizations, artists and residents, to 'reclaim' and defend the 'local banal public spaces' of residential streets for pedestrian and community enjoyment (Grannis 1998; Paddison and Sharpe 2007). For example, the UK national charity Living Streets seeks to make streets "attractive and enjoyable spaces in which to live, work, shop and play safely". Founded in 1929 as the 'Pedestrian Association', the 2001 name change reflects the move away from a reactive 'safety conscious' approach to traffic management towards a more proactive and holistic concern to cultivate 'sociability' and inclusion through everyday street social interactions. As we will see in Box 28.1, in the case study example of the Playing Out civic campaigning organization, ambitions to recreate the social life of the street and neighbourhood, as well as 'safe spaces' for children are often at odds with norms and cultures of parenting that prevent children's freedom to roam or engage in free-play. There are many other examples around the world of different motives and efforts to occupy the public spaces of the street. Probably the best known example is Occupy Wall Street which surfaced in New York before travelling the world as an idea for street reclaiming that captured the imagination as a spatial metaphor for challenging global social injustice (Penney and Dadas 2013).

Manuel Castells (2005) understands a culture of liveability in terms of the social-spatial connection of "local life, individuals, communes, and instrumental global flows through the sharing of public places" (2005: 60). Public space is understood as the key connector of human interaction and experience. Historically the form and use of public space has commanded considerable academic interest (Low 2000; Mitchell 2003). Lyn Lofland (1973: 34) points out that the pre-industrial city emphasized the multiplicity of uses to which public space could be put. The popular image is one of a bustling market-square in which traders and shoppers greet each other by name. This contrasts with the social isolation critics attribute to monotonous housing estates that lack local cultural identity and the drift to suburbanization that separates people from each other and from commercial districts and public amenities (Gehl and Svarre 2013). Post-war urban planning has been widely criticized for neglecting the humanistic significance of the neighbourhood landscape elements of ordinary streets, informal urbanism, the spaces between buildings and the transition or buffer spaces that often function as place-settings for community life. A notable exception is Jane Jacobs's 1961 classic 'anti-planning' thesis *The Life and Death of Great American Cities* (see also Sennett 1997) and, more recently, the counter-argument of Danish landscape architect Jan Gehl in his seminal text *Life Between Buildings: Using Public Space* (Gehl 2010).

Understanding liveability as a yearning for socio-spatial connection, it is constructive to draw on Sophie Watson's (2004: 210) notion of the street as a "space of democratic possibility". This emphasizes the co-constitutive functions of public space with public life or civil society. In the UK context, research undertaken by the Fabian Society highlights the crucial role of public green space, for instance, as providing 'places to be'

> where we can come together, build relationships and reverse society's long-term journey towards individualism and isolation. These places are under threat. Central government funding for local authorities has fallen by around 40 per cent, leaving councils without the means to adequately maintain facilities or engage with local people.
>
> *(Wallis 2015: 2)*

Sophie Watson argues that this social space of public life is not just about building social capital but also about creating and liberating spatial agency for imagining alternative possible futures. This justifies a turn away from looking at streets as fixed public spaces (auditing the quantity of benches or pedestrian access). Instead it suggests we need to understand and explore this landscape as a fluid dialogical space of inspiration and learning. The social learning may be to an extent scripted or engineered (as below in Box 28.1) or it may function through a more organic process of dreaming and enchantment (as below in Box 28.2, for the case study example of the LA Ecovillage, USA).

Public life, civil society and intimate encounter

Urban social theorist Henri Lefebvre argues that place-making is a social practice. The texture of everyday social settings and interactions operate largely at an unconscious level. This contrasts with the enduring faith that many planning practitioners and policy-makers hold in the power of physical design to change the social life of a community. In national policy terms, the late 1990s saw both Britain and the USA embark on an urban 'renaissance' which was replete with cultural and architectural motifs of village life inserted into the heart of the cosmopolitan city. This process appeared to capture a popular yearning for close-knit community affiliations, fondly remembered from snapshots of children playing in traffic-free streets in the 1950s and 1960s. A number of new social movements began to flourish; those associated with the 'post-material' values of slowness, simplicity, authenticity, community and conservation (movements spanning 'slow food', 'slow cities', 'transition towns', 'living streets' and 'voluntary simplicity'). At the same time, architects, planners and municipal leaders renewed their faith in the idea that positive attributes of community spirit could be cultivated by recreating 'traditional' urban residential neighbourhoods.

The social dimensions of public space introduced above are usefully explored with reference to what Georg Simmel has to say on the significance of co-present social interaction. Writing in the early twentieth century, his theorizing drew attention to a convivial scale of belonging that expanded upon the concept of *gemeinschaft* (close-knit community) previously introduced by his contemporary Ferdinand Tönnies. Simmel recognized an affective 'being in togetherness' to gemeinschaft, pointing to the universal occurrence in human development of a social pleasure in the physical company of others. He referred to this as *geselligkeit* or the 'play-form' of association. This play-form introduced novelty and disruption to otherwise routine exchanges. Pleasurable mutuality could be further deepened through social awareness and the questioning of taken-for-granted values (neighbours talking through differences in their upbringing, for instance). He observed that associations assumed greater depth when dialogue challenged taken-for-granted norms and values. To Simmel (1903) the virtue of geselligkeit is that engagement runs deep, beyond fleeting impressions. This recognizes the additional function of social learning as one of intersecting layers of understanding liveability.

Establishing a space of deep engagement, in a neighbourhood association or community cooperative, for instance, resonates with what Kittay (1999) and others have determined as 'love, care and solidarity'. It highlights concern not to romanticize public engagement without first asking who participates, for whose benefit (Jarvis 2015a). Activities such as attending public consultations, for instance, or promoting, campaigning and engaging in direct action inevitably involve unpaid 'work' as well as availability; just as someone can be 'present' but emotionally unavailable, the emotion-work of community associations can reproduce unequal divisions of labour or exclude certain population groups from the space of engagement. The emphasis placed at this scale of public life on 'social capital' and advanced inter-personal competencies also challenges the argument that community participation is sufficient as a mark of democratic development, "to encourage the inclusion of outsiders, to break down barriers created by wealth and privilege (or knowledge and motivation), or prevent those that are already better off and more dominant from flourishing at the expense of others" (Coote 2011: 85).

In practice, spatial and historical discontinuities highlight the paradox of 'localism' that is variously instrumental and involuntary. This is evident in the way that community groups and civic projects are increasingly expected in a climate of austerity to replace state welfare functions as reserve 'capacity' (Elwood 2004). In the Australian context, Argent (2005) describes this as a "neoliberal seduction". On the one hand the spatiality of localism is inherently attractive to rally popular support through "proximity, co-presence and reach". On the other hand, when public spending is cut, communities are often forced into driving local development as a kind of "mopping up" exercise (Argent 2005: 37).

From DIY to DIT

Just as 'liveability' can appear to be both self-evident and abstract, definitions of public life are similarly contested. How we understand public life differs according to whether the service involved functions within or outside formal institutions – of politics, religion or education, for instance, where this broadly describes the work of 'networking' – to be known to a lot of people or connect one group of actors to another. The approach taken by urban design consultants Jan Gehl and Birgitte Svarre (2013: 2) views public life ethnographically in terms of "everything we can go out and observe happening" in public space; the mundane human behaviour "that takes place between buildings, to and from school, on balconies, seated, standing, walking ... far more than just street theatre and café life". A more autonomous, ideological approach views public life on a continuum of voluntary service from grassroots activism, community organizing, informal urban development to being a 'good neighbour'.

There is a long history to the idea of ordinary people organizing into groups and associations, acting on a voluntary basis to solve local problems. This form of direct action goes by a variety of names, including 'self-help' or 'do it yourself' DIY community action or community organizing. This chapter traces a subtle but significant transition from emphasis on individual action (the 'self' of DIY) to cooperative collaboration in 'do-it-together' DIT development. From the previous definitions of public life, an argument can be made for shifting research and debate away from preoccupation with the motives of individual activists finding common ground with others 'in resistance' (such as Occupy Wall Street) to draw attention to what it takes for people to 'change the world' in ordinary ways. This cultivation of a public life is fundamentally about local empowerment: it is less about "the transfer of decision-making power from 'influential' sectors to those previously disadvantaged or 'other' sections of society, but about these 'others' taking control and initiating different or 'alternative' spatial processes" such as community-led housing or the transfer of public assets into mutual ownership or management (Schneider and Till 2009: 100). A shift in focus from the individual actor to activist groups working together on a shared endeavour highlights the 'soft skills' of association and collaboration that have received limited attention in urban studies.

The literature on movements of 'street reclaiming' such as Occupy Wall Street typically define civic activism in terms of DIY democracy. This describes the horizontal processes of leadership and consensus whereby individuals are 'making themselves up as they go along' as producers and consumers combined. As Ratto and Boler (2014: 5) observe,

> this self-creation can be seen in a positive light – for instance, as a reaction against the regulation of identity that can constitute the lived experience of a totalitarian government. However, it can also be understood as part of a hegemonic acceptance of the breaking apart and individualization of civil society.

It is therefore constructive to consider the emerging counter-trend of 'do-it-together' or DIT democracy. Richard Wolff (1998: 13) regards this as a socially progressive transition whereby

> creating your own direct environment with other people is the way to escape alienation and promote solidarity, respect and mutual support. If it is on the scale of a small village or a street, self-organization and direct action are the fundamentals of local action.

Further emphasis on renewed 'togetherness' is highlighted in two popular ethnographies that explore small self-organizing groups, including voluntary associations, church groups, clubs and civic societies (Hemming 2011; Sennett 2012). Both offer a timely counter-point to well-rehearsed concern for the decline in social trust and associational life over the past half century. Robert Putnam's (2000) thesis *Bowling Alone* epitomizes this concern. Rather than directly challenge accounts of increasing individualization and diminishing connections between people and between people and the places in which they live and work, Richard Sennett (2012) highlights the rich but fragile nature of these poorly understood associational relations. He argues that cooperation and collaboration need to be understood as craft skills that require people to understand and respond to one another in order to act together (Sennett 2012). In this sense the *process* of working together jointly is as significant as the goal or outcome: collaboration provides a catalyst of deeper levels of trust and cooperation, whether or not the group achieves their intended goal.

Envisioning: a utopian method

It is in order to highlight the place-based *process* of social learning and the 'play-form' of voluntary association that we turn now to a final linked concept: 'envisioning' liveability. This highlights an important distinction between the way that planners, politicians, policy-makers and practitioners typically conjure up an image of place through the development process, and how community groups imagine and realize what they want for their neighbourhood.

When asked to envision possibilities for a more equitable, just and ecologically sustainable urban future, David Harvey contends that most of what passes for city planning has been inspired by utopian modes of thought (Harvey 2000; MacLeod and Ward 2002: 153). Here, by contrast, envisioning is understood and conceived as a socio-cultural, socio-spatial project of shared *endeavour* as much as one that is rooted in a *vision* of a better alternative to mainstream options. This distinction is subtle but significant and it builds on an equally explicit distinction between an imagined utopia (destination) and utopian methods of thinking critically (journey) as motivation for transformation. In this sense, envisioning is about acting on a possibility to turn an idea into a reality where it is the process of this realization which is as transformational, if not more so, than the end result.

This discussion of concepts remains quite abstract. A more accessible way of illustrating what 'envisioning liveability' might look like in practice is suggested below; first in a poem, and then in two brief case study vignettes. Richard Delorenzi wrote the following poem while participating in a group activity for a community housing event in London on 29 June 2010. It was his inspiration for shared public space that was subsequently published in a Diggers and Dreamers handbook on the form of collaborative housing known as cohousing.

We will build this place – and they will play (a cohousing project)
My boy was born today, where will he play?
Will he play indoors on his own, all alone?

Will he play in the road, with the cars?
Will he play in the park, far away?
Let us get rid of the cars.
Let us know our neighbours.
Let us build a community together.
Let us talk in the streets, with no cars.
Let the children play.
Let the adults play.
Let the old people play.
Let the children play.

(Richard Delorenzi, Community Land Trusts Conference, 2010, quoted from Bunker et al. 2011: 149. https://richarddelorenzi.wordpress.com/2012/04/16/we-will-build-this-place-and-they-will-play-a-co-housing-project/)

Box 28.1 Case study example: Playing Out – whose street? Campaigning for children's right to 'play out'

Playing Out is a UK resident-led organization that echoes similar campaigns around the world intended to support people who want their children to be able to play freely in streets that are temporarily closed to traffic. The UK scheme suggests parallels with Donald Appleyard's advocacy of children's use of streets and sidewalks in Berkeley, California, in the 1960s and 1970s.

Figure 28.1 Let the children play: street closure for Big Lunch street party; children 'playing out', Newcastle upon Tyne, 2008

Source: photo: Helen Jarvis.

Playing Out began in Bristol on the modest scale of a single street: neighbouring mothers Alice Ferguson and Amy Rose shared their vision of making their residential street more 'liveable'. They recalled childhood memories of riding bikes and inventing games and lamented that, due to a high volume of traffic and lack of green spaces, their own children did not enjoy the same freedom. They had already set up a residents' association and this forum provided the necessary space to develop

Figure 28.2 Let the adults play: street closure for Big Lunch street party, Newcastle upon Tyne, 2008

Source: photo: Helen Jarvis.

friendships, trust and local knowledge required to generate widespread support for regular temporary road closures.

The first Playing Out session was organized to coincide with International Children's Day on 1 June 2009. It took several months to get permission to close the street from the relevant local authority because this required the full agreement of everyone living on the street. This necessarily involved a lot of talking and listening to each other's point of view on questions of whose street this was with respect to freedom and access. Residents were to be allowed car access during Playing Out sessions, escorted at walking speed by volunteer stewards, in effect creating a 'shared space' between cars and people. In the event, children of all ages played in the street together – on scooters and bikes, with balls and chalk or simply inventing games and talking together. Older residents and those less motivated initially to create child-friendly neighbourhoods learned to appreciate the 'buzz' that accompanied Playing Out sessions: there were impromptu picnics as outdoor seating provided a magnet for neighbours to spend time getting to know each other more deeply.

While Alice and Amy came up with the original proposal and galvanized other residents to share their enthusiasm, Playing Out would not have thrived and spread to other streets if it did not function as a 'do-it-together' project. It was not a trivial undertaking for neighbours to lift their gaze from their own household concerns to collaborate on a shared endeavour. But the hard work paid off. Residents on neighbouring streets wondered if they could also apply for temporary street closures and this way the idea travelled. Playing Out was formalized as a Community Interest Company in 2014, accompanied by the publication of a 'step-by-step' handbook and training sessions regularly delivered to resident groups around the UK (Ferguson and Rose 2014: 3–5).

Playing Out is a simple but effective example of how residents can act together to appropriate and 'domesticate' public space for new uses. It provides evidence of local people reclaiming access to the spaces in front of their homes, extending the reach of domestic space, metaphorically (if not literally) 'pulling down the fences' that typically contain and inhibit isolated, individual living arrangements. This temporary transformation is progressive because DIT planning introduces a space of deep engagement – what Simmel (1903) referred to as *geselligkeit*.

Box 28.2 Case study example: LA Ecovillage – a neighbourhood healing itself through cooperative development

On my 2014 visit to the inner city site of the LA Ecovillage (LAEV) I was initially struck by the unplanned appearance of an unruly mix of street art, arid edible planting (the area was suffering from a serious drought), home-made street furniture (including a miniature lending library) and a brightly painted permeable road surface at the intersections of Bimini Place and White House. There were numerous hand-drawn signs urging drivers to 'slow down' while murals encouraged creative play. Accompanied on this occasion by my young daughter I regarded this playful environment as 'out of place' relative to the highly regulated and materially uniform street layout. It was as is if the kids had taken charge, casting aside the conservative interventions of parents, teachers and planning guidance. Yet, further investigation and a tour of the area revealed an extraordinarily well-orchestrated diffusion of civic organizing and local enterprise in and around this two-block neighbourhood.

The LAEV project is located on the northern edge of an area that was caught up and seriously damaged in the 1992 'riots' when civil disobedience followed the acquittal of police officers who

Figure 28.3 LA Ecovillage, view of White House Place Learning Garden
Source: photo: Helen Jarvis.

were on trial for assaulting black man Rodney King captured in a videotape of his arrest. The neighbourhood is characterized by a mix of land uses including rental housing, commercial retail and light industrial zoning.

The project started as an initiative of the Cooperative Resources and Services project (CRSP), a non-profit organization founded in 1980 by activist Lois Arkin. According to Boyer (2015: 330):

> around 1983 Arkin began to envision a 'neighbourhood of coops' that would allow individuals to access multiple services in the same space. In 1996 CRSP purchased the rundown Bimini Apartment Building with loans from friends and relatives, inheriting a number of sitting tenants, retrofitting it over time as a collectively self-governed cooperative based on ecological values.

Today, in addition to the forty-eight unit ecovillage, 'do it together' developments co-produced:

> the Bimini Slough Ecological Park; the White House Place Learning Garden; the Bicycle Kitchen (and the CicLAvia festival, a tradition borrowed from Bogota, Columbia that temporarily closes down stretches of downtown streets for cyclists and pedestrians which has grown into a citywide social event held five times per year), a café and 'bulk food buying cooperative' (plus annual 'Eco-Maya' festival celebrating Mayan heritage amongst LA residents), a curb-side 'lending library', a healing centre and a neighbourhood orchard 'offering fruits and nuts for everyone who passes by'.
>
> *(Boyer 2015: 332; Bimini Place 'Our History' Youth Care signage)*

Figure 28.4 Open-air curb-side 'lending library' in front of LA Ecovillage
Source: photo: Helen Jarvis.

Each project demonstrates deep commitment to permanently affordable land and housing for lower-income households. In this sense, do-it-together for-all protects this neighbourhood from gentrification for a select few.

While Lois Arkin was particularly influential in the early envisioning process, from the outset she collaborated with activists sharing similar values. Over time the critical mass of LAEV comprised resident members (thirty-five 'intentional neighbours') working jointly with non-member neighbours and enlightened non-profit practitioners representing the city at large. The sheer momentum and diffusion of DIT collaborative practice distinguishes this retrofitted neighbourhood development from other community-led housing projects that I have visited around the world (Jarvis 2015a). Loosely connected to a global ecovillage movement, where the intention is to model positive solutions to global ecological crises, LAEV functions as a source of inspiration and innovation, most notably the 'soft architecture' involved in assembling multiple local cooperative shared space economies. LAEV is illustrative of the scope of possibilities for DIY and DIT street-level transformation.

Concluding remarks

By turning the attention to the banal and relatively unorganized spaces and practices of streets social interaction, this chapter finds evidence of local people, including children, claiming a place in the world; not through ownership but through shared endeavours and everyday habitual practice. Many of the terms and concepts used to describe this process of residential urban development, and what it takes for a neighbourhood to be 'liveable', invite simultaneously taken-for-granted and highly contested meanings. The simple term 'community' for instance is notoriously fuzzy and open-ended. It can refer to a neighbourhood or geographic association but it can also extend to common interests that transcend space and place, as with 'communities of interest'. Recognizing this problem and the way that politicians, planners and policy-makers often employ an imagined 'sense of community' and 'liveability' in rhetorical ways, this chapter calls for a new analytic approach.

The discussion and empirical examples highlight the significance of informal grassroots social spaces and DIT processes of local place-making. This is an important focus because we are witnessing a climate of local authority spending cuts in which many more public spaces and facilities are being transferred to management and delivery by volunteers. Discussion shows that people inhabit the neighbourhoods where they live and work in complex ways that involve many intangible qualities that are rooted in social connections – of inspiration, learning and wonder. These intangible qualities tend not to be recognized or valued by formal urban planning or policy-making.

Intangible qualities of association can be regarded as transformational because the active involvement of volunteers can empower them and the communities they are providing services for. At the same time, we must be careful not to romanticize the scale of the street, or the neighbourhood, as if participation at this local scale represents a distinct, homogenous whole, composed of people all speaking with a single voice. It is in order to challenge the shortcomings of narrow consultation that the analytic approach proposed here emphasizes the wider dialogical and evolving dynamics of envisioning.

References

Amin, A. (2004). Regions Unbound: Towards a New Politics of Place. *Geografiska Annaler: Series B: Human Geography*, 86(1), pp. 33–44.
Argent, N. (2005). The Neoliberal Seduction: Governing-at-a-Distance, Community Development and the Battle over Financial Services Provision in Australia. *Geographical Research*, 43(1), pp. 29–39.
Barrell, J. and Whitehouse, J. (2004). Home Zones – an Evolving Approach to Community Streets. *Municipal Engineer*, 157(4), pp. 257–265.
Bell, D. and Jayne, M., eds. (2006). *Small Cities: Urban Experience Beyond the Metropolis*. London: Routledge.
Boyer, R.H.W. (2015). Grassroots Innovation for Urban Sustainability: Comparing the Diffusion Pathways of Three Ecovillage Projects. *Environment and Planning*, 45 (2), pp. 320–337.
Buffel, T., Phillipson, C. and Scharf, T. (2012). Ageing in Urban Environments: Developing 'Age-Friendly' Cities. *Critical Social Policy*, 32(4), pp. 597–617.
Bunker, S., Coates, C., Field, M. and How, J., eds. (2011). *Cohousing in Britain: A Diggers and Dreamers Review*. London: D & D Publications.
Castells, M. (2005). Space of Flows, Space of Places: Material for Theory of Urbanism in the Information Age. In: S. Bishwapriya, ed. *Comparative Planning Cultures*. London: Routledge, pp. 45–63.
Coe, N.M. and Bunnell, T.G. (2003). Spatializing Knowledge Communities: Towards a Conceptualization of Transnational Innovation Networks. *Global Networks*, 3(4), pp. 437–456.
Cooper Marcus, C. (2002). Shared Outdoor Space and Community Life. *Places*, 15(2), pp. 32–41.
Coote, A. (2011). Big Society and the New Austerity. In: M. Stott, ed. *The Big Society Challenge*. London: Keystone Development Trust Publications, pp. 82–95.
Elwood, S. (2004). Partnerships and Participation: Reconfiguring Urban Governance in Different State Contexts. *Urban Geography*, 25(8), pp. 755–770.
Ferguson, A. and Rose, A. (2014). *How to Organise Playing out Sessions on Your Street – a Step-By-Step Manual*. Bristol: Playing Out.
Gehl, J. (2010). *Life Between Buildings: Using Public Space*. Copenhagen: Danish Architectural Press.
Gehl, J. and Svarre, B. (2013). *How to Study Public Life*. Washington, DC: Island Press.

Grannis, R. (1998). The Importance of Trivial Streets: Residential Streets and Residential Segregation. *American Journal of Sociology*, 103(6), pp. 1530–1564.
Harvey, D. (2000). *Spaces of Hope*. Edinburgh: Edinburgh University Press.
Harvey, D. (2009). *Social Justice and the City*. Athens: University of Georgia Press.
Hemming, H. (2011). *Together: How Small Groups Achieve Big Things*. London: John Murray.
Jacobs, J. (1961). *The Death and Life of Great American Cities*. Harmondsworth: Penguin.
Jarvis, H. (2015a). Community-led Housing and 'Slow' Opposition to Corporate Development: Citizen Participation as Common Ground? *Geography Compass*, 9(4), pp. 202–213.
Jarvis, H. (2015b). Towards a Deeper Understanding of the Social Architecture of Co-housing: Evidence from the UK, USA and Australia. *Urban Research and Practice*, 8(1), pp. 93–105.
Kittay, E.F. (1999). *Love's Labour*. New York: Routledge.
Lofland, L.H. (1973). *A World of Strangers: Order and Action in Urban Public Space*. Prospect Heights, IL: Waveland Press.
Low, S. (2000). *On the Plaza: The Politics of Public Space and Culture*. Austin: University of Texas Press.
Low, S. and Smith, N., eds. (2013). *The Politics of Public Space*. New York: Routledge.
MacLeod, G. and Ward, K. (2002). Spaces of Utopia and Dystopia: Landscaping the Contemporary City. *Geografiska Annaler: Series B, Human Geography*, 84(3–4), pp. 153–170.
Massey, D. (2005). *For Space*. London: Sage.
Mitchell, D. (1995). The End of Public Space? People's Park, Definitions of the Public, and Democracy. *Annals of the Association of American Geographers*, 85(1), pp. 108–133.
Mitchell, D. (2003). *The Right to the City: Social Justice and the Fight for Public Space*. New York: Guilford Press.
Paddison, R. and Sharp, J. (2007). Questioning the End of Public Space: Reclaiming Control of Local and Banal Spaces. *Scottish Geography Journal*, 123(2), pp. 87–106.
Penney, J. and Dadas, C. (2013). (Re) Tweeting in the Service of Protest: Digital Composition and Circulation in the Occupy Wall Street Movement. *New Media & Society*, 16(1), pp. 74–90.
Philips Liveable Cities Think Tank (2011). Liveable Cities-poster. http://thisbigcity.net/infographic-what-makes-a-liveable-city/ [accessed 20 April 2017].
Playing Out. www.playingout.net [accessed 3 November 2015].
Purcell, M. (2003). Citizenship and the Right to the Global City: Reimagining the Capitalist Working Order. *International Journal of Urban and Regional Research*, 27(3), pp. 564–590.
Putnam, R.D. (2000). *Bowling Alone: The Collapse and Revival of American Communities*. New York: Simon & Schuster.
Ratto, M. and Boler, M. (2014). *DIY Citizenship: Critical Making and Social Media*. Cambridge, MA: MIT Press.
Robinson, J. (2006). *Ordinary Cities: Between Modernity and Development*. London: Routledge.
Schneider, T. and Till, J. (2009). Beyond Discourse: Notes on Spatial Agency. *Footprint. Delft Architectural Theory Journal*, 4 (Spring), pp. 97–111.
Scott, J.C. (2012). *Two Cheers for Anarchism: Six Easy Pieces on Autonomy, Dignity, and Meaningful Work and Play*. New York: Princeton University Press.
Sennett, R. (1997). *The Uses of Disorder: Personal Identity and City Life*. New York and London: W.W. Norton.
Sennett, R. (2012). *Together: The Rituals, Pleasures and Politics of Co-operation*. London: Allen Lane.
Simmel, G. (1903). The Metropolis and Mental Life. *Individuality and Social Forms*. In: D.N. Levine, ed. (1971), *Selected Writings*. Chicago: University of Chicago Press, pp. 324–339.
Tait, M. and Jensen, O.B. (2007). Travelling Ideas, Power and Place: The Cases of Urban Villages and Business Improvement Districts. *International Planning Studies*, 12(2), pp. 107–127.
Van Vliet, W. (2011). Intergenerational Cities: A Framework for Policies and Programs. *Journal of Intergenerational Relationships*, 9(4), pp. 348–365.
Wallis, E. (2015). *Places to Be: Green Spaces for Active Citizenship*. London: Fabian Society.
Watson, S. (2004). Cultures of Democracy: Spaces of Democratic Possibility. In: C. Barnett and M. Low, eds. *Spaces of Democracy: Geographical Perspectives on Citizenship, Participation and Representation*. London: Sage, pp. 207–222.
Watson, S. (2006). *City Publics: The (Dis)enchantment of Urban Encounters*. London: Psychology Press.
Wolff, R. (1998). *Possible Urban Worlds: International Network of Urban Research and Action*. Zurich: Birkhauser.

29
INFRASTRUCTURAL CITIZENSHIP
Spaces of living in Cape Town, South Africa

Charlotte Lemanski

Introduction

The foundational premise of this chapter is that, while there has been widespread recognition that both infrastructure and citizenship are crucial for understanding the everyday spaces of life in the city, the connections between them are poorly understood and under-theorized. As a means to explore these connections this chapter deploys the phrase *infrastructural citizenship*. Whilst not the first to coin this phrase (see Justesen 2013; Shelton 2017), its meaning within this chapter centres on how citizens' everyday access to, and use of, infrastructure in the city affect, and are affected by, their citizenship identity and practice. Furthermore, the phrase provides the foundations for initiating critical connections between the two scholarly fields of urban infrastructure and citizenship.

The chapter starts by reviewing existing geographical debates on citizenship, demonstrating the distinction between citizenship practices and acts, and highlighting examples of how citizenship can be framed as both "insurgent" (Holston 2008) and "ordinary" (Staeheli et al. 2012). Second, the chapter analyses scholarship on infrastructure, particularly the contemporary shift towards conceptualizing infrastructure as socio-technical (Amin 2014; Graham and Marvin 2001) alongside an emerging recognition of the importance of infrastructure to the everyday lives of urban citizens (Graham and McFarlane 2015). Third, the chapter develops the notion of *infrastructural citizenship* as a conceptual lens through which to acknowledge the interconnected relationship between infrastructure and citizenship in spaces of urban living. Finally, the chapter uses a case study of state-subsidized housing in Cape Town to critically explore how public infrastructure connects with citizenship in contemporary South Africa. This empirical case is particularly pertinent because universal citizenship is still relatively new in South Africa, and has emerged in a context where post-apartheid urban politics have been framed around infrastructural provision.

Citizenship: legal practices and socio-political acts

Citizenship has long been a central concept within work on political, cultural and historical geography, exploring the relationship between citizens and the state (historically the nation-state). However, contemporary scholarship has shifted focus in terms of scale and scope. First, the scale of citizenship has both retracted and expanded, with recognition that citizenship functions at the local/urban scale as well as the global transnational scale, often resulting in different forms of citizenship at different scales (Marston and Mitchell 2004). Second, the scope of citizenship has shifted away from a static interpretation of citizenship through a legal statist framework of rights and responsibilities, to a dynamic recognition of the ways in which citizens exercise and demonstrate those rights and responsibilities, and thereby in turn produce new understandings and experiences of citizenship. In this context, citizenship is not merely a legal fact, but is also an everyday act that operates at multiple scales. This shift has brought renewed energy to debates exploring the processes through which citizens engage with and demonstrate their citizenship. Engin Isin and Greg Nielsen (2008) coined the phrase "acts of citizenship", highlighting the need to move attention away from those who determine the rules of citizenship (i.e. the state) or those who act out citizenship (i.e. citizens), to instead focus on the *acts* of citizenship themselves. "We propose to shift focus from the institution of citizenship and the citizen as individual agent to acts of citizenship – that is, collective or individual deeds that rupture social-historical patterns" (Isin and Nielson 2008: 2).

In other words, they argue that citizenship is less about the status of individuals or institutions, and more about the processes through which citizens engage with and demonstrate their citizenship. Consequently their focus is on citizenship actions rather than on citizenship status. Isin and Nielson (2008) further clarify their focus by distinguishing between everyday normalized and mundane *practices* of citizenship (e.g. voting, paying taxes), and *acts* of citizenship as non-repetitive and often confrontational actions taken by citizens to demand their citizenship rights. While other scholars use different labels to demonstrate citizenship acts, for example 'social citizenship' (Dwyer 2010; Taylor-Gooby 2008), there is a general consensus in the literature that citizenship is not just about the legalistic contract of rights between the state and its citizens, but includes the ways in which citizens demonstrate these rights, and consequently citizenship comprises a crucial space of everyday life in the contemporary city.

At the same time, the distinctions inherent in this binary, between citizenship as a legal fact or everyday act, are challenged by the concept of *ordinary citizenship*. Coined by Staeheli *et al.* (2012), ordinary citizenship argues that the complexities of citizenship are best understood as part of the legal system *and* everyday activities and encounters. Consequently, rather than consider citizenship as either a legal fact or an everyday act, Staeheli *et al.* (2012) employ the concept of ordinariness to consider both in tandem, using the example of US immigrants, a group excluded from formal citizenship, to demonstrate their argument.

> The conceptualisation of citizenship as ordinary trains our analytical lens on the ways in which laws and social norms are entwined with the routine practices and experiences of daily lives, as citizens – and other political subjects – negotiate exclusion and marginalisation.
>
> (Staeheli et al. 2012: 629)

Ordinary citizenship emphasizes the everyday and very normalized ways in which citizens demonstrate practices and acts of citizenship (in Isin and Nielson's terms) that provide legal and social order to their daily lives. This concept is particularly useful because it acknowledges the ways in which the very ordinariness of citizenship can be an implicit mechanism for exclusion, although arguably it could also be mobilized by marginalized people as a tool for political change. This resonates with James Holston's now famous text on *Insurgent Citizenship* (2008) exploring democratic citizenship in the urban peripheries of São Paulo. Holston argues that although Brazilian citizenship is universally inclusive in legal terms, it is inegalitarian in practice based on unequal access to land and services (i.e. infrastructure). Holston distinguishes between what he terms modernist citizenship, whereby citizenship is restricted to those who submit to state-led expectations and norms, and his notion of insurgent citizenship, whereby citizens challenge the state, often in violent and confrontational ways, for example illegally building houses and invading land as a means to claim their citizenship rights. This approach is pertinent because it highlights the ways in which while (modernist) citizenship can be exclusionary, (insurgent) citizenship can empower marginalized communities to take action against the state in order to secure a city that better reflects their interests and needs.

To conclude this section, within the contemporary focus on citizenship as incorporating both legal practices and socio-political acts, there is widespread recognition that citizenship plays a role in both marginalizing certain people (e.g. via legal frameworks and ordinary everyday acts) as well as being a tool for marginalized groups to protest (e.g. via insurgency or more 'ordinary' strategies) at multiple scales. Within this context there is an implicit assumption that those perceived as 'marginalized' are those with limited access to material goods and public infrastructure (often there are multiple marginalizations – e.g. immigrants lacking legal status, but weak access to infrastructure is typically connected), and yet the explicit connections between citizenship and infrastructure are rarely considered in depth. The next section provides an overview of contemporary scholarship on urban infrastructure, before the chapter turns to consider the concept of *infrastructural citizenship* as a means to explicitly connect these scholarly debates.

Infrastructure: socio-technical space of everyday life

While infrastructure has always been an implicit component of both urban and development geography subdisciplines, it has received a surge of interest from contemporary social scientists, recognizing that infrastructure is not a neutral technical matter to be addressed only by engineers and architects, but vital to social life. Within this contemporary perspective, infrastructure is conceptualized as more than just the physical manifestation of the city or the material means through which the urban is able to function. Certainly cities and urban dwellers need roads, electricity, water, sanitation and housing for example, which are all provided through the technology of road, pipes, cables and wires; but infrastructure is also inherently social in the way it is both produced and used. Within this context there is widespread recognition in the literature that a critical understanding of urban infrastructure requires a multi-disciplinary approach, spanning the physical and social sciences, to acknowledge the ways in which infrastructure represents a physical manifestation of a social *and* technological process, in terms of both construction and consumption.

Tracing the emergence of these socio-technical debates, Susan Star's (1999) 'Ethnography of Infrastructure' broke new ground by challenging existing conceptualizations of infrastructure as exclusively technical. Through her ethnographic fieldwork working alongside biologists, medics and computer scientists, she revealed the ways in which infrastructure is embedded in humans and their relationships. Within geography, Stephen Graham and Simon Marvin's now landmark *Splintering Urbanism* (2001) text has been at the forefront of new social science approaches to infrastructure. Graham and Marvin (2001) frame urban infrastructure as a socio-technical process, highlighting the ways in which the spaces of urban societal life (e.g. shopping, cooking, washing, sanitation) are made possible by technology, and consequently they reveal how the human body and technological machines are mutually dependent. Urban and cultural geographer Matthew Gandy (2004) uses the human biology concept of metabolism to unite the social and biophysical systems that underpin the city. Similarly, social anthropologist

AbdouMaliq Simone's (2004) conceptualization of *people as infrastructure* situates urban infrastructure not in roads and pipes, but in the social and economic collaborations that form between marginalized urban citizens.

Alongside the now widespread recognition that infrastructure is not merely technological, but also crucial to understanding contemporary urban societal life, comes increasing concern over unequal access to infrastructure. Graham and Marvin coined the phrase 'splintering urbanism' to describe the exclusionary and fragmentary effects of privatizing infrastructure provision. In the neoliberal era of privatization, the infrastructural division of the city is driven by market rationalities but has clear social implications; where water pipes bypass poor communities, or new train lines connect wealthy residential and commercial zones (but ignore poorer parts of the city), or where internet cables are prioritized in wealthy neighbourhoods. Consequently, urban splintering is both spatial and social, dividing the city into well-connected and under-connected zones, with societal consequences for poverty and inequality. So while historically infrastructure was viewed as neutral or technocratic, Graham and Marvin's approach reveals how networked infrastructure is a tool of social power that can extend and perpetuate inequality. Furthermore, this inequality is not merely about access to physical goods and services, but reveals broader processes of exclusion and marginalization that are intrinsically connected to citizens' rights.

Most recently there has emerged a shift towards exploring infrastructure from the perspective of citizens' everyday social and political lives. Graham and McFarlane's (2015) *Infrastructural Lives* focuses on the everyday scale as the means to illuminate citizens' perceptions and experiences of infrastructure. "While infrastructure debates have made important contributions to how we understand the 'supply-side' dimensions of infrastructure, there has been surprisingly little about how people produce, live with, contest, and are subjugated to or facilitated by infrastructure" (Graham and McFarlane 2015: 2).

The central goal of this everyday perspective is to explore how urban dwellers produce and manage their own infrastructure needs, in order to better understand the connections between urban life and infrastructure. Essentially, this is about exploring the relationship between infrastructure and everyday urbanism across a variety of empirical contexts, and consequently this perspective explicitly acknowledges infrastructure as a space of urban life. Similarly, urban geographer Ash Amin (2014) has analysed the sociality of infrastructure from the perspective of the poor, using a Brazilian case study to highlight the ways in which the poor themselves play a role in co-constructing infrastructure (e.g. water, electricity, shelter, sanitation), consequently highlighting how infrastructure is a crucial component of both individual and collective spaces of living and protest. This resonates with Simone's notion of *relational infrastructure* (2015), exploring low-income urban dwellers' incremental 'pay you go' strategies to improve and repair infrastructure as demonstrating people-led (rather than state-led) infrastructure. There seems a clear connection here, between three recent bodies of scholarship on infrastructure – Graham and McFarlane's (2015) *infrastructural lives*, Amin's (2014) *lively infrastructure* and Simone's (2015) *relational infrastructure* – and Holston's (2008) classic text on *insurgent citizenship*. All four explicitly recognize poor people's self-led strategies to provide their own infrastructural needs in the absence of adequate state provision, yet only Holston explicitly frames people-led infrastructure as a demonstration of urban citizenship. While Graham and McFarlane (2015) focus their everyday lens through questions related to violence, pacification and dispossession, with a special interest in waste, they acknowledge that this perspective is incomplete. Arguably, understanding infrastructure from the perspective of urban dwellers themselves has a clear connection to citizenship. While the physicality of inequality manifests in access to infrastructure, this is a material representation of unequal citizenship in the dual sense of both urban dwellers' access to citizenship rights as well as citizens' differentiated mobilization capacities. "Infrastructure is proposed as a gathering force and political intermediary of considerable significance in shaping the rights of the poor to the city and their capacity to claim those rights" (Amin 2014: 1).

The increasingly widespread conceptualization of infrastructure as both a technical and social element of institutional and everyday urban life has resulted in recognition of the role of infrastructure in other areas of urban life, such as urban violence (Rodgers and O'Neill 2012) and urban politics (McFarlane and Rutherford 2008), highlighting how broader processes of marginalization are operationalized through infrastructure in contemporary cities. This chapter argues the need for a more explicit connection between infrastructure and citizenship, using the phrase 'infrastructural citizenship' as the means for improved understanding of both concepts and practices, and the relationship between them.

Infrastructural citizenship

The connection between infrastructure and citizenship seems obvious, and indeed is often implicitly acknowledged in the literature. For example, there is widespread recognition that those with restricted citizenship rights (e.g. immigrants, homeless, slum-dwellers) suffer multiple marginalizations which typically include weak access to infrastructure such as water, housing, sanitation (e.g. Roy 2003; Staeheli et al. 2012). Similarly, the concept of *propertied citizenship* promotes an ideal of 'proper' citizenship, restricted to those whose legally entrenched access to infrastructure (via property ownership) gives them a vested stake in the nation-state that is physical, economic, political, social (Roy 2003). Indeed, Partha Chatterjee (2004) highlights the ways in which those with secure access to infrastructure (e.g. the middle class) use this advantage to stress their superior claim to citizenship (as law-abiding taxpayers) in order to further marginalize those with weak infrastructural status in the city. Furthermore, contemporary approaches to infrastructure as socio-technical highlight the ways in which infrastructure provides a physical representation of broader socio-political processes that implicitly include citizenship practices (e.g. "rights of the poor to the city", Amin 2014: 1) and acts (e.g. "capacity to claim those rights", Amin 2014: 1). And as a final example, the literature is rife with examples of citizenship acts that are not only focused on demanding improved access to infrastructure but frequently use infrastructure as a tool of protest (e.g. road blocks, gaining access to services illegally, self-built homes), such as Holston's insurgent citizenship (2008). However, despite these clear examples of the connections between infrastructure and citizenship, the relationship is rarely explicitly acknowledged or critically analysed, and is certainly under-theorized. In this chapter I propose the phrase *infrastructural citizenship* as a conceptual lens to tackle these oversights. Its definition is necessarily broad, focusing attention on the ways in which citizenship acts and practices are embodied in public infrastructure (and vice versa), in order to deepen understanding of the infrastructure-citizenship nexus in both theoretical and empirical terms. This is important because it explores potential connections between the infrastructural and civic nature of state-citizen relations.

This chapter is not the first to coin the phrase. Rune Justesen (2013) in a blog essay about *infrastructural citizenship* uses the concept to expose the active role of infrastructure in citizen-led forms of political mobilization. Framed around a case study of the Los Angeles port closures, Justesen (2013) argues that humans are not merely replaceable parts of the port system but are completely integral to this form of infrastructure, an observation dramatically revealed when the port was forced to shut down following strikes in 2012. Consequently, Justesen argues that infrastructure such as the port functions as a political space where active citizenship is demonstrated, for example via protests, strikes and occupations; and where citizens are fundamental to the effectiveness of infrastructure. While this example does indeed highlight the overlaps and connections between demonstrations of citizenship and sites of infrastructure, the emphasis on the port and its workers as the vessel for infrastructural citizenship is too narrow. While Justesen's infrastructural citizenship is restricted to those physically and economically positioned within infrastructure sites, this chapter argues that the concept has relevance for all citizens who (sometimes struggle to) access infrastructure.

Although not explicitly deploying the term infrastructural citizenship, Alex Wafer's (2012) research in South Africa on "discourses of infrastructure and citizenship" critically explores the relationship between the state's infrastructural capacity and imaginations of citizenship in everyday life within contemporary Soweto. Wafer (2012) dramatically highlights the centrality of infrastructure to citizenship in South Africa: from the ways in which infrastructure represented the power of the apartheid state, for example via urban segregation and the forced removal of non-citizens; to the post-apartheid government's agenda of service delivery as the mechanism for operationalizing universal citizenship; to the contemporary context of violent protests related to service delivery and illegal occupation of urban infrastructure as active demonstrations of citizenship. Wafer's (2012) Soweto case study reveals the ways in which the inherent connection between infrastructure and citizenship has changed over time, as he traces his involvement in the community from analysing protest-based citizenship (infrastructural in the sense that protests are framed around access to services) in the early 2000s, to witnessing more consumption-based citizenship in the 2010s (infrastructural in being framed around new shopping malls and materialism). Wafer's (2012) analysis is crucial in highlighting the embeddedness of the infrastructure-citizenship nexus in South Africa.

Most recently, Kyle Shelton's (2017) historical analysis of Houston residents' strategies of mobilization against highway construction in the 1970s uses infrastructural citizenship to highlight the ways in which the inert and technical materiality of infrastructure has political meaning. Shelton's rich historical analysis uses this case study to demonstrate how residents used infrastructure as a tool for political protest by arguing that their local streets and homes held equal value as major highways and urban regeneration projects, consequently using infrastructure as the means to assert their citizenship rights. In this context, the physicality of infrastructure took on a political and symbolic dimension as the location of citizenship practices and acts. Shelton uses this example to argue that citizen protests over infrastructure must not be sidelined as a minority or radical concern, but instead that infrastructure (both in a physical and symbolic sense) should be understood as central to urban citizenship. While embracing Shelton's call, and echoing the need to recognize the connections between infrastructure and citizenship, in this chapter I extend the concept beyond the protest-based forms of infrastructural citizenship that Justesen (2013), Shelton (2017) and Wafer (2012) all describe, to also consider forms of infrastructural citizenship that are not framed around radical or confrontational demonstrations but integral to everyday spaces of urban life.

Having introduced the concept of infrastructural citizenship, the chapter proceeds by exploring its utility within the empirical context of contemporary South Africa.

Infrastructural citizenship in Cape Town, South Africa

Within scholarship addressing contemporary urbanization in sub-Saharan Africa, there has been a marked focus on urban infrastructure, particularly in the context of rapid urbanization, hyper-fragmentation and low household incomes. African urbanist Edgar Pieterse (2014) argues that the expansion of slum urbanism has resulted in a polycrisis of extreme splintered urbanism with significant inequalities in both access to, and the quality of, infrastructure. More specifically, Pieterse and Hyman (2014) express concern regarding the extreme inequality represented in the current African model where the wealthy rely on private sector provision and the poor rely on self-construction. The theoretical implications of this urban political ecology have been situated within the African context by Lawhon et al. (2014), providing critical insights into the sociality and materiality of urban infrastructure in Africa, with a particular focus on the unequal distribution of resources. This city-scale approach is complemented by localized studies exploring the specific hurdles faced particularly by low-income urban-dwellers in accessing basic services and infrastructure such as electricity and sanitation, revealing the ways in which poor Africans circumvent state-infrastructure in order to independently construct or secure access (e.g. Baptista 2015; Jaglin 2014; Robins 2014). This emphasis on citizen-led infrastructure provision echoes trends identified elsewhere (e.g. Amin 2014; Graham and McFarlane 2015; Simone 2015).

Concurrently, scholarship on citizenship in urban Africa has received renewed attention, particularly in the context of democratization. Dorman et al. (2007) describe the ways in which citizenship has been re-framed in contemporary Africa around new forms of statehood and national belonging, in contrast to historic ethnic forms of belonging. They argue that state-based citizenship provides a framework for state-society relations. Building on this, Diouf and Fredericks (2014) define their *arts of citizenship* as the innovative and everyday ways in which African urban dwellers negotiate their citizenship rights and rewards, for example via involvement in churches, community groups and political organizations. They explore these arts in the context of infrastructures as a physical representation of the state and of broader processes of inclusion and exclusion, revealing "how delinations of citizenship are codified and contested in the built form of the city" (Diouf and Fredericks 2014: 6). In this sense, there is a clear connection between citizen-constructed infrastructure (e.g. house-building, electricity connections) and citizen-negotiated citizenship practices (e.g. associational life), particularly for low-income citizens. This is pertinent in revealing more about state-society relations in the context of infrastructural citizenship. Indeed, it seems obvious that infrastructure is not only one of the primary mediums through which citizens relate to the state (in terms of both expectations and physical realities), and also the primary way in which the state is physically represented at the local scale, and can play a role in creating and perpetuating (or potentially challenging and overcoming) marginalization.

Shifting attention specifically to South Africa, this provides a particularly interesting case study for exploring infrastructural citizenship because universal citizenship is only two decades old and has been accompanied by a raft of infrastructural-based policies. The post-apartheid government has prioritized an infrastructure-centric vision of citizenship, with the provision of housing and associated services (e.g. water, sanitation, electricity) implemented as the state's primary mechanism for overcoming past inequalities and implementing citizenship (Parnell and Pieterse 2010; Swilling 2006; von Schnitzler 2008; Wafer 2012). Apartheid South Africa provided a classic historic example of excluded citizenship. During apartheid citizenship was exclusively reserved for the white population, with other races designated citizens of their rural 'homeland' and therefore ineligible to own property, vote or move freely in South Africa's cities. This was clearly and explicitly not universal citizenship, and was inherently connected to infrastructure in terms of differentiated access. Furthermore, the collapse of apartheid in the late 1980s was in part a consequence of infrastructure-based protest, as black urbanization gradually threatened the apartheid system of urban segregation. In contrast, political change came more abruptly. On 27 April 1994, for the first time, all South Africans were legally considered full citizens with the right to vote in their new democracy. The new constitution and bill of rights came into force, with a particular emphasis on freedom and equality for all citizens. However, the spatial, social and economic legacies of apartheid remain, and in the contemporary era, mass protests over access to services (e.g. housing, electricity, water, jobs, education) have come to represent citizenship struggles for the urban poor. For while South Africa's progressive bill of rights gives full citizenship rights to all, access to services is still skewed according to income and spatial location, and thus access to infrastructure continues to play a major role in mediating perceptions and practices of citizenship.

The most significant infrastructural provision in the post-apartheid context is the construction of housing for low-income households. Since 1994 approximately three million state-subsidized fully serviced brick-built houses have been constructed and occupied (alongside full ownership rights). Despite criticisms of the housing policy in terms of the quality of construction, size of houses, location of settlements, and tendency to view beneficiaries as static recipients rather than active citizens, the scale of construction is impressive. For many housing beneficiaries there is significant symbolic association with becoming a citizen and receiving ownership of a fully serviced state-subsidized house located in a city in which they were previously considered a temporary sojourner. Based on fieldwork undertaken over a long period (2004–ongoing) in a single state-subsidized housing settlement in Cape Town, South Africa, I use a handful of residents' perceptions to indicate the potential utility of infrastructural citizenship as a lens through which to understand these experiences. This is necessarily indicative at this stage, as these ideas represent the emergence of a new research area on which further research is anticipated.

What is clear from over a decade of fieldwork in this state-subsidized housing settlement is that housing beneficiaries themselves demonstrate a strong connection between citizenship and infrastructure. In particular, residents frequently expressed a vivid association between the home that they now own and 'the end of the struggle'. In this sense, housing beneficiaries demonstrate explicit attachment to the physicality of the house and its associated infrastructure in terms of their identity as citizens of a post-apartheid South Africa and of Cape Town, a country and a city where they have potentially always lived, but have only recently received citizenship.

> Do you know how much this house is worth to me? Everything. I cannot give a price because this is the first and last house I have ever owned.
>
> (A.D. 3 September 2006)

> If I ever get the opportunity to buy somewhere bigger, I will never sell that house. I would rent it out because that house means so much to me, that's where all my dreams are. The first home I built for my family. It's a house that gave me strength … I won't just give it away, it's very close to my heart.
>
> (M.N. 14 September 2006)

> If a person asked me to sell my house they can go high and high, it doesn't matter. I won't sell it because I stay here … we struggled a lot and the Lord blessed me with this house.
>
> (Ry.M. 3 September 2006)

As these quotations demonstrate, the start of a new life in a democratic and free nation is emotionally and symbolically tied not merely to homeownership per se, but to the specific physical house and infrastructural services that embody this freedom. However, there are also dissenting views, with numerous residents expressing a more negative association with the physicality of their house and services.

> I was very unhappy when I moved in because we struggled for years and we got kak [shit]. You see the cracks in the wall, there's a lot, in the kitchen, bathroom and living room … They don't care about us here.
>
> (C.M. 16 September 2006)

> All the houses is falling apart – the doors, the windows, it all fall apart. They build it too quickly – there's a lot of cracks. I think they built in a month. They just wanted us to move out so the rich people can move in. They promised us a lot and none of it came.
>
> (L.T. 14 September 2006)

Although these latter quotations demonstrate less contentment with the material properties of the house, the relationship between the physical embodiment of a state-subsidized house and the sense of citizenship in the new South Africa remains prominent. While residents' perception of the quality of their house differed (although in fact, even those who highly valued their house still recognized its physical failings), the material provision of public housing and infrastructure is clearly framed around citizenship. While for some receiving a state-subsidized house embodies a sense of equal citizenship in post-apartheid South Africa, for others it highlights perceptions of ongoing marginalization as a citizen, particularly in comparison to the material wealth of other citizens.

In situating these empirical findings within contemporary literature exploring the connections between infrastructure and citizenship, a disjuncture emerges. While Amin (2014) and Holston (2008) both demonstrate how marginalized citizens construct their own forms of infrastructure, in this South African case study, the state has provided physical infrastructure in the form of housing and associated services. Consequently infrastructural citizenship emerges in the ways in which citizens relate to this state-provided infrastructure. Returning to Isin and Nielsen's (2008) distinction between citizenship practices and acts, this South African example demonstrates citizenship identity as rooted in the legality of access to infrastructure in the post-apartheid context of universal citizenship, with housing beneficiaries sufficiently confident in their new citizenship rights to criticize the quality and inequality of public infrastructure. This citizenship is rooted in legal *practices*, but demonstrated through public infrastructure as the physical representation of the state in citizens' everyday spaces of living. Although these quotations do not reveal the extent of citizenship *acts*, in terms of how citizens actively demonstrate their infrastructural citizenship, there is widespread evidence for example that housing beneficiaries demonstrate insurgent citizenship (Holston 2008) acts via building (illegal) informal structures in their backyards and accessing electricity extensions (see Lemanski 2009). Furthermore, it is evident that associational life as a form of arts (Diouf and Fredericks 2014) and/or ordinary (Staeheli et al. 2012) citizenship acts has been facilitated via the provision of infrastructure in terms of the spatial location of churches and sports groups, as well as providing the political focal point for community groups (see Lemanski 2006, 2008). Arguably then, this case study reveals how infrastructure both shapes citizenship practices and identities as well as determining citizens' capacity to engage in, as well as providing a focal point for, citizenship acts.

As this is an emerging area of enquiry, forthcoming research will explore the physical changes that citizens have made to their houses (e.g. extensions), methods of securing improved access to infrastructure (e.g. refuse collection, additional electricity lines, access to schooling and health care) as well as citizens' material aspirations for the future. These infrastructural acts will be analysed alongside residents' citizenship identity and practices in terms of involvement in everyday associational life, civic life and methods of protest, as a means to further explore the utility of infrastructural citizenship as a conceptual lens for analysing ongoing state-society relations.

Conclusion

This chapter has introduced the concept of infrastructural citizenship as a conceptual lens through which to connect scholarly debates on infrastructure and citizenship. This approach resonates with research situated elsewhere recognizing the ways in which citizens demonstrate their citizenships practices and acts (Isin and Nielson 2008) via the self-construction of infrastructure (e.g. Amin 2014; Graham and McFarlane 2015; Holston 2008; Simone 2015). Using a case study of state-subsidized housing in South Africa, the chapter has revealed the ways in which the legalities of citizenship practice, perceptions of citizenship identity and expressions of citizenships acts are all embedded in the physicality of public infrastructure as a representation of the state at the local scale. In this specific case study, housing beneficiaries express citizenship as embodied by their housing, with physical receipt of a house representing the start of a new identity as a South African citizen, accompanied by the confidence to criticize the state via infrastructure. This 'ordinary' approach to citizenship (Staeheli *et al.* 2012) allows the inclusion of a broad range of citizenship meanings, from everyday associational life to protest-based conceptualization of citizenship, in all cases highlighting the role of public infrastructure as the physical means through which citizens demonstrate their citizenship identity, practice and acts.

References

Amin, A. (2014). Lively Infrastructure. *Theory, Culture and Society*, 31(1), pp. 137–161.
Baptista, I. (2015). 'We Live on Estimates': Everyday Practices of Prepaid Electricity and the Urban Condition in Maputo, Mozambique. *International Journal of Urban and Regional Research*, 39(5), pp. 1004–1019.
Chatterjee, P. (2004). *The Politics of the Governed: Reflections on Popular Politics in Most of the World*. New York: Columbia University Press.
Diouf, M. and Fredericks, R., eds. (2014). *The Arts of Citizenship in African Cities: Infrastructures and Spaces of Belonging*. New York: Palgrave Macmillan.
Dorman, S., Hammett, D. and Nugent, P., eds. (2007). *Making Nations, Creating Strangers: States and Citizenship in Africa*. Leiden: Koninklijke Brill.
Dwyer, P. (2010). *Understanding Social Citizenship: Themes and Perspectives for Policy and Practice*. Bristol: Policy Press.
Gandy, M. (2004). Rethinking Urban Metabolism: Water, Space and the Modern City. *City*, 8(3), pp. 363–379.
Graham, S. and Marvin, S. (2001). *Splintering Urbanism: Networked Infrastructures, Technological Mobilities and the Urban Condition*. London: Routledge.
Graham, S. and McFarlane, C. (2015). *Infrastructural Lives: Urban Infrastructure in Context*. Abingdon: Routledge.
Holston, J. (2008). *Insurgent Citizenship: Disjunctions of Democracy and Modernity in* Brazil. Princeton, NJ: Princeton University Press.
Isin, E.F. and Nielsen, G.M., eds. (2008). *Acts of Citizenship*. London: Zed Books.
Jaglin, S. (2014). Regulating Service Delivery in Southern Cities: Rethinking Urban Heterogeneity. In: S. Parnell and S. Oldfield, eds. *The Routledge Handbook on Cities of the Global South*. London: Routledge.
Justesen, R. (2013). *Infrastructural Citizenship*. Available at: https://infrastructuralgossip.wordpress.com/2013/09/15/infrastructural-citizenship/ [accessed 21 January 2016].
Lawhon, M., Ernstson, H. and Silver, J. (2014). Provincializing Urban Political Ecology: Towards a Situated UPE through African Urbanism. *Antipode*, 46(2), pp. 497–516.
Lemanski, C. (2006). The Impact of Residential Desegregation on Social Integration: Evidence from a South African Neighbourhood. *Geoforum*, 37(3), pp. 417–435.
Lemanski, C. (2008). Houses without Community: Problems of Community (In)Capacity in a Low-Cost Housing Community in Cape Town, South Africa. *Environment and Urbanization*, 20(2), pp. 393–410.
Lemanski, C. (2009). Augmented Informality: South Africa's Backyard Dwellings as a By-product of Formal Housing Policies. *Habitat International*, 33(4), pp. 472–484.

Marston, S.A. and Mitchell, K (2004). Citizens and the State: Citizenship Formations in Space and Time. In: C. Barnett and M. Low, eds. *Spaces of Democracy: Geographical Perspectives on Citizenship, Participation and Representation.* London: Sage, pp. 93–112.

McFarlane, C. and Rutherford, J. (2008). Political Infrastructures: Governing and Experiencing the Fabric of the City. *International Journal of Urban and Regional Research,* 32(2), pp. 363–374.

Parnell, S. and Pieterse, E. (2010). The 'Right to the City': Institutional Imperatives of a Developmental State. *International Journal of Urban and Regional Research,* 34(1), pp. 146–162.

Pieterse, E. (2014). How Can We Transcend Slum Urbanism in Africa? UN-Habitat Global Urban Lecture Series. Available at: http://unhabitat.org/how-can-we-transcend-slum-urbanism-in-africa-edgar-pieterse-university-of-cape-town/ [accessed 21 January 2016].

Pieterse, E. and Hyman, K. (2014). Disjunctures between Urban Infrastructure, Finance and Affordability. In S. Parnell and S. Oldfield, eds. *The Routledge Handbook on Cities of the Global South.* London: Routledge.

Robins, S. (2014). Poo Wars as Matter out of Place: 'Toilets for Africa' in Cape Town. *Anthropology Today,* 30(1), pp. 1–3.

Rodgers, D. and O'Neill, B. (2012). Infrastructural Violence: Introduction to the Special Issue. *Ethnography* 13(4), pp. 401–412.

Roy, A. (2003). Paradigms of Propertied Citizenship: Transnational Techniques of Analysis. *Urban Affairs Review,* 38(4), pp. 463–491.

Shelton, K. (2017). Building a Better Houston: Highways, Neighbourhoods, and Infrastructural Citizenship in the 1970s. *Journal of Urban History,* 43(3), pp. 421–444.

Simone, A. (2004). People as Infrastructure: Intersecting Fragments in Johannesburg. *Public Culture,* 16(3), pp. 407–429.

Simone, A. (2015). Relational Infrastructures in Postcolonial Urban Worlds. In: S. Graham and C. McFarlane, eds. *Infrastructural Lives: Urban Infrastructure in Context.* London: Routledge, pp. 17–38.

Staeheli, L.A., Ehrkamp, P., Leitner, H. and Nagel, C.R. (2012). Dreaming of the Ordinary: Citizenship and the Complex Geographies of Daily Life. *Progress in Human Geography,* 36(5), pp. 627–643.

Star, S.L. (1999). The Ethnography of Infrastructure. *American Behavioral Scientist,* 43(3), pp. 377–391.

Swilling, M. (2006). Sustainability and Infrastructure Planning in South Africa: A Cape Town Case Study. *Environment and Urbanisation,* 18(23), pp. 23–50.

Taylor-Gooby, P. (2008). *Reframing Social Citizenship.* Oxford: Oxford University Press.

von Schnitzler, A. (2008). Citizenship Pre-paid: Water, Calculability and Techno-Politics in South Africa. *Journal of Southern African Studies,* 34(4), pp. 899–917.

Wafer, A. (2012). Discourses of Infrastructure and Citizenship in Post-Apartheid Soweto. *Urban Forum,* 23(2), pp. 233–243.

30
THE POLITICS OF URBAN AGRICULTURE

Sustainability, governance and contestation

Nathan McClintock, Christiana Miewald and Eugene McCann

Introduction

The Downtown Eastside Neighbourhood House (DTES NH) is a community services centre in a low-income neighbourhood of Vancouver, British Columbia. Like many organizations of its type, its programmes are based on a set of principles – an 'operating philosophy' – of inclusivity and activism. Unlike most, the neighbourhood house also articulates a separate, if related, 'food philosophy':

> Food is a key determinant of individual and community health – physical, mental, emotional and spiritual. We take every modest opportunity to remind DTES Residents of our Right to quality food. We use the offering of food to reflect back upon our neighbours their inherent dignity, deservedness and welcome within the DTES NH.
>
> *(Right To Food Zine 2016)*

Food, in its production, its consumption and its associated meanings, is clearly central to sustenance for the DTES NH, as it is for all of us. What their food philosophy makes clear, furthermore, is that food is also fundamentally political.

This chapter explores food production and, specifically, urban agriculture (UA) as a set of fundamentally political practices, both in terms of their role in neoliberal governance and 'sustainability' policy-making, and also as objects of contestation. We provide a brief overview of UA with a focus on the changing nature of urban food production in the Global North, then engage with UA's role in supporting food security, its contributions to environmental and social sustainability, as well as its entanglement in processes of gentrification. In particular, we use case studies from Portland, Oregon and Vancouver, British Columbia to highlight the contentious nature of UA in cities that explicitly frame their policy-making in terms of sustainability, resilience and 'greenness'.

While the practice of UA, which we broadly define as the production of food in cities, is as old as urbanization itself (Lawson 2005), it has enjoyed a striking resurgence in recent years in the Global North. This has been due in part to widespread assertions of its transformative contributions to food security and urban sustainability. Yet, the distribution of UA is socio-spatially uneven, benefiting some and excluding others. Therefore we ask, how do these differential – and inequitable – patterns both *arise from* and *contribute to* the fundamental tensions between economic growth, environmental regulation and social equity that define sustainability? And how do municipal policies mediate these processes? Can and do these UA policies open new spaces for more equitable models of sustainability? We will detail the role that both municipal policies and activist politics have played in shaping where and for whom UA is integrated into the urban landscape. In turn, we ask whether spaces of food production are part of a right to the city and if they can produce alternative visions of urban life and economic relationships. Our aim is to problematize the often-uncritical celebration of UA, highlight spaces of conflict within this growing movement and, at the same time, emphasize the social, health and environmental benefits of urban food production.

Approaching urban agriculture

Widely hailed as a key component of urban sustainability, and taking a prominent place in municipal planning efforts, UA is fast becoming entangled in the contradictions of sustainable urbanism in the Global North and is developing new forms and functions in the neoliberal city. New gardens are cropping up at a furious pace in a variety of types: residential gardens, community or allotment gardens, organizational gardens working for 'food justice' in so-called 'food deserts' (low-income areas with limited access to fresh produce) and market gardens and larger-scale urban farms that provide restaurants and residents with 'ultra-local' produce.

At the same time, UA is coming under scrutiny by community activists questioning who it is actually serving, raising concerns that this new wave of UA primarily caters and appeals to the affluent and opens the door to predominantly white gentrifiers (Crouch 2012; Markham 2014). In response, some UA advocates are organizing against gentrification and expanding their focus to broader struggles for social justice (Phat Beets Produce 2013; SFUAA 2013). Some are engaging in policy-making, bringing equity concerns to the fore. UA has, then, become a key site of political contestation over urban sustainability. It has become both a *driver of* and *reaction to* ongoing neoliberal urban transformations.

Early scholarship on UA tended to emphasize either its benefits or shortcomings, leading to a dichotomizing perspective on the practice as either radical or neoliberal. But these benefits and shortcomings actually co-exist as a function of UA's diverse motivations and forms (McClintock 2014). Often manifest at different scales, such contradictions cannot be fully understood without treating UA as a *process* operating within broader contradictory tensions of the 'uneven development' of the city, i.e. how flows of capital shape the city differentially for the benefit of ongoing accumulation of capital, regardless of impacts on urban residents (Smith 2008). This analytical lens brings into focus the relationship between political economic cycles of disinvestment and reinvestment (Hackworth 2007; Harvey 1989) and UA's emergence as a socio-spatial phenomenon that both reproduces and contests capitalist urbanization (McClintock 2014; Sbicca 2014).

Critical perspectives on urban agriculture

Urban agriculture's renaissance over the past decade has been accompanied by a groundswell of new UA organizations, projects, media attention and scholarship. Scholars have documented the multiple attributes of UA, including its provision of a suite of environmental and social benefits. These include: enhancing biodiversity; managing stormwater infiltration; improving nutritional and mental health; fostering community interactions and cohesion; mitigating urban food insecurity; and serving as an ethical alternative – albeit limited in scale – to the dominant industrial agri-food system (Barthel *et al.* 2015; Draper and Freedman 2010; Taylor and Lovell 2014).

Yet, critical food scholars in geography, sociology and anthropology have increasingly challenged UA's progressive potential on several grounds. One line of critique sheds light on how UA activists are inadvertently complicit in neoliberal restructuring, despite their radical intentions. Drawing on agrarian political economy, scholars demonstrate how alternative food networks, including UA and other "interstitial food spaces" (Galt *et al.* 2014), arise from political economic restructuring (Jarosz 2008). Some of these networks ultimately subsidize capital by shifting a portion of the responsibility for social reproduction onto volunteer-run groups, such as those organizing UA projects, since they replace services once provided by the welfare state (Allen and Guthman 2006; Rosol 2010). Further, many UA efforts ultimately instil a 'neoliberal governmentality' that encourages both personal responsibility for coping with economic restructuring, and market-based consumption-oriented approaches to food politics over collective action (Drake 2014; Pudup 2008; Weissman 2015).

A second relevant line of critique challenges 'the local' as a normative scale of intervention, and warns against reducing food justice to a spatial problem that can be easily ameliorated by constructing a garden or grocery store in a food desert (Shannon 2014). Falling into this 'local trap' prevents practitioners from addressing macro-scale structures mediating food access (Born and Purcell 2006), and the historical processes and contingencies that mediate access in particular neighbourhoods (Bedore 2013; McClintock 2011). Some scholars have therefore advocated for a more 'reflexive localism' (DuPuis and Goodman 2005; Levkoe 2011) that situates alternative food efforts within broader food systems and political economic contexts.

Finally, a third relevant critique draws on critical race theory to argue that alternative food networks (including UA) are often constructed as 'white spaces', where "bringing good food to others" (Guthman 2009) re-inscribes paternalistic, colonial patterns of oppression of people of colour. Disproportionate funding of white-led UA programmes further fuels this trend (Reynolds and Cohen 2016). Frequently, the motivations of well-meaning UA advocates who pursue work in communities with limited access to healthy food do not correspond to the expressed needs of community members themselves (Lyson 2014; Ramírez 2015; Slocum 2007). Some scholars parallel this critique by applying the concept of 'food sovereignty' to urban contexts in North America. Food sovereignty not only underscores resistance to the hegemony of the global agri-food system, but also the need for low-income communities of colour to determine what and how to eat (Bradley and Galt 2014; Roman-Alcalá 2015). How people frame UA – in terms of 'food justice', 'food sovereignty', 'community' or other frames – can differ greatly within cities and even within gardens.

While each of these lines of critique offers important insights into the politics and practice of UA, a discussion of UA from the perspective of urban theory in general and urban political theory in particular is only just beginning to gain traction (Eizenberg 2012; McClintock 2014; Purcell and Tyman 2014). Much of this discussion builds on a handful of foundational studies addressing tensions between UA and gardens in New York City (Schmelzkopf 2002; Smith and Kurtz 2003). It is crucial to foreground the political economic processes that shape the built environment and UA if we are to better differentiate UA's multiple and sometimes contradictory manifestations.

Neoliberal urbanization and the political economy of sustainability

Sustainability is never politically neutral. Popular narratives suggest that everyone benefits from the implementation of sustainable or 'green' infrastructure and amenities. Yet, these narratives tend to obscure both the spatial disparities in access to such benefits and the socio-economic inequities that are exacerbated by differential or uneven investment in this infrastructure. Geographic scholarship on 'eco-gentrification' emphasizes how investment in green infrastructure in the urban core – LEED buildings, bike lanes, walkable neighbourhoods, farmers markets, urban gardens, etc. – further entrenches inequity by serving some at the expense of others. Beneath a veil of environmentalist rhetoric, sustainable development fuels rising property values and rents, green spaces become commodified for consumption, and local residents are forced out of their neighbourhoods by more affluent newcomers seeking these amenities (Anguelovski 2016; Dooling 2009; Quastel 2009; Tretter 2013). Such processes are illustrative of the contradictions emerging from the 'urban sustainability fix' (While et al. 2004), the selective implementation by cities of environmental goals and values (often under the banner of sustainability) to legitimate and advance long-standing neoliberal entrepreneurial development strategies (Goodling et al. 2015; Temenos and McCann 2012; Walker 2015).

The geographic literature on neoliberal urban restructuring of the post-industrial Global North city (Hackworth 2007; Peck and Tickell 2002) can help us relate UA motivations and practices to (uneven) processes of urbanization. This literature explains how restructuring has reorganized urban space by examining how capital reinvestment in the urban core over the past few decades has transformed industrial 'wastelands' and inner-city 'ghettos' into new spaces of consumption, typified by the service economy, luxury condos, boutique shops, tourism, hotels and conference centres. A related body of work on gentrification examines how the return of capital has led to widespread displacement of low-income residents from the urban core (Blomley 2003; Lees 2012). For a key theorist of capital's return to the urban core, gentrification is a "consummate expression of neoliberal urbanism" (Smith 2002: 446).

Urban agriculture and the sustainable city

Since certain forms of UA – notably large community gardens, collective gardens run by non-profits, and commercial market farms on vacant lots – arise opportunistically in economically devalued urban areas (McClintock 2014), we must attend to how variegated processes of neoliberal development (Brenner et al. 2010) shape the form and distribution of UA and how UA, in turn, both contributes to and challenges these processes. In particular, we must critically examine the relationship between these processes and the recent resurgence of interest in UA – both large-scale and residential – in gentrifying areas. A useful starting point is to examine the sustainability fix employed by many cities in the neoliberal era. Investment in sustainability is a means of responding to both widespread environmental concern (particularly in politically progressive urban centres) and the entrepreneurial logic driving economic growth. Investment in green infrastructure appeals to green consumers and investors alike, both assuaging public anxiety and laying the groundwork for ongoing accumulation of capital. Mainstream sustainability goals thus tend to privilege profit-motivated development over equity (Temenos and McCann 2012).

Since it is such a broad and ambiguous concept with a generally positive connotation, sustainability is often shielded from critique and politicization. It frequently operates in a 'post-political' framework that focuses on technicalities (for example, how best to rework zoning ordinances to allow and encourage UA, or to draft building codes for green roofs) and, thus, downplays questions about rights and interests and, in turn, stifles collective mobilization for more radical change (Davidson and Iveson 2014; Swyngedouw 2009). Celebratory and post-political sustainability discourse frequently obscures the growing inequality in purportedly sustainable cities, as well as the historical processes of uneven development that make sustainability investments possible in the urban core (Goodling et al. 2015).

There is still significant work that needs to be done to clarify the particular dynamics linking UA and eco-gentrification, however. For instance, UA projects often take root in vacant lots in the urban core, places awaiting the next wave of redevelopment. As investment returns under banners of 'smart growth' or 'livability', UA sites either become coveted for their development potential and gardens are resultantly displaced alongside the residents who created them (Schmelzkopf 2002; Smith and Kurtz 2003). Or, these UA sites become internalized by the gentrification process and are marketed by developers as signs of middle-class conviviality in ways that add value to new, high-end construction projects (Quastel 2009). In cities where vacant and cheap land abounds (e.g. Detroit and other Rust Belt cities), 'land-grabbing' by nascent urban agribusiness has fuelled bitter debates over whose interests UA ultimately serves (Colasanti et al. 2012; Safransky 2014). Research in Portland, Oregon, for example, reveals that *residential* UA may also be enmeshed in the gentrification process, as young, educated newcomers, drawn to cities renowned for green amenities, move into cheaper neighbourhoods (McClintock et al. 2016). Furthermore, the motivations of practitioners vary along socio-economic lines, with sustainability and environmental concerns resonating more with more affluent and educated gardeners, suggesting a class-based 'eco-habitus' (Carfagna et al. 2014) that motivates them to garden. Urban agriculture thus appears to take on a particular socio-spatial form in the neoliberal city, and has become an increasingly important space of politics.

Urban agriculture governance and contestations in Portland, Oregon and Vancouver, British Columbia

We turn now to a case study of two cities – Portland and Vancouver – to begin to understand how such processes unfold. While Portland and Vancouver are both known for their green infrastructure, policies and lifestyles, they have different economies, social histories, demographics and regulatory frameworks. For instance, while the population of both municipalities is similar (594,000 and 603,000 respectively in 2011), Vancouver has nearly three times the population density and significantly higher property values. Further, while Portland is a majority white city with long-established African American and Latino communities, less than half of Vancouver's population is white and over one-third is of Asian descent. Vancouver's economy is more oriented towards the global market, and is a major hub of international resource trade and real estate investment (Berelowitz 2010). Portland, on the other hand, is "a regional city" (Abbott 2001: 40) historically tied to domestic markets, but has recently capitalized on its ability to export its models of sustainability planning and green economic development around the world (Temenos and McCann 2013).

Despite these fundamental differences, common regional, cross-border narratives have highlighted similarities in order to construct a shared identity, premised on: a common physical geography between the Cascade mountain range and the Pacific Ocean; a history of environmental activism that laid the foundation for innovations in sustainability and green infrastructure (from light rail and bike lanes to green buildings and farmers markets); and even an elite dream of economic integration (Sparke 2000). Both cities have topped various sustainability and livability rankings over the past decade, garnering worldwide attention and accolades. Municipal officials in both cities have begun to recognize the range of benefits that UA can provide and have taken interest in expanding its functional role in local food systems by adopting new ordinances and zoning updates in efforts to expand residential food production (McClintock and Simpson 2016). At the same time, sustainability's unevenness in both cities has served as a rallying point for food systems activism and other social justice struggles, particularly as food takes a more prominent role in the revitalization and, some would argue, gentrification of urban neighbourhoods (Burnett 2014).

Both cities use UA to further their sustainability goals in ways that both promote and manage urban food production. At the same time, in both cities, UA is increasingly squeezed out, both in terms of economics and space, as a result of a municipal commitment to increased building density. Furthermore, the two cities have been engaged in a relatively long-term relationship around policy-making, particularly in relationship to food. As we might expect in a neoliberal context, their learning engagements with each other and associated sharing and borrowing of certain policies and practices with other cities around sustainability are also tinged with a compulsion to compete (McCann 2013). Finally, in Vancouver and Portland, UA is unevenly implemented and embraced. Affluent neighbourhoods often reject an UA aesthetic that prioritizes food production over manicured lawns, which suggests a lesser concern with the environmental ethics of sustainability. At the same time, residents of low-income neighbourhoods often lack the necessary resources – notably land and time – to be involved in UA. Moreover, current models of individual plot ownership replicate hegemonic property relations that may be incommensurate with collective land stewardship models, which are often tied to culturally specific value systems. The ability of UA to go beyond the creation of 'feel good' projects to provide alternative models of urban existence is constrained by which projects are sanctioned and which are marginalized by city policies, land use constraints, and by the sometimes divergent goals and motivations of UA practitioners.

Despite the dominance of the sustainability narrative that undergirds UA in these two cities, there continue to be contestations around the application of UA that unfold along various lines, from the institutional to spatialized class and racial/ethnic divisions. First, while we might consider UA to be a relatively low-cost sustainability fix (compared to more structural interventions such as bioswales or green roofs) and one that is often celebrated for its health, social and environmental benefits, it has not been without detractors within cities themselves. For example, municipal staff in Portland and Vancouver were initially reluctant to allow food production in parks. This resistance came from an entrenched perspective within Vancouver's Parks Board and Portland's Department of Parks and Recreation that park space should be primarily used for recreation, not food production. Additionally, there was concern that by assigning plots to individuals and charging a fee, public space was essentially closed off to the public. People from both cities who were involved with developing community gardening programmes remember the resistance they faced. Portland's first director of the city's Community Gardening programme, explains,

> When the Parks [and Recreation department] originally started doing the community gardens it was definitely recreation, and a couple of years after that they said, "Well okay maybe we can be a part of this food access question", but they had to be dragged kicking and screaming to that because they were unsure about a programme that subsidizes individuals in what usually turns out to be [in] perpetuity once you get your plot.

Similarly, an early champion for UA in Vancouver, recalls,

> A lot of the Parks Board staff, especially those who are kinda more horticulturalists, were asking, "Why are food and gardens being talked about in the Parks Board? That should all be private, public land should only be open green space."

In Portland, this tension was largely resolved with the advent of the Community Gardens programme in 1975 but it took another two decades for the Vancouver Parks Board to institutionalize community gardens on city greenspace. Today, however, tensions remain despite the formalization of food production. Because parks (as well as school grounds) are some of the few remaining places in the city where open space is protected from market forces, questions over how public greenspace should be used figure centrally into activists' efforts to increase urban food production. However, the public has not welcomed the encroachment of food production onto park space in every instance. In Vancouver, one attempt by the Parks Board to turn an unused parking lot into a community garden was met with resistance by neighbourhood residents who were concerned about privatization of public space.

Beyond challenging gardening in public parks, some within municipal government simply question municipal involvement in questions of food production, more broadly. Resistance to UA was evident in a recent Vancouver mayoral election, when the City's UA initiatives, including a bylaw allowing backyard chickens and providing funding to a non-profit organization to encourage the growing of wheat in residential yards, were used, unsuccessfully in the end, to discredit the incumbent Vision Vancouver party. In particular, the opposition argued that food production is outside of the City's mandate. One member of the Vancouver Food Policy Council, a local government advisory board, explained,

> The opposing party to Vision Vancouver has used urban farming and community gardens and wheat being grown in the city as ways of trying to poke holes in the mayor's agenda. You know, just saying that that's not actual city business, and that's hippie fluff.

Second, such debates highlight the ways in which the politics of UA take shape along cultural and class divides within cities as well as related spatial divides. In both Portland and Vancouver, UA's visibility often corresponds to the concentration of a particular demographic group. Food production in front yards, for example, tends to occur more frequently in middle-income neighbourhoods where levels of education attainment are high (McClintock et al. 2016). Here, people may be more likely to challenge conventional landscape norms of the manicured front lawn and a garden relegated to the backyard where it is out of sight. Indeed, in green cities, where gardens are well entrenched in the sustainability imaginary, front yard gardens may instead signal adherence to a particular set of environmental values that may not resonate in the most affluent neighbourhoods where such practice might be seen as a transgression of social norms (Naylor 2012).

According to one urban farmer, for example, residents of the more affluent Westside of Vancouver tend to maintain this more traditional aesthetic of "manicured lawns" over one transgressed by "chickens and vegetables", whereas Eastside neighbourhoods are more likely to have visible food production, from front yard gardens to gardens on boulevards and traffic circles (on general east-west differences in Vancouver, see Proudfoot and McCann 2008). Community and school gardens follow a similar pattern in Vancouver. In the affluent Shaughnessy neighbourhood, for example, there are neither, whereas the lower-income, trendy neighbourhood of Grandview-Woodland has at least seventeen such gardens.

Residential gardening in Portland follows similar patterns. Front yard gardens are most common in neighbourhoods closest to the median income, in gentrifying and recently gentrified neighbourhoods (see Figure 30.1), whereas most gardens in both the affluent south-west neighbourhoods and low-income neighbourhoods of East Portland tend to be hidden from view. The motivation to garden also differs in line with this uneven socio-economic geography (on east-west differences in Portland, see Goodling et al. 2015). In East Portland, gardeners tend to be less affluent than the surrounding population and rely on their gardens for more of their produce, whereas gardeners living in inner Northeast and Southeast tend to be more affluent,

Figure 30.1 A front yard garden in a gentrifying neighbourhood of Portland, Oregon
Source: photo: Nathan McClintock.

have higher levels of educational attainment than their neighbours, and emphasize environmental sustainability to a greater extent (McClintock *et al.* 2016).

Third, contestations over how urban space should be used are often highly racialized. Increasingly, UA is viewed as a space of young, white, middle-class residents, with immigrants and people of colour frequently excluded from participation in more public forms of gardening because of language and other barriers (Reynolds and Cohen 2016). In Portland, many of the city's green amenities have become flashpoints over long histories of racial dispossession. Many African American residents see the arrival of green infrastructure such as bike lanes, bioswales and gardens as both a driver and result of the gentrification and displacement that has gutted the black community over the past decade, as well as the most recent example of top-down planning decisions imposed on the city's black community (Lubitow and Miller 2013). Given the proliferation of gardens within the broader context of wholesale gentrification of Portland's historically black Albina district, certain types of UA – especially collective and community gardens – are increasingly 'coded' or 'read' as white, a marker of the influx of white hipsters and the loss of African American culture. According to a leader of black business association in Albina, the emergence of community gardens was a "bad sign for the African American community. We always gardened. We always shared our gardens and our food. We didn't need 'community gardens', that's a white invention" (quoted in Hern 2016). Indeed, the way UA is currently implemented and managed is reflective of a certain notion of food production that may not reflect the diverse ways in which communities may connect with agriculture. In Vancouver, this process is less evidently racialized, but nevertheless cuts along class lines. Embodying the City of Vancouver's Greenest City claims, UA is entrenched in the green entrepreneurial logic undergirding municipal economic development strategies, which inevitably lead to gentrification and displacement as housing prices rise.

Contesting the sustainability fix?

As the cases of Portland and Vancouver demonstrate, understanding UA as a sustainability fix can help us to better recognize UA as a space of politics in the neoliberal city. Central to development of the green city, UA therefore enjoys a prominent role in municipal governance of sustainability. Food system localization is heralded as a central tenet of such policy work. But the belief that "fresh, local, and usually organic foods produced by small-scale farms [is] the alternative to the industrial food system and its environmental ills" (Bradley and Galt 2014: 172) has been critiqued for conflating local with ethical, for being exclusionary, and for failing to acknowledge structural inequalities that inhibit food access (Born and Purcell 2006; Ramírez 2015; Sbicca 2014).

Urban agriculture is a way of performing sustainability without addressing who actually benefits from the increasingly common municipal drive towards a neoliberal version of sustainability. While UA projects in both cities aim to address issues of sustainability, they are not necessarily designed to address issues of food security and inequality, especially when these projects are heavily managed by the City or constrained by funding parameters. Indeed, urban farms and community gardens often do not have an explicit social justice mandate, nor do equity concerns motivate all practitioners of UA. One urban farmer from Vancouver remarked, "I don't know that urban farming is so much an activist activity. Like, for a lot of people, it's small business, it's entrepreneurial." Both in Portland and Vancouver, community gardens and urban farms are viewed as spaces for community building and food production and are, largely, not sites from which to critique structural issues that contribute to food insecurity.

Other alternative UA models, such as social enterprises that do attempt to address issues like barriers to employment and social exclusion, are also entangled in the entrepreneurial logic of the sustainable city. For example, Sole Food Street Farms, which sells its produce at farmers markets and restaurants, has an explicit mandate to support low-income Downtown Eastside Vancouver residents through employment opportunities. Partially supported by the City through funding and access to land, the organization is a keystone within the wider food systems localization efforts. Its sprawling farm, located in the shadow of glittering apartment towers, two sports arenas and major highway viaducts, is emblematic of the Greenest City brand and the sustainability fix it supports (see Figure 30.2). Furthermore, by emphasizing the personal transformation through employment, the project, like many similar ones (Pudup 2008; Weissman 2015), unintentionally bolsters the neoliberal agenda emphasizing individual responsibility and the formation of 'useful' citizens, rather than challenging systems of economic oppression that result in their poverty and exclusion in the first place.

Nevertheless, UA does hold a radical potential for reimagining urban living or supporting alternatives to capitalism through the creation of alternative economies. In some instances, there are small-scale efforts at implementing bartering systems or work-shares for produce boxes, that point to the potential for UA to be subversive (see Galt et al. 2014). For example, Portland's Urban Farm Collective has land and water sharing agreements among its members. Volunteers are also paid in barter bucks called 'slugs' that can then be exchanged for produce from the farm. Such an arrangement is central to their effort, as noted on their website (www.urbanfarmcollective.com), to "strive to exit completely from the cash for goods, market economy" through their barter system. This system provides an alternative economic system that allows for greater participation of people in farming without being tied to a capitalist system.

Figure 30.2 Sole Food Street Farms, a large social enterprise located in downtown Vancouver, British Columbia

Source: photo: Nathan McClintock.

Embarking on UA using an explicit framework of food justice or social justice can also help transform – or at least push back against – dominant sustainability discourse. While sustainability is clear in its concern for *inter*generational equity, i.e. balancing the needs of the present with those of future generations, its concern with *intra*generational equity is less evident (Agyeman 2013). An explicit food justice orientation, on the other hand, clearly places concerns over structural inequities and oppression at the forefront of UA practice. Food justice, with its roots in the civil rights and environmental justice struggles of people of colour in the USA (Gottlieb and Joshi 2010), in general holds greater traction in Portland, but is increasingly informing the missions of Canadian UA organizations, as well.

As UA practitioners and activists increasingly engage a food justice framework, their insights have the potential to draw equity concerns into policy-making and planning conversations. That justice and equity will take centre stage in these efforts is not a foregone conclusion, however, given the lack of participation by people other than the 'usual suspects', described by the former director of the Vancouver Urban Farming Society (VUFS), as "young, white, twenty, thirty-something, educated, middle class" urban agriculturalists. She recognizes that attention to justice is "not something that as an overarching organization or community that I would say we've been as intentional about as we could be" (Interview, 18 June 2014). Activists must therefore be conscientious and continue not only to create these spaces for discussion, but also actively incorporate social justice into their organizational missions, goals and strategies.

As part of the neoliberal sustainability fix, urban agriculture has clearly become a flashpoint, where tensions between exchange value and multiple, often divergent use values come to a head along class and racial lines, as well as within and across municipal agencies. UA is a product of negotiation among various interests and institutions. At the same time, the outcomes of its negotiation, which are heavily influenced by dominant political-economic interests and hegemonic sustainability discourses, are nonetheless open and emergent to a certain degree. Urban agriculture is nothing if not thoroughly political.

Acknowledgements

Research presented here is supported by grants from the National Science Foundation (#1539750: 'Urban agriculture, policy-making, and sustainability') and Portland State University's Institute for Sustainable Solutions.

References

Abbott, C. (2001). *Greater Portland: Urban Life and Landscape in the Pacific Northwest*. Philadelphia: University of Pennsylvania Press.
Agyeman, J. (2013). *Introducing Just Sustainabilities: Policy, Planning, and Practice*. London: Zed Books.
Allen, P. and Guthman, J. (2006). From 'Old School' to 'Farm-to-School': Neoliberalization from the Ground Up. *Agriculture and Human Values*, 23(4), pp. 401–415.
Anguelovski, I. (2016). Healthy Food Stores, Greenlining and Food Gentrification: Contesting New Forms of Privilege, Displacement and Locally Unwanted Land Uses in Racially Mixed Neighborhoods. *International Journal of Urban and Regional Research*, 39(6), pp. 1209–1230.
Barthel, S., Parker, J. and Ernstson, H. (2015). Food and Green Space in Cities: A Resilience Lens on Gardens and Urban Environmental Movements. *Urban Studies*, 52(7), pp. 1321–1338.
Bedore, M. (2013). Geographies of Capital Formation and Rescaling: A Historical-Geographical Approach to the Food Desert Problem. *The Canadian Geographer/Le Géographe canadien*, 57(2), pp. 133–153.
Berelowitz, L. (2010). *Dream City: Vancouver and the Global Imagination*. Vancouver: Douglas & McIntyre.
Blomley, N. (2003). *Unsettling the City: Urban Land and the Politics of Property*. New York: Routledge.
Born, B. and Purcell, M. (2006). Avoiding the Local Trap: Scale and Food Systems in Planning Research. *Journal of Planning Education and Research*, 26(2), pp. 195–207.

Bradley, K. and Galt, R.E. (2014). Practicing Food Justice at Dig Deep Farms & Produce, East Bay Area, California: Self-Determination as a Guiding Value and Intersections with Foodie Logics. *Local Environment*, 19(2), pp. 172–186.

Brenner, N., Peck, J. and Theodore, N. (2010). Variegated Neoliberalism: Geographies, Modalities, Pathways. *Global Networks*, 10(2), pp. 182–222.

Burnett, K. (2014). Commodifying Poverty: Gentrification and Consumption in Vancouver's Downtown Eastside. *Urban Geography*, 35(2), pp. 157–176.

Carfagna, L.B., Dubois, E.A., Fitzmaurice, C., Ouimette, M.Y., Schor, J.B. and Willis, M. (2014). An Emerging Eco-Habitus: The Reconfiguration of High Cultural Capital Practices among Ethical Consumers. *Journal of Consumer Culture*, 14(2), pp. 158–178.

Colasanti, K., Hamm, M. and Litjens, C. (2012). The City as an 'Agricultural Powerhouse'? Perspectives on Expanding Urban Agriculture from Detroit, Michigan. *Urban Geography*, 33(3), pp. 348–369.

Crouch, P. (2012). Evolution or Gentrification: Do Urban Farms Lead to Higher Rents? *Grist*. Available at: http://grist.org/food/evolution-or-gentrification-do-urban-farms-lead-to-higher-rents [accessed 1 November 2012].

Davidson, M. and Iveson, K. (2014). Recovering the Politics of the City: From the 'Post-Political City' to a 'Method of Equality' for Critical Urban Geography. *Progress in Human Geography*, 38(5), pp. 543–559.

Dooling, S. (2009). Ecological Gentrification: A Research Agenda Exploring Justice in the City. *International Journal of Urban and Regional Research*, 33(3), pp. 621–639.

Drake, L. (2014). Governmentality in Urban Food Production? Following 'Community' from Intentions to Outcomes. *Urban Geography*, 35(2), pp. 177–196.

Draper, C. and Freedman, D. (2010). Review and Analysis of the Benefits, Purposes, and Motivations Associated with Community Gardening in the United States. *Journal of Community Practice*, 18(4), pp. 458–492.

DuPuis, E.M. and Goodman, D. (2005). Should We Go 'Home' to Eat?: Toward a Reflexive Politics of Localism. *Journal of Rural Studies*, 21(3), pp. 359–371.

Eizenberg, E. (2012). Actually Existing Commons: Three Moments of Space of Community Gardens in New York City. *Antipode*, 44(3), pp. 764–782.

Galt, R.E., Gray, L.C. and Hurley, P. (2014). Subversive and Interstitial Food Spaces: Transforming Selves, Societies, and Society–Environment Relations Through Urban Agriculture and Foraging. *Local Environment*, 19(2), pp. 133–146.

Goodling, E.K., Green, J. and McClintock, N. (2015). Uneven Development of the Sustainable City: Shifting Capital in Portland, Oregon. *Urban Geography*, 36(4), pp. 504–527.

Gottlieb, R. and Joshi, A. (2010). *Food Justice*. Cambridge, MA: MIT Press.

Guthman, J. (2009). Bringing Good Food to Others: Investigating the Subjects of Alternative Food Practices. *Cultural Geographies*, 15(4), pp. 431–447.

Hackworth, J. (2007). *The Neoliberal City: Governance, Ideology, and Development in American Urbanism*. Ithaca, NY: Cornell University Press.

Harvey, D. (1989). *The Urban Experience*. Baltimore: Johns Hopkins University Press.

Hern, M. (2016). *What a City is For, or the Curved Line of the Horizon, in Three Parts*. Cambridge, MA: MIT Press.

Jarosz, L. (2008). The City in the Country: Growing Alternative Food Networks in Metropolitan Areas. *Journal of Rural Studies*, 24(3), pp. 231–244.

Lawson, L.J. (2005). *City Bountiful: A Century of Community Gardening*. Berkeley: University of California Press.

Lees, L. (2012). The Geography of Gentrification: Thinking Through Comparative Urbanism. *Progress in Human Geography*, 36(2), pp. 155–171.

Levkoe, C.Z. (2011). Towards a Transformative Food Politics. *Local Environment*, 16(7), pp. 687–705.

Lubitow, A. and Miller, T.R. (2013). Contesting Sustainability: Bikes, Race, and Politics in Portlandia. *Environmental Justice*, 6(4), pp. 121–126.

Lyson, H.C. (2014). Social Structural Location and Vocabularies of Participation: Fostering a Collective Identity in Urban Agriculture Activism. *Rural Sociology*, 79(3), pp. 310–335.

Markham, L. (2014). Gentrification and the Urban Garden. *The New Yorker*. Available at: www.newyorker.com/business/currency/gentrification-and-the-urban-garden [accessed 23 August 2014].

McCann, E. (2013). Policy Boosterism, Policy Mobilities, and the Extrospective City. *Urban Geography*, 34(1), pp. 5–29.

McClintock, N. (2011). From Industrial Garden to Food Desert: Demarcated Devalution in the Flatlands of Oakland, California. In: A.H. Alkon and J. Agyeman, eds. *Cultivating Food Justice: Race, Class, and Sustainability*. Cambridge, MA: MIT Press, pp. 89–120.

McClintock, N. (2014). Radical, Reformist, and Garden-Variety Neoliberal: Coming to Terms with Urban Agriculture's Contradictions. *Local Environment*, 19(2), pp. 147–171.

McClintock, N. and Simpson, M. (2016). Cultivating in Cascadia: Urban Agriculture Policy and Practice in Portland, Seattle, and Vancouver. In: J. Dawson and A. Morales, eds. *Cities of Farmers: Problems, Possibilities and Processes of Producing Food in Cities*. Iowa City: University of Iowa Press, pp. 59–82.

McClintock, N., Mahmoudi, D., Simpson, M. *et al.* (2016). Socio-Spatial Differentiation in the Sustainable City: A Mixed-Methods Assessment of Residential Gardens in Metropolitan Portland, Oregon, USA. *Landscape and Urban Planning*, 148(1), pp. 1–16.

Naylor, L. (2012). Hired Gardens and the Question of Transgression: Lawns, Food Gardens and the Business of 'Alternative' Food Practice. *Cultural Geographies*, 19(4), pp. 483–504.

Peck, J. and Tickell, A. (2002). Neoliberalizing Space. *Antipode*, 34(3), pp. 380–404.

Phat Beets Produce (2013). *Statement of Gentrification*. Available at: www.phatbeetsproduce.org/full-statement-on-gentrification/ [accessed 12 December 2017].

Proudfoot, J. and McCann, E.J. (2008). At Street Level: Bureaucratic Practice in the Management of Urban Neighborhood Change. *Urban Geography*, 29(4), pp. 348–370.

Pudup, M. (2008). It Takes a Garden: Cultivating Citizen-Subjects in Organized Garden Projects. *Geoforum*, 39(3), pp. 1228–1240.

Purcell, M. and Tyman, S.K. (2014). Cultivating Food as a Right to the City. *Local Environment*, 20(10), pp. 1132–1147.

Quastel, N. (2009). Political Ecologies of Gentrification. *Urban Geography*, 30(7), pp. 694–725.

Ramírez, M.M. (2015). The Elusive Inclusive: Black Food Geographies and Racialized Food Spaces. *Antipode*, 47(3), pp. 748–769.

Reynolds, K. and Cohen, N. (2016). *Beyond the Kale: Urban Agriculture and Social Justice Activism in New York City*. Athens: University of Georgia Press.

Right To Food Zine. (2016). Available at: http://rtfzine.org/about/ [accessed 28 January 2016].

Roman-Alcalá, A. (2015). Broadening the Land Question in Food Sovereignty to Northern Settings: A Case Study of Occupy the Farm. *Globalizations*, 12(4), pp. 545–558.

Rosol, M. (2010). Public Participation in Post-Fordist Urban Green Space Governance: The Case of Community Gardens in Berlin. *International Journal of Urban and Regional Research*, 34(3), pp. 548–563.

Safransky, S. (2014). Greening the Urban Frontier: Race, Property, and Resettlement in Detroit. *Geoforum*, 56(2), pp. 237–248.

Sbicca, J. (2014). The Need to Feed: Urban Metabolic Struggles of Actually Existing Radical Projects. *Critical Sociology*, 40(6), pp. 817–834.

Schmelzkopf, K. (2002). Incommensurability, Land Use, and the Right to Space: Community Gardens in New York City. *Urban Geography*, 23(4), pp. 323–343.

SFUAA. (2013). Position on Gentrification. *San Francisco Urban Agriculture Alliance (SFUAA)*. Available at: www.sfuaa.org/position-on-gentrification.html [accessed 12 December 2017].

Shannon, J. (2014). Food Deserts: Governing Obesity in the Neoliberal City. *Progress in Human Geography*, 38(2), pp. 248–266.

Slocum, R. (2007). Whiteness, Space and Alternative Food Practices. *Geoforum*, 38(3), pp. 520–533.

Smith, C.M. and Kurtz, H.E. (2003). Community Gardens and Politics of Scale in New York City. *Geographical Review*, 93(2), pp. 193–212.

Smith, N. (2002). New Globalism, New Urbanism: Gentrification as Global Urban Strategy. *Antipode*, 34(3), pp. 427–450.

Smith, N. (2008). *Uneven Development: Nature, Capital, and the Production of Space*. Athens: University of Georgia Press.

Sparke, M. (2000). Excavating the Future in Cascadia: Geoeconomics and the Imagined Geographies of a Cross-Border Region. *BC Studies: The British Columbian Quarterly*, 127(1), pp. 5–44.

Swyngedouw, E. (2009). The Antinomies of the Postpolitical City: In Search of a Democratic Politics of Environmental Production. *International Journal of Urban and Regional Research*, 33(3), pp. 601–620.

Taylor, J.R. and Lovell, S.T. (2014). Urban Home Food Gardens in the Global North: Research Traditions and Future Directions. *Agriculture and Human Values*, 31(2), pp. 285–305.

Temenos, C. and McCann, E. (2012). The Local Politics of Policy Mobility: Learning, Persuasion, and the Production of a Municipal Sustainability Fix. *Environment and Planning A*, 44(6), pp. 1389–1406.

Temenos, C. and McCann, E. (2013). Geographies of Policy Mobilities. *Geography Compass*, 7(5), pp. 344–357.

Tretter, E. (2013). Sustainability and Neoliberal Urban Development: The Environment, Crime and the Remaking of Austin's Downtown. *Urban Studies*, 50(11), pp. 2222–2237.

Walker, S. (2015). Urban Agriculture and the Sustainability Fix in Vancouver and Detroit. *Urban Geography*, 37(2), pp. 1–20.

Weissman, E. (2015). Entrepreneurial Endeavors: (Re)producing Neoliberalization Through Urban Agriculture Youth Programming in Brooklyn, New York. *Environmental Education Research*, 21(3), pp. 351–364.

While, A., Jonas, A.E.G. and Gibbs, D. (2004). The Environment and the Entrepreneurial City: Searching for the Urban 'Sustainability Fix' in Manchester and Leeds. *International Journal of Urban and Regional Research*, 28(3), pp. 549–569.

31
RETROACTIVE UTOPIA
Class and the urbanization of self-management in Poland

Kacper Pobłocki

Introduction

There is a widespread hostility towards middle-class contentious politics in urban studies literature. Many researchers hold a tacit assumption that residents' movements are conservative and guided by narrow, parochial interests. A classic example of this is provided by Mike Davis in his *City of Quartz*. Homeowners' associations in southern California were by his account busy establishing "bourgeois utopias". Those entailed the "creation of racially and economically homogeneous residential enclaves glorifying the single-family home ... and defense of this suburban dream against unwanted development (industry, apartments, offices) as well as unwanted persons" (Davis 1990: 170). Most literature on American cities perceives middle-class activism as a battle to defend white privilege and property values in an increasingly non-majority-white urban milieu. The assumption that all forms of property-anchored politics is reactionary is perhaps the reason why the very first Anglo-Saxon book on the right to the city (Mitchell 2003) deems the homeless as the true revolutionary subject of urban politics. For authors such as David Harvey (2012: 130) or Neil Smith (1996: 3–29) this is still largely the case.

The same applies to the US residents' movement manifesto – *The Life and Death of Great American Cities* by Jane Jacobs. Although it is a radical text on many accounts (Berman 2010: 314–318; Taylor 2006), and while for many economists she became, as Paul Krugman (1995: 5) put it, "a patron saint of a new growth theory", very few urban scholars take Jacobs seriously. David Harvey (2011: 171) for example argued that after the glory days of sitcom suburbs were over,

> traditionalists increasingly rallied around the urbanist Jane Jacobs, who had very distinctive ideas as to what constituted a more fulfilling form of everyday life in the city. They sought to counter sprawling suburbanization and the brutal modernism of [Robert] Moses' large-scale projects with a different kind of urban aesthetic that focused on local neighbourhood development, historical preservation and, ultimately, reclamation and gentrification of older areas.

A careful read leaves no doubt that Jacobs was actually an outspoken critic of the very first portents of gentrification. "Cities need not 'bring back' their middle-class and carefully protect it", she argued, but instead, they should "grow" it (Jacobs 1961: 369). While Richard Florida and his acolytes managed to culturewash her work, Jacobs' idea of urban diversity was indeed distinctive. It was not, however, about "superficial architectural variety" (132) but about "a diversity of incomes" (374). Her urban utopia was not aesthetic or sentimental but sought to facilitate upward social mobility.

All this fails to appreciate a point consistently made by Ira Katznelson (1976, 1981, 1992) – American patterning of class/race and space is by and large idiosyncratic. It was the American "ghost acres" that allowed Britain to escape in the nineteenth century the trap of proto-industrialization and jumpstart the Industrial Revolution (Pomeranz 2000). United States became "the first country in the world ... to apply one simple ordering principle to the whole of its national territory" (Osterhammel 2014: 105), and the way its space was measured, sundered into pieces and then commodified, underpinned, as Andro Linklater (2002, 2014) brilliantly demonstrated, its capitalism (see also Smith 1991). This is why outside Anglo-Saxon countries the word property is *not* synonymous with real estate and the relationship between ownership, class and space follows a different logic. Yet, because most of the literature on urban activism in the Global South that emerged in the last decade follows in this wake, it focuses on denouncing for the many "bourgeois utopias" or middle-class "evil paradises" (e.g. Anjaria 2009; Baud and Nainan 2008; Bhan 2009; Ghertner 2012; Taguchi 2013) and contributes to turning an American predicament into an universal law.

There are three notable exceptions to this: the work of James Holston (2008) on Brazil, Asef Bayat (1997, 2013) on Iran and of Eyal Weizman (2012) on Israel. What singles them out, and also brings closer to the heritage of Jacobs, is their immersion in urban materiality. The right to the city for them is not an abstract "cry and demand" (Harvey 2012) but instead lurks in quotidian, collective practices. They describe how change is brought by a political redefinition of the material fabric of the city. Holston showed how the very practical and mundane owning of the urban infrastructure, known in Brazil as *autoconstrução*, became a fulcrum for making broader political claims and a vehicle for a fundamental redefinition of citizenship. Property, he shows, does not have to be reactionary. Bayat describes how daily struggles, patterns of street and sidewalk use amalgamate into the "encroachment of the ordinary" that led to major political victories for the urban "non-movements". Finally, Weizman showed how spatial tactics of Israeli activists engendered an "ordered chaos" and a six-dimensional space that makes the two-state solution practically impossible. My chapter follows this material tradition and describes different lineages and dynamics of the relationship between political contention, class and space, showing yet another case of the material and daily struggles for the right to the city.

Lineages of class politics

One appreciates the uniqueness of the American case immediately after looking at housing from a global perspective. Like in the United States, it is key for perpetuating inequality in Poland. While class disparities are usually calculated on the basis of incomes, if we factor in the property, then Poland's Gini coefficient rises significantly from 38.4 per cent to 57.9 per cent (zero is a situation of perfect equality and 100 per cent is perfect inequality). Today 10 per cent of most affluent households hold 37 per cent of all property, while the 20 per cent of poorest households own only a margin (0.1 per cent) of Poland's assets. Most of this is housing. While the average (median) Polish household owns assets worth 61,700 euro (which is half the EU average), only a fraction (2,014 euro) of this was constituted by financial assets (this is five times lower than the EU average). Not only are Polish households not plugged into the global financial system, but even the housing sector is only superficially financialized. Most households own their homes

(76.4 per cent as compared to the EU average of 60.1 per cent). But only 12.1 per cent of Polish households have a mortgage (Bańbuła and Żółkiewski 2015). This means that most of the property owned in Poland is old and was acquired during the country's largest building boom I called the "Long Sixties" (Pobłocki 2012a). It is clearly visible in Figure 31.1. Thus in order to understand Poland's property distribution and patterns of inequality today, we need to delve into the very moment Poland turned majority-urban.

While there is a direct relationship between income and education and one has to have a university degree to be considered 'middle class' (Domański 2012), if we look at the pattern of property ownership, a very different dynamic emerges. There has been a persistent housing shortage since the end of the Second World War, estimated at roughly 1.5 million units. Thus by EU standards nearly half (44.8 per cent) of Polish apartments are overcrowded. Yet, there are many households with a second home – which in most of the cases is rented out. And here lies the crucial statistic as far as class is concerned: 56.9 per cent of those who own a second house have a secondary (often vocational) education, 39.3 per cent have higher education and 3.8 per cent completed only a primary school. The curious fact that most of those who make money on property are not 'well' educated points to the critical peculiarity of the Polish 'middle' class. It is a vestige of the socialist class structure that defies categories coined on the basis of American or British experience. The propertied class in Poland is by and large the old socialist middle class. It was a broad category that encompassed skilled workers, office clerks and highly educated professionals – contrasted to people performing the most unskilled of tasks for which even secondary, vocational, education was not necessary. What is more, the mushrooming of urban activism all over Poland in the last decade can be seen as its late coming-of-age.

Another distinctive trait is that Poland never had its own 'bourgeois revolution' that forged the middle classes. The only equivalent of the French, American or the British 'Glorious Revolution' is represented by the Second World War. It was during the Nazi occupation that the ancient regime in Poland was wiped out. For centuries before that, cities in Poland were inhabited and shaped by Jews and Germans, while Poles dominated the countryside. The "revolutionary war" (Gross 1997) changed this. While looting of Jewish property was a Europe-wide phenomenon (Dean et al. 2007) in occupied Poland it gained a peculiar twist. The previously marginalized Polish lower-middle classes organized gruesome "golden harvests" (Gross and Grudzińska-Gross 2012) which became a way of claiming the so-called ex-Jewish property

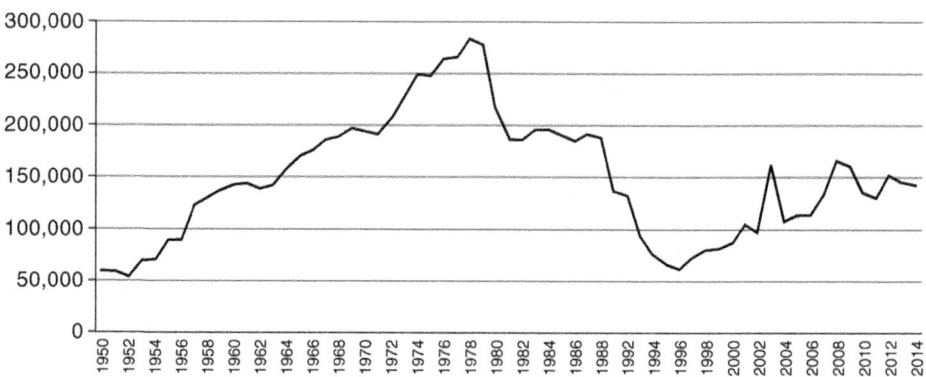

Figure 31.1 New construction in Poland, 1950–2014

(*mienie pożydowskie*). The antebellum nationalist right was well organized and led by local elites such as small shop owners, priests or school teachers. It was in the crucible of the Second World War that the Polish capitalist middle class emerged as a political subject.

As Leif Jerram (2013: 39–42) noted, while factories were in the first half of the twentieth century dominated by the left, European fascism was born from street fighting. In this sense, the fascists were the very first to claim the right to the city. This explains why killings did not stop with the war. As Jan Gross (2006) demonstrated, the Kielce Pogrom of 1946 was a struggle over both public and private property: law-abiding (lower) middle-class citizens in a provincial city went on a two-day-long killing spree in order to prevent Jews from returning to their homes as they now (or so they insisted) belonged to the Poles. Since Poland's borders were also significantly shifted to the West, the first post-war decade entailed a polonization not only of ex-Jewish but also ex-German (*poniemieckie*) property and cities. As it happened, this overlapped with 'building socialism'. It occurred not only in new, socialist towns (Kotkin 1995; Lebow 2013) but also in older cities and thus entailed an eradication of older, working-class urban cultures. This was documented brilliantly by Padraic Kenney (1997), who showed how communists established their grip over Łódź, a leading industrial hub and Poland's de facto capital between 1945 and 1950, by striking an alliance with peasant sons and daughters who just migrated to the city instead of the extant working classes.

The Łódź proletariat coalesced during the 1920s and 1930s from a unique hybrid of Polish, German and Jewish cultural material and was a child of a global upheaval. Between 1905 and 1911 peoples' revolutions broke out in countries (like Mexico, Iran, China, Russia or the Ottoman Empire) that together housed a quarter of the global population. Since they were the poor and 'backward' ones, this global moment (unlike the 1780s, 1840s or 1968 that revolutionized the West) did not stick in global memory. Yet, as Robert Blobaum (1995) showed, the 1905 revolution marked the birth of contentious politics in Poland. Only afterwards can we speak of mass party politics, urban protests or heated political contention between the right and the left. This urban/industrial environment, encompassing co-ops, self-organized cultural clubs and militant trade unions produced what Paul Mason described as "probably the most successful example of concentrated community organization in the history of the working class" (2010: 241). The very first wave of anti-communists strikes that shook Łódź between 1945 and 1947 was exactly this: an urban struggle over wages, production quotas and self-management at the workplace. This is why there were no pogroms in Łódź – politics at this moment confirmed to Jerram's dictum that the left fought for the workplace, while the right looted the streets. This was, however, soon to be over. And the socialist middle class that emerged later differed substantially from its capitalist counterpart.

The socialist middle class

While during the first post-war decade the power of the only discursively 'totalitarian' state was limited to the major industrial centres, after Poland turned majority-urban in 1965 building socialism gained a wholly new meaning. Major cities were officially closed for new migrants in 1956, and it was time to conquer the countryside and small towns. Thus while during the European 'urban miracle' that in Poland lasted, roughly, between the 1860s and 1960s, population growth occurred in the largest cities at the expense of the countryside and towns, after 1965 the focus was on second- and third-tier cities. As Stephen Collier (2011) demonstrated, this was a deliberate Soviet policy. Socialism was never municipal (as it was occasionally in the West) but universal. In 1931 one of the Bolshevik leaders "famously defined the socialist city as any settlement on the territory of the Soviet Union. The claim has often been taken to be a mindless tautology" (Collier 2011: 75) – but it meant that abolishing all spatial divides, the one between the town and the country, or between large metropolitan centres and small towns, was the true objective. The idea was that it did not matter where a socialist citizen lived – he or she was to be liberated from the constraints of place (Collier 2011: 21, 34). It was effectively Lefebvre's (2003) "urban revolution" realized.

This is why, with the exception of Moscow, there are no megacities in Eastern Europe. This is not a sign of putative "under-urbanization" (Szelényi 1996) but a consequence of socialist spatial justice. As Robert C. Allen calculated, if not for the Soviet demographic, educational and social policies, Russia's population would have exceeded a billion by now. Between 1928 and 1989, the population of the USSR rose by 70 per cent, compared to the three- to five-fold increases realized by countries (such as India or Pakistan) at a similar level of development in the 1920s. Furthermore, although the Cold War was an unequal fight, by 1970 standards of living in both East and West were already converging (Allen 2003: 116–120, 136–137). Only a handful of countries (notably Argentina, Uruguay and Chile) that were rich at the beginning of the twentieth century have fallen into the camp of the poor countries, and only a few countries that were poor in 1800 have joined the prosperous. The latter includes Japan, Taiwan, South Korea as well as Eastern Europe (Allen 2003: 6). The 'Soviet experiment' was many things but not a failure. By the 1970s erosion of the rural-urban divide was already visible. The urban scholar Marcin Czerwiński (1974: 6) had do doubts that "the social meaning [and] physical form of contemporary urbanization is on many levels radically different from everything hitherto seen in history". Urbanization, like socialism, was universal, and thus it did not matter that only 60 per cent of the Polish population lived in areas defined administratively as urban.

It was part and parcel of a larger effort. Also, income disparities were flattened. It ceased mattering if one worked as a university professor or a street sweeper – the salary was more or less the same. Because unemployment was eliminated, everybody had a basic income. Although housing was scarce, everybody had the right to an apartment. Regimes were largely popular and, as David Priestland put it, they "may not have created 'new socialist people', but they did create men and women with many socialist ideals". An independent opinion poll from 1980 showed that equality was the second most important value for Poles (after family), and that although "democracy was seen as valuable … it was less important than equality" (Priestland 2009: 511). Sociability became more important than careers. Friendship in the Soviet Union

> seems to have been taken much more seriously than in the West … 16 per cent of people met friends every day, 32 per cent once or several times a week, and 31 per cent several times a month. American single men, in contrast, met friends on average four times a month.
>
> *(Priestland 2009: 442)*

Most people living under socialism "were neither dissidents nor true believers", as Alexei Yurchak argued, and conducted a life "that was neither too activist or too oppositional, implying instead that this life was interesting, relatively free, full, creative" (2006: 118).

This does not mean that social frictions vanished, only that they moved to a different turf. Already the 1968 upheaval was no longer about self-management at the workplace but about distribution of the means for collective consumption (Pobłocki 2012a, 2012b). The 1970s saw three powerful workers' protests against a hike in … meat prices. These were both an expression of the urban 'moral economy' and part and parcel of global protests. The Global South, hard-pressed by the consequences of the oil-shocks, saw over 150 food riots only between 1976 and 1982 (Prashad 2012: 125). In Poland they gained a distinctive twist. In a society where everybody earned more or less the same, the stratifying factor was consumption and the family size. Since most men and women worked, households had similar incomes. Prices were centrally set and flat. Thus the amount of money per capita depended on how many children there were in a family. Working-class families that tended to have more children consumed more items such as cabbage, lard and dark bread, while families with two or one children had more butter, white bread and meat on the table. With over 70 per cent of incomes spent on food, no wonder this was a heated issue. The wiener schnitzel became the symbol of socialist and urban prosperity. Meat consumption in 1968 was already twice higher than before the war and stood at 68 kg (in the nineteenth century an average Pole ate 6 kg of meat annually). So having meat every day, not only on Sundays, became the way the working class translated all the lofty promises of socialism into their daily lives (Pobłocki 2010: 230).

In order to keep up with the surging demand, the government imported substantial amounts of grain (used for animal fodder) and even meat itself. When it tried, increasingly squeezed by a global recession, to increase the price of meat, it was met by a fierce resistance. Thus, as Jacek Kurczewski (1993: 10) noted, the Solidarity movement of 1980, in which more than ten million people participated, was not "a rebellion of people in despair, but a revolution of those whose hopes remained unfulfilled". This is why the strikes erupted in Gdańsk – a city with a highly skilled young labour force, that felt disenfranchised by the ruling elite and who thought, in the spirit of self-management, that they were competent enough to deliver the promise of socialism. "The cultural and advancement of millions of people", argued Kurczewski,

> led to the situation in which a new middle class was formed, blocked in its aspirations on the one hand by the close borders of the ruling class, and on the other by the misgovernment of the country and its economy.
>
> *(1993: 188)*

Solidarity was thus a 'self-managing trade union' that significantly broadened the idea of self-management also to the realm of consumption. Food rationing, introduced in 1980, was one of Solidarity's demands, and so were free Saturdays or lowering of the retirement age. Although what followed was the very first bankruptcy of a socialist economy that contributed significantly to shifting the scales in the Cold War in favour of the United States (Priestland 2009: 522), the Solidarity movement can be seen as the very pinnacle of this socialist middle class – just as the 1946 pogrom was perhaps the final expression of the capitalist middle class in Poland.

The birth of urban movements

The "Polish crisis" (Simatupang 1994) was a protracted one, and (as Figure 31.1 shows) both the 1980s and the 1990s were lost decades. They culminated in the 1998–2003 crisis, that for Poland became "the first crisis *of* capitalism itself rather than of the transition to capitalism" (Ost 2005: 166, original emphasis). The project of de-industrialization, initiated already in 1980, came to its definitive phase. The Polish domestic economy moved from one constituted by large companies to one dominated by sweatshops – that in 2013 employed over 56 per cent of the workforce. Because unemployment surged to 20 per cent in 2003, many multinationals saw Poland as a reservoir of cheap labour. Thus in the wake of its EU accession in 2004, Poland quickly became the "Mexico of Europe" (Turner 2008: 63). With scanty competition from domestic companies, many sectors became monopolized by multinationals. In some cases they simply took over the existing infrastructure – like the American agribusiness that bought out large parts of the former state-owned farms and turned them into profitable enterprises (Dunn 2003). In many others, new industries came in.

'White goods' manufacturing, that moved from North to South Europe in the 1960s, now shifted from Spain and Italy eastwards. Poland's main cities like Wrocław or Łódź re-industrialized in the 2000s and became major producers of fridges, washing machines and the like. Others, like Kraków or Warsaw, became places where call-centres or accounting services were outsourced to. Germany eclipsed Russia as Poland's main trading partner, and a (modest) building boom started in 2004. Roughly 90 per cent of money that went in the last decade into the building new office spaces, shopping malls and warehouses were provided by German pension funds, who thus recycled capital made from servicing the Chinese industrial revolution and producing luxury goods for the Chinese middle class. Residential construction was financed mainly by domestic money and the vast majority of it occurred in the 'villages' on the fringes of metropolitan areas. The already modest percentage of urban population fell below 60 per cent in the 2000s and thus Poland became further 'under-urbanized'.

This substantially changed forms of contention. While 1992 saw a hike in labour protests, with over 700,000 people on strike against the 'shock therapy' of de-industrialization, by 2000 the number of striking workers fell nearly to nil. The opening of the British labour market that accompanied Poland's EU accession, absorbed a further 1.1 million workers and potential labour discontents too. Trade union membership fell from 40 per cent to 14 per cent, and many of the new jobs were created on precarious terms, which today constitute over one-third of all job contracts. The so-called precariat organized itself only as late as 2014, so for the first decade of Poland's EU membership most political contention was urban. Already between 2000 and 2004 the number of urban protests (demonstrations or blockades) increased threefold. It then withered away only to return with vengeance in 2010 (Urbański 2014: 188–191). This was exactly the moment when urban movements entered the stage – as a grassroots response to the policies of the mid-2000s, of course with a structurally unavoidable time lag.

There are many factors that explain its emergence into which I cannot delve. One of the them was a literal flood of second-hand cars after 2004. While in the 1990s an automobile was still largely a luxury item, and in 2003 only 35,000 used cars crossed the German-Polish border, already a year later this surged to 823,000 units annually and has remained at such astounding levels ever since. With infrastructure and regulations wedded to the erstwhile, moderate rates of traffic, this was bound to generate tensions. Most of the new suburbanites who live in the 'rural' counties retain very strong ties with the cities – they commute not only to work, but also drive their children to city schools or go there to movies (Kajdanek 2012: 86). New suburbanization and the flood of used cars that was embraced by city populations clogged inner cities; the traffic jams extended further and further out. The worst off were the older, socialist suburbs that quickly turned into transit zones. Little wonder, then, that residents of exactly those areas were the first to realize universal car ownership was a bad idea. While the poorer and older inner-city residents were more inclined to embrace automobiles and would be often in agreement with the new suburbanites that the most burning issue was a lack of parking spaces, groups like *My-Poznaniacy*, founded in 2007/2008, would quickly start challenging the rationality of the modernization programme engendered by the early few years of Poland's EU membership.

My-Poznaniacy was an association of smaller residents' movements from many corners of Poznań. It became the bellwether of Polish urban movements after it earned nearly 10 per cent of votes in the municipal elections of 2010. If one looks at spatial distribution of their support (see Figure 31.2), *My-Poznaniacy*'s candidate for major (who, although on a different ticket, won in 2014) registered most votes in areas between the inner city and the new suburbs (which are located outside the official city limits). These are exactly the suburbs built for, or rather by,

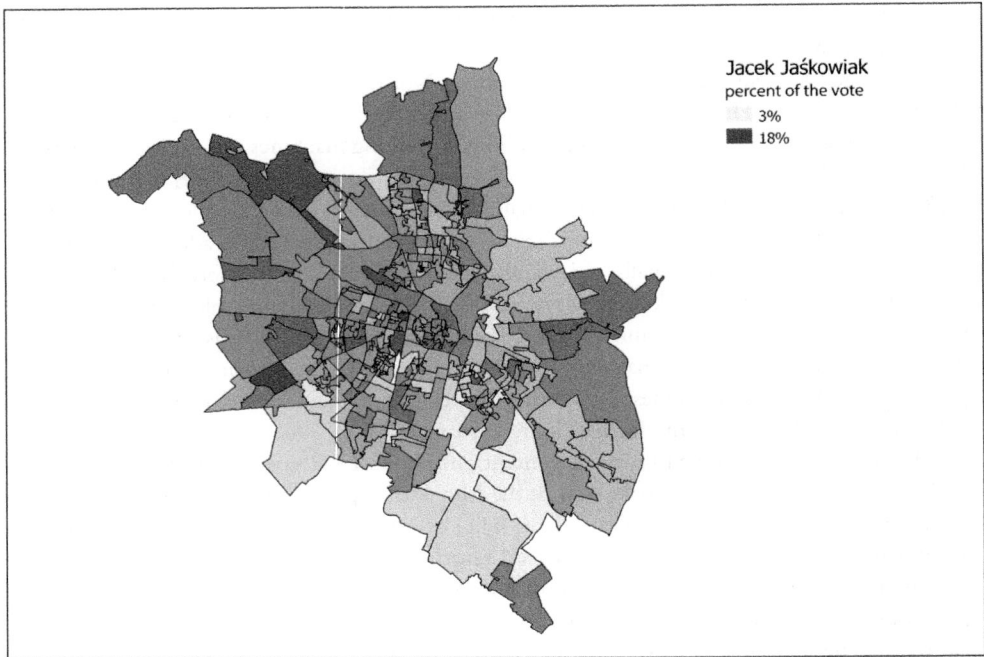

Figure 31.2 Spatial distribution of My-Poznaniacy support in 2010

the socialist middle class in the 1970s and 1980s. They were largely "unfinished utopias", to borrow Ketherine Lebow's (2013) apt phase – but in a more quotidian sense. The socialist state was notorious for building new apartments without the amenities. While for the high-rise estates these 'shortages' were about schools or retail stores, for low-rise areas the problems were even more dramatic. Although many of those areas had comprehensive plans, like in Poznań's Strzeszyn (see Figure 31.3), most of the amenities, including electricity, sewerage or telephone, had to be self-built by the residents themselves. Thus especially during the crisis-ridden 1980s, many grassroots committees were formed – exactly like in Brazilian cities described by James Holston (2008). Many of them were active during the political festival of 1988–1991, and during the 1990s were formally registered as community boards (*rady osiedla*).

At the same time, the idea of self-management (*samorządność*) – central for nearly a century of Polish politics, was urbanized. While during communism local administration either did not exist or, when it was established in the 1970s, was stymied by the state and party administrations, these institutions were allocated increasing power after 1989. Throughout most of the 1990s jobs at local governments, now called *samorząd*, were considered paltry and given to the least demanding echelons of party apparatuses. But once most state assets had been privatized and the post-2004 building boom took off, suddenly urban space became of substantial value. The new building boom was conspicuously chaotic. An entirely new legal structure ushered in in 2003 favoured scattered, fragmented and disjointed developments. Since developers cut corners, many preferred to construct new apartments in between those that already existed, so they could rely on the existing hook-ups. When residents of those socialist apartments and houses realized that the park in front of their windows was being built over, they started getting interested in how all this came about. It turned out the new legal framework meticulously excluded them from the decision-making process. But a local neighbourhood association precisely in one of

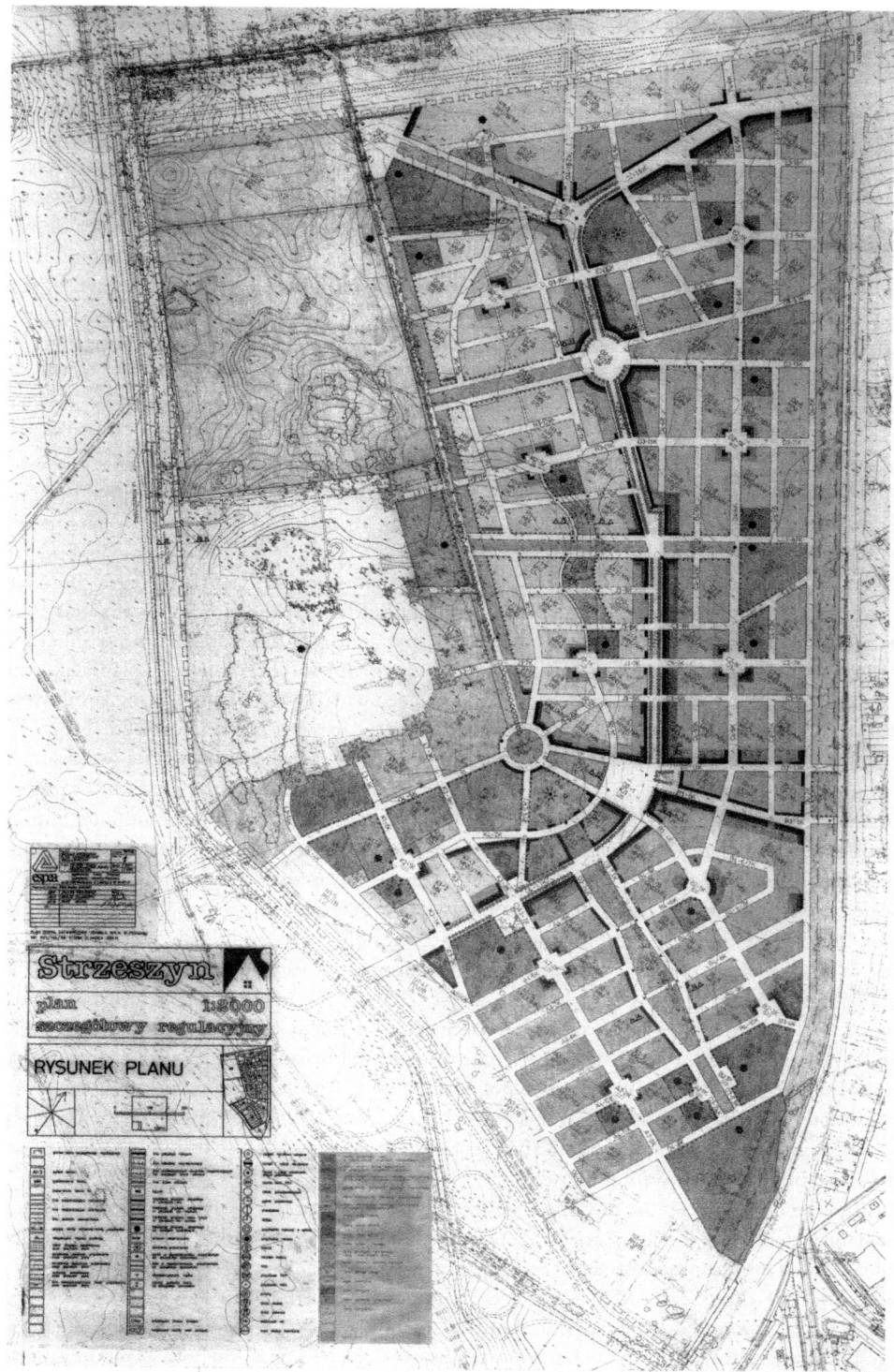

Figure 31.3 The layout for Strzeszyn

these areas found a loophole that allowed them to block the development (Mergler et al. 2013: 92). This discovery, made most probably for the very first time in 2007, gave residents substantial leverage and became the fulcrum of their political clout. This is also when the activists realized that community boards, dormant for nearly two decades, became a useful vehicle for voicing residents' demands.

In the process they discovered that although the country was a liberal democracy, the planning process was entirely undemocratic and intransparent. This autocratic practice stood in stark contrast to the democratic theory and rhetoric, and residents sought to act upon the idea of self-management in the place of residence. It is not surprising that Poznań – perhaps the only large city that was sidestepped in the re-industrialization of the 2000s – became the hotbed of this contention, as the city was mainly a recipient of the problems and not the benefits associated with new investments. The key concept that soon became the main idea pursued by all Polish urban activists, pundits and scholars alike, was that of *ład przestrzenny*. As the lawyer Hubert Izdebski argued,[1] it is broader than the Anglo-Saxon "spatial harmony" but narrower than "spatial governance" (and the German *Raumordung*). It should be not translated as a "spatial order" – one of the utopias of Western middle classes that protect their little "evil paradises" against any sorts of contamination. The Polish word for order – *porządek* – which has been an integral part of the right-wing vocabulary, was never used by urban activists. *Ład przestrzenny* belonged to a different political lineage. As one journalist put it, it is "something everybody in Poland has heard about but nobody has seen for a very long while" (Springer 2013: 9). What everybody saw with their naked eye was its very opposite – spatial chaos, i.e. a practice of ad hoc, uncoordinated and opportunistic investments that destroyed the spatial harmony inherited from the socialist building boom and its modernist principles. The modernist plans, like the layout for Strzeszyn (see Figure 31.3), that had only been partially realized under socialism, were actually the material proof what *ład przestrzenny* once was, or what could have been.

Ład przestrzenny was a what I call an "retroactive utopia", because during state socialism no such principle was ever codified in law. Nor was it in any way used by the communist propaganda – unlike for example *czyn społeczny*, i.e. voluntary work done for the community, which, alongside the derogatory term *samowola budowlana*, would be the Polish equivalents of the Brazilian *autoconstrução*. Spatial reality during state socialism was in fact substantially "chaotic", because it was also an outcome of overlapping and sometimes divergent policies (Jałowiecki 2010: 263). Now, because the principle of *ład przestrzenny* has been literally etched in space, even in its absence, it could be mobilized for generating a forward-looking, inclusive political agenda. To be sure, it was conservative – but it harked back to a very progressive political project. Without the knowledge of the spatial and social heritage it wanted to 'bring back', one could easily mistake it for a NIMBY-type of movement. But because it was socialist in a material and not discursive way, it could become politically potent without running into the danger of being accused of being 'nostalgic' for state socialism. While *czyn społeczny* was consistently mocked as an element of a failed political project, *ład przestrzenny* was free from such stigmas and thus could became a powerful political tool.

Although it came from the low-rise neighbourhoods that had the most extensive experience in self-management at the residential level, it was embraced by vast swathes of the (post)socialist middle class. Thus the most important campaign of 2010, one that landed one of *My-Poznaniacy*'s leaders the title of Poznań's Man of the Year a month before the elections, was a struggle to build a park – promised by socialist authorities already in the 1960s – in Rataje, one of the largest housing estates in Poland. This promise was not delivered upon by both communists and neoliberals, and residents thought it was in their capacity to compete the 'unfinished utopia' of socialist planning. While the communists established a pre-fab housing factory in lieu of the park, arguing it was only a 'temporary' solution and once the estate was competed the space would revert to its 'original' function, in 1994 it was sold for a song as a brownfield. Three years later a powerful local developer acquired it but for the next ten years a legal deadlock prevented him from building new apartments in there. Only in 2006 when he finally managed to get a building permit, the residents – who followed the litigation but did not participate in it – entered the conflict, and once *My-Poznaniacy* was formed it became their legal vehicle.

The principle of *ład przestrzenny*, that applied to high-rise Rataje in the very same way as to Strzeszyn and other low-rise areas, was the engine of this supra-local alliance. As Lebow (2013) insisted, residents' activism during the socialist era was inscribed in the planning philosophy, so this was actually not very surprising. But a movement that spanned both low-rise and high-rise areas with such ease would be unthinkable in the United States. For Harvey (2012: 138) the right to the city "has to be construed not as a right to that which already exists, but as a right to rebuild and re-create the city as a socialist body politic in a completely different image". The Polish case, just as the Brazilian one, shows that it does not have to be an abstract 'cry or demand' but a material practice, and that the extant fabric of the city, and especially the housing, can become a powerful vehicle for progressive politics. Sometimes, a (modest) urban revolution can be achieved not by erecting progressive spaces on clean slates but by enacting on a political potential locked in tangible spatial relations even if it is – like in this case – dormant for decades.

Note

1 He made this point at a conference on spatial governance held at the Polish Presidents' Office in the autumn of 2013.

References

Allen, R.C. (2003). *Farm to Factory: A Reinterpretation of the Soviet Industrial Revolution*. Princeton, NJ: Princeton University Press.
Anjaria, J.S. (2009). Guardians of the Bourgeois City: Citizenship, Public Space, and Middle-Class Activism in Mumbai. *City & Community*, 8(4), pp. 391–406.
Bańbuła, P. and Żółkiewski, Z. (2015). *Zasobność gospodarstw domowych w Polsce. Raport z badania pilotażowego 2014 r.* Warszawa: Narodowy Bank Polski.
Baud, I. and Nainan, N. (2008). 'Negotiated Spaces' for Representation in Mumbai: Ward Committees, Advanced Locality Management and the Politics of Middle-Class Activism. *Environment and Urbanization*, 20(2), pp. 483–499.
Bayat, A. (1997). *Street Politics: Poor People's Movements in Iran*. New York: Columbia University Press.
Bayat, A. (2013). *Life as Politics: How Ordinary People Change the Middle East*. Stanford, CA: Stanford University Press.
Berman, M. (2010). *All That Is Solid Melts into Air: The Experience of Modernity*. London: Verso.
Bhan, G. (2009). "This Is No Longer the City I Once Knew": Evictions, the Urban Poor and the Right to the City in Millennial Delhi. *Environment and Urbanization*, 21(1), pp. 127–142.
Blobaum, R. (1995). *Rewolucja: Russian Poland, 1904–1907*. Ithaca, NY: Cornell University Press.
Collier, S.J. (2011). *Post-Soviet Social: Neoliberalism, Social Modernity, Biopolitics*. Princeton, NJ: Princeton University Press.
Czerwiński, M. (1974). *Życie po miejsku*. Warszawa: PIW.
Davis, M. (1990). *City of Quartz: Excavating the Future in Los Angeles*. New York: Vintage Books.
Dean, M., Goschler, C. and Ther, P., eds. (2007). *Robbery and Restitution: The Conflict over Jewish Property in Europe*. New York: Berghahn Books.
Domański, H. (2012). *Polska klasa średnia*. Toruń: Wydawnictwo Naukowe UMK.
Dunn, E.C. (2003). Trojan Pig: Paradoxes of Food Safety Regulation. *Environment and Planning A*, 35(8), pp. 1493–1511.

Ghertner, D.A. (2012). Nuisance Talk and the Propriety of Property: Middle Class Discourses of a Slum-Free Delhi. *Antipode*, 44(4), pp. 1161–1187.

Gross, J.T. (1997). War as Revolution. In: N.M. Naimark and L.I. Gibianskii, eds. *The Establishment of Communist Regimes in Eastern Europe, 1944–1949*. Boulder, CO: Westview Press, pp. 17–40.

Gross, J.T. (2006). *Fear: Anti-Semitism in Poland After Auschwitz: An Essay in Historical Interpretation*. New York: Random House.

Gross, J.T. and Grudzińska-Gross, I. (2012). *Golden Harvest: Events at the Periphery of the Holocaust*. Oxford: Oxford University Press.

Harvey, D. (2011). *The Enigma of Capital*. London: Profile Books.

Harvey, D. (2012). *Rebel Cities: From the Right to the City to the Urban Revolution*. London: Verso Books.

Holston, J. (2008). *Insurgent Citizenship: Disjunctions of Democracy and Modernity in Brazil*. Princeton, NJ: Princeton University Press.

Jacobs, J. (1961). *The Death and Life of Great American Cities*. New York: Vintage Books.

Jałowiecki, B. (2010). *Społeczne wytwarzanie przestrzeni*. Warszawa: Scholar.

Jerram, L. (2013). *Streetlife: The Untold History of Europe's Twentieth Century*. Oxford: Oxford University Press.

Kajdanek, K. (2012). *Suburbanizacja po polsku*. Kraków: Nomos.

Katznelson, I. (1976). *Black Men, White Cities*. Chicago: University of Chicago Press.

Katznelson, I. (1981). *City Trenches: Urban Politics and the Patterning of Class in the United States*. New York: Pantheon Books.

Katznelson, I. (1992). *Marxism and the City*. Oxford: Clarendon Press.

Kenney, P. (1997). *Rebuilding Poland: Workers and Communists, 1945–1950*. Ithaca, NY: Cornell University Press.

Kotkin, S. (1995). *Magnetic Mountain: Stalinism as a Civilization*. Berkeley: University of California Press.

Krugman, P. (1995). *Development, Geography, and Economic Theory*. Cambridge, MA: MIT Press.

Kurczewski, J. (1993). *The Resurrection of Rights in Poland*. Oxford: Clarendon Press.

Lebow, K. (2013). *Unfinished Utopia: Nowa Huta, Stalinism, and Polish Society, 1949–56*. Ithaca, NY: Cornell University Press.

Lefebvre, H. (2003). *The Urban Revolution*. Minneapolis: University of Minnesota Press.

Linklater, A. (2002). *Measuring America: How an Untamed Wilderness Shaped the United States and Fulfilled the Promise of Democracy*. New York: Walker & Co.

Linklater, A. (2014). *Owning the Earth: The Transforming History of Land Ownership*. London: Bloomsbury.

Mason, P. (2010). *Live Working or Die Fighting: How the Working Class Went Global*. Chicago: Haymarket Books.

Mergler, L., Pobłocki, K. and Wudarski, M. (2013). *Anty-Bezradnik przestrzenny: prawo do miasta w działaniu*. Warszawa: Res Publica Nowa.

Mitchell, D. (2003). *The Right to the City: Social Justice and the Fight for Public Space*. New York: Guilford Press.

Ost, D. (2005). *The Defeat of Solidarity: Anger and Politics in Post-Communist Europe*. Ithaca, NY: Cornell University Press.

Osterhammel, J. (2014). *The Transformation of the World: A Global History of the Nineteenth Century*. Princeton, NJ: Princeton University Press.

Pobłocki, K. (2010). *The Cunning of Class: Urbanization of Inequality in Post-War Poland*. Budapest: Central European University, unpublished doctoral dissertation.

Pobłocki, K. (2012a). Knife in the Water: The Struggle over Collective Consumption in Urbanizing Poland. In: P. Bren and M. Neuburger, eds. *Communism Unwrapped: Consumption in Cold War Eastern Europe*. Oxford: Oxford University Press, pp. 68–90.

Pobłocki, K (2012b). Class, Space and the Geography of Poland's Champagne (Post)Socialism. In: M. Grubbauer and J. Kusiak, eds. *Chasing Warsaw: Socio-Material Dynamics of Urban Change since 1990*. Frankfurt: Campus, pp. 269–289.

Pomeranz, K. (2000). *The Great Divergence: China, Europe, and the Making of the Modern World Economy*. Princeton, NJ: Princeton University Press.

Prashad, V. (2012). *The Poorer Nations: A Possible History of the Global South*. London: Verso.

Priestland, D. (2009). *The Red Flag: Communism and the Making of the Modern World*. London: Penguin.

Simatupang, B. (1994). *The Polish Economic Crisis: Background, Causes, and Aftermath*. London: Routledge.

Smith, N. (1991). *Uneven Development: Nature, Capital, and the Production of Space*. Oxford: Blackwell.

Smith, N. (1996). *The New Urban Frontier: Gentrification and the Revanchist City*. London: Routledge.

Springer, F. (2013). *Wanna z kolumnadą: reportaże o polskiej przestrzeni*. Wołowiec: Czarne.
Szelényi, I. (1996). Cities Under Socialism – and After. In: G.D. Andrusz, M. Harloe and M.I. Szelényi, eds. *Cities After Socialism: Urban and Regional Change and Conflict in Post-Socialist Societies*. Oxford: Blackwell, pp. 286–317.
Taguchi, Y. (2013). Civic Sense and Cleanliness: Pedagogy and Aesthetics in Middle-Class Mumbai Activism. *Contemporary South Asia*, 21(2), pp. 89–101.
Taylor, P.J. (2006). Jane Jacobs (1916–2006): An Appreciation. *Environment and Planning A*, 38(11), pp. 1981–1992.
Turner, G. (2008). *The Credit Crunch: Housing Bubbles, Globalisation and the Worldwide Economic Crisis*. London: Pluto Press.
Urbański, J. (2014). *Prekariat i nowa walka klas: przeobrażenia współczesnej klasy pracowniczej i jej form walki*. Warszawa: Książka i Prasa.
Weizman, E. (2012). *Hollow Land: Israel's Architecture of Occupation*. London: Verso.
Yurchak, A. (2006). *Everything Was Forever, Until it Was No More: The Last Soviet Generation*. Princeton, NJ: Princeton University Press.

PART VII

Spaces of circulation

Andrew E.G. Jonas and Kevin Ward

This section examines an important yet often overlooked theme in the analysis of urban politics, namely, how cities and the spaces within them are interconnected through circulatory systems and flows. In recent years, concepts of mobility and flow have injected new life into critical thinking about the city and, indeed, space more generally. Urban scholars, such as Manuel Castells (2000) and Doreen Massey (1991), amongst others, have argued that seemingly 'urban' social and political processes are always constituted through socio-spatial relationships, flows and networks which can extend far beyond the boundaries of the city. Such relational thinking about the city aims to challenge the viewpoint that the underlying conditions shaping urban social formations and politics are to be found exclusively inside the city itself (Amin and Thrift 2002).

Relational approaches to the urban open up exciting new opportunities to examine urban politics from the vantage point of wider circulations and flows. Such circulations and flows involve not just the different circuits of capital operating through the built environment as described, for example, by David Harvey (1985); they can also refer to flows of material goods and resources, nature, transport, policies, ideas and political discourses. The analysis of such spaces of circulation, in turn, requires a different way of thinking about the underlying social configuration of different spaces of urban politics; namely, it puts causal priority on exposing and explaining the spatial movements, flows and networks that stretch far beyond urban political territory and often at a global scale. At the same time, however, it is also important not to ignore how such ostensibly 'global' circulations operate within and around particular urban places.

These six chapters explore a series of issues related to the spatial politics and counter-politics of circulation. They include analyses and discussions of different sorts of circulatory systems such as those associated with climate change, urban experiments, the e-economy, water management, transportation and social justice movements. Each chapter highlights the importance of seeing cities not as spatial containers of politics but rather as spaces of flows, negotiations, experimentations and contestations, which cut across and actively disrupt received notions of the 'urban' as having fixed jurisdictional limits. Collectively, chapters throw into critical perspective the limitations of seeing urban politics as always contained, negotiated and defined within formal political boundaries and fixed territories.

The chapter by Kevin Fox Gotham and Clare Cannon develops the concept of 'circulating risks' in order to illuminate the ways in which the growing extension, intensification and speeding up of social and economic activities contributes to spatialization of urban politics and climate change risks. It begins by describing the spatial and temporal components of climate change, and delineates the ways in which cross-border flows of capital and finance and increasing global connectivity expose cities to risk and vulnerability. It then highlights the centrality of relations of domination and subordination in the production of climate change risk, exposing their linkages to gender and social inequality. The authors proceed to describe the global circulation of climate change risk reduction measures – adaptation and mitigation – and discuss their impacts on cities and on the construction of spaces of urban policies. In doing so, they stress the importance of viewing cities as constituted by flows, circulations and cross-scale interactions.

Markus Hesse argues in his chapter that the digital economy is a catalyst for the circulation of people, goods and information, both material and virtual, affecting both urban and non-urban realms. He uses the empirical case of *Amazon.com* to discuss the digital arrangement of production, labour, distribution, advertising, purchasing or consumption as a means of creating particular spaces of politics. What once started as mail-order retail for books about twenty years ago can now be understood as an all-encompassing platform for performing a variety of socio-economic practices. Thanks to the peculiarities of disruptive innovation, and backed by venture capital and an aggressive market strategy, the 'system' has taken an almost totalitarian turn.

Hesse considers the implications of the e-economy for urban place, space and territory. These are manifold: first, the compartmentalized organization of value chains, physical distribution and labour relations accelerates the fragmentation of space-time-relations; second, the associated 'logics of dislocation' (Barnes 1996) help to create spaces of both agglomeration and peripherality, where contemporary notions of urban and rural are becoming further blurred; third, the abstract system's imperative not only steers the orchestration of the firm's networks, but also performs as a powerful agent of policy-making. Against this background, the chapter also explores the spaces of urban politics that are produced and reproduced by *Amazon.com*.

The next chapter is co-authored by James Evans, Harriet Bulkeley, Yuliya Voytenko, Kes McCormick and Steven Curtis. It examines the production and circulation of urban experiments through urban living laboratories (ULLs), a mode of governing that is rapidly growing in popularity as cities attempt to learn how to become more sustainable under conditions of economic, social and environmental uncertainty. ULLs can be buildings, neighbourhoods or districts within cities that are designated to host a multitude of social and technical experiments to inform urban policy, planning and, in particular, politics. ULLs require us to rethink how we approach urban politics as they reconfigure traditional processes of governing urban development in terms of approach, participation and scale. The promise of ULLs lies in their potential to transform top-down planning and development into local innovation processes that open up spaces for citizen-centric and place-based urban politics. In this way they are being championed as a new approach to solve long-running urban issues and transform under-developed urban areas into hubs of the new economy. Current evidence suggests that while participation is rather narrow they play a key role in providing legitimacy for alternatives, which in turn is driving their circulation as approach to urban governance.

In her chapter, Julie Cidell examines the Chicago Area Waterways System as a complex assemblage of water, infrastructure, fish, sewage and ships. A system of canals connects the Chicago River to the Great Lakes and Mississippi River watersheds, making it possible for water and its contents to traverse the subcontinental divide. The properties and capacities of fish, water, sewage and other elements of this assemblage have produced unexpected effects as elements interact with and act upon each other, illustrating the concept of distributed agency and making the attempted reassembly of this network fraught with difficulty.

A proposed 'ecological separation' plan to reduce the threat of the invasive species of Asian carp in the Chicago region would separate the non-human ecologies of the two watersheds while keeping their human ecologies connected, thus keeping the region safe for capital in addition to keeping the region's ecology safe from outside invasion. Considering the carp, Lake Michigan and the Chicago and Des Plaines Rivers as components in an assemblage offers a way to understand how the various properties and capacities of water have shaped and continue to shape the city of Chicago and the region around it and why this particular environmental problem is so intractable. Using this example, Cidell's chapter considers how the spaces of urban politics regarding water incorporate not only its consumption, but also its role as a medium for other kinds of circulations – wastes, freight, capital and biophysical life – and therefore the politics of deciding what should flow and what should be blocked.

With the rapid global diffusion of Uber and similar transportation network companies (TNCs), as well as global fascination with self-driving cars, tech mobility is likely going to be making greater claims on urban spaces around the world. Jason Henderson suggests the utility of examining tech mobility with greater critical scrutiny and especially its implications for the urban politics of mobility, i.e. political debates over what urban transportation modes are prioritized, and who decides which modes are important.

After describing some of the key characteristics of tech mobility, Henderson draws on a politics of mobility framework to consider how tech mobility is an irreducibly political project promoting progressive visions of urban liveability such as reduced car ownership, whilst also aligning decidedly with a neoliberal politics of privatization and deregulation. Tech mobility may have far-reaching consequences in shaping urban geographies of the future – not by transforming cities, but, rather, by intensifying and solidifying contemporary inequities and uneven geographies.

Struggles for social justice and recognition of difference often begin and coalesce in cities. The final chapter in this section – written by Helga Leitner and Sam Nowak – examines the making of multi-racial (counter-)publics engaged in urban social justice struggles, and their attempts to create egalitarian political spaces to enact a potentially radical democratic politics. Informed by the political philosophies of Nancy Fraser and Jacques Rancière, the authors examine the worker centre movement – a swell of grassroots organizations that have emerged during the past twenty-five years to provide services to, advocate on behalf of and empower immigrants and people of colour. These organizations have also waged successful multi-racial social justice campaigns, which this chapter examines in the context of Los Angeles.

Leitner and Nowak highlight four dimensions of the making of multi-racial publics through the Los Angeles worker centre movement as follows: (1) the enactment of egalitarian forms of governance and experimentation with new ways of organizing; (2) a sustained commitment to learning from each other and negotiating across racial and other lines of difference; (3) publicizing existing injustices and making demands on the state to rectify these; and (4) multiple spatialities in the making of these multi-racial counter-publics. The chapter reveals how an especially critical role was played by spaces of withdrawal (safe physical space of the worker centres) along with spaces of circulation (mobilities of people and ideas and networks of connectivity amongst worker centers). These different spaces shape – even as they are shaped by – the construction of multi-racial publics and egalitarian demands within the wider arena of urban politics.

References

Amin, A. and Thrift, N. (2002). *Cities: Reimagining the Urban*. Cambridge: Polity.
Barnes, T.J. (1996). *Logics of Dislocation: Models, Metaphors, and Meanings of Economic Space*. London: Guilford Press.
Castells, M. (2000). *The Network Society*. 2nd ed., Cambridge, MA: Blackwell.
Harvey, D. (1985). *The Urbanization of Capital*. Baltimore: Johns Hopkins University Press.
Massey, D. (1991). A Global Sense of Place. *Marxism Today*, 35(6), pp. 24–29.

32
CIRCULATING RISKS
Coastal cities and the spectre of climate change risk

Kevin Fox Gotham and Clare Cannon

Introduction

Climate change is an urgent problem for coastal regions already facing the challenges of urbanization, overcrowding, subsidence and nutrient loading. Global mean temperatures are on the rise, ice sheets and glaciers are rapidly melting, and sea levels are rising (Nicholls et al. 2007; for an overview, see Moser et al. 2012). Scientists expect sea-level rise to exacerbate existing risks such as coastal flooding, coastal land loss, rising water tables and drainage problems, and salinization of coastal environments (Blum and Roberts 2009; Gonzalez and Tornqvist 2006; IPCC 2014; Karl 2009; National Academy of Sciences 2003: 4; Stern 2007; USGCRP 2013). Apart from climate change, predicted land loss combined with increasing population growth in coastal areas could endanger regional economies, threaten sources of fresh water, and alter land cover patterns and ecosystem services (e.g. provisioning services such as food and water; regulating services such as flood and disease control; and cultural services such as spiritual, recreational and cultural benefits) (Millennium Ecosystem Assessment 2005). "The scope, severity, and pace of future climate change impacts are difficult to predict", according to the White House Council on Environmental Quality (2010: 6), but "coastal areas will need to prepare for rising sea levels and increased flooding." Adding to the challenge of responding to these impacts, climate-related changes will likely exacerbate the impacts of other non-climatic risks and stressors such as pervasive inequality, jurisdictional fragmentation, fiscal strains, ageing infrastructure, habitat destruction and pollution (IPCC 2014).

We have two arguments in this chapter. First, we argue that cross-scale circulations – of energy, people, policies and financing – are major sources and drivers of climate change risk and vulnerability in coastal cities. Second, we suggest that the increasing extension, intensification and velocity of social activities surrounding climate change risks express the proliferation of spaces in and through which the politics of the 'urban' are constructed and contested. Globally, 1.2 billion people (23 per cent of the world's population) live within 100 km of the coast, and 50 per cent are likely to do so by 2030 (UNESCO 2009). Around the world, coastal cities like Dhaka (Bangladesh), Jakarta (Indonesia), Manila (Philippines), Kolkata (India), Phnom Penh (Cambodia), Ho Chi Minh city (Vietnam), Shanghai (China), Bangkok (Thailand), Hong Kong (China), Kuala Lumpur (Malaysia) and Singapore, among others, are at risk not just because of rising sea levels but urban population growth (Hallegatte and Corfee-Morlot 2011; Nicholls and Cazenave 2010; O'Brien and Leichenko 2000).

It is tempting to argue that coastal cities face different levels of risk and vulnerability due to their geographical location, exposure to storms and susceptibility to sea-level rise. Yet risk, exposure and vulnerability are not the same thing. Risk is a general category that scholars define as the combination of people's *exposure* to a hazard and their social *vulnerability* (i.e. their capacity to anticipate, respond to and recover from damage) (for an overview, see Tierney 2014). The concept of vulnerability has become particularly important within natural hazard and disaster research (Cutter 2003; Pelling 2012) and researchers have used it as an analytical lens to explain how social context including policies, corporate investment patterns and socio-legal regulations shape risk. Researchers often describe vulnerability as having three components: exposure to a hazard, susceptibility to harm and adaptive capacity. This differentiated conceptualization usefully illustrates the relationship between human action and risk (i.e. certain policy and institutional actions can potentially alter exposure and/or vulnerability to a hazard) and reveals the socially and political constructed nature of risk definitions, risk assessments and risk analyses.

By cross-scale 'circulations' we mean influences, connections and interactions among institutions, government agencies and networks to facilitate the movement of information, people and resources across borders. We explain cross-scale circulations in terms of the extension, intensification and acceleration of government actions, interchanges between public and private actors, socio-economic activities, and flows and networks of investment and finance. 'Flows' refer to the movement of commodities, money, people and information across space and time, while 'networks' refer to patterned interactions among agents, organizations and activities. In this sense, cross-scale circulations involve increased interregional interconnectedness, a widening reach of networks of social activity, and the possibility that local events and actions (by individuals, corporations and governments) can have far-reaching and long-lasting consequences for urban politics.

As we argue, cities are not just backdrops or platforms against which cross-scale circulations unfold, but are central to the ways in which people and organizations interact to construct and produce climate change risks. Just as circulations are sources and drivers of climate change risks, climate change risk governance policies – e.g. mitigation and adaptation strategies – are both in motion and simultaneously fixed, or embedded in place. Changes in the extension, intensification and speed of circulations – e.g. in/out-migration of people, financial (dis)investment in cities, etc. – combined with changes in how people use and manage resources in cities are central to understanding the impacts and consequences of climate change risk. Our approach to examining climate change risks focuses attention on the intersection of gender and inequality and the ways in which they appear and reappear in matters of urban politics. In arguing that gender matters for understanding climate change risks, we make the case that the concept of circulations has implications for the gendering of urban policies.

Spatiality, temporality and climate change risk

Scholars have used the terms mobilities, interactions, exchange and re/de-territorialization to describe and understand the movement of commodities, money, capital and people across socio-spatial boundaries and jurisdictions (Brenner 1999; Sassen 2001). Moreover, using concepts such as time-space distanciation, time-space compression, glocalization, and other temporal and spatial metaphors, researchers have re-conceptualized notions of capital circulation and commodity exchange to provide insight into the changing geographies of trade, finance, retail, distribution and consumption (Giddens 2000; Harvey 1989; Wellman 2002). While much of the literature on forms of exchange and circulation has focused on the growing intensity, extension and increased velocity of flows of capital under conditions of globalization, there has been a turn more recently to consider forms of circulation that cut across cities and actively disrupt notions of 'the urban' as having fixed jurisdictional boundaries and delimited spatial borders. Against views of cities as containers of social action and relationships, recent critical interrogations by Scott and Storper (2015) draw attention to the importance of studying the "circulation space of the city" that is represented by infrastructures and arterial connections that facilitate intra-urban flows of people, goods and information. "There are many urban processes for which neither formal administrative boundaries nor the functional regions of cities would be the relevant scale for comparison", according to Robinson (2011: 14). Instead, processes that exceed a city's physical extent – circulations, mobilities and flows – offer rich potential for empirical analysis and theorization to challenge extant scholarship and provide novel and original insights into the nature of urban politics.

Spatially, the concept of circulation attends to the specific flows and mobilities at work in drawing people, ideas, activities into proximity and close relationships. If we apply this thinking to climate change risk, then we can analyse climate change risks as products of the sharply increased connectivity between places and accelerated and condensed modes of human interaction in today's globalized world. According to the Intergovernmental Panel on Climate Change (IPCC 2014), climate change risks to coastal cities include sea-level rise which could lead to loss of land and communities; increased frequency and destructiveness of storms which could undermine infrastructure systems, property and livelihoods through flooding; warming of glaciers which could reduce fresh water supplies and availability; changes in winter and summer energy demand which could lead to brownouts; and changes in the incidence of vector-borne diseases (for an overview, see Hunt and Watkiss 2011). Importantly, climate change risks are not objective forces that exist in an a priori fashion. Rather, climate change risks "coalesce with other stresses" including water scarcity, institutional and jurisdictional fragmentation, fiscal strains and ageing infrastructure, limited revenue streams for public-sector risk reduction action, and inflexible patterns of land-use (UNISDR 2004). Obviously, the severity of these different types of stress vary across cultural, political and socio-economic contexts (UN-Habitat 2003).

Viewing climate change in terms of circulation suggests that climate change impacts have spatial and temporal components that reflect and reinforce global-local relations and cross-scale interactions. Climate change is a de-spatialized, global phenomenon and, simultaneously, a spatialized, urban problem. On the one hand, climate change impacts appear to be remote, distant and extra-local phenomena in which global climate change processes – e.g. greenhouse gas (GHG) emissions, atmospheric conditions, global disruption, extreme events and international negotiations – unfold and develop albeit in an unpredictable and geographically uneven fashion. On the other hand, climate change impacts vary geographically with local consequences that reflect different local histories and geographies and have different implications for local economies and communities. Understanding the urban nature of climate change requires that we examine not only how climate change is framed and contested within cities, according to Bulkeley (2013: 4), but also how the politics of reducing the risks of climate change has come to be so closely connected to the imagination of urban futures.

Temporally, climate change has both short- and long-term traumatic components. Changes in the incidence and intensity of extreme weather events – storms and floods – take place in the context of long-term and chronic traumatic processes such as sea-level rise, ocean acidification and rising temperatures (Bulkeley 2013; Millennium Ecosystem Assessment 2005). At the same time, the short- and long-term traumatic components of global climate change are producing a complex geography in which the differential impacts of climate change are being superimposed on dissimilar vulnerabilities. To add further complexity to the picture, climate change is occurring in a rapidly changing world marked by ongoing processes of economic globalization, large-scale human migration, costal urbanization, subsidence and nutrient loading (O'Brien and Leichenko 2000). Government policies, socio-legal regulations and other human behaviours and actions that drive climate change have enduring global consequences with local impacts that are often unevenly distributed across environments and societies (Bagstad et al. 2007). In short, viewing climate change risk in terms of circulations suggests that climate impacts do not happen *to* cities per se (in a top-down fashion) but are fundamentally shaped and transformed *through* urban processes.

Gender and inequality in the city

The concept of circulating risks draws our attention to the impact of unequal power relations in shaping patterns of social mobility and access to social, political and economic resources. Researchers have documented that the impacts of global climate change will not be equally distributed around the world, and "many of the countries least responsible for the rise in greenhouse gases will be most likely to feel its impacts in changes in weather, sea levels, human health costs, and economic hardships" (Nagel et al. 2009: 17). One important predicted outcome of climate change is an accelerated circulation of people out of poorer regions and countries into more developed, less impacted areas, an environmental migration that has the potential to strain the resources and social fabrics of receiving societies and exacerbate tensions in sending communities. Variations in individual, community and national vulnerability to the impacts of climate change *across* cities are only part of the structure of inequality in global climate change. As the 2014 IPCC report notes, there is an unequal distribution of impacts and vulnerabilities to climate change *within* cities especially those associated with intense social class, race and ethnic, gender and age stratifications. Moreover, disadvantaged individuals and their communities – due to the socio-economic status, geography, racial and ethnic health disparities and lack of access to care – are likely to face greater susceptibility to such storms and floods associated with climate change (Nath and Behera 2011).

Women are one such disadvantaged group the IPCC (2014) notes as being particularly susceptible to increased risks associated with climate change. Gender is a mode of social stratification that affects the distribution of resources, wealth, power, as well as the extension of rights and entitlements (Dankelman 2010; Denton 2002; United Nations 1999). Women are by no means a homogenous group of persons; they experience risk differently in part due to the intersection of their social location (e.g. along axes of race, class, nationality, age, ability, etc.) (see Arora-Jonsson 2011; Khosla and Masaud 2010). Generally, theories of gender inequality elaborate on connections between women's political-economic power and environmental exploitation, gendered-based differences in the interpretation and perception of environment problems and concerns, and gender discrimination as a driver of socio-environmental vulnerability (e.g. Agarwal 1995; d'Eaubonne 1974; Gaard 2011; Mies 1999; Mies and Shiva 1993; Rocheleau et al. 1996; Ruether 1993; Warren 1990). The relative poverty of women in cities creates greater barriers in the face of environmental degradation, since women tend to experience poorer nutrition, limited health care and, in the case of divorced and widowed women, fewer sources of social support (see Mies and Shiva 1993).

Variants of feminist political ecology focus on the uneven distribution of access to and control over resources within the social, political and economic context. This approach, with a particular emphasis on family and community health, argues that gender is a critical variable in people's ability to sustain livelihoods and communities' ability to develop sustainably (Rocheleau et al. 1996). An overarching goal, then, is to explain local experiences of inequality within global environmental and economic change. More generally, feminist political ecology seeks to analyse the context in which gender interacts with class, race, culture and national identity to distribute hazard risks within the larger social-ecological environment (Rocheleau et al. 1996: 5). With increases in frequency and magnitude of climate-related catastrophes (such as hurricanes, typhoons, earthquakes, desertification, etc.), women experience exacerbated hardships and threats to their livelihoods and those of their families (Dankelman 2010; Davidson and Freudenburg 1996; Denton 2002; McCright 2010; Nelson et al. 2002; Terry 2009). Furthermore, women tend to face disproportionately higher obstacles to participate in decision-making, to articulate and communicate collective needs, and to access external resources to cope with climate change risks.

Scholars have identified three major mechanisms that increase women's already vulnerable social position in the wake of climate change, and which have implications for the urban as a space of politics. First, women are the largest group of farmers in the world and yet have very few, if any, rights to the land they cultivate. They also tend to have little control over the resources necessary for survival (Rocheleau et al. 1996). Along with increased disasters that threaten the lives and livelihoods of women, climate change brings increased desertification of land thus making rural agriculture livelihoods unsustainable. This reality, in part, drives rural to urban migration (see Dankelman 2010). Increased urbanization in turn drives climate change with its use of non-renewable energy, inadequate basic services and infrastructure, and urban sprawl (Khosla and Masaud 2010). Generating a feedback loop, greater climate change increases risks women face and their socio-environmental vulnerability. Scholars then must theorize and empirically examine the gendered nature of environmental rights and responsibilities. Here women have greater responsibility over the health of families and communities but more often than not do not have access to the rights over property and land use and resource decisions that would provide them an equal opportunity to address their responsibilities (Rocheleau et al. 1996).

Second, women's knowledge of environmental management and conservation typically are not included or valued in policy discussions or by policy-makers as well as those in positions of power to enact widespread policies (Dankelman 2010; Denton 2002; Nelson et al. 2002). As stewards of the land in their agricultural capacity, women have garnered intimate knowledge of ecosystems and their unique and diverse functions, yet this valuable knowledge is not included in policy discussions because the knowledge is deemed as non-quantitative, unscientific and unreliable (Mies and Shiva 1993). Oppositions to incorporating traditional ecological knowledge into policy discussions are embedded in the social and political realities of cities and regions, and are designed to reproduce a particular social order and rely on definitions of the 'environment' that are tailored to the ruling interests. The right to social protection, the definition of social justice and, in a broader sense, the definition of citizenship, are defined accordingly (Dankelman 2010; Rocheleau et al. 1996).

Third, women themselves are often not included in policy discussions at key levels of global society (e.g. United Nations, national governments, local magistrates, etc.). The exclusion of women in these conversations continues not only the power asymmetry that produces and supports patriarchal society, which creates social inequality, but also misses key perspectives women may contribute to discussions for the mitigation and adaptation of climate change's most perilous effects (Dankelman 2010; Terry 2009). Feminist scholars approach these issues by analysing gendered environmental politics and grassroots activism in which women's roles in struggles over natural resources and the environment redraw lines of identity and the nature of environmental problems (Rocheleau et al. 1996: 4).

As greater numbers of people move to urban areas, particularly to coastal cities, more and more have to confront risks caused by sea-level rise, greenhouse gas emissions and climate catastrophes, all consequences of global climate change (Khlosa and Masaud 2010). This circulation of ever-greater risks exacerbates women's already disadvantaged position due to power asymmetries and resulting social inequalities. Women, in this context, are not solely victims but often enact their own mitigation and adaptive strategies at the local and regional levels through their agricultural practices and urban organizing strategies (Arora-Jonsson 2011; Terry 2009). They are agents of change and, although not always, can have an ameliorating effect on global climate change (McKinney 2014). Moreover, although women as a class of persons experience social disadvantage due to patriarchal cultural and structural norms, they can exercise their own agency in both their local communities and when able, nationally (see McKinney 2014; Mies and Shiva 1993; Rocheleau et al. 1996; Terry 2009). In highlighting the centrality of women's social vulnerability to climate change risks, it is important to note there is nothing inherently vulnerable or virtuous, for that matter, in women's social location (Aron-Jonsson 2011). As a large percentage of those affected by climate change, and as a class of people who are often responsible for children and elders, it is important not only to research the ways gender inequality play out but also necessary to include women in decision-making and governance roles as coastal cities, particularly, and all regions, generally, debate and implement mitigation and adaptation tactics and strategies.

Climate change mitigation and adaptation

The concept of circulating risks draws attention to two paradoxical features of the relationship among urban politics and climate change risks. On the one hand, cross-scale circulations can enhance urban adaptive capacity through processes of coordination, inter- and intra-government collaboration, social learning, knowledge sharing and integration, trust building and conflict resolution (Folke et al. 2005). On the other hand, cross-scale circulations can limit capacity for social-ecological renewal, and reinforce patterns and processes of vulnerability and unsustainable development. Different cross-scale circulations can perpetuate social-ecological inequality, generate and exacerbate group struggles and antagonisms, and impede conflict resolution. The challenge is to identify and explain how and under what conditions increases in the extension, intensification and acceleration of cross-scale circulations can enhance risks for some cities rather than others, and lead to different individual, community and national capacity to respond to climate change causes and consequences.

Over the last decade, governments and non-government organizations have relied upon, and developed, new multi-level institutional arrangements to encourage cross-scale circulations among government agencies, private firms and non-profit organizations in order to maximize the effectiveness of climate change countermeasures. Scientists, researchers and policy-makers have developed several climate change risk governance strategies – mitigation and adaptation – to manage changes in urban water resources, agriculture, infrastructure and settlement issues, human health, tourism, transport, and energy usage and production. Climate change *adaptation* involves developing ways to protect people and places by reducing their vulnerability to climate impacts. For example, to protect against sea-level rise and increased flooding, communities might build seawalls or relocate buildings to higher ground. For coastal systems, adaptation can mean identifying, preparing and funding the construction of coastal wetlands restoration projects in order to reduce risks associated with hurricane storm surge, sea-level rise and land subsidence (Day et al. 2007; Turner and Boyer 1997). Coastal restoration techniques include marsh creation and restoration, shoreline protection, hydrologic restoration, use of dredged material, terracing, sediment trapping, vegetative planting, barrier island restoration, bank stabilization and wetland mitigation banking, among other techniques (Allison and Meselhe 2010). Climate change *mitigation* involves attempts to slow the process of global climate change, usually by lowering the level of greenhouse gases in the atmosphere. Mitigation can also mean using new green technologies and renewable energies, making older equipment more energy efficient, or changing management practices or consumer behaviour.

Policy models of climate change risk governance include provision of low-carbon infrastructures, services and goods; tax incentives, deductions and exemptions to encourage corporations to curb GHG emissions and adopt green technology; adoption of new land-use planning, codes, standards and zoning; and education campaigns directed at diverse audiences (Adger et al. 2003; Field 2012; Hallegatte 2009; Pachauri et al. 2014; Smit and Wandel 2006; Smith 2013). Many cities have developed sophisticated policy approaches based on measuring and monitoring GHGs, setting targets and developing action plans. The extent and specific type of climate change adaptation and mitigation have been constrained and enabled by cross-scale interactions, institutional capacity between cities and levels of the state, policy coordination, information sharing, availability of financing and the legacy of past public-private investments in the built environment. Importantly, human response to climate change risks cannot be confined to jurisdictionally defined city spaces or left to the actions of the state. Rather, climate adaptation and mitigation is governed and realized through the interrelations between global, national and local actors across state/non-state boundaries.

Effective climate change governance implies a historically unprecedented level of global cooperation among different levels of government and also among governments and non-governmental organizations. International organizations such as World Business Council for Sustainable Development, World Resources Institute, IPCC, Carbon Disclosure Project, United Nations Environment Program and the International Council for Local Environmental Initiatives have worked to create workable rules, timetables, exchanges, credits and subsidies through extensive policy experimentation and debate. Through their cross-scale communication and information sharing efforts, these organizations have expanded our understanding of the impact of land use and land changes on GHG emissions, effects of patterns of property and resource control on climate change risks, and identified the drivers and barriers for international climate change policy (for an overview, see Dunlap and Brulle 2015). Today, these organizations are engaged in extensive cross-border collaborations using communication and information technologies that make spatial and temporal barriers obsolete and political borders irrelevant. As a result, climate change risk reduction measures are global in terms of the reach of the networks of knowledge production associated with their formulation; ubiquitous in terms of their relevance and implementation; regionally differentiated in terms of the political cultures and modes of policy-making that shape them; and often deeply localized in terms of their politics and statutory content (IPCC 2014).

Circulations of environmental culture and cultural meanings systems are also important in shaping urban political responses to mitigate or adapt to global climate change. Social, cultural and political efforts by international organizations to encourage consumers to 'go green' represent concerted attempts to circulate information about the benefits, obstacles and feasibility of creating more sustainable consumption habits and lifestyle choices in a consumer society. International climate change organizations operate as transfer agents and carriers of policy information. The process of information transfer is both territorializing and de-territorializing as international organizations circulate progressive global models that are often adopted by local actors. Meyer et al. (1997) describe the emergence of the "world environmental regime" composed of environmental international non-governmental organizations (EINGOs) and other civil society groups that promote the universal adoption of environmental policies, programmes and standards. EINGOs diffuse norms, environmental cultures and policy models related to consumer-advocacy, eco-friendly consumption and green sustainability. Broadly, the rise and global spread of EINGOs offer opportunities to re-conceptualize cities as produced in relation to processes circulating across geographical scales while recognizing that cities provide the basing points for those wider processes.

Conclusion

Scientists and researchers predict that climate change will have a range of impacts in cities, from sea-level rise to increased destructiveness of extreme events (storms, heatwaves, droughts and flooding). However, predicting the specific impact that will occur in particular cities is fraught with scientific complexity, uncertainty and political conflict. The landscape of risk in a city is therefore not merely a matter of which assets and populations happen to be located in vulnerable sites. Rather, climate change risks are produced through the interaction between environmental and social processes so that the risks that poorer areas face have been historically and systematically produced through different circulations of people, capital and finance. In short, it is the effect of climate change combined with a range of other political and socio-economic processes operating across different infrastructural systems that produce urban climate change risks. While studies of future climate change impacts have focused on predicted levels of population growth and economic development, it is important to recognize that urban risk and vulnerability to climate change is affected by temporal and spatial dimensions of circulation through the underlying historical, political and economic dynamics of cities.

Whilst identifying risks is important, our examination suggests that we need to rethink the urban as a space of politics around climate change. In the realm of climate change politics, recent years have witnessed a movement from *municipal voluntarism* – a focus on the voluntary activities of government authorities as a means of enhancing capacity to address climate change – to *strategic urbanism* in which cities integrate climate change mitigation and adaptation measures with conventional socio-economic development strategies. Municipal authorities in New York, Tokyo and London, among other places, now partner with businesses, universities and civil society actors to engage in 'experimental' interventions often targeted at creating low-carbon urban infrastructures. Policy-makers around the world now are now considering a variety of new financing mechanisms and institutional structures – carbon taxes, carbon markets, carbon trading schemes, alternative energy development, and technologies (e.g. smart grids) – to reduce the risks of GHG emissions. While these new financing mechanisms and structures draw attention to the global context in which climate change operates, they are also designed to highlight the social construction of other scales, such as the 'urban' and the 'regional', as a means of drawing investment into urban political space and thereby reducing the negative risk consequences of climate change impacts. As this chapter demonstrates, global environmental concerns associated with climate change help to redefine the urban as a distinctive space for new political interventions and struggles surrounding risk and vulnerability to climate change impacts.

We believe that the concept of circulating risks can be a useful heuristic device to help scholars theorize and analyse the impacts and consequences of increased global connectivity on the production of climate change risks. Different forms and types of circulations – population growth and movement of people to coastal cities, globalization of financing, movement of capital and commodities across borders – are closely related to the production of climate change risk. Our heuristic device also draws attention to how climate change mitigation and adaptation strategies are characterized by a range of power relations associated with their adoption and execution, from top-down imposition by bureaucratic organizations to enthusiastic public acceptance or deep animosity and grassroots resistance. Moreover, our concept highlights the centrality of relations of domination and subordination in cross-border circulations in which power and inequality structure how and where resources are deployed to address and combat climate change risks. The recent and ongoing upsurge in economic, political and ecological globalization allow for more powerful and wealthier cities and nations to external portions of their climate change risk to less-powerful cities and poorer nations. In addition, the environmental and public health consequences of rising sea-levels, urban heat islands, and flood-related and heat-related diseases (e.g. malaria) disproportionately affect the poor cities and nations and vulnerable populations in all nations. Through our analysis of climate change risks, we have stressed the importance of viewing cities not as a passive medium and or receptacle of risk and vulnerability but as spaces of flows in which climate change risks are circulating and cutting across cities, a conceptualization that disrupts notions of the urban as having fixed jurisdictional boundaries.

References

Adger, W.N., Huq, S., Brown, K., Conway, D. and Hulme, M. (2003). Adaptation to Climate Change in the Developing World. *Progress in Development Studies*, 3(3), pp. 179–195.
Agarwal, B. (1995). *A Field of One's Own: Gender and Property in South Asia*. Cambridge: Cambridge University Press.
Allison, M.A. and Meselhe, E.A. (2010). The Use of Large Water and Sediment Diversions in the Lower Mississippi River (Louisiana) for Coastal Restoration. *Journal of Hydrology*, 387(3), pp. 346–360.
Arora-Jonsson, S. (2011). Virtue and Vulnerability: Discourses on Women, Gender and Climate Change. *Global Environmental Change*, 21, pp. 744–751.
Bagstad, K.J., Stapleton, K. and D'Agostino, J.R. (2007). Taxes, Subsidies, and Insurance as Drivers of United States Coastal Development. *Ecological Economics*, 63(2), pp. 285–298.
Blum, M.D. and Roberts, H.H. (2009). Drowning of the Mississippi Delta Due to Insufficient Sediment Supply and Global Sea-level Rise. *Nature Geoscience*, 2(7), pp. 488–491.
Brenner, N. (1999). Globalisation as Reterritorialisation: The Re-scaling of Urban Governance in the European Union. *Urban Studies*, 36(3), pp. 431–451.
Bulkeley, H. (2013). *Climate Change and Cities*. New York: Routledge.
Cutter, S.L. (2003). The Vulnerability of Science and the Science of Vulnerability. *Annals of the Association of American Geographers*, 93(1), pp. 1–12.
d'Eaubonne, F. (1974). *La Feminisme ou la Mort*. Paris: Pierre Horay.
Dankelman, I. (2010). *Gender and Climate Change: An Introduction*. London: Earthscan.
Davidson, D. and Freudenburg, W. (1996). Gender and Environmental Risk Concerns: A Review and Analysis of Available Research. *Environment and Behavior*, 28(3), pp. 302–339.
Day, J.W., Boesch, D.F., Clairain, E.J., Kemp, G.P., Laska, S.B., Mitsch, W.J., Orth, K., Mashriqui, H., Reed, D.J., Shabman, L. and Simenstad, C.A. (2007). Restoration of the Mississippi Delta: Lessons from Hurricanes Katrina and Rita. *Science*, 315(5819), pp. 1679–1684.
Denton, F. (2002). Climate Change Vulnerability, Impacts, and Adaptation: Why Does Gender Matter? *Gender and Development*, 10(2), pp. 10–20.
Dunlap, R.E. and Brulle, R.J., eds. (2015). *Climate Change and Society: Sociological Perspectives*. Oxford: Oxford University Press.
Field, C.B., ed. (2012). *Managing the Risks of Extreme Events and Disasters to Advance Climate Change Adaptation: Special Report of the Intergovernmental Panel on Climate Change*. Cambridge: Cambridge University Press.
Folke, C., Hahn, T., Olsson, P. and Norberg, J. (2005). Adaptive Governance of Social-Ecological Systems. *Annual Review of Environmental Resources*, 30, pp. 441–473.
Gaard, G. (2011). Ecofeminism Revisited: Rejecting Essentialism and Re-Placing Species in a Material Feminist Environmentalism. *Feminist Formations*, 23(2), pp. 26–53.
Giddens, A. (2000). *Runaway World: How Globalization Is Reshaping our Lives*. New York: Routledge.
González, J.L. and Tornqvist, T.E. (2006). Coastal Louisiana in Crisis: Subsidence or Sea Level Rise? *EOS, Transactions American Geophysical Union*, 87(45), pp. 493–498.

Hallegatte, S. (2009). Strategies to Adapt to an Uncertain Climate Change. *Global Environmental Change*, 19(2), pp. 240–247.

Hallegatte, S. and Corfee-Morlot, J. (2011). Understanding Climate Change Impacts, Vulnerability and Adaptation at City Scale: An Introduction. *Climatic Change*, 104(1), pp. 1–12.

Harvey, D. (1989). *The Condition of Postmodernity*. Oxford: Blackwell.

Hunt, A. and Watkiss, P. (2011). Climate Change Impacts and Adaptation in Cities: A Review of the Literature. *Climatic Change*, 104(1), pp. 13–49.

IPCC (Intergovernmental Panel on Climate Change). (2014). *Climate Change 2014: Synthesis Report*. Geneva: IPCC. Available at: www.ipcc.ch/pdf/assessment-report/ar4/syr/ar4_syr.pdf [accessed 30 May 2017].

Karl, T.R. (2009). *Global Climate Change Impacts in the United States*. Cambridge: Cambridge University Press.

Khosla, P. and Masaud, A. (2010). Cities, Climate Change and Gender: A Brief Overview. In: I. Dankelman, ed. *Gender and Climate Change: An Introduction*. London: Earthscan, pp. 78–96.

McCright, A. (2010). The Effects of Gender on Climate Change Knowledge and Concern in the American Public. *Population and Environment*, 32, pp. 66–87.

McKinney, L. (2014). Gender, Democracy, Development, and Overshoot: A Cross-National Analysis. *Population and Environment*, 36(2), pp. 193–218.

Meyer, J.W., Frank, D.J., Hironaka, A., Schofer, E. and Tuma, N.B. (1997). The Structuring of a World Environmental Regime, 1870–1990. *International Organization*, 51(4), pp. 623–651.

Mies, M. (1999). *World Accumulation and Patriarchy on a World Scale*. New York: Zed.

Mies, M. and Shiva, V. (1993). *Ecofeminism*. Nova Scotia, CA: Fernwood Publishing.

Millennium Ecosystem Assessment. (2005). *Ecosystems and Human Well-Being: Current State and Trends*. Washington, DC: Island Press. Available at: www.millenniumassessment.org/documents/document.356.aspx.pdf [accessed 30 May 2017].

Moser, S.C., Jeffress Williams, S. and Boesch, D.F. (2012). Wicked Challenges at Land's End: Managing Coastal Vulnerability Under Climate Change. *Annual Review of Environment and Resources*, 37, pp. 51–78.

Nagel, J., Dietz, T. and Broadbent, J., eds. (2009). *Workshop on Sociological Perspectives on Global Climate Change*. Arlington, VA: National Science Foundation.

Nath, P. and Behera, B. (2011). A Critical Review of Impact of and Adaptation to Climate Change in Developed and Developing Economies. *Environment, Development, and Sustainability*, 13, pp. 141–162.

National Academy of Sciences. (2003). *Grand Challenges in Environmental Sciences*. Washington, DC: National Academy Press.

Nelson, V., Meadows, K., Cannon, T., Morton, J. and Martin, A. (2002). Uncertain Predictions, Invisible Impacts, and the Need to Mainstream Gender in Climate Change Adaptations. *Gender and Development*, 10(2), pp. 51–59.

Nicholls, R.J. (2004). Coastal Flooding and Wetland Loss in the 21st Century: Changes Under the SRES Climate and Socio-Economic Scenarios. *Global Environmental Change*, 14, p. 69.

Nicholls, R.J. and Cazenave, A. (2010). Sea-Level Rise and Its Impact on Coastal Zones. *Science*, 18 June, 328(5985), pp. 1517–1520.

Nicholls, R.J., Wong, P.P., Burkett, V., Codignotto, J., Hay, J., McLean, R., Ragoonaden, S., Woodroffe, C.D., Abuodha, P.A.O., Arblaster, J. and Brown, B. (2007). Coastal Systems and Low-lying Areas. In: M.L. Parry, O.F. Canziani, J.P. Palutikof, P.J. van der Linden and C.E. Hanson, eds. *Climate Change 2007: Impacts, Adaptation and Vulnerability*. Cambridge: Cambridge University Press, pp. 315–356.

O'Brien, K. and Leichenko, R. (2000). Double Exposure: Assessing the Impacts of Climate Change Within the Context of Economic Globalization. *Global Environmental Change*, 10, pp. 221–232.

Pachauri, R.K., Allen, M.R., Barros, V.R., Broome, J., Cramer, W., Christ, R., Church, J.A., Clarke, L., Dahe, Q., Dasgupta, P. and Dubash, N.K. (2014). *Climate Change 2014: Synthesis Report*. Contribution of Working Groups I, II and III to the Fifth Assessment Report of the Intergovernmental Panel on Climate Change. Geneva: IPCC.

Pelling, M. (2012). *The Vulnerability of Cities: Natural Disasters and Social Resilience*. London: Earthscan.

Robinson, J. (2011). Cities in a World of Cities: The Comparative Gesture. *International Journal of Urban and Regional Research*, 35(1), pp. 1–23.

Rocheleau, D., Thomas-Slayter, B. and Wangari, E. (1996). *Feminist Political Ecology: Global Issues and Local Experiences*. New York and London: Routledge.

Ruether, R.R. (1993). Ecofeminism: Symbolic and Social Connections of Oppression of Women and the Domination of Nature. In: C. Adams, ed. *Ecofeminism and the Sacred*. New York: Continuum, pp. 13–23.

Sassen, S. (2001). *The Global City: New York, London, Tokyo.* 2nd ed., Princeton, NJ: Princeton University Press.

Scott, A.J. and Storper, M. (2015). The Nature of Cities: The Scope and Limits of Urban Theory. *International Journal of Urban and Regional Research*, 39(1), pp. 1–15.

Smit, B. and Wandel, J. (2006). Adaptation, Adaptive Capacity and Vulnerability. *Global Environmental Change*, 16(3), pp. 282–292.

Smith, K. (2013). *Environmental Hazards: Assessing Risk and Reducing Disaster.* London: Routledge.

Stern, N. (2007). *The Economics of Climate Change. The Stern Review.* Cambridge: Cambridge University Press.

Terry, G. (2009). *Climate Change and Gender Justice.* Oxford: Oxfam.

Tierney, K. (2014). *The Social Roots of Risk: Producing Disasters, Promoting Resilience.* Stanford, CA: Stanford University Press.

Turner, R.E. and Boyer, M.E. (1997). Mississippi River Diversions, Coastal Wetland Restoration/Creation and an Economy of Scale. *Ecological Engineering*, 8(2), pp. 117–128.

UNESCO (United Nations Educational, Scientific and Cultural Organization). (2009). *Water in a Changing World.* World Water Assessment Programme, the United Nations World Water Development Report 3. New York: United Nations. Available at: www.unesco.org/new/fileadmin/MULTIMEDIA/HQ/SC/pdf/WWDR3_Facts_and Figures.pdf [accessed 28 February 2013].

UN-Habitat. (2003). *Global Report on Human Settlements 2003: The Challenge of Slums.* New York: United Nations.

UNISDR (United Nations International Strategy for Disaster Reduction). (2004). *Terminology: Basic Terms of Disaster Risk Reduction.* Geneva: UNISDR.

United Nations. (1999). *World Survey on the Role of Women in Development.* New York: United Nations.

USGCRP (US Global Change Research Program). (2013). *National Climate Assessment.* Third Report, draft. Washington, DC: USGCRP. Available at: www.globalchange.gov/publications/reports [accessed 12 December 2017].

Warren, K.J. (1990). The Power and the Promise of Ecological Feminism. *Environmental Ethics*, 12(2), pp. 125–146.

Wellman, B. (2002). Little Boxes, Glocalization, and Networked Individualism. In: T. Ashida and M. Tanabe, eds. *Digital Cities II: Computational and Sociological Approaches.* Berlin and Heidelberg: Springer, pp. 10–25.

White House Council on Environmental Quality. (2010). *Progress Report of the Interagency Climate Change Adaptation Task Force: Recommended Actions in Support of a National Climate Change Adaptation Strategy.* United States Government. Washington, DC: US Government Printing Office.

33
THE LOGICS AND POLITICS OF CIRCULATION

Exploring the urban and non-urban spaces of *Amazon.com*

Markus Hesse

> We seek to be Earth's most customer-centric company.
>
> *Amazon.com (2015)*

> Mankind has taken once [*sic*] small step towards space tourism. On Monday, the *New Shepard* spacecraft performed a perfect vertical landing making it the first rocket to launch into space and land back on earth safely. The momentous occasion was a feat for Amazon boss Jeff Bezos who founded *Blue Origin*, the space company behind the vehicle.
>
> *Rajan (2015)*

Introduction

This chapter explores the contemporary digital world, discussing recent changes and their implications for the politics of urban spaces and contexts. It examines the case of the enterprise *Amazon.com*, thus exploring the urban and non-urban spaces that are being used, transformed or influenced by the digital and non-digital dimensions evolving from the firm's habits and practices. The two quotations above reveal not only the company's ambition in planetary regards, but may also be taken as a clue that its understanding of what *space* is about goes far beyond traditional territorial meanings – as *Amazon.com* not only plans to deliver parcels by air drones, but has already started to explore outer space for the purpose of introducing space-trips for its customers.

However, this chapter sticks to Planet Earth and aims to investigate the more ordinary but nevertheless peculiar notions of digital circulation in a spatial context. The geographies that were created or enabled to develop by technological innovation are fuzzy, manifold and also contradictory, as the growing body of literature confirms a broad variety of events, developments and associated consequences (see Zook 2008). While new information and communications technologies were initially discussed concerning the *impact* they were generating (Hesse 2002), their role is now becoming something rather *systemic* (Kitchin 2014). A related transformation has changed *Amazon.com*, which is explored as a case study here. What once started little more than twenty years ago as mail-order retail for books now appears as an unprecedented and all-encompassing platform for performing a broad range of socio-economic practices. It now looks as if *Amazon.com* is becoming comprehensive, if not totalitarian, in terms of the information, products and services supplied — given the power it has achieved in order to control the related data, supply chains and value-adding networks.

This chapter argues that the orchestration of the firm's networks produces an abstract system's imperative that leads to fragmentation and dislocation by means of digital and functional integration. The implications of this for place, space and territory are manifold, creating spaces of circulation both in urban and non-urban realms, which are most comprehensive in socio-economic terms. In the remainder of this chapter, I will present the conceptual background of this subject matter by addressing the implications of information and communication technologies for spatial development in general and the urban in particular. I will then illustrate my argument against the case of *Amazon.com* — that there is dislocation emerging as a consequence of disruption and digital integration. Finally, I will discuss what this could mean for situating the digital transformation of circulation in urban spatial contexts.

Conceptual background: circulation, the digital and the material world

The discourse on the relationship between virtual flows and the material world has evolved parallel to technological innovation and to a lesser degree to real-life application of the new systems. The role of new information and communication technologies in societies in general and geographies in particular represents a major topic of debate for some time now (see Graham and Marvin 1996; Warf 2013; Wheeler *et al.* 2000). One important aspect of the debate suggests that a certain dissolution of physical (or territorial) space into the virtual world is occurring, a narrative that has had a stunning tenacity over the last two decades. It started in the 1980s and accelerated in the mid-1990s, caused by the advent of the Internet and revealing a certain degree of speculation and exaggeration. Its specific vocabulary suggested the emergence of new urban spatial forms and processes as captured in terms such as "silicon landscapes" (Hall and Markusen 1985), "city of bits" (Mitchell 1995) or the "death of distance" (Cairncross 1997). In retrospect, these terms carry a somewhat territorialist, if not essentialist, notion. The related myths often unfolded into technological determinism (Graham and Marvin 1996: 80ff.), as a linear projection of changing framework conditions into prescribed urban futures, which were then translated into impact assessment and possible strategies of response.

While it seems plausible to understand technological change as a transformation of space-time relations from *destiny* to *opportunity*, as Kellerman and Paradiso (2007) suggest, its concrete spatial implications remain unclear. Technological change was clearly overestimated at the beginning of its lifecycle (that is, the onset of the digital revolution), given the belief that spatio-temporal patterns of humans and the associated structures would emancipate from the traditional imperatives of time and space. Meanwhile, it looks as if the evolution of the Internet, of Web-based social media and corporate services inhibit a degree of change that, though having taken time to evolve, can hardly be underestimated now.

That said, I will briefly turn to the urban question, the second conceptual background which is relevant here. It seems widely accepted that the traditional division of labour between urban and non-urban places has tended to dissolve, as urbanization of societies has become ubiquitous (Cloke 2011). Whilst this remains a general proposition, the picture appears rather mixed when linking the Information and Communication Technologies (ICT) discourse named above with the debate on urbanization. One the one hand, communication was initially imagined as undermining traditional patterns of spatial development and settlement structure, by eliminating friction and reducing the cost of overcoming distance quite substantially. Information flows were theorized as developing in three different ways: *complementarily*, by *adding to*, or *substituting for* material flows (Couclelis 2004). In so doing, their possible impact on the various geographies (places, flows) may be neither one-dimensional nor causal. Rather, they function as a catalyst of existing trends and developments, regardless whether the related setting is characterized as urban, semi-urban or rural.

This issue is further complicated by contemporary thoughts on how the urban and non-urban can be defined – seemingly ubiquitous, on the one hand, yet, as a consequence, less distinctive, on the other hand. The more cities and the urban are considered constructs rather than pre-given entities, spatially unbound, fuzzy and ambivalent, the more difficult it might be to detect clear interrelations between the urban and the digital. Given the emphasis placed on knowledge economies by recent academic work, there seems to be reason to assume that the widespread use of ICT gives significant support to core urban properties such as agglomeration, which Scott and Storper (2015: 8) theorize as "the spatial concentration of production and a multifaceted, circular, cumulative dynamic of clustering and sorting". However, one could also think about other forms of agglomeration, where big streams and bundles of data are nested in their own locational dynamics and mechanisms, and are becoming manifold forms of interaction with the material world (Kitchin 2014). The related consequences of these enhanced dynamics of flows for cities may be as mixed and ambivalent as it was the case with earlier forms of circulation, that is, transportation: material flows were considered to enable *and* to demolish the logics of agglomeration, transport thus being viewed as a *maker* and *breaker* of cities (Clark 1958).

Spider in the web: the rise and performance of *Amazon.com*

Amazon.com was founded in 1994 in Seattle, Washington, USA, by Jeffrey P. Bezos, currently president, CEO and chairman of the board, and also an important shareholder. First registered as an enterprise in Seattle, *Amazon.com* became fully operational in July 1995; however, since 1996 it has been registered in the State of Delaware. Starting from its home base in the United States (USA), it has now established suborganizations in countries as varied as Australia, Brazil, Canada, China, Czech Republic, France, Germany, India, Italy, Japan, Mexico, Poland, Spain and the UK, with physical (not only virtual) presence in many of these. According to its 2014 Annual Report, it serves customers in 200 countries worldwide, thus encompassing much of the globe. *Amazon.com* is now without a doubt one of the real giants of e-commerce, the digital economy and the Web. With regards to market value and turnover, various rankings list *Amazon.com* among the top five global corporations, close behind *Google*, the Chinese *Alibaba* and *Facebook*. On the revenue-based *Fortune 500* list of all corporations, it ranks twenty-ninth behind many superpowers of manufacturing, services or finance. It is estimated that it has captured as much as a 5 per cent market share of total global e-commerce, only second worldwide behind *Alibaba*. In 2014 *Amazon.com* realised net sales of about $89 billion, almost three times as much as in 2010; it employed 154,100 people worldwide by the end of 2014, and also has 15,000 robots in ten fulfilment centres across the USA (Amazon.com 2015).

The firm has gained this position in the digital economy – not yet dominance, but probably close to hegemony – by pursuing a constant path of innovation and improvement, failure and disruption. First, it took the book market by storm, targeting authors, readers, publishers and stores (and also printers) – a story that has been brilliantly conveyed in various books and media reports.[1] Competitive pricing, a huge catalogue and a service quality not known before were key ingredients of this strategy, and also the aggressive execution of competitive power. Second, *Amazon.com* expanded from books into trading consumer products of almost any kind. Third, it began offering services, particularly those that are based on online operations such as *Amazon Marketplace*, *Fulfilment by Amazon* and *Amazon Web Services* (the early incarnation of cloud-computing). In this regard, the firm's business operations could be understood as an ongoing experiment, an ever-lasting process of exploration, creative construction and associated destruction. Services that were successful internally (for the purpose of its own business) were also opened up to third parties, so infrastructure and services could be established as a platform for use by a variety of actors other than *Amazon.com* itself. Third-party platforms such as *Marketplace* and *Web Services* have proven to be ideal for expanding the business. The combination of these three steps, based on an unusual mix of innovation and economies of scale, inhibits those properties that make the firm's role both disruptive and successful (see Christensen 2006).

Today there are a huge variety of activities assembled under the roof of *Amazon.com*, alongside an enormously stretched value chain. The company now does much more than just delivering books and other consumer goods. It enables individuals and corporations to create value within a rather diversified portfolio: it operates gigantic server farms in order to provide a proper information flow for its own purpose and also for third parties; it produces hard devices for the end-consumer, such as phones, tablets, readers, TV devices; it operates delivery services from fulfilment centres to the customers' doors, which is at least partly (in major metro regions) offered as same-day delivery and delivery on Sundays, more recently including grocery delivery as well; it introduced a self-publishing platform for authors, which perfectly links the writers' communities of practice with the growing population of reviewers on *Amazon* websites; it deploys on-line marketplaces as a hugely successful sales platform which provides almost half of its fulfilment volumes; it has now also started to produce its own TV series and films with its *Amazon Studio*; and last but not least it is about to experiment with the operation of 'unmanned' vehicles such as robots (inside fulfilment centres) or air drones (outdoors), promising the next big leap in establishing highly routinized, profitable business operations.

The application of ICT was a fundamental requirement for successfully establishing and steering these complex systems. ICT sits at the heart of this corporate practice, as there are two elements of *Amazon*'s business operations that seem to be key to the company's success. These two are packed with high technology: (1) data centres for the management and collection of information of all kind; and (2) fulfilment centres for the seamless but (cost-) efficient flow of commodities to further points of sale or to the point of consumption. By using data centres to direct the information flow, and deploying fulfilment centres to navigate the consolidation and physical movement of consignments, *Amazon.com*'s network creates an entirely new geography of distribution that appears quite distinct from its predecessors. As the firm's customers are geographically dispersed, and given the corporation's ambition to serve these customers not only as perfectly as possible, but almost all over the Planet, the spatial implications of its business operations and the challenges for space-time management are obvious.

Implications for place, space and territory: towards the logics of dislocation

While avoiding the simplistic and one-directional underpinnings of the term *impact* when analysing the role of the digital economy for urban-regional developments, one could indeed assume particular *implications* that the business model of *Amazon.com* and others has on place, space and territory. Its physical imprint comprises real-life usage of digital devices and services of all kinds (including the related energy consumption and demand for space), the associated hardware and infrastructure needs, and also the sorting, bundling and delivery operations that are necessary to satisfy the customer. Key spatial implications are obvious from the firm's geographical coverage, which in turn is relevant for land use and circulation. Most concretely, these implications materialize in the newly established network of fulfilment or distribution centres (DCs) that serve as backbones of the company's network, and the associated commodity flows that are going into and out of these nodes towards the final destination. Meanwhile, *Amazon.com* operates 250 active distribution facilities worldwide, of which 156 are (in 2016) fulfilment centres and also an increasing number of hubs for *Amazon-Prime* customers (*Prime Now*-hubs) and grocery delivery (*Amazon Fresh*) (see MWPVL International 2016). While about two-thirds of the facilities were first established in the USA, foreign markets are increasingly served by an own network of DCs, as soon as the business on these markets offers sufficient critical mass to make it economically reasonable. Among these international markets, the UK and Germany stand out as most extensively covered.

The DCs vary in size, ranging from 50,000 to more than 1.2 million square feet, depending on their function and their role in the specific setting of the regional network. Land consumption is one factor here that characterizes all sites or spaces that are dedicated to circulation, massive trip generation enhanced by economies of scale being the other. All centres are equipped with conveyor belts and radio-frequency devices, and they are certainly up-to-date in terms of data processing. While staff work on two different ends of the supply chain, unloading incoming or loading outgoing consignments, both from and onto lorries, others stack the shelves or assemble orders for shipment. Increasingly, robots are being used for the later task. Thus the requirements of data management and automated operations are not only high, but they actually determine the configuration and layout of the sites – this being not necessarily an impact, but instead a concrete implication of the virtual for the material. For these reasons, fulfilment centres usually had to be newly built, and in most cases they have been developed on previous greenfield sites, with prime motorway access. In most cases existing depots or warehouses could not be used for this purpose due to the size of the outlets that were needed to provide a maximum of services under one roof. Trade-offs between land prices and accessibility were, and still are, most important for the locational decision-making (Hesse 2004), as most preferred localities were those in the vicinity of major metropolitan areas – as close to the labour market and to the customers as possible, albeit away from the dense, costly urban core. Traffic generation by massive distribution centres is an issue as well, creating local traffic jams, while it is controversially discussed whether the overall net balance in terms of the associated vehicle movements triggered by large-scale fulfilment centres is positive or negative (Jaller *et al.* 2015). To some extent, *Amazon.com* is also creating jobs in semi-peripheral regions, which provide access to metropolitan areas but avoid the costs and negative externalities of agglomeration.

Figure 33.1 shows that the company's outposts are distributed throughout Europe, revealing a sort of logical spatio-temporal pattern.[2] While physical distribution is located close to urban centres (albeit not too close), customer service centres are placed at the periphery in call-centres, almost offshore. The knowledge-intensive tech- and headquarter functions are situated in a handful of metropolitan or quasi-metropolitan areas. When components and finished goods travel around the globe, so do the associated services, thus it is difficult to make a strong statement on the urban- or rural-ness of certain activities in the entire value chain. When selecting the locations shown in Figure 33.1, population potential and thus the size of each market area were crucial considerations especially when placing fulfilment centres. These 'market-based' factors are always combined with the search for the seamless flow of commodities, thus bringing together site and situation. Other issues have also played a role here, as in some cases non-operational rationales turned out to be decisive as well, such as unions or taxation policies (see below).

This pattern of newly established hubs and DCs becomes the core 'space of circulation', as it were within a broader network of relations and entails various forms of *mobilities*. The first of these was the launch of the corporation and the decision to build up its own network of DCs, 'mobilizing' the sites as such (Hesse 2006), as it triggered the development of new logistical cathedrals, creating distribution centres that are spread, or 'distributed' (Cidell 2015), across

The logics and politics of circulation

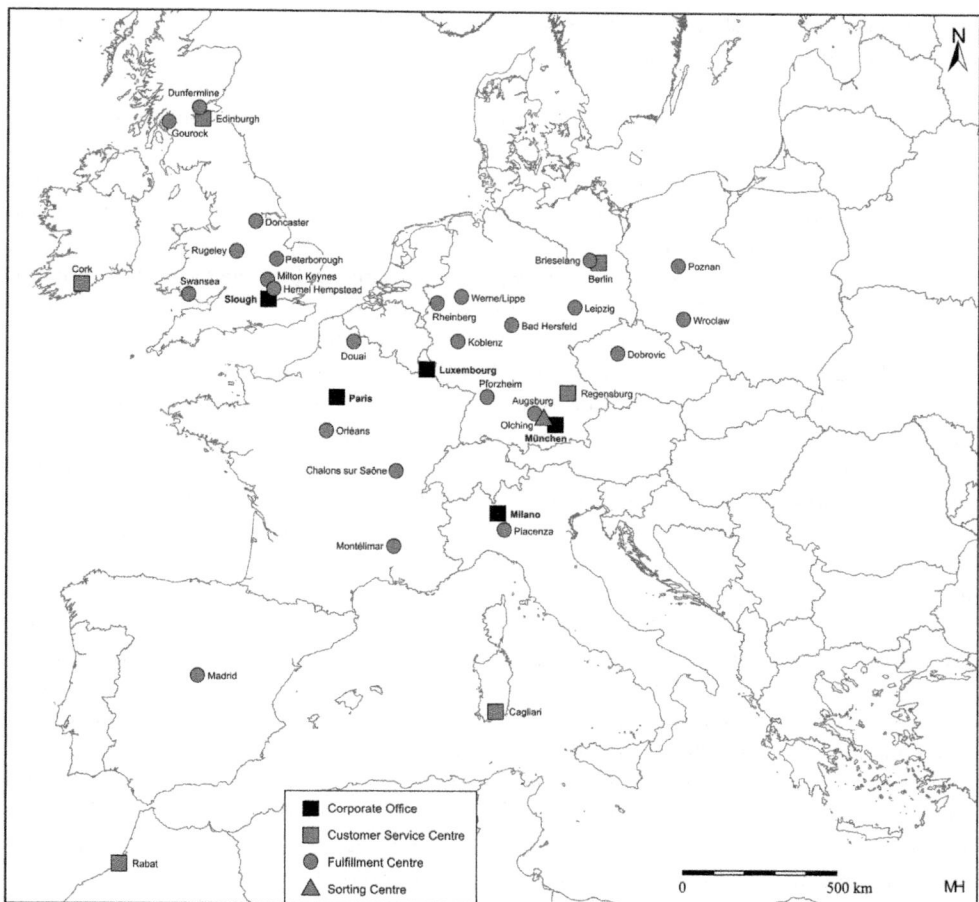

Figure 33.1 Amazon.com locations in Europe (as of March 2016). Map created by author based on various corporate sources; Cartography: Malte Helfer, UL.

territory. Second, this network enables a constant flow of vessels and vehicles, alongside the main routes that connects hubs and customers. Third, it triggers an efficient in-house flow of consignments, mostly on conveyor belts, operated by humans or robots. Fourth, these so-called pickers, that is, staff who gather ordered items from the shelves and stow them in a trolley walk quite a stretch on a typical workday. O'Connor quotes an *Amazon* official saying that "some of the positions in our fulfilment centres are indeed physically demanding, and some associates may log between seven and 15 miles walking per shift" (O'Connor 2013: 6).

Moreover, there is a regional development impact or implication visible from the evolution of *Amazon.com*, as all new distribution centres hire a handful of IT and middle management experts, as well as a certain range of 'fulfilment associates' who are in charge of the tasks named above. Such investments can easily reach one thousand or more employees at a single site, which is usually much appreciated by economic development officials, particularly in regions that suffer from de-industrialization or are peripheral (Hesse 2015). For obvious reasons, the company claims that it has created thousands of jobs in the USA, the UK, in Continental Europe and other parts of the *Amazon* universe, and still continues to do so. However, there are major concerns about labour conditions, salaries and regulatory circumstances, which not only reflect different corporate ethics, but also seem to echo the firm's unlimited ambition and rush for improvement. Quite impressively, O'Connor (2013) presents the small town of Rugeley, UK, a former coal-mining place in Staffordshire of about 22,000 people, which was chosen as the site of a 700,000 square feet fulfilment centre by *Amazon.com* in 2011. Certainly, some people welcomed the opportunity to re-enter the labour market, while others complain about the rigid ('Darwinian') corporate culture, the surveillance of employees and the constant tracking of their productivity. Temp staffing seems to be a rule not only for new appointments, but is often practised to enhance overall employee performance through competition, and minimum adult wages were reported to earn just £6.19 per hour. "It seemed like this was the town's chance to reinvent itself after decades of economic decline. But as they have had a taste of its 'jobs of the future', their excitement has died down" (O'Connor 2013: 2).

Finally, as important as its immediate locational pattern and thus the related spatial imprint of *Amazon.com* is the implication of emptiness caused by its competitive pressure. Since *Amazon* started two decades ago, it has probably contributed to putting thousands of independent shopkeeping enterprises out of business, beginning with indie bookstores, and continuing to threaten small retailers and department stores. It is hard to calculate how many jobs and stores have vanished as a result of the rising market shares and competitive power of e-commerce in general, and of *Amazon.com* in particular; while structural change usually comes in a package of various factors, it is difficult to hold *Amazon* responsible for all the related losses. However, it is obvious that the enormous growth of the firm has been, and still is, necessarily associated with the hollowing out of the traditional landscape of retail. This effect might be even enhanced when *Amazon* gets deeper into the operation of physical bookstores, and has to be taken into account when praising the many jobs created through *Amazon*'s various platforms.

The practices discussed here evolve from disruptive innovation, they are backed by venture capital and also by an overly aggressive market strategy. They lead to a comprehensive but compartmentalized organization of value chains, whose implications on place, space and territory can be interpreted in the context of a certain "logics of dislocation" (Barnes 1996) – the geographical manifestations of disruption and fragmentation in the economic process. According to Barnes (1996: 249ff.), the logic of dislocation includes: (1) the disruption of a previously established order; (2) the idea of displacement and the process of which an established body of ideas, people or things gives way to another; and (3) a sort of medical definition, pointing to the discontinuity of corporeal issues. While it was far too early to make any assumption on the concrete implications of virtual economies, trade and retail back in the mid-1990s, Barnes' claim can be perfectly alluded to the case of *Amazon.com*: the firm's practices provide a rather distinct set of how to do things, and they mean a fundamental break with the previously existing order, which particularly applies to spatial or locational terms. Also, it is a development that brings about spaces of circulation in both centres and peripheries. As a consequence, the development of the digital economy can no longer be properly associated with categories such as the urban or the rural.

Fostering continuous improvement through time-space compression

Competition in the digital business and e-commerce tends to be high and is driven by rivals both virtual and material. Thus the system created by *Amazon.com* is under constant pressure to improve, and the company has prepared for standing the competition by pursuing an attitude that apparently follows "customer obsession rather than competitor focus" (*Amazon.com* 2015: 2). Regardless whether the focus is indeed on the customer or on rival companies, the competitive imprint of the firm appears to be rather strong. Most notably, *Amazon.com* has cultivated a peculiar management of time and space in order to respond best to these difficult framework conditions. This particular management of time and space unfolds in various ways. First, expansion is envisaged not least in territorial terms: after arriving with physical operations in China and India, *Amazon* is now about to expand to further markets such as Australia, which until recently was served by fulfilment taking place in Europe (Germany), thus becoming almost planetary. Second, *Amazon.com* is about to extend its portfolio of products and services such as repair or maintenance, which can be ordered via the *Amazon Dash Button*. Also, it has been successfully experimenting with grocery delivery (*Amazon Fresh*) for some time now, an e-commerce service that turned out to be rather costly so far, but promises particular advantages given the firm's overall portfolio. While *Amazon Fresh* is already established in US cities such as Seattle, Los Angeles and San Francisco, the firm has plenty of other metropolitan areas on its screen, where customers will receive food and general merchandise deliveries at very short term.

Third, and most importantly, *Amazon.com* is accelerating supply chain processes, in order to get even closer to the customer, by annihilating space and compressing time. It now adds a completely new layer of distribution to the system, situated in between large-scale fulfilment centres and the customer. Small-scale sorting and delivery centres allow keeping the promise of same-day service or even shorter delivery time frames. Instead of shipping to customers in high-density urban areas from massive fulfilment outlets based at the periphery or in remote low-cost areas, the company will move towards a fast-track shipping strategy, thus streamlining the supply chain even further. For that purpose, it now operates about twenty-three smaller sortation centres only in the USA and a dozen currently emerging in the UK, with each facility employing 100–300 associates (MWPVL International 2016). The speed and reliability of this distribution strategy calls for a same day turnaround in most metro area cities of the USA, and probably also British conurbations like London, Manchester, Glasgow or Edinburgh. *Amazon lockers* for storage and brick-and-mortar outlets, of which the first opened in Seattle in 2015, will allow the company to accelerate business operations further and to get as close to the customer as possible.

Gaining more control over the supply chain – and independence from freight forwarding and parcel delivery partners – seems vital for the firm. For that purpose, *Amazon* is experimenting with taking over the last mile delivery in its own responsibility. Moreover, in March 2016 it announced its plan to lease twenty Boeing 747 freight carriers for long-haul shipments under its own regime. The next step in these regards will be overcoming the barriers and friction of physical space, such as water, mountains, traffic congestion or the dense urban fabric as such. For that purpose, *Amazon.com* has been introducing air cargo distribution with the help of 'unmanned' drones. These flying robots may be well suited to making deliveries in remote areas such as islands, or to speeding up distribution in high-density settlements such as Manhattan or other inner cities. While an agreement with air traffic authorities in various countries remains lacking, it can be expected that Amazon will eventually fly parcel drones on a regular basis given the firm's pattern of successfully pursuing time-space compression as a driver of its business model.

Conclusion: the spaces and politics of circulation of *Amazon.com*

The spaces of circulation as introduced in this chapter are heavily constructed on the grounds of a peculiar combination of virtual and physical means, where data management and logistics competencies unfold to develop a strikingly competitive business model. However, these innovations can only unfold against the background of a particular 'politics of circulation' – a practice that is pursued on various levels of engagement: within the firm, between *Amazon* and partners, and also by interacting with state authorities at both local and national levels. First, there is good reason to assume that the firm is governed internally rather rigidly, by top-down decision-making and control, in contrast to recent claims for inclusive corporate governance. The overall extent to which the firm's operations are data-driven allows for an absolute level of control to be exerted that was not known before. This may apply to staff at the management level in the same ways as to the pickers and packers in the fulfilment centres, since the respective performances and their output are data-driven and data-controlled.

Second, as the biggest e-commerce actor and one of the top data-handling providers, the company needs the support of many other service firms, subcontractors and the like. Also, by its internal processes and infrastructure for use by third parties, the owner and operator of the network has become a spider in the web. While much of the internal machinations of the firm remain unknown, we do know that the selling and purchasing activities of *Amazon Marketplace* are constantly tracked and assessed by the owner, which sometimes also triggers market intervention to the benefit of *Amazon*, and at the cost of the sales partner. Data management makes it easy to detect products or services that successfully fill a gap in a given market, which provides the firm with an opportunity to quickly develop its own product. This may reframe the balance between cooperation and competition with its partners (Ritala *et al.* 2014). Third, the massive expansion of its physical network, most notably the system of fulfilment and sorting centres as backbones of the delivery web, requires to be subject of local negotiation. Every single DC needs to conform to local building codes and planning ordinances. Hosting an *Amazon.com* fulfilment centre is most often perceived by municipal administrations as an opportunity, rather than a risk. For this reason, planning bodies tend to be receptive to building permit requests. However, even with the accelerating impact of municipal competition, planning and building a DC is a political-economic issue that is usually carefully planned and, almost always, controversial.

Fourth, negotiations are also prevalent at the national scale, taking place with governments and overarching regulatory bodies, such as nation-states or the European Union, in order to ensure the company's well-being. This particularly applies to the processes of 'tax optimization'. As sales tax rates can vary widely, for example between states of the USA, these differences have been directing investments to those areas that do not charge taxes on online businesses. The definition of *Amazon*'s business as *logistics* rather than *retail* makes a significant difference here as well. The issue of definition also applies to the minimum wage in European countries. In Germany, the battle between the services union *Verdi* and *Amazon* on minimum salaries has already led to a series of strikes at DCs. Among other issues, the dispute is rooted in the union's claim to classify fulfilment operations as retail, while the company insists on considering this as logistics work, which usually pays less. This labour conflict was most likely driving the company's decision to open up fulfilment centres in Poland and the Czech Republic, offering lower labour cost and a more comfortable political environment. Also, tax reasons were probably the drivers behind *Amazon.com* establishing its European headquarters in Luxembourg back in 2006, which was secretly negotiated between the two bodies. As the *Luxleaks* revealed, the Grand Duchy's taxation policy not only offered minimum obligations for the company as such, but were also supportive of downsizing profits due to 'patent'-spending between different subsidiaries of *Amazon.com*, based on discretely offered tax rulings by the government. Also, tax advantages for e-commerce earnings were calculated based on Luxembourg rates, even though the related operations mostly took place in other countries. This problem has been solved in the meantime with the re-territorialization of e-commerce sales taxes across Europe, and the efforts fostered by the Organisation for Economic Co-operation and Development to regulate base erosion and profit spending are expected to have a similar effect in the future.

To conclude, due to the peculiarities of disruptive innovation and backed by a combination of a huge amount of capital and an aggressive market strategy, *Amazon.com* is continuing to reshape the landscape of e-commerce quite significantly. In this regard, it is no longer situated in between *complementarity* and *cannibalism*, which were the major impacts initially expected from e-commerce or the Internet on existing patterns of communication and on space-time practices. Rather, with the broader implications of this business model for society, economy and territory, it is becoming most comprehensive if not all encompassing. After having successfully introduced a completely new structure for distributing goods and also offering services based on ICT, data management and fulfilment operations, *Amazon.com* is constantly approaching its vision of becoming almost planetary in scale. In so doing, it increasingly contributes to the blurring of boundaries between production, distribution and consumption.

In this context, it has already been argued that *Amazon.com* is neither a retail firm nor a logistics corporation, but a '*matrix* commerce' that has evolved from the liaison of virtual and physical activities (Williams 2012). In so doing, the system seems to have almost become totalitarian. It reveals a certain irony that, by disruptive innovation and exerting unlimited competitive pressure, the firm resembles to some extent the old, all-encompassing postal services of the pre-privatization era – when only one institution offered a comprehensive set of products and services, authorized by the government, and without competition. While certainly not suggesting that this would be a desirable future, given the poor performance often employed by the state-owned postal services, the idea that the marriage of big data and sophisticated distribution creates a new super-monopoly in private hands bears a fundamental concern for open societies: the machine is now operated, controlled and valorized by a private enterprise, fit for profit and subject to the relentless exploitation of resources, human and non-human, urban and non-urban. This evokes some fundamental concerns about the combination of technology and power (see, e.g. Ullrich 1977), which should be taken into account when further discussing the peculiarities of the relationship between the digital and the material worlds.

Last but not least, the system represented by *Amazon.com* inspires further reflections upon the role of the urban and non-urban as specific spaces of politics. In the first instance, e-commerce in general and *Amazon.com* in particular seem indifferent to territorial classifications such as the urban or the rural, while its real challenge and achievement is obviously overcoming if not annihilating the friction of space and time. E-commerce aims to be all-encompassing and wide-spread, eliminating borders of all kind, not only between different markets and sectors of economic activity, but also concerning territorial distinctions. As a consequence, it looks as if *Amazon.com* is somehow emblematic for the critical debate on the urban and rural in the context of what is known as planetary urbanization – thus considering the urban as such to be theoretically no longer distinctive and thus practically no longer relevant (see Angelo 2017). However, one could also argue that peripherality, understood as structural disadvantage reproduced by socio-economic development in a territorial context, is prevalent in both urban and more remote settings. This problem remains as a general challenge for urban policy and politics.

The problem was perfectly illustrated by a study released on *Bloomberg.com* in April 2016 (see Ingold and Soper 2016), revealing that the *Amazon-Prime* concept inhibits a certain sense of redlining, by refusing same-day delivery to economically disadvantaged areas. The study analysed the distribution of *Prime*-service according to the firm's website, and the ZIP-codes were correlated to population and area data provided by the *American Community Survey* (ACS). Based on this data assessment, it turned out that *Prime*-services were not being offered in certain parts of metropolitan areas, including urban districts such as South Atlanta, Georgia, The Bronx or parts of Queens in New York City, the primarily Black neighbourhood Roxbury in Boston,

Massachusetts, the South Side of Chicago, Illinois, and others – all areas where a large concentration of the non-White, immigrant and non-affluent part of the population can be found. Since these obvious blind spots on the company's services maps coincide with the social, racial or economic composition of the population, the claim was made that *Amazon.com* would potentially discriminate against these parts of the population.

In response to the public debate initiated by the *Bloomberg* study, the firm promised to offer *Prime*-services at least for three of the neighbourhoods not served before. Nevertheless, the case demonstrates that social inequalities tend to persist in the digital age, and when looking at the systemic – if not totalitarian – nature of matrix enterprises such as *Amazon.com*, it is hardly possible to identify the subject and target of political action. While the associated problems and conflicts are perceived to be local or urban, regulating the spaces and politics of circulation would require an abstract, multi-level power to act which is difficult to imagine.

Notes

1 See for example *The Everything Store* by Stone (2013), and also the lengthy magazine pieces by Wasserman in *The Nation* (2012), Brauck et al. in *Der Spiegel* (2013), O'Connor in the *Financial Times* (2013), Packer in the *New Yorker* (2014), Wohlsen in *Wired* (2014), Foer in the *New Republic* (2014), Hooper in the *Guardian* (2014), or Kantor and Streidfeld in the *New York Times* (2015).
2 It would be interesting to apply a comparable mapping exercise to the demand side, tracing the location of the firm's customers; this might reveal whether the aggregate customer-population is more urban, more rural, or simply indifferent to such categories. However, such data belong to corporate secrecy and are almost impossible to obtain.

References

Amazon.com. (2015). *2014 Annual Report*, 29 January. Available at: http://amazon.com [accessed 3 March 2016].
Angelo, H. (2017). From the City Lens Toward Urbanisation as a Way of Seeing: Country/City Binaries on an Urbanising Planet. *Urban Studies*, 54(1), pp. 158–178.
Barnes, T.J. (1996). *Logics of Dislocation: Models, Metaphors, and Meanings of Economic Space*. London: Guilford Press.
Brauck, M., Müller, M.U. and Schulz, T. (2013). Gnadenlos.com. *Der Spiegel*, 51, pp. 58–65.
Cairncross, F. (1997). *The Death of Distance: How the Communications Revolution Will Change Our Lives*. London: Orion Business Books.
Christensen, C.M. (2006). The Ongoing Process of Building a Theory of Disruption. *Journal of Product Innovation Management*, 23, pp. 39–55.
Cidell, J. (2015). Distribution Centres as Distributed Places: Mobility, Infrastructure and Truck Traffic. In: T. Birtchnell, S. Savitzky and J. Urry, eds. *Cargomobilities: Moving Materials in a Global Age*. New York and London: Routledge, pp. 17–34.
Clark, C. (1958). Transport: Maker and Breaker of Cities. *Town Planning Review*, 28(4), pp. 237–250.
Cloke, P. (2011). Urban-rural. In: J. Agnew and D. Livingstone, eds. *The Sage Handbook of Geographical Knowledge*. Los Angeles: Sage, pp. 563–570.
Couclelis, H. (2004). Pizza over the Internet: E-commerce, the Fragmentation of Activity, and the Tyranny of the Region. *Entrepreneurship and Regional Development*, 16, pp. 41–54.
Foer, F. (2014). Amazon Must Be Stopped. It's Too Big. It's Cannibalizing the Economy. It's Time for a Radical Plan. *The New Republic*, 9 October.
Graham, S. and Marvin, S. (1996). *Telecommunications and the City: Electronic Spaces, Urban Places*. London and New York: Routledge.
Hall, P.G. and Markusen, A. (1985). *Silicon Landscapes*. London: Allen and Unwin.
Hesse, M. (2002). Shipping News: The Implications of Electronic Commerce for Logistics and Freight Transport. *Resources, Conservation & Recycling*, 36(3), pp. 211–240.
Hesse, M. (2004). Land for Logistics: Locational Dynamics and Political Regulation of Distribution Centres and Freight Agglomerations. *Tijdschrift voor Economische en Sociale Geografie*, 95(2), pp. 162–173.

Hesse, M. (2006). Logistikimmobilien. Von der Mobilität der Waren zur Mobilisierung des Raumes. *DISP*, 42(4/167), pp. 43–51.

Hesse, M. (2015). Selling the Region as Hub: The Promises, Beliefs and Contradictions of Economic Development Strategies Attracting Logistics and Flows. In: J. Cidell and D. Prytherch, eds. *Transport, Mobility and the Production of Urban Space*. London and New York: Routledge, pp. 207–227.

Hooper, M. (2014). Amazon at 20: Billions, Bestsellers and Legal Battles. *Guardian*, 21 July.

Ingold, D. and Soper, S. (2016). Amazon Doesn't Consider the Race of Its Customers. Should It? *Bloomberg*, 21 April. Available at: www.bloomberg.com/graphics/2016-amazon-same-day/ [accessed 27 April 2016].

Jaller, M., Wang, X.C. and Holguin-Veras, J. (2015). Large Urban Freight Traffic Generators: Opportunities for City Logistics Initiatives. *Journal of Transport and Land Use*, 8(1), pp. 51–67.

Kantor, J. and Streidfeld, D. (2015). Inside Amazon: Wrestling Big Ideas in a Bruising Workplace. *New York Times*, 15 August. Available at: www.nytimes.com/2015/08/16/technology/inside-amazon-wrestling-big-ideas-in-a-bruising-workplace.html?_r=0 [accessed 30 May 2017].

Kellerman, A. and Paradiso, M. (2007). Geographical Location in the Information Age: From Destiny to Opportunity? *GeoJournal*, 70, pp. 195–211.

Kitchin, R. (2014). The Real-Time City? Big Data and Smart Urbanism. *GeoJournal*, 79(1), pp. 1–14.

Mitchell, B. (1995). *City of Bits: Space, Place and the Infobahn*. Cambridge, MA: MIT Press.

MWPVL International. (2016). Amazon Global Fulfillment Center Network. Available at: www.mwpvl.com/html/amazon_com.html) [accessed on 6 March 2016].

O'Connor, S. (2013). Amazon Unpacked. *Financial Times*, 8 February.

Packer, G. (2014). Cheap Words. Amazon is Good for Customers. But is it Good for Books? *The New Yorker*, 17 February.

Rajan, N. (2015). Amazon's Jeff Bezos Lands First Reusable Space Rocket. *Huffington Post UK*, Tech page, 25 November. Available at: www.huffingtonpost.co.uk/2015/11/25/amazon-jeff-bezos-lands-first-reusable-space-rocket_n_8645930.html [accessed 30 May 2017].

Ritala, P., Golnam, A. and Wegmann, A. (2014). Competition-based Business Models: The Case of Amazon.com. *Industrial Marketing Management*, 43(2), pp. 236–249.

Scott, A.J. and Storper, M. (2015). The Nature of Cities: The Scope and Limits of Urban Theory. *International Journal of Urban & Regional Research*, 39(1), pp. 1–15.

Stone, B. (2013). *The Everything Store: Jeff Bezos and the Age of Amazon*. New York: Bantam Books.

Ullrich, O. (1977). *Technik und Herrschaft*. Frankfurt/Main: Suhrkamp.

Warf, B. (2013). *Global Geographies of the Internet*. Heidelberg/Berlin: Springer.

Wasserman, S. (2012). The Amazon Effect. *The Nation*, 18 June.

Wheeler, J.O., Aoyama, Y. and Warf, B., eds. (2000). *Cities in the Telecommunications Age*. New York: Routledge.

Williams, A. (2012). Amazon Is Not a Commerce Company. Blogpost on *www.techcrunch.com*, 30 December. Available at: https://techcrunch.com/2012/12/30/amazon-is-not-a-commerce-company/ [accessed 26 January 2016].

Wohlsen, M. (2014). A Rare Peek Inside Amazon's Massive Wishfulfilling Machine. *Wired*, 16 June.

Zook, M. (2008). *The Geography of the Internet Industry: Venture Capital, Dot-coms, and Local Knowledge*. London: John Wiley & Sons.

34
CIRCULATING EXPERIMENTS
Urban living labs and the politics of sustainability

James Evans, Harriet Bulkeley, Yuliya Voytenko, Kes McCormick and Steven Curtis

Introduction

Urban experimentation has gained traction in cities all over the world as a way to find new, more sustainable ways to plan and develop cities. Interventions designed to address a diverse range of urban challenges bring innovative social and technical components together to learn by doing. Seen through this lens, the modernist planning that dominated the urban arena for much of the twentieth century seems to have given way to what we might term the *experimental city* – a condition where the urban both forms an arena for experimentation and is shaped by it (Evans et al. 2016). The appeal of urban experiments lies in their ability to be radical in ambition while limited in scope; ground-breaking rather than rule-breaking. Experimentation permits learning, which is increasingly identified as a necessary ingredient to 'scale up' solutions both within and between cities.

This chapter examines the production and circulation of urban experiments through urban living laboratories, a mode of governing the city and type of experimentation that is growing in popularity. Urban living labs (ULLs) constitute a form of experimental governance, where urban stakeholders develop and test new technologies and ways of living to address the challenges of climate change and urban sustainability (Bulkeley and Castán Broto 2012). ULLs can be considered both as an arena (geographically or institutionally bounded spaces), and as an approach for intentional collaborative experimentation of researchers, citizens, companies and local governments (Schliwa 2013). They are often located in socio-economically and environmentally challenged neighbourhoods and areas in order to facilitate regeneration (Curtis 2015; Voytenko et al. 2015). The rapid spread of ULLs is playing a key role in the production of more experimental approaches to cities, producing a multitude of experiments intended to inform urban policy, planning and, in particular, politics.

ULLs provide a window upon the processes through which cities are attempting to learn to become more sustainable. ULLs require us to rethink how we approach the urban as a space of politics as they reconfigure traditional processes of governing urban development in terms of approach, participation and scale. By adopting a user-led approach to co-creation, ULLs recognize the need to base urban development priorities and actions around the needs and desires of urban dwellers and others who use the city. This approach not only opens up processes of urban planning and development to a wider range of participants, but potentially empowers participants in new ways; notably entrepreneurs, private partners, NGOs, residents and users. In doing so, ULLs herald a new kind of urban politics that transforms top-down planning and development into local innovation processes staged in spaces such as buildings, neighbourhoods or districts. Despite the proliferation of ULLs across Europe, their ability to deliver a new more politically inclusive and innovative approach to urban development remains little more than an article of faith. Drawing on the results of a range of research projects, this chapter outlines the emergence of ULLs, before considering the political dimensions of this approach and their implications for cities and those that live in them.

The rise of urban experimentation

The rise of urban experimentation is related to a range of social, economic and environmental challenges facing cities today. The challenges of sustainability, climate change, development and security increasingly exceed the capacities of planning, a practice that has traditionally relied on at least a degree of certainty and the ability to separate out the components of urban problems in order to address the whole. Yet it also reflects the changing nature of urban governance and politics. Cities have at once become charged with a growing agenda of issues to address whilst also being subject to new forms of political rationality, many of which emphasize the need to move beyond a centralized state towards local forms of governing that work through partnership and participation.

The burgeoning realization that 'business as usual' will no longer do has prompted a search for alternative ways to organize, plan, manage and live in cities. Experimentation offers a way to do all of this, gaining traction in cities all over the world as a mode of governance to stimulate alternatives and steer change (Bulkeley and Castán Broto 2012). Policy-makers, designers, private companies, communities and NGOs are initiating innovation activities to trial alternative future visions of local economic development, social cohesion, environmental protection, creative sector expansion, policy evolution, service delivery, infrastructure provision, academic research, and so forth (Karvonen et al. 2014). In this context experimentation emerges as a means through which to orchestrate and govern the urban milieu in ways that are simultaneously innovative and creative while reframing distant targets and government policies into concrete actions that can be undertaken by a variety of urban stakeholders in specific places (Karvonen and van Heur 2014).

While assuming many forms, urban experimentation can be distinguished conceptually from conventional urban development and policy by an explicit emphasis on learning from real world interventions. Urban experimentation offers a framework within which to arrange instruments, materials and people in order to induce change in a controlled manner, and subsequently evaluate and learn from those changes (Karvonen and van Heur 2014). Universities and other knowledge-based organizations have become enrolled in urban experimentation to help demonstrate their wider societal value and create an impact of their activities, and in doing so can take advantage of the opportunities that such forms of collaboration provide (König 2013; Trencher et al. 2014). Understanding experiments as sites through which "particular urban infrastructure regimes ... are configured and challenged" (Bulkeley et al. 2014: 1477) resonates with current understandings that emphasize the relational and provisional aspects through which the city is comprised (Graham and McFarlane 2015). For example, Urban Political Ecology places an emphasis on flows of power and materials, socio-technical studies on the co-evolution of technology and society, and critical infrastructure studies on the ways in which urban institutions, techniques and artefacts are established, maintained and challenged (Monstadt 2009). The process of urban experimentation reworks the relationships between social and material networks in the context of existing economic, social and political trajectories.

Considering the political aspects of urban experimentation leads inexorably to the question of how an experiment or set of experiments drives wider transformation. Urban experiments are currently being deployed in a more widespread and explicitly recognized fashion than at any other time (Evans 2011). For instance, the Mayor of Bogotá, Enrique Peñalosa, is famous for transforming the city from one of the most dangerous and unpleasant places to live at the end of the 1990s into a global leader in sustainable urbanism today. What he calls the Bogotá 'experiment' started with a far smaller one, the *dia sin carro* or day without cars, which lit the fuse of the 'Happy City' movement that has taken hold globally as a new, human-centric approach to planning cities (Montgomery 2013). Numerous other urban approaches likewise share a commitment to changing the way in which we build, manage and live in cities through explicitly staging experiments. Smart cities, eco cities, low-carbon urbanism, urban adaptation, urban living labs and sustainable urban development all draw on the idea that experimentation can generate more liveable, prosperous and sustainable urban futures (de Jong et al. 2015). Experimentation forms a common thread running through otherwise disparate contemporary urban trends, from corporatized attempts to create smart, low-carbon cities to grassroots civic movements to make neighbourhoods more socially cohesive and environmentally benign. It is for this reason that urban experimentation has assumed such rapid prominence across a range of urban practice and academic thought.

In relation to the low-carbon agenda, numerous successful experiments have been established in cities over the last twenty years, leading funding bodies, policy-makers, charities, companies and communities to a mutual focus on how to translate discrete experiments into broader change. Cities are increasingly seeking to learn from experiences elsewhere when planning programmes of sustainable transition management, yet the contingencies of policy-learning arrangements in this field have not been explored in great depth. The policy mobilities approach rejects the notion of objective best-practice knowledge, or its neutral transfer between contexts (Clarke 2012). As policy knowledge is 'mobilized' and circulated across time and space, by different actors, through networks, and for particular purposes, its content and understanding has been observed to mutate (Peck 2011; Peck and Theodore 2010). When planning programmes of sustainable transition management cities are increasingly seeking to learn from experiences elsewhere; yet relatively little research exists considering the origins, developments or influences of policy knowledge in long-term sustainability planning and urban governance.

Laboratories come to town

There is no uniform definition of living labs (Schliwa 2013; Ståhlbröst 2008). Some scholars and organizations define them as partnerships between sectors (often between public, private and people) (European Commission 2015; Rösch and Kaltschmitt 1999) where universities play a key role (Evans and Karvonen 2011), while others look at living labs more in the light of pilot and demonstration projects, which function as supportive tools for private actors and industry, helping them commercialize their services, products and technology (Kommonen and Botero 2013; Reimer et al. 2012). The concept of a living laboratory first appeared in academic literature in the 1990s (Veeckman et al. 2013). William Mitchell, a professor at the Massachusetts Institute of Technology, is widely credited with first articulating the concept (Mensink et al. 2010). He sought to combine the growing use of sensory technology and computing power to stage experiments in the real world that could be monitored in a level of depth usually confined to the laboratory (Evans and Karvonen 2011; Mensink et al. 2010). The idea of rigorous testing in real world contexts resonated with numerous contemporary trends including the centrality of consumers to product development, shortened innovation iterations and the development of Information Communication Technology (ICT) capabilities (Ståhlbröst and Holst 2012). This early vision of the living lab has subsequently evolved into a broader mechanism to spur innovation and address social and environmental issues, being used to cover all kinds of testing and innovation in real world settings. When applied to cities this thesis holds that cooperation networks will form the backbone of future urban innovation (Schaffers et al. 2011), and that networking stakeholders within and between cities will lead to high knowledge spillover effects with low transaction costs (Cooke et al. 2002). Living labs can be employed by municipal or regional governments with limited budgets because they increase the efficiency of tacit and codified knowledge transfer between end-users and stakeholders (Mensink et al. 2010).

Living labs have proliferated very quickly in Europe due to substantial financial and policy backing from the European Commission. They are seen as a magic bullet to overcome the so-called European paradox whereby leadership in innovation and sustainability fails to translate into commercial success (Veeckman et al. 2013). The Lisbon Strategy outlined at the European Council in 2000 is motivated by the perceived difference in growth and innovation of the EU compared with the USA and Japan (Directorate General for Internal Policies 2010: 11). To increase technological capacity and innovation the EU sought the synchronization and coordination of ongoing and upcoming structural reform, building on the Luxembourg Process (European employment strategy), the Cardiff Process (integrating environmental issues in EU policy) and the Cologne Process (establishing social and macroeconomic dialogue). The European approach to innovation is distinctive for its emphasis on democratic processes and citizen engagement, and the harmonization of the European Union innovation system was set out in the Helsinki Manifesto, which states "that new, concrete measures are needed for turning the Lisbon Strategy into a living reality and making Europe more competitive and innovative in a human-centric way" (Finland's EU Presidency 2006: 1). The Manifesto outlines a seven-step roadmap that seeks to harmonize innovation pathways and financial eServices, harness ICT in enabling efficient working environment and increase EU interoperability of digital standards. However, most importantly, the Helsinki Manifesto called for the creation and financing of the European Network of Living Labs.

The European Network of Living Labs launched in 2006 with twenty living labs from fifteen Member States. The European Network of Living Labs calls itself an "international federation of benchmarked Living Labs in Europe and world-wide" (ENoLL 2015a), operating as a platform for the coordination of best practices while serving as a repository of economically viable and socially acceptable examples of living labs (European Commission 2014). ENoLL seeks to support living labs and offers services including communication and promotion, project development, brokering, policy and governance, and learning and education services (ENoLL 2015b). Along with this, members are afforded the right to use the 'Living Lab' label signalling proof of certification as a living lab by ENoLL (Curtis 2015). To date, the European Network of Living Labs has had nine membership calls, known as waves, with more than 370 living labs operating in a variety of sectors including ICT, energy, elderly care and health care, mobility and rural development. The European Network of Living Labs is the largest and most comprehensive living lab initiative in the world (Veeckman et al. 2013), playing a critical role in exporting the idea of living labs to cities.

Research grants from regional and national governments flow for those academics seeking to understand open innovation ecosystems, in particular, living labs. The European Commission's largest-ever funding scheme, Horizon 2020, actually specifies that projects must adopt a living lab approach to be considered for funding under many of the urban and innovation related calls. ULLs represent an approach to innovation that evolved in the ICT sector and has subsequently been applied to solve urban challenges. The rapid colonization of urban policy and funding discourses by living labs has generated considerable current interest in how they are being established in cities, and what their impacts are.

Urban living labs in practice

While emerging from the broader field of living labs, ULLs are physical places where stakeholders including public agencies, firms, universities and users come together to develop, deploy and validate new technologies, services, products, ways of living and modes of governance (Juujärvi and Pesso 2013). ULLs share the broader living lab approach to co-production that involves users in developing innovations in real world settings, but tend to take place in areas of cities that are designated in some way for experimenting with new approaches to urban planning and development. They also differ from living labs in their focus on 'urban' or 'civic' innovation, which strengthens the public elements of urban innovation and aims broadly at prompting some transformation in urban governance (Baccarne et al. 2014).

This is important, as previous research shows that overly technical approaches to deploying new technologies in cities fail to produce innovation or learning and can be easily co-opted by dominant economic interests. In their study of the Clean Urban Transport Europe Programme, which trialled green transport solutions in major European cities, Hodson and Marvin (2009) argue that these projects are little more than demonstrations of existing technologies and services, and that they did not engage local populations or context. Work conducted on the Oxford Road Corridor in Manchester, UK, has highlighted similar challenges concerning how to achieve social inclusion in ULL projects, and the de-politicization of urban governance that corporate-led partnerships and scientific modes of governance threaten (Evans 2011; Evans and Karvonen 2014). ULLs aspire to a quadruple helix model of partnership, whereby government, industry, the public and academia work together to generate innovative solutions (Voytenko et al. 2015).

An initial ULL typology (see Silver and Marvin 2016) differentiates between ULLs with various logics (i.e. post-carbon living, techno-oriented, knowledge producing and supporting economic growth), settings (i.e. different constellations of actors involved in designing and running ULL), focus (i.e. new technology, climate, building retrofit, food production, urban landscape, sustainability, low-carbon economy), activities (i.e. research, testing, training and education, R&D), and timeline (i.e. long term, temporary, uncertain). Certainly the association of ULLs with discourses of innovation has meant that they are often tied to the narratives of the potential for both low-carbon transitions and the growth of smart cities to provide new forms of urban economic development. ULLs are part of the proliferation of experimental approaches to governing climate change (Bulkeley et al. 2015), which tend to focus on the issues of how to foster low-carbon development. ULLs also offer a vehicle through which to enact and test smart city initiatives, providing platforms for co-production and innovation aimed at transforming urban governance (Baccarne et al. 2014).

The ownership of ULLs can be divided between three different forms: strategic, civic and grassroots. *Strategic* ULLs are led by government or large private actors, use urban as an arena for the pursuit of the interests of other actors, and tend to operate in the whole city area with multiple projects under one umbrella. An example of a strategic ULL is Renewable Wilhelmsburg Climate Protection Concept in Hamburg, Germany (see Table 34.1).

Civic ULLs are led by urban actors such as universities, cities and urban developers, focus on economic and sustainable urban development, are represented by either stand-alone projects or city-districts, and often have co-funding as central to a partnership model. An example of a civic/municipal ULL is Urban.Gro.Lab in Groningen, the Netherlands (see Table 34.2).

Table 34.1 Strategic urban living lab example

Renewable Wilhelmsburg Climate Protection Concept
Location: Hamburg, Germany
Lead partner: International Building Exhibition (IBA)
Partner types: Government (local), Government (national), business
Driver: Former industrial and port area in need of regeneration

Using the IBA showcase area, which extends from Veddel to Harburg's upriver port, this ULL illustrates how urban planners, architects, the public and planners can make creative use of energy savings, energy efficiency and renewable energy potential. By 2015, half of the electricity needs of all buildings on the Elbe islands will be produced on site by climate-friendly and renewable energy, whereas by 2025 the entire electricity need of the buildings will be produced on site. The coverage of the heat demand is aimed for 2050.

Table 34.2 Civic/municipal urban living lab example

Urban.Gro.Lab
Location: Groningen, the Netherlands
Lead partner: Rijksuniversiteit Groningen and Municipality of Groningen
Partner types: Local government (municipality), academia, consultancy, artists, cultural festival, business and entrepreneurs, NGOs
Driver: Innovation and improvement of urban environment

Urban.Gro.Lab is a creative breeding ground for people, questions, creativity, inspiration and technology in the city of Groningen. It is an initiative of the Municipality of Groningen and the Department of Planning of the University of Groningen. The city of Groningen is used by the urban living lab as a 'testing ground' for high quality applied research that focuses on current spatial and societal issues. The urban living lab aims to generate knowledge and inspiration for the liveable city of the future, with a specific emphasis on bringing science and practice together in a series of dynamic urban experiments.

Grassroots or organic ULLs are led by urban actors in civil society or not-for-profit actors, focus on a broad agenda of well-being and economy, often host micro-projects or single-issue projects and have limited budgets. An example of a grassroots ULL is Stapeln open maker-space in Malmö, Sweden (see Table 34.3).

While varying in terms of leadership, composition, focus and scale, ULLs are proliferating across cities in Europe as a new approach to address a range of classic urban challenges, ranging from the physical regeneration of derelict of industrial areas to social capacity building and environmental sustainability. As an approach to innovation that evolved in the ICT sector and has subsequently been applied to solve urban challenges, ULLs represent the intrusion of design thinking into the urban arena (Schliwa, forthcoming). The next section considers the political consequences of this for the ways in which cities are governed and lived in.

The political dimensions of ULLs

While not explicitly positioned as political spaces, ULLs clearly bring different combinations of urban stakeholders together to engage with urban policy and development at local scales. This section outlines three dimensions of ULLs that are influencing urban politics, loosely organized

Table 34.3 Grassroots urban living lab example

Stapeln Open Maker-Space
Location: Malmö, Sweden
Lead organization: NGO Stpln
Partner types: NGO, local government, university, SMEs and individual activists
Driver: Upskilling and social cohesion

Stapeln (STPLN) is a cultural house that hosts a co-working facility, a space for exhibitions and performances, and several do-it-yourself workshops for textile printing, sewing, knitting, carpentry, digital production, bicycle service and construction, and creative reuse/recycling. It is targeted at people of all ages active within arts, technology, innovation, design or crafts. In most cases, people may use STPLN for free, but in return pay with time and knowledge. The STPLN building is owned by the City of Malmö, which also provides basic financial support. STPLN offers new work and leisure opportunities for all Malmö citizens, which encourage more sustainable lifestyles, enhance social cohesion, allow for new ways of interaction, learning and exchange of skills.

around the themes of legitimacy, inclusivity and inequality associated with the devolution of urban problem solving to local scales.

First, ULLs produce new forms of political legitimacy based upon material demonstrations. For citizens, industry and politicians, being able to actually see a zero-carbon energy house, or engage with an innovative electric bike or sharing scheme can have powerful effects on their thinking and values. In this sense, 'seeing is believing' and the power of ULLs to demonstrate alternatives is in itself politically powerful. City leaders require a political mandate to act on urban sustainability challenges which can be strengthened by successful demonstrations. For example, cities that have successfully restricted car usage through congestion charging (London) or allowing odd and even number plated cars on alternate days (Delhi) have run trials first to increase political legitimacy for the idea. A key characteristic of ULLs is that they produce knowledge "in the real world" and "for the real world" (Evans and Karvonen 2014), and the ability of ULLs to bring alternatives to life in spaces such as buildings, neighbourhoods or districts is critical. Alongside the political effects of seeing and interacting with alternative solutions, ULLs also produce practical insights into the implementation and performance of new technologies and services. This kind of practical learning is intended to make it easier to adopt specific innovations and alternatives, by providing knowledge about the barriers and evidence about real world performance. In this sense, the examples given in the previous section play a political role in articulating innovations into place by overcoming practical challenges and fostering legitimacy.

In practice this is challenging to achieve. Bulkeley and Castán Broto (2012: 13) find that "experiments are often vested with particular interests and strategic purposes in the governing of the city", and experiments are frequently portrayed as beneficial to cities as a whole while sidestepping troubling issues about who is doing the experimenting, who is being experimented on, who is benefiting and who is being left out. ULLs carry implicit assumptions about what a city is and who should be involved in creating and managing it. Recent research by Curtis (2015) highlights that while ULLs are often established to develop innovative solutions in lower-income neighbourhoods, civic participation is dominated by relatively highly educated middle classes who have more time and motivation to engage around specific urban issues like for example cycling. Who is involved and what kinds of knowledge and competencies are prioritized by ULLs plays a key role in determining their goals, outcomes and ultimately broader social impact.

Challenges around inclusivity in turn generate questions of inequality. In substituting top-down planning and development for local innovation ULLs risk creating highly uneven political landscapes. In creating discrete and episodic experiments the ULL approach to urban development is inherently fragmentary in both a geographical and institutional sense. Cities and communities also vary in their ability to deliver this kind of approach, which is based around attracting competitive funding at some level, and subsequently produce 'successful' experiments. The focus of ULLs on enabling places to solve their own problems can be viewed as part of a broader neoliberalization of urban politics, whereby services and functions that used to be performed by the state are transferred to individuals and communities. In letting a thousand flowers bloom ULLs can create a vastly more varied palimpsest of urban politics, but this variation must not create greater social inequality. The need to experiment and the ability to experiment are not distributed evenly across cities; neither do they necessarily coincide.

Conclusions

Urban living labs constitute bounded spaces that enable intentional collaborative experimentation by researchers, citizens, companies and local governments. In urban governance terms they substitute top-down planning for local innovation processes staged in spaces such as buildings, neighbourhoods or districts. They are designed to facilitate real-time experimentation with social and/or technical alternatives in cities in order to test new solutions to urban problems and learn from them. In this sense, they open up spaces for new groups to engage in urban development, empowering non-state actors including companies, NGOs and communities. In practice though, participation tends to be relatively narrow, and ULLs often reflect the existing status quo of dominant actors and priorities.

The proliferation of ULLs reflects a broader shift towards a more experimental approach to urban development, whereby cities must put partnerships together and attract competitive funding to deliver projects. Their value to cities lies in the ability of ULLs to provide a political and geographical focus around which to bring partners and funding together. As a result, the ULL concept itself has become a highly sticky mode of urban governance, rapidly circulating around cities in Europe. Here, the medium is the message – form trumps content as cities strive to create the right conditions for innovation through the creation of ULLs. The design-led thinking that inspired the original living lab concept runs up against the inevitably political context in which cities are planned and developed. Current research shows that ULLs struggle to realize their potential to convene new groupings of stakeholders and novel approaches to cities, often falling back on familiar ways of doing things. This presents both a challenge and a major opportunity. In an era in which it is widely recognized that we need to reinvigorate urban politics in order to reshape our cities and societies in new and more sustainable ways, ULLs offer a crucible in which new, more collaborative and material, forms of urban politics can be invented.

References

Baccarne, B., Mechant, P., Schuurman, D., Colpaert, P., & De Marez, L. (2014). Urban Socio-Technical Innovations with and by Citizens. *Interdisciplinary Studies Journal*, 3(4), pp. 143–156.

Bulkeley, H. and Castán Broto, V. (2012). Government by Experiment? Global Cities and the Governing of Climate Change. *Transactions of the Institute of British Geographers*, 37, pp. 1–15.

Bulkeley, H., Castán Broto, V., Hodson, M. and Marvin, S. (2013). *Cities and Low-Carbon Transitions*. London: Routledge.

Bulkeley, H., Castán Broto, V. and Maassen, A. (2014). Low-carbon Transitions and the Reconfiguration of Urban Infrastructure. *Urban Studies*, 51, pp. 1471–1486.

Bulkeley, H., Castán Broto, V. and Edwards, G. (2015). *An Urban Politics of Climate Change: Experimentation and the Governing of Socio-Technical Transitions*. London: Routledge.

Clarke, N. (2012). Urban Policy Mobility, Anti-politics, and Histories of the Transnational Municipal Movement. *Progress in Human Geography*, 36(1), pp. 25–43.

Cooke, P., Davies, C. and Wilson, R. (2002). Innovation Advantages of Cities: From Knowledge to Equity in Five Basic Steps. *European Planning Studies*, 10(2), pp. 233–250.

Curtis, S. (2015). *Innovation and the Triple Bottom Line: Investigating Funding Mechanisms and Social Equity Issues of Living Labs for Sustainability*. MSc Thesis, Lund University. Available at: http://lup.lub.lu.se/luur/download?func=downloadFile&recordOId=8055265&fileOId=8055267 [accessed 7 July 2016].

de Jong, M., Joss, S., Schraven, D., Zhan, C. and Weijnen, M. (2015). Sustainable-Smart-Resilient-Low Carbon-Eco-Knowledge Cities: Making Sense of a Multitude of Concepts Promoting Sustainable Urbanization. *Journal of Cleaner Production*, 109, pp. 25–38.

Directorate General for Internal Policies. (2010). *The Lisbon Strategy 2000–2010: An Analysis and Evaluation of the Methods Used and Results Achieved*. Available at: www.europarl.europa.eu/document/activities/cont/201107/20110718ATT24270/20110718ATT24270EN.pdf [accessed 28 June 2016].

ENoLL. (2015a). The European Network of Living Labs – the First Step Towards a New Innovation System! *European Network of Living Labs*. Available at: www.openlivinglabs.eu/ [accessed 28 June 2016].

ENoLL. (2015b). ENoLL: About Us. What Is a Living Lab? *European Network of Living Labs*. Available at: www.openlivinglabs.eu/aboutus [accessed 28 June 2016].

European Commission. (2014). *Living Well, Within the Limits of Our Planet. 7th EAP – The New General Union Environment Action Programme to 2020*. Available at: http://ec.europa.eu/environment/pubs/pdf/factsheets/7eap/en.pdf [accessed 28 June 2016].

European Commission. (2015). Open and Participative Innovation. *European Commission: Digital Agenda for Europe. A Europe 2020 Initiative*. Available at: ec.europa.eu//digital-agenda/en/open-and-participative-innovation [accessed 28 June 2016].

Evans, J. (2011). Resilience, Ecology and Adaptation in the Experimental City. *Transactions of the Institute of British Geographers*, 36(2), pp. 223–237.

Evans, J. and Karvonen, A. (2011). Living Laboratories for Sustainability: Exploring the Politics and Epistemology of Urban Adaptation. In: H. Bulkeley, V. Castán Broto, M. Hodson and S. Marvin, eds. *Cities and Low Carbon Transitions*. London: Routledge, pp. 126–141.

Evans, J. and Karvonen, A. (2014). "Give Me a Laboratory and I Will Lower Your Carbon Footprint!" – Urban Laboratories and the Governance of Low-carbon Futures. *International Journal of Urban and Regional Research*, 38(2), pp. 413–430.

Evans, J., Karvonen, A. and Raven, R., eds. (2016). *The Experimental City*. London: Routledge.

Finland's EU Presidency. (2006, 20 November). *The Helsinki Manifesto*. Available at: http://elivinglab.org/files/Helsinki_Manifesto_201106.pdf [accessed 28 June 2016].

Graham, S. and McFarlane, C. (2015). *Infrastructural Lives: Urban Infrastructure in Context*. Abingdon: Routledge.

Hodson, M. and Marvin, S. (2009). Cities Mediating Technological Transitions: Understanding Visions, Intermediation and Consequences. *Technology Analysis & Strategic Management*, 21(4), pp. 515–534.

Juujärvi, S. and Pesso, K. (2013). Actor Roles in an Urban Living Lab: What Can We Learn from Suurpelto, Finland? *Technology Innovation Management Review*, 3(11), pp. 22–27.

Karvonen, A. and van Heur, B. (2014). Urban Laboratories: Experiments in Reworking Cities. *International Journal of Urban and Regional Research*, 38, pp. 379–392.

Karvonen, A., Evans, J. and van Heur, B. (2014). The Politics of Urban Experiments: Radical Change or Business as Usual? In: S. Marvin and M. Hodson, eds. *After Sustainable Cities*. London: Routledge, pp. 105–114.

Kommonen, K-H. and Botero, A. (2013). Are the Users Driving, and How Open Is Open? Experiences from Living Lab and User Driven Innovation Projects. *The Journal of Community Informatics*, 9(3). Available at: http://ci-journal.net/index.php/ciej/article/view/746/1026 [accessed 31 May 2017].

König, A. (2013). *Regenerative Sustainable Development of Universities and Cities: The Role of Living Laboratories*. London: Edward Elgar.

Mensink, W., Birrer, F.A. and Dutilleul, B. (2010). Unpacking European Living Labs: Analysing Innovation's Social Dimensions. *Central European Journal of Public Policy*, 1, pp. 60–85.

Monstadt, J. (2009). Conceptualizing the Political Ecology of Urban Infrastructures: Insights from Technology and Urban Studies. *Environment and Planning A*, 41, pp. 1924–1942.

Montgomery, C. (2013). *Happy City: Transforming Our Lives Through Urban Design*. London: Penguin.

Peck, J. (2011). Geographies of Policy: From Transfer-Diffusion to Mobility-Mutation. *Progress in Human Geography*, 35(6), pp. 773–797.

Peck, J. and Theodore, N. (2010). Mobilizing Policy: Models, Methods, and Mutations. *Geoforum*, 41(2), pp. 169–174.

Reimer, M., McCormick, K., Nilsson, E. and Arsenault, N. (2012). Advancing Sustainable Urban Transformation Through Living Labs: Looking to the Öresund Region. *International Conference on Sustainability Transitions*, pp. 29–31. Available at: http://climate-adapt.eea.europa.eu/metadata/publications/advancing-sustainable-urban-transformation-through-living-labs-looking-to-the-oresund-region [accessed 28 June 2016].

Rösch, C. and Kaltschmitt, M. (1999). Energy from Biomass: Do Non-technical Barriers Prevent an Increased Use? *Biomass and Bioenergy*, 16(5), pp. 347–356.

Schaffers, H., Komninos, N., Pallot, M., Trousse, B., Nilsson, M. and Oliveira, A. (2011). Smart Cities and the Future Internet: Towards Cooperation Frameworks for Open Innovation. *The Future Internet Assembly, Vol. 656*. Berlin/Heidelberg: Springer, pp. 431–446.

Schliwa, G. (2013). Exploring Living Labs Through Transition Management – Challenges and Opportunities for Sustainable Urban Transitions. *IIIEE Master thesis*. Available at: www.lunduniversity.lu.se/lup/publication/4091934 [accessed 28 June 2016].

Schliwa, G. (forthcoming) Smart cities by design? Interrogating design thinking for citizen participation' In: Kitchen, R. *The Right to the Smart City*, Bingley: Emerald.

Silver, J. and Marvin, S. (2016). The Urban Laboratory and Emerging Sites of Urban Experimentation. In: J. Evans, A. Karvonen and R. Raven, eds. *The Experimental City*. London: Routledge, pp. 47–60.

Ståhlbröst, A. (2008). Forming Future IT – The Living Lab Way of User Involvement. Luleå, Sweden: Luleå University of Technology. Available at: http://epubl.ltu.se/1402-1544/2008/62/LTU-DT-0862-SE.pdf [accessed 28 June 2016].

Ståhlbröst, A. and Holst, M. (2012). *The Living Lab Methodology Handbook*. Luleå, Sweden: Luleå University of Technology. Available at: www.ltu.se/cms_fs/1.101555!/file/LivingLabsMethodologyBook_web.pdf [accessed 28 June 2016].

Trencher, G., Bai, X., Evans, J., McCormick, K. and Yarime, M. (2014). University Partnerships for Co-designing and Co-producing Urban Sustainability: A Global Survey. *Global Environmental Change*, 28, pp. 153–165.

Veeckman, C., Schuurman, D., Leminen, S. and Westerlund, M. (2013). Linking Living Lab Characteristics and Their Outcomes: Towards a Conceptual Framework. *Technology Innovation Management Review*, 3(2), pp. 6–15.

Voytenko, Y., McCormick, K., Evans, J. and Schliwa, G. (2015). Urban Living Labs for Sustainability and Low Carbon Cities in Europe: Towards a Research Agenda. *Journal of Cleaner Production*, 123, pp. 45–54.

35
ASSEMBLING AND RE-ASSEMBLING ASIAN CARP

The Chicago Area Waterways System as a space of urban politics

Julie Cidell

Introduction

The assemblage of the Chicago Area Waterways System (CAWS), including the collection of invasive species known as Asian carp, offers a way to understand how the varied properties and capacities of water have shaped and continue to shape the City of Chicago and the region around it. In the process, the CAWS demonstrates how the city and region might be investigated as spaces of urban politics regarding what should be allowed to flow through the city and what should be blocked. The CAWS includes not just the Chicago River, whose flow has been reversed, but also a system of canals that unites the Great Lakes and Mississippi River watersheds. The water within the CAWS plays an important role in two different ways: as excess in times of heavy rainfall that needs to be carried away to prevent its unwanted presence in streets and basements; and as a means of circulating other substances, perhaps most importantly the sewage of the City of Chicago.

When considering the CAWS from the vantage point of spaces of urban politics it is important to reflect upon how the political ecology of urban water concerns not only its *consumption*, but also its role as a *medium* for other kinds of circulation: wastes, freight, capital and biophysical life. For example, economic interests rely on the CAWS for cheap transportation, including steel mills and chemical factories, claiming that the ability to carry barges and their cargo across the subcontinental divide is vital to their continued existence in the region. Nevertheless, closing or blocking off parts of this system, and re-assembling it to allow the passage of some elements but not others, is argued by politicians outside of Illinois to be the only permanent solution to keeping invasive carp out of the Great Lakes. The century-long struggle between Illinois and the rest of the Great Lakes states and provinces over the diversion of water from Lake Michigan further fuels this confrontation. Disentangling these multiple purposes which the waters of the CAWS serve is therefore a complex task, albeit an urgent one in policy terms. It also demonstrates different ways of approaching and conceptualizing the spaces of urban politics, namely considering the discursive and material elements of circulation in understanding what is and is not allowed to flow through the city.

The concept of assemblage has been employed by geographers and other urban researchers in a variety of fields from architecture (Edensor 2011) to urban policy (McCann 2011).

Developed out of work by Deleuze, Guattari and DeLanda, among others, assemblage is related to complex systems in that it explores heterogeneity and complexity, where the whole is something other than the sum of its parts (Braun 2006). Entities have different *properties* that stay constant despite the context (for example, trees take in carbon dioxide and give off oxygen), but also have different *capacities* to affect other entities depending on their context (for example, tree roots may break apart a sidewalk or hold together an eroding cliff) (DeLanda 2006). Assemblage theory therefore has the potential to help us understand the complex workings of urban environmental systems made up of human and non-human elements, including points of policy intervention and spaces of urban politics.

Asian carp themselves have been 'assembled' as a threat to regional human and non-human ecologies due to their voracious appetites that threaten to outcompete existing species, as well as their surprising habit of leaping from the water when startled, potentially endangering boaters and fishers (Colón 2014). Multiple species, each with their own properties and capacities, entered the Mississippi River watershed between ten and twenty years ago and have been making their way steadily northward ever since. Evidence of fish presence within or very near to the Great Lakes has been given through environmental DNA (eDNA) rather than the observation of fish bodies. The fluid property of water and its capacity to carry both eDNA, such as fish scales, and actual bodies of fish make it difficult to answer the vital question, 'Where are Asian carp?'

A proposed 'ecological separation' plan would dis-assemble the CAWS into two separate watersheds. Beyond the projected ecological benefits, this plan creates opportunities for new cycles of investment via reinvesting in obsolete infrastructure and riverfront neighbourhoods within the central city. At the same time, separation of the watersheds would reduce or eliminate restrictions on drawing drinking water from Lake Michigan, allowing suburbs to grow without limit. The reassembly of the CAWS may therefore have as much to do with continuing to enable the circulation of capital as it does with keeping fish from circulating through the region's waterways, demonstrating how ecological spaces are also spaces of urban politics.

This chapter begins with a brief introduction to the concept of the assemblage as used in geography and related fields, followed by a review of recent geographical work on the properties and capacities of water and fish. This is followed by a history of the Chicago Area Waterway System and Asian carp in the Midwest as assemblages of urban infrastructure and biological life. I consider not only how the carp-river-lake-sewage assemblage has been put together and maintained, but the work that is being proposed to break it apart and resistance to that proposal on the part of both humans and non-humans. I conclude with some thoughts on the properties and capacities of water as both a means and an object of circulation and how thinking of the CAWS as an assemblage changes our thinking about spaces of urban political intervention.

Geographical assemblages

The concept of the assemblage has recently been taken up by geographers with an underlying concern for how objects or concepts that we take for granted as stable or structural are in fact temporary fixities of heterogeneous elements. For example, "the electrical grid is better understood as a volatile mix of coal, sweat, electromagnetic fields, computer programs, electron streams, profit motives, heat, lifestyles, nuclear fuel, plastic, fantasies of mastery, static, legislation, water, economic theory, wire, and wood" (Bennett 2010: 25). One motivation for the development of the assemblage has been to go beyond actor-network theory to incorporate power, history and materiality into explanations of socio-environmental processes. As will be seen below, history and materiality matter very much to the current situation of the CAWS, along with the rapidly changing nature of the assemblage itself. Another motivation for the theory has been to identify potential sites of intervention, which is highly relevant to spaces of urban politics.

Broadly speaking, there are three ways in which the concept of assemblage has been put to use in urban geography and related fields. The first is as a means of description, employed by authors such as Ong and Collier (2004) or Sassen (2007) to explain how different elements come together into formations like 'globalization'. The second is as an ontology of how disparate elements relate to each other (McFarlane 2011b), or a theoretical concept that tries to explain where agency comes from in order to effect change (McFarlane 2011a; Vandergeest et al. 2015). This is the way in which assemblage is being deployed in this chapter. Finally, assemblage can also be a way of conceiving of the world, as a conceptualization or ethos of engagement (Anderson and McFarlane 2011; Ranganathan 2015) that produces new political opportunities and limits (Bennett 2010); however, this latter conceptualization is beyond the scope of the current chapter.

Assemblage emphasizes not only the characteristics of individual elements but their interactions (Bear 2013; McFarlane 2011a, 2011b). As we will see below, Asian carp are posed as a threat to the Great Lakes ecosystem because of their interactions with other species, from algae to humans, more than their behaviour in and of itself. Importantly, the assemblage as a whole may also have agency, as with Bennett's example of the electrical grid above: normally, it provides power to consumers (and profits to its owners, and pollution to downwind residents), but during the 2003 blackout in the eastern United States (USA) and Canada, it affected individual lives and socio-economic systems in a variety of unexpected ways. Recognizing this agency on the part of the assemblage necessarily complicates its politics, by emphasizing that non-humans, too, play a role in the spaces of urban politics.

While the formation and maintenance of assemblages has often been a subject of critical interest, there has been relatively little work on attempts to deliberately pry them apart or to remove certain elements. In fact, assemblage theory would tell us this is impossible, because relationships among elements of the assemblage remain even if the elements themselves are gone. To use Edensor's (2011) example of St Ann's Church in Manchester, the stone of the church remains weakened in places because of past exposure to soot, even if the improved air quality of Manchester means that soot is no longer in the atmosphere. Nevertheless, actors frequently seek to remove problematic elements of an assemblage in an effort to reduce those elements' effects. In the assemblage of the CAWS, two of the most problematic elements are water and fish, specifically because of their interactions with other elements, as discussed in the next section.

Circulating water, circulating fish

In recent times, human geographers have contributed to a significant body of research on urban water systems (e.g. Ekers and Loftus 2008; Gandy 2002, 2006; Kaïka 2005; Meehan 2014; Swyngedouw et al. 2002; Waley 2000) as well as wastewater, sewerage and runoff (Dimpfl and Moran 2014; Jones and Macdonald 2007; Karvonen 2011; McFarlane 2008; Ranganathan 2015; Young and Keil 2005). Much of this has focused on the provision of urban drinking water, including issues of water quality, privatization and unequal access (Bakker 2005; Jauhiainen and Pikner 2009; Loftus and McDonald 2001; Ranganathan and Balazs 2015; Swyngedouw 2004). Urban rivers, in particular, have been studied because of their position at the culture/nature interface, especially in places where they have been heavily engineered (Revill 2007). The

Thames, for example, was the site of conflict between the landed interests of millers and the mobile interests of barge captains, each of whom sought a different use value from the river and struggled to 'improve' the river accordingly (Oliver 2010, 2013). In a different context, the Los Angeles River was engineered into the confines of concrete channels to control periodic but devastating floods and keep the riverbanks available for development (Gumprecht 1999). More recently, however, the possibility exists to return to a naturalized river channel for the same purpose of flood control, relying on restored wetlands to absorb runoff (Gandy 2006).

Rivers and water are also known to their human users through the (sometimes delayed) effects they produce in other entities, like fish and flooded houses (Bull 2011; Waley and Åberg 2011; Walker et al. 2011; Woelfle-Erskine 2015). Though not necessarily framed in terms of assemblages, much of this more recent work nevertheless has a similar emphasis on the relations among heterogeneous entities and the unpredictable nature of the interactions between them.

However, this work has not generally considered water as a carrier of other things – except fish. Such work demonstrates how the bodies of fish themselves become sites of tension or conflict over what is 'natural' or 'human', either regarding the wild or farmed nature of a species (Bull 2011) or how well a 'foreign' fish has assimilated into the native waterscape (Colón 2014; Lavau 2011; Warren 2007; see also Barker 2010; Eskridge and Alderman 2010; and Head and Muir 2004 on invasive plant species). Interactions among humans, water and fish shape all three, sometimes in surprising ways (Bear 2013; Woelfle-Erskine 2015), which in turn could stimulate new ways of thinking about spaces of urban politics as we consider what species or objects are allowed to flow and which ones are not.

In other words, the study of water and urban waterways should not only be about the *commodification* of water or what happens when it is out of place, but its role as a *medium* for other kinds of circulation – wastes, freight and capital, as well as biophysical life – and as an actor or agent of change in its own right. Ever since Chicago was founded at a low point between the Mississippi River and Great Lakes watersheds, its existence has depended upon the flow of water and capital through its territory (Cooke and Lewis 2010; Cronon 1991). Understanding how this assemblage of water, industry and fish has come together is key to understanding how the threat of Asian carp has been constructed and how re-assembling the CAWS might reduce that threat. In the following discussion, I follow water and fish through the complex interconnections between them and the other entities within and around the CAWS, drawing out some implications for the analysis of spaces of urban politics.

The Chicago Area Waterways System

The possibility of linking the Great Lakes and Mississippi River watersheds has been floated ever since Louis Joliet and Father Père Marquette explored the Great Lakes region in the mid-1600s. Joliet noticed that it was only a six-mile portage across the evocatively named Mud Lake, and a canal could easily connect the two. In fact, in times of high water, the Des Plaines River, a tributary of the Illinois and thus also the Mississippi, overflowed into Mud Lake and hence also the Chicago River and Lake Michigan. By 1848, the Illinois and Michigan Canal was finished across the subcontinental divide, securing Chicago's future as an entrepôt between the East Coast and the Heartland (Cronon 1991) and spurring the first of many debates over the appropriate role of the federal government in approving and funding modifications to the region's waterways (Hill 2000).

As Chicago grew, its waste was deposited into the river and thus into the lake. Even offshore water intake pipes became contaminated during periods of high rain. The bodily wastes of thousands of Chicagoans began to affect the population through small outbreaks of cholera and typhoid leading to fears of larger epidemics, and through the stench of the river/sewer (Hill 2000). The water of the river and lake was not playing its desired role of carrying sewage outside the range of human consumption of drinking water.

While the use of rivers as sanitary infrastructure was hardly unique to Chicago (Keeling 2005; Platt 2004), the degree to which engineers remade the river by changing its direction of flow certainly was. The Sanitary and Ship Canal was dug partially in parallel and partially replacing the Illinois and Michigan Canal, totalling more excavated material than the Panama Canal (Hill 2000). A series of pumps and locks was installed so that once the canal was cut into the Des Plaines River, water could be persuaded to flow from Lake Michigan into the river and down through the canal into the Des Plaines, to the Illinois and then the Mississippi (Solzman 2006). Of course, the population downstream of Chicago was not enamoured of becoming drawn into this sewerage assemblage, and so they fought the plan in court. The State of Missouri planned on filing an injunction with the Supreme Court against the State of Illinois as construction was drawing to a close. On 16 January 1900, in the middle of the night before the suit was to be filed, the Board of Trustees opened the gates at Lockport, and the reversal began. Up until 2015, Chicago was the only major city in the USA that did not chemically disinfect its sewage. Instead, it was run through a series of mechanical and microbial filters before being pumped into the CAWS and further diluted (Hill 2000). Since the Environmental Protection Agency's declaration that Chicago was no longer exempt from the Clean Water Act, the State of Illinois has contributed funds for a chemical disinfection system (Hawthorne 2011).

As the name of the Sanitary and Ship Canal suggests, two kind of flows had to be balanced: fast-flowing water to wash away Chicago's sewage, and a navigable waterway whose current was not too fast for ships. Water was being asked to fulfill two potentially contradictory roles at once, leading to conflicts between the Sanitary District and the Federal Government over how much water could be diverted for sanitary purposes based on the speed of the resulting current, itself controlled by the locks and pumps near the lake. Later, the other Great Lakes states sued Chicago over the amount of water being diverted for drinking purposes (Cain 1974). A final agreement was reached in 1967, named the Chicago Diversion for the water that is diverted from the Great Lakes to the Mississippi River watershed because of the CAWS.

The Chicago Diversion has three separate components (see Figure 35.1 and Espey et al. 2004), all of which increase in volume as the population of the Chicago metropolitan area grows:

- direct flow through daily lock operations and larger releases to improve water quality by flushing wastewater downstream;
- stormwater runoff that passes through the built environment into the canals and river but would otherwise reach the lake; and
- drinking water from the lake that passes through human bodies and is filtered into the canals and river.

While water makes it possible for ships and sewage to travel between watersheds, it also enables non-humans to make the same journey. Scores of non-indigenous species have made their way from one watershed to the other, the zebra mussel being one of the most well-known. The most recent threat to the integrity of the Great Lakes ecosystem, due to its generalist property that gives it the capacity to outcompete many existing species, is the Asian carp. The assemblage of the CAWS therefore becomes an urban space that needs to be reconfigured to allow the flow of water and *some* of the things it carries – but not others.

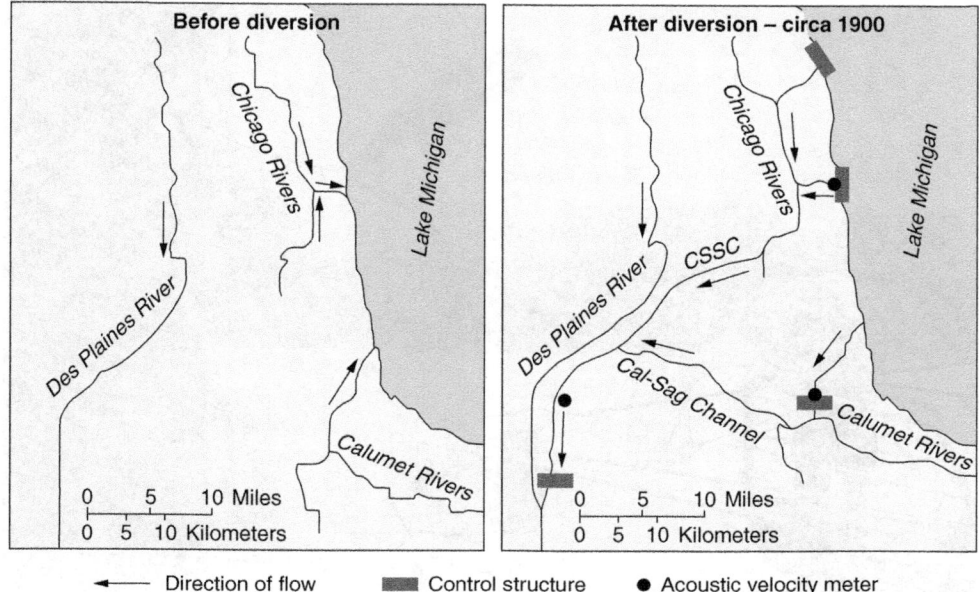

Figure 35.1 The Chicago Diversion as a result of reversing the flow of the Chicago River

Source: in the public domain as a U.S. Geological Survey document, available at https://pubs.usgs.gov/fs/FS-014-99/index.html.

Foreign fish

The 'Asian carp' is itself an assemblage, composed of two to five species of carp, including the bighead (*Hypophthalmichthys nobilis*) and silver (*Hypothalmichthys molitrix*) but sometimes also the grass (*Ctenopharyngodon idella*), black (*Mylopharyngodon piceus*) or common (*Cyprinus carpio*) carp. These species were imported from China in the 1970s to filter out plankton from aquaculture and retention ponds in the South. Sometime in the 1980s or 1990s, they are thought to have escaped via flooding into the Mississippi River (though see Colón 2014). They can grow up to a hundred pounds (45 kg) and can eat 40 per cent of their body weight in a day, and they have no natural predators on this continent.

The threat that Asian carp are said to pose is threefold. First is direct competition with native or existing species for food. As generalists, their capacity to interfere with already existing species changes with the availability of food (Sampson et al. 2009). Second is the possibility they could significantly reduce the amount of plankton in the Great Lakes, thus disrupting the entire food web. Since these carp prefer warm, fast-flowing water for spawning, which is the opposite of the waters of Lake Michigan, even the passage of individual fish into the Great Lakes does not mean that a population would become established (Cooke and Hill 2010). Nevertheless, river mouths emptying into the Great Lakes are at risk should Asian carp make their way *en masse* into that watershed (Cooke et al. 2009).

The third threat is to human lives and livelihoods. Many news stories on the Asian carp include an image of silver carp leaping out of the water when startled (see, for example, Matheny 2016 and Figure 35.2); videos posted on a variety of websites show the jumping fish in action. Boaters have been injured by 40 or 50-pound fish (roughly 20 kg) hitting them. Furthermore, fisheries may be in danger if native species are outcompeted; in fact, there are stretches of the

Figure 35.2 Asian carp leaping from the Wabash River in Ohio
Source: Steve Hillebrand/U.S. Fish and Wildlife Service Headquarters.

Illinois River where the majority of the biomass is already Asian carp. While there has been some small success with fishing commercially as part of the so-called 'invasivore' movement, the fine bone structure of the fish makes them more difficult to prepare than typical North American fish, and the potential for contamination from mercury and other heavy metals exists as well (Gorman 2010; Rogowski *et al.* 2009). It is therefore not only the behaviour of the fish, but such material elements as their bone structure and the heavy metals they unknowingly transport, that makes their interactions with humans fraught with difficulty and difficult to disassemble.

Despite the urgency of the threat, there is significant controversy over where exactly Asian carp are located. Single bighead carp have been found in Lake Erie in years past, and a single bighead was found in Lake Calumet in 2010 on the eastern side of the locks. In order to slow the spread of non-native species, electric barriers were previously installed by the US Army Corps of Engineers (USACE) at various points in the CAWS, and they seem to be affecting Asian carp as well. When the Cal-Sag Channel between Lake Michigan and the Sanitary and Ship Canal was dosed with rotenone (fish poison) in December 2009 while the electric barrier was off for maintenance, a single carp was found among 90 tons of dead fish. Another fish kill in May 2010 along the Little Calumet River, this time explicitly for the purpose of searching for the Asian carp, found no carp among 11,000 dead fish. In other words, so far the aquatic environment is safe – aside from the rotenone.

What *has* been found in Lake Michigan is eDNA, or environmental DNA. As forensic science uses human traces such as hair or skin to identify if an individual has been in a certain location, eDNA tests water for traces of a specific species that shed its cells through mucus, faeces or urine. There have been multiple positive readings for silver and bighead carp eDNA in the region, including the Calumet Harbor on Lake Michigan (Jerde *et al.* 2011). However, this DNA might merely have travelled in the water through the locks or via barge water after being expelled from a fish's body; it does not mean there are actual fish in the harbour (see Darling and Mahon 2011). In other words, although fish bodies have not been observed in the Great Lakes, Asian carp have already been placed in that watershed through eDNA.

Re-assembling the Chicago Area Waterways System

The waterway system that has carried away sewage from Chicago for over a century is now positioned at the cusp of a threat to the ecological integrity of the international Great Lakes ecosystem. This threat and the possible solutions are not new: former Mayor Daley convened a regional summit on Asian carp as early as 2004. Participants concluded that re-installing the subcontinental divide, re-reversing the Chicago River and closing the multiple sets of locks were advised. However, the costs of installing a fully modernized sewage treatment system in the place of the Sanitary and Ship Canal were deemed too high, not to mention the lost passageways for barges and their bulk cargo. The canals, microbes, locks and the water and its contents have become too closely intertwined to separate without spending billions of dollars, and maybe not even then. Thus even if the solutions are not new, treating CAWS as an assemblage requires a rethinking of the spaces for political intervention.

For example, the Natural Resources Defense Council (NRDC), the Alliance for the Great Lakes and the Great Lakes Fishery Commission are strong proponents of a re-reversal, or 'ecological separation', which "prohibits the movement or inter-basin transfer of aquatic organisms between the Mississippi and Great Lakes basins via the [CAWS]" (Brammeier et al. 2008: ii). Proponents argue that native species on both sides of the watershed divide have suffered invasions from the other. In other words, too much circulation is as bad as not enough, and the USACE should be ordered to close the locks.

However, assemblages – by their nature – refuse to break up completely. The locks are not watertight, so water and the fish it carries may be able to get through even if the locks were closed to ship traffic. The locks need to be occasionally opened to allow water to flow towards the lake to prevent it from flooding backyards and basements. Additionally, there are over twenty points throughout the region where the USACE has estimated that flooding could lead to a transfer of species between watersheds. The chemical, steel and food processing industries of south-east Chicago and north-east Indiana argue that the cheap transportation provided by the CAWS is the only reason they can remain in business. Furthermore, on a particularly ironic note, the water quality in some the canals of the CAWS is so low – 70 per cent effluent – that water has to be occasionally flushed in from Lake Michigan to increase the oxygen level to meet Federal Environmental Protection Agency (FEPA) standards (NRDC 2010). In fact, an alternative solution to the carp threat is to deliberately de-oxygenate a stretch of canal for a long enough distance that no plant or animal life could journey through it, a designated sacrifice zone in a region rife with unofficial ones that would remove the capacity of the water to support biological life.

Because of all of these forms of resistance to breaking apart the CAWS assemblage, human and non-human alike, the argument has come to be framed by the NRDC and others as *ecological* separation rather than complete physical or hydrological separation. Stopping one type of flow – water and the fish it carries – while maintaining other flows such as wastewater, floodwaters and barge traffic, therefore becomes the ambitious goal. In other words, the aim is to rework the assemblage of the CAWS to allow some flows to continue, such as sewage, barges and excessive or out of place water, while blocking other flows such as fish bodies. Ecological separation recognizes the complex interconnections in the CAWS assemblage and the difficulty or impossibility of breaking them, instead proposing to rebuild the assemblage as a different kind of system. It also partially recognizes the agency of non-humans, both animal and matter, in that water and fish are acknowledged as having the power to affect different interactions.

Most proposed political solutions, however, have focused not on a range of options, but on the hard barrier of hydrological separation. In January 2010, Michigan's Attorney General Mike Cox – who was running for the Republican nomination for governor – filed suit in the US Supreme Court to force the USACE to halt an existing study of the watersheds and close the locks.

At the same time, the Close All Routes and Prevent Asian Carp Today Act, or CARP ACT Act, was introduced in the US House and Senate with co-sponsors from Minnesota, Wisconsin, New York and Ohio, with a similar goal. The Supreme Court turned down Cox's case, and the bill never went beyond an initial reading. Then in June 2010, a single live carp was found in Lake Calumet, east of the locks. This time, legislation tucked into the federal transportation appropriations bill was successfully co-sponsored by Michigan and Illinois senators, directing the Secretary of the Army and the USACE to study physical separation of the watersheds in a sped-up timeframe. The USACE determined that it would not be possible under this shorter deadline to recommend the best option, to conduct a full environmental impact analysis of any one option, or to consider any possible non-CAWS pathways for carp to enter the Great Lakes (USACE 2014).

The sense of urgency in Congress's requirements is somewhat puzzling. Asian carp have not only been in the Illinois River since the early 1990s, but also they have been caught commercially since 1994. The properties of the fish have not changed in the meantime, and their capacities to effect change in the Great Lakes remain largely speculative. Furthermore, there are dozens of other, arguably more egregious, threats to the aquatic ecosystem that do not receive the same kind of attention, from industrial pollution to climate change. Carp are also making their way up the Mississippi and Ohio Rivers, and while state fish and wildlife officials are certainly aware of the threat, there is no federal legislation being proposed to construct barriers or close spillways across these waterways. Chicago's former Mayor Daley pointed out that dozens of other species have entered the Great Lakes and the Mississippi watershed through the St Laurence Seaway, and yet there is no regional outcry to close that important shipping channel for ecological purposes.

After ridiculing the idea of re-reversing the Chicago River, the former Mayor Daley suddenly came out in favour of exploring the idea in the fall of 2010 (Dardick 2010). Perhaps after his failure to attract the Olympics to Chicago and his subsequent announcement not to seek re-election, Daley was searching for a lasting legacy. But there is also the issue that the amount of water that can be diverted from Lake Michigan into the Mississippi River watershed is capped by Supreme Court decree in the Chicago Diversion. As discussed above, there are three pathways by which water that is 'naturally' destined for the Great Lakes watershed flows down the Mississippi instead. As growing suburbs deplete their own aquifers and seek to tap into Lake Michigan water, there is a very real threat of running up against the federally mandated limit of how much water can be transferred over the subcontinental divide. The Chicago Metropolitan Agency for Planning found that with current usage rates and projected population growth, water withdrawals from Lake Michigan would reach the maximum amounts allowed by 2050 (CMAP 2010). Reduced precipitation as a result of climate change will further exacerbate the situation (Changnon 1994).

Re-reversing the river, on the other hand, would eliminate the Chicago Diversion and therefore the cap on water use. Stormwater and drinking water that becomes filtered wastewater would return to Lake Michigan instead of flowing away via the Des Plaines and Illinois Rivers, thus allowing growing suburbs to take water from the lake almost without limit. At the same time, the FEPA's recent decision to require cleanup of the Chicago River through the heart of the city to permit swimming could open up miles of riverfront or canalfront within the city for redevelopment. Considering the broader population of the Chicago region as part of the carp-river-lake-sewage assemblage, not only in terms of contributing to the sewage but in taking water from the lake, reminds us that capital always has a role to play in environmental issues (Baldwin 2009). The sudden interest in reconfiguring the CAWS may have as much to do with keeping the region safe for flows of capital circulation via suburbanization and urban redevelopment as it does with keeping the Great Lakes' ecology intact.

Conclusion

The glaciated landscape of north-eastern Illinois blurs the boundaries between watersheds, making the flow of water between them relatively easy. Into this flat landscape, the canals, locks and pumps of the CAWS have persuaded water to flow in a specific direction and to dilute and carry with it the human wastes of the region. Once that relationship between water and sewage was established, other entities such as barges and bulk goods entered the assemblage, with fixed infrastructure such as factories and ports being placed as well. Entering the watershed from a more distant location, new fish species have changed the ecology of parts of the region and could potentially have even broader impacts, were they to cross the watershed divide and thrive in the waters of the Great Lakes.

In describing how the disparate elements of water, sewage, carp, barges and boaters have been drawn together across time and space and maintained in certain configurations, this chapter has argued for considering water as not only an object of circulation itself, whether for drinking purposes or as rainwater runoff, but as a medium of circulation for other substances. Determining what should or should not be allowed to circulate, and where, is vital to the spaces of urban politics.

The concept of distributed agency is relevant here as well, as it helps us identify spaces for political intervention. Are Arkansas or Mississippi aquaculturists responsible for importing Asian carp in the first place? Is the City of Chicago responsible for breaching the subcontinental divide at the turn of the twentieth century? Are the carp responsible for seeking out new food sources and relatively warm, flowing water, forcing other species out of the way? Are new techniques of eDNA analysis responsible for placing fish in the Great Lakes even if fish bodies have not been observed there? Is water responsible for carrying invasive species in addition to sewage but also failing to stay where it should? The answer is 'yes', because all of these elements have played a role in producing this particular socio-environmental assemblage, even if uneven power relations determine whose role is larger and longer-lasting and where the ultimate blame should be placed. In this case, water might well be the element with the most power overall, since its property of flowing downhill has been shown to eventually win out over watershed boundaries and flood control systems alike. Phrased another way, the lively materiality of water will likely override attempts by humans to control its agency.

While Bennett (2010) argues that we cannot fairly place blame because of the distributed agency of an assemblage, making any kind of material change to the water infrastructure of the Chicago region requires the placing of responsibility in order to determine who will bear the cost. Identifying Chicago as the most culpable jurisdiction puts the fiscal burden on that city and on the USACE as the maintainers of the locks, which is what elected officials in the rest of the Great Lakes states are trying to do. Untangling the various elements of the assemblage instead offers the opportunity for alternative entry points and alternative urban politics. For example, identifying growing suburbs and their demands for more drinking water and increased capacity to absorb stormwater runoff and sewage would force those jurisdictions to reduce consumption or contribute to the cost of a re-assembled waterway system that would chemically treat wastewater instead of relying on water's capacity to dilute and transport wastes. Identifying water itself with its dual role as object and medium of circulation would lead to a more nuanced understanding of urban environmental systems and, perhaps, a more balanced solution to retaining the positive properties and capacities of the CAWS while minimizing its troublesome features. In any case, the emergent properties of this assemblage mean that attempts to break it apart or re-assemble it are likely to be unsuccessful for political, biological, economic and hydrological reasons. Nevertheless, determining what can and cannot flow through Chicago will continue to be a key space of urban politics and one worthy of analysis in other cities and regions as well.

References

Anderson, B. and McFarlane, C. (2011). Assemblage and Geography. *Area*, 43(2), pp. 124–127.
Bakker, K. (2005). Neoliberalizing Nature? Market Environmentalism in Water Supply in England and Wales. *Annals of the Association of American Geographers*, 95(3), pp. 542–565.
Baldwin, A. (2009). Carbon Nullius and Racial Rule: Race, Nature and the Cultural Politics of Forest Carbon in Canada. *Antipode*, 41(2), pp. 231–255.
Barker, K. (2010). Biosecure Citizenship: Politicising Symbiotic Associations and the Construction of Biological Threat. *Transactions of the Institute of British Geographers*, 35(2), pp. 350–363.
Bear, C. (2013). Assembling the Sea: Materiality, Movement and Regulatory Practices in the Cardigan Bay Scallop Fishery. *Cultural Geographies*, 20(1), pp. 21–41.
Bennett, J. (2010). *Vibrant Matter: A Political Ecology of Things*. Durham, NC: Duke University Press.
Brammeier, J., Polls, I. and Mackey, S. (2008). *Preliminary Feasibility of Ecological Separation of the Mississippi River and the Great Lakes to Prevent the Transfer of Aquatic Invasive Species*. Executive Summary of report prepared for the Great Lakes Fishery Commission. Available at: www.glfc.org/research/reports/Brammeier_ecological%20separation.htm [accessed 31 May 2017].
Braun, B. (2006). Environmental Issues: Global Natures in the Space of Assemblage. *Progress in Human Geography*, 30(5), pp. 644–654.
Bull, J. (2011). Encountering Fish, Flows, and Waterscapes Through Angling. *Environment and Planning A*, 43, pp. 2267–2284.
Cain, L. (1974). Unfouling the Public's Nest: Chicago's Sanitary Diversion of Lake Michigan Water. *Technology and Culture*, 15(4), pp. 594–613.
Changnon, S., ed. (1994). *The Lake Michigan Diversion at Chicago and Urban Drought: Past, Present and Future Regional Impacts and Responses to Global Climate Change*. Final report, National Oceanic and Atmospheric Administration (NOAA) contract 50WCNR306047.
CMAP (Chicago Metropolitan Agency for Planning). (2010). *Water 2050: Northeastern Illinois Regional Water Supply/Demand Plan*. Available at: www.cmap.illinois.gov/regional-water-supply-planning [accessed 22 September 2016].
Colón, S. (2014). Occupying Nature: Fishing for Meaning in the Asian Carp. *Transforming Anthropology*, 22(1), pp. 24–30.
Cooke, S. and Hill, W. (2010). Can Filter-feeding Asian Carp Invade the Laurentian Great Lakes? A Bioenergetic Modeling Exercise. *Freshwater Biology*, 55(10), pp. 2138–2152.
Cooke, J. and Lewis, R. (2010). The Nature of Circulation: The Urban Political Ecology of Chicago's Michigan Avenue Bridge. *Urban Geography*, 31(3), pp. 348–368.
Cooke, S., Hill, W. and Meyer, K. (2009). Feeding at Different Plankton Densities Alters Invasive Bighead Carp (*Hypophthalmichtys Nobilis*) Growth and Zooplankton Species Composition. *Hydrobiologia*, 625(1), pp. 185–193.
Cronon, W. (1991). *Nature's Metropolis: Chicago and the Great West*. New York: W.W. Norton and Company.
Dardick, H. (2010). Daley Wants to Turn Around River's Flow. *Chicago Tribune*, 10 September. Available at: www.chicagotribune.com/news/elections/ct-met-daley-0911-20100910,0,5972863.story [accessed 9 September 2016].

Darling, J. and Mahon, A. (2011). From Molecules to Management: Adopting DNA-based Methods for Monitoring Biological Invasions in Aquatic Environments. *Environmental Research*, 11(7), pp. 978–988.

DeLanda, M. (2006). *New Philosophy of Society: Assemblage Theory and Social Complexity.* New York: Continuum.

Dimpfl, M. and Moran, S. (2014). Waste Matters: Compost, Domestic Practice, and the Transformation of Alternative Toilet Cultures around Skaneateles Lake, New York. *Environment and Planning D*, 32(4), pp. 721–738.

Edensor, T. (2011). Entangled Agencies, Material Networks and Repair in a Building Assemblage: The Mutable Stone of St Ann's Church, Manchester. *Transactions of the Institute of British Geographers*, 36, pp. 238–252.

Ekers, M. and Loftus, A. (2008). The Power of Water: Developing Dialogues Between Foucault and Gramsci. *Environment and Planning D*, 26(4), pp. 698–718.

Eskridge, A. and Alderman, D. (2010). Alien Invaders, Plant Thugs, and the Southern Curse: Framing Kudzu as Environmental Other Through Discourses of Fear. *Southeastern Geographer*, 50(1), pp. 110–129.

Espey, W., Jr, Melching, C. and Mades, D. (2004). *Lake Michigan Diversion – Findings of the Fifth Technical Committee for Review of Diversion Flow Measurements and Accounting Procedures.* Report prepared for the US Army Corps of Engineers Chicago District, Chicago.

Gandy, M. (2002). *Concrete and Clay: Reworking Nature in New York.* Cambridge, MA: MIT Press.

Gandy, M. (2006). Riparian Anomie: Reflections on the Los Angeles River. *Landscape Research*, 31(2), pp. 135–145.

Gorman, J. (2010). A Diet for an Invaded Planet: Invasive Species. *New York Times*, 31 December. Available at: www.nytimes.com/2011/01/02/weekinreview/02gorman.html [accessed 9 September 2016].

Gumprecht, B. (1999). *The Los Angeles River: Its Life, Death, and Possible Rebirth.* Baltimore: Johns Hopkins University Press.

Hawthorne, M. (2011). U.S. Demands Chicago River Cleanup: Stretches Must Be Safe Enough for Swimming, EPA Orders. *Chicago Tribune*, 12 May. Available at: www.chicagotribune.com/news/local/ct-met-chicago-river-swimming-20110512,0,7855236.story [accessed 9 September 2016].

Head, L. and Muir, P. (2004). Nativeness, Invasiveness, and Nation in Australian Plants. *The Geographical Review*, 94(2), pp. 199–217.

Hill, L. (2000). *The Chicago River: A Natural and Unnatural History.* Chicago: Lake Claremont Press.

Jauhiainen, J. and Pikner, T. (2009). Narva-Ivangorod: Integrating and Disintegrating Transboundary Water Networks and Infrastructure. *Journal of Baltic Studies*, 40(3), pp. 415–436.

Jerde, C., Mahon, A., Chadderton, W. and Lodge, D. (2011). 'Sight-unseen' Detection of Rare Aquatic Species Using Environmental DNA. *Conservation Letters*, 4(2), pp. 150–157.

Jones, P. and Macdonald, N. (2007). Making Space for Unruly Water: Sustainable Drainage Systems and the Disciplining of Surface Runoff. *Geoforum*, 38(3), pp. 534–544.

Kaïka, M. (2005). *City of Flows: Modernity, Nature, and the City.* New York: Routledge.

Karvonen, A. (2011). *Politics of Urban Runoff: Nature, Technology, and the Sustainable City.* Cambridge, MA: MIT Press.

Keeling, A. (2005). Urban Waste Sinks as a Natural Resource: The Case of the Fraser River. *Urban History Review*, 34(1), pp. 58–70.

Lavau, S. (2011). The Nature/s of Belonging: Performing an Authentic Australian River. *Ethnos*, 76(1), pp. 41–64.

Loftus, A. and McDonald, D. (2001). Of Liquid Dreams: A Political Ecology of Water Privatization in Buenos Aires. *Environment and Urbanization*, 13(2), pp. 179–199.

Matheny, K. (2016). Invasive Asian Carp Less than 50 Miles from Lake Michigan. *Detroit Free Press*, 22 December. Available at: www.freep.com/story/news/local/michigan/2016/12/22/asian-carp-great-lakes-michigan/93970746/ [accessed 31 May 2017].

McCann, E. (2011). Veritable Inventions: Cities, Policies and Assemblage. *Area*, 43(2), pp. 143–147.

McFarlane, C. (2008). Governing the Contaminated City: Infrastructure and Sanitation in Colonial and Post-colonial Bombay. *International Journal of Urban and Regional Research*, 32(2), pp. 415–425.

McFarlane, C. (2011a). Assemblage and Critical Urbanism. *City*, 15(2), pp. 204–224.

McFarlane, C. (2011b). The City as Assemblage: Dwelling and Urban Space. *Environment and Planning D*, 29, pp. 649–671.

Meehan, K. (2014). Tool-power: Water Infrastructure as Wellsprings of State Power. *Geoforum*, 57, pp. 215–224.

NRDC (Natural Resources Defense Council). (2010). *Rebuilding Chicago's Stormwater and Wastewater Systems for the 21st Century: Understanding Hydrologic Conditions in the Region.* Technical Report. Available at: http://docs.nrdc.org/water/files/wat_10102001a.pdf [accessed 9 September 2016].

Oliver, S. (2010). Navigability and the Improvement of the River Thames, 1605–1815. *The Geographical Journal*, 176(2), pp. 164–177.

Oliver, S. (2013). Liquid Materialities in the Landscape of the Thames: Mills and Weirs from the Eighth Century to the Nineteenth Century. *Area*, 45(2), pp. 223–229.

Ong, A. and Collier, S. (2004). *Global Assemblages: Technology, Politics, and Ethics as Anthropological Problems.* Malden, MA: Blackwell Publishing.

Platt, H. (2004). 'Clever Microbes': Bacteriology and Sanitary Technology in Manchester and Chicago During the Progressive Age. *Osiris*, 2nd series, 19, pp. 149–166.

Ranganathan, M. (2015). Storm Drains as Assemblages: The Political Ecology of Flood Risk in Postcolonial Bangalore. *Antipode*, 47(5), pp. 1300–1320.

Ranganathan, M. and Balazs, C. (2015). Water Marginalization at the Urban Fringe: Environmental Justice and Urban Political Ecology Across the North-South Divide. *Urban Geography*, 36(3), pp. 403–423.

Revill, G. (2007). William Jessop and the River Trent: Mobility, Engineering and the Landscape of Eighteenth-century 'Improvement'. *Transactions of the Institute of British Geographers*, 32(2), pp. 201–216.

Rogowski, D., Soucek, D., Levengood, J., Johnson, S., Chick, J., Dettmers, J., Pegg, M. and Epifanio, J. (2009). Contaminant Concentrations in Asian Carps, Invasive Species in the Mississippi and Illinois Rivers. *Environmental Monitoring and Assessment*, 157(1–4), pp. 211–222.

Sampson, S., Chick, J. and Pegg, M. (2009). Diet Overlap Among Two Asian Carp and Three Native Fishes in Backwater Lakes on the Illinois and Mississippi Rivers. *Biological Invasions*, 11(3), pp. 438–496.

Sassen, S. (2007). *Territory, Authority, Rights: From Medieval to Global Assemblages.* Princeton, NJ: Princeton University Press.

Solzman, D. (2006). *The Chicago River: An Illustrated Guide to the River and its Waterways.* 2nd ed., Chicago: University of Chicago Press.

Swyngedouw, E. (2004). *Social Power and the Urbanization of Water: Flows of Power.* New York: Oxford University Press.

Swyngedouw, E., Kaika, M. and Castro, E. (2002). Urban Water: A Political Ecology Perspective. *Built Environment*, 28(2), pp. 124–137.

USACE (United States Army Corps of Engineers). (2014). *The GLMRIS Report: Great Lakes and Mississippi River Interbasin Study.* Chicago: USACE.

Vandergeest, P., Ponte, S. and Bush, S. (2015). Assembling Sustainable Territories: Space, Subjects, Objects, and Expertise in Seafood Certification. *Environment and Planning A*, 47, pp. 1907–1925.

Waley, P. (2000). What's a River Without Fish? Symbol, Space and Ecosystem in the Waterways of Japan. In: C. Philo and C. Wilbert, eds. *Animal Spaces, Beastly Places: New Geographies of Human-Animal Relations.* London: Routledge, pp. 159–181.

Waley, P. and Åberg, E. (2011). Finding Space for Flowing Water in Japan's Densely Populated Landscapes. *Environment and Planning A*, 43, pp. 2321–2336.

Walker, G., Whittle, R., Medd, W. and Walker, M. (2011). Assembling the Flood: Producing Spaces of Bad Water in the City of Hull. *Environment and Planning A*, 43, pp. 2304–2320.

Warren, C. (2007). Perspectives on the 'Alien' Versus 'Native' Species Debate: A Critique of Concepts, Language and Practice. *Progress in Human Geography*, 31(4), pp. 427–446.

Woelfle-Erskine, C. (2015). Thinking with Salmon About Rain Tanks: Commons as Intra-actions. *Local Environment*, 20(5), pp. 581–599.

Young, D. and Keil, R. (2005). Urinetown or Morainetown? Debates on the Reregulation of the Urban Water Regime in Toronto. *Capitalism, Nature, Socialism*, 16(2), pp. 61–84.

36
GOOGLE BUSES AND UBER CARS
The politics of tech mobility and the future of urban liveability

Jason Henderson

Introduction

Since the mid-2000s large technology companies based in the San Francisco Bay Area have superimposed a new 'tech mobility' over the existing transportation palimpsest of the metropolis. Fusing smart phone apps and nanotechnology with conventional buses and cars, companies such as Google and Uber have introduced distinctive configurations and flows of urban travel which to varying degrees are spreading, or have potential to spread, globally. Outfitted with wireless service, ample seating and legroom, laptop plug-ins, and accessed via employer-provided digital IDs and smart phones, privately run, third party-contracted 'tech' buses ply the freeways between San Francisco and Silicon Valley, transporting technology workers to huge suburban corporate campuses such as Apple (Cupertino), Facebook (Menlo Park) and Google (Mountain View). New pre-arranged, on-demand car services like Uber and Lyft use global positioning systems for navigation and tracking, while smart phones connect drivers to potential customers who have downloaded an 'app' that allows 'e-hailing', cashless payment and driver ratings. After disembarking from their tech buses, these car services ferry youthful 'creatives' between apartments, condos and trendy nightlife venues. Meanwhile Google, Uber, Lyft and an array of Silicon Valley tech companies have made allegiance with traditional automobile manufacturers, and through nanotech and software, are poised to transform future automobility with fleets of self-driving cars.

With the rapid global diffusion of Uber and similar transportation network companies (TNCs), as well as global fascination with self-driving cars, tech mobility is likely going to be making greater claims on urban spaces around the world. It is therefore useful to consider tech mobility with greater critical scrutiny and especially its implications for the urban politics of mobility, i.e. political debates over what urban transportation modes are prioritized, and who decides which modes are important (Henderson 2013). In this chapter, after describing some of the key characteristics of tech mobility, I draw on a politics of mobility framework to consider how tech mobility is an irreducibly political project promoting progressive visions of urban liveability such as reduced car ownership, whilst also aligning decidedly with a neoliberal politics of privatization and deregulation.

Tech mobility's liveability emphasis imagines urban futures comprised of car parking stalls converted to housing, streets retrofitted to bikeways and parks, and renewable energy and electrification making future urban transport carbon-free. Yet at the same time tech mobility also envisions transportation shaped by pricing and markets rather than by regulation and collective action, consistent with the broader agenda of the privatization of space and market-based pricing of public access to space. This 'neoliberal liveability' is a profit-oriented, de-regulated organization of transport that may result in an abandoned public realm, intensify precariousness in labour and austere governance, all the while installing a highly automated, optimally surveilled urban transport system in erstwhile liveable but exclusive urban cores. Tech mobility may have far-reaching consequences in shaping urban geographies of the future – not by transforming cities, but, rather, by intensifying and solidifying contemporary inequities and uneven geographies.

Tech mobility

On the surface, tech mobility takes conventional vehicles like buses and cars, which it retrofits with new 'smart' technologies like Global Positioning Systems (GPS) and wireless mobile phones, and then fuses these technologies into new ways of organizing labour and passengers alike. For example, 'Google buses' – the slightly pejorative moniker for the Bay Area's elite private bus system – are in fact traditional bulky, over-the-road coaches harkening back to a Greyhound or Casino tour bus. The innovation, however, is in the organization of the Google-bus system and access to it. The buses are not 'common carriers' whereby anyone can access them, making it a very different arrangement from conventional public transit. Instead, the system operates as an elite mobile space, heavily monitored and deploying smart technologies for matching long-distance commuters with routes as part of traffic reduction mitigations for sprawling Silicon Valley corporate campuses such as Apple, Facebook and Google.

TNCs, on the other hand, involve conventional cars owned by ordinary people but acting as taxis. Through wireless technology, geolocation devices, payment apps and rating apps, smart phones are the key transportation innovation; both drivers and passengers must have one in their possession. Rides are mostly local and shorter in length (one to five miles) and are pre-arranged, sometimes almost instantly, and this differs from traditional taxis, which a customer can hail on the street. A key to TNCs is reliance on non-professional drivers with casual, flexible schedules rather than full-time taxi employees. Traditional social interaction is also eliminated with TNCs – passengers and drivers no longer discuss or handle the fares, which are paid on credit cards using the app. Considered incredibly convenient by many tech workers, business travellers and young urban professionals around the world, TNCs are also rife with labour conflict born out of the broader trend of an insecure, part-time, underemployed, precariat (Harvey 2012).

Self-driving cars promise to end this labour strife for TNCs by ending the need for human operators, a welcome outcome for the most ardent tech mobility enthusiast. Instead of humans, cars will be driven by microscopic sensors, cameras, remote sensing technology, three-dimensional map navigation tools, and other nanotechnologies. A commercially viable fully automated self-driving car is probably over a decade away, but there are many degrees on a continuum of automation that are quickly unfolding, including 'connected cars', collision avoidance, automated parking, and cars that can self-drive but have an override function so that humans can take over operation (NHSTA 2013).

While private tech buses are limited to the Bay Area and other high-tech urban centres like Seattle, political pushback in San Francisco reveals tensions over tech mobility's political project of neoliberal liveability. This has implications for future politics of mobility in many cities around the world where public transit is inadequate and private interests might try to circumvent political gridlock. TNCs have expanded rapidly globally, from San Francisco, New York and Paris, to Sao Paulo and Mexico City, Beijing and Delhi, but there has also been considerable political pushback with similar implications for future urban liveability. And while there is debate about when self-driving cars will proliferate (some suggest the mid-2020s, others suggest longer – see NHTSA 2013 and McKinsey & Company 2016), the increasing amount of attention and resources dedicated to self-driving cars, and the distinctive emphasis on urban liveability, warrants consideration for how they will impact the politics of mobility. Of particular confusion in the promotion of self-driving cars are the decades of 'in-between' when human-operated cars are interspersed with self-driving cars.

Tech mobility and urban liveability

Responding to critics and political opponents, and in defence of the new configurations and flows tech mobility produces, boosters of tech mobility have invoked politically progressive discourses about urban liveability and made liveability part of their political project. Notably, tech CEOs, government officials and planners, and various business and non-profit organizations allied with tech mobility claim that private buses, TNCs and self-driving cars will reduce car dependency. Here tech mobility is politically aligned with urban-based liveability movements such as proponents of mass transit and cycling, along with the sustainable streets media (such as *Streetsblog* and *CityLab*, in the USA), which share a desire to see car dependency ended, and to reconfigure urban space into denser, transit-oriented, walkable built forms – development patterns in the USA associated with smart growth or the new urbanism.

For example, in the Bay Area large suburban tech companies, business groups and many (but not all) local elected officials and transportation managers insist that thousands of San Francisco residents who work in suburban tech have intentionally chosen a car-free lifestyle and that the buses make it possible for them to achieve this and also work in Silicon Valley, which is 35–45 miles south of the city. They point to one survey of tech bus riders which suggest that almost half of the private bus clientele would likely drive alone if the service were not available (SFMTA 2015a). The same survey of over 500 tech bus riders also showed that 45 per cent of those surveyed did not own a car. At contentious public hearings over the legality of the buses and their impact on the city, tech workers repeat that they would buy a car and drive if it were not for the tech buses.

Uber and Lyft also claim TNCs reduce the need for car ownership in cities, and that rather than compete, their services complement walking, cycling and transit. Many sustainable-streets advocates initially expressed enthusiasm about TNCs because of the potential for car-free lifestyle and less driving. The CEO of Lyft sums up the liveability discourse arguing that the TNCs' vision is to use smart phone apps to fill empty seats in cars, reducing traffic and greenhouse gas (GHG) emissions, and predicting car ownership will plummet (Bradley 2015). The academic Transportation Research Board (TRB) is especially, albeit speculatively, enthusiastic that TNCs will reduce car ownership, free up urban land now tied up in parking provision, and bring dramatic and transformative reductions in vehicle miles travelled (VMT) (TRB 2015: 19, 43) TRB envisions a positive feedback loop as TNCs and smart phones enable 'on the fly coordination' for fleet-based, fixed route ridesharing and first and last mile service to transit stations (Thompson 2016; TRB 2015).

Self-driving cars round-out tech mobility's urban liveability discourse figuratively and, to the more boosterish, in totality. The rationale begins with safety, and the expectation that self-driving cars will obey all traffic laws – thus ending speeding, reckless driving, road rage and drunk driving. The safety impact makes it possible to produce smaller and more efficient cars, with a concomitant narrowing of streets and freeway lanes, less urban space consumed for parking, and more space for pedestrians, cyclists and denser urban infill. There will be highway capacity surplus, thus opening opportunities to remove urban freeways that take up valuable land. The loudest boosters claim that self-driving cars will end automobile ownership altogether as people will just hail a car when needed. Less utopian but still significant, one academic forecast suggests that, in the USA at least, household car ownership could be reduced by 43 per cent as cars are shared rather than each household member having an individual car (Schoettle and Sivak 2015).

To be sure, most of this enthusiasm is speculative and not without contrary evidence suggesting that the benefits to urban liveability are overstated. Critics of tech buses in San Francisco argue that the reduced driving and car ownership arguments lack clear supporting evidence and present a very narrow lens into regional mobility. They especially argue that many tech workers would forgo living so far from work and instead move closer to their jobs if they had to drive (the average tech bus commute is 47 miles each way). Hinting at this, a UC Berkeley working paper (also based on surveying tech bus riders) concluded that up to 40 per cent of tech workers would move closer to work if the buses disappeared, and many more would switch to public transit (Dai and Weinzimmer 2015). A notable quote from that survey states that "I chose to live in San Francisco because of my employer-provided commute shuttles. I would otherwise have lived in [the South Bay], because I don't have a car and who the hell wants to drive that much anyway" (Dai and Weinzimmer 2015: 25).

Moreover, many advocates in San Francisco's 'right to the city' movement, such as tenants' rights, social service providers and racial justice organizations, have argued that regional VMT (and GHGs) is likely increasing because the tech buses contribute to gentrification and displacement, pushing lower- and middle-income households into car-dependent suburbs and away from transit-optimal urban cores.[1] They posit that a thorough environmental study (which was the focus of litigation in 2015 and 2016) would reveal a more negative impact on regional liveability and equity than what tech bus defenders present. Yet the evidence and data is muddled, in part because no environmental study has yet to be conducted, and meanwhile the tech companies that contract out the commuter tech shuttles have been resistant to share data on their employees, vehicle occupancy, routing or travel times, although the city of San Francisco has introduced a programme that involves data collection – but only voluntary participation (SFMTA 2015a).

TNCs are also resistant to sharing their vast records of travel behaviour, citing privacy and competition, and so their claims about car ownership impacts of TNCs are also questionable. TRB points out that while car ownership could potentially be reduced, TNCs may also be leading to more congestion, more emissions and escalating mobility – more energy-intensive induced travel (TRB 2015). 'Dead mileage' and cruising may be more widespread than understood, especially if, in expensive housing markets like San Francisco, the TNC drivers reside in outlying, but more affordable districts. Moreover, there is evidence that rather than reducing car use, TNCs may actually be increasing car-based travel as people who might have previously taken a local bus, walked or biked now might use TNCs (Rayle et al. 2016). Coupled with Uber's aggressive emphasis on luxury sport utility vehicles (SUVs) and drawing business travellers away from taxis might mean net VMT and emissions are higher with TNCs. In San Francisco this is especially relevant as the city required the taxi industry to convert to hybrid cars, while the unregulated TNCs encourage less-fuel-efficient luxury vehicles.

The liveability arguments for self-driving cars are both loud and utopian, but also speculative because commercial deployment is years away. The central question is: will self–driving cars simply extend the traditional model of household car dependency, albeit with more efficient household car use? Or, will self-driving cars end automobile ownership altogether, with households hailing cars for things like 'last-mile' trips but otherwise organizing daily life around transit, walking and cycling?

One future scenario suggests that self-driving cars may reduce household car ownership in the USA but increase VMT at the same time. Households that currently have two or more cars might shift to a single shared car, but that shared car would actually drive more as it shuttles (often empty) back and forth chauffeuring the household members throughout the day (Schoettle and Sivak 2015). In that scenario the single vehicle is driven 75 per cent more than previously, and average US household VMT remains exceptionally high compared to VMT in the USA today. Additionally, children, the disabled and elderly would more easily drive, likely increasing net VMT and demand for cars.

Self-driving cars may also generate a 'rebound effect' whereby people find that using an automated car is so easy they don't mind longer trips or congestion because they can be entertained in the vehicle. To that end, an automotive industry white paper suggests future rates of car ownership might stabilize in some wealthier urban cores with high levels of urban liveability, but that throughout the world car ownership will continue to rise 2 per cent annually into the 2050s (McKinsey & Company 2016). Globally, one out of three cars might be autonomous but in 2050 105 million cars would be sold annually (higher than today), propelled by mass motorization in developing countries coupled with higher replacement rates for efficiently utilized self-driven cars.

The material and energy resources needed to prop up this global-scale expansion of automobility, even if electrified and supplied with renewable energy, might undermine GHG reduction policies and counteract urban liveability. Worse, the intense focus on a future of driverless cars means there could be less political support for urban liveability as people who might have previously seen merit in transit or bicycling now hopefully await a driverless car future. In a contemporary political climate of public sector austerity and fiscal discipline, and corollary inadequate levels of public investment in transportation infrastructure, coupled with an already deeply skewed political bias in favour of roads over transit, self-driving cars could sap what public funding exists for transit and other liveable infrastructure. Given these pitfalls, and turning to tech mobility's neoliberal political project, instead of a utopian future, tech mobility might simply accentuate already unsustainable and inequitable urban conditions.

Neoliberal liveability

Sceptics of tech mobility's future impact on car ownership and VMT might conclude that tech mobility has co-opted or greenwashed liveability, but it is more complicated than that. Tech mobility needs the armature of liveability because liveability is the site of the knowledge economy, especially for young, highly educated tech workers or what commentators such as Richard Florida (2005) call 'creatives'. Dense, walkable urban spaces provide the close human interaction that the knowledge economy needs (Newman and Kenworthy 2015).

This post-industrial 'shop floor' includes physical infrastructure improvements such as pedestrian plazas and bicycle lanes, which are amenities but also the necessary foundations for a tech-ecosystem and production spaces (Stehlin 2015). Tech hubs like San Francisco embrace liveability not just because it reduces driving and GHGs, but because liveable urban configurations are "enabling workers to participate more fully in a spatial milieu that drives innovation, and in particular have access to the younger, more dynamic firms" (Stehlin 2015: 451).

Developing this liveability-knowledge nexus further, Newman and Kenworthy (2015) posit that spatial efficiencies of walking, cycling and high-quality rail transit enable concentration of the digital, knowledge and service economies, and argue that "walkability is enabling the growth of new knowledge-related jobs" (2015: 58) and is why young tech workers flock to older walkable urban forms like San Francisco's Mission District or Haight Ashbury.

Yet both dwelling and working in liveable urban configuration is not always possible, especially when the largest tech corporations such as Apple, Google and Facebook seek vast amounts of land for their campuses, and that kind of land is hard to assemble and develop due to regulations and higher land costs in older urban cores. In this case the 'shop floors' on the tech campuses are internally walkable, bikeable, and contain all manner of green urbanism, but erstwhile are located in gulfs of unattractive, lower density, car-oriented sprawl.

The internet search engine and software giant Google epitomizes this geography of tech mobility. The 'Googleplex', Google's headquarters, is 35 miles South of San Francisco in Mountain View, and has upwards of 20,000 employees in a vast verdant campus setting next to the 101 Freeway. Internally the campus is outfitted with colourful Google bikes, walking paths and other accoutrements of the tech shop floor, but thousands of Google's young creatives reside in San Francisco, in neighbourhoods like the Mission, Noe Valley and the Haight. To bridge the gulf between bucolic and insular production sites and exciting, dense, liveable urbanity, San Francisco tech companies have implemented a de facto regional private bus system. Yet beyond urban form and urban design, what kinds of urban political spaces might be revealed through tech mobility?

Over the long term, an erstwhile pro-transit constituency among tech workers and wealthy professionals might favour environmental policies and urban liveability, yet also, through tech buses and TNCs, might be seceding from the public and in due time may embrace policies that starve public transit as it resists taxes and fees that fund public transit. In this vein, tech mobility's leading promoters in the industry, such as the CEO of Uber and executives at Google, often invoke a libertarian, self-reliance discourse whereby the consumers of tech mobility – tech bus commuters, Uber riders and dreamers of self-driving cars – are giving up on collective political solutions to urban and labour issues and instead rely increasingly on privatized networks; private transit and private 'shared' cars like Uber Pool or Lyft Line. This libertarian self-reliance path invokes environmental arguments along the way, but with no personal sacrifice and less willingness towards empowering government (Hill 2015: 219). In sum, tech mobility gives up on the public good and relies on privatized networks that accentuate uneven urban geographies.

The uneven geography of neoliberal tech mobility intensifies as its impacts on urban and regional planning begin to manifest themselves. In the case of San Francisco's tech buses, public planning is discarded in favour of a privatized solution. Rather than engage in public solutions, such as adequately funding public transit, and aggressively supporting land use policies that enable housing built proximate to tech campus, tech mobility circumvents difficult but democratic local and regional government stalemates such as that found in San Francisco's regional land use and transportation planning process. While the parallel commuter rail, which also links San Francisco to Silicon Valley, languishes in crowding, delay and political opposition that stifles improvement, private buses sidestep the public stalemate. Tech workers might not only shirk responsibility, but also they may not share the burden or feel the impacts of crumbling public systems.

Tech buses also indirectly accommodate conservative land use politics in the Bay Area's wealthier low-density suburbs. In enclaves like Cupertino (Apple), Menlo Park (Facebook) and Mountain View (Google), politically conservative homeowners and local governments object to proposals for higher-density housing and other liveable mixed use developments. These jurisdictions welcome the tax revenue generated by commercial development, but avoid paying for the social services (schools, health care) needed for residents in new housing. In San Francisco's Silicon Valley, transit-connected, walkable communities have been made all but illegal by decades of local initiatives and interventions often led by conservative middle and upper class, anti-density, anti-tax homeowners in suburban localities. In fact, paralleling the Bay Area's tech boom beginning around 2011, a vitriolic Tea Party-style conservative politics dampened *Plan Bay Area*, a progressive-leaning regional plan meant to encourage suburban densification in order to reduce car dependency and GHGs. This resulted in a weak regional housing plan with an underfunded and lacklustre transit vision (Frick 2013; Henderson 2013).

Ironically, these same low-density suburbs mandate transportation demand management schemes, or TDM, which requires that the massive tech firms mitigate the traffic impacts of their campuses. Tech campuses are required to provide private buses as part of the conditions of approval and as part of caps on car trips. The City of Mountain View, home to Google, requires no more than 45 per cent of commute trips in single-occupant vehicles, and Menlo Park can impose penalties and fees on Facebook if it exceeds a locally mandated vehicle trip cap. Furthermore, the tech bus systems enable the tech campus to reduce parking provision, providing lower development costs and maximum utilization of land.

Trip caps, development agreements and local, suburban environmental mitigations meant that in 2015 the City of San Francisco expected a 41 per cent increase in the growth of private tech buses on its city streets (SFMTA 2015a: 18). The private bus system is expected to expand as all of the Bay Area's public transit agencies strain under declining revenues, deferred maintenance and deep federal and state shortfalls in funding. This foreshadows a potential transit future in which a premium system serves the wealthy in first-class coaches – and in premium liveable neighbourhoods with on-demand TNC service – and a dilapidated, economy-class system serves the lower classes that are gentrified out of the core.

Tech mobility and labour

The uneven geography and planning impacts of neoliberal tech mobility also fit well with the broad ambit of the neoliberal project against organized labour. For example, by outsourcing the tech buses to third-party contractors, tech firms are insulated from direct labour negotiations and capital investment, while circumventing hard-fought labour rules and wages found in the public sector transit systems. Of particular relevance is the split shift, or paying bus operators during midday during off-peak commute hours. All public transit systems have the conundrum of idled workers during off-peak operations, but standby is necessary and inevitable in order to have professional, well-trained and healthy operators. Decades of labour struggle and negation in the last century resolved the conundrum in favour of organized workers, but this does mean transit can be expansive to operate.

Tech mobility's neoliberal vision means privatized transit constitutes a resolution to the muddled labour union work rules like those found in public transit. For the tech bus system, drivers are flexible, work part time and are largely unorganized although the Teamsters Union has made a small dent. In late 2015, after lobbying by Teamsters and other labour groups, the San Francisco Board of Supervisors (similar to a city council) mandated 'labour harmony' between the bus drivers and the private bus firms, but the language of the local law is ambiguous and the city cannot legally ban a private bus company from existing if it violates the policy.

The labour question permeates into the TNC system in similar ways to the private bus system. As TNCs utilize new strategies to access labour through tech innovation like crowdsourcing, tech mobility also bridges the broader effort of firms to optimize expenses by paying workers only when they perform a task. Similar to the tech bus driver's split shift conundrum, TNC driver labour value is only in exact minutes of work, logged onto a computer or smart phone, and there is no compensation during bathroom breaks or for rest, which is a core aspect of twentieth-century labour organizing (Hill 2015). This means that a TNC driver is only working when carrying a passenger, not while looking for a passenger. Drivers then have to work longer hours, take fewer breaks and could compromise safety as this can contribute to reckless behaviour and fatigue.

For TNC drivers, the conundrum is that they must be available on demand, but they are only demanded sporadically. This hyper-market efficiency leads to less humane work conditions and yet is exacerbated by a 'speeding-up' of workers as TNCs firms like Uber cut rates paid to drivers and then insist that drivers can make up the slack by working longer hours, or working more quickly. Moreover, the lack of regulation and professionalism required by TNC firms means that there is a low bar of entry to take on the job. This and the promise of easy money attracts lower-income workers who, in precarious labour markets, have little other choice.

TNC firms like Uber and Lyft counter that they are simply attracting everyday people who might have spare time or who are underemployed. In this vein tech mobility is cast as a new 'savior of the middle class' because TNCs arguably boost incomes in middle-class households that have yet to fully recover from the 2008 global financial crisis (Said 2015). The extra income for dislocated workers is said come with highly desired flexibility, letting drivers work on their own terms, being 'their own boss' according to Uber. As Uber positions itself as fighting for all underemployed American workers, it invokes hip new micro-entrepreneur discourses like 'micro-business' and a 'do-it-yourself (DIY)' economy of liberated (and libertarian) workers.

Yet this arrangement requires a depressed, uneven labour market to thrive. It requires precariousness, uncertainty, lack of security and scarcely waged or unwaged labour that is also the bedrock of the new tech piecemeal task work based online such as Task Rabbit (Hill 2015). In the USA, at least until labour litigation is concluded, TNC drivers are '1099' employees, a category used by the Internal Revenue Service to describe independent contractors and small business owners. An army of underemployed, low-paid, desperate workers, many among the most vulnerable, and who are marginally attached and discouraged, shifting from job to job, or displaced by globalization and automation, end up flocking to TNCs to serve elite urban core tech workers and their fellow creatives.

Consider the mechanics of price surging by Uber. When demand for TNC services are high, such as after large sporting events, festivals, or in vibrant, liveable entertainment districts like the Mission in San Francisco, the logic of surge pricing is that high consumer demand is met by raising prices, thus attracting more TNC drivers to the event or geographic locality. While outfits such as TRB (2015) suggest surge pricing is a 'best practice' because it lures drivers to an area where there is high demand, the Uber model also requires too many drivers in order to swamp competition – a reserved driver pool. Focusing on the on-demand wait time for costumers, feeding the consumers' sense of entitlement and speed, also means high numbers of vehicles roaming streets "chasing the surge" (Hill 2015: 93). Meanwhile during non-surge periods, there can be a shortage of TNCs as drivers, as their own boss, wait for the opportune moment to make more money during the surge.

TNC's boast of middle-class wages also comes before acknowledging the costs of vehicle maintenance, car payments, insurance and the depreciation cost of the car, which accelerates with more use. Liability is pushed onto workers, including responsibility for crashes, sexual assaults, background checks and other insurance, while the TNC firm is exposed to no risk, even as there are muddy lines between when a TNC car is being operated commercially rather than as private auto.

The situation of TNC workers, like tech bus operators, has not gone uncontested. Labour organizations such as the Teamsters in the USA have helped organize TNC drivers, and drivers have themselves formed unions in cities like Los Angeles and Seattle. A national taxi association has organized drivers in Austin, Philadelphia and San Francisco, and an important class-action lawsuit filed in 2015 may shift the burden of liability and compensation to the TNC firms. A progressive-majority city council in Seattle has cleared the path for legal unionization in that city, and this is the first ordinance of this kind in the USA. And, at the time of writing, in California the US Federal Court will soon determine if a TNC driver is an employee or independent contractor. This decision is expected to have a key impact on the future of TNCs because it could mean that TNC firms must pay unemployment, workers' compensation, minimum wage and overtime, as well as reimbursement for car expenses, smart phone bills and tips that are currently paid to Uber as a percentage of fares.

Considering uneven geographies of gentrification and displacement in San Francisco and nearby Oakland, liveable urban cores for the wealthier stratum may become surrounded by 'car cities' (referring to Newman and Kenworthy 2015), where the tech bus and TNC drivers dwell because they cannot afford the urban core. When forecasting for self-driving cars is plotted out, the map is one of a collection of elite global urban cores where TNCs are automated, while the peripheries of the core have conventional cars organized around unsustainable sprawl (McKinsey & Company 2016). As a handful of elite global cities wire themselves for automated cars (and soak up billions of dollars of public investment as well as the earth's remaining fossil fuel and material resources) the vast majority of the planet (including in the USA) continue to choke on annual sales of conventional cars of around 105 million, up from 85 million in 2015 (McKinsey & Company 2016).

Conclusion: contesting the political project of tech mobility

There is no doubt that the products designed by the employees of large technology companies, from cancer-fighting drugs to search engines and social networking platforms, and extremely useful smart phone apps that make everyday life easier for working class and elite alike, are here to stay and have many benefits to society. In the field of transport, there is ample hope and enthusiasm that tech mobility, motivated by these kinds of innovations, may indeed make cities more liveable and greatly reduce GHG emissions worldwide. Yet peeling back the hip, exciting new sheen of the latest app, there is also a very poignant political project that may in fact magnify and deepen already tenuous urban inequities and instability.

Tech mobility's champions maintain a pronounced neoliberal ideology that, in the words of one critic, "ascribes magical powers to technology and the private sector" (Eskow 2015). A prominent national libertarian-conservative political operative in the USA, Grover Norquist, has expressed hope that Uber may help the Republican Party finally gain control of urban strongholds (Hill 2015). Tech firms such as Google, Uber and Facebook have hired professional political strategists to manage policy and communications teams, ostensibly in preparation for political backlash from labour unions and right-to-the-city advocates concerned about gentrification and displacement and the broader uneven geographies of tech mobility (Carr 2015).

Ultimately the goals of the neoliberal-strand of tech mobility, depriving government of taxes, reducing citizen participation and diminishing confidence in public solutions to urban problems such as traffic, pollution and transport finance, mean that the beneficiaries of tech mobility withdraw into privatized, wired, liveable but electronically gated urban cores or enclaves.

Moreover, all indications are that tech mobility will continue to be cast as liveable, but also, ironically, very much centred on automobility. In early 2016 General Motors invested

$500 million into Lyft, and Ford was rumoured to be seeking partnership with Google, signifying that tech firms and automobile manufacturers are positioning themselves to propel automobility with self-driving cars and new variations of TNCs. In considering urban futures, it is worthwhile to remember that automobility has historically been important to gentrification in US cities, especially for the first and second waves of gentrification in many eastern and southern cities. As white middle-class professionals re-urbanized into lower-income minority residential areas, fixed-up devalued homes and admired the charm and potential walkability of urban cores, they nonetheless continued to drive almost everywhere out of fear (real and perceived) of crime on the neighbourhood streets. This is a form of social and spatial secession through automobility that tech mobility will likely perpetuate. The cocoon of automobility often enables gentrification because the gentry is sealed off while moving through a part of the city perceived as unsafe – which happens to be their own gentrifying neighbourhood.

Political pushback against the neoliberal liveability–tech mobility nexus is emerging, as hinted in labour disputes over private tech buses and TNCs. Beyond labour organizing of drivers, right-to-the-city activists have mobilized in San Francisco to contest the very meaning of who the city is for. Between 2014 and early 2016 an array of opponents of tech buses organized protests and filed litigation questioning the legality of tech bus operations on San Francisco's city streets. Affordable housing and social justice advocates began protesting the buses in 2014, including some high-profile blockades and media spectacles that drew national and international attention to both the buses and San Francisco's affordable housing crisis. Between 2014 and 2016 housing, labour and environmental advocates routinely challenged the buses at public hearings in San Francisco and filed litigation against the city of San Francisco's effort to manage the private buses.

The City of San Francisco, in response to protests and in an effort to legitimize and manage the rapidly growing private system, initiated a pilot programme in 2015 that was made permanent in early 2016. This 'Commuter Shuttle Program Policy' directs the private buses to certain city bus stops, and certain city streets, while levying a 'stop-event fee' every time a private bus uses a public bus stop (SFMTA 2015a). However, participation in the Program is voluntary, and critics argue the fee rate is exceptionally low, while the entire Program is technically illegal because California law bans private uses of public bus stops. Moreover, litigants argued that the City failed to conduct a required environmental review of the Program, and thus many of the environmental arguments in favour of the buses (less congestion, reduced GHGs) are dubious.

Yet, the City states, on behalf of tech mobility enthusiasts: "there is no demonstrable evidence of physical displacement" nor "a demonstrative causative link between commuter shuttles and housing demand" (San Francisco Planning Department 2015: 24). Tech workers already demand the housing regardless of the shuttles. Without the tech buses, they'd drive to Silicon Valley. At a public hearing in (SFMTA 2015b), a dozen tech workers reiterated "we love the [liveable] community and we are part of it" … and that the buses "allow us to live in the city we love". Speakers said they loved walking, biking and loved Muni (Municipal Railway, San Francisco's public transit system), and that "without the shuttles we'd have to drive, or buy a car".

That might very well be true, but it begs the question, can there be the possibility for a progressive urban politics of tech mobility that fuses the benefits of technology and innovation (including direct service commuter buses) but without the neoliberal hue? Can tech mobility's liveability project be separated from the neoliberal political project and especially bring about labour peace? Can tech mobility be harnessed in ways that, instead of propping up the car system and metropolitan inequity, be targeted at redoubling the efforts of expanding bicycle and transit spaces, densifying suburbs and overlaying that with a social housing strategy that ensures an equitable urban future? These questions, and the divergent progressive versus neoliberal politics of tech mobility is arguably going to be one of the greatest urban political debates of the next few decades.

Note

1 Here I invoke Harvey's (2012) notion of the right to the city as a challenge to neoliberal urban policies, specifically that market forces and private property should not triumph over other human rights such as adequate housing, health and access to urban political power.

References

Bradley, R. (2015). Lyft's Search for a New Mode of Transport. *MIT Technology Review*, 118(6), pp. 49–53.

Carr, P.B. (2015). Noone is Telling You the Real Story Behind Uber's Latest Layoffs. *Pando*, 15 December. Available at: https://pando.com/2015/12/14/no-one-telling-you-real-story-behind-ubers-policycomm/ [accessed 10 July 2016].

Dai, D. and Weinzimmer, D. (2015). Riding First Class: Impacts of Silicon Valley Tech Shuttles on Regional Mobility, Commute Mode, and Residential Location Choice. Metropolitan Transportation Commission Planner's Breakfast, Oakland, CA, 5 March.

Eskow, R. (2015). The Sharing Economy Is a Lie: Uber, Ayn Rand and the Truth About Tech and Libertarians. *Salon*, 1 February. Available at: www.salon.com/2015/02/01/the_sharing_economy_is_a_lie_uber_ayn_rand_and_the_truth_about_tech_and_libertarians/ [accessed 10 July 2016].

Florida, R. (2005). *Cities and the Creative Class*. New York: Routledge.

Frick, K.T. (2013). The Actions of Discontent: Tea Party and Property Rights Activists Pushing Back Against Regional Planning. *Journal of the American Planning Association*, 79(3), pp. 190–200.

Harvey, D. (2012). *Rebel Cities: From the Right to the City to the Urban Revolution*. New York: Verso.

Henderson, J. (2013). *Street Fight: The Politics of Mobility in San Francisco*. Amherst: University of Massachusetts Press.

Hill, S. (2015). *Raw Deal: How the 'Uber Economy' and Runaway Capitalism Are Screwing the American Workers*. New York: St Martin's Press.

McKinsey & Company. (2016). *Automotive Revolution: Perspective Towards 2030*. Frankfurt: McKinsey & Company.

Newman, P. and Kenworthy, J. (2015). *The End of Automobile Dependence: How Cities Are Moving Beyond Car Based Planning*. Washington, DC: Island Press.

NHTSA (National Highway Traffic Safety Administration). (2013). *Preliminary Statement of Policy Concerning Automated Vehicles*. Washington, DC: NHSTA.

Rayle, L., Dai, D., Chan, N., Cervero, R. and Shaheen, S. (2016). Just a Better Taxi? A Survey-based Comparison of Taxis, Transit, and Ridesourcing Services in San Francisco. *Transport Policy*, 45(1), pp. 168–178.

Said, C. (2015). Airbnb, Uber: "We Are the Saviors of the Middle Class". *San Francisco Chronicle*, 10 November. Available at: www.sfchronicle.com/business/article/Airbnb-Uber-We-are-the-saviors-of-the-middle-6620729.php [accessed 10 July 2016].

SFMTA (San Francisco Municipal Transportation Agency). (2015a). *Commuter Shuttle Program Policy Staff Report to SFMTA Board of Directors*, 17 November. San Francisco: San Francisco Municipal Transportation Agency.

SFMTA (San Francisco Municipal Transportation Agency). (2015b). *Hearing on the Adoption of the Commuter Shuttle Program*, 17 November. Available at: http://sanfrancisco.granicus.com/MediaPlayer.php?view_id=55&clip_id=24143 [accessed 10 July 2016].

San Francisco Planning Department. (2015). SFMTA Commuter Shuttle Program: Certificate of Determination Exemption from Environmental Review. San Francisco Planning Department. Available at: www.sfmta.com/sites/default/files/projects/2015/Commuter%20Shuttle%20Program%20Certificate%20of%20Exemption%20from%20Environmental%20Review.pdf. [accessed 10 July 2016].

Schoettle, B. and Sivak, M. (2015). *Potential Impact of Self-Driving Vehicles on Household Vehicle Ownership and Usage*. Ann Arbor: University of Michigan Transportation Research Institute.

Stehlin, J. (2015). The Post-Industrial 'Shop Floor': Emerging Forms of Gentrification in San Francisco's Innovation Economy. *Antipode*, 48(2), pp. 474–493.

Thompson, C. (2016). No Parking Here. *Mother Jones*, January/February. Available at: www.motherjones.com/environment/2016/01/future-parking-self-driving-cars [accessed 10 July 2016].

TRB (Transportation Research Board). (2015). *TRB Special Report 319: Between Public and Private Mobility: Examining the Rise of Technology-Enabled Transportation Services.* Committee for Review of Innovative Urban Mobility Services; Transportation Research Board. Washington, DC: National Academies of Sciences, Engineering, and Medicine. Available at: www.trb.org/Main/Blurbs/173511.aspx [accessed 10 July 2016].

37
MAKING MULTI-RACIAL COUNTER-PUBLICS
Towards egalitarian spaces in urban politics

Helga Leitner and Samuel Nowak

Introduction

Struggles for social justice often begin and coalesce in cities. This chapter is concerned with the making of progressive multi-racial publics engaged in urban social justice struggles, and their attempts to create egalitarian political spaces to enact a potentially radical democratic politics. The making of multi-racial publics has multiple dimensions. First, it involves publicizing and making visible existing injustices and discrimination, and making demands on the state to rectify these. Second, it involves the creation and enactment of alternative forms of governance, and experimentation with new ways of organizing. Third, it requires a sustained commitment to learning from each other and negotiating across racial and other lines of difference, and creating spaces where such negotiations are possible and encouraged rather than foreclosed. Last, but not least, multiple spatialities, including macro- and micro-geographic contexts, the construction and configuration of physical spaces and extra-local connectivities and mobilities, shape and are shaped by the construction of multi-racial publics.

We examine these questions through the case study of Los Angeles, a center of immigration with considerable racial and cultural diversity but also dramatic inequalities in wealth and well-being. Specifically, we examine the role of worker centers,[1] grassroots organizations that have emerged during the past twenty-five years to provide services to and advocate on behalf of immigrants and people of colour, also waging successful multi-racial social justice campaigns. Most recently, the Los Angeles $15 minimum wage campaign has enhanced the visibility and influence of participating grassroots organizations within the urban political landscape. Drawing upon a study of the worker center movement in Los Angeles conducted in the summer of 2015, which included interviews with leaders within the movement, we first summarize arguments advanced about the making of (counter)publics and the construction of egalitarian political spaces. We then turn to our case study, starting with a brief discussion of the space-time of the worker centers, followed by a discussion of the different acts towards constructing egalitarian multi-racial publics. Our discussion highlights the following themes in respect of spaces of urban politics: flat modalities of governance and organizing; learning about the Other and negotiating across social and spatial differences; and spaces of publicity – the making of the $15 minimum wage campaign, a recent successful cross-racial networked collaboration publicizing injustice and demanding remedies. We conclude by reflecting on the multiple spatialities of worker centers – especially the importance of spaces of withdrawal (safe physical spaces of the worker centers) and spaces of circulation (networks of connectivity and mobilities of people and ideas) – which we suggest both shape and are shaped by the construction of multi-racial publics and egalitarian political spaces.

The making of publics and egalitarian political spaces

Conceptually, this chapter is concerned with how grassroots movements challenge undemocratic forms of authority and discrimination and attempt to create egalitarian political spaces, and the role of spatialities in this process. We begin with two political philosophers whose ideas have been important in debating these issues: Nancy Fraser and Jacques Rancière. We bring them in conversation with one another, recognizing that both are concerned with reflecting and elaborating on the conditions of possibility for radical emancipatory action and equality as a presupposition rather than a goal of political action. However, there also are distinct differences between them in terms of terminology and in their understanding of politics – especially their conceptions of the relationship between the state and publics. In recent years, Jacques Rancière has been a far more popular and influential philosopher than Nancy Fraser among critical urban and political geographers working on social movements, social justice struggles and radical democracy. Scholars such as Erik Swyngedouw (2009, 2014), Mark Purcell (2014), Mark Davidson and Kurt Iveson (2014, 2015), have drawn on Rancière and other radical democratic theorists (such as Laclau and Mouffe), extending their thinking to the urban scale.

Rancière is fundamentally concerned with working towards radical democracy through a theory of egalitarian political action. For him, equality is not an outcome to be achieved, but is based on the assumption that each and every one is equal: to be taken seriously as a valid partner, with "no particular group of people better able to, or born to, rule" (see Iveson 2016: 11). In this view, democracy is distinct from the rule of the best (meritocracy), the rule of experts (technocracy), the rule of the wealthiest (plutocracy), and so on. It is an alternative to those "governments of paternity, age, wealth, force and science, which prevail in families, tribes, workshops and schools and put themselves forward as models for the construction of larger and more complex human communities" (Rancière 2006: 45). Democracy, therefore, must continually be reworked through demands for equality. Importantly, a demand for equality is not merely one for inclusion made on the part of the excluded. Rather, it is a refusal of an existing social order that already presumes inclusion and equality (i.e. citizenship and the bourgeois public sphere) through a demand that "they be taken into account not as subordinates with a limited (or no) part to play in society, but as *equals*" (Davidson and Iveson 2014: 139, original emphasis).

Issues of equality and the notion of publics as important elements of and arenas for democratic politics have been undermined in the neoliberal era. First, has been the neoliberal emphasis on individual responsibility and rights; second, this way of thinking appropriates the notion of the 'public', also reducing it to public-private partnerships in which 'public' generally stands in for government and 'private' for private business. Yet this does not mean that the making of publics has ceased to exist, both on the ground and as an object of theorization. One of the most influential and debated theoretical explications of the public sphere is due to Jürgen Habermas (1989), who conceives of the bourgeois public sphere as a single overarching arena where rationality and equality are the bases of deliberation (Calhoun 1993). His conception has come under heavy criticism, however, especially for its assumption of equal participation. As Fraser (1990) has pointed out, social differences among participants – both socio-economic inequalities and socio-cultural differences – cannot be bracketed because they are never eliminated. They continue to pervade deliberations in many different guises – such as differences in the linguistic repertoire of different social groups. Thus, while conceding value to the concept of the public sphere, Fraser argues that there is no a single public sphere. Rather, multiple public spheres exist and should be acknowledged, especially in socially and culturally diverse capitalist societies.

Within this frame of multiplicity, counter-publics are proposed as overt alternatives to the hegemonic public sphere (Fraser 1990). Counter-publics constitute "parallel discursive arenas where members of subordinated social groups invent and circulate counter-discourses" (1990: 67). They allow individuals to come together to voice their grievances, needs and desires within particular venues, and to develop skills and strategies to communicate them to a wider public (Sziarto and Leitner 2010). As Fraser puts it, "on the one hand, they function as spaces of withdrawal and regroupment; on the other hand, they function as bases and training grounds for agitational activities toward wider publics" (Fraser 1990: 124). She notes that "members of subordinated social groups – women, workers, peoples of color, and gays and lesbians – have repeatedly found it advantageous to constitute alternative publics" (1990: 67). Such counter-publics contest and publicize discrepancies between the norm of liberal democracy's claims to equality, freedom and citizenship, and the reality whereby women, racialized groups and sexual minorities are subject to discrimination, oppression and exclusion.

An important element in the making of counter-publics is questioning the boundary between the public and private spheres, and who participates in defining matters of 'public' or 'private'. As Fraser (1990) argues, the drawing of a boundary separating the public and private spheres is already a political act: it defines what kinds of issues can be debated, what behaviours belong and do not belong, and who is recognized in the hegemonic public sphere (Sziarto and Leitner 2010). For example, in an era of (neo)liberal governance, with its focus on autonomous individuals responsible for their own well-being, notions of social responsibility and issues of social justice and care are increasingly expelled from political and public debate, and as guidelines for public behaviour. Challenging this banishment, counter-publics reinsert these issues and behaviours that have been relegated to the private sphere back into publics, re-politicizing them.

In contrast to much on the scholarly literature on publics and counter-publics (Warner 2002), which focuses on their rhetorical strategies in seeking publicity, we focus on the making and enacting of publics. Examining the processual aspects of public making reveals that they are not completely free of hegemonic relations produced by the dominant power, nor are they simply in an agonistic relationship with the hegemonic public sphere and the state. Following Fraser (1990), we argue that the making of publics involves not only creating spaces of refuge for developing an oppositional or alternative politics, *but also* engaging with the state and multiple publics. Invoking multiple publics means framing the public sphere as a "structured setting where cultural and ideological contest or negotiation among a variety of publics takes place" (Eley 1993: 306).

Further, cultural and political differences exist *within, as well as between* publics, as we demonstrate below for multi-racial publics. Radical democratic theorists such as Rancière and Laclau and Mouffe suggest that such differences, disagreements and associated conflict should be welcome, seeing agonistic struggle among different positions as the very substance of democratic politics (Purcell 2014: 170). While acknowledging that conflict and contestations are part and parcel of democratic politics, we contend that differences and disagreements have both strengths and vulnerabilities. Socially and culturally diverse publics provide opportunities to learn from one another, benefiting from the distinct backgrounds, experiences and perspectives of their members. At the same time, intra-group social inequalities and socio-cultural and ideological differences also pose challenges, which necessitate openly acknowledging differences and devising ways for negotiating across them. Failure to do so may foreclose effective collective action and even lead to the demise of a public.

Making publics and enacting egalitarian political spaces also entail multiple spatialities and temporalities. Fraser and Rancière conceive of these in very different ways, but do not fully explore them. Drawing on socio-spatial theory, we extend their conceptions, showing how physical and virtual spaces, networks and mobilities both help shape and are shaped by the construction of multi-racial publics and egalitarian political spaces (Leitner et al. 2008). Fraser sees physical spaces – coffee shops, community centers, union halls, sports fields, etc. – as crucial sites for coming together, congregating and withdrawing from the eye of the dominant power. Our study confirms the importance for the construction of egalitarian political spaces of physical spaces that are sheltered from the dominant public sphere and the surveillance of the state (Sziarto and Leitner 2010; see also Brown 2008).

Conversely, designated public spaces – streets, squares, public buildings – become centers for gaining publicity for existing injustices, making demands and proposing alternatives. Responding to increasing surveillance by the state and business interests, seeking to monitor and restrict political expression in designated public spaces, virtual public spaces have become an increasingly important means to gain publicity in recent decades (Clough and Vanderbeck 2006; Graham and Aurigi 1997; Mitchell 2003). The other crucial spatiality in the construction of publics are spaces of circulation – the mobilities of people and ideas and the web of local- and extra-local networks, connecting them with other publics. As we demonstrate below, such networks are important sources of information, inspiration and support, operating both through face-to-face interaction as well as in virtual space through diverse communications technologies.

Whereas Fraser and urban geographers have focused on the role of spatialities in the making of publics and social movements, Rancière's conceptualization of a politics of equality highlights another aspect of the spatio-temporality of contentious politics. He argues that a politics of equality has no single, proper place and time, but remains open to the emergent spaces and times of politics in the city (Davidson and Iveson 2015). Politics does not emerge from particular places, but in practice. As Rancière (1999: 30) puts it: "What makes an action political is not its object or the place where it is carried out, but solely its form, the form in which confirmation of equality is inscribed in the setting up of a dispute." As we read it, he is not arguing that politics is everywhere and always, but that any privileging of certain sites as the proper space of politics a priori is itself an act of ordering, one that presupposes what should count as progressive political action – and where. Avoiding such privileging implies that the spatio-temporalities of an egalitarian politics become a matter of empirical investigation.

The space-time of worker centers

Los Angeles's history and culture, as well as its economic dynamism, have been deeply shaped by its ethnic and racial diversity. With the origin of immigrants changing and diversifying in the second part of the twentieth century, so too has the city's ethnic and racial diversity. Since the late 1960s, immigration to Los Angeles has been dominated by migrants from Mexico and other Latin American countries, as well as Asian migrants (e.g. from China, Korea, Vietnam and Thailand), transforming Los Angeles into a metropolitan area with a majority Spanish-speaking and a substantial Asian population. This racial and ethnic diversity intersects with dramatic inequalities in wealth, manifest in distinct ethno-racial geographies. For many immigrants of colour, everyday lived experiences of exploitation, discrimination, poverty, racism and fear of deportation have been powerful triggers to seek collaboration among communities of colour and become active in the immigrant rights movement (Brodkin 2007; Milkman 2006; Pulido 2006, 2007; Widener 2008).

Cross-cultural and cross-racial organizing has a long history in Los Angeles. During the last quarter of a century new forms of multi-racial politics, new sites of struggles and new alliances have been forged (Nicholls 2003; Widener 2008). Examples include social justice non-profit organizations, such as worker centers (operating outside labour unions), and collaborative multi-racial alliances and campaigns, such as the most recent successful $15 minimum wage campaign. While such campaigns might be single issue, these grassroots social justice organizations engage in multiple struggles that range from living wages and benefits, to wage theft,[2] to low-income housing, public health, transportation, police brutality, immigration reform, and so forth.

Worker centers in Los Angeles, California, have experienced some of the most rapid growth and greatest density in American cities (Fine 2016).[3] Broadly, contemporary worker centers tend towards three spheres of action: service, organizing and advocacy (Fine 2006). First, most centers provide such services as legal assistance, English as a second language (ESL) classes, and on occasion affordable housing. Most also connect their members with other service agencies. Second, worker centers organize low-wage workers, mostly immigrants and minorities of colour, for political mobilization, placing a high emphasis on popular education and leadership development (Fine 2006: 13; Theodore 2015). Third, they advocate for worker and immigrant rights and environmental justice through diverse strategies and tactics, including direct action and anti-racism campaigns. In short, while they are called *worker* centers, their actions and campaigns far exceed the sphere of production to focus on issues such as education, immigrant rights, environmental justice, housing and policing. Each organization deploys its own unique combination of these, underlying its distinctive character. Further, worker centers in Los Angeles are organized around a particular industry (e.g. The Garment Worker Center), around race/ethnicity (e.g. The Black Worker Center), or are place/community based (e.g. Korean Immigrant Worker Alliance (KIWA)). Yet there are no hard and fast lines of difference between these different types.

Much of the scholarship on worker centers has focused on their role and importance in the American labour movement (Fine 2006; Milkman 2006; Narro 2005; Theodore 2015). In this chapter we focus on them as spaces for enacting an egalitarian politics: a flat and flexible governance structure that is open to experimentation; learning about and negotiating across social differences; and engaging in multi-racial campaigns prosecuted through active local and extra-local networking. Further, the physical spaces of the worker centers have been crucial for enacting egalitarian politics, serving as spaces of withdrawal from the dominant public sphere, where experiments with a flattened governance structure become possible, where popular education is pursued, where members can express themselves and their identities without fear, and where members learn to negotiate across their differences.

Alternative modalities of governance and organizing

One of the stated goals of worker centers is the enactment of a flat governance structure that provides equal voice in decision-making processes that identify the needs of workers and work to address them. As Alexandra (the director of KIWA) put it, this distinguishes them from the hierarchy and guild system that characterizes unions. Worker center leaders expressed a strong commitment to democratic decision-making amongst staff and workers, facilitated in their opinion in part by their relatively small scale. As Alexandra notes,

I'd like to say that we are so democratic because we are so righteous, but if I had a staff of 10,000, well there probably would be a lot more hierarchy. We probably wouldn't sit around building consensus with all 10,000 people.

But as one worker center director pointed out, decision-making processes through a flat governance structure are difficult and lengthy even at that small scale, requiring patience and endurance. Center directors describe this as something to work through by using workshops, classes and training to facilitate decision-making amongst workers. Embracing a flat governance structure is also in line with centers' ideals of popular education and leadership development, sometimes explicitly identifying with the pedagogy of Paulo Freire and other leaders of the popular education movement. The institutional infrastructure of worker centers also facilitates this philosophy and pedagogy. For instance, meetings of KIWA, which is embedded in a diverse ethno-racial neighbourhood, are simultaneously translated into Korean, Spanish and English to allow for equal participation across their member base.

While worker centers embrace flat governance, this does not mean the absence of structures governing behaviour. For example, the Instituto de Educacion Popular del Sur de California (IDEPSCA) worker center and the day labourer centers have developed codes of ethics prohibiting discriminatory language in an effort to create a space of mutual respect for difference amongst their members. Maegan, director of IDEPSCA, recounts a situation when these rules were enforced as a church donor spoke to the center's members:

> They were singing and then they started talking about how horrible gay marriage was, and how horrible the Supreme Court decision was, and I was like, "you can't do that here. That's not what's allowed." [It was difficult] because some of the workers who do identify as evangelical Christians were kind of like "well, that's how I feel". I said, "that's great but not here". You know we have rules, we have code of ethics, and non-discrimination and non-hateful language is a part of that.

This exemplifies how rules can be crucial for negotiating social difference amongst members, including enacting a politics of equality that does not reinforce heteronormativity.

The less hierarchical structure and smaller scale of worker centers has enabled them to evolve into what Janice Fine (2011: 606) calls "organizing laboratories", capable of developing and experimenting with innovative governance and organizing strategies. At the Pilipino Worker Center (PWC), for instance, the leadership has tried to accommodate the transportation challenges faced by their workers through experimenting with a small-circle model, facilitated by digital technologies. Members are divided by neighbourhood into about fifty groups, with an appointed leader who regularly attends not only circle meetings, but also meetings with leadership at the center. Aquilina explains how this model helps meet their workers' needs, builds leadership and organizes workers:

> In this workforce – where they are live-in workers – transportation is an issue, but they all have their cell phones, so how can we create spaces for them to be engaged, involved, to exercise leadership, even if they can't be here physically present? ... We're trying to think about how we are creating these spaces that include building community, building care, and building action within our membership, and that things can be more dispersed.

To ease the challenge of organizing and communicating at great distances across the city, the small groups use digital technologies to coordinate within themselves, with their organizers and for action. They utilize video and phone conferencing for meetings of the group leaders, and group texting for ongoing dialogue between members and organizers. This dispersed model serves an organizational function, but also provides a socio-technical infrastructure for collective action. Aquilina explains:

> It was really hard to get people to do call-ins, for example, to legislators because, even if they have the capacity to do that, it's very intimidating. But when you have your group and then your group leader does it and then texts to the rest of the group, "OK, I just called!" then the next person says "OK, I just called!" then it makes it a collective action, instead of being isolated. So it's, how do you create those structures so people get out of isolation?

In short, the flexibility of worker center governance enabled the PWC to experiment with the digitized small-circle model to accommodate workers' needs and experiment with new strategies of organizing. This is just one example of worker centers' innovative organizing strategies, often leveraging digital technologies.

Spaces of negotiation and learning

Beyond being spaces of experimentation in organizing models, worker centers also facilitate being together in difference. They are sites of daily learning about the Other and negotiations across differences, problematizing and addressing the hard conversations about prejudice and stereotypes, racism, sexism, religious intolerance and other forms of dominance, rather than avoiding them. There are numerous challenges in accomplishing this. As Maegan commented, a major challenge is to begin naming the various -isms and calling out individuals who engage in racist and/or sexist talk:

> I think some of the biggest challenges that we've had is just – there has *not* been a lot of discussion about the racism and anti-blackness coming from lots of the immigrant workers themselves. And that's only just been something that we've started to talk about, especially in the midst of everything that's happened in the last few years.

Maegan went on to say that when racial stereotypes, prejudice and sexism came up in conversation and practices they felt compelled to make it visible and start conversations about where this is coming from, through organizing a popular education workshop on race and gender:

> I think they're a good starting place for conversation. ... I don't think people have ever named these things ... It hasn't been named. Or, it's said – I think a lot of the time what happens too – is it's said and it sometimes might have been ignored in the past. Like "oh, that's just how they are". And it's like "no, wait a minute. Why?" So be like, "no, wait a minute, why'd you say that?" And not in a way that shames, not in a way that we're saying "how dare you?" because we're working together here. Just trying to break that down and understand why. So, I think for us, it's not the *end;* it's just the starting point because these things have never been spoken about. Never.

Maegan notes that these conversations are not an achieved end point, but rather are a place from which to begin the process of naming racism, sexism, homophobia and inequalities. This parallels Rancièrian thinking, whereby gender and racial equality are not the end point but a normative assumption from which to proceed – in this case, through a disruption of prejudiced beliefs through conversation. Indeed, like other center directors Maegan identified conversations as the most important ingredient to moving forward together across difference:

> Lots of talking. Our meetings can be very long for that reason ... We really end up talking through a lot of it ... I think we throw all these terms [queer, LGBTQ, for example] around and we think everyone is on the same page about it. And we're not! So I think having lots of conversations and understanding that we may not always agree in the end, but at least in terms of our policies we have strict policies about non-discrimination and respect, so we're not going to call names, we're not going to use those terms because we also have to understand that there are some queer-identified staff members, there are some staff members who are not Christian. So, how do we – and we talk about it.

Here, Maegan identifies a very important aspect of negotiating across differences. It is not enough to have conversations about stereotypes, prejudice and discrimination, and to learn about the Other. A set of non-negotiable rules of conduct governing interactions with the Other are also necessary – for example, the banning of name calling, using derogatory terms and harassment. While these are not necessarily clearly specified in writing and made visible at the worker centers, the day labour programmes run by some of the worker centers post rules that are clearly visible to everybody entering the day labour space. Posters spell out rules of conduct, alongside job distribution rules, participants' rights and responsibilities, personal appearance, in order to "build a healthy and safe environment, to promote transparency, justice and equality, to prevent conflict and promote friendship, and project an image of order and professionalism to the community" (Pasadena Community Job Center Internal Rules 2016, courtesy of Nik Theodore).

Alexandra reminds us of another mundane obstacle in communicating with the Other and negotiating across differences: language barriers and cultural mores. In her experience, this often poses a greater challenge in the 'private' physical space of the worker center than in the public spaces of marches and protests:

> But then, on the other hand, the challenge for us is that people speak different languages, they eat different food, they sometimes have different religions, they have different cultural practices. And *that* part is hard. So the easy part is going to May Day and marching together. The hard part is eating pizza afterwards because the Koreans don't really want to eat pizza and you sit with people whose language you can actually understand. ... So it's those non-formal spaces that are actually more challenging in some ways. I mean everybody is willing and makes an effort, but when you literally don't speak the other person's language and there is no translation devices and no simultaneous translation going on, you know.

Here, Alexandra articulates a common opinion expressed by center directors: that the close social and spatial proximity of worker centers incites, requires and facilitates negotiations across differences. On the one hand, the spatial closeness of the worker center requires its members to acknowledge, come to terms with and address differences among them. On the other hand, their shared social experience of discrimination of oppression facilitates such negotiations.

Spatialities of making social justice campaigns

If the space of worker centers is one of 'withdrawal' and the formation of new (counter) publics and politics, then their campaigns for social justice are the outward 'agitational' dimension of worker centers – spaces of publicity (Fraser 1990). Campaigns range widely, from those primarily driven by a single center on a single issue to coalitions of multiple worker centers and other social justice organizations making demands on the local or national state. The recent $15 minimum wage campaign for the City of Los Angeles is an example of the latter, and required ongoing coordination across different actors and groups within the progressive community in the city, in addition to public displays of protest. As we suggest above, political action entails many spatialities in and beyond the city. The physical space of the centers, which serve as spaces of withdrawal, are dialectically related to the spaces of circulation – the mobilities of people and ideas, as well as local and extra-local networks with other publics (Leitner et al. 2008). The minimum wage campaign reveals these spatialities and how they make visible the ongoing social justice work of centers and their making of multi-racial publics.

In May 2015, Los Angeles became the largest city in the United States to pass a $15 minimum wage, following the precedent set by cities like Seattle and San Francisco. A key actor in the movement behind this was the Raise the Wage Campaign, a coalition of labour groups, faith-based organizations, worker centers, day labourer centers, labour unions, and other advocacy and community organizations that fronted the campaign. Organizationally, the Raise the Wage Campaign formed a steering committee consisting of twenty organizations that took the lead in the campaign. Every organization represented on the steering committee, several of them worker centers, committed to contributing a full-time organizer or in-kind contributions to the campaign, signing a memorandum of understanding (MOU) to work towards the collaboratively developed set of demands. These included the $15 minimum wage but also, no less importantly, a city ordinance against wage theft and an office of enforcement within city government. These last two demands were eventually passed alongside the $15 minimum wage ordinance.

Worker centers were instrumental for building momentum within the movement, especially for the wage theft and enforcement provisions. The Los Angeles Coalition Against Wage Theft, whose leadership included each of the worker center directors we interviewed, worked closely alongside Raise the Wage. Wage theft has been a pervasive problem amongst the low-wage workers who make up the membership of worker centers. Indeed, Los Angeles has the highest rates of wage theft in the country, with workers from low-wage industries losing more than $26.2 million per week, mostly from minimum wage violations (Milkman et al. 2010). As Maegan sees it, "Everybody else can talk about where it [the wage theft ordinance] came from. It came from worker centers, and specifically it came from *jornaleros* [labourers]."

Yet neither the origin nor the success of this campaign was straightforward. Just as negotiation across social difference is a challenge within the space of worker centers, the solidarity required for campaigns also needs to be actively constructed. As Pulido (2006: 144) notes, referencing earlier coalitional politics in Los Angeles, "connections between various groups are not inevitable or automatic but must be articulated". In other words, it takes social and political work to bring together different groups, and to move forward when differences are encountered in how to approach a common goal, or when competing claims for recognition come into conflict. This is the work of making multi-racial publics.

For example, there was intense discussion amongst worker centers over the meaning of wage theft and therefore the coalition's demands. For the Black Worker Center, the incredible rates of job discrimination against African Americans in Los Angeles – what Lola calls the "black jobs crisis" – constituted a form of wage theft. The campaign against wage theft then becomes a moment for political action, but also a claim for equality. Lola explains:

> The minimum wage campaign ... [is] a really good example of our coming to the table to fight for raising the floor for workers, but educating that table about what's happening to black workers and how this policy can be improved to address something like discrimination and saying that ... that discrimination is wage theft. And therefore we need to enforce and protect workers against discrimination, as well as against wage violations.

For the Black Worker Center, the political momentum behind the wage theft ordinance was an opportunity to fight discrimination against African American workers in the Los Angeles region, but also, crucially, a moment to force the progressive labour movement in Los Angeles to recognize the plight of black workers. This claim brought with it its own set of issues, however. Rosemarie describes the challenges of negotiating competing claims for recognition and understandings what constitutes wage theft:

> There was an issue when the Black Worker Center started participating in our space. There was a push to change our messaging to "being unemployed is the ultimate form of wage theft, or being denied". And there was a lot of resistance against that because we were like "that might be true for your membership, but our membership got – one person got $67,000 stolen from them. That's pretty significant". So for them, that's the ultimate form of wage theft is working for free.

Rosemarie continues, describing how the Coalition sought resolution:

> Honestly, it was a lot of meeting collectively, a lot of talking, challenging each other. ... I think what we came to in the end is ... [saying] "this train is moving and we have every intention of getting that piece of the policy passed. So even if it's not in the forefront, you have to know and trust us, that in every place we can, we are inserting that anti-discrimination language" ... and the Black Worker Center still came out to every single action. And we are lifting up the anti-discrimination policy.

Rosemarie underlines the importance of learning and trust for moving past differences towards a common goal, but also the collective commitment to common objectives that was enshrined in the MOU. The MOU, which makes each organization on the steering committee accountable to work in good faith towards passing the full package agreed upon, is itself of course the outcome of intense negotiations across differences.

While the $15 minimum wage campaign erupted in moments of protest and high local public visibility, it was, of course, in no way limited to those spaces and times. Intra-metropolitan and extra-metropolitan networks of actors, ideas and policies have been equally important, providing a crucial foundation of inspiration and support for worker centers in all their campaigns for social justice. Locally, worker centers benefited from both formal and informal relationships among their respective leaders. Through working collaboratively on coalitions, the worker center leaders we spoke to had developed informal, personal relationships that support their organizing. Speaking of leadership for the Coalition Against Wage Theft, Rosemarie explains:

> With the women that I've worked with, we genuinely care about each other in more than just the work sense. And that's with ... all of the worker centers. ... Because we

are trying to lift each other up, all at the same time. We all work with marginalised bases. We can't just build power for carwash workers, and not think about the garment workers, not think about the black workers in South L.A., who are unemployed. And I think that since all of us in the higher positions feel that way, we're able to do that together.

Rosemarie connects these informal relationships with a solidarity that extends across making the multi-racial public. She feels that the leaders' care for one another translates into a collective identity, whereby one worker center cannot make claims for equality without the others. Aquilina reinforces how networks connecting worker centers have strengthened collaborations and collective action: "we've been able to develop pretty strong networks and collaboration with other worker centers, which it seems is not necessarily the case in other places". Rosemarie and Aquilina suggest the importance of informal relations of care in the making of multi-racial publics.

Beyond these informal relationships, worker center leaders also have sought to develop a formal institutional network through a Los Angeles Federation of Worker Centers to share resources, coordinate action and build power within the regional labour movement for marginalized populations that often are considered 'unorganizable' by labour unions (Fine 2016; Milkman 2006). As Walter Nicholls (2003: 892) has argued, such formal and informal networks constitute a kind of "organizational infrastructure" that can "serve as the relational grounds for the elaboration of more institutionalized forms of coordinated interactions". Indeed, these types of "coordinated interactions" grow as worker centers gain more political clout and prominence through victories like the wage theft ordinance: The American Federation of Labor and Congress of Industrial Organizations (AFL-CIO), the largest federation of unions in the United States, now maintains a full-time worker center director position that coordinates between organized labour and worker centers, and worker centers have begun to federate nationally (Fine 2007). These intra-metropolitan networks articulate with extra-local circulations of ideas and people. Both in their modalities of governance and campaigns strategies, worker centers have cultivated extensive relationships with other worker centers and national and international progressive movements, thereby opening up new spaces of circulation connecting urban politics to the wider labour movement. Particularly for the $15 minimum wage campaign, worker center leaders – and the coalitions of which they were a part – learned from activists and organizations from the successful Seattle and the Bay Area $15 minimum wage movements. For instance, The Raise the Wage campaign only implemented the MOU after organizers learned how an MOU had been usefully leveraged in Oakland.

Worker center networks and mobilities also circulate beyond national borders. Both KIWA and PWC leaders were influenced and inspired by popular movements in Korea and the Philippines, respectively. These international networks of inspiration and learning fold into local contexts and networks of ideas and people: as KIWA was inspired by the Korean democracy movement, so the PWC was influenced and supported in its formation by KIWA. Similarly, the PWC drew on international organizing models in implementing their small circles, assembling different elements that would work in the Los Angeles context: "[We started studying] these more, decentralized organizations, and pulling different elements from that, looking at domestic worker organizations abroad, like Kenyan domestic workers' union, South Africa and other places, and pulling in different elements and seeing what would work."

Conclusion

In this chapter we have highlighted the making of multi-racial publics and their construction of egalitarian political spaces that attempt to enact a potentially radical democratic politics. Using the worker centers in Los Angeles as a case study, we show how they have been creating and enacting a flat and flexible governance structure that is open to experimentation, where everybody has a voice in decisions affecting their lives, and people, rather than the government, establish rules; where popular education, learning and negotiating across social and cultural differences can and does happen; and people engage in networking across differences in order to collectively prosecute successful social justice campaigns. Two spatialities have been vital for the construction of these egalitarian political spaces and new ways of doing things: first, the physical spaces of the worker centers, which serve as spaces of withdrawal, to some extent sheltered from the dominant public sphere; and second, the spaces of circulation – the mobilities of ideas and people and the real and virtual local and non-local networks of actors. The latter provide a crucial foundation of inspiration and support for worker centers and their campaigns for social justice. Los Angeles's worker centers benefited from both formal and informal relationships between local leaders and members of grassroots organizations, NGOs and politicians within Los Angeles, the USA and beyond. Through cultivating such extensive relationships, worker centers also have been opening up new spaces of circulation connecting urban politics to the wider labour movement, the state political landscape, international social justice movements.

In sum, worker centers work towards constructing egalitarian political spaces that could indeed be transformative of urban politics. We use 'towards' deliberately; not to suggest that these spaces are a priori more democratic and equal, but that they enact a processual method of equality – a presumption of equality that is verified through the ongoing practice of being together in difference. There are practical limitations and challenges to this process. First, there are the mundane issues of maintaining these alternative spaces. Strapped for resources, both financially and in terms of labour power, staff and members are overworked and faced with near-constant financial insecurity. Second, as we have shown, negotiations across lines of difference constitute both a great source of inspiration and solidarity but also a potential problematic for the multi-racial public. Insofar as breakdowns in learning and exchange occur, effective political action may be foreclosed. These challenges put strain on the potential of worker centers for alternative urban politics. As shown through the case of the Los Angeles $15 minimum wage campaign and wage theft ordinance, however, these progressive small-scale publics also show the transformational potential of grassroots political involvement. Worker centers were instrumental in the $15 minimum wage in San Francisco and Los Angeles and, no less importantly, ordinances against wage theft have already opened the door for such initiatives in other cities. Furthermore, the State of California has passed state-wide legislation to raise the minimum wage to $15 by 2022. While a direct causal line is impossible to establish, and altogether beside the point, we suggest that worker centers as alternative, egalitarian spaces have been influential in Los Angeles urban politics, and may indeed be transformative on the way to broader social change.

Notes

1 American spelling is used for 'worker centers' throughout this chapter.
2 Wage theft is the practice of employers not paying workers their full wage by violating minimum wage laws, stealing tips, withholding overtime pay, or forcing workers to clock out and continue working.

3 The first wave of contemporary worker centers emerged in the US South during the late 1970s and 1980s in response to economic restructuring and the decline of manufacturing, but also racial and gender discrimination in union representation (Fine 2006: 11). The number of worker centers bourgeoned in the 1990s and 2000s, growing from approximately 155 in 2007 to over 225 in 2016 (Fine 2016). Many of these centers were organized around immigrant groups from Latin America and Asia that relocated to the United States during this time period.

References

Brodkin, K. (2007). *Making Democracy Matter: Identity and Activism in Los Angeles*. New Brunswick, NJ: Rutgers University Press.

Brown, M. (2008). Working Political Geography Through Social Movement Theory: The Case of Gay and Lesbian Seattle. In: K. Cox, M. Low and J. Robinson, eds. *The SAGE Handbook of Political Geography*. London: Sage, pp. 285–304.

Calhoun, C. (1993). Introduction: Habermas and the Public Sphere. In: C. Calhoun, ed. *Habermas and the Public Sphere*. Cambridge, MA: MIT Press, pp. 1–48.

Clough, N. and Vanderbeck, R.M. (2006). Managing Politics and Consumption in Business Improvement Districts: The Geographies of Political Activism on Burlington, Vermont's Church Street Marketplace. *Urban Studies*, 43(12), pp. 2261–2284.

Davidson, M. and Iveson, K. (2014). Occupations, Mediations, Subjectifications: Fabricating Politics. *Space and Polity*, 18(2), pp. 137–152.

Davidson, M. and Iveson, K. (2015). Recovering the Politics of the City: From the 'Post-political City' to a 'Method of Equality' for Critical Urban Geography. *Progress in Human Geography*, 39(5), pp. 543–559.

Eley, G. (1993). Nations, Publics, and Political Cultures: Placing Habermas in the Nineteenth Century. In: C. Calhoun, ed. *Habermas and the Public Sphere*. Cambridge, MA: MIT Press, pp. 289–339.

Fine, J. (2006). *Worker Centers: Organizing Communities at the Edge of the Dream*. Ithaca, NY: Cornell University Press.

Fine, J. (2007). A Marriage Made in Heaven? Mismatches and Misunderstandings Between Worker Centres and Unions. *British Journal of Industrial Relations*, 45(2), pp. 335–360.

Fine, J. (2011). New Forms to Settle Old Scores: Updating the Worker Center Story in the US. *Relations industrielles/Industrial Relations*, 66(4), pp. 604–630.

Fine, J. (2016, 29 January). *Protecting Immigrant Workers: New Strategies for Strengthening Labor Standards Enforcement*. Institute for Research on Labor and Employment, UCLA. Los Angeles: UCLA.

Fraser, N. (1990). Rethinking the Public Sphere: A Contribution to the Critique of Actually Existing Democracy. *Social Text*, 25/26, pp. 56–80.

Graham, S. and Aurigi, A. (1997). Urbanising Cyberspace? *City*, 2(7), pp. 18–39.

Habermas, J. (1962 trans. 1989). *The Structural Transformation of the Public Sphere: An Inquiry into a Category of Bourgeois Society*. Cambridge: Polity.

Iveson, K. (2016). *Placing Encounters*. Unpublished manuscript.

Leitner, H., Sheppard, E.S. and Sziarto, K. (2008). The Spatialities of Contentious Politics. *Transactions of the Institute of British Geographers*, 33(2), pp. 157–172.

Milkman, R. (2006). *L.A. Story: Immigrant Workers and the Future of the U.S. Labor Movement*. New York: Sage.

Milkman, R., González, A.L., Narro, V., Bernhardt, A., Theodore, N., Heckathorn, D., Auer, M., DeFilippis, J., Perelshteyn, J., Polson, D. and Spiller, M. (2010). *Wage Theft and Workplace Violations in Los Angeles: The Failure of Employment and Labor Law for Low-Wage Workers*. Institute for Research on Labor and Employment, UCLA. Los Angeles: UCLA.

Mitchell, D. (2003). *The Right to the City: Social Justice and Fight for Public Space*. New York: Guilford Press.

Narro, V. (2005). Impacting Next Wave Organizing: Creative Campaign Strategies of the Los Angeles Worker Centers. *New York Law School Law Review*, 50, pp. 465–513.

Nicholls, W. (2003). Forging a 'New' Organizational Infrastructure for Los Angeles' Progressive Community. *International Journal of Urban and Regional Research*, 27(4), pp. 881–896.

Pulido, L. (2006). *Black, Brown, Yellow, and Left*. Berkeley: University of California Press.

Pulido, L. (2007). A Day Without Immigrants: The Racial and Class Politics of Immigrant Exclusion. *Antipode*, 39(1), pp. 1–7.

Purcell, M. (2014). Rancière and Revolution. *Space and Polity*, 18(2), pp. 168–181.
Rancière, J. (1999). *Disagreement: Politics and Philosophy*. Minneapolis: University of Minnesota Press.
Rancière, J. (2006). *Hatred of Democracy*. London: Verso.
Swyngedouw, E. (2009). The Antinomies of the Postpolitical City: In Search of a Democratic Politics of Environmental Production. *International Journal of Urban and Regional Research*, 33(3), pp. 601–620.
Swyngedouw, E. (2014). Where is the Political? Insurgent Mobilisations and the Incipient 'Return of the Political'. *Space and Polity*, 18(2), pp. 122–136.
Sziarto, K. and Leitner, H. (2010). Immigrants Riding for Justice: Space-time and Emotions in the Construction of a Counter Public. *Political Geography*, 29(7), pp. 381–391.
Theodore, N. (2015). Generative Work: Day Labourers' Freirean Praxis. *Urban Studies*, 52(11), pp. 2035–2050.
Warner, M. (2002). Publics and Counterpublics. *Public Culture*, 14(1), pp. 49–90.
Widener, D. (2008). Another City is Possible: Interethnic Organizing in Contemporary Los Angeles. *Race/Ethnicity: Multidisciplinary Global Contexts*, 1(2), pp. 189–219.

PART VIII

Spaces of identity

Byron Miller and David Wilson

The notion of 'spaces of identity' foregrounds the role of the subject and how affective bonds, norms and affiliations are formed – through the agency of individual subjects, through the agency of collectives, and through the agency of others who seek to define particular collectivities, often in unfavourable ways, for their own benefit. While the notion of identity may foreground the individual, its formation is an innately social process. Conceptions and categories of identity frequently become concretized in institutional and social structures, producing structural patterns of both privilege and oppression. In an increasingly urban world the formation of identities, the exercise of privilege and oppression on the basis of group identities, and the contestation of those power relations are increasingly urban processes.

'Identity politics' have become a central dimension of modern political contestation. Socially constructed 'race', gender, sexual orientation and class identities are now commonly understood to be fluid, intersectional, multiple and positional, rather than singular and stable. While oppression is often practised against those defined in some way as 'other', the city can offer hope as a refuge for difference. As Iris Marion Young asserts, "the unoppressive city is ... defined as openness to unassimilated otherness" (1986: 234).

Identity politics are directly linked to the "struggle for recognition". As Nancy Fraser argues,

> demands for 'recognition of difference' fuel struggles of groups mobilized under the banners of nationality, ethnicity, 'race', gender and sexuality. [But this], of course, is not the whole story. Struggles for recognition occur in a world of exacerbated material inequality – in income and property ownership; in access to paid work, education, health care and leisure time; but also more starkly in caloric intake and exposure to environmental toxicity, hence in life expectancy...
>
> *(1995: 68)*

Fraser compellingly argues that questions of recognition and material redistribution are inextricably intertwined and that the political and policy strategies taken in one realm have inevitable implications for the other.

An overriding theme of the six chapters on 'spaces of identity' is that identity politics are simultaneously struggles for recognition and struggles over material resources. These struggles, moreover, are fluid and situated in a wide range of different contexts, producing not only differing material outcomes but also different constructions of identity. The contingency of outcomes is in many ways shaped by the contingencies of the city.

In 'A city of migrants: migration and urban identity politics', Virginie Mamadouh highlights the central role of migration in the growth of cities and the ways in which migrant politics and urban politics overlap. Migrant populations may be very diverse, not simply because of a multiplicity of countries of origin, but also due to a variety of transnational and translocal networks and the different ways in which these networks may articulate with social and political circumstances in the city of landing. Networks may entail not only flows of information, resources and communication, but also differences in political organization, political opportunity structures, legal frameworks and policies. Migrant struggles for recognition and resources are often met by anti-immigrant mobilizations, usually by groups that perceive themselves as powerless and express their frustrations through racist and nationalist populism. These political conflicts are bound up with issues of redistribution, which implies a politics of scale involving both the nation-state and, in the European context, the EU. Multi-scalar political interventions tend to follow one of two logics, cultural (targeted to a group) or territorial (targeted to a place), each with its own implications for urban politics.

In 'Class, territory and politics in the American city', Kevin Cox addresses two basic forms of identity construction: class and territory. Cox focuses specifically on the process of suburbanization in the United States and how it has produced a metropolitan scale fracturing of the working class. Struggles among different factions of the working class have come to be "organized along territorial lines – this or that neighbourhood or even independent suburb – but representing wider social strata and their interests". In the context of a highly decentralized state and the promotion of homeownership, urban politics often revolve around rent-seeking and property values. With an emphasis on spatially fixed property capital, it is no surprise to see a territorialized class politics – a politics of exclusion aimed at lower-status 'others' – centred on the living place.

Carolina Sternberg's 'Urban middle-class shifting sensibilities in neoliberal Buenos Aires' displays some disturbing commonalities with Kevin Cox's analysis, although focused more on the construction of class identity and status. Sternberg describes the transformation of the 'sensibilities' of Buenos Aires' middle class in the wake of the 2001 economic crisis, the adoption of neoliberal policies and subsequent policies adopted to help the worst-off. Prior to the economic crisis Buenos Aires had a reputation as a very inclusive and socially integrated city with good opportunities for upward social mobility. That changed dramatically in the aftermath of the 2001 crisis as neoliberal policies inculcated values of individualism and responsibilization. Middle-class optimism about the future was replaced by a sense of decline, further exacerbated by policies adopted in 2007 to help the poor and working classes. As the very poorest became better off, the anxieties, resentment and fear of the declining middle class were exacerbated. In an attempt to salvage their class status, the middle classes pursued a revanchist politics against the least well-off, pushing squatters, immigrants and the homeless to both the geographical and social peripheries. The middle class sought to preserve their own social identity and status by excluding those they believed to be their social inferiors.

In 'Compassionate capitalism: tax breaks, tech companies and the transformation of San Francisco', Lauren Alfrey and France Winddance Twine describe a different politics of social exclusion. Their analysis focuses on the City of San Francisco's strategy for 'revitalizing' the racially and ethnically diverse low-income Tenderloin neighbourhood in the middle of San Francisco's downtown. The City's strategy has been to offer payroll tax breaks to tech firms, rationalized as a means of providing job opportunities for the poor residents of the neighbourhood. Alfrey and Twine find that the poor residents of the Tenderloin lack the skills and training to obtain jobs with high tech firms and that the tax breaks provided have done little to improve the economic circumstances or social inclusion of Tenderloin residents. Despite the co-presence of tech workers and the poor and marginalized in the same neighbourhood, "they live in separate worlds segregated socially, economically, and racially".

Fran Tonkiss's 'Gendering urban protest: politics, bodies and space', focuses on women's mobilizations in public spaces in actions aimed to challenge gendered forms of oppression. She does not wish to "treat gender as a given demographic marker, but rather to think about how gendered identities and engagements are produced through political mobilization, public expression and spatial figuration". Tonkiss discusses a variety of women's protests around the world, stressing not only the range of intersecting social and political issues addressed, but also the power of women's bodies in space as a protest repertoire. The very presence of women in public space, in many contexts, can be a challenge to structures of power and authority. Gendered identity is actively produced through protest, even as protest challenges gendered identities.

Finally, in 'Queering urban politics and ecologies', Will McKeithen, Larry Knopp and Michael Brown provide a wide-ranging historical account of the evolution of development of urban political ontologies and their relationship to conceptions of queer politics. They begin their account with a discussion of Ray Pahl's work of the 1970s, which saw urban politics as concerned with "allocation, access, rent, and property". Allocation and access, of course, strongly parallel Nancy Fraser's notions of redistribution and recognition. It was within this framework that the politics of "sexual minorities (lesbians, gays, bisexuals, trans*, and otherwise 'queer' folk)" were initially conceptualized. The issue of the territoriality of queer identity politics has been an important one, both in the defensive strategies of oppressed sexual minorities and in the successful electoral campaigns of politicians such as Harvey Milk. Clearly access to the local state apparatus could be spatialized, as could the allocation of (state-provided) collective consumption goods. Early work on queer urban politics had, nonetheless, relatively narrow conceptions of the arenas, structures and identities of politics, often based in heteronormative assumptions. More recent work, based on explicitly relational ontologies, has stressed the fluidity, positionality, plurality and intersectionality of identity construction, emphasizing the "ways in which [identities] are articulated with each other in the lived experiences of actually existing, embodied social actors". These intersectional approaches have given rise to an anti-identitarian politics as well as a broad rethinking of an urban politics based on fixed boundaries and categories. Another boundary now in question is that between human and non-human worlds. McKeithen, Knopp and Brown argue that queer politics have always hinged on questions of inclusion and that "queer theory offers urban political inquiry a new way of thinking about urban politics altogether: queer urban ecology".

We often fail to recognize the boundaries of our categories and how those boundaries inscribe power relationships. Understanding the world of urban politics relationally rather than categorically, and as always situated, opens up new ways of understanding both who we are and how we might challenge inequities in power.

References

Fraser, N. (1995). From Redistribution to Recognition? Dilemmas of Justice in a 'Post-Socialist' Age. *New Left Review*, 212, July–August, pp. 68–93.

Young, I.M. (1986). The Ideal of Community and the Politics of Difference. *Social Theory and Practice*, 12(1), pp. 1–26.

38
A CITY OF MIGRANTS
Migration and urban identity politics

Virginie Mamadouh

Introduction

Cities have been spaces of identity politics in many ways. No doubt that in the past decade migration has been a key component of these identity politics in most cities across the world. This chapter introduces briefly the relation between migration and the city before discussing three topical dimensions of identity politics linked to migration. The first foregrounds the political agency of immigrants. The second reflects, in contrast, collective mobilizations against immigrants. The last deals with urban policies.

The chapter is mostly informed by the experiences of Western European cities (especially in the Netherlands, France, Belgium, Germany and the United Kingdom) and, to a lesser extent, Northern American cities. Similar experiences are found in South African cities with their Southern African immigrants, and in the Gulf states with migrants from Southern and South-East Asia to the Gulf States, notwithstanding the specific characteristics of their immigrants, dynamics of their cities and features of their political arenas. Likewise the focus here is on larger cities, but smaller towns can also be important sites of re-bordering migration and urban politics (see Gilbert (2009) for the thorough analysis of two Northern American examples of anti-immigration municipal measures or Coleman (2012) for local state enforcement of border policies in the US Southwest).

Cities and migrants

The relationship between cities and migration is complex. If we conceive of the urban as characterized by density and diversity (Guillaume in Lévy and Lussault 2003: 949), cities are linked to movement and migration in seemingly contradictory ways. Cities are specific forms of socio-spatial organization that can be seen as devices that create co-presence (Lussault in Lévy and Lussault 2003: 988) reducing distance, and movement needed to overcome it, to almost nothing. However, this very concentration of diverse people – the co-presence of people originating from and connected to different places elsewhere – rests ultimately on the migration of people into the city.

The chapter focuses on the impact of a specific kind of migration on a specific aspect of urban politics, namely international migration (as opposed to migration from smaller town and rural surroundings) and identity politics (as opposed to the politics of land use and spatial planning and material redistribution issues). It should be stressed, however, that other types of migration to cities have similar impacts, especially in times of enhanced urbanization or rural flight (metaphorically but appropriately called exodus, or *l'exode rural* in French). This is especially true when the new urban dwellers come from far away and do not share many cultural traits (language, religion, customs...) with their place of landing or have different traditions of political rights and social entitlements: think of Irish migrants moving to Liverpool in the nineteenth century, Southern Spaniards in Barcelona, Southern Italians in Milan, emancipated Virginian former slaves in Chicago in the early twentieth century, Anatolians in Istanbul in the late twentieth century or rural Chinese in booming megacities, handicapped by the *hukou* system, in the twenty-first century. Likewise identity politics – a politics foregrounding cultural (self and ascribed) identification over other interrelations – is only part of multi-dimensional urban politics, never detached from material struggles over the appropriation of urban space and places in the city.

Generally speaking, (larger) cities are the gateways through which foreign immigrants (i.e. migrants from abroad) enter their new country of residence. They provide shelter to very diverse groups (in social, cultural and/or economic terms) and function in the best case as a social elevator and integration machine and in the worst case as a trap in which immigrants with little social, cultural and/or economic capital are stuck in deprived inner-city or suburban neighbourhoods, segregated from the mainstream of society. As analysed early in the twentieth century by the Chicago School, cities have often been the mechanisms by which immigrants are integrated into their new country of residence. Not surprisingly, larger cities have often been associated with immigration; this should not obscure the fact that international migration can be geared towards rural areas, especially where labour-intensive agriculture depends on seasonal labourers from abroad or where states have created reception centres for asylum seekers in rural locations. Migrant politics need not be urban. Nevertheless, there is a strong overlap between migrant politics and urban politics.

The migrants and the urban grassroots

(Im)migrants display great diversity and are often only perceived as a group because they have their origins outside the territory of the state. They need not have more in common than this common lack of local roots but usually share other social, cultural or political characteristics. They can be foreigners or have obtained full citizenship. Sometimes descendants of migrants are classified as immigrants, even if they were born in the country and have full citizenship.

Since migrants are very diverse, grassroots mobilizations of migrants are very diverse too, depending on their specific background and the characteristics of the urban locality in which they mobilize (or not). Typically the main issue is whether migrants participate in collective action *as migrants* with other migrants, or if they mobilize in broader grassroots movements. To some extent, the participation of immigrants in broader urban movements, housing and neighbourhood associations, public transport users action groups, urban gardeners' organizations or trade unions could be interpreted as an indicator of their social and political integration into the host society, while separate actions groups could be read as the result of (self-) exclusion. This is, however, too simplistic. Participation in broader grassroots movements from systematically marginalized positions (e.g. if they are welcomed as rank and file but excluded from pivotal positions in these organizations) would not be a very convincing case of integration. Fully separate mobilization could be seen as a demonstration of the mastery of the rules of the local political game and prove to be a valuable foundation for leaders and bridge builders from within the community. More likely, neither is fully realized and both dynamics can potentially enable and disable individual migrants.

Grievances may be related to different tactics. When the focal grievances are specific to the migrant experience (e.g. pertaining to relations with police, residence permits, work permits, family reunification, formal and informal exclusion in the housing or the labour market, lack of support for traditional customs one wishes to continue in the new country of residence, etc.), mobilization is likely to be among migrants from the same country of origin. When the focal grievances are common to many urbanites (e.g. living conditions in the city such as housing shortages, unemployment, poor education, poor or expensive public transport, air pollution, noise nuisances, health hazards, corruption, tax raises, the closure of health services or sport facilities…) mobilization is likely to be based on other affinities and feelings of belonging and solidarity.

Beyond grievances, the specific characteristics of the migrant population also affect emerging tactics. Resources that migrants are able (or unable) to mobilize can set them apart from other urban residents. Typically guest workers and irregular migrants have limited resources to activate for collective action, both in terms of language skills, legal skills and schooling, and in terms of money, time and contacts in the local state, local political parties and civil society organizations. Some immigrants, especially expats, knowledge workers, diplomats, academics, students and asylum seekers may be better equipped, from the standpoint of knowledge and finances. Even if they lack local language skills and local contacts, they can hire professional mediators to defend their interests.

Some migrant characteristics may be advantageous: a language unfamiliar to the police, informal networks based in hometown connections, and strong social control inside the community could become invaluable resources for organizing into separate organizations. Hometown associations were common, especially in the 1960s and the 1970s, when labour migration was expected to be temporarily (Çaglar 2006, 2007). The translocal and trans*national* character of these hometown associations was sometimes perceived as undermining national politics and national sovereignty, both in the country of origin and in the country of residence, and the association of migrant politics with transnationalism and globalization had an impact far beyond the city of residence (Østergaard-Nielsen 2003).

Possibly the most important factor influencing the way migrants organize (separately or not from other urbanites) is the political opportunity structure. Depending on the civic and political rights they have been assigned, migrants can be empowered actors in local politics or might only mobilize in clandestine associations. For long periods of the twentieth century Western European states limited the rights of foreigners to be politically active and to fund and run political organizations, a policy meant to limit diaspora politics among exiles and prevent them from interfering with the political regime in their state of origin, upholding respect of other states' sovereignty.

Provisions in the country of origin (especially when still the country of citizenship) can also severely limit the opportunities of emigrants to be politically active while living abroad, a provision generally meant to prevent emigrants from organizing to intervene in domestic politics and to suppress 'foreign' expectations about democratic rights, political activism and political culture in the event they return home. The states of origin can, however, promote migrant self-organization as a way to control their nationals, to make sure they are encouraged to maintain strong ties with their country of origin so that cultural habits and nostalgic memories are kept high, children are taught their heritage language, remittances are sent to extended family, investment is made 'at home', and the country's interests are promoted in the new country of residence. These organizations are important instruments through which governments can reach their nationals; they can also be used to lobby or to pressure the government of the country of residence, when the migrant group is sizeable. Polarization and division can also be easily created in the country of residence. For example, after the attempted coup of 15 July 2016 in Turkey, both Erdogan supporters and Gülen proponents reported fears of serious clashes among members of the Turkish diaspora in the Netherlands and both complained that the Dutch state would not protect them sufficiently against threats from the other camp.

Nowadays many countries like the Netherlands and Sweden have granted foreign legal residents' voting rights for local elections. Moreover citizens of the European Union, living in a country other than where they hold citizenship, have voting rights for local and European elections where they reside. This influences the local electoral dynamics in cities with a large migrant population.

Differences in organization strategy echo to some extent the political organization and culture of the country of origin and therefore differs among groups of different origin in the same city. Differences are also quite noticeable among the receiving political cultures of the countries of reception. The Netherlands, for example, originally strongly encouraged religious organization as a way to organize guest workers arriving from Muslim countries, as it seemed to fit the confessional logics of Dutch society that had been 'pillarized' into segregated politico-denominational pillars for most of the twentieth century as the result of the pacification of struggles between Protestants, Catholics and Socialists. This strategy was adopted despite the fact that those consociational arrangements were steadily losing ground with the secularization of the 1960s and that religion-based services like schools, hospitals, universities, public broadcasts, etc. were increasingly anachronisms in a largely atheist society. France, by contrast, foregrounded a republican ideal to elude any kind of communitarian organization that would interfere in the direct relation between the state and the individual. A similar contrast can be found in Belgium between Flanders (with consociational arrangements between Catholics and Socialists) and Wallonia (more akin to France). As a result, migrant organizations in Flanders often use communitarian labels (religious, national or ethnic) while their counterparts in Wallonia prefer general labels (anti-racism, anti-discrimination, inclusion...).

This political opportunity structure is much more than the assemblage of formal rules about citizenship, human, civic, political and social rights, and access to legal protection through supranational legal frameworks enforced by the European Court and/or the Court of Justice of the European Union. It is complemented by informal rules and practices of inclusion and exclusion performed in social interaction by non-state actors in all walks of daily life. These informal rules and practices can vary quite a lot among cities, even within the same state with consistent immigration and citizenship laws. Local governments have their own agency in the translation and application of national integration policies. Contrasting Amsterdam and The Hague, Hoekstra (2015) reveals quite different conceptions of diversity and citizenship in the two cities located only 60 km apart, within the same centralized welfare state.

Local governments, local political parties and civil society organizations can be more or less inclined to welcome migrants into their ranks, to support their demands and to negotiate alliances with them. Depending on attitudes, it might be easier to develop separate organizations and then enter alliances with other local actors, or to participate in existing organizations. In the Netherlands, for example, there are significant differences between Amsterdam and Rotterdam that reflect broader differences in local political culture (more diverse leftist movements in the first, a stronghold of the Labour party in the second), different socio-economic dynamics (more diverse sectors with finance and services in the first, industrial and port activities in the second) and different migration flows (more diverse, economically and culturally, in the first). Similarly, in their comparative studies of minority politics and immigrant rights activism in Los Angeles,

Paris and Amsterdam since the early 1970s, Nicholls and Uitermark (2013, 2017) document the specificity of local processes. For example, they analyse contradictory trajectories (from multicultural clientelism to class empowerment through a broad alliance for social justice in Los Angeles; from class empowerment to division and religious containment in Amsterdam), explaining them through differing cultural narratives and frames and differing local state capacities to co-opt minority groups (Nicholls and Uitermark 2013).

This brings us to another dimension of collective action beyond grievances, resources and opportunities – namely the ability to frame and reframe societal issues. In this regard the very figure of the migrant should be problematized (see Nail 2015). The foregrounding of a migrant identity – at the expense of other identities – is a relatively recent phenomenon, even if some can argue that the tensions between mobile and sedentary people are recurrent in the history of humankind (as ancient myths in many cultures remind us). In many Western European states, the category of *migrant* is taken for granted, although the group it invokes may be vague and include locally born descendants of migrants, even when only one of their parents was an immigrant. The term *migrant* is often used in combination with other labels that allow the foregrounding or backgrounding of different features (foreigners, strangers, minorities, autochthony/allochthony). While many migrant issues are framed in an essentialized and ethnicized way, labels are disputed and contested and shift over time. In many Western European countries, we have seen a shift from national labels referring to states of origin (Turks, Moroccans, Pakistani) to religious labels referring to religion of origin (Muslims, Hindus).

Migrants' descendants (inappropriately called second generation migrants), with access to the national language and to the education system, are better able to produce new framings that place themselves at greater distance from their parents' country and culture of origin. These new framings have evolved in the context of a common experience of discrimination, racism and colonialism, but also stress belonging and agency in the country of residence. These framings and experiences have given rise to diverse movements ranging from *SOS Racisme* (SOS Racism) in the early 1980s, and later *les indigènes de la république* (the Indigenous of the Republic – indigenous meaning here the indigenous people of (North) Africa in the French colonial empire) in France, *Afro-Deutsch* (Afro-German) in Germany, the *Arabisch-Europese Liga* (Arab-European League AEL) in Belgium and the Netherlands, and *Black Lives Matter* in the UK, to name only a few initiatives (see El Tayeb's work on postnationality and race (El Tayeb 2008) and what she calls the queering of ethnicity in postnational Europe (El Tayeb 2011)). Others do not cultivate their origins but choose instead to join new transnational Islamist political movements that only superficially relate to the cultures of their parents and grandparents.

It is worth noting that some recent mobilizations insist on the identity of migrants as mobile persons – almost devoid of other identities. After the 2004 and 2007 EU enlargements encompassing most former Eastern European countries, mobile EU citizens (especially Roma from Central and South Eastern Europe) were frequently confronted with attempts of local governments to prevent them from camping within their jurisdiction. Irregular migrants and asylum seekers often claim space for themselves. Near the entrance of the Eurotunnel under the Channel near Calais, France, spontaneous settlements of migrants aiming at entering the United Kingdom evolved into an autonomous city known as *the Jungle* (until its displacement in 2016) (see Davies and Isakjee 2015). Asylum seekers whose applications were rejected have often tried to get by in the city, organizing to make themselves visible and claim their right to the city. In Amsterdam *Wij zijn hier/We are here* is such a group. Asylum seekers in Hamburg, united by the common experience of a traumatic crossing of the Mediterranean, did the same under the banner *Lampedusa in Hamburg*.

Anti-immigrant mobilizations in cities

Anti-immigration activism must also be considered to understand the impact of migration on identity politics: immigration has become a major issue in urban politics. Urban mobilizations can target immigrants directly, or contest specific changes in the city, in certain neighbourhoods, and/or in public spaces associated with the presence of migrants. Questions about territory, belonging and a symbolic sense of disruption are related to (perceived and expected) clashes between newcomers and those already established in a certain place.

Anti-immigrant mobilizations pursue a politics of spatial (in)justice that does not focus on the precarious position of migrants in the city. Instead they frame the presence of migrants as a nuisance in certain neighbourhoods, and as an extra burden for other disadvantaged urban residents. Socio-economic issues are sometimes explicitly articulated – when narratives highlight intensified competition for lower-paid and precarious jobs, for poor but cheap housing or for limited and mediocre public services (education, health, welfare...). Often these issues are labelled as cultural, stressing the impossibility of living together due to incongruous beliefs, customs and practices of newcomers. Everyday irritations with visible, audible, smellable or otherwise tangible expressions of different cultures become part of anti-migrant political framings. Specific social and cultural facilities for migrants such as mosques are often perceived as neighbourhood liabilities – comparable to a highway or a waste facility, and targeted by NIMBY (not-in-my-backyard) mobilizations of would-be neighbours for the noise and traffic that they might generate (see Uitermark and Gielen 2010 about the role of the media in such a struggle around the establishment of a large purpose-built mosque in Amsterdam and the role of the discursive power of opponents and proponents, but for other interpretations about the same struggle see Sunier 2006 or van Westerloo 2006.)

When they resist social changes induced by the arrival of newcomers, these movements often explicitly criticize the social groups that welcome immigration at the macro level: economic and political elites that support(ed) the import of cheap labour and/or political and cultural elites that support(ed) cultural diversity, very few of whom live and work with migrants on a daily basis. In other words, these mobilizations express the sense of powerlessness of those who perceive immigration as a problem imposed upon them in the interest of others. Moreover, they tend to blame newcomers for not blending in: migrants – unlike established residents – have presumably made a choice to migrate and are expected to adapt to their new place of residence – regardless of the actual circumstances in which migrants made their 'choice'.

While tensions between newcomers and established inhabitants of a place may be inevitable, there are many ways to frame them and to resolve them. Migration has become the single most significant political issue in Western European cities in the past decades, but the nature of the issue has changed over time. In the 1960s and 1970s, trade unions, communist and socialist movements, labour parties and urban movements in Western Europe were inclined to support foreign workers' rights, empowering both groups at the same time: they aimed at both improving the lives of foreign workers and preventing them from being used to undermine the collective bargaining position of national workers. Once guest workers settled, became citizens or obtained voting rights in local elections, labour and socialist parties were also keen to mobilize them in electoral politics.

Later on, mobilization aimed at the social and political exclusion of immigrants and eventually their departure became more common. These anti-immigrants mobilizations reflect different shades of nativism and nationalism, but they share the naturalizing of an 'us versus them' competitive binary, essentialist identities and an inclination towards analogies between migration and military invasions or natural disasters (flood, tsunami). These movements can be labelled *identitaires* (identitarian movements) to stress their focus on fixed notions of identity to be defended against newcomers (as if identities were not always in flux and as if migration was the only cause of cultural change).

These anti-immigrant movements are at least as diverse as those of migrant populations (and the distinction between the two is not as sharp as expected as it is not uncommon to see former migrants metaphorically 'closing the door' to newer newcomers). They can also be characterized according to their particular grievances, resources, opportunity structures and framings of problems and solutions. Such mobilizations often distinguish between different categories of migrants, the newcomers that can and do adapt and those who do not fit because they are 'too different', or between deserving and undeserving migrants ('genuine refugees' versus 'gold diggers'). Anti-immigrant groups are sometimes overtly racist (although white supremacy is a marginal ideology in Europe, while cultural superiority is often taken for granted), at times anti-Semitic, sometimes conservative Catholic (against the more progressive current in the Catholic Church), and oftentimes anti-clerical and anti-religious, opposing in particular political Islamism. Local urban identities are sometimes more attractive and 'politically correct' alternatives to nationalism as a basis for resisting globalization and Europeanization and the migration flows that accompany them. More recently, nationalism has been revived through new parties targeting both migrants and the European Union, while promoting national sovereignty (for example the UK Independence Party UKIP in the UK) and opposition to immigration and migrants. These were key issues in British cities in the campaign for Brexit in the June 2016 referendum.

Such movements and parties generally have a strong populist character, expressing opposition to political and cultural elites and established political parties. Labour parties have been particularly affected – locally and nationally – as they have often been seen as the political parties that have been the most ardent defenders of migrants' rights and interests, at the expense of natives'. Their success mobilizing the new migrant electorates is now threatened by severe backlashes (see the Dutch pun regarding the acronym of the Labour Party PvdA as *Partij van de allochtonen* instead of *Partij van de Arbeid-*, literally Party of the Allochtones – the label used as blanket etiquette for foreign-born residents – instead of Party of the Labour).

Spatially, mobilizations against migrants focus on the defence of specific territories. Narratives of 'invasion' to be resisted have been framed at different scales (Mamadouh 2012), calling for action to stop migration and its consequences for the demographic makeup of certain neighbourhoods, cities, countries or even complete continents. A particular struggle pertains to the use of, and practices in, public space. The dress codes and public activities of certain migrant groups have been perceived as changing the character of public space and affecting others' freedom to use, and feel comfortable in, that space. The French case has been most politicized around the use of Islamist headscarves (see for example Hancock 2014). Typically, groups that have benefited from the freedom and tolerance of urban life – such as women and LGBTIQ people (LGBTIQ standing for Lesbian, Gay, Bisexual, Transsexual, Intersex and Questioning) – see their individual freedom contested directly or indirectly when newcomers wish to organize public space according to norms that do not allow for the overt expression of gendered and sexualized identities. Incidental aggressions against women and LGBTIQ people by migrants or their descendants are in turn politicized by anti-migrant movements that claim equal rights for women and homosexuals (and by extension other queer sexual identities) as cultural heritage – no matter how recently these have been achieved. This kind of homonationalism, 'pinkwashing' xenophobia, is only one of the expressions of the disintegration of a broad alliance for diversity among diverse minorities (ethnic and sexual) that emerged in the 1970s in Western cities. But this should not hide real concerns with women's and LGBTIQ rights that the state should guarantee, nor with issues of cultural relativism in criminal law cases.

Urban politics from above: urban policies, migration policies and diversity

The anti-immigrant grassroots has been associated with revanchist urbanism, especially when it has been able to capture local governments (a function of both their level of support and the electoral system). Comparing the reconquest of the (inner) city by capital and middle class originally described in the USA to Western Europe, but stressing the specificities of revanchism in Western European welfare states (financed largely by the central state), Uitermark and Duyvendak (2008) discuss the confrontational discursive politics that emerged in Rotterdam in the early 2000s with the electoral success of a populist party that replaced the reformist and emancipatory policies of the Labour Party. Cities have become laboratories for the management of urban marginality. Uitermark (2014) reviews urban policies and integration agendas in six Western European countries, including the Netherlands and the United Kingdom, as attempts to manage the urban marginality that results from neoliberal globalization (both the deregulation of the economy and international migration). In his view, heavy government intervention in deprived urban neighbourhoods is the direct successor of the national welfare policies of the Fordist economy. In any event urban policies are not only local, but involve redistribution mechanisms at higher scale levels. Many central states have developed urban policies (France, the UK, the Netherlands) and at the supranational level the European Commission has designed urban policies (for an early analysis of EU policies for social inclusion see Samers 1998).

Again, the diversity of urban identity politics should be emphasized. Broadly speaking we can contrast two models: one following a cultural logic and one a territorial logic. In the Netherlands, for example, policies target minority ethnic groups that the government has identified as needing extra care (Turks, Moroccans, Surinamese, Antilleans); in France, territorial policies target neighbourhoods that the government has identified as needing extra care. These approaches reflect different ways of thinking about urban problems and urban identities, different ways of collecting information (registration of ethnicity/birth place, etc.) and different political dynamics: the first foments ethnic prejudice (e.g. the 'Moroccans') and ethnic mobilization, the second territorial prejudice (the 'banlieues') and territorial mobilization. In the Netherlands unsatisfactory results with group targeting led to experimentation with territorial policies and the selection of priority neighbourhoods, while in France there have been pleas for policies to assist specific minorities. Both ethnic and territorial policies combined seem to generate particularly explosive situations (on *het Zwart Beraad*, literally the black caucus, in local politics Amsterdam Zuidoost, see Dukes (2007); on riots in the French *banlieues*, see Dikeç 2007a and 2007b).

In the EU urban interventions are always multiscalar. National urban policies (or EU urban policies for that matter) are eventually implemented by local authorities. More recently the retreat of the national state has been 'compensated' by giving local authorities more say and more responsibilities (but generally not proportionally more financial means) to tackle a broad range of urban and social issues. This resonates with a broader movement that frames mayors as pragmatic and efficient managers who can provide better solutions to social problems than national states (e.g. Barber (2013) and his global parliament of mayors). The European Commission has long developed urban policies to target urban problems and is trying to involve cities more systematically in policy-making to make sure that its policies are 'urban proof' (codified in the Urban Agenda of the EU which was eventually adopted in 2016 after years of negotiations).

Local authorities react to national policies when they impact local populations, and sometimes even to the national policies of neighbouring countries. Denmark's restrictive family formation policies for Danish citizens with a foreign partner, combined with free moment in the EU, have pushed binational families to settle in Malmö, on the Swedish side across the Öresund, even when family members work in Copenhagen. In some other countries, local authorities have tried to compensate for ostensibly inadequate national policies though local policing and disciplining. In the summer of 2016, in the aftermath of the terrorist attack and mass murder claimed by the Islamic State of Iraq and the Levant (ISIL) in Nice on Bastille Day (the French National Day), several coastal municipalities in the Alpes-Maritimes department on the French Côte d'Azur, including Nice, banned the burkini bathing dress. This was done in the absence of national legislation, with municipalities claiming they were trying to prevent civil unrest on the beaches. Their argumentation was eventually rejected by the highest administrative court, the *Conseil d'État*, due to the policy's disproportional impact on the fundamental freedom of individuals.

Cities have not hesitated to act against national immigration policies or at least to refrain from facilitating their implementation. In 1979 Los Angeles proclaimed itself a sanctuary city and it has been followed by many US and Canadian cities. Sanctuary cities have adopted local ordinances to prevent police officers and local civil servants from checking the immigration status of people, therefore not collaborating with immigration authorities' attempts to expel immigrants. In the 2010s, Dutch cities have defied national state policies regarding asylum seekers whose asylum request had been rejected. These migrants were expelled from reception centres and supposed to leave the country, but did not (for a variety of reasons) instead becoming irregular residents in Dutch cities. Although the central state forbade aid to these irregular residents, cities have developed minimal reception policies at the local level (the so-called bed-bath-bread measure, see Versteegh 2016) to alleviate dramatic situations.

The so-called 2015 European migration crisis prompted some European cities to act directly to deal with the uncoordinated arrival of migrants. Barcelona (run by Barcelona en Comú, a citizens' platform, since 2015) declared itself a 'refugee city' and in March 2016 reached direct agreements with Athens and Lesbos in Greece and Lampedusa in Italy. In April 2016 the Mayors Conferences of European Capital Cities similarly discussed collaboration for the reception of migrants under the leadership of Athens and Amsterdam. These cities are also key players in the Urban Agenda for the European Union pilot partnership on migration and asylum, a collaborative effort of Member States, the European Commission, and cities to improve EU policies and their implementation at the local level.

Next to such practical policies, many local authorities have sought to frame urban identity as diverse. This framing can be seen as the reinvention of multiculturalism as urban citizenship (Uitermark et al. 2005), foregrounding the city as the site of negotiation of multicultural interactions and stressing the fluid, multifaceted dynamics of urban identities (as opposed to rigid national ones).

Beyond urban identity politics, diversity has become an instrument for city branding (such as I AMsterdam), invoked to boost one's score in Richard Florida's creative city framework, an influential model among local politicians and policy-makers trying to make their city competitive in the global economy. This strategy includes the promotion of ethnic entrepreneurship, niche shops, Chinatowns and the incorporation of newcomers in local cultural events. The latter include the adoption of new local traditions created by migrants such as the Summer Carnival in Rotterdam (initiated in 1984 by migrants from Latin America and Cape Verde wanting to have organized street parades) or the Kwaku Summer Festival in Amsterdam (created in 1975 by the Surinamese community and named after the day of the abolition of slavery in the Dutch colony of Suriname in 1863), as well as the incorporation of migrants in local traditions (such as the 'election' of a Syrian 'Prince Carnival' – the shadow mayor of the city during Carnival – in Maastricht in 2016), and new mega-events (such as ethnically branded boats at the Canal Parade of the annual Amsterdam gay pride celebration).

Conclusion

Despite the problems of discrimination and contention over multiculturalism, the diversity produced through migration is now widely acknowledged as a reality. It has been branded as "super diversity", a term coined by Vertovec (2007) to stress the growing diversity in the demographic make-up of cities. Others speak of "hyper diversity", "extreme diversity" (Crul 2015) or "deep diversity" (Kraus 2008 and 2011). They stress the diversity of groups and the diversity within groups in the city and the fact that many cities like Brussels or Amsterdam have actually become majority-minority cities (Crul 2015): cities where there is no ethnic majority, not even the main national group of the state. The ascendancy of migrants to higher local office is often seen as emblematic for these diverse cities: Ahmed Aboutaleb was appointed mayor of Rotterdam in 2009, Anne Hidalgo elected mayor of Paris in 2014, and Sadiq Khan elected mayor of London in 2016 (although he was born in London his parents' foreign origins have been widely foregrounded). *Our City*, the enchanting documentary of (Italian born Brusseler) Maria Tarantino (2015), is another example of a new hyper-diverse local identity that celebrates the presence and languages of the many minorities in Brussels, a majority-minority city (tellingly the original title is in English, not in French and/or in Flemish).

We should not romanticize these multicultural encounters as migrant groups often seem to live in parallel worlds. Nor should we be oblivious to the conflicts of these extremely diverse cities, nor to the *mal être* of urbanites longing for an imaginary culturally homogenous city (be it informed by a nostalgic representation of the same city fifty years ago or by a dystopian religious fundamentalism). It could be argued that most urban residents are well-acquainted with diversity and negotiate their way through the identity politics of the city. In that case, a major challenge of urban identity politics in the 2020s may be the management of the clash between super-diverse cities where multiculturalism, multilingualism, multiracialism is celebrated or contested but always lived, and their rural surroundings where this diversity is observed from a distance and considered an imminent threat to local homogenous enclaves. In the Dutch case, for example, the neo-nationalist PVV was electorally very strong in Volendam, a town with hardly any migrants, located nearby Amsterdam. Perhaps we should revisit the photography of Ingrid Pollard, portraying her out-of-place-ness as a Black Briton in the rural English landscape (Woods 2005) and develop understandings of urban identity politics that take into account relations not only within the city, but between the city and the many (rural) places around it.

References

Barber, B. (2013). *If Mayors Ruled the World: Dysfunctional Nations, Rising Cities*. New Haven, CT: Yale University Press.
Çaglar, A. (2006). Hometown Associations, the Rescaling of State Spatiality and Migrant Grassroots Transnationalism. *Global Networks*, 6(1), pp. 1–22.
Çaglar, A. (2007). Rescaling Cities, Cultural Diversity and Transnationalism: Migrants of Mardin and Essen. *Ethnic and Racial Studies*, 30(6), pp. 1070–1095.
Coleman, M. (2012). The 'Local' Migration State: The Site-Specific Devolution of Immigration Enforcement in the U.S. South. *Law & Policy*, 34(2), pp. 159–191.
Crul, M. (2015). Super-diversity vs. Assimilation: How Complex Diversity in Majority–Minority Cities Challenges the Assumptions of Assimilation. *Journal of Ethnic and Migration Studies*, 42(1), pp. 54–68.
Davies, T. and Isakjee, A. (2015). Geography, Migration and Abandonment in the Calais Refugee Camp. *Political Geography*, 49, pp. 93–95.
Dikeç, M. (2007a). *Badlands of the Republic: Space, Politics and Urban Policy*. Oxford: Blackwell.
Dikeç, M. (2007b). Revolting Geographies: Urban Unrest in France, 1: 1190–1206. *Geography Compass*, 1, pp. 1190–1206.

Dukes, T. (2007). *Place, Positioning and European Urban Policy Discourse: Examples of Politics of Scale in 'Brussels' and the Netherlands*. PhD Thesis. University of Amsterdam.

El-Tayeb, F. (2008). 'The Birth of a European Public': Migration, Postnationality, and Race in the Uniting of Europe. *American Quarterly*, 60(3), pp. 649–670.

El-Tayeb, F. (2011). *European Others: Queering Ethnicity in Postnational Europe*. Minneapolis and London: University of Minnesota Press.

Gilbert, L. (2009). Immigration as Local Politics: Re-bordering Immigration and Multiculturalism Through Deterrence and Incapacitation. *International Journal of Urban and Regional Research*, 33(1), pp. 26–42.

Hancock, C. (2014). 'The Republic Is Lived with an Uncovered Face' (and a Skirt): (Un)dressing French Citizens. *Gender, Place & Culture*, 22(7), pp. 1023–1040.

Hoekstra, M. (2015). Diverse Cities and Good Citizenship: How Local Governments in the Netherlands Recast National Integration Discourse. *Ethnic and Racial Studies*, 38(10), pp. 1798–1814.

Kraus, P.A. (2008). *A Union of Diversity: Language, Identity and Polity-building in Europe*. Cambridge: Cambridge University Press.

Kraus, P.A. (2011). The Multilingual City: The Cases of Helsinki and Barcelona. *Nordic Journal of Migration Research*, 1(1), pp. 25–36.

Lévy, J. and Lussault, M., eds. (2003). *Dictionnaire de la géographie et de l'espace des sociétés*. Paris: Belin.

Mamadouh, V. (2012). The Scaling of the 'Invasion': A Geopolitics of Immigration Narratives in France and The Netherlands. *Geopolitics*, 17(2), pp. 377–401.

Nail, T. (2015). *The Figure of the Migrant*. Stanford, CA: Stanford University Press.

Nicholls, W. and Uitermark, J. (2013). Post-multicultural Cities: A Comparison of Minority Politics in Amsterdam and Los Angeles, 1970–2010. *Journal of Ethnic and Migration Studies*, 39(10), pp. 1555–1575.

Nicholls, W.J. and Uitermark, J. (2017). *Cities and Social Movements: Immigrant Rights Activism in the United States, France, and the Netherlands, 1970–2015*. Chichester: Wiley Blackwell.

Nicholls, W., Miller, B. and Beaumont, J., eds. (2013). *Spaces of Contention: Spatialities and Social Movements*. Aldershot: Ashgate.

Østergaard-Nielsen, E. (2003). The Politics of Migrants' Transnational Political Practices. *The International Migration Review*, 37(3), pp. 760–786.

Samers, M. (1998). Immigration, 'Ethnic Minorities', and 'Social Exclusion' in the European Union: A Critical Perspective. *Geoforum*, 29(2), pp. 123–144.

Sunier, T. (2006). The Western Mosque: Space in Physical Place. *ISIM Review*, 18, pp. 22–23.

Tarantino, M. (2015). *Our City*. Brussels: Wildundomesticated. www.ourcityfilm.com/.

Uitermark, J. (2014). Integration and Control: The Governing of Urban Marginality in Western Europe. *International Journal of Urban and Regional Research*, 38(4), pp. 1148–1436.

Uitermark, J. and Duyvendak, J.W. (2008). Civilising the City: Populism and Revanchist Urbanism in Rotterdam. *Urban Studies*, 45(7), pp. 1485–1504.

Uitermark, J. and Gielen, A.-J. (2010). Islam in the Spotlight: The Mediatisation of Politics in an Amsterdam Neighbourhood. *Urban Studies*, 47(6), pp. 1325–1342.

Uitermark, J., Rossi, U. and van Houtum, H. (2005). Reinventing Multiculturalism: Urban Citizenship and the Negotiation of Ethnic Diversity in Amsterdam. *International Journal of Urban and Regional Research*, 29(3), pp. 622–640.

van Westerloo, G. (2004). Revolutie bij de zwartekousenmoslims. *M-Magazine NRC Handelsblad*, pp. 11–22.

Versteegh, L. (2016). About Bed, Bath and Bread: Municipalities as the Last Resort for Rejected Asylum Seekers. In: V. Mamadouh and A. van Wageningen, eds. *Urban Europe: Fifty Tales of the City*. Amsterdam: Amsterdam University Press, pp. 363–368.

Vertovec, S. (2007). Super-diversity and its Implications. *Ethnic and Racial Studies*, 30(6), pp. 1024–1054.

Woods, M. (2005). *Rural Geography: Processes, Responses and Experiences in Rural Restructuring*. London: Sage.

39
CLASS, TERRITORY AND POLITICS IN THE AMERICAN CITY

Kevin R. Cox

Introduction

In the United States, the two decades following the Second World War are commonly regarded as the period when suburbanization first seized the public imagination. They also happened to coincide, for the most part, with a stiff political reaction against the progressive elements of the New Deal. Led by big business and marked by major assaults on labour, not least the Taft-Hartley Act of 1947, and then bolstered by McCarthyism, the way was paved for a Republican hegemony. In some minds the two facts of suburbanization and a shift to the right were connected. This generated a debate among political scientists and particularly those interested in voting behaviour. Did increasing suburbanization, therefore, imply a long-term tilt to the advantage of the Republicans? Or could any contextual effect be disregarded in favour of one that emphasized the socially selective nature of suburbanization, so that explanation for the Republican resurgence had to be found elsewhere?

This is mentioned at the outset to emphasize the close relation that was seen, at least by some, between what was happening in urban areas, a new socio-spatial nexus perhaps, to the national level. This book is about urban politics and there are certainly logics of social change and of power relations distinct to urban areas. But these have not only fed into developments at the national level; the federal government in turn has been implicated in creating that context within which seemingly distinct urban processes would unfold.

For subsequent to the Second World War, politics in urban areas *did* change and not necessarily in the immediately partisan ways that interested the political scientists. Rather what is concomitant of the – admittedly expansive – suburbanization of the period, and which continues down to the present day, is what some have tried to grasp through the idea of a politics of turf: a highly territorialized politics focusing on the activities of neighbourhood organizations and resident groups as they try, through various exclusionary and inclusionary tactics, and not just in the suburbs, to structure their living place environments. In fact the term 'politics of turf' does not quite grasp what has been going on since class relations in the loose sense of social stratification have been absolutely constitutive. Rather what has unfolded over the last seventy years has been more a territorialized class politics, and one with very serious distributional effects in urban areas: matters of the distribution of real income (Harvey 1973: Chapter 2).

It is this particular politics that is the present focus: how it came about, what it implies for people's identities, both in urban areas and nationally; how, that is, it has tended to nurture support for a broader socially conservative alliance. At the same time, it has been underpinned by developments, including government policies, at that same more all-embracing scale.

In thinking about urban politics, including its spatiality, I take the centrality of the accumulation process as axiomatic. The question then becomes one of how it has unfolded over space, the contradictions that it has generated as a result of its spatial integument, and the subsequent social tensions. Harvey (1985a) has written about this but largely in the context of what is usually referred to as the inter-urban: a focus emphasizing the importance for urban politics of competition between urban areas, the class relations underlying that competition, rather than what occurs within. What he has emphasized is the contradiction between capital's necessary fixity in urban areas and the general indifference of capital in its more mobile forms to that fixedness: bolstering economic bases here, undermining them elsewhere and calling forth strategies to structure that movement to local advantage. Cross-class coalitions of capital and fractions of the working class with their own immobilities and which see themselves as also threatened by, say, plant closures, or simply seduced by the promise of belonging to a major league city, come together around programmes of defending a particular inter-urban geography; even deepening it to their own advantage.

But in many respects these relations, tensions and tendencies are mirrored by processes *within* urban areas. There is the same sort of business-orchestrated competition over space, incorporating factions of the working class, and designed to ensure that value continues to flow through particular areas and neighbourhoods. Typically it has been developers with anxieties about rent appropriation, that have taken the lead, though sometimes in alliance with local governments which have their own particular interests in appropriating rents through the property tax. The struggle between the older, inner suburbs and the outer ones that have the advantage of access to the large swathes of undeveloped land favoured by the developers is exemplary. In an alternative riff on this theme, the owners of older shopping centres see their anchor stores departing for newer ones further out and do battle to the extent that they can, drawing on the support of local residents who see them as a stabilizing force, a symbolic barrier to the *sans-culottes* beyond, and who will willingly join in the campaign to frustrate the competition, as in opposing tax breaks or infrastructural improvements that the city government might offer the upstarts. A developer planning a major shopping centre may also join forces with neighbourhood groups to stymie another centre which is ahead in the planning game. It is certainly not the case, though, that shopping centres are always part of the story. Developers of long-stay residential projects will ally with their recent homebuyers to beat off the proposals of another developer which threaten values already in place.

Cross-class coalitions like these are by no means always assured. There is often conflict between developers and residents around who will pay for the social costs of development: expansion of local schools, improvements to local highways, expanded water and sewer provision – anything, in other words, that developers will habitually try to push off on to somebody else. There is also what is, in effect, a metro-wide struggle among different factions of the working class, organized seemingly along territorialized lines – this or that neighbourhood or even independent suburb – but representing wider social strata and their interests. This is that massive majority of the population living off wages, salaries or pensions, as opposed to property income, but divided along stratum lines: something so clearly evident in the struggles to avoid displacement by gentrification, or to exclude the housing projects from which the less-well-to-do are likely to benefit – smaller houses, apartments or, dread the thought, public housing.

This chapter is about understanding how this politics came to be. What did the world have to be like for it to be possible? In part it has been a question of social changes of quite striking diversity: not just the expansion of homeownership that has given so many fits about what new developments will do for their property values, but changes in types of employment, and family structure, including gender; and even the expansion of automobile ownership that has allowed the creation of a metropolitan housing market and made that social mix that is so repellent to some, a possibility. But social change aside, social changes that could not necessarily be anticipated, the real estate industry, that branch of capital responsible for the provision of all manner of premises including housing, has been massively involved in providing the conditions for this politics. And it is no coincidence that the explosion of residential exclusion and defence of turf corresponds to the maturation of property capital; its perfected separation from capital's industrial, commercial and financial branches. Here though we are dealing with something more systemic than conjunctural. Again, as with the various social changes that have been significant, attention has to turn to developments that have been much more national than urban. The urban has been made by changes beyond its boundaries. But by the same token, what has transpired in urban areas has had implications for national politics: its own contribution, more specifically, to the great moving-right show.

Territorializing the class politics of the living place

After the mid-1960s, then, the metropolitan area seemed to be composed of a mélange of service and economic areas, varied in size, which existed to foster pursuit by individuals of personal goals and objectives. Each of these areas competed for economic resources, power and 'top-notch' citizens, and the competition not only pitted Cincinnati against its suburbs but also big city neighbourhood against big city neighbourhood, suburb against suburb, and neighbourhoods within a particular suburb against one another. Each of these communities, in other words, comprised a community of advocacy. And the larger units, such as Forest Park or Cincinnati, constituted a community of advocacy made up of smaller communities of advocacy.

(Miller 1981: 239)

The most important motivation contributed by the socio-cultural system in advanced capitalist societies consists of syndromes of civil and familial-vocational privatism ... Familial-vocational privatism ... consists in a family orientation with developed interests in consumption and leisure on the one hand, and in a career orientation suitable to status competition on the other. This privatism thus corresponds to the structures of educational and occupational systems that are regulated by competition through achievement.

(Habermas 1973: 75)

In his book on neighbourhood politics in the Cincinnati area, Miller does not make reference to Habermas's familial-vocational privatism, nor even to Young and Willmott's (1975) suburbanizing 'symmetrical family' though, and despite the European provenance of these ideas, he might reasonably have done so. And if there is an error in Miller's description it is that it is insufficiently attuned to the obverse of the 'competition for "top-notch" citizens', which was and remains, of course to keep out those who aren't 'top-notch'. It would not be till 1976 that exclusionary politics in itself was given some lengthy academic attention (Danielson 1976). There is, though, empirical evidence of an impressive post-war surge of resident opposition to multi-family projects, to re-zonings to high density as well as to re-zonings for non-residential use (Cox 1984: 289–292). Equally if not more significant is the way social science attention shifted to questions raised by the jurisdictional fragmentation so closely associated with the competition that Miller refers to and with exclusionary policies. The economists tried to explain it through a Candide-like salve bookended by Tiebout's public choice approach in 1956 and Wingo's (1973) likening of local governments to clubs. Sociology and political science tended to be more critical. Newton (1975) provided a stinging rebuttal to the economists. Researching intra-metropolitan inequalities became something of a cottage industry and the political scientist Oliver Williams (1971) was the first to provide a synoptic view linking them to fragmentation and exclusion.

Even so, analysis remained incomplete. In particular it lacked any sort of historical depth, linking what was happening to broader social changes of the sort that Habermas attributes to 'advanced capitalist societies'. One starting point is the change in how people related to their living places: from the living place as community, a highly particularized notion of something of a non-substitutable sort; to a position in which it was viewed as more of a commodity – as composed of various separable attributes like local schools, exclusion of commercial uses, a particular neighbourhood aesthetic,[1] and social environment which could be compared across places, given a price and traded. This would then become the object of attention of the home value studies that proliferated from the 1970s on, assigning so much explanatory significance to different attributes.

Living places as self-referential, inward-looking communities could and did emerge around different nucleating and isolating forces, sometimes overlapping and sometimes reinforcing one another. Limits to personal mobility were an important condition for occupational communities clustering around places of employment. Sharing a common origin from elsewhere and, as chain migration entered in, speaking a common foreign language, could do the trick. In both instances, the occupational community and the immigrant one, inter-marriage would add the cement of expanded kinship ties. People worked and lived alongside each other, they worshipped together, they rented houses from each other, and they even circulated their savings through the neighbourhood through a savings and loan exclusive to it.

From the 1950s on the breakup of these intensely face-to-face, interactive, communities starts attracting academic attention. Much of this was British. A landmark study was that of Bethnal Green in East London by Young and Willmott (1957) tracing the movement of younger married couples outward to suburban housing developments and the social changes that that seemed to set in motion: like a different way of valuing people, not so much in the round but more in terms of their outward display of possession. At the same time, and simply in virtue of dispersal, family structure started to shift: away from the strongly kin-focused environment of East London towards a more inward looking nuclear form.

Some of these themes were echoed in a later (1965) study of family change in South Wales by Rosser and Harris. Focusing on a community of tin-plate factory workers on the edge of Swansea they documented the breakup of what they call 'the cohesive society' and its displacement by 'the mobile society'. This advances the argument a little since it moves beyond the traces of spatial determinism evident in Young and Willmott and shifts attention to broader social change. In their book they document the dispersal of a younger generation from a community in which kin and neighbour had worked alongside each other, helped one another at times of unemployment and need, and, though they don't remark on it, one structured along the lines of a traditional gender division of labour – as had been the case in Bethnal Green. The change is both locational and occupational as the young move into a more far-flung marriage market, take up more white-collar sorts of work, marry from within the wider metropolitan area giving multiple meanings to the idea of a 'mobile society' that carries in its wake the dissolution of the 'cohesive' one.

There are no truly comparable studies for the American case but plenty of supportive evidence for similar sorts of processes. The fact of the occupational community, what John Bodnar would call the 'enclave community' is well documented (Bodnar 1982). Kin lived alongside kin to a remarkable degree. The effects of urban renewal in the 1950s and 1960s would then focus attention on the disintegration of some of these communities, recorded most notably in Herbert Gans' *The Urban Villagers*. John Alt (1976), recalling Young and Willmott's *Symmetrical Family*, then carried that breakup forward to a world of family-based, privatized consumption built around family vacations, home improvement and the television; a theme that is also apparent in Maurice Stein's (1960) survey of American community studies.[2]

How, therefore, might we understand what led to this change and brought about new connections to the living place that would be so pregnant in their implications for urban politics? The conjuncture of forces at work turns out to be quite complex. Not all of what came together moved in tandem. Habermas's 'familial-vocational privatism', like Young and Willmott's 'symmetrical family' is an ideal type, necessarily overgeneralizing and lacking analytical depth. What one can say, though, is that the changes were latent in the concrete trajectory of capitalist development: expanding out from the workplace therefore to affect how people lived their lives outside it.

In first place has been the deepening of not just the social division of labour but also the technical. The balance between mental and manual forms of labour, in the conventional senses, has changed. Job hierarchies have become more elongated. The probabilities of 'moving up' have increased. Sectoral change and the growth of state employment and of skills-intensive forms of manufacturing have had similar effects. Labour demand has shifted in the direction of employment requiring some sort of formal qualification, at one time a high school diploma and now a degree. Meanwhile the diversification of jobs, the increased locational discretion of firms in newer sectors, has helped spell the demise of the occupational community.

Family structure has been transformed: a more nuclear, child-centred family, distancing itself from grandparents both literally and metaphorically, and with gender roles somewhat less segregated than they used to be.[3] The meanings of the child and of gender roles both, have changed. In place of the child as the little worker of the nineteenth century, working from an early age to supplement the family budget has emerged the child as, in effect, an object of parental consumption: to be delighted in and to be shown off in public; even to be treated vicariously as a vehicle for their own frustrated ambitions or as potentially their *oeuvre*. Zelizer (1985) has linked this change to the late nineteenth century campaign of middle-class philanthropists to eliminate child labour; something that would then be reinforced, and over working-class protests, by the introduction of mandatory schooling and minimum school-leaving ages. Children now became expensive, paving the way for the widespread adoption of contraception and the transition to smaller families so apparent in the earlier years of this century. In other words, at least one of the conditions for reducing domestic labour and allowing women, still subject to gender norms, to take up wage work, was being put in place. Changes in occupational structure, towards employment less demanding of the classical 'masculine' attributes would lend further impetus to this and, in effect, hoist the masculinists keen to keep women out of the workplace, on their own petard.

Those changes in the character of occupations, the possibilities of upward mobility, would then combine with child centredness to create the career orientation that Habermas referred to. Parents wanted what was 'best' for their children and this has in significant part been defined in terms of their future life chances: putting them in a position where they can indeed, like their parents in *their* turn, engage effectively with "structures of educational and occupational systems that are regulated by competition through achievement" (Habermas 1973: 75).

This has occurred against a background of family-centred consumption: certainly the car, which initially was indeed the 'family' car, but most notably the privately owned home. This has made its own contribution to family-vocational privatism: a symbol of occupational achievement, an expression of individual enterprise, therefore, and of the middle-class norms of independence and providence. But this particular ingredient of familial-privatism, complementing other components in different ways, has proved central to the idea of the living place as commodity; as something to be bought and possibly sold and certainly a store of value for the family savings.

Home purchase provides access to a status-endowing address, a favoured school district, access to open space, a particular ambience. But it also exposes owners to the risk of devaluation. The disintegration of living place as community and an associated increase in personal mobility releases people for the process of what would come to be called, significantly, 'residential choice': something to be pandered to, even engineered, by the expansion of a speculative housing industry. From the standpoint of those who have already made 'the choice' the new developments promised by this turn of events can threaten values in place. New development can be contested and almost certainly will be. The house can clearly be sold but, even while new development is being opposed, this will likely, in virtue of the uncertainty, be at a lower value than otherwise. Homeownership, in other words, brings with it the dangers of fixity; the house can't be moved in order to take advantage of better market conditions elsewhere! And even when the decision is made to sell, the transaction costs of realtor fees, and the more intangible ones of putting the house in shape for buyers and looking for a house elsewhere add to the immobilizing effect. In other words: the living place now becomes something to be defended. As a commodity subject to devaluation, it assumes a new meaning.

It was in consequence of this that the newly developing suburbs on the urban periphery became battlegrounds in the 1950s, 1960s and 1970s. Meanwhile those making the move, embracing the living place as through-and-through commodified, left behind the decaying remnants of living places redolent of a different era; that of their parents who were brought up in a world populated by communities which by today's standards were more closely knit. What would hasten the final demise would be the emergence of the post-industrial city and renewed interest in living there on the part of people with money.

Gentrification produced a different sort of relation between the social strata. In the suburbs it was the better off trying to keep out the lesser mortals who threatened to come and live in tract housing or, so-called 'starter' homes, or worse yet, apartments. In the inner city, though, roles were, at least initially, reversed. Now it was those of lesser means trying to keep out the more affluent. The living place in question was to some degree seen in commodity terms: relatively low rents close to downtown sources of employment. But bonds of community might also be present, originating in some shared place of origin. Cybriwsky (1978) showed quite early on just how strong these ties could be as well as the depth of the resentment at the arrival of the early gentrifiers. But usually resistance has been a losing cause. Once incursions were made, the gentrifiers would take steps to further the process[4] and as real estate values turned to their advantage, to expel the recalcitrant; and then by a blanket re-zoning to single family[5] granted by a city equally keen to promote 'neighbourhood improvement' or 'renewal', try to ensure that they would not return.

In sum, this is a politics of the living place that deepens the uneven development of the urban area, chasing away the *démunis* into the older housing, so long as the gentrifiers have yet to discover it, the weaker school districts and the blue-collar suburbs: the neighbourhoods that get to be defined as 'gritty'. For the very poor there is a different sort of fixity; instead of one that is chosen, it is imposed. They are trapped: consigned to parts of the city that they wish they could escape. Interest in any form of action to improve conditions is weak and confined to a core of stranded homeowners possibly supplemented by urban 'pioneers', struggling against brownfield contamination, housing abandonment, the homeless shelters expelled from a tarted-up downtown (Mair 1986) and drive-by shootings. Meanwhile people move around a few blocks at a time in search of something that is a marginal improvement.

This is an uneven development that gets reproduced, in part quite effortlessly and partly by the expenditure of political muscle. The effortless bit is a politics of difference that is utterly taken for granted: a division between those who 'made it' through their own hard work and those who could have done but in virtue of personal shortcomings, didn't. The absurdity of this, and the fallacy of composition aside (can everyone be a lawyer?) is obvious. It is both self-congratulatory and dishonest and wreaks appalling psychological damage on those who are, at least in terms of the accepted norms of argument, defenceless. It is a politics, though, that has institutional, as well as discursive underpinnings and this is where the effort is required. The benefits yielded by a particular living place, benefits that will show up among other things in property values, have their conditions. Not the least of these is the quite extraordinary jurisdictional fragmentation of metropolitan areas in the USA: a division into local governments and most crucially of all for Habermas's familial-vocational privatism, school districts.

This is something that has to be defended, and it is. In most states inequalities of per pupil spending between metropolitan school districts achieve levels of a quite alarming magnitude: among the three major urban areas[6] of Ohio only for Cincinnati does the difference between the smallest to the largest spending district fall below a factor of two. Aside from the single case of Vermont, school financing reform has met stubborn resistance from those in the high spending districts that stand to lose their privileges; reform, ironically, tabbed as 'Robin Hoodism'. Suburban exclusionism has clearly had more than a simply fiscal rationale, though. It has also been explicitly socially exclusionary and schools have been at the centre of those anxieties.

All of this might seem to suggest that what has been discussed here is a politics of social stratification projected onto space and independent of other tendencies in the capitalist form of society in which it is embedded. This would be a major mistake. Just because resident organizations and even school districts have taken the lead does not mean that other interests of a relatively non-territorialized nature have not been in play, as we will now see.

Metropolitan housing markets, property capital and the search for class monopoly rents

In Harvey's geopolitics of capitalism, the contradiction between fixity and mobility looms large. It is a contradiction apparent in the current discussion of a territorialized class politics of the living place. It is, though, a class politics in a limited sense of the term 'class': class as a distributional category, so more a matter of stratification. What is missing is class in the sense of a production relation: the relations that people enter into with one another in order to produce. These are capitalism's more fundamental class relations: between the owners of productive capital on the one hand and the mass of the working class on the other, for sure, but also those different claimants for what productive capital is able to appropriate from labour, including the developers and other agents of property capital. In other words, in coming to grips with the historicity of this territorialized class politics, while familial-vocational privatism has been one element of the story, itself embedded in highly mediated ways in the trajectory of capitalist development, there has been another condition; namely, the emergence of property capital as the dominant moment in urban housing markets.

After the Second World War housing provision was transformed. Prior to that most houses for single-family occupation – the term itself is evocative – were custom built on lots purchased by the future homeowners, or even self-built. The large-scale, speculatively built development lay for the most part in the future. Anxieties during the depression about boosting domestic demand would start to change things by lending urgency to the creation of the institutional underpinnings of a mass market for housing. At the centre of this effort was the Federal Housing Administration or FHA, established in 1934. The FHA is most noted for the introduction of mortgage insurance, to which is widely attributed the loosening of the housing finance spigot: reduction in the risk of mortgage loans made lower-interest and longer-term loans possible. Homeownership, after a decline in the depths of the depression, did indeed start to take off. Less well known is the fact that the FHA, through its mortgage insurance policies, also gave incentives to stiffening up local zoning rules. Whether or not a house qualified for mortgage insurance depended on a variety of criteria. One of these was local zoning: how resistant were zoning designations to change? This was something that an incipient class of speculative developers, organized through NAREB or the National Association of Real Estate Boards, had pushed for and in fact they got to write the rules: rules intended to protect investments in their long-term projects (Weiss 1987).

In other words, the developers were in it from the start. In fact the push to expand home ownership had been under way since the 1920s. This started in 1922 with the federally initiated 'Own Your Own Home' campaign which then morphed into 'Better Homes in America, Inc.': a coalition of bankers, real estate development companies, builders and building supply firms scattered across the USA in over 7,000 local chapters pushing for government support for the development of houses for owner occupancy (Hayden 2001). In the post-war period, the fate of any tenure other than private and particularly private homeownership, would then be sealed by the outcome of a very different war – that on public housing; a war which drew unabashedly on the anti-communist hysteria of the time (Parson 1982, 2007).

In other words, an essential aspect of the fixing of the more nuclearized families moving out of the older rental housing in the city is to be laid at the door of the developers, the housing industry and the real estate lobby more generally. The advantages of homeownership were constructed, for the most part, from the bottom up by those who stood to benefit directly from it.[7] The working classes collaborated but things could have been otherwise, as they were in other of Habermas's 'advanced capitalist societies': homeownership in France and Germany was for a long time not the preferred tenure and in Germany that remains the case. But then, once people were effectively immobilized in space and given an asset to defend against devaluation, property capitals would became a crucial aspect of the subsequent politics, sometimes collaborating with residents, sometimes provoking them through what was being planned, and all in the interest of what Harvey (1974) called class monopoly rents.

What emerged in full clarity after 1945 was in many respects symbolized by the much publicized Levittown: a highly competitive home building industry, often integrated upstream with land purchase and development, bent on speculative construction, and therefore committed to a combination of cost cutting[8] and product innovation. The big money though has been in creating the conditions for enhanced rents: the clever design of developments so as to (e.g.) internalize its externalities: building and retaining a shopping centre or creating a golf club; or adding to the desirability of lots by putting them around a lake or indeed the golf course. This has meant an increased commitment to scale. The pull has been towards developing on such a large scale as to form the character of the suburb as a whole, particularly its schools. The patience demanded of long-stay projects has been of the essence, creating part of that island-like structure of advantage that David Harvey saw as key to class monopoly rent, though the sheer distinctiveness and cutting-edge novelty of developers has been another.

But while housing developments, pushed forward by the speculative nature of the business, have become bigger, that has been a problem for those who got there earlier. In part this has been a matter of the subsequent congestion that all have to confront and who then get charged for cleaning up the mess. Schools have been a central issue, often requiring new construction which then has to be paid for through increases in the property taxes, and not just those of the new arrivals. This in turn has given impetus to large lot zoning so as to control the numbers of the newcomers and this in turn has imparted a social bias to construction.

Not that this social bias is entirely a matter of minimizing the cost to the residential taxpayer. Quite to the contrary, for there is always the worry that the new houses may be priced so as to appeal to buyers who, to use Zane Miller's expression, are not quite 'top notch'. And of course the temptations from the developer's point of view are there. The demand coming from the 'top notch' is limited. On the other hand, developments that cater to the less so in a highly rated, even celebrated, school district can be an attractive source of rents. And to the extent that exclusionary forms of zoning become more widespread as a defence against 'them', so the attractions to the developer of a re-zoning, the increase in land values resulting from the proverbial stroke of the pen, become all the more attractive. There is no formal market in re-zonings but the demand for them as a result of the gains to be made and the willingness to 'supply' them by a notoriously corrupt officialdom means that there is an informal one. As a result new value continues to flow through the suburbs, while threatening to undermine values already in place and provoking battles around the shape that new development should assume, and even referenda to throw out the re-zonings.

None of this, though, could have happened without corresponding changes in financing homeownership. It wasn't just the real estate industry promoting 'Own Your Own Home'; banks were in it as well. The rise of a speculative home building industry was conditional on the availability of finance and not just for purchasers but also for itself; given the huge amounts of money that can go into a development, developers are inevitably highly leveraged. As the size of developments increased, so these needs became more and more pressing. So while FHA mortgage insurance did indeed make it easier for people to buy, less widely recognized is the way in which it also facilitated the financing of the developments themselves. From 1934 on, in addition to sheltering bank risk when making mortgage loans, the FHA also insured the loans that the banks and savings and loans made to the developers (Hayden 2001) and many of the savings and loans had in fact been created by builders and developers: nice work if you can get it. After the war the financial markets for the development industry would be increasingly deepened as banks and savings and loans sold their developer loans and mortgages on to secondary markets. Not least, this released urban housing markets from the limits of local saving.[9]

Concomitant with these changes has been the enhanced mobility of people that came with the automobile. This furthered the breakup of the living place as a community and set in motion the creation of a housing market on a metropolitan scale, which in turn would greatly increase not only the possibilities of class monopoly rents, but also the tensions around the living place as a commodity.[10] Initially the attention of the speculative developer focused on the urban periphery, soaking up customers from both the centre and from outside. The pressures from the centre were in part due to population growth, notably in the post-war period, of African Americans; and in part a result of the demolition of housing to make way for so-called 'urban renewal' and freeways. This then put pressure on an inner ring of older suburban development going back to prewar years, generating a set of studies of what was called 'neighbourhood change': something facilitated in its turn by a complex of conditions including the seductions of newer housing, the decline of mass transit and the shift to an automobile-centred way of life but which would also stimulate rearguard actions on the part of those about to be displaced (Ginsberg 1975; Leven et al. 1976; Molotch 1973; Wolf and Lebeaux 1969). Recently, gentrification has absorbed much of the critical attention formerly accorded exclusionary zoning on the urban periphery[11] completing the disintegration of the living place as community that had started to take shape in the 1930s.

Concluding comments

There are several points about the approach adopted in this chapter that merit emphasis. The first is the power of Harvey's (1985a, 1985b) vision of the geopolitics of capitalist development. Central to this was the drive to accumulate, to create ever wider spaces through which capital could circulate in search of enhanced profits, and then the subsequent contradictions as it ran up against earlier investments in fixed capital that stood to be devalued. The result was a geopolitics of simultaneous competition over space and of pressures on the working class that could then generate struggles among the latter. In the instance reviewed here all these elements are in play. On the one hand, the working classes experienced a new fixity in space: in part a result of broader social changes induced by capital's developmental trajectory, but also of property capital's insistence that the dominant housing tenure should be that of homeownership. On the other the emergence of a metropolitan-wide housing market, the result of speculative development searching out new niches, spatial and otherwise, and the revolution in personal mobility ensured that those fixities would be challenged and in ways that would create tensions among different factions of the working class.

The second point concerns the way the discussion here has been limited to the American case. It is true that much of the above, taken at a sufficient level of generality, finds strong echoes in other of the advanced capitalist societies. There is evidence there of the same exclusion and resistance to new development and to gentrification, similar sorts of anxieties about schools and property values, and a dominance of housing provision by speculative developers. The USA, though, remains quite different; even peculiar (Cox 2016). This is a complex area to which complete justice cannot be done in the space available here. In part it is a matter of a very different state structure: in particular a high degree of decentralization of formal power. From the standpoint of the territorialized class politics of space in urban areas, what is most relevant here is the quite extraordinary jurisdictional fragmentation of metropolitan areas into entities with both powers and needs to influence the development process: powers over zoning, including the power to call a referendum over re-zoning ordinances, which can be used by resident groups bent on exclusion; but also their own need to raise revenues, which can affect their position on appropriate forms of development.

This radical fragmentation of state power in the USA is in part an expression of a quite different social formation: still a capitalist one but formed discursively in distinct ways. The absence of a strong labour movement has long been recognized. Rather there have been pressures towards the fragmentation of the working class. This has been propelled in significant part by notions of individual upward mobility and enterprise, and the blame and shame heaped upon people to the extent that they are unable to move up: from inner city, to inner suburb, from inner suburb further out, from tract housing, to something more individualized. And so on. To them are denied the moral resources of an inward-looking, self-referential working class. Of course, the working class in Western Europe is not what it was; but the embers remain and the sort of 'socialist' label given public housing in the USA, as the big developers swept away its challenge in the 1950s, hardly on the horizon. The balance of political forces remains somewhat different therefore, giving residential politics its own *differentiae specifica* even while similarities remain.

Finally, there is the way in which new working-class sensibilities induced by new relations to the living place have informed politics at the national level. In part this has been about the rise of homeownership and the sense of belonging – deceptively – to the propertied classes. But more has been at stake than that. At the same time as American political scientists were investigating the relationship between suburbanization and a shift in partisan sentiment towards the Republican Party, in Great Britain, the question was being asked *Must Labour Lose?* (Abrams *et al.* 1960) and housing was once again part of the picture but not just private housing; the public housing estate was also being fingered as the culprit for shifting political loyalties: not just the breakup of the old working- communities like those studied in Bethnal Green by Young and Willmott (1957) or Rosser and Harris in Swansea (1965) but also the incorporation of the dispersed into new residential environments – ones that fragmented and induced, as Young and Willmott suggested, a status-driven aspirationism which would, as Abrams *et al.* suggested, create fertile ground for that displacement of the political centre of gravity that has been under way ever since.

Notes

1 The 'social class aesthetic' defined by Werthman *et al.* (1965): winding, tree-lined streets, equipped with numerous cul-de-sacs and bordered by single-family homes to create a park-like atmosphere.
2 Stein echoes some of the themes of British sociology though dating the changes a little earlier. He sees traces of them in the Lynds' 1929 *Middletown*:

> the essential fact remained constant – that life plans, whatever they may have been, had to be reoriented around the pursuit of status through money and commodity display. Thus the willingness that the second generation of urban Americans of every nationality showed to move to 'more modern' areas of second settlement rested upon this commitment.
>
> *(p. 54)*

3 Compare Young and Willmott in discussing what they term the symmetrical family:

> The young couples had had to, or have chosen to, move to new houses and away from relatives. Husbands have often been the most willing because it meant that their wives could be wrestled away from the influence of their Mums.
>
> *(p. 93)*

4 The formation of a residents' association is often the first step, followed by guided tours for the sake of the bankers in an effort to lift any redlining restrictions on lending to those wanting to buy into the area, and by pressure on a typically all-too-willing city council for some 'ye oldeing' of the streetscape as in cobbles and turn-of-the century gas lamps.
5 A lot of older single-family housing in the United States is located in areas zoned for apartments. One can only speculate on the motivation behind this curious conjunction. Certainly in the 1920s and 1930s homes were viewed in part as safety nets; not just the safety net of having a roof over one's head at times of unemployment but of having space that could be rented out to provide an income.
6 As defined by the central county of the metropolitan area.
7 Compare Dolores Hayden:

> Vast American suburbs of the post-World War II era were shaped by legislative processes reflecting the power of the real estate, banking and construction sectors, and the relative weakness of the planning and design professions. Despite the fact that FHA programs were effectively a developer subsidy, they were presented as assistance to the American consumer.
>
> *(2004: 151)*

8 This has included the increasing use of pre-fabricated components of the house and, in so-called tract housing, the simulation of assembly line techniques, where the workers move to the structure in process rather than the other way around.
9 This was particularly important since financial agencies were barred from inter-state and in some states, inter-county branching until the laws started to be relaxed in the 1970s. This was not, though, a limit for insurance companies which after 1945 and the relaxation of state legislation, became increasingly involved in financing developments as well as mortgages. For a case study see Hanchett (2000).
10 This is ongoing as through the realtor practice of multiple listings and less intentionally through the diffusion by the newspapers of school district rankings; something resulting in turn from the practice of state-wide testing.
11 Drawing on the Google Ngram viewer, references to exclusionary zoning take off very rapidly after the mid-1960s and then decline quite sharply after 1978. 'Gentrification' takes off about 1970 and has continued to increase at a rapid rate ever since.

References

Abrams, M., Rose, R. and Hiden, R. (1960). *Must Labour Lose?* Harmondsworth, Middlesex: Penguin.
Alt, J. (1976). Beyond Class: The Decline of Industrial Labor and Leisure. *Telos*, (28), pp. 55–80.
Bodnar, J. (1982). *Workers' World: Kinship, Community and Protest in an Industrial Society, 1900–1949*. Baltimore: Johns Hopkins University Press.
Cox, K.R. (1984). Social Change, Turf Politics, and the Concept of Turf Politics. Chapter 12 in A. Kirby, P. Knox and S. Pinch, eds. *Public Service Provision and Urban Development*. Beckenham, Kent: Croom Helm.
Cox, K.R. (2016). *The Politics of Urban and Regional Development and the American Exception*. Syracuse, NY: Syracuse University Press.
Cybriwsky, R.A. (1978). Social Aspects of Neighborhood Change. *Annals of the Association of American Geographers*, 68(1), pp. 17–33.
Danielson, M.N. (1976). *The Politics of Exclusion*. New York: Columbia University Press.
Gans, H. (1982). *The Urban Villagers*. Glencoe, IL: Free Press.
Ginsberg, Y. (1975). *Jews in a Changing Neighborhood: The Study of Mattapan*. Glencoe, IL: Free Press.
Habermas, J. (1973). *Legitimation Crisis*. Boston: Beacon Press.
Hanchett, T.W. (2000). Financing Suburbia: Prudential Insurance and the Post-World War II Transformation of the American City. *Journal of Urban History*, 26(3), pp. 312–328.
Harvey, D. (1973). *Social Justice and the City*. Baltimore: Johns Hopkins University Press.
Harvey, D. (1974). Class-Monopoly Rent, Finance Capital and the Urban Revolution. *Regional Studies*, (8), pp. 239–255.
Harvey, D. (1985a). The Place of Urban Politics in the Geography of Uneven Capitalist Development. Chapter 6 in *The Urbanization of Capital*. Oxford: Basil Blackwell.
Harvey, D. (1985b). The Geopolitics of Capitalism. Chapter 7 in D. Gregory and J. Urry, eds. *Social Relations and Spatial Structures*. London: Macmillan.
Hayden, D. (2001). Revisiting the Sitcom Suburbs. *Land Lines*, 13(2), pp. 1–3.
Hayden, D. (2004). *Building Suburbia: Green Fields and Urban Growth, 1820–2000*. New York: Pantheon.
Leven, C., Little, J., Nourse, H. and Reed, R. (1976). *Neighborhood Change: The Dynamics of Urban Decay*. New York: Praeger.
Mair, A. (1986). The Homeless and the Post-industrial City. *Political Geography Quarterly*, 5(4), pp. 351–368.
Miller, Z. (1981). *Suburb: Neighborhood and Community in Forest Park, Ohio, 1935–1976*. Knoxville: University of Tennessee Press.
Molotch, H. (1973). *Managed Integration: Dilemmas of Doing Good in the City*. Berkeley and Los Angeles: University of California Press.
Newton, K. (1975). American Urban Politics: Social Class, Political Structure and Public Goods. *Urban Affairs Quarterly*, 11, pp. 241–264.
Parson, D. (1982). The Development of Redevelopment: Public Housing and Urban Renewal in Los Angeles. *International Journal of Urban and Regional Research*, 6(3), pp. 393–413.
Parson, D. (2007). The Decline of Public Housing and the Politics of the Red Scare: The Significance of the Los Angeles Public Housing War. *Journal of Urban History*, 33(3), pp. 400–417.

Rosser, C. and Harris, C. (1965). *The Family and Social Change: A Study of Family and Kinship in a South Wales Town*. London: Routledge and Kegan Paul.
Stein, M.R. (1960). *The Eclipse of Community*. Princeton, NJ: Princeton University Press.
Tiebout, C. (1956). A Pure Theory of Local Public Expenditures. *Journal of Political Economy*, 64(4), pp. 16–24.
Weiss, M. (1987). *The Rise of the Community Builders*. New York: Columbia University Press.
Werthman, C., Mandel, J. and Dienstfrey, T. (1965). *Planning the Purchase Decision: Why People Buy in Planned Communities*. University of California at Berkeley: Center for Planning and Development Research, Institute of Urban and Regional Development (Preprint 10).
Williams, O.P. (1971). *Metropolitan Political Analysis*. Glencoe, IL: Free Press.
Wingo, L. (1973). The Quality of Life: A Micro-economic Definition. *Urban Studies*, 10(1), pp. 3–18.
Wolf, E.P. and Lebeaux, C. (1969). *Change and Renewal in an Urban Community*. New York: Praeger.
Young, M. and Willmott, P. (1957). *Family and Kinship in East London*. London: Routledge and Kegan Paul.
Young, M. and Willmott, P. (1975). *The Symmetrical Family*. New York: Pantheon.
Zelizer, V. (1985). *Pricing the Priceless Child*. New York: Basic Books.

40
URBAN MIDDLE-CLASS SHIFTING SENSIBILITIES IN NEOLIBERAL BUENOS AIRES

Carolina Sternberg

Introduction

This study examines the interconnections between the shifting sensibilities of middle-class identities in Buenos Aires, i.e. their encompassing sentiments, moralities, perceptions, expectations, and their significant role in mediating the process of neoliberalization in this city, particularly from the early 2000s to the present.

Drawing from interviews conducted during Spring 2014 with journalists, scholars and grassroots organizations on their perceptions of middle-class identities in Buenos Aires, and secondary source analysis from 2001 to the present, I examine the transformation of middle-class sensibilities, from cultivating a progressive and inclusive city, to favouring all-encompassing policies directed at increasing segregation and exclusiveness. This is illustrated through discourse analysis that reflects socially resentful, revanchist and fearful attitudes towards the poor and working-class citizens of Buenos Aires.

Closely following cultural historians' and anthropologists' critical approaches to the term 'middle class' (Defonline 2014; Furbank 2005; Heiman 2012; Parker and Walker 2013; Visacovsky 2009, 2010), I am interested in examining what discourses have been ascribed to them in Buenos Aires to constitute them as a coherent group. Here, I am trying to comprehend the social uses of these delimitations, in other words, how these discourses participate in the construction of identities and social practices (Lamont and Molnar 2002; Visacovsky 2009; Visacovsky and Garguin 2009). Throughout my narrative, I engage with journalists', scholars' and neighbourhood organizations' accounts that have attributed a specific character and identity to the Buenos Aires middle class, from the 2001 economic crisis to the present. This study also draws from content analysis of a variety of scholarly studies, blogs, documents, reports and dailies that all together offer significant renditions of the middle-class *porteña* (porteño/a is popular term used to refer to Buenos Aires's residents).

I argue that the convergence of the economic meltdown experienced in Argentina in 2001, in particular in Buenos Aires, with what I term the shifting sensibilities of the urban middle class, have profoundly contributed to the success and sustainability of an ongoing and ascendant neoliberal governance since 2007. As I discuss later, urban middle-class *porteños* have increasingly shown a lack of interest in, and support for, urban redistributive policies.[1] Furthermore, a profound neoliberal ethos has been shaping the *porteños*' social relationships, moving them to support the neoliberal regime and resist progressive movements supported by the marginalized, poor and working classes. With this claim, I am not belittling cross-class inclusive politics that have existed and still exist between the poor and the middle class, as was persuasively revealed in the work by Farías (2016) and Lawson *et al.* (2015). However, I question the celebrated egalitarian politics that is often-claimed to characterize Argentina in the post-crisis period, conceptualized by many as the 'post-neoliberal' period.

First, I find that Buenos Aires' successful neoliberal initiatives are anchored in what many have characterized as middle-class fears and resentment towards those who are said not to belong in the city's urban, economic and social spaces. These sentiments, I argue, emerged with the onset of Argentina's 2001 crisis. In parallel, as decades passed, they became profoundly articulated with neoliberal principles invoked by the local government. Second, I examine how neoliberal governance reinforces middle-class fears and resentments through repressive policies, and associated laws that allow the expulsion of people from parks, paths and city squares. Third, I propose that these policies have created the conditions for middle-class re-appropriation of public spaces that were said to be 'occupied' by the poor, immigrants and working classes.

Cultural approach to the term middle-class

There is a wide consensus among social science scholars that the term middle class is a slippery one, as its boundaries are contested and forever evolving over time and place (Adamovsky 2009; Farías 2016; Furbank 2005; Heiman 2012; Lamont and Molnar 2002; Parker and Walker 2013; Visacovsky and Garguin 2009). I closely follow the approaches from critical cultural historians and anthropologists to the term 'class' insofar as classes are "more ideas and discourses rather than objective things ... They provide a short-hand that people use to make sense of reality" (Parker and Walker 2013: 16). From this perspective, historian Adamovsky (2009) has argued against the existence of a middle class in Argentina.[2] Adamovsky asserts that a series of articulated elements – economic, cultural, racial and a specific understanding of Argentina's national values – characterize the identity of the Argentine middle class. In Adamovsky's words: "[The middle class in Argentina] exists as an identity ... a compound of representations articulated across time ... these are precisely what are being displayed when people identify themselves as part of the middle class" (Defonline 2014).

In this work, I examine the publicly circulated narratives of the 'middle class' in their own time and place by engaging with journalists', scholars' and neighbourhood organizations' accounts which, in their own terms, attribute a specific character and identity to the urban middle class in Buenos Aires. I focus on the period from the aftermath of the crisis to the present.

It is important to keep in mind some important aspects about the city of Buenos Aires and its *porteños*. Since the late nineteenth century, Buenos Aires has constituted a prosperous and vibrant city to which working-class immigrants from southern Europe flocked to look for economic opportunities. They were able to rapidly integrate into the social fabric of the city. The ideas of 'having expectations' and 'gaining social mobility', fundamentally through public education, were widely embraced by the six million foreign immigrants who poured into Argentina between the years of 1880 and 1930, a time in which Buenos Aires evolved from a small town to a large growing city and a site of opportunity. Today, there is a wide consensus among Argentinian historians, sociologists and journalists that the notion of social mobility through public education has constituted one of the key identifiable values of the middle class in Argentina (Moreno 2001).

Of course, social mobility did not develop without political and social antagonisms among the different social strata that composed Buenos Aires at that time (see Adamovsky 2009; Garguín 2012), but as decades passed, it became more clear that the city would gain a positive reputation for being one of the most socially integrated and inclusive cities within the country and the Latin American region (Germani 1942; Guano 2002; Massuh 2014; Wortman 2003). In addition, despite frequent cycles of economic hardships and high inflation, Argentina has long been known for having a more substantial middle class than elsewhere in Latin America (Germani 1962). In fact, until the early 1980s, the city of Buenos Aires represented the primary location of Argentina's middle class, home to about 70 per cent of it (Minujín and Kessler 1995).

The national context and the fall of the middle class: 1990s–2001

In the 1990s, President Carlos Menem adopted a neoliberal development model for Argentina. A key feature was the 1991 Convertibility Law which stabilized the Argentine peso by making pesos fully exchangeable for US dollars. For many in the middle class, convertibility made it easier to travel abroad, to purchase imported products, to get credit, and to save and withdraw their money in US dollars. The general perception of affluence in society that convertibility engendered, as well as the dominant neoliberal ethos (Hackworth 2007) of the time (which fostered a belief in self-reliance, entrepreneurship and personal responsibility[3]), reinforced the idea among those who became poor during the 1990s that their hardship was an isolated incident for which they were personally responsible (González Bombal and Svampa 2001). Under such circumstances, they overwhelmingly responded privately and favoured coping strategies rather than protest.

However, the Convertibility Law proved to be unsustainable and by 2002 the national picture changed dramatically. Argentina had been in recession for three years and poverty had become rampant, having risen from 29 to 53 per cent since October 2000 (González Bombal and Svampa 2001). Disenchantment with the political system grew due to politicians' inability to resolve the country's economic woes, coupled with a series of political scandals. Following the election of President Fernando De La Rua in 1999 worries about Argentina's weakening economic fundamentals led to a run on the banks as both Argentine and foreign investors began to withdraw their dollar savings. In December 2001, as it became evident that those banks lacked the dollar reserves to cover withdrawals, Minister of Finance Domingo Cavallo announced a near-complete freeze on all bank accounts; Argentines would only be allowed to withdraw small amounts of their savings at a time, and only in pesos. This policy, which came to be known as the *corralito* (or playpen), was the spark that set off the December riots, known as the Cacerolazo or Argentinazo, forcing De La Rua to resign. The end of the convertibility, the devaluation of the peso and the forced conversion of US dollar savings into peso accounts followed the corralito (Kessler and Di Virgilio 2008).

Economic crisis ensued in December 2001 and in early 2002 the national economy nosedived. The gross domestic product sank by more than 16 per cent during this period, while manufacturing output plunged 10 per cent. A total of 20 per cent were unemployed, while an additional 23 per cent were underemployed (Minujín and Anguita 2004). One in five Argentinians at that time lived in severe poverty, which also meant hunger and struggles to obtain decent housing.

By the end of the crisis, the middle class had seen their economic situation drastically deteriorated. If in 1974 Argentina's middle class received almost 42 per cent of national income, by the end of 2001 it received only 33 per cent. In contrast, the "upper class increased their share of national income by ten percent" (Barros 2005: 35). More starkly, nearly seven out of ten Argentines were of the middle class in 1974; in 2001, before the Cacerolazo, five out of ten were middle class (Barros 2005), and in 2005, only four out of ten.[4]

The 2001 crisis: reverberations in the city of Buenos Aires

The impacts of the 2001 crisis have been extensively analysed in scholarly articles and widely covered in the media (see Barros 2005; Gomez Fulao 2011). In this short section I will only point to the publicly circulated narratives of the impacts and reactions to the economic and political crisis in Buenos Aires in 2001.

In the city of Buenos Aires the effects of the political and economic crisis severely shocked every working-class neighbourhood and left almost every member of this class in a situation of despair and economic vulnerability. Early on, the major media focused on recording the 'fall of the middle class'. This narrative quickly invaded all arenas of Buenos Aires' public and private life. In fact, as a prominent journalist had noted, the *porteño* middle class became fully aware that the ground under their feet was crumbling (Barros 2005). The loss of their previous lifestyle was everyday more real, everyday more tangible. As many have argued, the middle class in Buenos Aires was decimated (it fell to 15 per cent in 2001), as were expectations for and aspirations to a prosperous future.

Accompanying this new economic reality, disillusionment and a sense of declining social progress and social mobility pervaded middle-class circles. The expression "we hit the bottom" (interview with Lorena, a 40-year-old representative of the neighbourhood organization 'Buenos Aires Para Palermo', in 2014) became widely used, embodying a profound sense of eroding middle-class consumer power. Furthermore, the middle class confronted the possibility, and in many cases the reality, of falling into poverty – an acute, menacing and disturbing situation for many. As one newspaper eloquently summarized in a headline, the middle class was confronting "The fear of not being [part of this class anymore]" (Moreno 2001).

National economic recover and a new political scenario in Buenos Aires

The 2001–2003 period of shocking economic and social conditions spawned the election of Nestor Kirchner (2003–2007). Immediately after taking office in 2003, he initiated a significant roll-back of neoliberal policies and pushed for Argentina's international financial autonomy. Remarkably, in matter of few years, the Argentine economy showed signs of recovery driven, for the most part, by an influx of cash due to an unprecedented export boom (Kessler and Virgilio 2008). This recovery was accompanied by a series of welfare policies including higher rates of employment, increasing consumption levels and all-encompassing social aid programmes benefiting the country's low-income and working-class households.

The economic recovery of the first Kirchner term (2003–2007) reached both the middle and working classes in Argentina. By 2007–2008, macroeconomic indexes seemed to indicate that the economic crisis was past and the middle class was recovering. Interestingly, a study conducted by Pew Research Centre showed a substantial expansion of the middle-class population across Argentina, from 15 per cent in 2001 to 32 per cent in 2011. Similar increases were registered for Buenos Aires (Buenos Aires Herald 2015).

In the context of the promising national economic scenario, Mauricio Macri led in the first round of Buenos Aires' 2007 mayoral election against rival Daniel Filmus. Macri's programme proposed an expansive and vibrant business climate, an innovative transportation system and neighbourhood revitalization to make the city more attractive for global capital investment.

At the opposite end of the political spectrum, Filmus proposed policies to economically improve under-resourced neighbourhoods and to increase investment in public education and housing in order to make the city more inclusive and socially integrated. Many anticipated the victory of Macri, and he won the second round election in December 2007 with 61 per cent of the vote. But this victory seemed to also expose something new: the image of social inclusiveness that had once been central to Buenos Aires' reputation was no longer in the political lexicon. In fact, as the next sections will discuss, Macri consolidated neoliberal governance in Buenos Aires after winning another term in 2011.

New middle-class sensibilities fifteen years after 2001[5]

Today, more than fifteen years since the crisis erupted, two prominent and convergent narratives reverberate throughout my interviews with prominent journalists, scholars and representatives of neighbourhood organizations in the city of Buenos Aires. On a superficial level, these accounts echo and replay former discourses and perceptions that widely circulated characterizing the urban middle class at the apex of the economic crisis, especially in Buenos Aires. Yet, some of these accounts also exposed something *new and deeper*. They now seem infused with *new sensibilities* that suggest new perceptions, sentiments and moralities, new constructions of Buenos Aires' middle class.

There is a distinctive and overarching narrative characterized by a shared sentiment of a 'bygone era', replaced by limited expectations and broken aspirations. Permeating this narrative is a sense of "eroding long-term consumptive realities" that now seem to live, dwell, and colour the middle-class world. When I interviewed Lorena, she firmly communicated these sentiments:

> What they [the middle class] are seeking today are ways of getting by, not thinking about the future of their kids. [The middle class] doesn't ask for long-term solutions. Immediate things need to be solved today.

Laura, a 60-year-old long-time activist in the Asociación del Barrio Pueyrredón, communicated a sense of transitory and ephemeral times when referring to middle-class expectations today:

> [It is] what you can access in the short-term that today gives [the middle class] some peace.

Lorena and Laura's statements deeply resonate with common sentiments after the economic crisis erupted in 2001. Many newspaper accounts, even thirteen years later, recall a sentiment of middle-class disillusionment spurred by the loss of upward social mobility:

> [As] never before, the days in 2001 exhibited a social fabric dangerously broken ... a country where [there] still exists an imaginary tied to a social mobility, the reality strikes back ... if the old expectations of having a good and stable job [haven't] been broken yet, these are now all a very hard goal to achieve.
>
> *(Fernández Irusta 2014: 4)*

Yet other narratives point in a different direction, revealing still other new sensibilities. Despite the substantial economic recovery of many in the middle class since 2007, an ongoing and unsettling sentiment among middle-class *porteños* has become apparent, prompted by the increasing upward social mobility of low-income and working classes across the country, and particularly in Buenos Aires. In a city where many low-income and working-class settlers progressively improved their standards of living as a result of federal redistributive policies, Buenos Aires' traditional middle class began to fear that it might fall to the level of the lower classes. A journalist from a famous local daily eloquently summarized this point: "the Buenos Aires middle class does not want to yield its position in the social pyramid" (Moreno 2001).

To be clear, many events of the past represent attempts by the middle class *porteña* to socially distance itself from the lower classes (see Guano 2002, 2003). Yet, post-2001, these steps became more vigorous, inviting a variety of reactions. Mario, a representative from a neighbourhood association in the centre of Buenos Aires, characterized Buenos Aires' middle class, post-2001, this way:

> What pisses the middle class off is that the 'morochos' [referring to the working class in Buenos Aires] rub their shoulders with them [the members of the middle class].

Mario makes two significant points. First, the term *morochos* is a derogatory category primarily used by the upper classes towards the working class during the 1940/1950s when Perón was president of Argentina (1945–1955). Today this category is still used in some *porteños*' circles. Second, this term, intentionally used by Mario, illustrates middle-class anxiety over the possibility of social mixing and the erasure of material and symbolic class boundaries with the working-class *porteña*.

Similar desires to maintain class distinctions were expressed during my interview with Nicolas, a young student and representative of a grassroots organization located in the southwest of the city, when discussing the post-crisis middle-class *porteña*. According to Nicolas, despite being historically recognized for taking pride in sending their kids to public schools (Guano 2003), today's Buenos Aires middle class suffers from a mix of fear, racial anxiety and frustrated class aspirations:

> [I think] there's a micro-fascism going on … The middle class that re-launched after the [2001] crisis is more fascist. … They don't like to be sending their kids to public schools when they see that kids are [socially] mixed, I mean … when middle-class parents see their kids mixed with kids coming from working-class families like the janitors' or *cartoneros*' (garbage pickers) families … I've heard of a woman who moved her kids from a public to a private school even if the private school was unaffordable for her.

Shortly after this interview, I came across the following editorial echoing Nicolas' perceptions: "There's [public] criticism that the middle-class has shifted to the right, for the most part in the city of Buenos Aires, since Macri won the election" (San Martin 2013) (note: The interviewee is referring to the victory of Mauricio Macri, re-elected as city mayor in 2011 with 64 per cent of the vote).

These accounts indicate new middle-class *porteña* sensibilities towards the working class and poor, sensibilities that became more vigorous during the transition from the 2001 crisis to the economic recovery. Signs of class distinction (Bourdieu 1984), such as public schooling, acquired considerable semiotic force during this time, grounded in middle-class moral logics and a sense of righteousness (see Centner 2010). Narratives also seem to reveal the decline of the socially inclusive imaginary that used to be integral to the reputation of the city and its residents. Fragmentation and exclusiveness are now pervasive.

In short, we see a new class culture due to the confluence of new processes. The 2001 crisis broke the social fabric of the city and transformed social classes and moralities in critical ways (Fernández Irusta 2014), aided by the expanded social policies of the Kirchners. In unison, these conditions inadvertently deepened the insecurity, resentment and class resistance of Buenos Aires' traditional middle class.

Next, I illustrate the series of conservative unpopular policies advanced in Buenos Aires, primarily anchored and shaped by the aforementioned urban middle-class changing moralities around the possibility of social mixing or closing inter-class boundaries.

Ascendant neoliberal governance in Buenos Aires

As mentioned earlier, the city of Buenos Aires has historically had a reputation as one of the most progressive and socially integrated cities, not only in Argentina but also throughout South America. However, since the adoption of neoliberalism in the early 1990s, Buenos Aires began to radically alter its character, becoming increasingly neoliberal and revanchist (see Sternberg 2012).

In the 1990s the unfolding of structural adjustment policies across the country re-sculpted Buenos Aires' urban governance to be more business-oriented and supportive of real-estate capital's drive to gentrify neighbourhoods, a dramatic departure from its traditional role of ensuring equity in redevelopment. Today, austerity policies and a neoliberal ethos are ascendant.

When Mauricio Macri, leader of the PRO (Propuesta Republicana) party, took office as mayor of Buenos Aires in 2007 he not only accelerated neoliberalization, he also shifted toward revanchist policies and rhetoric[6] (Aalbers 2010; Kunkel and Mayer 2012; MacLeod 2002; Smith 1996). In the eight years of his administration, Macri systematically attacked and harassed the homeless, profoundly cut public spending in education, health and public housing, and systematically neglected declining neighbourhoods that were previously considered priorities for urban renewal (Sternberg 2015). His successor in office, Horacio Rodriguez Larreta, has reinforced Macri's political agenda. In 2010 an estimated 12,000 people were reported to be homeless, along with 173,721 households (i.e. more than half million residents) living in critically substandard housing (CESO 2015). Paradoxically, according to a study published by journalist Gabriela Massuh in 2014, about 30 per cent of the total housing stock in the city is vacant, the equivalent of 400,000 units. Massuh points out that these numbers reveal a "process of real estate speculation led by the local government ... [that] is sweeping out ... the social coexistence of different social sectors" (Pertot 2014a, my translation). She later added that "Buenos Aires is losing its character as a diverse city where middle, upper and lower classes used to live together in the same neighbourhood" (Pertot 2014a, my translation).

From the beginning of his administration, Macri embarked on a concerted geographical campaign to remove the homeless from central gathering places, bridges, tunnels, empty lots, settlements and gentrifying neighbourhoods. To this end, the local government created a state body – UCEP (*Unidad de Custodia del Espacio Público*). During its three years of operation, this state body assaulted, evicted and killed homeless persons occupying public land with apparent impunity, until reports of brutality skyrocketed, which resulted in the closing of the UCEP (Cartoceti 2011).

These policies were coupled with quintessential neoliberal financial strategies: sell off city properties for short-term gain with revenue used to attract corporate and real-estate investment, changes in zoning ordinances, land donations, and a series of tax abatements to favour real-estate speculation.

During his second administration, Macri continued implementing significant unpopular policies, and as result, drew increasing criticism from the Buenos Aires public, grassroots organizations, housing activists, working-class citizens, and legislators. Despite this, mayoral elections in July 2015 returned the PRO neoliberal government to power for four more years, this time under the leadership of Horacio Rodriguez Larreta.

Today, a business-oriented city is set to continue its revanchist policies. In 2014 Minister of Economic Development, Mr Cabrera, anticipating a victory of the PRO in the 2015 mayoral elections, expressed his strategic plan for the city: "Our objective for 2015 is to continue developing tools that will allow the unfolding of all the entrepreneurial potential of the city" (Pertot 2014b, my translation).

Neoliberal governance in Buenos Aires has strategically mobilized the resentment and insecurity of the middle class to undergird unpopular and regressive policies and programmes. How can such revanchist governance continue undaunted in its political projects?

Buenos Aires' neoliberal governance: mobilizing new middle-class sensibilities

Argentinian journalists and scholars (see Vommaro et al. 2015) have attributed support for revanchist governance among Buenos Aires' middle class to their strong reaction against the populist and welfare policies adopted during the twelve-year rule of the two Kirchner administrations.[7] More precisely, Vommaro et al. suggest that the unprecedented acceptance of broadly unpopular, austere and revanchist policies among Buenos Aires' middle class put in place since 2007 by Macri and successors, converged with their overt rejection of the Kirchners' populist national agenda.

However, looking more deeply to the aftermath of the 2001 crisis and the shift in middle-class sensibilities, I am prompted to offer the following explanation: more than an overt political rejection of the Kirchners, this revanchist turn is the result of confluence of new processes and factors: the 2001 crisis transformed social moralities in a critical way, triggering class anxieties and uncertainties about the boundaries of inclusion and exclusion in the imagined future of the city.

The idea of governance through fear and anxiety is not entirely new. Buenos Aires' post-2001 governance perfectly extols the deepened insecurity, resentment and fear of the middle class which has bred support for a neoliberal planning ideal including efforts to re-sculpt downtowns, advance gentrification, and push squatters, immigrants and the homeless further to the geographical and political peripheries. A central driver is the middle class's fear of sharing public spaces such as parks, schools and hospitals with working-class individuals, poor families and/or immigrants.

The events of December 2010 epitomize the use of fear as a governance strategy. A violent protest broke out in early December 2010 leaving three people dead in one of the largest parks of Buenos Aires, the *Parque Indoamericano*. The local government forcibly removed hundreds of families to 'take back' the park that was illegally occupied by local working-class residents and Bolivian families. This episode triggered visceral anti-immigrant and anti-social sentiments among many in the urban middle class. According to a prominent independent daily, "The violent speech given by Macri [in response to the evictions] twitched the entire city … fascism has begun to grow in the middle class … this [episode] illustrates the rejection toward the poor" (Bruschtein 2010). Buenos Aires media pointed out that the violent evictions executed by the local government were celebrated by the middle-class *porteños* who claimed that Bolivians and the poor had "taken over some of their valuable public space" (Mills 2011). Cristian Ritondo, chief of the legislature for the PRO, claimed in an interview with the daily *Pagina12* in December 2010:

> The PRO and the neighbours strongly support our eviction policy ... If we allow this to happen [referring to the people who occupied the parks], tomorrow they will come to take *Parque Las Heras, Parque Pereyra, Parque Centenario o Parque Chacabuco.*
> (Pertot 2010, my translation)

The final eviction of immigrants and the working class from *Parque Indoamericano* cemented the stern anti-homeless, anti-immigrant and anti-social ethos of neoliberal and revanchist Buenos Aires.

Conclusions

Buenos Aires' neoliberal governance is anchored in middle-class fear and resentment, and in opposition to sharing urban, economic and social spaces with the poor, immigrants and working classes. These sentiments emerged during Argentina's 2001 economic crisis. In the aftermath of the 2001 crisis repressive laws and policies were adopted to push people out of Buenos Aires' parks, paths and city squares. Paradoxically, policies that facilitate middle-class re-appropriation of public spaces only serve to reinforce middle-class fears and anxieties.

With the rise of austerity policies and a neoliberal ethos, Buenos Aires' middle class seems to be more receptive to conservative, racist and opportunist overtures than it was the decade preceding 2001. What can we learn about middle-class identities and values through an examination of the experiences of Buenos Aires' middle class?

First, this study sheds light on what is not evident in income bracket categories: narratives expressing tension and/or resentment in cultural, moral, aesthetic and visceral terms. Narrative accounts lay bare middle-class anxieties exacerbated by the social advancement of the working classes and the poor, advancement manifest in the 'taking over' of public spaces. Shifts in class boundaries and relative class position, I argue, contributed to the articulation of a neoliberal ethos and the adoption of neoliberal policies in the city.

Second, discourses are central to the construction of identities and social practices; middle-class identities are seen to manifest themselves in the everyday politics of the city, stretching beyond its boundaries. These identities also continue to vigorously structure the spatial relations and social milieus of Buenos Aires.

Third, there is a regulative relationship between the changing material conditions undergirding middle-class life and middle-class citizens' common-sense understandings of everyday life that both influence, and are influenced by, politically organized efforts. A profound neoliberal ethos has been shaping the *porteños*' social outlook: they have become supporters of a neoliberal regime, acting as a conservative barrier to progressive movements organized by the marginalized, poor and working classes.

It would be fruitful to continue rethinking and interrogating the role of the middle class, traditionally conceived, as an agent of urban social change, in particular in South America where it has become embroiled in inter-class disputes that affect the nature of urban interventions, especially their distributive implications. Additionally, it could also be worthwhile to consider the processes by which dramatic events, such as economic and political crisis and/or other acute material and social circumstances, shape class identities, sentiments and relational experiences with other classes.

Finally, all scales of government-federal, state and local – as this study illustrates – will continue to provide fruitful sites from which to unearth micro-processes of meaning-making and understandings of how discursive and material practices shape the anxieties, aesthetics and moral logics of everyday life.

Notes

1 It is important to note other interpretations of this phenomenon: many authors have strongly emphasized the collectivist approach that emerged in the aftermath of Argentina's 2001 crisis, wherein groups of middle-class *porteños* built cross-class alliances and reworked class politics around poverty and inequality through cooperatives, community and grassroots organizations (Massey 2012; Sautu 2001; Svampa 2005). Yet, other scholars have certain caution to claim the continuity of these more collective class identities after the crisis (Gago and Stzulwark 2009; Villalon 2007).
2 Very briefly Adamovsky (2009) explains that it was only during the 1950s that the middle class consolidated itself as a class insofar its perception, political participation and living conditions were aligned, regardless of its growing number. Following the author, this is the decade when the middle class formed as the main oppositional class to Perónism.
3 See Adamovsky 2009; Brown 2003; Garguin 2012; Germani 1981; Grimson 2007; Guano 2002; Sautu 2001.
4 Note that the characteristics of their earnings vary greatly.
5 All quotes in this section were translated from Spanish to English by the author. All names of individuals are pseudonyms.
6 Urban revanchism expresses a race/class terror felt by the middle class, a vicious reaction against minorities, the working class, homeless people, the unemployed (women, gays and lesbians), immigrants (Smith 1996: 211).
7 President Nestor Kirchner maintained his power from 2003 to 2007. He died of a heart attack in 2010. Cristina de Kirchner, his wife, succeeded him. She was democratically elected for two consecutive terms as president (2007–2015).

References

Aalbers, M. (2010). The Revanchist Renewal of Yesterday's City of Tomorrow. *Antipode*, 43, pp. 1697–1724.
Adamovsky, E. (2009). *Historia de la Clase Media Argentina. Apogeo y Decadencia de Una Ilusión*. Buenos Aires: Planeta.
Barros, R. (2005). *Fuimos. Aventuras y desventuras de la clase media*. Buenos Aires: Aguilar.
Bourdieu, P. (1984). *Distinction: A Social Critique of the Judgement of Taste*. Cambridge, MA: Harvard University Press.
Brown, W. (2003). Neoliberalism and the End of Liberal Democracy. *Theory and Event*, 7. Available at: muse.jhu.edu.
Bruschtein, L. (2010). Macri Vainilla. *Pagina12*, 11 December. Available at: www.pagina12.com.ar [accessed 10 June 2014].
Buenos Aires Herald. (2015). Argentina Leads Middle-Class Ranking. *Buenos Aires Herald*, 23 July. Available at: http://buenosairesherald.com/article/194618/argentina-leads-middleclass-ranking [accessed 10 January 2018].
Cartoceti, C. (2011). Macri Investigated for UCEP's Abuse of Poor. *The Argentina Independent*, 7 February.
Centner, R. (2010). Spatializing Distinction in the Cities of the Global South: Volatile Terrains of Morality and Citizenship. *Political Power and Social Theory*, (21), pp. 281–298.
Centro de Estudios Económicos y Sociales Scalabrini Ortiz (CESO). (2015). La situación habitacional en la ciudad de Buenos Aires. *Informe CABA*, nro 1, Febrero.
Defonline. (2014). La clase media no es una clase social, sino una identidad. Interview with Ezequiel Adamovsky, *Defonline*, 3 March. Available at: www.defonline.com.ar [accessed 8 March 2017].
Farías, M. (2016). Women's Magazines and Socioeconomic Change: Para Ti, Identity and Politics in Urban Argentina. *Gender, Place and Culture: A Journal of Feminist Geography*, 23(5), pp. 607–623.
Fernández Irusta, D. (2014). Los 'sin techo' que nadie ve. La clase media y el sueño perdido de la casa propia enfoque. *La Nación*, 9 July. Available at: www.lanacion.com.ar [accessed 9 September 2014].
Furbank, P.N. (2005). *Un Placer Inconfesable o la Idea de Clase Social*. Buenos Aires: Paidós.
Gago, V. and Sztulwark, D. (2009). Notes on Postneoliberalism in Argentina. *Development Dialogue*, 51, pp. 181–190.
Garguin, E. (2012). Los argentinos descendemos de los barcos: The Racial Articulation of Middle-Class Identity in Argentina (1920–1960). In R. López and B. Weinstein, eds. *The Making of the Middle Class: Toward a Transnational History*. Durham, NC and London: Duke University Press.

Germani, G. (1942). La clase media en la ciudad de Buenos Aires: Estudio preliminar. *Boletín del Investigaciones del Instituto de Sociología* 1, Facultad de Filosofía y Letras, Universidad de Buenos Aires, pp. 105–126.

Germani, G. (1962). Estrategia para estimular la movilidad social. *Desarrollo Económico*, 1(3), pp. 59–96.

Germani, G. (1981). La clase media en la ciudad de Buenos Aires: Estudio preliminar. *Desarrollo Económico*, 81(21), pp. 109–127.

Gomez Fulao, J.C. (2011). *Las Dos Miradas de la Crisis: Aceptación o Resistencia. La Clase Profesional Argentina en la Crisis de 2001*. Buenos Aires: Biblos.

González Bombal, I. and Svampa, M. (2001). *Movilidad Social Ascendente y Descendente en las Clases Medias Argentinas: Un Estudio Comparativo*. Serie Documentos De Trabajo 3. Buenos Aires, Siempro, Secretaría de Tercera Edad y Acción Social, Ministerio de Desarrollo Social y Medio Ambiente.

Grimson, A. (2007). *Cultura y Neoliberalismo*. Buenos Aires: CLACSO.

Guano, E. (2002). Spectacles of Modernity: Transnational Imagination and Local Hegemonies in Neoliberal Buenos Aires. *Cultural Anthropology*, 17, pp. 181–209.

Guano, E. (2003). A Color for the Modern Nation: The Discourse on Education, Class, and Race in the Porteño Opposition to Neoliberalism. *Journal of Latin American Anthropology*, 8(1), pp. 148–171.

Hackworth, J. (2007). *The Neoliberal City: Governance, Ideology, and Development in American Urbanism*. Ithaca, NY: Cornell University Press.

Heiman, R. (2012). Gate Expectations: Discursive Displacement of the 'Old Middle Class' in an American Suburb. In: R. Heiman, C. Freeman and M. Liechty, eds. *The Global Middle Classes: Theorizing Through Ethnography*. School for Advanced Research. Santa Fe: New Mexico.

Kessler, G. and Di Virgilio, M. (2008). La nueva pobreza urbana: Dinámica global, regional y Argentina en las ultimas dos décadas. *Revista de la CEPAL*, 95, pp. 31–50.

Kunkel, J. and Mayer, M., eds. (2012). *Neoliberal Urbanism and its Contestations*. London: Palgrave Macmillan.

Lamont, M. and Molnar, V. (2002). The Study of Boundaries in the Social Sciences. *Annual Review of Sociology*, 28, pp. 167–195.

Lawson, V., Elwood, S., Canevaro, S. and Viotti, N. (2015). 'The Poor Are Us': Middle-Class Poverty Politics in Buenos Aires and Seattle. *Environment and Planning A: Economy and Space*, 47(9), pp. 1873–1891.

MacLeod, G. (2002). From Entrepreneurialism to a Revanchist City? On the Spatial Injustices of Glasgow's Renaissance. *Antipode*, 34, pp. 602–624.

Massey, D. (2012). Learning from Latin America. *Soundings*, 50, pp. 131–141.

Massuh, G. (2014). *El Robo de Buenos Aires*. Buenos Aires: Sudamericana.

Mills, N. (2011). Political Fighting Continues Over Evictions. *The Argentina Independent*, 15 March.

Minujin, A. and Anguita, E. (2004). *La Clase Media, Seducida y Abandonada*. Buenos Aires: Edhasa.

Minujin, A. and Kessler, G. (1995). *La Nueva Pobreza en la Argentina*. Buenos Aires: Editorial Planeta.

Moreno, L. (2001). El aguante de la clase media. El temor de ya no ser. *Clarín Digital*, 5 May. Available at: www.defonline.com.ar [accessed 8 March 2017].

Parker, D.S. and Walker, L.E., eds. (2013). *Latin America's Middle Class: Unsettled Debates and New Histories*. Lanham, MD: Lexington Books.

Pertot, W. (2010). La política de Macri es no dar la cara. *Página 12*, 8 December. Available at: www.pagina12.com.ar [accessed 8 March 2017].

Pertot, W. (2014a). Echan a la clase media de la ciudad. Interview with Gabriela Massuh, *Página12*, 12 December. Available at: www.pagina12.com.ar [accessed 8 March 2017].

Pertot, W. (2014b). De eso se ocupa el mercado. Interview with Francisco Cabrera, *Página12*, 11 November. Available at: www.pagina12.com.ar [accessed 8 March 2017).

San Martin, R. (2013). El Kirchnerismo no tiene el arraigo emotivo del peronismo en los sectores populares. Interview with Enfoques Adamovsky, *La Nación*, 27 January. Available at: www.lanacion.com.ar [accessed 9 September 2014].

Sautú, R. (2001). *La Gente Sabe. Interpretaciones de la Clase Media Acerca de la Libertad, la Igualdad, el Éxito y la Justicia*. Buenos Aires: Lumiere.

Smith, N. (1996). *The New Urban Frontier: Gentrification and the Revanchist City*. New York: Routledge.

Sternberg, C. (2012). *The Dynamics of Neoliberal Contingency: Neoliberal Redevelopment Governances in Chicago and Buenos Aires*. Doctoral dissertation, Department of Geography, University of Illinois at Urbana–Champaign. Available at: www.ideals.illinois.edu/bitstream/handle/2142/31099/Sternberg_Carolina.pdf? sequence=1 [accessed 14 December 2017].

Sternberg, C. (2015). Urban Sustainability, Neoliberal Urban Governance, and the Possibilities of 'Greening' Buenos Aires. In: D. Wilson, ed. *The Politics of the Urban Sustainability Concept*. Champaign–Urbana, IL: Common Ground Press.

Svampa, M. (2005). *La Sociedad Excluyente. Argentina bajo el Signo del Neoliberalismo*. Buenos Aires: Taurus.

Villalon, R. (2007). Neoliberalism, Corruption, and Legacies of Contention: Argentina's Social Movements, 1993–2006. *Latin American Perspectives*, 34(2), pp. 139–155.

Visacovsky, S.E. (2009). Imágenes de la 'clase media' en la prensa escrita Argentina durante la llamada 'crisis del 2001–2002'. In: S.E. Visacovsky and E. Garguin, eds. *Moralidades, Economías e Identidades de Clase Media. Estudios históricos y etnográficos*. Buenos Aires: Antropofagia.

Visacovsky, S.E. (2010). *Hasta la Próxima Crisis. Historia Cíclica, Virtudes Genealógicas y la Identidad de Clase Media Entre los Afectados por la Debacle Financiera en la Argentina (2001–2002)*. Documentos de Trabajo-División de Historia 68, División de Historia del Centro de Investigación y Docencia Económicas. México DF. Available at: www.cide.edu/publicaciones/status/dts/DTH%2068.pdf [accessed 14 December 2017].

Visacovsky, S.E. and Garguin, E. (2009). Introducción. In: S.E. Visacovsky and E. Garguin, eds. *Moralidades, Economías e Identidades de Clase Media. Estudios Históricos y Etnográficos*. Buenos Aires: Antropofagia.

Vommaro, G., Morressi, S. and Belloti, A. (2015). *Mundo PRO. Anatomia de un Partido Fabricado Para Ganar*. Buenos Aires: Planeta.

Wortman, A., ed. (2003). *Pensar las Clases Medias. Consumos Culturales y Estilos de Vida Urbanos en la Argentina de los Noventa*. Buenos Aires: La Crujía ediciones.

41

COMPASSIONATE CAPITALISM

Tax breaks, tech companies and the transformation of San Francisco

Lauren Alfrey and France Winddance Twine

Fidgety loiterers frequent the streets of the Tenderloin, a fickle San Francisco neighbourhood not necessarily suited to certain palates. Raw to the core, the Tenderloin is an abrasively honest neighbourhood in the heart of the city. Cheap hole-in-the-wall eateries and paling historic buildings provide the backdrop to this neighbourhood's cast of sidewalk occupants. Recognizing the Tenderloin for its rich diversity and architectural potential, San Franciscans are working to redefine their dubious heartland.

This description appears on *Airbnb*, a global website where one can list or rent apartments, homes, and condominiums for short-term residential use.[1] In the section of the website titled, 'the community says', guests create one-word descriptions, known as tags, for fellow travellers. Airbnb descriptions offer a partial glimpse into how middle- and upper middle-class residents and tourists understand the Tenderloin as a neighbourhood that remains racially diverse and is home to some of the poorest residents of the city.

Like many neighbourhoods of San Francisco, the Tenderloin is being transformed by the recent arrival of technology companies. Tags provided by Airbnb visitors, including 'gritty', 'inner city', 'urban' and 'emerging', illustrate the abiding tensions. Once a multi-ethnic and economically mixed district in San Francisco, the Tenderloin has undergone a rapid transition in the last decade, from an entertainment district with hotels, theatres and single room occupancy (SROs) hotels that house the city's poorest residents to a vein in the nation's most concentrated technology hub.

Since 2011, at least twenty-seven companies, including Twitter and Zendesk, have received tax breaks to open up offices in the Tenderloin or the abutting area, known as mid-market.[2] As a neighbourhood that provides centralized social welfare, and medical and religious services for the city's poor and houseless, the Tenderloin now faces the challenges of neoliberalism from three aligned forces: the Mayor's office, commercial developers and the technology industry – arguably the most powerful industry in the early twenty-first century.

In this chapter, we provide a case study that examines how increasing racial and class inequalities are interpreted and refashioned in the new economy. First, we consider the ideological, material and symbolic impacts of tech companies in the mid-market district of San Francisco. How do technology workers make sense of their status as members of what can best be characterized as a new 'labour aristocracy', against the stark poverty and struggles of the city's unhoused and long-term residents? To address this question, we draw upon interviews with twenty tech workers employed at four technology firms receiving payroll tax benefits. We consider how these workers make sense of their presence in this 'abrasively honest' neighbourhood in which long-term residents are physically proximate, but socially a world apart.

Second, we examine a policy that drew tech firms to the area, known as the 'Twitter tax breaks', and the philanthropic efforts that accompanied this arrangement. These tax breaks were introduced by Mayor Ed Lee's administration in 2011 to promote economic development in neighbourhoods adjacent to tech growth. Our analysis compares the Twitter tax breaks to the public-private partnerships of the late twentieth century, known as empowerment or enterprise zone policies. Although both policies were designed to uplift impoverished urban communities, the ideological undercurrents of the Twitter tax breaks appear distinct. They represent a particular version of neoliberal redevelopment that we call *compassionate capitalism*.

Compassionate capitalism

In our interviews with technology workers, we identified a gap between utopian and meritocratic discourses about the promise of technology and the realities of the residents of impoverished communities where technology workers were employed. This is a central feature of what we call *compassionate capitalism*. Compassionate capitalism refers to the philanthropic commitments that accompany corporate redevelopment efforts in underserved neighbourhoods. This type of capitalism consists of both symbolic and discursive manoeuvres. It is characterized by contradictions of extreme income inequality, neoliberalism and race-evasion. In the case of the Twitter tax breaks, technology corporations and their employees are transforming the racial, class and ethnic demographics of neighbourhoods, while actively engaging in charity campaigns that minimize the optics of gentrification.

Social segregation and the new economy

Residential segregation by race, class and national origin has long been analysed as a central dimension of economic inequality, social mobility, urban poverty, achievement gaps in education and health disparities. Hundreds of years of *de jure* racial segregation, that is state-sanctioned legal discrimination in the United States housing, has been followed by ongoing de facto segregation. This form of residential segregation structures the childhood friendships, adult social networks, intimate relationships and family formation. People excluded from highly resourced neighbourhoods experience negative impacts on their quality of life including education, health and housing (Frankenberg 1993; Lipsitz 1998/2006, 2011; Massey and Denton 1993; Oliver and Shapiro 2006; Steinbugler 2012).

In *Seeking Spatial Justice*, urban theorist Edward Soja (2010) explains how spatial geographies can be sites of injustice and discrimination. Soja argues for a new spatial consciousness that addresses the ways in which space both shapes and is shaped by social inequality. As Soja writes, "The geographies in which we live can intensify and sustain our exploitation as workers, support oppressive forms of cultural and political domination based on race, gender, and nationality, and aggravate all forms of discrimination and justice." Without recognition of this fact, we do not perceive the ways that spaces are filled with "politics, ideology, and other forces" that shape our daily lives and engage us in struggles of justice and injustice (Soja 2010: 19).

Drawing upon the French philosopher Henri Lefebvre's concept of the 'right to the city', Soja calls on social scientists to shift their attention to urban space as a site of production, reproduction and contestation of inequalities generated from all arenas of social life. For Soja, the concept of *social reproduction* is key to understanding the mechanisms that drive the unequal distribution of vital public services and the privileging of private property for an area's elite. Boundaries of a space can be redrawn to "to serve both positive and negative purposes, giving greater or lesser representation to certain population groups on a kind of sliding scale of inequality" (Soja 2010: 38). While this point is essential, one limitation of Soja's analysis is that he does not consider the impacts of spatial inequality on different racial-ethnic groups. Blacks and Native Americans, for example, are the most segregated racial and ethnic groups in the United States.

The arrival of technology companies and their employees to San Francisco has changed the racial and class demographics of the city as a whole, and of the Tenderloin, in particular. These new residents are mostly White and Asian (Pepitone 2011), largely male (Cagle 2014), frequently under the age of thirty-five (Cushing 2013) and earning an average of over $100,000 annually (Cagle 2014). Since 2000, the number of Whites in the Tenderloin has increased by 2.5 per cent and the number of Asians by 8 per cent, while the Black population has decreased by 6 per cent, and the Native American population has been cut in half.[3]

Although they share the same sidewalks, long-term residents live in a socially segregated and stigmatized world that is unfamiliar to the average tech worker. Office jobs in the tech industry typically require a four-year degree and years of experience.[4] In the twenty-first century, white-collar positions such as those in tech are increasingly secured through personal referrals, a system based on the social connections a job candidate shares with prospective employers. Evidence shows that hiring in the tech industry is frequently based on a candidate's 'cultural fit' (Alfrey and Twine 2017). This is determined by the interests, tastes, life experiences and educational credentials shared between a job candidate and the individuals who evaluate them. Cultural fit, in other words, is built from exclusive forms of residential and educational segregation that are spatially organized by race and class (Alfrey and Twine 2017). In new economy work organizations, social networks have become the principal means by which workers are hired (Williams et al. 2012). In the tech industry, these networks are often composed of one's childhood, college and professional peers (Twine and Alfrey, forthcoming).

Technology companies in San Francisco

San Francisco has the highest high tech related job growth of any city in the United States. Between 2012 and 2014, the number of high tech jobs in the city grew by 51 per cent. With a population of 852,000 people, San Francisco ranks thirteenth in population size among US cities. The San Francisco Bay Area is currently home to the software industry's leading players including Facebook, Google, Twitter, Yelp, Uber, Slack, Yammer, Airbnb, Salesforce, Pinterest, StumbleUpon and Zendesk. In 2013 alone, San Francisco created 8,000 new tech jobs (Shinal 2014). Foreign tech start-ups looking to establish operations in the United States often consider San Francisco, exclusively, for its European flavour and proximity to so much talent and so many investors (Swartz and Martin 2012).

Aided by payroll tax incentive programmes, social media and internet application companies such as Facebook, Google, Twitter, Yelp and Yammer have signed long-term leases in downtown San Francisco. Many of their employees have moved to the vibrant urban landscape of the city, where they can walk or bike to bars, cafes, restaurants, theatres and public parks near their offices. Those who live, but do not work, in the city can travel via private company shuttle buses to Silicon Valley.

These forces have generated a dramatic shift in the class and racial demographics of the city as a whole. Long-term residents of all racial and ethnic backgrounds, who are working-class or lower middle-class renters, are being priced out of San Francisco entirely. As reporters Jon Swartz and Scott Martin write, "affordable office leases relative to the old Valley, favourable business taxes, and the allure of a major city [are] transforming the most vibrant and multiethnic neighborhoods in the city" (Swartz and Martin 2012).

In 2007, Google was one of the first tech companies to open offices in a downtown San Francisco building, which housed over 900 employees. In 2014, Google opened a second San Francisco office for 200 employees in the Mission District, the city's historically Latino neighbourhood (King 2014). In 2014, San Francisco began offering six-year exemptions from its payroll taxes to attract technology firms and encourage their long-term investment in the mid-market area. Companies such as Twitter were relieved of paying city payroll taxes in exchange for inhabiting buildings in a zone targeted for redevelopment. City planners argued that these 'new residents', who can buy and rent commercial and residential property at higher prices, would funnel money into new businesses through their consumption, create demands for entry-level jobs and boost overall tax revenue to provide a 'rising tide for all boats'.

San Francisco has since become one of the most 'tech friendly' cities in the United States. In 2012, Mayor Ed Lee, the first Chinese American mayor of San Francisco, did not mince words when describing his administration's commitment to

> ensuring San Francisco remains a centre for tech and the innovation capital of the world ... top creative and innovative talent wants to live in a vibrant, transit-friendly, global city that offers access to not only great jobs but also great food, entertainment and culture.
>
> (Swartz and Martin 2012)

History and demographics of the Tenderloin

According to historians Rob Waters and Wade Hudson (1998), the Tenderloin got its name from a similar red-light district located in in New York City. As reported in the *New York Times*, the New York neighbourhood was named as such by a police inspector appointed in the late 1800s, who commented that moving from a relatively obscure precinct to one of the area's most notorious, "I have had chuck for a long time, and now I am going to eat Tenderloin."[5]

Oblong in shape, the Tenderloin neighbourhood of San Francisco is bounded on the West by the Western Addition, to the North by Nob Hill, and to the immediate West by the hotels and department stores of Union Square (Groth 1994: 113). Its boundaries have shifted over time relative to surrounding areas. For example, some San Franciscans include the area of Polk Street from Eddy to Hayes within its border, while others refer to this area separately as 'Civic Centre'. Despite border discrepancies to the West and the North, Market Street is the clear demarcation of where the Tenderloin ends on its Southern side. This is where downtown mid-Market begins, the area now under redevelopment by tech companies receiving payroll tax benefits.

Once the centre for travelling sailors and miners in the late nineteenth century, after the 1906 earthquake and subsequent fires, the district developed into a rooming house area offering cheap lodging, food and entertainment for a male and immigrant labour force (Groth 1994). It functioned as a safe haven for gay men abandoned by families and friends during the 1930s–1960s. In 1968, before the now famous Stonewall riots in New York, it was a site for one of the first gay protests. According to Michael Bronski, the police attempted to "eject rowdy customers" at the Compton Cafeteria (Bronski 2011: 209). A sidewalk plaque now commemorates this event.

Within its boundaries is the Theater District, once home to the famous Black Hawk jazz club, and Fantasy Records where Dave Brubeck, Miles Davis and Thelonious Monk all recorded live albums in the late 1950s and early 1960s.[6] The Tenderloin remained overwhelmingly White, working class and populated by single men or couples as late as the 1970s (Groth 1994: 113). Following the Vietnam War, the neighbourhood welcomed large numbers of refugees and their families from South-East Asia (Bronski 1994). As historians Rob Waters and Wade Hudson (1998) report, two events in the late 1970s – the influx of thousands of South-East Asian immigrants, and a rising tide of neighbourhood activism and community organizing – made the Tenderloin one of the most diverse neighbourhoods in the city. Over time, the infusion of funds from local foundations and religious charities, such as St Anthony's Church, created a vital set of social services and a rare neighbourhood where low-income people could afford to live.

During the past three decades, as San Francisco's tech population has exploded, and housing costs have skyrocketed, the Tenderloin has persisted as the last neighbourhood to provide centrally located social services. St Anthony's Foundation, located on Golden Gate Avenue, continues to offer daily services that include daily meals, a computer lab, medical clinic and addiction recovery, clothing and jobs training programmes to the impoverished. Today, the Tenderloin has the densest population of any neighbourhood in the city, as well as the highest proportion of families and children. Median household income for Tenderloin residents was just over $15,000 in 2015, about one-sixth of the median household income of the city as a whole.[7] As policing practices and commercial redevelopment have pushed more and more of the city's dispossessed into its territory, the Tenderloin now functions as an over-crowded ghetto for the city's most underserved – home to thousands of poor Whites, Blacks, Latinos and South-East Asians of all ages, as well as refugees, street hustlers, drug dealers and the city's large homeless population (Gowan 2010). (Table 41.1) Approximately 28,000 people currently live in the forty-square-block area in single-room occupancy hotels and apartments.[8]

As ethnographer Teresa Gowan (2010) highlights, having a skid row style neighbourhood in the centre of one of the nation's wealthiest cities is rare. While other cities have poor and even depopulated neighbourhoods near the city centre, the Tenderloin *is* the middle of San Francisco's downtown. It touches all power centres of the city – running up to City Hall and the federal building on one side, the luxury hotels and upscale department stores of Union Square on another, and the financial district, Westfield Shopping Center, and the city's most used BART station for tourists and workers on yet another side. Thus, the Tenderloin forces a daily

Table 41.1 2010 racial demographics for the Tenderloin vs San Francisco

	Tenderloin		San Francisco	
	Number	%	Number	%
Total population	31,176		805,235	
White	14,147	45.4	390,387	48.5
Asian	7,922	25.4	267,915	33.3
Hispanic or Latino (of any race)	5,893	18.9	121,774	15.1
US Black or African American	4,343	13.9	48,870	6.1
Two or more races	1,469	4.7	37,659	4.7
Other race	2,866	9.2	53,021	6.6
American Indian and Alaska Native	306	1.0	4,024	0.5
Native Hawaiian or other Pacific Islander	123	0.4	3,359	0.3

Source: US Census 2010.

confrontation between the haves – tourists and well-paid finance and tech workers – and the have-nots of the city. This reality makes the ever-growing income and housing inequalities of San Francisco particularly difficult to ignore, and the Tenderloin has become a most persistent site of ire for business leaders, real estate developers, tourists and local politicians hoping to solidify San Francisco as a premiere destination for cosmopolitan elites.

Payroll tax breaks and the corporate 'revitalization' of the neighbourhood

In 2011, Mayor Ed Lee proposed a plan to redevelop the mid-Market area. To remove barriers to redevelopment, Lee encouraged high tech business to buy or lease abandoned or underused buildings adjacent to and in the neighbourhood. Known as the 'payroll tax incentive', or the 'Twitter tax breaks' this arrangement gave at least twenty-seven companies like Twitter, Dolby, Dropbox, Yelp, Yammer and Zendesk a six-to-eight-year payroll tax break on new employees hired for opening offices or keeping offices in the mid-market and Tenderloin Districts (Gordon 2011). The plan exempts eligible companies from 1.5 per cent payroll expense tax (or 'payroll tax') that San Francisco levies on all businesses that operate in the city and whose annual payroll expense exceeds $250,000 (Gordon 2011). Tax breaks include the salaries of new employees, as well as their bonuses and stock options, the latter of which can be a substantial part of a new employee's benefits package at a company of increasing market value. As of 2014, these policies had cost the City of San Francisco over $56 million in lost revenue.[9]

City planners argued that this legislation would lead to "higher job growth and property values in the area, and, in turn, would increase sales, hotel, utility user, property, and transfer tax revenues".[10] This is a trickle-down economic argument for job growth: it presupposes that by luring major corporations into the city, well-paid employees will spend money in surrounding neighbourhoods, creating entry-level jobs and generating income. As Mayor Lee similarly forecasted in 2011, when the programme was first proposed: "Central Market and the Tenderloin have been burdened with high vacancies and blight for decades and ... the payroll tax exclusion is a powerful tool that will help us bring in much-needed jobs, services and retail" (Gordon 2011).

Opponents view this agreement as an act of corporate welfare, one that reduces revenue to fund the very city services that the poor and houseless so desperately need. They suggest that any service sector jobs that might be created do little to correct for the vast disparities in skills and life experience that would allow a resident of the Tenderloin, for example, to compete for a job at the very tech companies hiring in their neighbourhood. In short, jobs being created by tech companies are far from reach for long-term residents in need of stable employment and affordable housing.

Empowerment and enterprise zones

We see the Central Market Street and Tenderloin Area Payroll Tax Exclusion as a variant of enterprise or empowerment zones. These policies have been well-criticized for using tax revenues to subsidize business location in a depressed area without stimulating a genuine process of long-term economic development that can survive the expiration of the subsidy (SF Controller's Office 2011: 2).

Formerly known as 'enterprise zones', empowerment zones had bipartisan appeal and were embraced by Democrats and Republicans (Kasinitz and Rosenberg 1996: 182). With empowerment zones, it is assumed that bringing business into the area will help the job prospects of poor residents who live there (Kasinitz and Rosenberg 1996: 182). According to Kasinitz and Rosenberg, "Since 1981, 35 states and the District of Columbia have introduced some version of an enterprise zone program, with modest results at best" (1996: 182). In her law review of enterprise zones and new market tax credits, Jennifer Forbes (2006) shows that there is little evidence to demonstrate success with these programmes. For example, in the city of Baltimore: among residents who participated in job training classes, only slightly more than half were working two years later (186). Said another way by Kasinitz and Rosenberg:

> The idea sounds like simple common sense: use tax breaks and other incentives to encourage private employers to locate in poor areas, bringing job opportunities to local residents … [However] the primary reason for ghetto unemployment is not the lack of nearby jobs but the absence of social networks that provide entry into the job market.
>
> (Kasinitz and Rosenberg 1993: 1)

Trickle-down economics are largely ineffective for people who are working class or working poor. A such, some advocates argue that long-term subsidized or free housing would be more useful to residents of the Tenderloin, who lack the educational credentials and social connections necessary to secure a job in the new economy, and in one of the world's most competitive industries. The argument that the mere presence of a corporation will generate jobs for the people living in the area is particularly unlikely in the tech sector – where our interviewees reported that software engineers need approximately ten years of training or the equivalent in computer programming to compete for jobs, and the average office worker must have a four-year degree and years of experience. This is particularly consequential for the 14 per cent of the population of the Tenderloin who are over the age of sixty-five.[11] It is unclear how advocates of tech company tax breaks imagine a future for long-term residents under these conditions, where the gap in credentialling and education between the average worker at a tech company and the average resident of the Tenderloin is so enormous.

Community benefits agreements

As part of the tax break arrangement, businesses applying for the Central Market Street and Tenderloin Area Payroll Expense Tax Exclusion with payroll of greater than $1 million annually must enter a Community Benefit Agreement (CBA) with the City Administrator. As of 2015, six tech companies – 21 Tech, One Kings Lane, Twitter, Yammer, Zendesk and Zoosk – have CBAs with the City Administrator. These beneficiaries have committed their employees to volunteering a certain number of hours in the community each month. Companies like Zendesk have even hired employees to oversee philanthropic efforts and to measure their impacts in the local community.

To date, CBA activities have included staffing local soup kitchens, volunteering at local schools and assisting with community gardens. Other interventions, designed to 'bridge the tech divide', have included the addition of a computer lab at St Anthony's church and the development of a Yelp style social services application known as 'Link-SF', which we discuss in the following section. Despite the well-intentioned goals of CBA agreements, there is scant evidence to demonstrate occupational mobility benefits for residents of the Tenderloin. In contrast, our interviews show that these philanthropic programmes appear to provide the greatest benefit to tech workers themselves, in part by retaining them within a spatial environment that sits in stark contrast to Silicon Valley.

Being a good neighbour: technological solutions to social problems

Early reports suggest that these community benefits have had marginal impacts on community members. Instead, the goal of assisting the community appears to blunt the optics of gentrification, and to present tech companies as 'good neighbours'. One example includes the Zendesk non-profit initiative. Known as *The Zendesk Neighbor Foundation*, it distributes some portion of $1 million for 'corporate social responsibility' initiatives. As described in a company press release, the fund allows Zendesk to donate to "outstanding organizations" that include: charity water, St Anthony's Foundation, Dry July, COSMIC and Temple Street University Children's Hospital. Company representatives claim:

> This new fundraising model will help us support organizations addressing issues of poverty, homelessness, and healthcare; improving education and promoting gender equality; workforce development; and technical literacy. It is these pillars that support the Zendesk Neighbor Foundation ... The Foundation will provide financial and strategic support to not only organizations in the community around our San Francisco headquarters, but also to the company's 10 other offices in 10 countries worldwide. By doing this, we can extend the community involvement that began in our headquarters to the neighbourhoods around the world in which we're rooted.

In addition to these efforts, engineers at the same company conceived of and designed a Yelp style application service, known as 'LinkSF', that allows residents to identify services such as shelter, medical assistance, computer labs and food shelters in their area (Wagner 2014). There is little evidence showing that this technological tool has changed the conditions of the area's struggling residents in meaningful ways.

In a city where there are approximately 6,500 homeless individuals, a report used to support LinkSF suggested that 40–45 per cent of them have cellphones. However, closer investigation reveals that this 40–45 per cent of cellphone users are merely those who use St Anthony's computer lab. Said another way, of the group of people technically savvy enough to seek assistance in a computer lab, less than half have a cellphone that is enabled for the LinkSF application (Here & Now 2014; Wagner 2014).

The software engineer who helped design the application and served as a spokesperson with the press stated that the underlying goal was not larger user numbers, but rather meaningful user experiences:

> A wild success for us might be 50, 100 or 200 people using the [site] ... The goal [is] having the providers and the referral services reach out and teach the service-seekers about this stuff and how to use it on their own terms.

Soon dubbed the 'homeless app', LinkSF received mixed reviews in the press. What is perhaps more striking than the limits of this technological solution is the logic that guided its creation. This story illustrates a particular kind of race and class *habitus* that typifies the technology sector.[12] A daily problem faced by a tech worker – such as how to make decisions about which restaurant or hotel to patron – is categorically distinct from individuals who are unhoused and lacking in safe residences, private bathrooms, showers and regular health care. A homeless services rating app is both ineffective for solving structural problems and highly neoliberal in its most basic premise – that homeless people lacking adequate services is a problem of 'choice' rather than the consequence of inequality and disinvestment.

CBAs function to introduce a new class of 'compassionate' capitalist managers to the city's poor and marginalized whilst upholding a myth of how they acquired such occupational power. While tech workers and Tenderloin residents share some limited public spaces, they live in separate worlds segregated socially, economically and racially.[13] In an industry that is predominantly White and Asian, in which company founders often hire former classmates and colleagues within similar backgrounds and educational credentials, it is difficult to imagine how the development of software applications and the construction of computer labs can uplift poor and unhoused communities that do not share a racial and class location with local technology workers. Without a recognition and interruption of the non-meritocratic channels of access enjoyed by the typical tech worker, long-term residents of the Tenderloin are likely to remain excluded from the region's tech boom.

Tech volunteerism in the Tenderloin: what is the real impact?

In this section, we examine how tech workers make sense of their presence in the Tenderloin, including their perspectives on the occupational responsibilities committed by their company's community benefits agreements. We begin with Sumi, a 35-year-old Chinese American interface designer who works at a start-up company receiving payroll tax benefits. Sumi earns between $100,000 and $150,000 annually. Her first tech industry job was in marketing. When asked to describe her current workplace, Sumi noted that it is "very comfortable". Her team typically arrives at the office around 10am, and her company offers free catered dinners every evening. Her compensation package includes a membership at a high-end gym, and she often exercises on her way into work. Sumi and her husband own a two-bedroom apartment north of the Panhandle, a hip neighbourhood near Golden Gate Park with highly desirable restaurants and bars.

Sumi is also the daughter of immigrants from Hong Kong who are landlords in the Tenderloin. She grew up in the Mission District, which is a formerly working-class, economically mixed and predominantly Latino neighbourhood that welcomed immigrants from Central America, Mexico and Brazil. This district has since become a premiere neighbourhood of San Francisco with restaurants, bars, mid-century furniture stores and concert venues used by Asian and Anglo-American tech workers.

When asked what she thought about her company's location in the Tenderloin, Sumi responded:

> Well, my parents co-owned a building in the Tenderloin with the rest of my family ... I remember going there as a child to pick up rent with my Mom there. And it was really scary! And even as a teenager going downtown, I remember people asking me if they could walk to the Asian Art Museum from the downtown area ... I would always advise them NOT to. I would advise them to take a bus or take Muni underground ... because I wouldn't walk that stretch myself. And now I'm totally comfortable doing that, I can go into the Tenderloin and have lunch! And I see lots of people I know doing the same thing. ... Growing up here, I think that's really great.

When asked about the tech industry and economic inequality in the neighbourhood, Sumi replied:

> I think it's something that would happen naturally anyway ... I think these things happen, and this is just one flavour of that happening. That's not to say I don't care about people who have lived here wanting to continue to live here. I mean, I was one of those people.

> I have friends who had to move to the East Bay [Oakland, Berkeley, Richmond] and I think that's natural. You know, if you want a better deal, you want an actual house, there's only so much in San Francisco ... no one deserves to stay here *just because*.

In this excerpt, Sumi rejects a 'right to the city' claimed by other less economically advantaged long-term residents of San Francisco. She views the forces remaking the Tenderloin as "natural" rather than as the consequences of targeted political and economic arrangements favouring individuals of her same occupational class. Her observations do not account for the structural advantages that gave her the opportunity to remain in an increasingly unaffordable city while others left. Her family was a rare group of immigrants that could afford property in San Francisco, an asset that accumulated over time and provided upward mobility to live in more expensive neighbourhoods. Sumi was also lucky to secure a job in the tech industry early in her career, before jobs in the industry reached a competitive fever-pitch. In short, being a property owner in one of the world's most expensive cities, with a six-figure salary in a highly competitive industry, is far from typical for a person in in their mid-thirties. Sumi's reality represents a rare combination of timing, job market advantage and family resources.

Dane is a 29-year-old Filipino-American who grew up in a middle-class suburban area outside San Francisco. He attended a private high school and, later, a private college in Silicon Valley. At the time of his interview, Dane was working in a customer support role at one of the companies receiving payroll tax benefits. He was referred for his job through a friend employed at the same company.

When asked about his company's mandated CBA volunteer activities, Dane initially expressed scepticism about their effectiveness, saying, "In terms of my volunteering in the Tenderloin, I'm not totally sold on how much of a positive effect it's actually going to have."

However, when asked whether these programmes do more harm than good for Tenderloin residents, Dane re-fashioned his argument:

> I think it's totally fine. [Name of his company] actually does a good job of participating and volunteering. I can't say the same about other companies that are signed on ... like Zoosk and Twitter ... [name of his company] has been praised for what they've actually accomplished ... they've followed through. Whereas other companies have been highly criticized for their lack of volunteering.

Here Dane uses the number of volunteer hours as a measure of the programme's success, comparing the total number of hours volunteered by employees at his company with those of other companies receiving payroll tax breaks in the area. When asked to clarify the impact of his company's volunteer efforts on members of the community, Dane struggled:

> I've done kitchen duty at one of the shelters ... and one of the best experiences I had was volunteering at St Anthony's busing tables. I had this stereotypical view of what homeless people looked like, and when I got there, I would say at least one out of every three people looked like regular people. ... If I saw them on the street, I would have never thought they were homeless. Being Filipino, I saw a bunch of Filipinos that were my parents' age, that could have easily been my parents. ... I've volunteered in other countries, and I think it's always hard to think about how much of an impact you're having. Whether it matters, or whether you made a difference. *But I think the most important part is that you at least know that you contributed at some level, and helped out ... for me, it's the motivation to do [good] rather than the actual impact.*

Dane's claim that "motivation to do good" is more important than "actual impact" is telling. He believes that the intent of tech workers is of a higher order than the impact on displaced and poor residents of the Tenderloin. Said another way, tech workers are the meaningful beneficiaries of these initiatives. Dane's interview reveals a kind of race-and-class innocence,[14] protecting 'people like him' from feeling responsible for the increasing displacement and inequality that their presence is driving.

Erin is a 32-year-old European-American, and works in customer support for a software services start-up receiving payroll tax benefits. She grew up in a suburb of Silicon Valley and attended a private Catholic High School. In 2005, she graduated from the University of California. At the time of her interview, Erin was living with her boyfriend, who also works in the tech industry as a software engineer. Like Dane, Erin was referred for her current job through a friend.

Erin distinguishes working in the Tenderloin from the Google campus in Palo Alto. She argues that being immersed in an urban area with social problems has benefits for young privileged workers who are otherwise geographically segregated from people that are poor and unhoused. In her analysis:

> I think it's actually a good arrangement now that I've been here. I think it's good for people in the tech industry, especially, myself included, a lot of the young people that are coming straight from school ... It's like – it forces you to be part of the community and to learn more and to come face-to-face with some of the realities of market ... I've come to embrace it more. And also, the fact that we're encouraged to volunteer, I think, is a good thing because I just haven't had the time to in the past ... now, you almost have an excuse to do something.

She continued:

> [Volunteering] was kind of hard because you're faced with the reality, you're not in the best part of town and all that stuff ... it helps to put faces to people and realize that everyone is not like you ... so you know, it's good to see that, but it's not easy all the time. Um, it just makes you aware of what you have.

Here we see that Erin sympathizes with impoverished and homelessness residents of the Tenderloin. However, in her analysis of CBAs, she makes it clear that she and members of her occupational class are the primary targets of volunteer initiatives. In Erin's words, proximity to the poor and unhoused makes her 'aware of what she has'. Her interactions with Tenderloin residents function as a vehicle for self-reflection and enhance her sense of gratitude. Yet these feelings are not accompanied by a sense of responsibility, or a concern for social injustice. Rather, Erin's recognition that not "everyone is not like you" is a race-and-power-evasive discourse that elides the fact that most tech employees are members of racially dominant groups and come from middle- and upper-class backgrounds. Erin and her colleagues secured jobs in one of the world's most coveted industries often through personal relationships and college alumni networks composed of people from similar race and class backgrounds. They had citizenship or H1-B visa status. The typical Tenderloin resident, on the other hand, is more likely to be impoverished, to be an undocumented immigrant and to be Black American or Latino. While tech workers and Tenderloin residents may share a physical environment that overlaps spatially, their everyday lives are highly segregated socially.

Conclusion

The tech workers we interviewed varied in how they evaluated the merits of the CBA agreements. Rather, the evidence suggests that these initiatives allow technology companies to acquire reputational capital by being perceived as good neighbours. This is a symbolic move, rather than an economic intervention, that does not allow Tenderloin residents to rent or buy affordable housing and remain in their neighbourhood. Initiatives designed to provide computer training, while a necessary resource for poor and working-class residents lacking access to a computer, neglects the significant role that class privilege, racial inequality and social networks play in how tech workers secure their employment in this industry. In this way, community benefits agreements present a type of compassionate capitalism that appears to primarily benefit technology workers. Though these policies are well-intentioned in their design, they ultimately fail to challenge structural inequality in job markets and ongoing racism in housing. Worse, in light of charitable initiatives, they may function to create another form of advantage for the area's new elites, helping to retain them as workers in neighbourhoods where their presence simultaneously drives displacement.

Our interviews with tech workers, combined with our analysis of Community Benefits Agreements (CBAs), suggest that these efforts constitute emotionally rewarding social initiatives for tech companies and their workers that do little to address the structural inequalities responsible for the ongoing marginalization of the poorest residents of the city. When asked, "How have the CBA efforts of your company positively impacted the community?" the tech workers we interviewed struggled for evidence, and instead described how they benefited from such efforts. They detailed the ways in which the opportunity to volunteer made them feel good about themselves and enabled them to feel more 'connected to' or 'safe' in the neighbourhood in which they now worked. We see this as evidence of compassionate capitalism, a set of symbolic and discursive manoeuvres that provide area elites with a heightened sense of comfort with the homeless, mentally ill, impoverished and racially excluded, without being accountable to the community or bearing responsibility for their role in ongoing gentrification and resident displacement. CBAS function to uphold a class-based and racialized spatial and social hierarchy of private property holders and occupational elites, driving out long-term residents who were unable to accumulate the resources to purchase housing in their home city

In summary, we believe that CBAs reinforce social and economic inequality. They provide a tattered safety net under the banner of progress to economically marginalized residents without the kind of interventions that would allow those same individuals to stay. Rather than assist current residents financially and supporting, in Edward Soja's words, their 'right' to remain in San Francisco, CBAs reinforce a mythology. The myth is that knowledge of basic skills can allow residents to secure white-collar office jobs that typically require four-year college degrees, prior relevant experience and personal contacts. This is fundamentally disconnected from the reality of how a typical tech worker secures employment. A digital 'gold rush' is driving a new wave of internal migrants from other regions of the United States as well as international migrants from China, India, Russia and Western Europe, to accept well-paid technology jobs, while San Francisco residents outside their social class receive few meaningful benefits from this new industry.

Notes

1 *Airbnb* is an online global platform designed to assist people who wish to list, find and rent local homes including single-family detached houses, apartments, condominiums and lofts among others.
2 Source: City and Country of San Francisco, Office of the Controller, 'Review of the Impact of Central Market Payroll Tax Exclusion', October 2014.

3 Source: www.sfdph.org/dph/files/mtgsGrps/FoodSecTaskFrc/docs/AChangingLandscape-FoodSecurity intheTenderloin.pdf.
4 Source: http://sfpublicpress.org/news/2013-11/twitter-other-tech-companies-get-sf-tax-breaks-but-show-little-progress-hiring-in-neighborhood.
5 Source: http://query.nytimes.com/mem/archive-free/pdf?res=9901E5D9143AE433A25755C2A9659C946696D6CF.
6 Source: http://content.time.com/time/magazine/article/0,9171,825838,00.html.
7 Source: http://www.google.com/url?hl=en&q=www.city-data.com/neighborhood/Tenderloin-San-Francisco-CA.html&source=gmail&ust=1480977451606000&usg=AFQjCNFbJIuqqy2DsOFqzTrm25EvfzZu1Q.
8 Source: www.pbs.org/newshour/bb/san-francisco-working-class-neighborhood-left-behind.
9 Source: www.sfchronicle.com/bayarea/article/S-F-tax-day-protest-marches-on-Twitter-5405393.php.
10 Source: http://sfcontroller.org/sites/default/files/FileCentre/Documents/1865-PDF108.pdf.
11 Source: US Census 2010.
12 Pierre Bourdieu defines habitus as "structuring structures": the dispositions and tastes that are cultivated primarily through one's class position and interactions with other members of their class. Although these dispositions are often unconscious, they are practiced in a way that constructs and reproduces class difference in everyday life (Bourdieu 1984).
13 Racialized and impoverished groups occupy fundamentally disparate social networks, even when they live in the same geographic area (Bonilla-Silva 2014; Frankenberg 1993; Kasinitz and Rosenberg 1993, 1996).
14 In *Racing for Innocence*, sociologist Jennifer Pierce (2012) presents the concept of racial innocence, used by White lawyers who oppose affirmative action programmes as the way in which they can "disavow accountability for racist practices and, at the same time, practice racially exclusionary behavior" (p. 9).

References

Alfrey, L. and Twine, F.W. (2017). Gender-fluid Geek Girls: Negotiating Inequality in the Tech Industry. *Gender and Society*, 31(1), pp. 1–23.

Bonilla-Silva, E. (2014). *Racism without Racists: Color-Blind Racism and the Persistence of Racial Inequality in America*. 4th edn, New York: Rowman & Littlefield.

Bourdieu, P. (1984). *Distinction: A Social Critique on the Judgement of Taste*. Cambridge, MA: Harvard University Press.

Bronski, M. (2011). *A Queer History of the United States*. Boston, MA: Beacon Press.

Cagle, S. (2014). San Francisco's Class War, by the Numbers. *The Nib*, 12 February. Available at: https://medium.com/the-nib/700c51a43a4 [accessed 13 February 2014].

Cushing, E. (2013). The Bacon Wrapped Economy. *East Bay Express*, 20 March. Available at: www.eastbayexpress.com/oakland/the-bacon-wrapped-economy/Content?oid=3494301&storyPage=4 [accessed 8 February 2014].

Forbes, J. (2006). Using Economic Development Programs as Tools for Urban Revitalization: A Comparison of Empowerment Zones and New Markets Tax Credits. *Illinois Law Review*, 2006(1), pp. 177–204.

Frankenberg, R. (1993). *White Women, Race Matters: The Social Construction of Whiteness*. Minneapolis: University of Minnesota Press.

Gordon, R. (2011). Twitter Will Get Payroll Tax Break to Stay in SF. *San Francisco Chronicle*, 6 April. Available at: www.sfgate.com/news/article/Twitter-will-get-payroll-tax-break-to-stay-in-S-F-2375948.php [accessed 4 January 2015].

Gowan, T. (2010). *Hobos, Hustlers, and Backsliders: Homeless in San Francisco*. Minneapolis: University of Minnesota Press.

Groth, P. (1994). *Living Downtown: The History of Residential Hotels in the United States*. Berkeley: University of California Press.

Here & Now. (2014). Homeless in San Francisco? There's an App for That. *WBUR Boston Here & Now*, 26 March. Available at: http://hereandnow.wbur.org/2014/03/26/homeless-app-sf [accessed 8 April 2014].

Kasinitz, P. and Rosenberg, J. (1993). Why Enterprise Zones Won't Work: Lessons from a Brooklyn Neighborhood. *The City Journal*, 3. www.city-journal.org/html/why-enterprise-zones-will-not-work-12612.html [accessed 14 December 2017].

Kasinitz, P. and Rosenberg, J. (1996). Missing the Connection: Social Isolation and Employment on the Brooklyn Waterfront. *Social Problems*, 43, pp. 180–196.

King, J. (2014). Google May Now House Startups in San Francisco's Historic Latino Community. *Colorlines*, 17 February. Available at: http://colorlines.com/archives/2014/02google_opens_office_in_san_franciscos_historic_latino_community.html [accessed 17 February 2014].

Lipsitz, G. (1998/2006). *The Possessive Investment in Whiteness: How White People Profit from Identity Politics*. Philadelphia, PA: Temple University Press.

Lipsitz, G. (2011). *How Racism Takes Place*. Philadelphia, PA: Temple University Press.

Massey, D.S. and Denton, N.A. (1993). *American Apartheid: Segregation and the Making of the Underclass*. Cambridge, MA: Harvard University Press.

Oliver, M. and Shapiro, T. (2006). *Black Wealth/White Wealth: A New Perspective on Racial Inequality*. New York: Routledge.

Pepitone, J. (2011). Silicon Valley Keeps its Diversity Data Secret. *cnnmoney.com*, 9 November. Available at: http://money.cnn.com/2011/11/09/technology/diversity_silicon_valley/ [accessed 12 February 2014].

Pierce, J.L. (2012). *Racing for Innocence: Whiteness, Gender, and the Backlash against Affirmative Action*. Stanford, CA: Stanford University Press.

SF Controller's Office. (2011). *Payroll Expense Tax Exclusion in Central Market Street and Tenderloin Area: Economic Impact Report*. Available at: http://sfcontroller.org/sites/default/files/FileCenter/Documents/1865-PDF108.pdf [accessed 12 January 2018].

Shinal, J. (2014). It's Boom Time in San Francisco: Population, Jobs Are Growing. *USA Today*, 11 February. Available at: www.usatoday.com/story/tech/columnist/shinal/2014/02/11/san-francisco-is-tech-hub-grew-during-recession/5398931/ [accessed 3 December 2016].

Soja, E. (2010). *Seeking Spatial Justice*. Minneapolis: University of Minnesota Press.

Steinbugler, A. (2012). *Beyond Loving: Intimate Racework in Gay, Straight, and Lesbian Relationships*. New York: Oxford University Press.

Swartz, J. and Martin, S. (2012). San Francisco Is Fast Becoming a New Technology Hub. *usatoday.com*, 24 July 24. Available at: http://usatoday30.usatoday.com/money/industries/technology/story/2012-07-18/new-tech-hub-san-francisco/56441430/1 [accessed 8 February 2014].

Twine, F.W. and Alfrey, L. (forthcoming). *Geek Girls: Race, Class, and Power in the Tech Industry*. Cambridge: Cambridge University Press.

Wagner, K. (2014). The Key to Helping San Francisco's Homeless: Cellphones. *Mashable*, 28 February. Available at: http://mashable.com/2014/02/28/homeless-mobile-link-sf/#6TIrLoirGkqc [accessed 10 February 2014].

Waters, R. and Husdon, W. (1998). The Tenderloin: What Makes a Neighborhood? In: J. Brook, C. Carlsson and N.J. Peters, eds. *Reclaiming San Francisco: History, Politics, Culture*. San Francisco: City Lights Publishers, pp. 301–316.

Williams, C.L., Muller, C. and Kilanski, K. (2012). Gendered Organizations in the New Economy. *Gender & Society*, 26, pp. 549–573.

42
GENDERING URBAN PROTEST
Politics, bodies and space

Fran Tonkiss

Acts of protest are among the most immediate ways in which spaces of urban politics are made manifest. Street marches, demonstrations, vigils and occupations turn everyday urban places into sites of politics by concentrating diffuse and often distant issues of power in condensed public spaces. This works both through direct confrontation at emblematic sites of authority (the protest at the parliament building, ministry, central bank or barracks), and by making visible and audible operations of power which take place elsewhere – in dispersed or remote sites of military conflict, economic exploitation or environmental destruction, or in the private spaces of the home. The discussion that follows considers the politics of urban protest in regard to gender issues, identities and relations. My interest is in urban mobilizations that focus directly on gender themes as well as those which address 'other' political concerns through the figuration of women's bodies and the articulation of women's voices. This is not to suggest that political protest is only gendered when it is led by women or centres on matters of concern to them. Rather the point is to explore how women's mobilizations activate urban space in ways which make gender relations and identities visible, public and political.

Histories of public mobilization have been heavily masculinized, and both formal and routine rights to access and occupy public space remain inequitably distributed along gender lines in different political and cultural contexts; whether through force of law, the coercive weight of social norms, or uneven dynamics of gender interaction, autonomy and vulnerability. Spatial inequities around the gendering of bodies are particularly acute in respect of transgender, queer and non-binary subjects whose access to everyday spaces, public anonymity and urban amenity is compromised or proscribed in numerous ways (see Doan 2010). The production and protection of gendered spaces has been an important element of women's urban politics, creating and maintaining spaces of refuge, community centres, independent workplaces, businesses and leisure spaces, crèches and health facilities (see, for example, Spain 2015). Alongside these more embedded spaces of gender politics, women's movements of various kinds have taken to public streets and squares in a politics of protest, witness and memory. It is still the case, however – and across a range of national and urban settings – that women are less likely than men to take part in political protests and tend to engage in less 'confrontational' tactics of dissent, with men typically more involved in demonstrations, occupations and pickets (see Dodson 2015). Within protest movements, moreover, there persists a tendency for women to do the less visible and more mundane work of organization, so that the division of labour within such movements often can be seen to reproduce conventional gender roles and demarcations (Boler *et al.* 2014; Dodson 2015: 379–380; Sasson-Levy and Rapoport 2003: 392–395).

The urban politics of protest has been very prominent in recent years, with resistance and opposition to political repression, financial crisis, debt and inequality finding expression in street-based demonstrations and occupations in a number of cities. The involvement of women has been marked in these situations, with the presence and profile of women a feature of mobilizations in Middle Eastern and North African cities during 2011's Arab Spring (Al-Ali 2012; Hafez 2014; Khalil 2015; Morsy 2014), the *Indignados* 15M movement in Spain (Gámez Fuentes 2015), Occupy encampments in numerous urban sites (Hurwitz and Taylor 2018; Maharawal 2016), and the 2013–2014 Euromaidan protests in Kiev (Khromeychuk 2016; Martsenyuk 2016). While such movements raised gender injustice as part of broader political agendas, the position of women and the handling of gender issues in these settings were often sharp points of contention. Media representations of urban insurgencies drew on the symbolic power of individual women who were subject to police and military violence, as seen in the Green Movement protests in Tehran following Iran's 2009 elections, the 2013 Gezi Park demonstrations in Istanbul, or Cairo's Tahrir Square (see Al-Ali 2012; Arat 2013; Hafez 2014). Inside these spaces of protest, meanwhile, problems of marginalization and exclusion, homophobia and transphobia, sexual harassment and violence were highlighted in the Gezi Park, Tahrir Square, 15M and Occupy encampments, and during Hong Kong's 2014 Umbrella Movement protests (see Bhattacharjya *et al.* 2013; Deb 2015; Gámez Fuentes 2015; Hurwitz and Taylor 2018; Maharawal 2016; Wang *et al.* 2017).

In these respects, women's participation in urban protest is common, persistent and often still compromised. Rather than focusing on when and how gender becomes a problem – or women become a symbol – within social and political movements, my concern here is with when women take the lead in the streets and spaces of public protest. In what follows, I consider mobilizations that address more general politics of social and economic justice in terms of their implications for the lives and livelihoods of women and girls, as well as interventions which focus more explicitly on a gendered politics of the body, violence and autonomy. In both contexts, there is a critical interplay between gender as a domain of politics – configuring a complex set of political issues, problems and conflicts – and gender as a mode of political subjectivity, agency and expression. The point is not to treat gender as a given demographic marker but rather to think about how gendered identities and engagements are produced through political mobilization, public assertion and spatial figuration. Gender protests are animated both by a politics of solidarity – where partisan commitments are made based on degrees of commonality – and by a politics of intersection, where connections and causes are premised on the recognition of difference. In an extended manner, focusing on women's mobilizations offers insights into the active constitution of urban politics. The discussion closes by reflecting on what gendered movements might say about spatial interventions, the constitution of publics and the figuration of dissent – the *where*, *who* and *how* of urban protest – as claims to and in space are made by diverse bodies and discordant voices.

Gendering protest

On 21 January 2017, some half a million people gathered in Washington, DC to take part in a political demonstration. On the previous day, a rather smaller crowd had convened in the same city to witness the inauguration of Donald J. Trump as the forty-fifth president of the United States. The Women's March on Washington was mobilized in response to a figure whose litany of routine sexism on mainstream and social media formed a background to the more recent release of a 2005 recording of his claims about the opportunities for casual sexual assault afforded to him by his status as a 'star'. But if women were the largest, they were hardly the only group of people in the USA who had been the object of demeaning, disrespectful or demonizing statements on the part of the man who was now President, or would be vulnerable to the legal and policy moves he had promised to make while in office. The demonstration highlighted women's rights, but also the rights of minorities, immigration reform, racial equality, social justice, health care and the environment. It took its place in a long history of mass protest and gendered politics, including notably in the US case the Million Woman March led by African American activists in Philadelphia in October 1997 and the Million Man March in Washington of October 1995, led by the National African American Leadership Summit and the Nation of Islam. The latter is less commonly cited within a history of gendered protest, but its articulation of race politics in the USA centred on the representation, treatment and experience of black men. Mobilizations such as these open up linkages between various politics of gender – women's reproductive rights, sexual harassment and gender-based violence in 2017; self-determination and empowerment for African American women and girls in 1997; or the policing, incarceration and cultural stereotyping of black men in 1995 – and wider issues of inequality, exclusion and exploitation that cut across gender lines in manifold ways. This connective politics work both through a *politics of intersection* which stresses how gender identities and inequities are shaped and striated by factors such as race, class, age, religion, sexuality or immigration status; and via a *politics of solidarity* which allows individuals and groups to make common cause with others whose formation and experience may be very different and distant from their own.

In contemporary US cities the gendering of urban protest serves to underline the differentiated citizenships of women or minority men in a broader setting of legal and political equality. In situations where women's formal and substantive rights are more clearly restricted in relation to those of men, women's claims to public space are especially striking. But in spite of the variations in women's access to political rights and practical spatial freedoms, there remains a particularity around urban manifestations in which women lead or form the majority of the crowd, such that gendered protest retains a symbolic charge in the spaces of quite divergent cities. The modern history of women's urban mobilizations ranges from the Women's March on Versailles in October 1789 incited by women from the markets of central Paris or the revolutionary protests led by women workers in Petrograd in February 1917, to the many manifestations for women's suffrage in the early twentieth century, the Women's March in Pretoria against South Africa's apartheid pass laws in 1956 and international demonstrations in support of Women's Liberation from the 1970s. Some of these historical mobilizations had gender issues at their core – women's electoral enfranchisement or social and economic rights – but others involved women marching or rioting against poverty and hunger, racial injustice and repressive authority.

These mobilizations animate spaces of protest via the figuration of gendered bodies. The political affect of the declarative female body – in gestures of resistance, refusal, testimony or memory – links repertoires of gendered protest in diverse urban locations. The international anti-war movement Women in Black began in Jerusalem in 1988 during the first Intifada as groups of Israeli women made clear their opposition to the occupation of Palestine. Their protest took the form of silent vigils, as women clad in black stood their ground at street intersections holding stop-sign placards reading 'Stop the Occupation' (Blumen and Halevi 2009; Sasson-Levy and Rapoport 2003; see also Baum 2006). The Women in Black vigils took place weekly, early on Friday afternoons, and – having spread to other Israeli cities – became the model for an international network of women's protests against conflict, militarism and war.

Such tactics mediate domestic and military politics, everyday geographies and spaces of exception, through the figure of the protesting body. These displacements between symbolically gendered spaces, and between women's embodied protests and the more abstract body politic, resonate in a range of political contexts. Traditions of protest in Latin American cities, for instance, have mobilized the female body in actions which draw on conventional gender roles and speak back to a political order in doing so. In December 1971, the March of the Empty Pots saw women protest in the streets of Santiago de Chile against the Popular Unity government of Salvador Allende (Pieper Mooney 2007; Power 2002; Townsend 1993); in March 2014, women in Venezuelan cities turned out in a similar protest against the Maduro government. The *cacerolazo* protests – so named after the pots and pans that the protesters banged as they marched, and signifying the inflation and basic shortages that motivated them – take the kitchen to the street to emphasize the impact of economic policies on family livelihoods and domestic economies. While these moments of direct action can be read as bourgeois protests against left-wing governments (and therefore consistent with prevailing assumptions about women's political conservatism), their class complexion is not straightforward (see Townsend 1993), and since the early 1970s *cacerolazo* has been deployed in a number of cities against governments of left and right, both in street demonstrations and in protests that allow dispersed crowds to make political noise from inside their homes under conditions of curfew or over-policing.

An especially powerful instance of the figuration of the female body in urban space is the movement of the *Madres de Plaza Mayo* in Argentina. These networks of women came together in the late 1970s, demanding the recognition and return of children who had been 'disappeared' under the 1976–1983 military junta (see Bosco 2004, 2006; Taylor 2001). Given the restrictions on public gathering, political protest and human rights activism under the dictatorship, the symbolic and figurational centrepiece of the mothers' movement was their circling of the *Plaza Mayo*, the civic space in downtown Buenos Aires framed by the presidential palace, the city's cathedral, the colonial *cabildo* and other public and historic buildings. Silent, circling, bearing pictures of their vanished children, the mothers' wordless comportment offered a kind of political 'voice' in circumstances of censorship, repression and denial (Fabj 1993). The public mobilization of the figure of the mother runs across spatial and political settings in a connective politics which draws out the intersections between the 'private troubles' of the family and public issues of injustice, oppression and inequity. The Mothers of the Movement became a compelling public face of the Black Lives Matter protests which emerged after 2013, as women who had lost children to police brutality and gun violence in US cities mediated that larger political message through the testimony of those to whom these individual lives mattered most of all. In embodying the impact on families, older women stood in for the lives of younger men and boys in the urban encounter between male police and young black men. At the same time, the Say Her Name campaign spoke and stood up on police violence against African American women, underscoring the complex intersections of race and gender in the policing of black populations in the USA.

A leading edge of the politics of austerity in Britain has been sharpened by the Focus E15 group of young mothers who organized against the welfare cuts and eviction notices which would force them out of their supported social housing in East London in the shadow of the ongoing development of the city's 2012 Olympics sites. The women's mobilization in the face of threats to the security of their own and their small children's housing was one front in a movement of opposition to welfare retrenchment, benefit cuts, gentrification and social cleansing in an increasingly unequal city. Such a politics – like the 'purple tide' mobilized in anti-austerity protests first in Malaga and then in other Spanish cities and focusing on the roll-back of gender equity policies (Gámez Fuentes 2015: 363) – sits in a wider international context in which the politics of austerity has hit particularly hard on women's incomes, services and protections (see Karamessini and Rubery 2014). The Focus E15 activists drew on a variety of spatial tactics including occupations, demonstrations and lobbying in the more formal sites of government (see Watt 2016). In the politics of homelessness and shelter deprivation in cities such as London, the most obvious 'public' figure is often that of the rough sleeper, a largely male subpopulation; the Focus E15 group brings into the street, and into public view, the precarious housing condition of women and children in insecure spaces of temporary or inadequate accommodation.

These versions of 'militant motherhood' articulate 'private' troubles of bereavement, dispossession and displacement through public claims, presence and occupation (see Pieper Mooney 2007). Imogen Tyler (2013: 212) has argued that "maternal protests are an important site of political revolt against dehumanising and disenfranchising forms of subjugation to sovereign regimes of power" whose operations are often covert or concealed; Tyler's own account begins with the activism of detained asylum-seekers with babies born in Britain. Such actions embody, spatialize and gender politics in numerous senses: subverting and domesticating public space with the tools of the kitchen, performing private loss and bearing personal witness in places of public commemoration, challenging eviction and displacement through homely occupation, or simply through the still surprising spectacle of the female body as public actor. They also concentrate a more distributed politics: the Argentinean mothers organized across the country, over decades and via different political strategies; as well as connecting with other militant mothers and a broader politics of solidarity in Latin America and internationally (see Pieper Mooney 2007). Black Lives Matter began as a set of interventions on social media by three women activists: it now has chapters throughout the United States and a network of international groups, as well as engaging in more specific acts of urban protest.

Some of these movements and interventions focus on the politics of the city – issues which are concentrated in or heightened by (even if not exclusive to) urban environments: encounters between police and minorities in racially divided cities, urban violence and vulnerability, tenure insecurity and shelter discrimination in cities of deepening inequality. Others address politics which extend over larger geographies, cut across or get stuck at national borders, are circulated through media and information networks, or are sequestered within the spaces of the home, the cell or the camp. In these cases the spatial infrastructure of political protest is afforded by the streets and squares of the city as a platform for connection, solidarity and demonstration. The politics of the street and the square *publicizes* the domestic politics of the home, makes evident patterns of violence which rely on suppression, concealment and collusion, and constructs the suffering of specific individuals and families in collective terms.

The politics of gender in space

This relationship between the urban as the object of politics and the city as a site of political mobilization is paralleled by the interactions between gender as a matter of politics and gender as a mode of political expression. Given enduring and systematic inequities along gender lines – including unequal access to formal political power, representation and voice – the organization of women and girls is always 'about' gender to some degree. But political manifestations which focus directly on gender issues – from reproductive rights to gender diversity and sexual self-determination, gender-based violence, or women's legal and economic rights – make the formal and informal regulation of gendered bodies central to a figurative politics in space. Politics of this kind brings into public view issues and injustices which occur elsewhere and often out-of-sight, or which compromise women's and girls' access to the very spaces in which urban protests are enacted. Freedom of movement is protected under the Universal Declaration of Human Rights, yet women's mobility in space is impeded in various ways and diverse environments by law or regulation, by cultural norms and familial coercions, by the threat or experience of harassment and violence. Movements seeking to Reclaim the Night have campaigned over several decades to assert women's access to public space and free movement and to resist sexual violence and harassment. Such contestations over the gendered body – its autonomy, its representation and its regulation – make the body in space both the object and the subject of political mobilization. More generally, the politics of women's protest positions the bearer of rights as an embodied subject, and shows how often violations of legal and civil rights take the form of corporal harms or prohibitions.

The call to take back the night and take to the street echoes in recent movements which make public claims to space as part of an argument for the integrity and autonomy of the gendered body. The SlutWalk protests which originated in Toronto in 2011 and have spurred sister marches in numerous other cities responded to the formal and informal policing of women's bodies in public by insisting on the right to move (and to dress) freely without fear of harassment, abuse or assault (Borah and Nandi 2012; Carr 2013; Gwynne 2013; Mendes 2015). The direct action politics of groups such as Pussy Riot and Femen (Channell 2014) play on the sexualized body to protest controls over women's bodies and sexualities, whether in relation to cultures of religious conservatism, sex trafficking or attacks on reproductive rights. Movements of this kind are part of a history of protest which uses riotous bodies and perverse performance to politicize gender injustice, violence and sexual marginalization, with key antecedents in LGBTQ movements in the 1980s and 1990s such as ACT UP, Queer Nation and the Lesbian Avengers (see Rand 2013; Shepard 2010).

This is a highly visible mode of women's direct action, using the spectacular valency of unruly, and often undressed, female bodies (on women's naked protests, see Tyler 2013; Veneracion-Rallonza 2014). But other sorts of direct action are based on the offence attached to women doing very mundane things in public – breastfeeding flash mobs in cities from Budapest to Fuzhou, Hong Kong, Houston or London, or Saudi women driving cars in the streets of Riyadh and Jeddah in defiance of arrest and detention. Much gendered protest, after all, is premised on the simple provocation of women walking down the street. As Baum (2006: 564) writes of her experience as a Woman in Black: "It seems like a very minor kind of action." Yet the effect of these embodied protests stems in large part from the spatial disorder represented by women claiming and occupying public space; while the abuse protesters such as the Women in Black have been subject to in the street suggests that the voices of even silent women can be heard as too loud.

These forms of protest rely on tactics of situated embodiment and spatial presence, but their wider impact stems from their dissemination and mediation through both mainstream representations and activist social media networks. Social media have become increasingly critical to social movement politics generally, but are particularly salient in circumstances where women's spatial freedoms are most constrained. Given the risks of arrest run by women participating in the Right-to-Drive campaigns in Saudi Arabia, the street-based action was covert while the protest effectively 'took place' via social media uploads and messaging (Agarwal et al. 2012). And while social media networks are used to organize and publicize protest actions, and to make visible women's involvement in occupations and demonstrations (al-Natour 2012; Newsom and Lengel 2012), they also configure a mediated space of gendered protest in themselves, with networks such as *Hollaback!* in the USA, *Harassmap* in Egypt and *Women Shoufouch* in Morocco documenting and challenging street-based harassment of women and girls (Keller et al. 2016;

Skalli 2014). As the politics of the street articulates and connects dispersed spaces of inequity, exploitation and injury, so networks of social media create larger spheres of communication and information which link many 'minor' instances of abuse and assault across an extended geography of streets, squares, buses and trains in which women and girls encounter casual denigration and physical violation.

Conclusion

Gendering spatial politics raises certain issues which are especially acute in relation to women's protests and might also help us to think more generally about manifestations in urban space. There are three themes to highlight here: the means by which direct interventions constitute and concentrate politics in space; the 'publics' formed by gendered protesters as contingent collectivities; and the role of figuration in articulating political claims. Put simply, these themes speak to the *where*, *who* and *how* of urban politics, as spaces are re-made, collectivities are put together and political subjects are configured.

The political effect of interventions in space – forms of direct action which derail the usual order of urban streets, squares, buildings or parks – stems in part from the re-purposing of everyday and elite spaces in the city (from the road surface to the assembly building) not as sites of transit, commerce, leisure or government but as sites of dissent, opposition and occupation. The transformation of space through repertoires of political action takes on a particular charge given the persistent and peculiar transgression represented by the female body in public (cf. Cresswell 1996). Women's and girls' access to public and urban spaces remains highly uneven across national and cultural settings, but in no city in the world is a scene in which females predominate in an urban crowd a usual or unremarkable thing. The personal risks taken by women drivers in Riyadh or anti-rape activists in Delhi may be much greater than those of SlutWalkers in Toronto, but the claims made in these places call out the legal and cultural policing of women in public, and the various forms in which women's public 'modesty' is taken to be a matter of male concern. Some of the political actions considered in the preceding discussion seek to make their political points through spectacle and shock value, but a great deal of women's protest is premised on the simple visual and spatial challenge of women taking to the street on their own or on equal terms. Women's interventions in public spaces unsettle the norms around which everyday environments are ordered, the routine rights, entitlements and presumptions of different actors, and legitimized modes of occupation and use.

Women's actions in urban space, second, might help us to think more critically about the manner in which political publics are constituted. An important strand of social and political theory has challenged the historical and conventional terms in which the 'public' has been understood in liberal societies, and the exclusions entailed in both thought and practice along lines of class, gender, sexuality, race, culture, legal status and so on (see, for key examples, Fraser 1990; Warner 2002). Women's mobilizations and claims to political representation have been crucial in contesting and transforming normative conceptions of how – and by whom – publics are made. But 'women' also represent an extremely disparate category within which experiences, interests and attitudes vary sharply within and between national and urban contexts. Gendered protest movements have often reproduced established lines of power, as the priorities and voices of white, middle-class, straight and cisgender women take precedence, and those who are privileged in these respects may be less vulnerable to hostility or brutality from police, the wider public and, indeed, other protesters (for these arguments explored, for example, in relation to the SlutWalk movement, see Borah and Nandi 2012; Carr 2013; Dow and Wood 2014; Gwynne 2013; Reger 2015). The criticisms within women's protest movements around such exclusions have been important in promoting more diverse forms of political activism, and also offer insights into the fraught formation of political publics more generally. Women's mobilizations across urban and transnational spaces involve strategies of identity and intersectionality in varying degrees; where identity works not through assumed commonalities but on the basis of imaginative commitments, and connections are premised on the recognition rather than the erasure of differences.

The politics of gendered protest, finally, is instrumental in thinking about the role of figuration in spatial politics. The differentiation of gendered bodies in public space is an argument for taking both difference and bodies seriously in repertoires of political protest. Political expression is conventionally understood as a question of 'voice' and the noise of the crowd and the textuality of movement slogans bring political discourse to urban streets. But gendered protests signally deploy the body as a means of expression and communication. The body, that is, becomes a kind of voice. Political claims may be articulated in figurative ways; opposition, dissent and refusal enunciated through stubborn occupation and silent vigil. This is not to deny that women's protests have some great lines ("I can't believe I still have to protest this fucking shit" as the sign held by at least one older women read during the anti-Trump protests of January 2017), but – as in the tactics of queer protest politics and disability rights movements – the ordinary and the more staged spectacles of bodies which are marked by difference, and which take those differences into the skirmishes of the streets, speak to power in eloquent and emphatic terms.

Street-based protest, demonstration and occupation have been a keynote of urban politics in recent years. Such protests have been gendered in critical respects, not only in their composition but in terms of how gender politics, identities and solidarities are worked out through the act of protesting. Women's and girls' mobilizations in urban space call out issues of gender-based violence, discrimination and sexist stereotyping, but are also crucial in challenging political and policy regimes which impact most severely on women's services, incomes and family lives. From demonstrations against rape and its legal treatment in Delhi to the international SlutWalk movement, from the millions of women worldwide marching against Trump to crack teams of Femen activists, from breastfeeding flash mobs taking their 'lactivism' to public and commercial spaces in Hong Kong or Houston to clandestine networks of women driving cars in the streets of Riyadh: these urban interventions form part of a history of women's spatial acts which contest constraints on women's movement, freedoms, expression and embodiment; challenge gendered legal regimes, political processes and cultural norms; point up the gendered disparities in substantive rights to the city; and make clear the ways in which social and economic inequalities continue to be organized – and lived – along gendered lines.

References

Agarwal, N., Lim, M. and Wigand, R.T. (2012). Online Collective Action and the Role of Social Media in Mobilizing Opinions: A Case Study on Women's Right-to-Drive Campaigns in Saudi Arabia. In: C.G. Reddick and S.K. Aikins, eds. *Web 2.0 Technologies and Democratic Governance*. New York: Springer Science + Business Media, pp. 99–123.

Al-Ali, N. (2012). Gendering the Arab Spring. *Middle East Journal of Culture and Communication*, 5, pp. 26–31.

al-Natour, M. (2012). The Role of Women in the Egyptian 25th January Revolution. *Journal of International Women's Studies*, 13(5), pp. 59–76.

Arat, Y. (2013). Violence, Resistance, and Gezi Park. *International Journal of Middle East Studies*, 45(4), pp. 807–809.

Baum, D. (2006). Women in Black and Men in Pink: Protesting against the Israeli Occupation. *Social Identities*, 12(5), pp. 563–574.

Bhattacharjya, M., Birchall, J., Caro, P., Kelleher, D. and Sahasranaman, V. (2013). Why Gender Matters in Activism: Feminism and Social Justice Movements. *Gender & Development*, 21(2), pp. 277–293.

Blumen, O. and Halevi, S. (2009). Staging Peace through a Gendered Demonstration: Women in Black in Haifa, Israel. *Annals of the Association of American Geographers*, 99(5), pp. 977–985.

Boler, M., Macdonald, A., Nitsou, C. and Harris, A. (2014). Connective Labor and Social Media: Women's Roles in the 'Leaderless' Occupy Movement. *Convergence: The International Journal of Research on New Media Technologies*, 20(4), pp. 438–460.

Borah, R. and Nandi, S. (2012). Reclaiming the Feminist Politics of 'SlutWalk'. *International Feminist Journal of Politics*, 14(3), pp. 415–421.

Bosco, F.J. (2004). Human Rights Politics and Scaled Performances of Memory: Conflicts among the Madres de Plaza de Mayo in Argentina. *Social and Cultural Geography*, 5(3), pp. 381–402.

Bosco, F.J. (2006). The *Madres de Plaza de Mayo* and Three Decades of Human Rights Activism: Embeddedness, Emotions, and Social Movements. *Annals of the Association of American Geographers*, 96(2), pp. 342–365.

Carr, J.L. (2013). The SlutWalk Movement: A Study in Transnational Feminist Activism. *Journal of Feminist Scholarship*, 4 (Spring). www.jfsonline.org/issue4/articles/carr/.

Channell, E. (2014). Is Sextremism the New Feminism? Perspectives from Pussy Riot and Femen. *Nationalities Papers*, 42(4), pp. 611–614.

Cresswell, T. (1996). *In Place/Out of Place: Geography, Ideology, and Transgression*. Minneapolis: University of Minnesota Press.

Deb, B. (2015). Transnational Feminist-Queer-Racial Common Fronts: Who Is the 99%? In: T.A. Comer, ed. *What Comes after Occupy? The Regional Politics of Resistance*. Newcastle upon Tyne: Cambridge Scholars Publishing, pp. 62–80.

Doan, P. (2010). The Tyranny of Gendered Space: Reflections from Beyond the Gender Dichotomy. *Gender, Place and Culture*, 17(5), pp. 635–654.

Dodson, K. (2015). Gendered Activism: A Cross-National View on Gender Differences in Protest Activity. *Social Currents*, 2(4), pp. 377–392.

Dow, B.J. and Wood, J.T. (2014). Repeating History and Learning from it: What Can Slutwalks Teach Us about Feminism? *Women's Studies in Communication*, 37, pp. 22–43.

Fabj, V. (1993). Motherhood as Political Voice: The Rhetoric of the Mothers of Plaza de Mayo. *Communication Studies*, 44(1), pp. 1–18.

Fraser, N. (1990). Re-thinking the Public Sphere: A Contribution to the Critique of Actually-Existing Democracy. *Social Text*, 25(6), pp. 56–80.

Gámez Fuentes, M.J. (2015). Feminisms and the 15M Movement in Spain: Between Frames of Recognition and Contexts of Action. *Social Movement Studies*, 14(3), pp. 359–365.

Gwynne, J. (2013). Slutwalk, Feminist Activism and the Foreign Body in Singapore. *Journal of Contemporary Asia*, 43(1), pp. 173–185.

Hafez, S. (2014). The Revolution Shall Not Pass Through Women's Bodies: Egypt, Uprising and Gender Politics. *The Journal of North African Studies*, 19(2), pp. 172–185.

Hurwitz, H.M. and Taylor, V. (2018). Women Occupying Wall Street: Gender Conflict and Feminist Mobilization. In: L.A. Banasak and H.J. McCammon, eds. *100 Years of the Nineteenth Amendment: An Appraisal of Women's Political Activism*. New York: Oxford University Press, pp. 334–355.

Karamessini, M. and Rubery, J., eds. (2014). *Women and Austerity: The Economic Crisis and the Future for Gender Equality*. Abingdon: Routledge.

Keller, J., Mendes, K. and Ringrose, J. (2016). Speaking 'Unspeakable Things': Documenting Digital Feminist Responses to Rape Culture. *Journal of Gender Studies*. DOI: 10.1080/09589236.2016.1211511.

Khalil, A., ed. (2015). *Gender, Women and the Arab Spring*. New York: Routledge.

Khromeychuk, O. (2016). Negotiating Protest Spaces on the Maidan: A Gender Perspective. *Journal of Soviet and Post-Soviet Politics and Society*, 2(1), pp. 9–47.

Maharawal, M.M. (2016). Occupy Movements. In: N. Naples et al., eds. *The Wiley Blackwell Encyclopedia of Gender and Sexuality Studies*. Malden, MA: Wiley-Blackwell.

Martsenyuk, T. (2016). Sexuality and Revolution in post-Soviet Ukraine: Human Rights for the LGBT Community in the Euromaidan Protests of 2013–2014. *Journal of Soviet and Post-Soviet Politics and Society*, 2(1), pp. 49–74.

Mendes, K. (2015). *SlutWalk*. Basingstoke: Palgrave Macmillan.

Morsy, M. (2014). Egyptian Women and the 25th of January Revolution: Presence and Absence. *The Journal of North African Studies*, 19(2), pp. 211–229.

Newsom, V.A. and Lengel, L. (2012). Arab Women, Social Media, and the Arab Spring: Applying the Framework of Digital Reflexivity to Analyze Gender and Online Activism. *Journal of International Women's Studies*, 13(5), pp. 31–45.

Pieper Mooney, J.W. (2007). *Militant Motherhood* Re-visited: Women's Participation and Political Power in Argentina and Chile. *Geography Compass*, 5(3), pp. 975–994.

Power, M. (2002). *Right-Wing Women in Chile: Feminine Power and the Struggle against Allende, 1964–1973*. University Park: Penn State University Press.

Rand, E.J. (2013). An Appetite for Activism: The Lesbian Avengers and the Queer Politics of Visibility. *Women's Studies in Communication*, 36(2), pp. 121–141.

Reger, J. (2015). The Story of a SlutWalk: Sexuality, Race, and Generational Divisions in Contemporary Feminist Activism. *Journal of Contemporary Ethnography*, 44(1), pp. 84–112.

Sasson-Levy, O. and Rapoport, T. (2003). Body, Gender, and Knowledge in Protest Movements: The Israeli Case. *Gender and Society*, 17(3), pp. 379–403.

Shepard, B. (2010). *Queer Political Performance and Protest: Play, Pleasure and Social Movement*. New York: Routledge.

Skalli, L.H. (2014). Young Women and Social Media against Sexual Harassment in North Africa. *The Journal of North African Studies*, 19(2), pp. 244–258.

Spain, D. (2015). *Constructive Feminism: Building Women's Rights into the City*. Ithaca, NY: Cornell University Press.

Taylor, D. (2001). Making a Spectacle: The Mothers of the Plaza de Mayo. *Journal of the Association for Research on Mothering*, 3(2), pp. 97–109.

Townsend, C. (1993). Refusing to Travel *La Via Chilena*: Working-class Women in Allende's Chile. *Journal of Women's History*, 4(3), pp. 43–63.

Tyler, I. (2013). Naked Protest: The Maternal Politics of Citizenship and Revolt. *Citizenship Studies*, 17(2), pp. 211–226.

Veneracion-Rallonza, M.L. (2014). Women's Naked Body Protests and the Performance of Resistance: Femen and Meira Paibi Protests against Rape. *Philippine Political Science Journal*, 35(2), pp. 251–268.

Wang, K.J., St John, H.R. and Wong, M.Y.E. (2017). Touching a Nerve: A Discussion on Hong Kong's Umbrella Movement. In: G. Brown, A. Feigenbaum, F. Frenzel and P. McCurdy, eds. *Protest Camps in International Context: Spaces, Infrastructures and Media of Resistance*. Bristol: Policy Press, pp. 109–134.

Warner, M. (2002). *Publics and Counter-Publics*. Brooklyn, NY: Zone Books.

Watt, P. (2016). A Nomadic War Machine in the Metropolis: En/countering London's 21st-century Housing Crisis with Focus E15. *City*, 20(2), pp. 297–320.

43
QUEERING URBAN POLITICS AND ECOLOGIES

Will McKeithen, Larry Knopp and Michael Brown

"The urban political process is basically concerned with allocation, access, rent, and property" (Pahl 1975: 296; quoted in Pratt 1982: 481). Forty years ago such a nomothetic claim deftly summarized the interdisciplinary study of urban politics. It focused attention on the distributional issues of politics in the city, and prompted questions about both human agency and social structure in shaping them. Williams (1975) interpreted this coming-together of factors as a kind of urban political ecology: where the city was more interactional than territorially defined. The city was a location that intensely brought together competing interests, which in turn reflected structures of the possible. "The process of forming locational arrangements and rules for governing changes in them forms a logical central focus for urban political analysis" (Williams 1975: 131). And certainly this concise summary still captures key elements and research agendas in the field.

But nearly half a century later, both empirical and theoretical challenges have rendered these definitions reductionist. Pahl and Williams wrote at a time when public choice theory, Weberian sociology and Marxist political economy narrowly debated the ken of politics in the city. Since then, of course, less essentializing and more modest forms of theorizing prompt our quest for less exclusionary definitions of social phenomena like 'urban politics'. And the increased visibility and presence of sexual minorities (lesbians, gays, bisexuals, trans★ and otherwise 'queer' folk) in cities – have also been part of what Williams would call urban political ecology, although that term has also been resignified (see below). So our purpose in this chapter is to consider how sexual minorities have been considered in urban political geography, and how a queer urban political ecology might be imagined presently and for future work. For queer theory demands attention to the foreclosures, ignorances and, yes, closetings of others that need attention.

The chapter proceeds in the following steps. First, we chronicle how sexuality and space studies have empirically enriched Pahl's work on its own terms. Next we queer the spaces of urban politics, following central intellectual trends of queer theory. Specifically we problematize the ontologies implicit and explicit in urban political geography. Finally, we draw on recent queer theory and post-humanist interventions to reimagine an urban political ecology that attends to sexualities and intimacies excessive to the human subject.

Urban political ontologies

Allocation and access are closely related, of course. Allocation deals with the distributive question of who gets what, where, when and why, while access often refers to a place at the table, or to venues of decision-making. Territorial strategies of political empowerment in cities often were struck in order to secure or increase resource allocation from the state. Access to political capital could also be gained by jurisdictional creation. Through a nuanced account of coalition-building, for instance, Forest (1995) chronicled West Hollywood, California's creation as a negotiation between gays and elderly residents for greater autonomy and participation in local politics.

Work on AIDS activism could also be interpreted as being about access to social and health services, particularly on the parts of gay men. By the early 1990s such access was literally vital in the context of imminent death and complicated by (what we would now call) neoliberal state restructuring that included welfare retrenchment, and the rise of the voluntary sector in service provision. Brown's (1997) work on Vancouver stressed how state-centric notions of urban political geography occlude important spaces of politics that lie at the interstices between the local state, civil society and the home. Geltmaker's (1997) account of ACT UP Los Angeles and the Queer Nation emphasizes the power of direct action and confrontation to secure not just access to resources, but also politics of recognition that queer folk are part of the political community in which resources are allocated in the first place.

Electoral geography also has framed allocation questions for LBGTQ people. Bob Bailey's (1999) work explored the emergence of lesbian and gay voting blocs in US cities through the 1980 and 1990s, while case studies such as Krahulik's (2007) show how transformative such coalitions can be in local politics. Through a series of analyses of gay-rights initiatives, scholars have stressed support as a function of social class, education level, and other progressive markers (e.g. O'Reilly and Webster 1998). Race and immigration have a complicated indirect relation to support for local gay-rights initiatives (De Leon 1993; Brown et al. 2005; Chapman et al. 2007).

With respect to access, the strategic territoriality of gay men provided a rich empirics for late twentieth century urban political inquiry. The move from at-large to district elections in San Francisco during the mid-1970s, for example, allowed Harvey Milk, the first openly gay city supervisor in the USA, to join the Board of Supervisors. And while there was already a woman on the board, the first avowed feminist, Carol Ruth Silver, also was elected through this new electoral geography, as was Gordon J. Lau the first Chinese-American elected there. For Castells, the 'making of gay territory' through spatial concentration, housing revitalization and the move to district elections all augmented the political access of gays (or at least gay white men) in the city. Similar arguments have been made with respect to upper-class gay white men in Toronto's Village (Nash 2006) and the 'pink pound' in Manchester's Gay Village in the 1990s (Quilley 1997), where access was functional to capitalist accumulation in a context of urban economic restructuring.

Access could also be spatialized within the local state apparatus itself. Lesbians and gays working as urban managers or gatekeepers could make collective consumption provision more equitable. Andrucki and Elder (2007) and Brown (1994), for example, discuss the role of activists in the operation of non-profit and state entities as a means of increasing the equitable distribution of government services. Knopp (1987), meanwhile, discusses the role of activists in affecting a change in police practices around the enforcement of morals laws. However, access also refers to claiming urban space itself. The complexity of lesbians' quests for community space through the 1970s, for example, testifies to the creativity and flexibility of women searching for empowerment through access to community (Enke 2007; Podmore 2006, 2013; Valentine 1993). Enke's and Podmore's works in particular show how lesbians' limited but creative access to urban public space provided important moments of resistance to heteronormativity and patriarchy. At a more micro scale, Valentine stressed how heteronormativity structured access to both public and private space in the city. An article by Ki Namaste (1996), meanwhile, was one of the earliest pieces that chronicled and charted violence against trans* people across the city, especially in public space. And in a moving testimony, Doan (2010) has charted difficulties of access throughout her own lifecourse.

The emphasis on *rent* and *property* reflects a preoccupation with the politics and economics of land and housing as they intersect with various kinds of sexual politics in the city. Knopp's early work (1990a, 1990b) on gentrification in the Marigny neighbourhood of New Orleans, for example, explicitly linked the politics of sexuality and gay community development to the political economy of land, housing and labour markets. He traces the shift to a service economy in New Orleans, early failures of dominant market actors (banks, mortgage lenders) to recognize the economic development potential of gay consumers, and the ways in which this led to the commodification of local gay culture and gentrification by a cross-class coalition of gay workers and property developers. Anne Marie Bouthillette (1997) contrasts the economic and cultural dynamics of housing provision in the gay-male-identified West End and lesbian-identified Grandview-Woodland neighbourhoods in Vancouver to explain the shifting spatial boundaries of sexualized urban spaces in that city. Glen Elder's work on Cape Town, South Africa (2005a, 2005b) offers a slightly different angle on the roles of rent and property in the sexual politics of the city. He details the redevelopment of formerly 'whites only' neighbourhoods into hot spots for the consumption of a globalizing Western gay male culture. From still another angle, Muller (2007) examines spaces of lesbian women's basketball fan culture in two US cities in light of the larger role of professional sports (and women's sports in particular) in the redevelopment of central cities in the early 2000s.

More recently, scholars have investigated the shifting nature of these urban spaces in light of broader contemporary cultural, political and economic change. Nash (2013), for example, emphasizes generational issues – specifically the rise of an anti-identitarian 'post-gay' sensibility among many young people – in what she argues is a decline in the importance of gay neighbourhoods to gay political empowerment. Coming from a more voluntarist perspective, Ghaziani (2014) emphasizes changes in the nature of consumer demand to argue that gay neighbourhoods, while not disappearing, are both shifting their physical locations and becoming less easily distinguishable from more integrated 'gay friendly' neighbourhoods. Doan and Higgins (2011), meanwhile, focus on the role of gentrification and related planning initiatives in erasing an existing gay neighbourhood in Atlanta, rather than in facilitating its growth. They document the role of increasing housing costs and hostility towards gay businesses in the displacement and dispersal of many of the neighbourhood's gay residents.

While by no means exhaustive, this review charts the way sexual minorities were incorporated into classic questions of urban political geography. Their participation in the ecology of city politics – as Williams would chart it – revealed the literature around it to be heteronormative. But it also reflected an increasing and powerful presence in the city of these people.

Queering urban political ontologies

The ontologies at work here are diverse and to varying degrees implicit or explicit. Broadly speaking they can be viewed as coalescing around the following three types: *arenas* in which politics 'take place' (e.g. city, workplace, neighbourhood, community, etc.); *structures* (including processes, mechanisms and scale) that condition what is possible and impossible; and *identities* (both human and non-human, including social groups, commodities, capital, places, landscapes, and other allegedly material inputs to and outcomes of urban politics). The particular ways in which these have been conceptualized depends on the theoretical and philosophical commitments of those doing the conceptualization. Public choice theorists, for example, have tended to construct these ontologies around *idealist* conceptions of the way the world works. Here what is key are individuals (or groups of individuals) exercising a certain kind of *creative agency* or *will* (Buchanan and Tullock 1962; Downs 1957; Olson 1965). Weberians, by contrast, have tended towards more *positivist* ontologies, in which an 'objective' reality is more or less discernible via observation, measurement and methods of *critical rationalism* (Saunders 1981). And political-economists of various stripes have tended towards more *relational* ontologies in which *structural interests* and *social power*, discernible through methods like *historical materialism*, are driving forces (Harvey 1985; Soja 1980).

Notwithstanding these philosophical differences, the Pahlian consensus privileged certain ontologies and ontological imaginations over others. The relevant *arenas* of urban politics were more or less discrete and included various apparatuses of the state (particularly at the local scale), civil society entities that replicated, filled gaps in or contested these state apparatuses; the built environment (including, importantly, distributional infrastructures); and markets (especially property and labour markets). Capitalism was by far the most important *structural* force and as such was seen as being primarily about the production and distribution of wealth and, more broadly, the realization of class interests. And the relevant *identities* involved were particular social groups (most notably classes, but sometimes also 'races' and 'ethnicities'), particularly meaningful sites (i.e. 'places'), and various place-based 'communities'. Overlaying all of these were hierarchical ontologies of power in which wealth and resources were centralized and distributed strategically via networks established by elites (whether they be markets, physical infrastructures or cultural networks). Speaking even more broadly, the spatial ontologies of the Pahlian imagination tended to be conceived of in relatively fixed and material terms. That is to say, they were imagined as fairly stable (albeit changeable in their particulars), bounded and relatively easy to detect through either direct or indirect observation and/or measurement.

As we have seen, feminist and LGBT scholars working within (or in concert with) this tradition enriched it empirically by broadening the list of relevant arenas, structures and identities involved in urban politics. Sites such as the home were shown to be important politically, both as loci of capitalist (re)production and of resistances to it (Valentine 1993). Patriarchy, heterosexism and heteropatriarchy, meanwhile, became recognized as key structural forces operating at multiple spatial scales and, often, in concert with (or even as part of) capitalism (Foord and Gregson, 1986; Grier and Walton 1987; Johnson 1987; Knopp 1992; Knopp and Lauria 1987; McDowell 1986; Peake 1993). And of course gender and sexuality were now seen as constituting key identities to be contended with theoretically as well as empirically, along with various related spatial ontologies (e.g. neighbourhood, community and even market). The subsequent poststructural turn within the social sciences, however, has led to a deeper critique that strikes at the heart of the Pahlian tradition's implicit ontological imagination.

First, there has emerged a powerful critique of 'identity' and 'identity politics'. In this imagination (which draws, in part, on feminist standpoint theory – see Harding 1986; Hartsock 1983, 2004), social categories are conceived not just as 'socially constructed' (Berger and Luckmann 1967) but as fluid, hybrid and contradictory (Bachetta 2002; Butler 1990; Haraway 1991). Thus the term 'identity' itself is often eschewed in favour of 'positionality' or other terms more evocative of situatedness, context and contingency (Bachetta 2002; Oswin 2006; Puar 2002). Consequently, there has also emerged an emphasis on *pluralizing* identities and positionalities, such that masculinities replace masculinity, femininities replace femininity, homosexualities replace homosexuality, and so forth.

A closely related development emphasizes the *intersectional* dimension of positionality. Intersectional approaches stress the specificity of social actors' experiences *in situ* (Collins 1998; Crenshaw 1995; McWhorter 2009; Oswin 2008; Valentine 2007), thus directing analysis away from the more abstract and general characteristics of 'racisms', 'heteropatriarchies', etc. towards the various (including contradictory) ways in which these are articulated with each other in the lived experiences of actually existing, embodied social actors. While the limits of such an approach have begun to be debated (e.g. Brown 2012; Puar 2007), there is little doubt that intersectional approaches (as well as anti-identitarianism more broadly) open up the 'who' implicit in Pahl's formulation to a deeply critical deconstruction (some of the more radical implications of this opening up are explored in the third section below).

In terms of the 'where' implicit in Pahl's formulation, poststructural and related developments have shifted attention away from an implicitly Cartesian and modernist spatial imagination towards one that is more relational, topological, fluid, hybrid and 'open'. While the specific theoretical and philosophical influences here are diverse and by no means widely agreed upon, they include such frameworks as assemblage theory, non-representational theory, actor-network theory and affect theory (Knopp 2004; Massey 2005; Thien 2005; Thrift 1999; Whatmore 1999). The consequence for many queer, feminist and sexuality scholars is a radical rethinking of the spaces of (urban) politics: the state-civil society distinction, for instance, is reimagined as a fuzzy, fluid and multi-scalar set of relations that is as obfuscatory and ideological as it is 'real' (Brown 1997), while 'home', 'neighbourhood' and 'community' are recast as similarly unstable, multi-scalar, relational and contradictory. These reimaginings lead to a recognition that radical urban politics of movement and displacement (Knopp and Brown 2003; Waitt and Gorman-Murray 2011), anti-communitarianism (Joseph 2002), and 'unassimilated otherness' (Young 1990) are perhaps more theoretically *and* politically apposite – particularly for queers (including gender-queer folks) – than Pahlian place-based, identity and interest-based politics. They also lead to a reimagining of the 'power geometries' implicit in structuralist and other modernist notions of scale (Massey 2005), place, placelessness and movement (Knopp 2004), such that multi-scalar relationships, vectors, movements (of meanings as well as of 'things'), absence, and even ephemerality have as much ontological and political 'weight' as the more traditional spaces of Pahl's formulation.

Indeed, the 'structures' of urban politics themselves – most notably 'capitalism' but also 'patriarchy', 'racism', 'ableism' (among other 'big' structures) and even everyday bureaucracies – have been deconstructed by scholars seeking to foreground issues of gender, sexuality, desire (from a perspective that emphasizes the contingency, fluidity, hybridity, etc. discussed above) alongside issues of class, race, ethnicity, ability/disability, etc. Thus 'capitalism' is reimagined as not only always racialized, gendered, sexualized, etc. (albeit contingently), but as *in*complete, *dis*continuous, partial and rife with not just contradictions but opportunities for direct resistance (Gibson-Graham 1996). The result has been queer urban scholarship that reimagines urban politics as every bit as 'cultural' as 'material' (in a way that foregrounds issues of gender, sexuality and desire, while simultaneously disrupting that cultural/material divide (Bain and Nash 2006; Nash 2013; Nash and Bain 2007), that radically rethinks the *spaces* of urban politics (e.g. Bell and Binnie 2004; Bell et al. 1994; Oswin 2008), and that gives new ontological 'weight' not just to hybridity and fluidity but to flows, absences/placelessness and passings/ephemerality (Knopp 2004).

Possible futures: the queer ecologies of urban politics

These interventions have radically reconceptualized the arenas, structures and identities of urban politics. Arenas become multiple and interconnected across site, time and scale. Structures become contingent and interlocking. Identities become unfixed, fractured and always in flux between states of being and becoming. What Williams called 'urban political ecology' is far more complicated in its interactions than even he imagined. Yet, scholars continue to take Williams' ecological imagination at face value, imagining these interactions solely within the realm of human activity, including non-human actors only in the form of institutional structures or capital markets. In thinking about how scholars might build upon these legacies to further queer urban politics and the thinking around them, we interrogate this notion of 'ecology' to resignify it as an interactive, interdependent and more-than-human urban space brimming with animals, plants and biophysical forces generally termed 'nature'. Ecology is more than just a metaphor.

To think the urban as a more-than-human space, place and process has itself been considered a rather queer idea. Cities, particularly those situated within Western industrial and post-industrial contexts, have been imagined largely in opposition to or as refuge from the natural and the non-human. Over the past two decades, however, the field of political ecology – which foregrounds issues of political economy, socio-natural change and human/non-human relations – has considered urban politics and processes (Keil 2003; Swyngedouw and Heynen 2003; Wolch et al. 2002). Yet, both political ecology as a whole and its urban strains remain relatively disengaged from queer theory. So how might political ecology be queered, and what might a queer urban political ecology look like?

Queer scholars have redefined the spaces or 'arenas' of urban politics, opening up the conceptual and empirical spaces of home, neighbourhood, 'public' and 'private' and body. Queering urban political ecologies extends these insights to consider the ways gendered and sexualized political processes and struggles circulate through, manifest within, and mutually constitute urban natures. As Sandilands (2005: 7) writes, "many modern formations of natural space ... [have been] organized by prevalent assumptions about sexuality, and especially a move to institutionalize heterosexuality". While Sandilands' example – US National Parks campgrounds that modelled heteronuclear suburban planning design – demonstrates the diffusion of urban processes beyond urban forms, this naturalization (and its contestation) can be seen in urban natures, as well. For instance, the British and North American Garden Cities and City Beautiful movements of the late nineteenth and early twentieth centuries sought to instil urban social order through an orderly and harmonious natural environment. This order was heteronormative, as well as classist, capitalist and white supremacist. Frederick Law Olmsted and his imitators sculpted their urban parks within this same imagination as idealized sites for bourgeois family strolls or heterosexual trysts. While this heteronormalization of urban park space has induced heavy surveillance and policing of queer and commercial sexualities, these same parks have also provided spatial resources for gay men's sexual communion and likely done the same for other sexual-political subcultures. The political possibilities of these queer urban ecologies unfold in ways that expand 'the political' beyond the human. For instance, Gandy (2012) traces the ways in which abandoned and liminal spaces of urban nature (e.g. an overgrown London cemetery) nurture unruly subjects (humans, plants, insects), queer relationships (gay cruising, ecophilic fucking) and what he calls 'heterotopic' more-than-human alliances to protect and cultivate spaces that sustain multiple forms of 'queer' life deemed 'unproductive' by capitalist and heterosexist logics.

Queering urban political ecologies also reconceptualizes the 'structures' of urban politics. While formations of gender and sexuality remain central, other intersecting structures of speciesism, ecophobia and the capitalist production of urban nature come into view. This intervention echoes feminist political ecologists who argue that recently popular conversations around 'intersectional' identity and subject-formation might also consider the environment and non-human natures (Hawkins and Ojeda 2011). For instance, Patrick (2014) examines the redevelopment of New York City's High Line, a 1.45-mile-long park constructed on the remains of an abandoned elevated railway, and the ways its 'revitalization' depended upon the eviction of the 'unruly' weed *Ailanthus altissima* and a homonormative narrative of gentrification that displaced issues of race, class and gender displacement. Patrick's account demonstrates the structural and metaphorical complicities between 'green' and gay gentrification. Moreover, Patrick gestures to an 'ecological queerness' that values an ethical responsibility towards 'unruly', 'queer' and 'out of place' bodies, whether human or non-human. In so doing, he hopes to "provide political power and ethical grounding for human inhabitants of cities who are opposed to or impacted by gentrification" (2014: 921).

Queering urban political ecologies also brings attention to different subjects and 'identities' within urban politics. This is more than a question of turning to previously unexamined objects of study, however important that work might be, or of crafting a new queer orthodoxy. Rather, queering urban politics means tracking the relational processes – of love, sexuality, labour, desire – that produce and reproduce the bodies, subjects, forms and intimacies of urban life. In turn, queering urban political ecologies extends this process-based ontological imagination to the human–non-human relationships with which urban spaces are brimming. Pet-keeping represents perhaps the most culturally apparent of these, at least in post-industrial contexts. Indeed, in Seattle, a city that has seen rapid neoliberal restructuring and a city that we the authors call home, pets seem to be everywhere, albeit unevenly across economic, social and legal landscapes. Dog parks, pet adoption and obedience classes become sites to reproduce and contest social inequalities, habituated class norms and class status (Bourdieu 1984). The urban politics of pet-keeping are animated not only by questions of resource distribution or class habitus, however, but also relations of kinship, gendered care, emotion and normative sexuality. For example, Nast (2006) crafts a queer political economic approach to trace the parallel trends witnessed in post-industrial societies of increased consumption in the pet industry, psychosocial alienation and increased emotional investment in non-human companions. McKeithen (2017) examines the pop culture figure of the 'crazy cat lady' to understand how idealized notions of pet ownership intersect with cultural norms of heterosexual kinship and bourgeois domesticity. Everyday practices of homemaking become fraught sites in the urban politics of intimacy in which some human–animal relationships prove proper, normal or innocuously cute and others become too excessive, too inhuman or too queer. Across these cases, urban politics are not just populated with LGBT identity groups vying for resources or recognition. They are also filled with a multitude of queer relationships, including those that blur the bounds of species.

Lastly, this attention to more-than-human relationships and their queer valences also forces an attention to the hybrid and more-than-human agencies of urban politics. Urban political ecologists have long highlighted the forceful effects of non-human natures and their socionatural, networked productions of urban life. Queer geographers have recently highlighted the ways in which sexualities, sexual practices and desire are themselves more-than-human effects emerging from the entanglements of human bodies, inanimate objects, non-human natures, smells and built environments (Brown 2008; Gandy 2012). Queering urban political ecologies means not only rethinking the taken-for-granted arenas, structures and identities of urban politics but also appreciating the hybrid, differentiated and more-than-human agencies that produce and suffuse them. This approach takes seriously the phrase 'sexual chemistry'.

These gestures towards queering urban political ecologies should not be contained to conventional urban natures (e.g. parks) but should be extended in multiple directions. For instance, this queer ecological approach could be extended to consider: other domestic spaces; the place-based and intersectional machinations of speciesism; the ways in which 'gender', 'sexuality', 'queer' and 'nature' intertwine across sites spatially and temporally outside the Global North or urban modernity; the social and biological reproduction of more-than-human urban ecologies; the novel urban socio-natures of the Anthropocene; and the agonistic politics of care and violence within more-than-human alliances. While this field remains inchoate and thus any and all directions might be further queerly understood, we offer these and other spaces of urban politics as suggestions for future study.

Conclusion

We close this chapter with three points. First, the history of urban political geography's consideration of LGBTQ folks can be read quite conventionally. As marginalized and closeted individuals, they were not on mainstream scholars' radar when debates over the nature of urban politics were happening in the 1970s and 1980s. When recognized, they could handily be incorporated into the competitive ecology of city politics: as interest groups or voting blocs seeking allocation and access, or as capitalists or consumers in a post-industrial urban economy that restructured urban housing and labour markets. Their challenge, however, was not simply empirical. Thus our second point is that queer *theory*, which emerged alongside and, arguably, as part of the 'coming out' of LGBTQ people into mainstream culture (including the academy) prompted a critical interrogation of the very premises (especially ontologies) of the political as distributional, state-centred and identitarian. Finally – and this is our third point – the challenge posed by queer theory offers urban political inquiry a new way of thinking about urban politics altogether: queer urban ecology. By recognizing that the political struggles of gender and sexual minorities have, on the one hand, hinged upon questions of inclusion within or refusal of a liberal, humanist framework and, on the other hand, always entangled non-humans, queer scholars expand our imaginations of politics in the city.

References

Andrucki, M.J. and Elder, G.S. (2007). Locating the State in Queer Space: GLBT Non-profit Organizations in Vermont, USA. *Social and Cultural Geography*, 8(1), pp. 89–104.

Bacchetta, P. (2002). Rescaling Transnational 'Queerdom': Lesbian and 'Lesbian' Identitary–Positionalities in Delhi in the 1980s. *Antipode*, 34(5), pp. 947–973.

Bailey, R.W. (1999). *Gay Politics, Urban Politics: Identity and Economics in the Urban Setting*. New York: Columbia University Press.

Bain, A.L. and Nash, C.J. (2006). Undressing the Researcher: Feminism, Embodiment and Sexuality at a Queer Bathhouse Event. *Area*, 38(1), pp. 99–106.

Bell, D. and Binnie, J. (2004). Authenticating Queer Space: Citizenship, Urbanism and Governance. *Urban Studies*, 41(9), pp. 1807–1820.

Bell, D., Binnie, J., Cream, J. and Valentine, G. (1994). All Hyped Up and No Place to Go. *Gender, Place and Culture*, 1(1), pp. 31–47.

Berger, P. and Luckmann, T. (1967). *The Social Construction of Reality: A Treatise on the Sociology of Education*. London: Penguin.

Bourdieu, P. (1984). *Distinction: A Social Critique of the Judgement of Taste*. Cambridge, MA: Harvard University Press.

Bouthillette, A.M. (1997). Queer and Gendered Housing: A Tale of Two Neighbourhoods in Vancouver. In: G.B. Ingram, A.-M. Bouthillette and Y. Retter, eds. *Queers in Space: Communities, Public Places, Sites of Resistance*. Seattle, WA: Bay Press, pp. 213–232.

Brown, G. (2008). Ceramics, Clothing and Other Bodies: Affective Geographies of Homoerotic Cruising Encounters. *Social and Cultural Geography*, 9(8), pp. 915–932.

Brown, M.P. (1994). The Work of City Politics: Citizenship through Employment in the Local Response to AIDS. *Environment and Planning A*, 26(6), pp. 873–894.

Brown, M.P. (1997). *Replacing Citizenship: AIDS Activism and Radical Democracy*. New York: Guilford Press.

Brown, M.P. (2012). Gender and Sexuality I: Intersectional Anxieties. *Progress in Human Geography*, 36(4), pp. 541–550.

Brown, M., Knopp, L. and Morrill, R. (2005). The Culture Wars and Urban Electoral Politics: Sexuality, Race, and Class in Tacoma, Washington. *Political Geography*, 24(3), pp. 267–291.

Buchanan, J.M. and Tullock, G. (1962). *The Calculus of Consent* (Vol. 3). Ann Arbor: University of Michigan Press.

Butler, J. (1990). *Gender Trouble: Feminism and the Subversion of Identity*. New York: Routledge.

Chapman, T., Lieb, J. and Webster, G. (2007). Race, the Creative Class, and Political Geographies of Same-sex Marriage in Georgia. *The Southeastern Geographer*, 47(1), pp. 27–54.

Collins, P.H. (1998). It's All in the Family: Intersections of Gender, Race, and Nation. *Hypatia*, 13(3), pp. 62–82.

Crenshaw, K. (1995). *Critical Race Theory: The Key Writings that Formed the Movement*. New York: The New Press.

De Leon, R. (1993). *Left Coast City*. Lawrence: University of Kansas Press.

Doan, P. (2010). The Tyranny of Gendered Spaces – Reflections from Beyond the Gender Dichotomy. *Gender, Place and Culture*, 17(5), pp. 635–654.

Doan, P. and Higgins, H. (2011). The Demise of Queer Space? Resurgent Gentrification and the Assimilation of LGBT Neighborhoods. *Journal of Planning Education and Research*, 31, pp. 6–25.

Downs, A. (1957). An Economic Theory of Political Action in a Democracy. *The Journal of Political Economy*, 65(2), pp. 135–150.

Elder, G.S. (2005a). Love for Sale: Marketing Gay Male P/leisure Space in Contemporary Cape Town, South Africa. In: L. Nelson and J. Seager, eds. *A Companion to Feminist Geography*. Malden, MA: Blackwell Publishing, pp. 578–589.

Elder, G.S. (2005b). Somewhere, Over the Rainbow: Cape Town, South Africa, as a 'Gay Destination'. In: L. Ouzgane and R. Morrell, eds. *African Masculinities*. New York: Palgrave Macmillan, pp. 43–59.

Enke, A. (2007). *Finding the Movement: Sexuality, Contested Space, and Feminist Activism, Radical Perspectives*. Durham, NC: Duke University Press.

Foord, J. and Gregson, N. (1986). Patriarchy: Towards a Reconceptualisation. *Antipode*, 18(2), pp. 186–211.

Forest, B. (1995). West Hollywood as Symbol: The Significance of Place in the Construction of a Gay Identity. *Environment and Planning D: Society and Space*, 13(2), pp. 133–157.

Gandy, M. (2012). Queer Ecology: Nature, Sexuality, and Heterotopic Alliances. *Environment and Planning D: Society and Space*, 30(4), pp. 727–747.

Geltmaker, T. (1997). The Queer Nation Acts Up: Health Care, Politics, and Sexual Diversity in the County of Angels, 1990–92. In: G.B. Ingram, A.-M. Bouthillette and Y. Retter, eds. *Queers in Space: Communities, Public Places, Sites of Resistance*. Seattle, WA: Bay Press, pp. 233–274.

Ghaziani, A. (2014). *There Goes the Gayborhood?* Princeton studies in cultural sociology. Princeton, NJ: Princeton University Press.

Gibson-Graham, J.K. (1996). *The End of Capitalism (as We Knew It): A Feminist Critique of Political Economy, with a New Introduction*. Minneapolis: University of Minnesota Press.

Grier, J. and Walton, J. (1987). Some Problems with Reconceptualising Patriarchy. *Antipode*, 19(1), pp. 54–58.

Haraway, D. (1991). *Simians, Cyborgs, and Women: The Reinvention of Women*. London and New York: Routledge.

Harding, S.G. (1986). *The Science Question in Feminism*. Ithaca, NY: Cornell University Press.

Hartsock, N.C. (1983). The Feminist Standpoint: Developing the Ground for a Specifically Feminist Historical Materialism. In: S. Harding and M.B. Hintikka, eds. *Discovering Reality*. Dordrecht, Holland: D. Reidel Publishing Company, pp. 283–310.

Hartsock, N.C. (2004). Feminist Standpoint Theory Revisited: Truth or Justice? In: S. Harding, ed. *The Feminist Standpoint Theory Reader: Intellectual and Political Controversies*. New York: Routledge, pp. 243–246.

Harvey, D. (1985). *The Urbanisation of Capital: Studies in the History and Theory of Capitalist Urbanisation*. Baltimore: Johns Hopkins University Press.

Hawkins, R. and Ojeda, D. (2011). Gender and Environment: Critical Tradition and New Challenges. *Environment and Planning D: Society and Space*, 29(2), pp. 237–253.

Johnson, L. (1987). Realist Perspectives: Patriarchy and Feminist Challenges in Geography. *Antipode*, 19(2), pp. 210–215.

Joseph, M. (2002). *Against the Romance of Community*. Minneapolis: University of Minnesota Press.

Keil, R. (2003). Urban Political Ecology 1. *Urban Geography*, 24(8), pp. 723–738.

Knopp, L. (1987). Social Theory, Social Movements and Public Policy: Recent Accomplishments of the Gay and Lesbian Movements in Minneapolis, Minnesota. *International Journal of Urban and Regional Research*, 11(2), pp. 243–261.

Knopp, L. (1990a). Some Theoretical Implications of Gay Involvement in an Urban Land Market. *Political Geography Quarterly*, 9(4), pp. 337–352.

Knopp, L. (1990b). Exploiting the Rent Gap: The Theoretical Significance of Using Illegal Appraisal Schemes to Encourage Gentrification in New Orleans. *Urban Geography*, 11(1), pp. 48–64.

Knopp, L. (1992). Sexuality and the Spatial Dynamics of Capitalism. *Environment and Planning D: Society and Space*, 10(6), pp. 651–669.

Knopp, L. (2004). Ontologies of Place, Placelessness, and Movement: Queer Quests for Identity and their Impacts on Contemporary Geographic Thought. *Gender, Place and Culture*, 11(1), pp. 121–134.

Knopp, L. and Brown, M. (2003). Queer Diffusions. *Environment and Planning D: Society and Space*, 21(4), pp. 409–424.

Knopp, L. and Lauria, M. (1987). Gender Relations as a Particular Form of Social Relations. *Antipode*, 19(1), pp. 48–53.

Krahulik, K.C. (2007). *Provincetown: From Pilgrim Landing to Gay Resort*. New York: NYU Press.

Massey, D. (2005). *For Space*. London: Sage.

McDowell, L. (1986). Beyond Patriarchy: A Class-based Explanation of Women's Subordination. *Antipode*, 18(3), pp. 311–321.

McKeithen, W. (2017). Queer Ecologies of Home: Heteronormativity, Speciesism, and the Strange Intimacies of Crazy Cat Ladies. *Gender, Place and Culture*, 24(1), pp. 122–134.

McWhorter, L. (2009). *Racism and Sexual Oppression in Anglo-America: A Genealogy*. Bloomington: Indiana University Press.

Muller, T.K. (2007). The Contested Terrain of the Women's National Basketball Association Arena. In: C.C. Aitchison, ed. *Sport and Gender Identities: Masculinities, Femininities, and Sexualities*. New York: Routledge, pp. 37–52.

Namaste, K. (1996). Genderbashing: Sexuality, Gender, and the Regulation of Public Space. *Environment and Planning D: Society and Space*, 14(2), pp. 221–240.

Nash, C.J. (2006). Toronto's Gay Village (1962–1982). *The Canadian Geographer*, 50, pp. 1–16.

Nash, C.J. (2013). The Age of the 'Post-mo'? Toronto's Gay Village and a New Generation. *Geoforum*, 49, pp. 243–252.

Nash, C.J. and Bain, A. (2007). Reclaiming Raunch? Spatializing Queer Identities at Toronto Women's Bathhouse Events. *Social and Cultural Geography*, 8(1), pp. 47–62.

Nast, H. (2006). Critical Pet Studies? *Antipode*, 38(5), pp. 894–906.

Olson, M. (1965). *The Logic of Collective Action*. Cambridge, MA: Harvard University Press.

O'Reilly, K. and Webster, G.R. (1998). A Sociodemographic and Partisan Analysis of Voting in Three Anti-gay Rights Referenda in Oregon. *The Professional Geographer*, 50(4), pp. 498–515.

Oswin, N. (2006). Decentering Queer Globalization: Diffusion and the 'Global Gay'. *Environment and Planning D: Society and Space*, 24(5), pp. 777–790.

Oswin, N. (2008). Critical Geographies and the Uses of Sexuality: Deconstructing Queer Space. *Progress in Human Geography*, 32(1), pp. 89–103.

Pahl, R.E. (1975). *Whose City? And Further Essays on Urban Society* (Vol. 208). Harmondsworth: Penguin Books.

Patrick, D.J. (2014). The Matter of Displacement: A Queer Urban Ecology of New York City's High Line. *Social and Cultural Geography*, 15(8), pp. 920–941.

Peake, L. (1993). 'Race' and Sexuality: Challenging the Patriarchal Structuring of Urban Social Space. *Environment and Planning D: Society and Space*, 11(4), pp. 415–432.

Podmore, J.A. (2006). Gone 'Underground'? Lesbian Visibility and the Consolidation of Queer Space in Montréal. *Social and Cultural Geography*, 7(4), pp. 595–625.

Podmore, J.A. (2013). Lesbians as Village 'Queers': The Transformation of Montreal's Lesbian Nightlife in the 1990s. *ACME: An International E-Journal for Critical Geographies*, 12(2), pp. 220–249.

Pratt, G. (1982). Class Analysis and Urban Domestic Property: A Critical Reexamination. *International Journal of Urban and Regional Research*, 6(4), pp. 481–502.

Puar, J. (2002). A Transnational Feminist Critique of Queer Tourism. *Antipode*, 34(5), pp. 935–946.

Puar, J. (2007). *Terrorist Assemblages: Homonationalism in Queer Times*. Durham, NC: Duke University Press.

Quilley, S. (1997). Constructing Manchester's 'New Urban Village': Gay Space in the Entrepreneurial City. In: G.B. Ingram, A.M. Bouthillette and Y. Retter, eds. *Queers in Space: Communities, Public Places, Sites of Resistance*. Seattle, WA: Bay Press, pp. 275–292.

Sandilands, C. (2005). Unnatural Passions? Notes Toward a Queer Ecology. *Invisible Culture*, 9, pp. 1–31.

Saunders, P. (1981). *Social Theory and the Urban Question*. New York: Routledge.

Soja, E.W. (1980). The Socio-Spatial Dialectic. *Annals of the Association of American Geographers*, 70(2), pp. 207–225.

Swyngedouw, E. and Heynen, N.C. (2003). Urban Political Ecology, Justice and the Politics of Scale. *Antipode*, 35(5), pp. 898–918.

Thien, D. (2005). After or Beyond Feeling? A Consideration of Affect and Emotion in Geography. *Area*, 37(4), pp. 450–454.

Thrift, N. (1999). Steps to an Ecology of Place. In: D. Massey, J. Allen and P. Sarre, eds. *Human Geography Today*. Malden, MA: Blackwell Publishers, pp. 295–322.

Valentine, G. (1993). (Hetero) Sexing Space: Lesbian Perceptions and Experiences of Everyday Spaces. *Environment and Planning D: Society and Space*, 11(4), pp. 395–413.

Valentine, G. (2007). Theorizing and Researching Intersectionality: A Challenge for Feminist Geography. *The Professional Geographer*, 59(1), pp. 10–21.

Waitt, G. and Gorman-Murray, A. (2011). It's About Time You Came Out: Sexualities, Mobility and Home. *Antipode*, 43(4), pp. 1380–1403.

Whatmore, S. (1999). Hybrid Geographies: Rethinking the 'Human' in Human Geography. In: D. Massey, J. Allen and P. Sarre, eds. *Human Geography Today*. Malden, MA: Blackwell Publishers, pp. 22–39.

Williams, O.P. (1975). Urban Politics as Political Ecology. In: K. Young, ed. *Essays on the Study of Urban Politics*. New York: Palgrave Macmillan, pp. 106–132.

Wolch, J., Pincetl, S. and Pulido, L. (2002). Urban Nature and the Nature of Urbanism. In: M. Dear, ed. *From Chicago to LA: Making Sense of Urban Theory*. Thousand Oaks, CA: Sage Publications, pp. 367–402.

Young, I.M. (1990). *Justice and the Politics of Difference*. Princeton, NJ: Princeton University Press.

PART IX

Spaces of utopia and dystopia

Andrew E.G. Jonas and David Wilson

Generally attributed to the writings of Sir Thomas More, utopia is an ideal place or a kind of 'heaven on earth'. Dystopia, on the other hand, is an imaginary place, which is deemed undesirable or 'hellish'. Utopian urban imaginaries have provided all sorts of possibilities for the bourgeois to accumulate power and wealth in the capitalist city (Fishman 2002). By way of contrast, George Orwell in *Animal Farm* painted a dystopian portrait of Stalinism which drew upon a distinctly non-urban imaginary (Orwell 1945). Meanwhile, critical urban theory continues to be enlivened by occasional discussions of utopia and its social possibilities (Pinder 2002).

Contemporary political struggles around urban space are often motivated by a hopeful imaginary of the city of the future or its antithesis (Harvey 2000). The five chapters in this final section consider how urban politics are fought around conflicting, competing and, at times, complementary utopian and/or dystopian urban imaginaries. They examine issues such as dystopias of urban abandonment and ruin, ideas of the utopian model city, the hopes and fears associated with immigration, tensions between mobility and immobility in urban political imaginaries, and debates around climate change. More generally the chapters reflect on the tensions between seeing urban politics as static and idealistic and those process-orientated approaches that make reference to conflicting urban imaginaries, whether utopian or dystopian.

In his chapter, Ulf Strohmayer traces the work of creative destruction with the urban fabric. Using case studies from New York, Berlin and Paris, the chapter analyses the respective fortunes of railway-connected spaces in the aftermath of their original infrastructural uses. Despite their relatively comparable origins, these three spaces have evolved in markedly different forms and guises since their use as railways has ceased. Deploying a terminology gleaned from the utopian-dystopian distinction, the chapter dissects the importance of local politics, practices and longer-term ambitions embodied by these sites; it also makes visible the dependency of urban politics upon planning and creativity.

As a normative planning concept, the 'European City' model has sought a reconciliation between economic growth, social justice and participation, as well as ecological responsibility in urban planning. Samuel Mössner and Rob Krueger argue that the normative force of the 'European City' model has led planners, policy-makers and academics in cities throughout the world to search for best practices for implementing these goals. But how successful is the utopian 'European City' idea in practice?

The authors revisit the contested relationship between market logics and growth orientation in urban planning, on the one hand, and social justice and the communicative planning culture, on the other. Freiburg, Germany is among a few cities that are celebrated for their ostensibly successfully efforts to implement social, economic and ecological policies. The authors interrogate critically Freiburg's efforts of bringing together a coherent sustainable city agenda and argue that the city serves as an interesting example to investigate empirically how well the competitive, entrepreneurial city, which is guided by economic imperatives, reconciles the tension between market logics and social equity.

Drawing on the work of Agamben (1995) and an expanded view of his states-of-exception concept, the chapter by Stuart C. Aitken and Jasmine Arpagian argues that young ethnic minorities are particularly vulnerable to a revanchism that is hidden under state policies of seeming democracy and justice. It highlights the experiences of Roma people in two urban areas in Slovenia and Romania. The Roma are a collection of diverse Romani peoples who are often lumped together as travellers or are marginalized through their traditional nomadism and encampments in rural villages and occupation of substandard urban areas. The Romani peoples are found mostly in Europe and the Americas, and they are traditionally not tied to national ethnicities and cultures, although many consider their origins emanate from Northern India over a thousand years ago.

The authors consider the Roma primarily from their urban, but also national, political subjectivities. They conclude that the nature of the modern *homo sacer* is exemplified by the problematic dystopian imaginaries that circle some Roma peoples, and that these imaginaries are often codified spatially by local council and state policies.

In her chapter, Theresa Enright examines the utopian and dystopian imaginaries of the mobile city. In particular, she considers the use of public art in the revitalization of Union Station in Toronto, Canada, where cultural programming has been essential to the transformation of the historic rail terminal into a multimodal 'mobility hub' and potent symbol of the desired twenty-first century networked global city. On the one hand, art is mobilized in pursuit of the sustainable, safe, liveable and world class – i.e. 'good' – city. On the other hand, art reveals the elusive and uneven nature of the hypermobile urban fantasy and exposes the destructive patterns of neoliberal urbanization of which it is a part. The installation *Zones of Immersion* by artist Stuart Reid is used as an entry point from which to view the dialectics of utopia and dystopia in the late capitalist city, the functions of artistic representations in transit-led urbanism and the competing aesthetics of contemporary transportation.

It has been argued that the politics of climate change is no longer confined to the jurisdictionally defined policy space of the city. Instead such politics have a deeply embedded relationality – a multi-level governance rationale – to the politics and policies of other spaces of politics be these the city-region, region, national or global, and even more unexpectedly, at more proximate territorial scales of city districts, wards and even communities. The final chapter by Andrew Kythreotis explores this spatial relationality in the context of the architecture of global climate regime (knowledge, politics and policy) and how such a regime is based upon certain scientific truths and technological (ir)rationalities that have their provenance in positivist schools of thought, which make it difficult to neatly elide global climate change politics and policy with more constructivist and reflexive urban imaginaries.

Furthermore, the chapter considers the extent to which in the debate about climate change the global and urban scales can or should be conflated. Efforts to 'short-circuit' the link between the urban and global scales horizontally reinforces a sanitized form of dystopic urbanity in the way that cities politically respond and plan for climate impacts. The chapter therefore questions whether it is materially beneficial to imagine a utopian urban climate politics grounded in relational understandings of the urban.

References

Agamben, G. (1995). *Homo Sacer: Sovereign Power and Bare Life*. Stanford, CA: Stanford University Press.
Fishman, R. (2002). Bourgeois Utopias: Visions of Suburbia. In: S. Fainstein and S. Campbell, eds. *Readings in Urban Theory*. Oxford: Blackwell, pp. 21–31.
Harvey, D. (2000). *Spaces of Hope*. Oxford: Blackwell.
Orwell, G. (1945). *Animal Farm*. London: Secker and Warburg.
Pinder, D. (2002). In Defence of Utopian Urbanism: Imagining Cities after the 'End of Utopia'. *Geografiska Annaler: Series B, Human Geography*, 84, pp. 229–241.

44
DYSTOPIAN DYNAMICS AT WORK
The creative validation of urban space

Ulf Strohmayer

Introduction

Urban politics do not take place within a vacuum. Rather, their ambitions, instruments and practices are shaped by, and in turn shape, past articulations of politics. The result is a layering of sorts, where past expressions of social, economic and cultural processes rub shoulders with current materializations, processes and customs, a layering that is often referred to as forming a "palimpsest" comprising urban forms, morphologies and urban politics (Swyngedouw and Heynen 2010: 79). Describing an iterative, cumulative process that sees urban discourses and related practices interact in a stratified manner, palimpsests also account for the co-existence of erasure and survival of such articulations in urban landscapes everywhere. Engaging with such landscapes becomes the work of memory (Huyssen 2003) and invites the recognition of different materialities contributing to past, present and future urban politics and identity construction (Crang 1996; Cupers 2005; Graham 2009).

And yet, as this chapter will argue, the image of the palimpsest leads astray. It suggests an all-too comfortable synchronicity of erased or disappeared parts of the urban condition with presently existing ones, when it is their convoluted co-existence that arguably requires analytical labour. The survival of some and the disappearance of other such articulations, in other words, have everything to do with the themes explored in the present volume: they come about as a result of larger historical processes, are tied in with urban politics and involve aesthetic considerations alongside economic factors. This chapter addresses some of these articulations with the help of case studies arising from within the world of urban railway spaces – especially urban spaces associated with disused railways. Its aim is to broaden our approach to urban development and politics and to add to the vocabulary of urban change. The chapter will use three recently re-designated spaces in major metropolitan areas in the developed world to illustrate and analyse palimpsest-infused urban processes while thinking about contextual forces in play that help us account for presences and absences in each of these places. What emerge in each of these cases are spatial configurations that are by definition always temporary, differential valorization of space and as such come about as a result of political processes.

The existence of palimpsest-like qualities attaching to urban environments should come as a surprise to no-one. As sites of economic exchange and agglomeration, cities have been prone to change more than non-urban spatial arrangements arguably since urban forms of life emerged some five thousand years ago (Crüsemann *et al.* 2013). Modern cities have considerably accelerated this tendency by being embedded in capitalist processes of accumulation, production and exchange. Key here is the tendency of capitalism to work through processes of 'creative destruction' that valorize and invalidate spatial configurations according to their usefulness for the rate of profit to increase. Joseph Schumpeter's deployment of Marx's earlier insight relating to the tendency of capitalist economic systems creatively to engage with its material and immaterial foundations (Schumpeter 1942; see also Harvey 2010: 46) has rightly been accorded prime status amongst the many tools available to urban analysts; in the context of concrete urban morphologies, this process becomes open to sensual practices most pointedly in the form of urban ruins: neglected spaces of whatever kind that no longer form part of larger valuation processes. Notwithstanding historically recurrent romantic infatuations with the idea and material presence of 'ruins' as providing a guide to the present notwithstanding (Dubin 2010: esp. 1–10), 'ruins' embody a 'dis-ordering' or disruptive quality that renders them important in the context of discussions of utopian and dystopian urban spaces, allowing for a recasting of the utopian-dystopian differentiation within urban environments.

At their core, 'utopian' and 'dystopian' spaces express the idea that spatial arrangements can articulate qualities inherent in society and thereby open them up to sensory perception. In addition, such spaces embody distinct ordering principles in that they are capable of expressing binary and perhaps dialectically arranged distinctions: from the 'utopian' qualities attaching to suburban configurations, 'smart urbanism' (Marvin *et al.* 2016) or 'green futures' (Bradley and Hedrén 2014) to 'dystopian' everyday experiences emanating from urban ghettos (McLeod and Ward 2002), what is desirable and what is not appears open to observation, critique, creative inspiration and (alternative forms of) knowledge. Common to many such deployments of 'dystopian' places, is that customarily they are placed alongside utopian ones in a mirror-like constellation. Logically and linguistically, this makes sense as the two concepts are not just clearly related but almost answer to each other in a manner akin to a Hegelian dialectic. In concrete urban settings, however, such neat juxtapositions or constellations hardly ever occur. Instead, even a non-analytical, synchronous glance at cities will reveal the co-existence of dystopian and utopian elements within any urban setting. Forgetting for a moment that what classifies as being either dys- or u-topian emerges within culturally shared but historically contingent contexts – for instance, cemeteries were not always regarded as 'frightful' places (Laqueur 2015) – we can establish a link between spaces that fulfil a clearly identifiable function within society and their attributed utopian qualities.

Related to this are a number of qualitative differentiations that remain powerful no less for the lack of a clear definition attaching to the terms 'utopia' and 'dystopia' (see Baeten 2002 for an overview). Beyond those aesthetic qualities associated with the terms that render a particular place either attractive or unattractive, think of the differentiation between optimism and pessimism, between hope and despair or the disparity between serene and disaster-ravaged environments, which are all conjured up by the headline concepts of this section of the *Handbook*. Cast in this manner, the utopian/dystopian distinction is a highly affective differentiation that works through and with human perception; little wonder, therefore, that one key aspect to the distinction between utopian or 'good' and dystopian or 'bad' spaces is their ability to be compatible with notions of the home and of dwelling (Kraftl 2007). Positioned between the two, a home is on the verge of becoming uncanny, a site in a horror movie.

And yet (again), as this contribution will argue, it is not 'the home' or indeed the built environment and its classical focus on buildings that is central to creative destruction, with the ruin as a memento of dysfunctionality. Rather, as a highly cyclical 'abandonment' of past and potential value, the ruin becomes a temporal fragment, a reminder that the relationship between utopia and dystopia is a flexible, malleable one that depends on economic cycles, cultural context and aesthetic values. No surprise therefore that the literature on 'ruins' is rather inexhaustible. From Simmel's 1907 essay (Simmel 1965) to Sebald's (2001) ruminations, from Benjamin's interpretation of Parisian arcades (Benjamin 2002) to contemporary equations of ruins to waste and to numerous webpages engaging in aspects of 'ruinophila' (Boym 2011), never are we far from an appreciation of the metaphorical and epistemological value of the ruin in analytical processes.

Moreover, like the image of the palimpsest, the notion of the 'ruin' has a tendency to lead us astray. Its implied materiality of a dis-used and dilapidated and thus at least potentially dystopian building runs aground when placed in the context of contemporary heritage designations, a general appreciation of memory and present-days attempts to extract from the built environment a capacity for continuity. A lot has changed, in other words, since Walter Benjamin employed the ruinous state of many Parisian Arcades in his 1920s and 1930s attempt to interpret them as key sites in the ongoing capacity of capitalism to transform urban environments, as ruins of past articulations of capitalism's seductive power (Benjamin 2002): we appreciate, conserve and sell ruins in today's ever more inclusive marketization of just about everything. Ruins, furthermore, come in many shapes and guises; as exemplars of 'creative destruction' it is thus perhaps equally valid, if not outright preferable, to focus one's attention on that part of the built environment that is less prone to preservative desires and practices. Infrastructures fit that bill.

Infrastructural spaces are everywhere but are mostly taken for granted unless or until they malfunction. They provide the glue, rather than the nodes, required for urban life to function: they link, fill, connect and facilitate. Think of streets, sewers and electric power cables; think, too, of mailboxes, subways and hospitals. In the context of this chapter, think of railway infrastructures and their associated uses and how these change over time. Intimately associated with the nineteenth century phenomenon of space-time compression (Harvey 1989), railway spaces radically altered the fabric of urban and national spaces by transforming the temporal dimension of everyday spatial relationships. Nationally, this resulted in a far-reaching alteration of rural-urban relationships and the 'conquest' of hitherto inaccessible parts of national territories; within cities, it planted stations at what were then the peripheries of urban territories; infrastructural facilitators that would soon be swallowed by cities growing at often remarkable rates. This chapter will explore three pieces of railway infrastructures in major Western cities with a view to analyse key transformative processes at the utopian/dystopian interface. The three spaces have been selected to allow a wide set of practices and results to become apparent within broadly comparable set of conditioning circumstances.

Railway infrastructures: Paris

While images of major Parisian railway termini – from the Gare d'Austerlitz to the one named after Lyon and beyond – will readily come to mind to many familiar with the French capital, not many visitors will promptly associate material realities or practices associated with the tracks that criss-cross the city from these stations outward. Unless one travels on these railway lines, they do not become part of the urban experience for a majority of tourists visiting the French capital. What is true for currently operative railway infrastructures also holds true for abandoned stretches of infrastructure: even frequent travellers to Paris will be hard-pressed to locate the *Petit Ceinture* ring railroad encircling the French capital within its 1860 (and still current) administrative configuration. Nor, for that matter, will most Parisians know much about a space that in 1926 was still deemed to be an integral part of everyday life in Paris, as evidenced by its inclusion in André Sauvage's renowned documentary film *Études sur Paris*. As a piece of pre-automobile urban infrastructure, the *Ceinture* served a function roughly comparable to that fulfilled by the *Péripherique* motorway today: it delivered goods and people across the city along a non-concentric, radially aligned trajectory. Of course, if the *péripherique* has become the symbol of individualized stop-and-go mobility, the *Ceinture* would be included amongst an older mode of public transportation now largely supplanted in Paris by buses and the Métro. Constructed between 1852 and 1867 and in use until 1934 to connect railway stations within Paris and to allow for the movement of goods and the deployment of military personnel and materials within the newly constructed Mur ('Wall') de Thiers, the *Ceinture* (the 'petit' or 'small' setting it apart from another ring railway beyond Paris proper), has largely lain dormant in its abandoned state since then. But for a short stretch of the line in the 15th and 16th arrondissements, where the tracks have been converted to form a 'green' walking environment, the infrastructure remains intact and recognizable as pieces of railway infrastructures (in fact, until well into the 1990s, some of the tracks were regularly used for the movement of goods between the tracks radially extending from the various railway stations). Which is not to say that today the tracks have become mere ruins: quite the contrary indeed. As anyone who has travelled the line or parts thereof (see Strohmayer 2012) can testify, the *Ceinture* and those lands immediately adjacent to it abound with mostly informal uses across the various arrondissements: from transgressive, sometime spontaneous and mostly non-permanent practises (Figure 44.1) to the deployment of space for artists' studios and other creative uses (Figure 44.2), from the planting of vegetables on

Figure 44.1 The Ceinture near the Rue de Crimée (19th arrondissement)
Source: photo by Ulf Strohmayer.

Figure 44.2 Studios below the Ceinture near the Rue de l'Argonne (19th arrondissement)
Source: photo by Ulf Strohmayer.

adjacent lands (Figure 44.3) to squatting on the *Ceinture* (Figure 44.4) – even though the tracks remain mostly intact, the practices engendered by the space that was (and to a large extent still is) the *Ceinture* vary dramatically. Most of these post-railway infrastructural uses emerged organically and remain part of the Parisian landscape through informally tolerated politics.

One could easily add further examples to this list of alternative uses. The conversion, by former students of the École des Beaux-Arts, of the former Charonne train station along the *Ceinture* in the 20th arrondissement into a café and concert venue still resonates within Paris; as does the transformation of parts of the tracks into a 'Sentier Nature' or nature path in the 15th and 16th arrondissements (Foster 2010) or the transformation of another station at Porte de Clignancourt into a cultural-culinary space baptised '*La REcyclerie*'. The list could go on but the point is already clear: on and along the *Ceinture*, abandonment has given way to a ruinous state which, in turn, gave way to a plethora of subsequent uses. These latter owe everything to politics in more than one sense: toleration is part of the answer and involves everything from the tacit acceptance of squatter rights to the issuance of temporally limited, legally codified rights. Equally important and no less political is the continued preserve of the land as state or semi-state property through the SNCF (Société Nationale des Chemins de Fer Français), the French state-owned railway company, as was the fact that the space occupied by the *Ceinture* was consciously kept open in anticipation of future uses. These latter centred on the replacement of the substandard system of bus routes (the three 'PC' buses) circulating around the twenty arrondissements on the Boulevards des Maréchauds. Since the decision, taken early in the new millennium, to replace the buses *in situ* with newly constructed tramlines (operating since 2006 and not yet

Figure 44.3 Urban gardening adjacent to the Ceinture below the Rue du Ruisseau ('Les Jardins du Rou‑
isseau', 18th arrondissement)

Source: photo by Ulf Strohmayer.

Figure 44.4 Squatting along the Ceinture in the Parc des Buttes Chaumont (19th arrondissement)
Source: photo by Ulf Strohmayer.

completed), this rationale has faded but was replaced by a more robust mechanism in 2006 through the signing of a framework protocol by the SNCF and the City of Paris encouraging cultural and related uses, which was prolonged and amended in 2015. Crucially, all of these forms of urban planning leave open the possibility of re-converting the space of the *Ceinture* back to railway-related activities, thus embedding any changes in land-use within the principle of reversibility.

In the context of the present chapter, the continuing transformation of the *Ceinture* thus demonstrates the entanglement of urban politics with dys- and u-topian visions that came about as a result of a disused and largely ruinous piece of infrastructure; planning here could perhaps best be conceptualized in the form of a Foucauldian *dispositif* (Strohmayer 2016) given its embedding in multiple structures, agencies and individual ambitions. Most of the *Ceinture* at present still remains a rather ramshackle affair, its realm cut off from the city with the help of high, if run-down, fences. However, precisely because of these dystopian allusions, alternative articulations abound that embody many a utopian vision.

Railway infrastructures: New York

In marked contrast to the infrastructure associated with the *Ceinture*, New York's High Line has become thoroughly reintegrated into urban modes of circulation and mobility, albeit in a transformed manner. Conceived in 1929 as an above-ground condition of freight-related mobility, the High Line was increasingly rendered obsolete by political decisions to invest in inter-state truck-based mobilities from the 1960s onward. Its use as a piece of railway infrastructure ceased in 1980 and it lay dormant before it was gradually reintegrated into official uses as a public space from 2004 onwards and officially opened in 2009. Since then, the park has successfully been extended twice northwards from its initial starting point at Gansevoort Street (between W12th and 13th Street) towards 34th Street, comprising a stretch of some 2.33 km.

The role model for the High Line conversion project was arguably the *Promenade Plantée* which used the infrastructural space of the old Vincennes railway in the 12th arrondissement of Paris. Inaugurated in 1993, this transformation of derelict and arguably dystopian infrastructural space into a sustainable, pedestrian urban utopia, transformed a liminal, bordering space into a genuinely linking piece of infrastructure. Considerably more diverse and capable of attracting tourists than its Parisian role model, the High Line has become a veritable success story attracting "over six million visitors per year" (Ascher and Uffer 2015: 228) and at the time of writing is being copied in Seattle, Chicago, Atlanta and other US cities (Brown 2013).

Given its above-ground material existence, the materiality that was the High Line from 1980 onwards was at the same time both less threatened and more vulnerable than, say, the *Ceinture* was in Paris: less threatened because being above ground, it did not technically occupy any valuable land; and more vulnerable because it was easier to dismantle. As such, the High Line qualified for the designation 'ruin' in a more tangible sense – if only for a limited time. It was saved from destruction as a result of the actions initially of a local resident who was followed later by a non-governmental neighbourhood group called 'Friends of the High Line'. According to their own website, this group refers to itself as "the caretakers of the High Line" (see www.thehighline.org/about, last accessed 31 May 2017).

If thus the survival of the railroad-generated infrastructure owes everything to grassroots-inspired forms of local activism, the conversion of the High Line into the extended park landscape open to visitors and residents alike came about more as a result of high calibre networking and fundraising activities employing art (especially photography) and influence to ensure that the conversion would happen and become a success. Such deployment of highly modern means to achieve a desired result finds its expression in Robert Hammond's 'Ted talk' about the High Line in which he uses the economic value and impact of the project to explain its genesis and success – and is rather revealing for that reason alone (see www.ted.com/talks/robert_hammond_building_a_park_in_the_sky?language=en#t-282349, last accessed 31 May 2017). With the backing of financiers, architects, fashion designers and eventually that of Major Bloomberg through the creation, in 2005, of a re-zoned 'West Chelsea Special District' (see www1.nyc.gov/assets/planning/download/pdf/zoning/zoning-text/art09c08.pdf, last accessed 31 May 2017), the purported raising of initially $150 million in support of the conversion will surprise no-one; nor will the fact that subsequent developments along the park were entirely in keeping with the capital-saturated redevelopment of the High Line park itself. From the construction of the re-located, Renzo Piano-designed Whitney Museum (inaugurated in 2015) to the construction of the Standard High Line Hotel towering over the park or the naming of parts of the park after benefactors, the High Line embodies a model of urban regeneration markedly different from that of the *Ceinture*: in New York, the ruinous, dystopian state of a continuous development has been thoroughly incorporated into the fabric of the city again and has delivered the anticipated results. Interestingly, as Shi (2013) observed, the erasure of past materialities here includes not just the replacement of an unordered and 'wild' space with a considerably more ordered and controlled environment but extends to the substitution of organically emergent flora and fauna through 'more appropriate' replacement plants and attached animal life. As concerns the results, these have recently been described as a form of "uncontrolled capital accumulation" (Morenas 2013: 301; for a factual assessment of the impact of the High Line on house prices in adjacent urban areas, see Levere 2014), with "[m]ore than 2,500 new residential units, 1,000 hotel rooms and over 500,000 square feet of office and art gallery space" (Shevory 2011) having developed in the first couple of years since the opening of the park, resulting in "$100 million in property tax increase" in 2010 alone (Levere 2014: 15, figure 6). Initiating processes of gentrification in the meatpacking district, a hithertofore "largely ignored" (Ascher and Uffer 2015: 224) part of Manhattan, those responsible for the development have thus arguably created a visually more homogenous environment at the cost of displacing established businesses and social forms of cohesion.

It would, of course, be disingenuous to attribute processes of gentrification to one particular development alone. As elsewhere, the effects of the ruinous transformation that is the High Line have to be placed and understood within broader contexts. The fact, for instance, that the surrounding Meatpacking District exhibits a considerable heritage value by being comprised of historical buildings with a rich sediment of architectural buildings while "[a]t the same time, the area hosted an atypical number of parking lots, taxi garages, and gas stations" (Ascher and Uffer 2015: 226) – allowing for both the creative reinvention of vacant space and ruins alike through market-led mechanisms.

Railway infrastructures: Berlin

Our third exemplary spatial transformation associated with railway infrastructures comes in the form of the *Park am Gleisdreieck* bordering Berlin's Schöneberg and Kreuzberg districts. In these districts – formerly in West Berlin – a considerable amount of infrastructural space was effectively abandoned and neglected as a result of highly idiosyncratic contextual configurations. The park's original spatial articulation came in the form of an extensive piece of railway infrastructure ordering the approach of trains to the Anhalter Bahnhof just south of Potsdamer Platz in Berlin while also being a station for freight in its own right; the resulting space was heralded by the novelist Joseph Roth in 1924 as a "'landscape' that begot an iron mask" (quoted in Scharnowski 2006: 80). The specific situation of Berlin in the post-war configuration of Germany – indeed of Europe – placed this extensive site of some 26 hectares in a prolonged situation of abeyance. Not only was much of the built environment and technical infrastructure destroyed during the final days of the Second World War, but also the specific status of Berlin, the division of Germany and highly idiosyncratic issues attaching to the ownership of land and buildings rendered the space occupied by railway infrastructures a complex and muddled affair. Effectively cut off from its erstwhile hinterland, the tracks still occupying the site were no longer used by inter-city related forms of traffic but by the S-Bahn or local railway network. Operating across the entirety of Berlin and legally in East German ownerships (the German Democratic Republic or GDR having been the legal successor to the state-owner pre-war Deutsche Reichsbahn), trains, tracks and railway-related infrastructures were owned by the GDR while the land on which these resided belonged to West Berlin. This state lasted until 1984 when West Berlin took over the operation of the S-Bahn on its territory; unification of Germany in 1989 finally brought the convoluted situation to an end. The situated context invoked just now matters since it afforded post-unification Berlin with yet another rather large chunk of land to fill through the joint mechanisms of planning, artistic happenings, individual and collective engagements and urban politics. It also created "green parallel worlds" (Kowarik 2015: 216), a sizeable urban fallow, where "nature" developed biotopes that could not develop or flourish elsewhere.

When unification happened, the site that was to become the present-day park was an urban wasteland of some proportions. Initially (in 1965) the area was designated to become part of a West Berlin North–South motorway known as the *Westtangente*, which led to considerable forms of civic resistance from 1973 onwards and well into the 1980s. Anchorage provided by the *Deutsches Technikmuseum* (the German Museum of technology) by occupying the North-Easternmost corner of the site (and including two former railroad roundhouses) since 1982 became instrumental in initiating change and did a lot to make the site appear ripe for development. The kind of development that happened subsequently owed everything to the interaction between scientists (ecologists for most part), citizen's initiatives or *Bürgerinitiativen* and official planners (Kowarik 2015: 212) that has since become part of the 'Berlin mix'. Planning started in 1997 and from 2006 onwards the park was opened up to the public in three distinct successive phases between 2011 and 2014. In its present configuration, it includes children's play areas, wilderness areas, sports facilities, community gardens, a centrally located café and remnants of older railway-related heritage, including signal boxes, tracks and nineteen old iron bridges spanning the Yorkstrasse in the Southern part of the Park. Crucially from a functional point of view is the integration of cycle tracks into the park's layout and design through its incorporation of the Berlin-to-Leipzig trans-regional cycle path, announced in letters in the ground, as well as the linking ambitions towards neighbouring parks and pathways that are the expressed ambition of the Berlin *Landschaftsplanung* (landscape planning; see www.stadtentwicklung.berlin.de/umwelt/berlin_move/de/hauptwege/index.shtml, last accessed 31 May 2017). Cycling, skateboarding, jogging and walking here enter into an entirely urbane and hodological (or pathway-related) contrast with criss-crossing above-ground subway routes and the open but sunken tracks of the high-speed ICE (Inter City Express) trains on the way to and from the Hauptbahnhof towards the north of the park. In a direct way, then, this is a hybrid space, not entirely abandoned by railway-connected activities while venturing towards something new; it is both *in* and *of* the city that surrounds it.

As in New York, the *Park am Gleisdreieck* has initiated change in the adjacent urban fabric, especially in the Schöneberg district, where the construction of high-end apartment blocks has already impacted on the neighbouring immigrant population through the conversion of rental to speculative accommodation (Rada 2015: 274). But in contrast to New York, this process was largely set in motion through the extended construction of new high-class apartment blocks adjacent to the western side of the Park. Equally telling is the fact that on its eastern or Kreuzberg side, the municipality of Berlin had agreed to accord favourable terms to a co-op (the 'Möckernkiez Genossenschaft') to advance the construction of an inclusive, cross-generational housing project comprising 460 apartments within the area of the Park properly speaking (see www.moeckernkiez.de/wp-content/uploads/moeckernkiez-flyer.pdf, last accessed 31 May 2017). The inclusive, participatory construction of the Park itself thus finds its equivalence in the decidedly mixed and non-hegemonic approach taken by all relevant stakeholders towards housing.

Conclusion

Railway-inspired transformations at the nexus between dys- and u-topian imaginations and practices are but one example of how the creative destruction of spaces within capitalist urban economies unfolds. The re-designation of formerly abandoned spaces more generally for uses as park landscapes for instance was not restricted to railway infrastructures. The developments of the Parisian Parks at *Buttes Chaumont* in the 1860s (Strohmayer 2006) or at *Bercy* in the 1980s (Strohmayer 2013) can perhaps serve as reminders of such changes in recent historical memory in that they clearly form part of a general economy of change within urban environments. In attributing such change to the latent tendency creatively to dismantle and reassemble – or to 'deconstruct' – spaces, we are merely attempting to discover an underlying reason within the temporal and spatial rhythms that make urban environments what they are.

In its most basic sense, navigating the relationship between utopian and dystopian spaces is the domain of urban planning; in fact, what urban planning is all about is accentuating, articulating and materializing visions of society within space. The examples discussed in this chapter bear witness of this fact. But they also articulate the sheer breadth of planning ambitions and practices within relatively comparable urban settings while reminding us of the paradox at work in relationally intertwined dys- and u-topian settings, the "paradox that fragments of the industrial past can be preserved only if they are willing to relinquish their uses" (Lopate 2011). The logic expressed here arguably owes everything to Schumpeter's 'creative destruction' invoked further up in this chapter and extends it to become part of a highly dynamic, situated and historically contingent settings. What is perhaps surprising and worthy of note is that comparable forms of decline with relation to dystopian railway spaces lead to markedly different forms of creativity and reconstructive élan in the exemplary spaces we touched on above: in Paris, New York, Berlin and elsewhere, places are made and remade through a historically resonant and constant interplay between materiality, politics and identity (Massey 2005). The fact furthermore that the conversion of dystopian spaces first into heritage (or at least 'preserve-worthy') spaces that are subsequently reintegrated into the urban economy has become prevalent in many Western cities – in fact: has become a stated goal of urban politics – can be read a sign of 'narcissism' (Lopate 2011) but most certainly is fully embedded in the move away from productive urban environments towards more leisure- and image-orientated ones. The High Line, as indicated above, arguably owes its existence to a series of ruin-documenting photographs produced between 2000 and the transformation of the line that were used widely from their publication in *The New Yorker* to advertise the eerie beauty of the stretched-out ruin (Sternfeld 2012).

Finally, it is perhaps incumbent on a chapter using terms like 'making' and 'remaking' quite liberally to think in closing about the nature of these 'creative' acts. Through our opening, critical invocation of the notion of the 'palimpsest' we pointed towards the centrality of the 'presence'/'absence' dialectic, which ought not to be misunderstood as a categorical, dichotomic relationship. Instead, it is perhaps best characterized as the constant possibility (and necessity) of urban forms of maintenance and repair (see Graham and Thrift 2007). In this light, 'breakdown and failure' are no longer seen as 'atypical' but become "the means by which societies learn and learn to re-produce" (Graham and Thrift 2007: 5). This implies that there is no 'additive' quality attaching to the interaction of 'utopian' and 'dystopian' spaces. There is, in other words, a lot of contingency, chance and local historical idiosyncrasies that intervene and will determine the possibility of spaces to re-emerge within validated contexts.

References

Ascher, K. and Uffer, S. (2015). The High Line Effect, CTBUH (Council on Tall Buildings and Urban Habitat). Research Paper presented at the CTBUH conference in New York. Available at: http://global.ctbuh.org/resources/papers/download/2463-the-high-line-effect.pdf [accessed 20 July 2016].

Baeten, G. (2002). Western Utopianism/Dystopianism and the Political Mediocrity of Critical Urban Research. *Geografiska Annaler: Series B Human Geography*, 84(3–4), pp. 143–152.

Benjamin, W. (2002). *The Arcades Project*, trans. H. Eiland and K. McLoughlin. New York: Belknap Press.

Boym, S. (2011). Ruinophilia: Appreciation of Ruins. Blog entry in the Atlas of Transformation. Available at: http://monumenttotransformation.org/atlas-of-transformation/html/r/ruinophilia/ruinophilia-appreciation-of-ruins-svetlana-boym.html [accessed 20 July 2016].

Bradley, K. and Hedrén, J., eds. (2014). *Green Utopianism: Perspectives, Politics, and Micro-practices*. New York: Routledge.

Brown, R. (2013). Now Atlanta Is Turning Old Tracks Green. *New York Times*, 14 February. Available at: www.nytimes.com/2013/02/15/us/beltline-provides-new-life-to-railroad-tracks-in-atlanta.html [accessed 20 July 2016].

Crang, M. (1996). Envisioning Urban Histories: Bristol as Palimpsest, Postcards, and Snapshots. *Environment and Planning A*, 28, pp. 429–452.

Crüsemann, N., van Ess, M., Hilgert, M. and Salje, B. (2013). *Uruk: 5000 Jahre Megacity*. Petersberg: Michael Imhof Verlag.

Cupers, J. (2005). Towards a Nomadic Geography: Rethinking Space and Identity for the Potentials of Progressive Politics in the Contemporary City. *International Journal of Urban and Regional Research*, 29(4), pp. 729–739.

Dubin, N. (2010). *Future and Ruins:. Eighteenth Century Paris and the Art of Hubert Robert*. Los Angeles: The Getty Research Institute.

Foster, J. (2010). Off Track, in Nature: Constructing Ecology on Old Rail Lines in Paris and New York. *Nature and Culture*, 5(3), pp. 316–337.

Graham, M. (2009). Neogeography and the Palimpsest of Place: Web 2.0 and the Construction of a Virtual Earth. *Tijdschrift voor Economische en Sociale Geografie*, 101(4), pp. 422–436.

Graham, S. and Thrift, N. (2007). Out of Order: Understanding Repair and Maintenance. *Theory, Culture and Society*, 24(3), pp. 1–25.

Harvey, D. (1989). *The Condition of Postmodernity*. Cambridge: Blackwell.

Harvey, D. (2010). *The Enigma of Capital and the Crises of Capitalism*. London: Profile.

Huyssen, A. (2003). *Present Past: Urban Palimpsests and the Politics of Memory*. Stanford, CA: Stanford University Press.

Kowarik, I. (2015) Gleisdreieck: Wie Urbane Wildnis im Neuen Park Möglich Wurde. In: A. Lichtenstein and F. Mameli, eds. *Gleisdreieck / Park Life Berlin*. Bielefeld: Transcript Verlag, pp. 210–221.

Kraftl, P. (2007). Utopia, Performativity and the Unhomely. *Environment and Planning D: Society and Space*, 25(1), pp. 120–143.

Laqueur, W. (2015). *The Work of the Dead: A Cultural History of Mortal Remains*. Princeton, NJ: Princeton University Press.

Levere, M. (2014). The High Line Park and Timing of Capitalization of Public Goods. Working Paper, University of California San Diego. Available at: http://econweb.ucsd.edu/~mlevere/pdfs/highline_paper.pdf [accessed 20 July 2016].

Lopate, P. (2011) Above the Grade: On the High Line. *Places Journal*, November. Available at: https://placesjournal.org/article/above-grade-on-the-high-line/?gclid=Cj0KEQjwtaexBRCohZOAoOPL88oBEiQAr96eSPimUhkcco1mQRvkLUHyahFc5k7FJf5lVO-A3YKkLX8aAjCE8P8HAQ [accessed 20 July 2016].

Marvin, S., Luque-Ayala, A. and McFarlane, C., eds. (2016). *Smart Urbanism: Utopian Vision or False Dawn?* Oxford: Routledge.

Massey, D. (2005). *For Space*. London: Sage.

McLeod, G. and Ward, K. (2002). Spaces of Utopia and Dystopia: Landscaping the Contemporary City. *Geografiska Annaler B: Human Geography*, 84(3–4), pp. 153–170.

Morenas, L. (2013). A Critique of the High Line: Landscape Urbanism in the Global South. In: A. Duany and E. Talen, eds. *Landscape Urbanism and its Discontents: Dissimulating the Sustainable City*. Gabriola Island, BC: New Society Publishers, pp. 293–304.

Rada, U. (2015). Unter Beobachtung – Eine Halbzeitbilanz des Parks am Gleisdreieck. In: A. Lichtenstein and F. Mameli, eds. *Gleisdreieck / Park Life Berlin*. Bielefeld: Transcript Verlag, pp. 272–277.

Scharnowski, B. (2006). Berlin ist Schön, Berlin ist Groß. Feuilletonische Blicke auf Berlin: Alfred Kerr, Robert Walser, Joseph Roth und Bernard von Bretano. In: M. Harder and A. Hille, eds. *Weltfabrik Berlin. Eine Metropole als Sujet der Literatur*. Würzburg: Königshausen and Neumann, pp. 67–82.

Schumpeter, J. (1942). *Capitalism, Socialism and Democracy*. New York: Harper and Brothers.

Sebald, W.G. (2001). *Austerlitz*. London: Penguin.

Shevory, K. (2011). Cities See the Other Side of the Tracks. *New York Times*, 2 August. Available at: www.nytimes.com/2011/08/03/realestate/commercial/cities-see-another-side-to-old-tracks.html?_r=0 [accessed 20 July 2016].

Shi, Y. (2013). Colonizing the Urban Wilds: Invader or Pioneer? MA thesis in Landscape Architecture, The Ohio State University. Available at: https://etd.ohiolink.edu/rws_etd/document/get/osu1366333944/inline [accessed 20 July 2016].

Simmel, G. (1965). The Ruin. In: K. Wolff, ed. *Essays on Sociology, Philosophy and Aesthetics*. New York: Harper & Row, pp. 259–266.
Sternfeld, J. (2012). *Walking the High Line*. Göttingen: Steidl.
Strohmayer, U. (2006). Urban Design and Civic Spaces: Nature at the Parc des Buttes-Chaumont in Paris. *Cultural Geographies*, 13(4), pp. 557–576.
Strohmayer, U. (2012). Performing Marginal Space: Film, Topology and the Petite Ceinture in Paris (with Jipé Corre providing art work). *Liminalities*, 8(4), pp. 1–16.
Strohmayer, U. (2013). Non-events and Their Legacies: Parisian Heritage and the Olympics That Never Were. *International Journal of Heritage Studies*, 19(2), pp. 186–202.
Strohmayer, U. (2016). Planning in/for/with the Public. *Urban Planning*, 1(1), pp. 1–4.
Swyngedouw, E. and Heynen, N. (2010). Urban Political Ecology, Justice and the Politics of Scale. In: G. Bridge and S. Watson, eds. *The Blackwell City Reader*. London: Blackwell, pp. 79–85.

45
DECONSTRUCTING MODERN UTOPIAS
Sustainable urbanism, participation and profit in the 'European City'

Samuel Mössner and Rob Krueger

Introduction

In this chapter we examine the European City model and its potential to produce more sustainable urban areas/cities. However, we also show the problematic relationship between social justice and economic growth that is inherent to the European City model. By drawing on the example of sustainable urban development in Freiburg, Germany, internationally recognized as a successful implementation of the European City model, we show that in practice social justice is subjugated to economic growth. It is the aim of this chapter to critically question urban models in general and the European City model in particular.

It has been well documented that a new modality of urban governance, aimed at a significant reduction in the state's responsibility for local welfare measures and service provision, began to emerge in the 1980s and flourish in the 1990s. The shift away from the welfare state and the dismantling of the Keynesian geopolitical and geo-economic system (Brenner 1997) was accompanied by the rise of a 'new urban politics' that centred on private sector interests and associated investments in urban areas (Cox 1993). In many countries, central and federal governments devolved powers to local governments, including municipal (urban) governments. While providing more autonomy for local authorities, the shifting terrain of governance also meant significant fiscal constraints for cities (Mayer 2013). Local governments mobilized their resources "in the scramble for rewards in an increasingly competitive free market" (Hall and Hubbard 1996: 154). Cities desperately searched for finding new ways and instruments to revitalize urban infrastructures, including housing. City governments adopted an 'entrepreneurial' ethos where even the smallest cities around the globe hired marketing strategists to help them position themselves, to find their unique selling point, or brand, for the marketplace. Cities became growing enterprises.

At the same time, in Europe and North America cities produce new forms of privatized urbanism that follows a globalized pattern of growth-oriented urbanism: indoor ski halls and urban entertainment centres, Disneyfied inner cities, new waterfront redevelopments, too numerous to count, became flagship symbols for this "imitative redevelopment" (Krueger and Buckingham 2012: 487), one that celebrity planner Charles Landry once quipped as the symbol of the "age of aquariums" (quoted in Krueger and Buckingham 2012: 487). Nevertheless, there were critics of the market-led, *growth-mandated urbanism*, especially in Europe. Scholars, activists and also some planners argued that market-led developments could not redistribute wealth appropriately, compromising the promise of the city: a more egalitarian form of distribution (Fainstein 2010). In her book, *The Just City*, Susan Fainstein focused her critique on how "deindustrialisation and globalisation had dramatically changed the fortunes of cities in the United States and Western Europe causing leaders to respond by entering into intense competition for private investment" (Fainstein 2010: 1). Others took up the cause and explored ways to reconcile urban economic growth and social justice. Or, in other words, they sought to bring back social justice to urban planning and development. In the course of their searches, Fainstein and others revisited the ideal of planning the European City.

The European City has been heralded as the "ideal type for cities, particularly with regard to overcoming social exclusion and segregation but also with regard to supposed positive relations between ecology and density, as well as competitiveness and – though vaguely defined – social and spatial coherence" (Novy and Mayer 2009: 105). It is a conception of the city that draws on the historical roots of approaching the urban and constructs a coherent line of development over time (Häußermann 2001). Specifically, it entails a vivid *urbanity* that differs from deserted suburban landscapes (Kunzmann 2011), supports individualism over the conformity of the urban (Kaelble 2001) and prefers non-geometrical forms and architectural aesthetics. The European Cities central pillar is its synthesis of market logics with social justice. Here, a growth orientation is a precondition for local democracy and emancipation. The European City model therefore promises a way of reconciling capitalist modes of accumulation with social values. It bears "normative force that can guide political action in the present context of globalisation and … integration"; it appears as the European utopia that is "often ideal-typically opposed to its North-American counterpart, the 'American City'" (Novy and Mayer 2009: 105).

By the 1990s, the European City model emerged as a political programme when severe economic and political restructuring took place all over Europe and, especially, in post-1989 Germany. Indeed, under German presidency of the European Union (EU), Germany sought to have the idyllic construction of the city codified into EU policy. In 2007, at the end of the German presidency, the European cities signed the *Leipzig Charter on Sustainable European Cities* (Commission of the European Communities 2007). The Leipzig Charter aimed to encourage more sustainable policies and demands for new forms of urban governance that facilitate urban sustainability. The charter holds important the city's need to be socially sustainable, and to foster local democracy. A key measure of these goals was how well cities achieved far-reaching forms of citizen's participation – what Selle and others call the culture of communicative planning (Fainstein 2010; Selle 2007). As a normative concept, the European City model sought reconciliation between economic growth, social justice and participation, as well as ecological responsibility in urban planning. The normative force of the European City model has led planners, policy-makers and academics to search for *best practices* for implementing these goals. Freiburg, Germany is among a few cities that are celebrated for their ostensibly successfully efforts to implement social, economic and ecological policies. As far back as the 1990s, Freiburg's development reads as a success story that is ecologically progressive, economically sound and socially just, or – in the words of planner Sir Peter Hall – Freiburg is the city "that does it all" (Hall 2014: 248). But how successful is the idyllic European City concept in practice?

In this chapter, we examine the implementation of the European City model to expose the contested relationship between market logics and growth orientation in urban planning, on the one hand, and social justice and the communicative planning culture, on the other. The European City model has been held up as a set of norms that others should mimic. We explore one such model, the city of Freiburg, Germany. In particular, we interrogate Freiburg's efforts of bringing together a coherent sustainable city agenda. Thus Freiburg serves as an interesting example to investigate empirically how the competitive, entrepreneurial city, which is guided by economic imperatives, reconciles the tension between market logics and social equity.

The European City as a modern utopia?

Before exploring the case of Freiburg, we develop the idea of the European City further. The European City gained political relevance at the end of the 1980s when the fall of the Iron Curtain led to the geopolitical disintegration and, subsequently, the re-assembly of territories – especially those in former socialist states – across Europe. Urbanists, policy-makers and researchers referred to the *originality* of the European City in order to counteract the 'imported threats' of urban policies from North America. In particular, the European City model provided an antidote for the paradigm of private-sector-driven growth that commodified the urban space (or sphere) and subjugated public infrastructure investments and social services under the paradigm of growth. Instead, the European City model aimed to reconcile these aspects into a coherent social and economic urban policy framework. The policy framework started out of a specific *ex-post* interpretation of the historical conditions of European urbanization. The European cities are reputed to obtain certain characteristics that are supposedly different from those of other agglomerations elsewhere. Walter Siebel, an influential German thinker and planner, identified six core characteristics of the European City as follows (see Siebel 2012):

1 *Presence of history*: Keeping cultural memories alive to show distinction among the variegated European community. He also believes that the European City is the cradle from which modern society 'learned to walk'.
2 *The hope of emancipation*: Means freedom from nature's constraints. Empowers the middle class to shift from the closed circles of the household economy to a free market exchange; the city from a feudalistic dependency to democratic self-administration.
3 *Urban lifestyles*: Means that there is a gap between private and public spheres. The private sphere is a place of intimacy, corporality and emotionality. The public sphere is a place that is a stylized presentation of self.
4 *Gestalt* (with a focus on architectural form): The urban is the container and symbol of history. The urban lifestyle and hope for emancipation that connect the European City are embedded in the built environment.
5 *The city as a political subject*: The European City is capable of acting independently in a democratically legitimated subject of its own development.
6 *Urban development equals growth*: since the nineteenth century the European City has been characterized and defined in terms of growth: concerning population, expansion but also in terms of capitalist economic growth.

These characteristics hint towards the problematic reconciliation of market orientation, on the one hand, and a claim for democratic emancipation, on the other.

The European City as a political approach

Siebel's ideas and the six characteristics of the European City can be seen in the Leipzig Charter, which emerged from the Lisbon Strategy in 2000 – an action and development plan aiming to reconcile economic competitiveness and social cohesion in Europe (Commission of the European Communities 2007). According to the Leipzig Charter, cities are "valuable and irreplaceable economic, social and cultural assets" (Commission of the European Communities 2007: 1). They are at centre stage for sustainable development as well as "engines of social progress and economic growth" (Commission of the European Communities 2007: 1). Both concepts – the Leipzig Charter and the conceptualization of the European City – incorporate a profoundly market-oriented understanding for delivering social services and new sources of revenue via economic growth.

As a consequence of this shift in urban governance, new instruments and tools for urban planning were needed that were suitable to introduce managerial styles to urban administrations. New public management, public-private partnerships and urban development contracts are just a few notions that symbolize the reform from the *codifying state* to the *cooperating state*. In Germany, the quantity of *urban development contracts*, a specific planning tool regulating the cooperation between the state and markets on a market-oriented contractual basis, has significantly increased during the last years. Originally only intended as a more flexible tool helping to administrate and regulate development and growth in the former German Democratic Republic, urban development contracts have since then systematically replaced traditional sovereign acts all across the country (Jochum 2004: 214).

The transformation of urban administrations and the profound market orientation creates a vacuum when it comes to implementing social equity and justice in urban development. During the 1990s and early 2000s, it became evident that the promising tools and instruments have largely ignored this page. Despite generally taking into question these tools, they were flanked by other instruments promising to reintroduce social aspects into urban planning and to bring back a certain level of social responsibility that has been blown to the wind before in favour of flexibility, profit maximization and time-saving.

The dismantling of traditional hierarchical tools and the implementation of entrepreneurial instruments to govern the city is mitigated by far-reaching modes of citizen participation with the aim to contribute to the level of social justice in urban planning. This approach, however, ignores that social justice does not exist per se, but there are various nuances and facets of social justice that can mutually co-exist. As social justice is not a physical condition but socially and differently constructed, there is not the *one* form of *just* participation but different forms of participation with different relations to social justice. Specifically, the question is who is included and targeted and who is not? What aspects of urban planning are placed for discussion and which are not? And which results from a participatory planning process are accepted and which are not?

This latter point is realized in the planning process as some groups can link their notions of 'justice' to economic imperatives, while others cannot. Consequently, the embeddedness of market logics in city planning creates a significant 'blind spot' in planning, even when such planning goals and practices argue that they are in line with concepts of social justice (Krueger and Buckingham 2012). The conceptualization of the European City holds economic growth as a core component. Given the ideological relationship between growth, planning and the economy, it is no wonder that the other core aspects are subjugated in the practice of creating the European City.

Urban development in the European City: green/just/profitable?

Drawing on empirical material gathered from Freiburg, a city hailed for its implementation of social justice, we now show that within the intellectual boundaries of the European City model only a specific interpretation of social justice is practised and thinkable that integrates and conforms with the logics of markets and economic growth and disintegrates social groups as soon as their political ideas conflict with these logics.

Greening Freiburg

Freiburg i. Br., Germany, is a middle-sized and compact city in the southern upper Rhine valley. Many foreign visitors – tourists and academics in search of the 'best practice' – are often delighted by its medieval structure, historic architecture and compactness, even though most of this had been bombed and destroyed during the Second World War. Today, Freiburg is heralded as a paramount example for implementing the idea of the European City as set out in the Leipzig Charter, being awarded as the European City of the Year 2010 by the Academy of Urbanism in London (see: www.academyofurbanism.org.uk/awards/, last accessed 20 July 2016). The city has attracted other awards of a similar nature. In 2004, for example, Freiburg was awarded the 'future-oriented city' by the Deutsche Umwelthilfe e.V. (German Environmental Action) and in 2010, the city won the German sustainability award. Finally, in 2012, Freiburg became part of Sustainia 100, an elite club of innovative sustainable organizations or municipalities. In each instance, its participatory practices were hailed as important pillars for Freiburg's success.

Moreover, these awards contribute to the attention Freiburg has garnered in international circles of policy-makers, planners and academics for its 'successful' implementation of social, economic and ecological policies since the 1990s (Broaddus 2010; Carter 2011; Hamiduddin 2015; Kronsell 2013; Medearis and Daseking 2012; Späth and Rohracher 2011; Zuh 2008). In particular, Freiburg became famous for its efforts to bring together the latest innovations in building technology, strict energy standards and an innovative sustainable land-use management (Scheurer and Newman 2009: 3), which has provided the city not only a good level of economic prosperity but also an "outstanding living quality" (Rohracher and Späth 2014: 8). Freiburg's urban development reads as a success story that incorporates all necessary elements outlined in normative manuals about urban sustainability (Roseland 1997). Indeed, the city is considered a forerunner for a sustainable urban development; a place that is ecologically progressive, economically sound and socially just.

In 2010, the City announced its 'Freiburg Charter for Sustainable Urbanism' (Stadt Freiburg i. Br. and Academy of Urbanism 2010). At this point, the City paid specific attention to political "participation in life-long learning processes" as well as transparency (Stadt Freiburg i. Br. and Academy of Urbanism 2010: 8). Freiburg's Lord Mayor, Dieter Salomon, has often repeated that the 'real' driving force behind Freiburg's sustainable agenda is an engaged and active citizenry and has emphasized the importance of citizen participation. Many scholars acknowledge the City's efforts to cultivate citizen participation that goes beyond the level of participation institutionalized in the German federal building code and other sectoral planning frameworks. Central to these arguments is the development of the neighbourhood Vauban that is considered a 'best practice' for sustainable urbanism and important pillar for Freiburg's international competitiveness. The international literature unveils the successful reconciliation of economic growth with the strengthening of social values and justice. Freiburg is considered an inclusive city; it includes all citizens in urban planning by means of broad participatory practices.

A just Vauban?

Within the Freiburg experience, the eco-neighbourhood of Vauban has become an important demonstration project, showing how making and planning cities which are socially inclusive and economically viable is possible. As a distinct community within the city, Vauban first emerged soon after French troops withdrew from Freiburg in the early 1990s. The barracks vacated by the French as well as the surrounding green-field sites situated in the south-west of the city were squatted by people living alternative lifestyles. During this time, planners developed a master plan for the area which defined it as a residential neighbourhood. While the city planners were still thinking in traditional categories of planning and giving priority to large investors and developers, the reality in the meantime was made differently, and by social activists in Vauban. For the latter, the housing shortage was alarming and affordable housing was urgently needed, particularly for a younger, and ecologically conscious middle class. They claimed Vauban as their *promised land*, their window of opportunity, to realize *their* imagination of living (Frey 2011).

Being ecologically conscious, economically successful and socially homogeneous, Vauban residents initiated their work with a high level of political power, despite significant resistance from parts of the city administration and some politicians. Indeed, Vauban's self-styled 'community developers' eventually influenced the master-plan for their purposes. The strong influence and political pressure of this new 'eco middle class' was successful in lobbying the city to deviate from the standard procedures and accepted rather innovative and non-standard approaches for Vauban.

As a consequence of what became a politicalized process, the City planned Vauban very differently. Vauban was developed by groups of individuals working on small-scale projects rather than large-scale developers. The City gave priority to smaller groups, associations and individuals who were able to fund their projects, as well as develop a vision for the neighbourhood. So, while a small part of Vauban was developed in traditional ways with a single architect and investor/developer, the majority of the community was sold to homeowner groups. Such homeowner groups consisted of ten or more individual prospective homeowners, who gathered together during an early phase of planning and started developing and constructing their own buildings. In other words, homeowner groups (*Baugruppen*, in German) created the neighbourhood in their collective imagination before it physically existed. The homeowner groups were paramount in bringing innovative and sustainable lifestyles to Vauban, and played a crucial role in participating in the development of Vauban.

Unlike housing associations that pool financial resources, homeowner groups are individual owners with individual financial capacities. They gather together with others interested in this type of housing and, along with an architectural plan and design, attempt to realize *their* image of housing. Here the *Baugruppen* in effect act as developers and cut out the middleperson, namely, the professional developer. The *Baugruppen* represent an alternative to traditional models of housing development because of their focus on saving costs by developing cooperations with fellow owners, sharing a common vision for living together and as a consequence, creating a community – or in the case of Vauban an association of communities – before the construction of the building commences. Many *Baugruppen* were founded upon pre-existing networks of people. Among them were thinkers, leaders and lobbyists for innovations in green energies and technologies, individuals who had nevertheless been socially connected for some time. These strong networks and social ties enabled the *Baugruppen* and future residents to influence urban planning successfully during nearly all phases of Vauban.

In Vauban, moreover, the formation of homeowner groups was sometimes initiated by the future homeowners themselves; at other times, however, the lead was taken by an architect. In Freiburg, in particular, some architects specialized in working together with different homeowner groups. Most homeowner groups started with a vague idea of what the building was supposed to look like, mostly in terms of innovative ecological techniques that were seminal at that time and therefore could not be realized in standardized ways. The high personal engagement for constructing the building and its surrounding living environments was complemented by several initiatives, working groups and communities that dealt with the more comprehensive approach to sustainable urbanism in daily life. The city planners in Freiburg became aware of this critical potential and encouraged their formation in Vauban, setting up an organization known as Forum Vauban.

Crucial to the success of homeowners' groups in Vauban was the foundation of a neighbourhood association in 1995. The institutionalization of participation by the Forum Vauban was made possible by the "enormous potential of engagement and creativity" of the residents of Vauban, but also by the financial basis coverage (Sperling 2013: 4). In the Spring of that year, the City administration officially gave Forum Vauban the responsibility for facilitating cooperation with the city administration regarding the development of Vauban (Sperling 2013). By institutionally embedding the Forum Vauban in the planning processes the City gained a reliable contact in the community for the city planners. The Forum Vauban promoted, amongst other plans, a car-reduced neighbourhood, compact urban development, short distances between living and working, public transportation, affordable housing for private homeowners, ecological and sustainable technologies, and a dynamic neighbourhood life that included an organic supermarket cooperative, which is still in operation today. These features were all part of the European City model which assuaged city planners.

Profitable Vauban

Vauban's success as a 'sustainable' neighbourhood, which heavily relies on innovative sustainable technologies and policies, is also framed by an imaginary of social tolerance and a 'colourful' social structure. The high level of participation of its inhabitants during the planning was accompanied by an attitude of being open-minded when it comes to the inclusion and participation of different social groups. A section in the book *Freiburg, Green City*, a propaganda piece funded by the City's marketers, shows a squatter collective that had settled at the entrance to Vauban in 2009. This collective, named *Kommando Rhino*, occupied the area in Vauban in order to live there with their trucks and trailers, peacefully co-existing with the local population of Vauban and surroundings for the next couple of years.

Squatter collectives can also be found in other sustainable urban developments, such as in the French Quarter in Tubingen or in Christiania, Copenhagen. At first glance, squatter collectives seem to be as revolutionary in promoting different ways of life as the *Baugruppen* were with their dedication for innovative forms of constructing housing. And consequently, *Kommando Rhino* symbolized the Vauban's spirit to resist the mainstream, and to counter current traditional forms of housing. *Kommando Rhino* was a powerful symbol for inclusiveness and diversity supporting rather than threatening economic stability. Indeed, the collective was touted as an important pillar of Freiburg's marketing strategy and was repeatedly mentioned in publications and shown off to international observers (Frey 2011). *Kommando Rhino* was well integrated and contributed to the vivid urbanity Vauban stands for. Its members organized a weekly open-air cinema, documented the press review and pictures about Vauban at the entrance to the camp and organized a café. Sometimes there were even public concerts and exhibitions. *Kommando Rhino* lived alongside Vauban's *Baugruppen* for several years.

The very difference that distinguishes *Kommando Rhino* from the *Baugruppen*, is their fundamental critique of capitalist growth, which is implicit of their form of living. They reject growth orientation and market mechanisms as the ordering structure and, through their lifestyle, offer an alternative to established forms of top-down, representative democracy. Squatter collectives, like *Kommando Rhino*, stand for self-organization, reducing material consumption and rejection of immobilities of contemporary life. In contrast, the *Baugruppen* offer a different way of organizing *within* the system of real estate and while they realized and implemented a new form of *doing* real estate, they never took into question the logics of capitalist exploitation of the housing market. In recent years, Vauban has experienced annual rent increases of about 10 per cent (Amt für Liegenschaften und Wohnungswesen der Stadt Freiburg 2013).

The virtuous living arrangements between these alternative groups came to an end in 2010. Residents of Vauban felt an earthquake-like jolt when rumours circulated that the neighbourhood was to be sold to an international investor. After the confusion which often follows a real or perceived disaster abated, it became clear that no international investor would move in to take Vauban; rather it was the City of Freiburg itself which would unveil a development strategy for the neighbourhood. Their first project was a hotel – a green hotel, of course. The planning process that followed faced severe criticism for its lack of citizen participation. Planners and the *Freiburger Stadtbau*, the city-owned housing company, ignored ideas and concrete plans put forward by an active group of residents from Vauban and, in late 2010, rejected the idea to include *Kommando Rhino* in making these plans. The City of Freiburg finally communicated its vision for the hotel; they offered a green, "socially inclusive" project – one that employed handicapped people (Mössner 2016). The site for construction was on the ground underneath the *Kommando Rhino* community. *Kommando Rhino* was displaced from Vauban and no alternative site was proposed for their resettlement. The central argument proffered by the city was that the area was always earmarked for development, without exception. Indeed, the vision of the European City had run its course: urban development now equated with growth. The value *Kommando Rhino* brought to the community was trumped by the economic value the hotel would bring.

The chimera of utopia: friction and the European City model

In her book *The Just City*, Susan Fainstein identifies "equity", "democracy", "recognition" and "diversity" as fundamental cornerstones that guarantee – yet do not define – social justice (Fainstein 2010: 23). In her critique of the communicative planning approach, Fainstein warns that social justice should not be conflated with ostensibly *consensual* decision-making processes that built upon mutual understanding (Fainstein 2010: 26). Consequently, not every form of participation will lead to a socially just city but might in fact indicate its opposite. The case of Freiburg/Vauban illustrates this point very well. Indeed, it calls into question the European City model. While in theory the goal of the European City model is to contain the contradictions of urban change, it itself is a chimera, an illusion.

Vauban's homeowner groups successfully participated and in and shaped Freiburg's neighbourhood planning process, if only within its own territorial boundaries. They fought for the inclusion of sustainable aspects into urban planning and have paved the path for Freiburg's new sustainability agenda. In only a few years, they fundamentally changed Freiburg's political orientation and reached a remarkable outcome. By doing so, they drew upon resources that facilitated their access to the planning process. The inhabitants of Vauban could rely on a strong social network that was built upon, and stabilized by, a remarkably high level of educational and cultural capital. This allowed them to speak the same language as the planners and politicians they were collaborating with, targeting the technocratic approach underlying urban planning thus far. Many inhabitants of Vauban showed a profound and impressive technical understanding of sustainability implementations and a powerful social network that helped to develop Vauban as part of the Green City branding in order to intrude the logics of planning. From this perspective, the successful participation – widely heralded in many publications and by the city government and planners often for the benefit of international audiences – is obvious and rather unsurprising. It is important to understand, however, that they never radically questioned the economic logics of urban development and its consequences for social inequalities, but rather tried to reconcile different interests in consensual way. Economic prosperity was brought in line with innovative ecological techniques and the idea of a social homogeneity. Moreover, there was a private market to support these wants.

Final thoughts

Before we judge and celebrate the European City model as 'sustainable' or 'just' we have to look beyond the short-term rhetoric of politicians, city boosters and well-meaning planners. In this chapter we chose two different examples from the City of Freiburg, which is often held up as an exemplar of the utopian ideal of the European City in which the access to participation occurred differently and, consequently, participation was practised in diverse ways and with different impacts. The first example refers to a successful participation process of the middle class and the second one shows how participation ended up in the specific moment when competing interests entered the agenda. The successful branding of Vauban was built upon the social aspects of sustainable development. Tolerance, equity and diversity were considered important pillars to maintain economic prosperity. A very effective way to prove the co-existence of these values to the local as well as an international audience is the demonstration of tolerance towards diverse forms of living. The displacement of *Kommando Rhino* stands as a symbol for those who seek to challenge or *cannot challenge* the logics of capitalism. Although unplanned, it was a strategic move to demonstrate the capacity of capitalism – distortedly presented here as tolerance – to integrate outsiders. In the very moment, when *Kommando Rhino*'s concept of society competed with the capitalist concept, the City made a clear and brutal decision to fix the problem spatially, in other words, to displace *Kommando Rhino*.

References

Amt für Liegenschaften und Wohnungswesen der Stadt Freiburg. (2013). *Mietspiegel 2013–2014*. Freiburg: Stadt Freiburg.
Brenner, N. (1997). Die Restrukturierung Staatlichen Raums: Stadt- und Regionalplanung in der BRD 1960–1990. *PROKLA-Zeitschrift für kritische Sozialwissenschaft*, 109(4), pp. 1–21.
Broaddus, A. (2010). Tale of Two Ecosuburbs in Freiburg, Germany. *Transportation Research Record: Journal of the Transportation Research Board*, 2187, pp. 114–122.
Carter, J.G. (2011). Climate Change Adaptation in European Cities. *Current Opinion in Environmental Sustainability*, 3(3), pp. 193–198.
Commission of the European Communities. (2007). *Leipzig Charter on Sustainable European Cities/Leipzig Charta zur Nachhaltigen Europäischen Stadt*. Informationen zur Raumentwicklung, (2010), 4, pp. 315–319.
Cox, K.R. (1993). The Local and the Global in the New Urban Politics: A Critical View. *Environment and Planning D*, 11(4), pp. 433–448.
Fainstein, S. (2010). *The Just City*. Ithaca, NY: Cornell University Press.
Frey, W. (2011). *Freiburg Green City: Wege zu einer Nachhaltigen Stadtentwicklung*. Freiburg: Herder Verlag.
Hall, P. (2014). *Good Cities, Better Lives: How Europe Discovered the Lost Art of Urbanism*. New York and London: Routledge.
Hall, T. and Hubbard, P. (1996). The Entrepreneurial City: New Urban Politics, New Urban Geographies? *Progress in Human Geography*, 20(2), pp. 153–174.
Hamiduddin, I. (2015). Social Sustainability, Residential Design and Demographic Balance: Neighbourhood Planning Strategies in Freiburg, Germany. *Town Planning Review*, 86(1), pp. 29–52.
Häußermann, H. (2001). Die Europäische Stadt. *Leviathan*, 29(2), pp. 237–255.
Jochum, H. (2004). *Verwaltungsverfahrensrecht und Verwaltungsprozeßrecht. Die Normative Konnexität von Verwaltungsverfahrens- und Verwaltungsprozeßrecht und die Steuerung des Materiellen Verwaltungsrechts*. Tübingen: Mohr Siebeck.

Kaelble, H. (2001). Die Besonderheiten der Europäischen Stadt im 20. Jahrhundert. *Leviathan*, 29(2), pp. 256–274.

Kronsell, A. (2013). Legitimacy for Climate Policies: Politics and Participation in the Green City of Freiburg. *Local Environment*, 18(8), pp. 965–982.

Krueger, R. and Buckingham, S. (2012). Towards a 'Consensual' Urban Politics? Creative Planning, Urban Sustainability and Regional Development. *International Journal of Urban and Regional Research*, 36(3), pp. 486–503.

Kunzmann, K. (2011). Die Europäische Stadt in Europa und Anderswo. In: O. Frey and F. Koch, eds. *Die Zukunft der Europäischen Stadt*. Wiesbaden: VS Verlag für Sozialwissenschaften, pp. 36–54.

Mayer, M. (2013). Urbane soziale Bewegungen in der Neoliberalisierenden Stadt. *Suburban – Zeitschrift für Kritische Stadtforschung*, 1(1), pp. 155–168.

Medearis, D. and Daseking, W. (2012). Freiburg, Germany: Germany's Eco-capital. In: T. Beatley, ed. *Green Cities of Europe: Global Lessons on Green Urbanism*. Washington, DC: Island Press, pp. 65–82.

Mössner, S. (2016). Urban Development in Freiburg, Germany – Sustainable and Neoliberal? *Die Erde*, 146(2–3), pp. 189–193.

Novy, J. and Mayer, M. (2009). As 'Just' as it Gets? The European City in the 'Just City' Discourse. In: P. Marcuse, J. Connolly, J. Novy, I. Olivo, C. Potter and J. Steil, eds. *Searching for the Just City: Debates in Urban Theory and Practice*. New York and London: Routledge, pp. 103–119.

Rohracher, H. and Späth, P. (2014). The Interplay of Urban Energy Policy and Socio-technical Transitions: The Eco-cities of Graz and Freiburg in Retrospect. *Urban Studies*, 51(7), pp. 1415–1431.

Roseland, M. (1997). Dimensions of the Eco-city. *Cities*, 14(4), pp. 197–202.

Scheurer, J. and Newman, P. (2009). Vauban: A European Model Bridging the Green and Brown Agendas. Case study prepared for *Revisiting Urban Planning: Global Report on Human Settlements*. United Nations. Available at: www.unhabitat.org/grhs/2009 [accessed 20 July 2016].

Selle, K. (2007). Stadtentwicklung und Bürgerbeteiligung – Auf dem Weg zu einer Kommunikativen Planungskultur? Alltägliche Probleme, neue Herausforderungen. *Informationen zur Raumentwicklung*, 1, pp. 63–71.

Siebel, W. (2012). Die Europäische Stadt. In: F. Eckardt, ed. *Handbuch Stadtsoziologie*. Wiesbaden: Springer VS, pp. 201–211.

Späth, P. and Rohracher, H. (2011). The 'Eco-cities' Freiburg and Graz: The Social Dynamics of Pioneering Urban Energy and Climate Governance. In: H. Bulkeley, V. Castan Broto, M. Hodson and S. Marvin, eds. *Cities and Low Carbon Transitions*. New York and London: Routledge, pp. 88–106.

Sperling, C. (2013). Keine Wirkung ohne Risiko. *Planung neu denken – PND*, 2–3, pp. 1–10.

Stadt Freiburg i. Br. and Academy of Urbanism. (2010). *Freiburg Charter of Sustainable Urbanism*. Freiburg: authors.

Zuh, M. (2008). *Kontinuität und Wandel Städtebaulicher Leitbilder. Von der Moderne zur Nachhaltigkeit. Aufgezeigt am Beispiel Freiburg und Shanghai*. Darmstadt: TU Darmstadt.

46
DYSTOPIAN SPACES AND ROMA IMAGINARIES
The case of young Roma in Slovenia and Romania

Stuart C. Aitken and Jasmine Arpagian

> The Roma doesn't want to be civilized.
> (Arbër, personal interview, Maribor, 24 March 2014)

Introduction

If power struggles are often motivated by fearful or hopeful imaginaries of urban space, they are also predicated by imaginaries of peoples. People perceived as disorderly or fearsome because they do not fit prevailing urban sensibilities are commonly shunned or ostracized from urban culture and politics, and marginalized to peripheral but also politically overlooked, under-resourced, inadequate and, ultimately, dystopian spaces. Pulling from an obscure Roman law, Giorgio Agamben (1995, 2001) theorizes these people as akin to the ancient *homo sacer*, or sacred man, who could be killed with impunity as long as the process was not sacrificial (the *homo sacer* was less than human, and even ritual sacrifice was too good for him as it might offend the gods). Denied any kind of political status (*bios politikos*), the *homo sacer* was left with bare life (*zoē politikos*). With the rise of democracy in the Greek *polis*, Agamben proposes that the idea of the *homo sacer* increasingly made little sense, but with modernity and events like the holocaust, hunger strikes, food banks and so forth, bare life is back and biopolitics are now a foundational part of the state apparatus. And, with the modern state comes spaces – from Nazi concentration camps to Abu Ghraib and Guantanamo Bay, and contemporary refugee camps – to circumscribe bare life. Agamben elaborates the function of these spaces and bare life in modern politics through "the state of exception" (1995: 9) whereby citizens are protected by the sovereign rule of law and certain outsiders (the modern *homo sacer*) are purposively excluded.

In this chapter, we suggest that the nature of the modern *homo sacer* is exemplified by the problematic dystopian imaginaries that circle some Roma peoples, and that these imaginaries are often codified spatially by local council and state policies. The Roma are a collection of diverse Romani peoples who are often lumped together as travellers or are marginalized through their traditional nomandism and encampments in rural villages and occupation of substandard urban areas. The Romani peoples are found mostly in Europe and the Americas, and they are traditionally not tied to national ethnicities and cultures, although many consider their origins emanate from Northern India over a thousand years ago. Here, we consider the Roma primarily from their urban, but also national, political subjectivities. Drawing on the work of Agamben and an expanded view of his states-of-exception concept, as well as Aihwa Ong's (2006) work on contemporary neoliberal governance, we argue that young ethnic minorities are particularly vulnerable to a revanchism that is hidden under state policies of seeming democracy and justice. And, drawing on recent work on youth geographies and citizenship (Aitken 2015; Mitchell forthcoming), we argue further that it is from young people's actions (often galvanized by non-governmental organizations or NGOs and other political constituencies) that hope arises. In what follows, we examine the contexts of urban Roma youth denied full citizenship rights for a variety of reasons, and how that denial contextualizes itself in the urban spaces of Maribor in Slovenia, and Cluj-Napoca in Romania (see Figure 46.1). The chapter draws upon ongoing fieldwork – in the form of interviews and observation-bolstering archival and news media research – which began in Slovenia in Autumn 2013 and in Romania in Autumn 2015.

The place of Roma peoples in urban imaginaries and the construction of their sometimes dystopian spaces is crucial for understanding neoliberal urban politics and governance practices as they relate to the disenfranchisement of exceptional populations. How are states of exception constructed around minority peoples and how do they push back? What are the implications for larger urban transformations and the role of neoliberal economics and governance? In addition, at a time of inward-looking European nationalisms and the potential unravelling of the European Union (and the erosion of the freedom of movement upon which the pan-State was founded), it is important to study the implications for a highly mobile population that

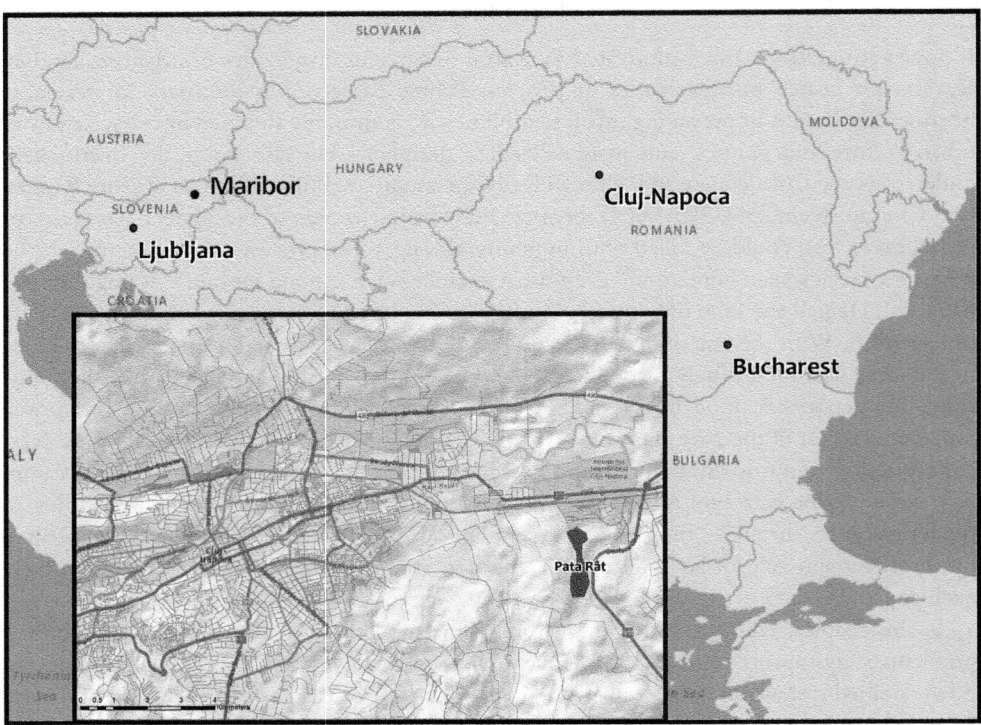

Figure 46.1 Study sites at Maribor and Cluj-Napoca (inset visualizes the peripheral location of Pata Rat)

Sources: Environmental Systems Research Institute (ESRI). Light Grey Canvas Map. ArcGIS Online; SPAREX Project: Field sites in SPAREX Project. Google My Maps. Available at: www.google.com/maps/d/viewer?mid=16 grzmqmmcEXuUImN8kYbzyqYzuY&hl=en&ll=0%2C0&z=8 [accessed 2 June 2017].

nonetheless is connected to very specific kinds of urban spaces. What happens when that movement is curtailed? What are the implications for urban Roma communities? And, perhaps of greatest import from what we are interested in here, what kind of action propels young Roma onto the streets and how do they affect change?

Roma youth imaginaries, mobilities, citizenship and space

Even when I became a Slovenian citizen, I still consider myself Roma and I will stick to that until I die ... Then again, as soon as I got Slovenian citizenship, things got better. I still keep pictures of places where we lived and what we had to go through and I just want to show it to people so they can see what we, as a family, had to overcome.

(Isni, personal interview, Maribor, 24 March 2014)

Issues of freedom, belonging, mobility, community and citizenship play into dystopian urban spaces in complex and contradictory ways. We encountered examples in Slovenia and Romania of Roma youth feeling connected to the places they live and protesting past evictions. As the epigram at the start of this section suggests, and flouting the usual 'travellers' imaginary' (Shubin and Swanson 2010), there is a strong sense of place amongst many of Roma that we talked to, with a stated gratefulness to their countries of citizenship in addition to the Roma communities within which they reside (Moise 2011). Controversy, then, surrounds notions of travelling and Roma place-attachment, with concerns about inclusivity and belonging sometimes outweighing ideas about exclusivity and mobility. For Roma youth, a focus on inclusivity and belonging brings to light the importance of their rights to space as well as their rights to movement and mobility. The right for young people to occupy urban spaces and move freely within them is one of the least understood and discussed rights but it is one of the most important because when young people gain the right to create and re-create their own spaces they simultaneously gain the propensity to create and re-create themselves (Aitken 2015: 143). Many Roma youth live in a state of exception to the degree that they are part of an underclass without freedom of movement or full citizen rights; their rights to space and mobility are ignored, and they are trapped in dystopian spaces despite feelings of belonging. The issues of entrapment and exclusivity/exception on the one hand, and mobility and citizenship on the other, play off against each other in the case of Slovenian Roma living in and around Maribor. The issues are also important for the Romanian examples, which we elaborate in the second part of the chapter through a very clear context of spatialized dystopia.

In what follows, we begin with the Maribor Roma community because it exemplifies contradictions between such issues as entrapment, exclusivity, mobility and citizenship, and it also elaborates an activist Roma community that moves some way from dystopian ideas through increased literacy and recent gains in cultural capital. What is less clear from the Maribor example is how far their gains in cultural capital are also a move away from neoliberal exceptionalism. For the most part, our discussion of Slovenian Roma parallels the Romanian examples that follow, with one important politico-legal exception. When Slovenia gained independence in 1991, 25,671 people from ethnic minority communities (primarily Serbs, Croats and Bosnians, but also Roma) were removed from the permanent residential register, and effectively lost legal status. This became known as 'erasure'. Unlike Slovenian Roma, Romanian Roma are not victims of deliberate state erasure although many similarly lack identity documents for a variety of reasons (e.g. inability to prove permanent residence because of informal housing, medically unassisted childbirths and illiteracy) (World Bank 2014).

The degree to which a neoliberal state-of-exception, whether by government fiat or duplicity, contextualizes Roma youth and the spaces within which they live and from which they travel is very much tied to how they see themselves and how much they push back against dystopian imaginaries. In the examples that follow, we elaborate the connections between bare life and hopeful futures, and tie these connections to young people's mobility, exclusivity, entrapment and citizenship rights.

"This new generation likes to work": Roma youth in Maribor

The Slovenian population of Roma is approximately 20,000: they mostly live in the rural areas of Prekmurje and Dolenjska, which represent respectively the most eastern and western extremes of the country. The largest urban concentration of Slovenian Roma is in Maribor, which is located in the eastern part of the country close to the Prekmurje region. In addition, there are several villages around Maribor whose inhabitants are exclusively Roma. Isni lives in Maribor, and perhaps because of his urban context ideas of mobility, connection and exclusivity permeate a narrative of hope:

> I know this fairy tale. What happens in your life, it's a fairy tale. My parents are from Kosovo. Before they came here, in the time of Yugoslavia, it wasn't so important if you were Bosnian, Slovenian or Roma, Macedonian. They accepted you. If you were working as a cleaner or miner, you were a man. My parents were valued here in Maribor because they were hard working people … They went to Kosovo for holiday. They treated them as they came from a big country. Not as from Slovenia, but as they would come from Germany, Italy or America. It was like that.
> *(Isni, personal interview, Maribor, 24 March 2014)*

Like Isni's parents, most Roma who live in Maribor come from Kosovska Mitrovica. Before the Balkans war more than 20,000 Roma lived in this area, but during the 1970s and 1980s they were attracted to the employment of a rapidly industrializing Maribor. Maribor is the second largest city in Slovenia with nearly 100,000 inhabitants. It is the major focus of heavy industry in the country, and its location in the eastern part of the country offers important trade with Austria and Germany. When Slovenia gained independence in 1991 and many Roma lost their permanent residential status, there was significant loss of mobility; without official papers they could be deported at any time and if they left in search of work would likely not be allowed re-entry into the country (Aitken 2016). Whereas an industrializing Maribor welcomed labour from the south, the tone changed in the 1990s when the new neoliberal Slovenian economy declined into recession. A combination of right-wing nationalism and the European sovereign debt crisis changed the perspective of Slovenians:

> "Hey you, the erased! Hey you, cigan! Go back where you're coming from!" Even if we were from here … I was two years old when we moved to Slovenia. All my childhood, everything I had, it's here. And they didn't recognize this.
> *(Isni, personal interview, Maribor, 24 March 2014)*

Isni describes how as a child in Maribor he felt like he was part of a normal family living in a nice apartment surrounded by a respected Roma community. What might be described as dystopia descended on Isni's family when they returned to an independent Slovenia after spending the summer in Kosova. It took six months of travel through the war-torn Balkans to get back home; Isni was twelve years old. On returning, they found another family living in their apartment and his parents were barred from resuming their former employment. Roma unemployment started climbing from this time and today in Maribor is estimated at 97 per cent.

> We went to the municipality and they said to us that we have no rights to get any kind of status in Maribor. From that day, we were living from day-to-day. We were collecting glass bottles from trash so we could live.
> (Isni, personal interview, Maribor, 24 March 2014)

Due to the Slovenian Alien Act of 1991, which created the political process of erasure, Isni was barred from going to school and his father was unable to obtain health care when he contracted tuberculosis. From thirteen to twenty-three years of age Isni was barred from subsidized housing, and lived in the apartments of Roma friends and acquaintances, often in deplorable conditions:

> Believe me, I still see things how and where we were sleeping and living. For more than ten years ... We were destroyed. You had no right [to space], you had troubles with every cop that stopped you, no one trusted you because you had no papers ... I was walking around the market [looking for] leftovers ... so we could survive. You couldn't get a job. You couldn't get support from the state. Nothing.
> (Isni, personal interview, Maribor, 24 March 2014)

When Isni was sixteen, his first daughter was born, and over the next eight years he and his wife had five more children. He made a living picking out things to sell from the trash:

> Glass bottles, paper, iron, everything that was useful. I was walking on the streets, collecting things ... It was sad when I came home and my child wanted milk or food. Because you didn't have no support, no one that would help you to take care of your child. I will never forget this. It's nothing for me ... that I was erased, I can take this ... but my children were suffering too.

The 1993 Slovenian Law on Local Self-Government afforded Roma people specific judicial protection; in Maribor this translated to at least one place reserved on the City Council for someone of Roma descent but it was soon clear that this representation did not necessarily help Roma families. To offset this, in 1996 the Roma Association of Maribor was created to help deal with specific issues of housing and children's health and illiteracy. Arbër, the vice-president of the Roma Association, told us: "We wanted to connect with local and national institutions, to arrange our living conditions and to create a better life for today and tomorrow." The Roma Association of Maribor is connected with the larger Council of Roma, which serves the Republic of Slovenia (*Svet Romske Skupnosti*), which in turn is directly connected to state government and the Roma Union of Slovenia (*Zveza Romov Slovenije*). Arbër has represented the needs of Roma for over three decades and part of his current job is to work on behalf of erased Roma. Illiteracy exacerbates contexts of erasure, he tells us, because many erased Roma cannot follow official documentation that offers a path to citizenship. Primary education in Slovenia is free, but there are issues such as lack of clothes and financial resources, illiteracy and weak knowledge of the Slovenian language, which prevent Roma children from attending school. Between 40 and 70 per cent of Roma children in the area attend elementary school, with the higher figure representing urban neighbourhoods. The Roma Association is responsible for opening access to municipal and state rights, but Arbër is at pains to note how difficult it is to maintain the rights of children to education:

> There were problems with the people who were erased. They didn't even know that they were erased. Because of the erasure they had problems also with children. What kind of problems? If you were erased, you had no right to get status. If you hadn't got a status, you couldn't sign your child to the kindergarten, which is the most important thing. Even if Roma could go to primary school later on, they had big problems with language. These troubles were serious and many Roma could not finish the primary school ... They signed them to ... school for disabled. But our children shouldn't go to this sort of school.
>
> *(Arbër, personal interview, Maribor, 24 March 2014)*

In 2007, Slovenia was the first European country to pass a law establishing Roma Community Councils in areas where there are a significant number of Roma people. The Roma Community Council is in charge of Roma rights, but it is not at all clear the extent of their mandate and their leverage with local authorities: "Even today it's not clear to me who has responsible for such acting, the state or municipality", notes Isni, and then adds with clear (at least to us) Agambenian emphasis: "Why they were treating us like slaves? Even slaves had some rights."

Slovenian Roma are part of the current European refugee crisis to the degree that lack of education sends them abroad in search of work. With Slovenia, Austria and Switzerland looking at closing their borders, and the Schengen Treaty (enacted to open borders and facilitate free movement for all Europeans) in question, many Eastern European Roma fear their loss of mobility to the west in search of work. With the so-called European refugee crisis continuing unabated, there are further concerns that Roma peoples get lumped in amongst Middle Eastern peoples as undesirable. Arbër makes the point that erasure exacerbates this issue:

> Recently a lot of Roma is migrating outside of Slovenia. If I ask myself why: some of them lost their residential rights [with erasure] and they were evicted. For some evictions Roma are guilty because they didn't pay in time or they didn't use a chance to pay by instalments, but many of them just didn't have enough money to pay the rent. Many of them was without work and the money they received from the social service wasn't enough to pay the debts. That's why many Roma left Slovenia [for] different parts of Europe...
>
> *(Arbër, personal interview, Maribor, 24 March 2014)*

Arbër suggests a need for more secure employment in Maribor, and is buoyed by a Roma-themed restaurant that opened in 2014, which he says not only creates a source of employment for Roma youth but also detracts from negative Roma imaginary. The Maribor Roma-owned, -themed, and -run restaurant is the first of its kind in Europe but its establishment was controversial. The choice of the building to house the restaurant was blocked by the city council for what Mayor Fistravec called "obvious xenophobic" motives (Cain 2013). Although supporting the venture, Fistravec's description of the motives as xenophobic speaks volumes about how life-long Roma residents in Maribor are perceived.

> Roma have rich culture and specific cuisine. Our food is quality because [of] our ancestry ... More important is that Roma works there and people are interested how Roma will work, what kind of food they will serve ... I introduced this point in front

of the state: If there were more chances for employment [there would be] less discrimination ... Through such projects many Roma people could get a job. For example, now we open restaurant and eighteen people will be employed there, eighteen! If the state would support organizations we could evolve more projects. In this projects many Roma could be employed in civil engineering, dressmaking ... The new generations that have finished [primary] school and also higher grades are without work and would like to work. My generation can't work anymore but this new generation likes to work.

(Arbër, personal interview, Maribor, 24 March 2014)

The Roma-themed restaurant might be considered an example of a state of exception turning into an exceptional opportunity for ethnic constructivism and Roma boosterism, but it may equally be considered a capitulation to a neoliberal economic model that requires marginal peoples to struggle to create their own opportunities at a time when the state has little regard for their welfare. Engin Isin (2008: 16) argues that "processes of 'globalisation', 'neoliberalisation' and 'post-modernisation' ... produce new, if not paradoxical, subjects of law and action, new subjectivities and identities, sites of struggle and new scales of identification". This raises questions about how much these new subjectivities are akin to Agamben's *homo sacer*.

Criticism is levelled at Agamben's unwillingness to distinguish between different forms of camps – or forms of exception not contained within camps – that could perhaps dislodge the notion of a clear connection between death camps and the plight of contemporary excluded people and their dystopian spaces (Owens 2009: 575). Ong (2006) argues that a loosening of Agamben's teleology enables consideration of the foundational relationship of the nation-state to other forms of exclusion. She notes that the extraordinary malleability of neoliberalism as a form of governance enables the re-engineering of political spaces and populations through exceptions beyond camps. It is debatable whether this process is entirely new, but global market forces and the "neoliberal logic" of "emerging states", she points out, "reconfigures the territory of citizenship" through "new economic possibilities, spaces, and political constellations for governing the (national) population" (Ong 2006: 75). Ong argues that neoliberal economies help govern what Mooney (1998) refers to as disorderly people in disorderly places. Isni's example suggests the draconian ways that Roma youth in Maribor were disadvantaged in the 1990s and early 2000s in terms of education, health, and housing. The dystopian spaces of Maribor are part of Roma youth's circulation through low-income housing and the city's wastelands. The question of new subjectivities is raised in the context of Arbër's proclamation that a new generation of Roma are willing to work in ways that their forebears were not. Is this a sign of hope or an example of Ong's neoliberal governance? In the next section, we consider this conundrum from the perspective of Romanian Roma, and a very specific dystopian space.

Working and living at the city dump: Roma youth in Pata Rat

For every sign of satisfaction amongst Roma people about their living conditions, there are an equal number of examples of dystopian concern. We now turn to a seemingly clear example of a spatialized dystopia in Romania. Pata Rat is an almost exclusively Roma urban settlement (Berescu 2006) in the outskirts of Cluj-Napoca, Romania's fourth-largest city. Pata Rat is considered a ghetto in public and professional imaginaries with the disorderliness of its Roma residents contributing to the community's deprivations and precariousness (Dohotaru 2013; Rat 2013). Like in Maribor, municipal interest in infrastructure and maintenance is ambivalent in many urban areas but it is particularly noticeable in spaces occupied by marginalized peoples. Pata Rat's infamy derives from its location near and on a recently closed landfill. Three hundred families and 1,500 individuals form Pata Rat's communities of Dallas, Cantonului, Coastei and families living directly on the dump (Pata-Cluj 2015a; Vincze 2013). Nearly half of the families (42 per cent) moved to Pata Rat after several rounds of evictions and relocations (Rat 2012, 2013). Roma families living in the forests outside Cluj were relocated to Pata Rat in 2003 (Rat 2013). A more prominent wave of migration occurred in December 2010 when families living on Coastei Street were evicted and forcefully moved to the periphery. Pata Rat is also a voluntary relocation site for individuals who can no longer afford high city rents and, while the landfill was still operational, for trash pickers. Dallas, one of the four communities in Pata Rat, formed in the 1990s when families moved to collect and sell recyclable waste (Tonk et al. 2014). The landfill's closure in July 2015 sent over 700 individuals into an economic crisis (Pata Cluj 2015a, 2015b). Some families left Pata Rat to camp in Cluj's forested outskirts and commute to the city in search of garbage.

Living conditions in Pata Rat are generally improvised, inconvenient and unsanitary. Half of the housing facilities are barracks assembled from plastic and timber (Tonk et al. 2014). Of the households, 43 per cent do not have toilets; only 24 per cent have electricity; 65 per cent use a public tap as a water source and 14 per cent get water from a spring near the landfill (Tonk et al. 2014). Soil samples from the landfill show a concentration of lead double national averages (Mionel 2012). There is a high frequency of acute and chronic illness: over a quarter of working-age residents suffer from ill health or a disability (Tonk et al. 2014). Doctors visiting the community report higher rates of dysentery, hepatitis and seasonal respiratory viruses among children (Moise 2011). Proximity to the landfill and unsanitary living conditions contribute to these health problems.

> [Roma] are poor. They don't even want to go to school anymore because they are made fun of. They're told all the time that they smell. But in what conditions could they bathe, if their parents live in Pata and work on the dump?
> *(Miki, interviewed in Dohotaru, 2015)*

In Pata Rat, children rummage through the waste to help their families collect recyclable material. For the popular television programme *Romania, I love you!*, a child demonstrates how she carefully picks through garbage to earn twenty lei a day: "I collect cans, metal, glass bottles, these things ... and I sell them and make money!" (Leonte 2009). Half of the children aged between seven and fourteen do not attend school for reasons similar to those in Maribor, including lack of resources and discrimination (Dohotaru 2015; Rat 2012).

As with Maribor, hope is hinged on a new generation of Roma who are better educated and willing to work. With support from the ROMEDIN (Socio-educational Services for Roma Inclusion) and a variety of Pata-Cluj projects, many children from Pata Rat attend schools in the city. In the 2014–2015 academic year, upwards of 270 Roma children enrolled. Following the enrolment of children from Pata Rat, the primary school closest to the community supports an 80 per cent Roma student body (ROMEDIN 2014). Unfortunately, this incipient segregation is further aggravated by a case of 'white flight', with a significant number of non-Roma parents removing their children from the school.

Also like Maribor, notions of movement and mobility are significant. Many arrived at Pata Rat following forced eviction and relocation. That said, young people are generally optimistic about movement away from waste collection in Pata Rat towards economic and social security:

I like school, I want to learn how to read and write. When I grow up I want to become an auto-mechanic, to repair cars, I don't want to work on the landfill.
(First-grader Flaviu, interviewed in Magradean 2015)

Routine movement from the periphery to the centre privileges students with mobility. They travel daily between spaces of exception and exclusion (home on the landfill) to spaces of citizenship and inclusion (schools in the city). These contradictions and tensions suggest at best an ambivalent relation between Roma communities and urban policy-making. Our final set of examples put forward a cautiously hopeful outcome from Roma youth agency and activism.

Recognizing their marginalized spatial and social status, Roma formerly living on Coastei Street formed the Community Association of Roma from Coastei hoping to redress violations of their human rights. In 2010, in the freezing temperatures of mid-December, residents from Coastei Street were evicted from the centre of Cluj and involuntarily relocated to Pata Rat (Rat 2013). The Roma 'colony' of legal and improvised shelters had to be removed because it was a hundred metres from the offices of a major phone company looking to expand its operations in the city (Veioza Arte 2011). More than seventy families were forced to leave their homes (Creta et al. 2013). Children from Coastei Street lamented the move:

> We had two parks. It was very nice; we would ride our bicycles at night. Here we have nothing, no playing field, no park. Only trash.
> *(Veioza Arte 2011)*

In collaboration with other local organizations and activists, the Community Association of Roma from Coastei organized demonstrations such as "I'm Rom – I want to live with dignity" and "Employment is not a luxury. It is a right!" (Dohotaru 2013). Young community members participated in the activism. Public debates and discussions included a performance of 'gypsy dance' by Pata Rat's children. These protests were actions against dystopian lived spaces with a focus on substandard housing near the landfill, under- and unemployment of working-age youth and young children not enrolled in schools. The protests motivated hope for permanently leaving the landfill and returning to the city. To a large degree the context of the activism focused on the best interests of Roma youth, which actively incorporated young people's ideas and energy.

A best-interests imaginary for Roma youth?

In 2013 and 2015, Maribor and Cluj-Napoca respectively were chosen as the European Youth Capital for that year. Youth Capitals spotlight uniqueness, diversity and youth empowerment programmes, as well as draw the attention of international media and youth-related conference organizers. It is telling that while the Maribor final report on activities in 2013 focused on education, housing and labour, there is no specific mention of issues that might arise for Roma Youth (Maribor City Council 2014). More disturbingly, the follow-up discussion and website created for the event focuses exclusively on entertainment, commerce and consumption (see: www.mb2013.si/, last accessed 31 March 2016). Among the Cluj-Napoca events in 2015, Phiren Amenca, an international network of Roma and non-Roma volunteers and organizations, hosted the 'So Keres Europa?!' (What's Up Europe?!) workshop series (Phiren Amenca 2015). About 400 young Roma and non-Roma from fifteen countries participated in the week-long event to stamp out anti-Roma racism. Phiren Amenca organized the protest "City for the people, not for profit" to commemorate the anniversary of the Coastei evictions. Marchers, mostly young and including Pata Rat residents, demanded that "we are people and want all rights" and "we are Rom and we have dignity" (Phiren Amenca 2015).

What remains to be seen is whether actions by various pro-Roma organizations, and even social business enterprises aiming to create jobs, are indeed inspired by a best-interests imaginary for Roma youth. We recognize and understand the critical role Roma communities play in directing NGO programmes but ask who could and should determine what is in the best interest of Roma youth – the municipal authorities, NGOs who have their own interests or the community itself? The activism we described educates and trains young Roma to make these decisions but does it move far enough from the disciplined neoliberal subjectivities about which Ong (2006) warns?

Conclusion

The specific demands for rights and dignity in the spaces of exception we describe in this chapter highlight how a problematical imaginary lands on young Roma people and their day-to-day contexts. The importance of the right to create and re-create themselves and their spaces is in the best interests of Roma youth and, as a consequence, the focus on spatial rights is not only about occupying spaces that are suitable for access to housing, livelihoods and education, but also the right to stay put as well as rights of movement and mobility in safe and secure ways. The application of new non-dystopian imaginaries by Roma peoples works only in accordance with the group's power and position in society and the roles that are attributed to them (Clark et al. 2005). For new Roma youth imaginaries to work, the ideas of culture, geography and history must be more than a dystopian tool that lands a universal notion of who Roma are in a particular locale, and it must be immune from the machinations of state politics and neoliberal imagineering. The translation of a larger Roma imaginary on particular locations is fraught with problems because the creation of imaginaries takes place not only between local, state and international institutions, but they also play out between different spaces Roma youth occupy. In each of these spaces, and between all these institutions, different forms of translation occur. That said, our ending examples suggest that out of the bare life and spaces of exception occupied by young people it is possible to translate a message of hope.

That young marginalized people get to create and re-create themselves as they create and re-create their city is an important right to the degree that it emboldens a new model of governance and practice, which is progressive and potentially utopian rather than reactive and more-than-likely dystopian. To enact this kind of governance with urban poor and marginalized peoples is to push against the states of exception proffered by neoliberal governance. And to the degree that neoliberal economics is now roundly criticized as unworkable (including by the International Monetary Fund that heretofore denied its existence, Chakrabortty 2016), perhaps it is time for a brave new step that lets loose young, exceptional peoples and their creativities.

References

Agamben, G. (1995). *Homo Sacer: Sovereign Power and Bare Life*. Stanford, CA: Stanford University Press.
Agamben, G. (2001). *Means without End: Notes on Politics*. Minneapolis: University of Minnesota Press.
Aitken, S.C. (2015). Children's Rights: A Geographical Perspective. In: W. Vandenhole, E. Desmet, D. Reynaert and S. Lambrechts, eds. *The International Handbook of Children's Rights: Disciplinary and Critical Approaches*. New York and London: Routledge, pp. 131–146.
Aitken, S.C. (2016). Locked-in-Place: Young People's Immobilities and the Slovenian Erasure. *Annals of the Association of American Geographers*, 106(2), pp. 358–365.

Berescu, C., Celac, M., Ciobanu, O. and Manolache, C. (2006). *Housing and Extreme Poverty: The Case of Roma Communities*. Bucharest: Ion Mincu University Press.

Cain, P. (2013). Roma Restaurant Project Sours in Slovenia. *BBC News*, 10 November. Available at: www.bbc.com/news/world-europe-24874578 [accessed 2 September 2015].

Chakrabortty, A. (2016). You're Witnessing the Death of Neoliberalism – From Within. *Guardian Weekly*, 31 May–7 June, Economics, p. 48.

Clark, R., Reilly, M. and Wheeler, J. (2005). Living Rights: Reflections from Women's Movements about Gender and Rights in Practice. *IDS Bulletin*, 36, pp. 76–81.

Creta, E., Greta, E. and Stancu, F. (2013). *Memoriu. Propunere de Parteneriat Catre Primaria Cluj-Napoca* [Memorandum]. Cluj-Napoca: Asociatia Comunitara a Romilor din Coastei.

Dohotaru, A. (2013). Antropologie Performativă. Cazul Ghetoului Pata Rât (Partea I). *Gazeta de Artă Politică*, 10 June. Available at: http://artapolitica.ro/?p=776 [accessed 19 February 2016].

Dohotaru, A. (2015). Celălalt. 3 Zile și 3 Nopți în Pata Cluj. *Social East*, 12 February. Available at: www.socialeast.ro/celalalt-3-zile-si-3-nopti-in-pata-cluj/ [accessed 19 February 2016].

Isin, E.F. (2008). Theorizing Acts of Citizenship. In: E.F. Isin and G.M. Nielsen, eds. *Acts of Citizenship*. London and New York: Zed Books, pp. 1–18.

Leonte, C. (2009). Drama Locuitorilor de la Pata Rat. *Romania te Iubesc*. Stirile ProTV. Available at: http://romaniateiubesc.stirileprotv.ro/stiri/romania-te-iubesc/romania-te-iubesc-drama-locuitorilor-de-la-pata-rat.html [accessed 19 February 2016].

Magradean, V. (2015). Început de an Scolar și Pentru Elevii Romi de la Pata Rât, Șansa Lor de a Scăpa de pe Rampă. *Mediafax*, 14 September. Available at: www.mediafax.ro/social/reportaj-inceput-de-an-scolar-si-pentru-elevii-romi-de-la-pata-rat-sansa-lor-de-a-scapa-de-pe-rampa-foto-14712010 [accessed 19 February 2016].

Maribor City Council. (2014). *Final Report: European YouthC, Maribor 2013*. Available at: www.europeanyouthcapital.org/wp-content/uploads/2013/05/Final-report-Maribor.pdf [accessed 18 February 2016].

Mionel, V. (2012). Pata Care Devine din ce în ce Mai Vizibilă. Cazul Segregării Entice din Ghetoul Pata Rât, Cluj-Napoca. *Sfera Politicii*, (168), pp. 53–62.

Mitchell, K. (forthcoming). Changing the Subject: Education and the Constitution of Youth in the Neoliberal Era. In: T. Skelton and S.C. Aitken, eds. *Establishing Geographies of Children and Young People*. Springer Major Reference Work: Springer Publishing Company.

Moise, C. (2011). Pe-Un Picior de Iad. *In Premiera cu Carmen Avram*. Antena 3, 20 November. Available at: www.youtube.com/watch?v=h_YFqoKlWdM [accessed 19 February 2016].

Mooney, G. (1998). Urban 'Disorders'. In: S. Pile, G. Mooney and C. Brook, eds. *Unruly Cities?: Order/Disorder*. New York and London: Routledge, pp. 50–83.

Ong, Aihwa (2006). *Neoliberalism as Exception: Mutations in Citizenship and Sovereignty*. Durham, NC: Duke University Press.

Owens, P. (2009). Reclaiming 'Bare Life?': Against Agamben on Refugees. *International Relations*, 23, pp. 567–582.

Pata-Cluj. (2015a). *What Is Pata-Cluj*. 4 December. Available at: www.patacluj.ro/what-is-pata-cluj/ [accessed 19 February 2016].

Pata-Cluj. (2015b). *Buletin Informativ*, June. Available at: www.patacluj.ro/wp-content/uploads/2015/08/PataCluj-news-6-iulie2015_RO_f.pdf [accessed 19 February 2016].

Phiren Amenca. (2015). Mission and Principles. *Network*, 10 December. Available at: http://phirenamenca.eu/network/mission-principles/ [accessed 19 February 2016].

Rat, C. (2012). *Participatory Assessment of the Social Situation of the Pata-Rat and Cantonului Area, Cluj-Napoca*. UNDP.

Rat, C. (2013). Bare Peripheries: State Retrenchment and Population Profiling in Segregated Roma Settlements from Romania. *Studia Universitatis Babes-Bolyai-Sociologia*, (2), pp. 155–174.

ROMEDIN. (2014). *Buletin Informativ Nr. 3*, 3 December. Available at: www.desire-ro.eu/wp-content/uploads/Bul_info_3_dec2014_ROMEDIN.pdf [accessed 19 February 2016].

Shubin, S. and Swanson, K. (2010). "I'm an Imaginary Figure": Unravelling the Mobility and Marginalization of Scottish Gypsy Travellers. *Geoforum*, 41, pp. 919–928.

Tonk, G., Adorjani, J. and Lacatus, O.B. (2014). *Coordinated Interventions for Combating Marginalization and for Inclusive Development Targeting Inclusively but Not Exclusively the Vulnerable Roma Through Desegregation and Resettlement of the Pata Rat Area Using the Leverage of EstF*. Draft Outline to the De-segregation/Resettlement Action Plan for Pata Rat 2014–2023. UNDP.

Veioza Arte. (2011). *Prea-fericitii din Groapa de Gunoi.* Available at: https://vimeo.com/32605959 [accessed 19 February 2016].

Vincze, E. (2013). Socio-Spatial Marginality of Roma as Form of Intersectional Injustice. *Studia Universitatis Babes-Bolyai-Sociologia,* (2), pp. 217–242.

World Bank. (2014). *Achieving Roma Inclusion in Romania – What Does it Take?* World Bank. Available at: www.worldbank.org/content/dam/Worldbank/document/eca/romania/Summary%20Report%20RomanianAchievingRoma%20Inclusion%20EN.pdf [accessed 19 February 2016].

47
MOBILE FUTURES
Urban revitalization and the aesthetics of transportation

Theresa Enright

Introduction

In the spring of 2015 a fresh controversy came to dominate discussions among the beleaguered commuters of Toronto's public transit system. This time, the problem was not congested vehicles, service breakdowns, ageing infrastructure, cuts to vital routes, transit deserts, funding shortfalls, rising fares, poor customer service, construction delays, stalled masterplans or undemocratic governance, but instead the content of a new art installation on the subway platform at Union Station, the city's most historic and largest commuter hub. The piece in question – *Zones of Immersion* by artist Stuart Reid – defied the public expectation that art should be a whimsical and soothing distraction from the transit system's problems. Instead, it depicted the subway as an alienating place and forced riders to confront the often bleak reality of urban travel. A *Toronto Star* article suggested that the installation was "depressing", "creepy" and "gloomy", and a reminder of the "miserable" experience of daily commuting (Spurr 2015); while one commentator on a local Reddit forum complained that the new art resembled "scribblings from a post-apocalyptic wasteland" (bipolar_sky_fairy 2015). These accusations were all the more notable given that the artwork was poised to be the centrepiece of campaigns to improve the brand association and journey experience of the Toronto Transit Commission (TTC), to modernize the rapid transit network, to revitalize Union Station and to develop the surrounding urban neighbourhood. Indeed, in stark contrast to the dystopian interpretation of many commuters, the TTC repeatedly celebrated the installation as a vital element in their utopian visions of urban mobility.

Zones of Immersion was commissioned as part of the TTC's corporate revisioning and as an element in an ambitious renovation of Union Station. The sleek artwork complements the clean white tile and streamlined signage of a revamped subway platform and enhances the visual environment, combatting what a local architectural critic has described as the drab "public toilet aesthetic" of the existing network (Hume 2015b). It is part of a broad effort to recast Union Station as an engaged cultural space and inviting destination point, improving the attractiveness of the busy central city interchange to commercial and tourist activity. It is thus a key element in beautifying the deteriorating building, positioning the station at the heart of the twenty-first century city, and transforming the experience of urban commuting along the supposedly frictionless networks of global transport.

Zones of Immersion supports the imaginary of the elite mobile city at the same time it paradoxically shatters it. Consisting of 170 metres of floor-to-ceiling panels of silver-stained, enamelled, engraved and laminated glass that are embedded along the central divide between the subway's two platforms, the installation offers a candid portrait of the more prosaic aspects of urban travel. The panels display approximately sixty enlarged images based on real-time observational sketches of people riding the subway, as well as poems inspired by commuting. They represent workers, students, families, tourists, and even the homeless in quotidian patterns of mobility. Bearing familiar countenances of the underground, these figures have despondent and melancholic expressions, testaments to the tedium and alienation of everyday life. At the same time, their bodily gestures – arms crossed, heads bowed, shoulders hunched – display the weary bodily comportments of urban existence under conditions of neoliberalism. Every few metres magnified anonymous faces with sunken eyes gaze daringly out at their real world counterparts. Indeed, the observer cannot help but be drawn in by the image and its dynamic composition. Forming a long translucent screen, the panels force commuters to see themselves "figuratively and literally mirrored in this subjective rendering of the very experience they are objectively part of at the moment of their viewing" (Reid, quoted in Canadian Architect 2015). In this sense, the installation captures the elegant rhythms and aleatory encounters of urban transit, while reflecting the transient and spectral crowd back onto itself.

According to Reid (2015), the panels enact "the push and pull of darkness and light" that define the subway experience. Whereas glass is a common design feature in metro stations to illuminate dim underground spaces, it here refracts a split image of urban mobility. The installation is characterized fundamentally by ambiguity. While its form and function are tied to the pursuit of Toronto as a world-class creative hub and to burnishing a brilliant future of pleasant and unfettered mobility, the content, in contrast, depicts the grim reality of moving and living in the city. The controversy over the installation and the divergent urban imaginaries of which it is a part signal a highly political aesthetics of metromobility.[1] The installation is thus a useful place from which to consider the networked utopias and dystopias of the contemporary global city and the role of cultural production in realizing and contesting these worlds. In this chapter, I use the installation to problematize the development of Union Station, and to consider in turn the contested aesthetics of the twenty-first century mobile city in circulation globally.

In what follows, I first consider the changing spatial contours of urban utopias in order to argue that the ideal city today is expressed not as a bounded territorial unit, but as a complex relational matrix that finds its most perfect expression in infrastructural networks. I then demonstrate how the desired networked utopia (as a 'good-place' and 'non-place') is pursued through transit art and the creation of a global city aesthetic. In the final sections of the chapter, I situate this paradigm within broader urban development dynamics of the late capitalist city and, using the example of Toronto, consider the political stakes of imagined mobile urban futures.

Imagining networked utopias

Urban utopias consider the possibilities of ideal forms of collective life and the organization and proper operation of the well-ordered city. While mass transit only rarely features directly in these urban imaginaries (Timms *et al.* 2014), since the emergence of the first metro systems in the late nineteenth century the art and design of the underground transit network has been deeply connected to representations and transformations of the city at large (Bownes *et al.* 2012; Ovenden 2009, 2013). The architecture, interior styles, typeface, signage, art, advertising, station layout and system maps shape how people move through and understand the city, contribute to urban culture and civic values, and influence the material constitution of the built environment. The famous palatial architecture of the Moscow metro, for example, was part of Stalin's "immense iconography of power" and was essential to cementing the modern city as well as selling the Soviet dream (Buck-Morss 1995: 22). Representations on and of mobility systems reflect and produce urban worlds and, considered in context, station design and artworks thus offer important insights on how the ideal city is imagined as well as the priorities orienting city growth and development. This chapter focuses on how metro art – defined after Ström (1994) as design styles as well as temporary and permanent installations and cultural performances – creates an aesthetic of mobility appropriate to the networked postmodern metropolis.

Borrowing from Maria Kaika's (2011: 968, original emphasis) work on architecture, it argues that metro art can be understood as utopian insofar as it is involved in "the narrativisation of the desires of elites during a given era, and as a key component in instituting a society's *radical imaginary* during moments of change". Just as architectural icons exemplify the aspirations of societies, establish collective identities and purposes, and are "embodiments of myths and wish images for the future" (Castoriadis 1987; Kaika 2011: 970), so too can cultural products such as art and design create a symbolic economy of the desirable future urban order. While there is an increasing conflation between art, architecture and design in metro construction today, taken together, this multiplicitous aesthetic realm is an important site of articulation for cities that are regionally and globally connected, entrepreneurial, revanchist, creative and liveable.

In the transition of cities away from the automobile and towards more multimodal mobilities, art and design are positioned as key instruments in representing and cementing a desired transit future. Private cars and automobile landscapes are no longer poised to lead urbanism and save the city, but are rather being replaced by more collective and sustainable mobility modes and by the purported seamless networks of subways, ports, logistical centres, airports and high speed trains (ARUP 2015; Siemens 2011). For city boosters, transport planners and global architects, these global networks – of which urban rail is an important part – are poised to redeem and redefine urban society in the twenty-first century. To realize this imaginary of the good city, fantastical post-industrial hubs of "apparent hypermobility" (Urry 2007: 149) are cast in monumental stations of glass and steel. These are then connected through the premium, smart, deterritorialized, flexible and smooth assemblages of global transport, information and communications infrastructure (see also Castells 1996). Because of its promise of speed and its ability to transform land use and value, the metro in particular is poised to lead this mobility transition and rebranding exercise. If metromobility promises a wholesale reorientation of urban everyday life, new environmental practices, social relations, spatial forms and patterns of economic growth, then metro art provides a means to imagine these dynamic links between transit, urbanity, progress and capital. Comprising a non-topographical set of relations, connections and flows, the ideals of metromobility also imply a new spin on the traditional interpretation of utopia as being both the 'good-place' (*eu-topia*) and 'no-place' (*ou-topia*).

Transit networks represent cities in their optimal and degenerate forms and the mobilities they offer determine whether one is "entering the gates of heaven or hell" (Urry 2007: 187). Transit infrastructures and their artistic elements express utopian hopes for a better urban world, yet they also express their dystopian correlates. They reveal the structural ambiguities of the present and, through the counter-image of the desired urban future, make us aware of the limits of existing situations and trajectories (Jameson 2005). Whether the delirious and homogenized "non-places" of supermodernity (Augé 2008); the "splintered" networks of premium and subordinate spaces (Graham and Marvin 2001); or the messy conflict-ridden spaces of public interaction that characterize mobile worlds (Sennett 1992; Urry 2007); transit systems signal the problems and challenges of speed, acceleration and circulation in global "spaces of flow" (Castells

1996). They are also crucial to the subjective bifurcation of populations into the "kinetic elite" and the "kinetic underclass" (Cresswell 2006; Urry 2007). Contemporary mobility systems articulate the promises of freedom, progress and prosperity, but they are also central bearers of inequality and domination (Bauman 1998; Cresswell 2006, 2010; Massey 2013).

The aesthetics of metromobility express this fundamental duality of claims to modern urbanity. On the one hand, transit art, design and architecture participate in selling the dream of smooth, global circulation. In this sense they ensure that the built form of urban networks promotes the flourishing of accumulation (Tafuri 1976). On the other hand, insofar as transit representations can reveal the antinomies of urban life and the radical distance between the experience and the ideal of urban movement (Jameson 2005; Timms et al. 2014), they have a vital critical function that reveals the possibility of a radically different (albeit not always progressive) future.

Metro art is a particularly salient site from which to view these utopian dynamics as it is at the forefront of post-industrial urban transformation. The turn to arts and cultural programmes in transportation infrastructure coincides in particular with a dramatic increase in modernization, expansion and creative efforts whereby urban rail systems are revived and redeveloped (Ström 1994). Transport agencies integrate art in a variety of ways into urban rail cars, stations and multimodal terminals through permanent and temporary, object, event or process oriented public art projects. Poetry in motion competitions, train designs, sculptures, historical dioramas, tile mosaics, interactive performances, musical concerts, graphic arts, installations, photography exhibitions and iconic architecture are the most common of the various experiments. In virtually any large metropolitan city today art is a crucial element of mass transit and urban development and is essential to the development of new urban imaginaries.

Cities around the world vary considerably in terms of the styles and designs of metro art and their use in articulating transit utopias. Situated conflicts over the aesthetics of metromobility undoubtedly reflect the particular stakeholder dynamics, cultural politics and built environment of a given city. Yet there are also common patterns and trends that shape the global context in which these battles take place. While there is great variation in the content and function of transit art worldwide, increasingly transit art has become implicated in economic development priorities. Most policy justifications come down to increasing ridership through transforming the meaning of the transit network and improving the brand identity of the transit agency and the city at large through associated changes in the morphologies of city space. By creating a distinctive locality, projecting an aura of quality, improving liveability and demonstrating global power and prestige, metro art facilitates attracting investment, increasing land values and making cities more competitive.

Mobilizing aesthetics of the global city

Insofar as the aesthetics of mobility constructs a future image of the city in order to capitalize upon it through anchoring global capital, the revival of metro art is inseparable from the neoliberalization of urban governance, policy-making and planning. For Ström (1994: 7), the profound reinvestments in art and the "aesthetic potential" of metros in the 1980s and 1990s can be explained as attempts by municipal governments to address the problems associated with the collapse of urban welfare and the defunding of public services. The global urban crisis precipitated by structural adjustment turned many subways into "danger zones", beset by mechanical and service failures, overrun by gangs, the homeless and the unemployed, and associated with both the stigmatized, hollowed-out inner-city ghettoes and the degenerate suburbs. The New York City metro was notoriously depicted throughout the late 1970s and early 1980s as such a gritty urban wasteland. Armed with a technocratic view that public art could alleviate social problems, officially sanctioned aesthetic projects were a way to combat rampant graffiti while sanitizing the image of a crumbling public asset. These relatively cost-effective steps could thereby stand in for more systematic upgradings of vital infrastructure.

The rhetoric of safety and cleanliness which dominated arts campaigns in the 1980s and early 1990s evolved in the decades that followed towards a glorification of the cultural value of transit systems themselves (Ström 1994). More recent works seek to position the upgraded metro as a more active vehicle of urban renewal and as a key symbol of the attractive and liveable city. Two main features, both tied to these development mandates, seem to dominate rationalities of transit art today. First, metro art is employed in landmark place-making and marketing to distinguish stations and to mark the unique character of the surrounding urban area. Second, metro art generates a metropolitan aesthetic, whereby gleaming new infrastructures cement the exclusive urban forms expected of aspiring global cities and mark the arrival of a metropolis to creative, prosperous, world-class status. Both of these functions are features of marketized patterns of optimization and growth. However, insofar as it is involved in rendering these utopian scenarios, metro art can have a third critical role: to point out the exclusions and unevenness constitutive of the hypermobile ideal and to question the viability and desirability of the neoliberal "dreamworlds" (Davis and Monk 2008) of which it is a part. Sometimes, as in the Toronto case, these three features overlap in competing and contradictory patterns.

Historically, one of the most common aims for transit art is to "landmark" significant places, buildings or historical sites (Ström 1994). As underground stations are often mere generic stops in disorienting subterranean labyrinths, art has long served a differentiating function, punctuating the homogenous and repetitive circuit. In the contemporary conjuncture, this landmark function of design is also inseparable from a broader programme of urban regeneration whereby the installation and performance of public art in transit stations is an active "place-making" activity (Kearns and Philo 1993; Kwon 2002). Not only does metro art today commemorate historic events or improve the legibility of the underground, but it also reframes the experience and meaning of transit stations and vehicles so as to reshape the built environment above ground. Metro arts are never autonomous, but are part of the material relations of urban identity, property and production. Indeed, like public art more generally, since the 1980s metro art has been justified not solely in purely aesthetic terms but as a mode of urban regeneration (Deutsche 1996; Hall and Robertson 2001; Miles 1997; Phillips 1988; Sharp et al. 2005). In resymbolizing transit spaces as productive assets, art can thus "help secure consent to redevelopment and to the restructuring that make up the historical form of late capitalist urbanism" (Deutsche 1996: 10). The use of iconic works by star architects ('starchitecture') (e.g. Foster + Partners' design of the Bilbao metro or the Jeddah transit system) and site-specific installations by blue-chip artists (e.g. Turner Prize-winning artists Douglas Gordon and Richard Wright whose work will be featured at the flagship station of the London Crossrail, Tottenham Court Road) effectively make stations more than places to pass through. This competitive landmarking not only contributes charismatic atmosphere to a place, but plays a promotional role in marketing, attracting investments, contributing to tourism, catalysing surrounding developments, adding to land values and thus closing rent gaps. A shift in the aesthetic qualities of transit can deeply change the urban fabric, especially by altering development of zones surrounding stations, revitalizing hollow city centres and building polycentric suburbs of the integrated urban-region.

Beyond revalorizing individual landmarks, however, the realty and rebranding emphasis of art is achieved at the urban regional scale through recasting the entire metro network as a metonym for the good (i.e. utopian) city. More than a means to move commuters from A to B, transit systems are key regulators of the contemporary global city and are vital to urban image making. As architect Peter Blake has stated, "a metropolis without a metro is like a church without a steeple" (quoted in Ström 1994: 46). By signifying metro networks as both necessary and appealing spaces art projects a particular image of the future that is sold to users, at the same time that it actively transforms the political economy of the city through bringing it into line with global city norms. Improving the attractiveness of the metro and, by extension, the city, is frequently linked in this respect to major planning initiatives and mega events.[2]

The use of metro art and design to improve the image of the city also dovetails with the redefinition of once-industrial cities as cultural hubs and the reorientation of urban policy towards alluring the coveted creative class (Florida 2004). Indeed, upscaling of the metro through public art is frequently pursued in conjunction with the promotion of other cultural industries, festivals and institutions. The new metro line in Dusseldorf, for example, has a network-wide design and arts scheme that is promoted by the City's Cultural department as way to advertise the Kunstakademie Düsseldorf (Arts Academy) and as means to position Dusseldorf globally as a high-ranking node of the avant-garde (Dunmall 2016). Also exemplary in this respect is London's Crossrail which has an official "artist in residence" and a designated "Culture Line" featuring "an integrated exhibition of site-specific public artworks by internationally renowned artists, which are sympathetic to the Crossrail railway and the location of the relevant station[s]" (Crossrail 2016). Advised by Futurecity, a cultural and place-making agency, the Crossrail Art board encourages corporate funders to invest in the network as "a once in a generation opportunity to associate their brand with an iconic piece of London infrastructure and some of the world's most famed galleries and artists" (McDougall 2014). Here the formation of a world-class railway requires that its buildings have an aesthetic quality reflecting the ambition and prestige of the entire urban project.

Importantly, it is not merely any metro, but only a beautiful and luxury metro associated with comfort, safety, cleanliness, speed, advanced technology and high-end culture that defines world-class urbanity. In order to look the part of a world-class city, a distinctive and exuberant 'metro style' now dominates over functional aesthetics. Indeed, from Dubai to Paris to Beijing, metro stations and networks around the world feature remarkably similar schemas of exposed concrete structures, glass skylights and ornamentation, stainless steel passageways and furniture, bright white wall and floor coverings, cavernous interiors, intelligent wayfinding and, of course, glitzy art. These airy and bright aesthetic norms work against the crowding, darkness and disorientation historically associated with subway systems as well as the bland and institutional stylings of the post-war era. In this sense, the transformations in metro art and design are an example of how aesthetic norms and visual appearance, more than social policy, are key to the production of the imagined global city and world class rule (Ghertner 2015).

This is not to say that other non-commodified futures of the mobile city are not imagined. Indeed, transit is not merely a site of local economic competitiveness, but also of social reproduction. As such, transit art and design is frequently used as a forum for community engagement and participation (Bertsche 2013; Breitbart and Worden 1994), as a way to highlight sustainability and the socio-nature of urban assemblages (Valley Metro 2008), as a means to integrate populations and democratize culture (Cox 2015), and as a vector for the values of solidarity and collectivism (Kingsbury forthcoming). However, even when art has the potential to produce counter-hegemonic images and discourses of the urban future, it is questionable to what extent these experiments succeed because of ongoing conflicts between social sustainability and economic development. Transit designs are not entirely appropriated or captured in the dominant neoliberal idea, but transit schemes worldwide are embedded within powerful spatial processes of capitalist urbanization that continuously work to undermine alternatives. Moreover, because of the corporate patronage and bureaucratic planning apparatuses that produce metro art, it, like public art more generally, is structurally constrained to produce primarily "minimum risk art" that does not offend, nor enlarge discourse or debate (Phillips 1988) and merely acts to boost culture-led place promotion (McCarthy 2006).

Even within the utopian paradigm of a competitively conceived mobile and networked global city, however, there are subversions and countervailing tendencies. Notably, insofar as these utopias promise an ideal future they necessarily counterpose themselves to the present. In revealing the impassable abyss between the myth of elite, frictionless and accelerated mobility, and the often disagreeable and sclerotic realities of moving, utopian representations contain a critical kernel. Critical distance enables one to view the processes of urban development and decline as being dialectically linked and to see, for example, the complex interdependencies that bring together the cosmopolitan and the homeless in the same space of a train. The dystopian other emerges most remarkably when aesthetics can expose non-conventional perspectives, "estrange the taken for granted" and "interrupt space and time" (Pinder 2002: 229). In this sense utopias can challenge the status quo and bring to light conflicts and social divisions they were meant to conceal. They can thus prompt us to reconsider the role mobility plays in the differentiation of society, and the pervasive alienation of everyday life. The abyss between experience and image also raises questions about the desirability of increasing speed and acceleration, of who has the ability to move seamlessly and who is denied this freedom, of free versus coerced mobilities, and of the kind of public created through a politics of propinquity. In this critical aspect, metro art can point out a radical difference from the present as well as the antinomies of transit-oriented futures (Jameson 2005). This is especially true when the ideal future depictions are hollow branding exercises that substitute an image for the material relations necessary to achieve them.

Art's role in illuminating conflicts over urban restructuring is also intensified with what Ström (1994: 131) classifies as imagery of "transportation within the transportation system", in which the content of the artistic representations features the very mobility systems of which they are a part. These self-reflective pieces not only demand an interrogation of the urban forms in which they are embedded, but also the content of their cultural representations. 'Transit art about transit' can estrange the familiar experiences of commuting, and in so doing, disrupt the status quo assumptions about movement through the city. Representations of mobility within mobility systems can be very poignant social commentaries offering explicit and immanent criticisms of the possibilities and limits of urban relations and revealing the multiple tensions of utopian imaginaries.

Toronto's transit vision

In Toronto it is possible to see these multivalent dynamics at work in examining the role of art within utopian discourses of transport planning. Transit is a highly contested point of convergence for many aspects of Toronto's neoliberal development. The challenges are numerous but there is a widespread frustration at the chronic lack of adequate and secure funding (following the downloading of transit responsibility from the province), the poor quality of transit operations and persistent stalling on the construction of new projects. These have resulted in expensive and rising ridership costs; a high fare box ratio that puts the financial burden of transit operation on users; an outdated network rife with 'transit deserts', particularly in the inner suburbs; inaccessible stations and vehicles for those with reduced mobility; chronic congestion; and frequent service delays (Boudreau et al. 2009; Hertel et al. 2015; TTCriders 2016). For many across the city riding the TTC is an expensive, uncomfortable and time-consuming burden of necessity.

In this context, it is somewhat surprising that the TTC is investing heavily in art and design, allocating 0.5 per cent of the budget for each new station for artwork and special finishes. Today art is integral to the network brand and imageability even while large-scale systemic transformations remain elusive. Indeed, transforming the aesthetics of metromobility in order to improve the 'customer service' of the network and sell the asset to investors, riders and tourists is a priority. Not only is art and design a touchstone of the most recent subway extensions (e.g. the Sheppard line completed in 2002 and the Spadina extension currently under construction), but the TTC has involved itself in multiple ways with the city's creative industries and with policies oriented to reinvent Toronto as a "global cultural capital" (Boudreau et al. 2009; City of Toronto 2003). Since 2007 the Art in Transit programme sponsored by Pattison media has brought cultural programming to stations and vehicles across the transit network. In 2014, Pattison Outdoor and the TTC Art for Commuters programme cooperated to install digital displays on subway platforms across the city in what they described as a "uniquely successful mingling of business and the arts" (Art in Transit 2016).[3] These screens serve as a venue for filmmakers, video artists, photographers and animators to display their work (Pattison partners with the Toronto Urban Film Festival and Contact Toronto photography festival), while also enabling traditional advertising, thus blurring the lines between aesthetic and commercial aims.

Zones of Immersion is one, albeit a flagship, element of this broader artistic investment and the new function for art in pursuing the competitive world-class city. Indeed, returning to the specifics of the installation, it is clear that the work has imagineering and urban development functions that far exceed its content. Reid's installation was completed as part of a dramatic C$640 million overhaul of the beaux arts Union Station heritage site which began in 2006. The changes to the TTC platform are just one part of the wider endeavour to expand the capacity and quality of movement throughout the station and beautify the transit node and its surroundings.[4] Since acquiring Union Station in 2000, the City of Toronto continues to own and manage the station and has implemented an aggressive agenda "that will make Union Station, one of Toronto's crown jewels, spectacular again" (City of Toronto 2016). Through coordination with the TTC, Metrolinx (a provincial agency overseeing transport for the Greater Toronto region) and private developers, the City of Toronto aims to transform the station into a rail hub for the twenty-first century that is both modern and user friendly.[5] In the words of former mayor, David Miller, "By 2015 what Torontonians are going to see is the foremost transportation hub in Canada completely revitalized … Picture one of the great stations in the United States of America [and] Europe and combine them" (quoted in Lewington 2009).

In addition to the *Zones of Immersion* installation on the TTC platform, another major installation, *fLux*, by Belgian artists LAb[au] is planned for a passenger walkway through a busy pedestrian corridor. The historic Great Hall is set to host visiting exhibitions, as well as "gaming competitions and hack-a thons", and it will become a major stop on Toronto's Nuit Blanche (Canadian Broadcasting Company (CBC) 2016). In order to draw non-commuters to the space, the station has planned summer block parties and concerts, and is touting its new food court as a 'must-visit' culinary wonderland. These features are all coordinated and curated by the station's new 'event producer'. Moreover, to aid with the reconstruction and to finance the infusion of cultural programming, TD Bank was named Union Station's Founding Sponsor in 2014. The private-public partnership with TD "is founded on a city-building vision, with a plan for more pop-up cultural events and artistic programming in the future" (Osmington Inc. CEO Lawrence Zucker, quoted in Novakovic 2015). Across all of these features, the arts-led transformation is justified explicitly in economic terms. Indeed, the Toronto City Council justified its contribution, saying: "By revitalising of the Station [sic], we're also investing in the City. A strong and rejuvenated transportation hub will bring more funding to the City through extra leasing and increased commuter traffic" (City of Toronto 2016). Not only does the renovation aim to address problems of coordination and flow, but the facelift is an attempt to turn the site into "a major destination for shopping, dining and visiting" (City of Toronto 2016).

More broadly the revitalization of Union Station is the keystone of the regional Metrolinx Big Move initiative, and is one of Waterfront Toronto's priority projects. The renovations were an impetus for the creation of a new privately operated airport connection, the UP (Union Pearson) Express and they were also leveraged as selling points in advance of the Pan American Games in 2015 and as a launching pad for a potential 2024 Olympic bid. A major aspect of upgrading the station is to convince residents of Toronto as well as investors and tourists that the city's transit network is a state-of-the-art amenity fit for the world-class city it aspires to become. Despite ongoing obstacles to improving the network, arts investments here *appear* to be a bold new direction for a city that is repeatedly criticized for lack of ambition and action (Hume 2015a; Levine 2014; Lorinc 2012).[6] Far beyond the functional needs of how to keep a growing city moving, the renovations involve rebranding the TTC and repositioning Toronto in the global order to meet the imagined needs of the twenty-first century future.

Despite the fact that the mobility future imagined by arts-led renovation in Toronto is based on the provision of premium networks to bring the cosmopolitan elite to and through Toronto, the content of *Zones of Immersion* draws attention to the more mundane realities of a transit system under stress. It does not depict an abstract cartography of worldliness and access, but offers instead a view from within the messy space of the train car. This field of vision – both a representation and reflection – necessarily includes those typically left out of imaginaries of the global city, including homeless, poor, working-class and otherwise discontented travellers. With candour unexpected of metro art, the installation questions the future in which urban commuting is pleasant and the spectacle reigns uninterrupted. More often than not, the installation is merely glanced at momentarily in a state of distraction and anxiety. Yet this fecund if fleeting critical opening is nevertheless important in generating meanings and significations of mobility that challenge the status quo.

Conclusion

This chapter has argued that utopian and dystopian scenarios of urban transport wrought through metro art are key to negotiating contemporary visions of the good city and the good society. To understand and evaluate how an aesthetics of metromobility functions, it must be contextualized in relation to its surrounding urban environment, its process of production and implications on planning, policy and everyday life. Using the case of Toronto, this chapter has shown in particular how metro art is used to reframe the relation between Toronto's transit network and the city, and to recast the image of Toronto globally. It also demonstrates how metro art articulates and symbolizes the idealized city as a non-space of world-class transit and how aesthetics revalorize these networks in line with the priorities of the aspiring global city.

More broadly, the chapter claims that utopian visions of transit are important elements in the complex and internally conflicting cultural economy of late capitalism. They are frequently used as tools of economic assessment that rely on claims to future urban value thus supporting the entrenchment of market relations. However, they may also place the present in critical perspective. As *Zones of Immersion* reveals, transit art and design may register simultaneously the dreams of unimpeded urban movement and the persistent frictions that give these dreams traction.

Notes

1 By metromobility I refer to the physical infrastructures of rapid mass transit; to the social, economic, cultural and political systems which they compose; and to the interdependence of these systems with the production of the metropolis.
2 Both Rio de Janeiro (2016) and Paris (2024), for example, used their metro extensions and renovations to strengthen their respective Olympic bids.
3 Arts for Commuters was founded in 2007 as a curatorial collective that brings art to the TTC network. Its primary partner has been Pattison, but it has also collaborated with the National Film Board, Coach House Books, ImagineNATIVE Film and Media Arts Festival, Workman Arts and *Spacing* Magazine.
4 Union Station houses commuter bus, commuter and passenger rail, subway, bicycle and pedestrian transport. It is located in the central business district of Toronto, close to the waterfront and bordering the Bay Street financial corridor. Although the station was once a monument to the Canadian national railway, recent decades have seen the building itself deteriorate, transport service suffer and the neighbourhood become a hollow and sterile environment outside of commuter rush hours.
5 The revitalization comprises a number of concurrent projects including a new GO (regional rail) concourse, a revitalized exterior and public plaza on Front Street, a rebuilt Bay Street concourse, indoor bicycle parking and extensive new commercial and retail offerings. The endeavour also involves the restoration and preservation of many of Union Station's heritage elements, a renovated VIA (national rail) concourse, the creation of a new PATH underground pedestrian system link and new connections to tourist destinations such as the Air Canada Centre and Maple Leaf Gardens.
6 Originally to be completed in 2014, the Union Station makeover is behind schedule and over budget. Due to its exorbitantly high fares, the UP Express ran well below capacity throughout its first year of operation. The farce of Toronto's transit dream, however, was perhaps most notably laid bare when it was revealed in 2016 that the new C$248 million train shed at Union Station would not actually fit the trains it was meant to house in the future.

References

Art in Transit. (2016). *About*. Available at: www.artintransit.ca/about [accessed 26 July 2016].
ARUP. (2015). *The Future of Rail 2050*. London: ARUP. Available at: www.arup.com/homepage_future_of_rail [accessed 26 July 2016].
Augé, M. (2008). *Non-Places: An Introduction to Supermodernity*, trans. John Howe. London and New York: Verso.
Bauman, Z. (1998). *Globalization: The Human Consequences*. New York: Columbia University Press.
Bertsche, L. (2013). Public Art and the Central Corridor: Place Promotion and Creative Placemaking. *Cities in the 21st Century*, 3(1), p. 1.
bipolar_sky_fairy. (2015). Is the New Public Art at Union Station Depressing? *Reddit Toronto*. Available at: www.reddit.com/r/toronto/comments/35abbg/is_the_new_public_art_at_union_station_depressing/ [accessed 26 July 2016].
Boudreau, J.A., Keil, R. and Young, D. (2009). *Changing Toronto: Governing Urban Neoliberalism*. Toronto: University of Toronto Press.
Bownes, D., Green, O. and Mullins, S. (2012). *Underground: How the Tube Shaped London*. London: Penguin Books.
Breitbart, M.M. and Worden, P. (1994). Boston – Creating a Sense of Purpose: Public Art and Boston's Orange Line [art and the Transit Experience]. *Places*, 9(2), pp. 80–86.
Buck-Morss, S. (1995). The City as Dreamworld and Catastrophe. *October*, 73, pp. 3–26.
Canadian Architect. (2015). *Zones of Immersion*. Unveiled at Toronto's Union Station. *Canadian Architect*, 30 June. Available at: www.canadianarchitect.com/architecture/zones-of-immersion-unveiled-at-torontos-union-station/1003728126/ [accessed 26 July 2016].
Canadian Broadcasting Company (CBC). (2016). Season 1 Episode 4. *Disrupting Design*, 27 March. Available at: www.cbc.ca/player/play/2685965488 [accessed 26 July 2016].
Castells, M. (1996). *The Rise of the Network Society: The Information Age: Economy, Society, and Culture*, Vol. 1. New York: John Wiley & Sons.
Castoriadis, C. (1987). *The Imaginary Institution of Society*. Cambridge: Polity Press.

City of Toronto. (2003). *Culture Plan for the Creative City*. Toronto: City of Toronto. Available at: www1.toronto.ca/city_of_toronto/economic_development__culture/cultural_services/cultural_affairs/initiatives/files/pdf/creativecity-2003.pdf [accessed 26 July 2016].

City of Toronto. (2016). *About Union Station*. Toronto: City of Toronto. Available at: www1.toronto.ca/wps/portal/contentonly?vgnextoid=5e05962c8c3f0410VgnVCM10000071d60f89RCRD&vgnextchannel=dfacd50749604510VgnVCM10000071d60f89RCRD [accessed 26 July 2016].

Cox, D. (2015). A Tour of the Stockholm Metro – the World's Longest Art Gallery. *Guardian*, 20 October.

Cresswell, T. (2006). *On the Move: Mobility in the Modern Western World*. London: Taylor & Francis.

Cresswell, T. (2010). Towards a Politics of Mobility. *Environment and Planning D: Society and Space*, 28(1), pp. 17–31.

Crossrail. (2016). *Crossrail's Art Programme: Integrating Art and Infrastructure*. Available at: www.crossrail.co.uk/sustainability/art-on-crossrail/ [accessed 26 July 2016].

Davis, M. and Monk, D.B., eds. (2008). *Evil Paradises: Dreamworlds of Neoliberalism*. New York: The New Press.

Deutsche, R. (1996). *Evictions: Art and Spatial Politics*. Cambridge, MA: MIT Press.

Dunmall, G. (2016). Ad-free Art on the Underground: Düsseldorf's 'Pure' New Metro Line. *Guardian*, 19 February.

Florida, R. (2004). *The Rise of the Creative Class*. New York: Basic Books.

Ghertner, D.A. (2015). *Rule by Aesthetics: World-class City Making in Delhi*. New York: Oxford University Press.

Graham, S. and Marvin, S. (2001). *Splintering Urbanism: Networked Infrastructures, Technological Mobilities and the Urban Condition*. London: Routledge.

Hall, T. and Robertson, I. (2001). Public Art and Urban Regeneration: Advocacy, Claims and Critical Debates. *Landscape Research*, 26(1), pp. 5–26.

Hertel, S., Keil, R. and Collens, M. (2015). *Switching Tracks: Towards Transit Equity in the Greater Toronto and Hamilton Area*. Commissioned by Metrolinx. Available at: http://suburbs.apps01.yorku.ca/wp-content/uploads/2015/03/Switching-Tracks_9-March-2015.pdf [accessed 26 July 2016].

Hume, C. (2015a). Toronto Transit Planning Lacks Vision. *Toronto Star*, 9 November.

Hume, C. (2015b). Union Station Artwork an Exercise in Artistic Transparency. *Toronto Star*, 31 March.

Jameson, F. (2005). *Archaeologies of the Future: The Desire Called Utopia and Other Science Fictions*. New York: Verso.

Kaika, M. (2011). Autistic Architecture: The Fall of the Icon and the Rise of the Serial Object of Architecture. *Environment and Planning D: Society and Space*, 29(6), pp. 968–992.

Kearns, G. and Philo, C., eds. (1993). *Selling Places: The City as Cultural Capital, Past and Present*. New York: Pergamon.

Kingsbury, D. (forthcoming). Infrastructure and Insurrection: The Caracas Metro and the 'Right to the City' in Venezuela. *Latin American Research Review*.

Kwon, M. (2002). *One Place after Another: Site-specific Art and Locational Identity*. Cambridge, MA: MIT Press.

Levine, A. (2014). *Toronto: Biography of a City*. Toronto: Douglas & McIntyre.

Lewington, J. (2009). Private Partners Named in Union Station Reno. *Globe and Mail*, 2 December. Available at: www.theglobeandmail.com/news/national/private-partners-named-in-union-station-reno/article1205371/ [accessed 26 July 2016].

Lorinc, J. (2012). How Toronto Lost Its Groove. *The Walrus*, 12 September. Available at: http://thewalrus.ca/how-toronto-lost-its-groove/ [accessed 26 July 2016].

Massey, D. (2013). *Space, Place and Gender*. New York: John Wiley & Sons.

McCarthy, J. (2006). Regeneration of Cultural Quarters: Public Art for Place Image or Place Identity? *Journal of Urban Design*, 11(2), pp. 243–262.

McDougall, H. (2014). Two Turner Prize Winners Commissioned for Crossrail Art Programme at Tottenham Court Road Station. *Crossrail News*, 22 December. Available at: www.crossrail.co.uk/news/articles/two-turner-prize-winners-commissioned-for-crossrail-art-programme-at-tottenham-court-road-station [accessed 26 July 2016].

Miles, M. (1997). *Art, Space and the City: Public Art and Urban Futures*. London: Routledge.

Novakovic, S. (2015). Union Station Taking Shape with New Retailers and TD Partnership. *urbantoronto.ca*, 21 December. Available at: http://urbantoronto.ca/news/2015/12/union-station-taking-shape-new-retailers-and-td-partnership [accessed 26 July 2016].

Ovenden, M. (2009). *Paris Underground: The Maps, Stations, and Design of the Métro*. New York: Penguin Books.
Ovenden, M. (2013). *London Underground by Design*. London: Penguin Books.
Phillips, P. (1988). Out of Order: The Public Art Machine. *Artforum*, December, pp. 92–96.
Pinder, D. (2002). In Defence of Utopian Urbanism: Imagining Cities After the 'End of Utopia'. *Geografiska Annaler: Series B, Human Geography*, 84(3–4), pp. 229–241.
Reid, S. (2015). *Union Station Project*. Video, 16 October. Available at: www.stuartreid.ca/zones-of-immersion-video/ [accessed 26 July 2016].
Sennett, R. (1992). *The Conscience of the Eye: The Design and Social Life of Cities*. New York: WW Norton & Company.
Sharp, J., Pollock, V. and Paddison, R. (2005). Just Art for a Just City: Public Art and Social Inclusion in Urban Regeneration. *Urban Studies*, 42(5–6), pp. 1001–1023.
Siemens. (2011). *Hubs of the Future: An Integrated Mobility Network for Passengers and Freight*. Munich: Siemens AG. Available at: www.mobility.siemens.com/mobility/global/SiteCollectionDocuments/en/integrated-mobility/future-of-hubs/hubs-of-the-future-en.pdf [accessed 26 July 2016].
Spurr, B. (2015). Is the New Public Art at Union Station Depressing? *Toronto Star*, 7 May.
Ström, M. (1994). *Metro-Art in the Metro-polis*. Paris: ACR edition.
Tafuri, M. (1976). *Architecture and Utopia: Design and Capitalist Development*. Cambridge, MA: MIT Press.
Timms, P., Tight, M. and Watling, D. (2014). Imagineering Mobility: Constructing Utopias for Future Urban Transport. *Environment and Planning A*, 46(1), pp. 78–93.
TTCriders. (2016). *Our Work*. Available at: www.ttcriders.ca/ [accessed 26 July 2016].
Urry, J. (2007). *Mobilities*. London: Polity.
Valley Metro. (2008). *Metro Art 2008*. Available at: www.valleymetro.org/images/uploads/lightrail_publications/METRO-Art-Book.pdf [accessed 26 July 2016].

48
REIMAGINING THE URBAN AS A DYSTOPIC RESILIENT SPACE

Scalar materialities in climate knowledge, planning and politics

Andrew Kythreotis

Nature has spoken.
  (Tagline from the 2004 movie, The Day After Tomorrow, dir. Roland Emmerich)

City growth has caused climate change, but that growth is also what's going to get us out of it.
  (Matthew Kahn, UCLA economist, Climatopolis, 2010)

Introduction

In the 2004 post-apocalyptic Hollywood blockbuster movie *The Day after Tomorrow* (dir. Roland Emmerich), one of the lead characters, a climatologist named Professor Jack Hall played by Denis Quaid, gets into an argument with the Vice-President of the USA regarding the immediate need for action on climate change to save future generations. The Vice-President argued that the price to pay for the Kyoto Agreement (remember, this was 2004) would reach billions of dollars and Hall replied that the costs of doing nothing would be even higher. Fast-forward nearly a decade and the 2013 Canadian science fiction horror film *The Colony* (dir. Jeff Renfroe) is set in a backdrop where technological development saves the human race through human-built weather machines that control the warming of the planet. These machines break down, causing incessant snowfall, and leading ironically to plummeting global temperatures.

The above examples are merely representations of (present and future) global climate change from the silver screen. However, they are not too far removed from contemporary popular political and lay debate on the way global climate change is seen to affect cities. Debate about the impact of global climate change on cities often seems to lurch from one extreme – the apocalyptic (e.g. cities engulfed by rising sea levels) – to another, the utopian (e.g. cities will be able to solve protracted issues like climate change because they are economic growth machines, as the quote from Matthew Kahn at the start of this chapter intimates). It is true that cities, especially in the developing world, have borne the brunt of climate impacts and are now faced with having to plan differently than previously. Both the frequency and intensity of climate impacts are largely due to urbanization, and are compounded by inadequate planning measures coupled with the increased exposure of rising urban populations to impacts like flooding.

Nevertheless, many cities are now paying increased attention to adapting to such impacts and attempting to implement autonomous measures that are aimed at protecting their burgeoning populations from the ravages of climate-related events exacerbated by anthropogenic interference. Such measures are designed to increase the 'resilience' of cities. For example, the Rockefeller Foundation has pioneered the foundation of the Resilient Cities Network (Rockefeller Foundation 2016) which consists of a hundred member cities from developed and developing countries all working towards greater urban resilience. The Resilient Cities Network defines urban resilience as "the capacity of individuals, communities, institutions, businesses, and systems within a city to survive, adapt, and grow no matter what kinds of chronic stresses and acute shocks they experience" (Rockefeller Foundation 2016). There are many other similar examples of city initiatives that promote greater urban mitigation, adaptation and, increasingly, resilience.

Arguably, resilience is now the new buzzword for cities, in part, because it systemically encapsulates the economic, social and environmental governance fabric of cities in a more holistic fashion than, for example, climate adaptation. Climate adaptation (vis-à-vis climate mitigation) has been viewed as an environmental 'add-on' to urban politics and policies (Betsill and Bulkeley 2007; Granberg and Elander 2007). Furthermore, adaptation to long-term climate change does not fit neatly into the short-term lifecycles of local urban planning (Wilson 2006). Yet this is slowly changing in light of recent extreme weather affecting cities on a drastic scale and due to the fact that many political responses developed at spatial scales beyond the urban remain reactive (Porter et al. 2015; Tompkins et al. 2010). Conversely, the notion of the 'resilient city' clearly elides to complex systemic networks of connectivity and flows that can successfully (re)organize the vulnerable, cultural and physical components of a city (Batty 2008; Desouza and Flanery 2013). Additionally, the term resilience can also play to the sympathies of economic development and private sector investment interests that otherwise align with cities as 'economic growth machines' whilst at the same time appeasing the climate adaptation policy purists (Kythreotis and Bristow 2017). Hence, there is great mass appeal (political and economic) in the idea of the 'resilient city'.

What this chapter aims to explore is the long-term implications of these burgeoning resilient cities initiatives for the space of urban politics. It will do this in two distinct, yet interrelated ways. First, such initiatives can potentially rupture the individuality of each city, and thereby suppress alternative urban political imaginaries, by producing and reflecting a monolithic resilience model; one that is assumed to be best practice for climate planning in all cities. Second, and with the advent of resistance to such a monolithic resilient city model agenda, the production of a multi-spatial urban policies might be engendered around scales other than the urban and also other 'urban imaginaries' associated with planning for climate change. Whilst these other urban imaginaries are materially grounded and played out in the formal space(s) of the city, how resilient cities are produced is not neatly elided with urban limits.

For example, it has been argued that the (urban) politics of climate change is no longer confined to the jurisdictional spaces of cities. Such politics have a deeply embedded relationality – a multi-level governance rationale – to the politics and policies of other spaces of politics be these the city-region, region, nation or the global, or even at more proximate territorial scales of city districts, wards and communities as the 100 Resilient Cities definition of urban resilience above clearly illustrates (Bulkeley 2012; Bulkeley and Betsill 2005; Burton et al. 2007; Corfee-Morlot et al. 2011; Hunt and Watkiss 2011). I explore this spatial relationality in the context of the architecture of the global climate regime (knowledge, politics and policy) and how such a regime is based upon certain scientific truths and technological (ir)rationalities that have their provenance in positivist schools of thought, which make it difficult to neatly elide global climate change politics and policy with more constructivist and reflexive urban imaginaries. I also consider the extent to which in the debate about climate change the global and urban scales can or should be conflated: whether "[T]he urban politics of the twenty-first century will be both a local politics and a global politics" as Allan Cochrane (1999: 123) envisaged. I consider whether any potential disconnect or what I call 'short-circuiting' between the urban and global scales horizontally reinforces a sanitized form of dystopic urbanity in the way that cities politically respond and plan for climate impacts. I conclude by questioning whether it is materially beneficial to imagine a utopian urban climate politics based upon a relational turn.

Urban imaginaries and the politics of climate change

The debate about climate change has been fuelled by a variety of imaginaries, ranging from the utopian to the dystopian. On the face of it, attaining a climate-resilient city would for a city planner or policy-maker represent a policy panacea or, in other words, a future vision of an urban utopia. If a city were fully climate resilient it would ideally be able to absorb future disturbances (e.g. a climate-related event like flooding for instance) and reorganize itself in a way that maintains or even improves its overall function and structure after the disturbance in question. This means the city could cope and function (at the very least) just as before the environmental disturbance had occurred; there would be limited or no human or financial costs incurred directly attributable to the event. However, cities and urban areas can never be truly resilient because it is impossible to predict the uncertain nature of extreme climate events; climate events like flooding have a fluid, unpredictable and discursive nature that cannot be engineered and managed precisely through planning measures and associated technologies.

The December 2015 flooding in the City of Carlisle, Cumbria, UK, exemplifies this. In 2010, nearly £40 million was spent on a 1-in-200-year flood alleviation scheme designed to protect 3,000 homes in and around Carlisle. Yet just five years on, a UK record pluvial event in Cumbria of nearly 350 mm of rainfall in a 24-hour period highlighted the frailties of technological engineering to engender a future of comprehensive urban resilience to flooding. It is important to note that technological engineering does protect cities from flood events, but this protection is only temporary because there will always be an extreme event that exceeds the planned protection levels. Whilst in hindsight such extreme events could be construed as a case of maladaptation, it does highlight the limitations of technology in adaptation planning when faced with extreme climate events.

Most well-planned and -engineered cities are susceptible to other non-climatic variables like human decision-making (Bell *et al.* 2014; Kythreotis *et al.* 2013). This, I would argue, is the very reason why cities tend to cast their gaze to other cities for inspiration when it comes to designing policies for adaptation and resilience in the hope that there is some scope for co-mutual planning. But this in turn involves constructing a different space of urban politics; a politics operating between cities rather within a city. In identifying this 'between space' of urban politics, Cochrane (2011) and Ward (2011) recognize that urban policies found within cities are often assembled elsewhere beyond the territorial limits of the city itself. For urban climate adaptation planning the 'elsewhere' in question seems to be defined through the best practices of other cities, especially in relation to urban resilience. However, such horizontally worked urban 'otherness' can only mitigate impacts specific to particular urban areas. An alternative – arguably even more geographically nuanced yet relational – way of thinking about the interchanging of climate planning and politics between cities suggests that such practice(s) necessarily involve theoretical concepts that can transcend the limits of the local, cities, the urban, etc.

So when I speak of 'other' urban imaginaries in the context of climate change I mean that cities tend to mobilize imagined ideas of what 'other' cities in the world experience and then bring such best practice to their own climate planning, policies and implementation processes. Such a practice would certainly (re)position the idea that the city is a homogenous political space capable of self-engineered resilience, much like the human-built weather machines in *The Colony*. It can be argued that 'best practice' forms the main aim of the 100 Resilient Cities Network, which "aims not only to help individual cities become more resilient, but will facilitate the building of a global practice of resilience among governments, NGOs, the private sector, and individual citizens" (Rockefeller Foundation 2016). In this quote lies a danger of not actually acknowledging and celebrating the diversity of individual cities but rather engendering the production of climate planning practices based upon what Castells (1998: 26) referred to as the

> world centered on the welfare state, on rigid zoning, in the belief of models of metropolitan growth, on the predictability of social patterns … on the long term benefits of economic growth without social or environmental constraints and on the view of the world from patriarchalism as a way of life.

Rather, urban climate planning in cities should be based upon the heterogeneous and relational contexts that define them.

Ward (2011) suggests that we study cities using a comparative relational approach rather than through the lens of categorized, Cartesian spatial fixity. In relation to planning for future climate impacts, one has to consider whether it is practically possible to model one city on another given the social, cultural, political and even economic diversity and distinctions that are evident between all cities. If we were to eliminate the idea of categorized spatial fixity in any analysis of how cities respond to climate impacts, then it is imperative that our analytical thinking extends to an analysis of the ontological nature and construction of (climate) knowledge. This is because such constructions are inherently spatial in the way they legitimize a specific form of urban politics and planning.

Since the inception of a global governance climate regime under the United Nations, constructions of climate knowledge have for too long relied on more powerful epistemic actors that legitimize a reductionist climate knowledge based on the natural and physical sciences, often because such science speaks the language of (local) decision-makers (Turnhout *et al.* 2016). The result of this is the uni-directional flow of knowledge from the global down to the local scale. Yet the thrust of recent urban theory suggests that cities require iterative examination as fluidic forms of contested scalar politics (Smith and Kurtz 2010), "scalar manoeuvring" (Kythreotis and Jonas 2012) or "scalecraft" (Fraser 2010) in which the political[1] agencies and knowledge(s) of diverse sets of actors across different spatial scales of governance are dynamically utilized. In particular, city planners are key to this success because they are able to multi-directionally draw from particular knowledges between and across scales, both horizontally and vertically in order to continually (re)shape urban climate planning and policy.

Although the actions and agencies of urban managers should be central to the process of knowledge accumulation and dissemination, it is not just the actions of these actors that end with harmonious and rational political decisions. It is rather the fact that the relationality of scale(s) (e.g. Howitt 1998) enables an end political decision or outcome to be reached using multiple geographies of different disciplinary knowledges. In this sense, scales as relation are able to materially and constructively absorb different political viewpoints so as to engender a specific desired urban outcome. If we transpose this thought to the current normative resilient cities discourse, we could certainly see how such a discourse can potentially endanger more progressive and constructivist post-structural, multi-spatially networked accounts of climate politics within cities and how, in turn, different urban constructions of climate planning knowledge can be effectively 'upscaled' to inform global constructions of knowledge.

The resilient cities discourse might even be doing the notion of 'the urban' an injustice by assuming that a monolithic, scientifically and technologically inspired reductionism is adequate enough for successful future climate urban planning practice. Writing in *Fortune* magazine in 1958, Jane Jacobs argued, "There is no logic that can be superimposed on the city; people make it, and it is to them, not buildings, that we must fit our plans" (Jacobs 1958: 127). Accepting Jacobs's argument about the need to promote a diversity of spaces in cities as a pre-given, we could actually argue that sharing best practice for climate planning could be counter-productive to an individual city's future planning needs because they reflect the realities of other cities back into the individual city – they are 'urban imaginaries'. Hence, the 'logic' of best practice in climate planning can in hindsight turn out to be illogical at best, and planning dystopia at worst, with such imagined ideas not necessarily resulting in well-adapted or climate-resilient cities of the future, and in some cases, maladapted cities.

New York is a prime example of this. Since the turn of the millennium, the city has invested heavily, both economically and politically, in a number of high-profile climate adaptation initiatives (Rosenzweig *et al.* 2007; Solecki 2012). But then came Hurricane Sandy in October 2012. What was previously thought a planning model of an autonomously well-adapted World City was immediately disproved. The main elephant in the room when it comes to how cities plan for extreme climate-related hazards will always be ensconced within the uncertainty and unpredictability of that given hazard. This is constantly in tension with the Promethean-inspired effect that humanity believes technological development and progress can completely eliminate any urban vulnerability. The immediate reaction of the Mayor's Office in New York City after Sandy was to become one of the first members of the 100 Resilient City Network. This is all well and good insofar as in the aftermath of such climatic devastation the city needs to be actively seen as planning 'bigger and better' for future climate risks. This is usually achieved through the silver bullet default of technological dependence and innovation, e.g. larger flood engineering defences. After all the process of becoming resilient should involve learning from current technological limitations and inadequacies.

New York's response to Sandy is the classic example of Downs's (1972) issue-attention cycle which explains how certain large (e.g. climate-related) events strongly affect public attitudes which are then lurched into the policy domain for immediate attention. One reason why cities like New York prescriptively (re)act like this can be attributed to the political fear of uncertainty and unpredictability brought by extreme climate-related events like hurricanes and floods. Such uncertainty and unpredictability is in tension with the established tenets of traditional city planning practices from the past 200 years or so, which in New York's case arguably owe their provenance to the city's planned grid system of 1811, which was (is) based on a standardized and rigorous application of orthogonal mathematical planning to a highly differentiated and complex natural topography. Yet this tension is not just a result of historic political actions inside urban territory, but also refers to how knowledge and ideas of how to instil such urban climate resilience are framed 'upscale'. In this sense, the urban imaginaries that I have described are not solely a result of the horizontal spatial relationships that go on between (other) cities in terms of best practice, but are rather shot through by a top-down vertical spatiality that makes engendering a utopian urban climate politics of space far more difficult and complex. McCann and Ward (2011) make this very point when they argue how cities are a hub in which spatial networks circulate via the importing of external influence. Accordingly localized (e.g. urban) political struggles have to reinterpret and even implement externally imported norms, knowledge, agendas and policies.

From the global to the urban; but not back again

An urban theoretical sensibility requires adjustments to how knowledge on climate change is generated and circulates. The *modus operandi* and architecture of the global climate regime – how it derives its knowledge, undergoes a particular politics based on certain (scientific) norms, and how these norms are then transformed into policies – are integral to how cities plan for and respond to climate change. The global climate regime was originally constructed around an explicit assumption that science would form the rationale basis of decision-making and policies where truth and knowledge would speak to power through a linear model of expertise (Jasanoff and Wynne 1998) (see Figure 48.1).

Such forms of knowledge are originally based on positivist empirical observations of natural facts (and in particular 'absolute truths') that could be used as evidence to prove a particular theory. This form of knowledge was (still is) deemed the essence of true science. The evolution of positivism since Comte has undergone many nuanced changes that have been integral to understanding how the global climate regime has evolved. In particular, the Popperian view that theories should rather be disproved (empirical falsification) using natural facts as forms of evidence and the Kuhnian notion that science undergoes non-linear paradigm shifts defined by a consensus of a particular scientific community. Significantly, the structural underpinning that has remained constant throughout the historical evolution of positivism is the Comtian interdependent hierarchical ordering of the disciplines where mathematics assumed the bottom place of this hierarchy but ironically was regarded as the greatest of all the disciplines because it allowed humans to explore and understand basic natural laws and absolute truths. This classification in particular has had an institutionalizing influence in the way certain powerful epistemic actors at the international scale understand, construct and reify a reductionist climate knowledge (Turnhout et al. 2016), and how this knowledge is spatially redistributed, or in McCann and Ward's (2011) words "imported", across geographical space(s) to (re)produce a certain form of climate change politics.

A utopian climate policy world would be characterized by a consensus between our planners and policy-makers regarding the use of *all* forms of scientific knowledge – the social, humanities, arts as well as the natural – and this would feed neatly into appropriate planning processes and responses at the urban level. We would have a Huxleyan 'brave new world' underpinned by co-mutual and co-productive planning and policy across geographical scales. Yet historically

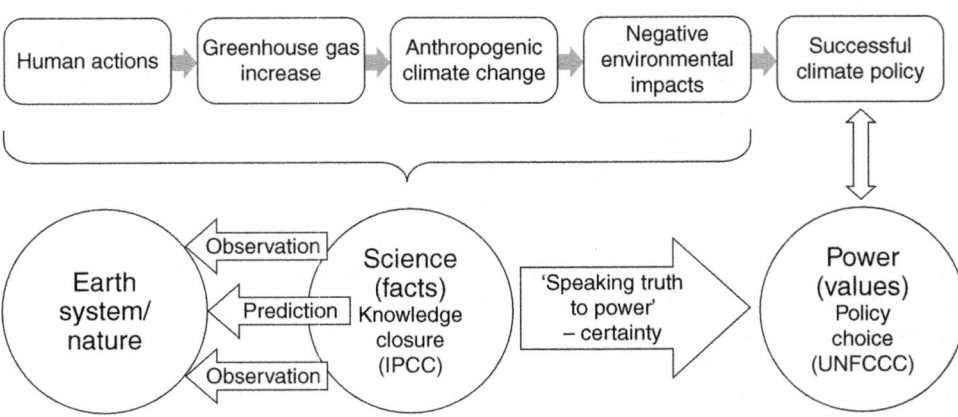

Figure 48.1 The linear model of (climate) expertise

it is only the absolute truths and natural laws derived from mathematics, physics and chemistry that have been politically legitimized as truth linearly speaking to power and policy. Yet because society interacts with nature and the physical world through a progressive social evolution of the theological, metaphysical and then positive, as Comte argued, we are left with an omnipresent sociology of climate change. Our ideas of climate and how it changes the environment around us are socially constructed and based upon complex social and physical associations. Jasanoff and Wynne (1998) have argued how the social sciences have been underplayed and underexplored vis-à-vis the natural sciences. They challenge the idea that "global problems such as climate change exist in a world that can be unproblematically accessed through direct observation" (Jasanoff and Wynne 1998: 3). Rather "environmental issues of global scale (or indeed any scale) emerge from an interplay of scientific discovery and description with other political, economic and social forces" (Jasanoff and Wynne 1998: 4). Hence, because anthropogenic climate change is socially constructed, there needs to be a degree of reflexivity in how society understands and creates new knowledges to respond to the physical and social impacts of climate – and these social constructions should, in practice, have a multi-directional influence across and between different geographical scales.

Science and Technology Studies (STS) has been an important disciplinary lens from which to examine the relationship between climate knowledge (science), society and politics. Yet a theoretical and/or empirical examination into the ways in which, or whether at all, this tripartite can continually (re)organize and (re)configure across geographical space has not. The idea of the linear model of expertise as represented in Figure 48.1 has been used to explain the unidirectional and closed nature of climate science (based, e.g. on the observations and predictions of the Intergovernmental Panel on Climate Change or IPCC) as the 'trigger' that leads to policy formulation under the United Nations Framework Convention on Climate Change (UNFCCC) and Kyoto Protocol (Beck 2011; Jasanoff and Wynne 1998). In essence, the linear model holds that more scientific research based on the natural sciences and absolute truths will lead to better certainty, which leads to better policy. The STS literature argues that this is far too simplistic and reductionist as a model. The relationship between science (knowledge closure) and power (policy choice) is characterized by a far more complex and multi-directional movement of knowledge uncertainty, resources, policies and planning ideas (see Figure 48.2), even though the science-policy relationship continually remains to be legitimized by scientific reductionism and socio-technical interventionism. It is this legitimization that ultimately proves far more

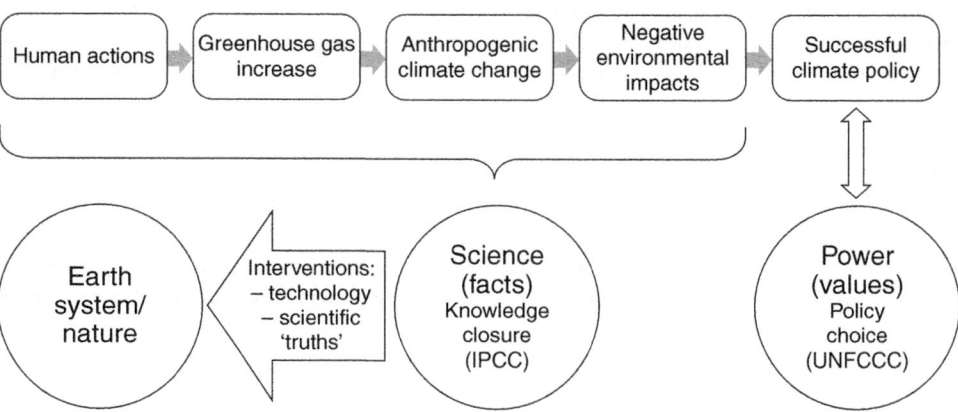

Figure 48.2 The multilinear model of (climate) expertise

destructive to eliding global climate change politics with a post-modern urban planning rationale in cities.

Beck (2011) and Dessai *et al.* (2009) have argued how the IPCC still externally portrays itself to policy bodies as using the linear model of expertise where absolute scientific truths dominate policy formulation around scenario-based assessments. This places certain limits on adaptation decision-making at the urban level. Since its inception the IPCC has monopolized itself as the paramount international advisory institution to policy-makers at different scales of political governance (Bolin 2007). This has resulted in the IPCC being continually critiqued both in terms of how it represents and normalizes scientific knowledge and influences policy around future scenarios and modelling (Edwards 2001; Hulme and Mahony 2010; Tol 2011) and how it governs itself as an institution (Beck *et al.* 2014; Carraro *et al.* 2015; Edwards and Schneider 2001; Hulme and Mahony 2010; Petersen *et al.* 2015). The IPCC has taken a reductionist scientific approach in the way it reports the latest knowledge, which is dominated by the physical properties and behaviour of greenhouse gases (GHGs) in the atmosphere and how these are modelled in terms of future temperature projections that should not exceed the two degrees centigrade thermostat threshold. It has been argued that modelling our future atmosphere based on certain projection scenarios of human GHG emissions in order to limit future temperature increase is "good science" (Norton and Suppe 2001). The linear model of expertise normalizes climate modelling and subsequent temperature thresholds as an indisputable and infallible epistemic form of knowledge to be used in UNFCCC policy-making as if there is no other alternative knowledge that can be used; what Hulme (2011: 249) describes as "universal predictors of future social performance and human destiny".

This deterministic approach of the IPCC has led to the 'Ultimate Objective' of the UNFCCC of stabilizing GHG concentrations through highly interventionist, technocratic mechanisms. The most obvious is the Kyoto Protocol, which established three 'flexibility mechanisms' to try and tackle the global mitigation of GHGs – Emission Trading Schemes; Joint Implementation; and the Clean Development Mechanism – all of which are based upon the main absolute scientific truth heralded by the IPCC (and UNFCCC) that lowering GHG emissions (particularly carbon dioxide) because they have physical properties that warm the earth, will result in decreased global temperatures. Such technocratic assumptions are also embedded into the governance structures of the UNFCCC, with permanent bodies like the Subsidiary Body for Scientific and Technological Advice having a key role advising on scientific and technological matters relating to the Convention and the Kyoto Protocol in relation to these self-prescribed nominal numbers. Many other influential UNFCCC governance bodies are continually coming into existence and are driven by this reductionist and technocratic knowledge, like the Technology Executive Committee which was established in 2010 to oversee the policy development and transfer of climate technologies between countries.

But what does this mean for the implementation of distinct spatially determined policy on climate in our cities? If the linear model still holds true in the way that climate policy is solely determined and legitimized through the natural sciences, forming a particular form of epistemic power amongst global policy-governance actors (Turnhout *et al.* 2016), then this model will also continue to create differential power inequalities between the urban and global scales, where the freedoms and reflexivity of urban politics more generally to plan for climate impacts are continually eroded.

Conclusion

In conclusion, the reification of the linear model approach to climate knowledge-policy at the global level results in our cities being transformed into dystopic, sanitized urban spaces. This not only short-circuits a post-modernistic form of urban planning and politics governed by networks and flows, as heralded by Jacobs and Castells, but also acts to disable such a reflexive and alternative urban politics and knowledge feeding (and neatly eliding) back into the global sphere and iteratively building on our policy knowledge base that will enable us to more readily tackle the effects of climate more innovatively. Rather such knowledge is transferred horizontally to other cities through sharing of best practice around ideas of what should constitute an 'imagined' resilient city (e.g. being able to promote adaptation actions that can cope within the limits of two degrees centigrade average global temperature rise). This is predicated on the assumption that what works for one city is good enough for the other. Yet this only makes future urban climate planning a more temporally abstract distant threat (Luers and Sklar 2014). However, some shared best practice can be beneficial, e.g. best practice that is dependent on reductionist technical knowledge like the methods and properties of materials used for engineering flood defences, but there remains the caveat of considering human agency in political and policy decisions. Not all knowledges are interchangeable between cities. The short-circuiting of knowledge flows from the city up to the international scale only acts to further sanitize the notion of the urban into dystopic imaginaries through the false panacea of horizontally shared best practice, as Figure 48.3 illustrates.

This chapter has considered the efficacy of the idea of planning the climatically resilient, well-adapted city. By introducing the reductionist way in which the global climate regime operates, I have argued that far from the local and the global being elided as spatial scales where multi-level governance enshrines and legitimizes the multi-directional flow of knowledges, there is rather a more interesting spatial dynamic at play which in turn opens up opportunities for rethinking the space of urban politics in relation to climate change and similar 'global' policy issues. The global is certainly in the urban (local), but the urban remains limited by current

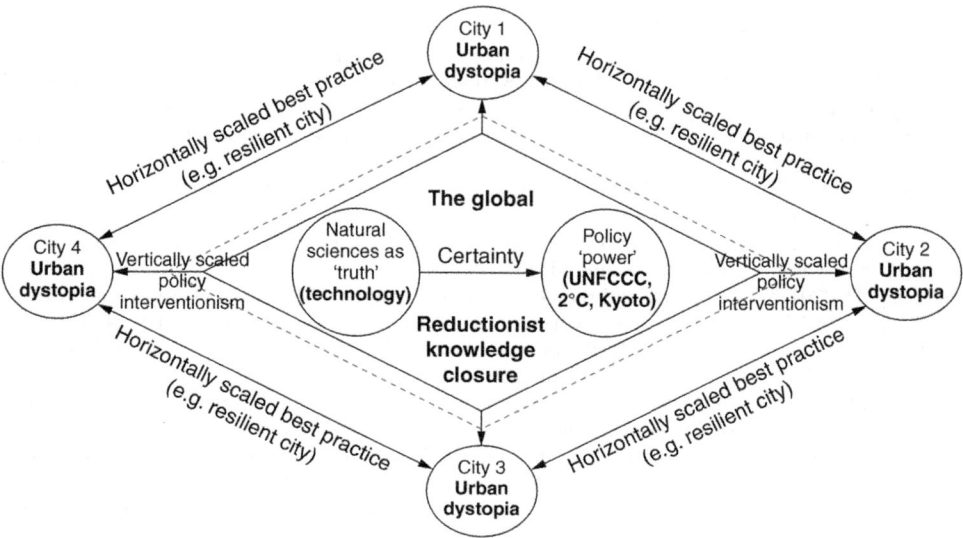

Figure 48.3 Urban dystopias: the short-circuiting of knowledge flows from the city to the international scale

scientific and planning practices in how it can promote climate resilience back up to the global scale. This uni-directional vertical flow of knowledge results in a horizontal but nested reworking of planning practice across different cities with a singular, globally defined form of urban planning best practice based on reductionist forms of resilience knowledge. It is this very process that creates such dystopic urban imaginaries in planning for climate change, limiting the ability of different knowledges to be reworked upscale so as to produce more reflexive knowledges that can then be iteratively reworked vertically and horizontally across geographical space. In this sense, maybe we are witnessing a reduction in empirical diversity in relation to the multi-level governance rationale as a solution to effectively tackling the impacts of climate change. It is too premature to imagine an elusive yet more utopian climate policy fix that is predicated upon the intertwining of policy and political scales in more effectively planning for climate change. In the context of climate adaptation planning, we are yet to fully witness a *fait accompli* empirical turn in relational urban geography where urban politics would be both a local politics and global politics of the twenty-first century as desired by Cochrane (1999). Rather, relational urban geography remains skewed towards a theoretical normativity based on the global still coercively pushing upon, and dictating, the urban.

Note

1 Political in this sense is not just defined as the art or theory of government, but also how state and non-state actors construct ideas and arguments (around climate change) based on different philosophies, cultures, economics, political systems and government policy.

References

Batty, M. (2008). The Size, Scale, and Shape of Cities. *Science*, 319(5864), pp. 769–771.
Beck, S. (2011). Moving Beyond the Linear Model of Expertise? IPCC and the Test of Adaptation. *Regional Environmental Change*, 11(2), pp. 297–306.
Beck, S., Borie, M., Chilvers, J., Esguerra, A., Heubach, K., Hulme, M., Lidskog, R., Lövbrand, E., Marquard, E., Miller, C., Nadim, T., Nesshoever, C., Settele, J., Turnhout, E., Vasileiadou, E. and Goerg, C. (2014). Towards a Reflexive Turn in the Governance of Global Environmental Expertise. The Cases of the IPCC and the IPBES. *GAIA – Ecological Perspectives for Science and Society*, 23(2), pp. 80–87.
Bell, J., Saunders, M.I., Leon, J.X., Mills, M., Kythreotis, A., Phinn, S., Mumby, P.J., Lovelock, C.E., Hoegh-Guldberg, O. and Morrison, T.H. (2014). Maps, Laws and Planning Policy: Working with Biophysical and Spatial Uncertainty in the Case of Sea Level Rise. *Environmental Science & Policy*, 44, pp. 247–257.
Betsill, M. and Bulkeley, H. (2007). Looking Back and Thinking Ahead: A Decade of Cities and Climate Change Research. *Local Environment*, 12(5), pp. 447–456.
Bolin, B. (2007). *A History of the Science and Politics of Climate Change: The Role of the Intergovernmental Panel on Climate Change*. Cambridge: Cambridge University Press.
Bulkeley, H. (2012). Governance and the Geography of Authority: Modalities of Authorisation and the Transnational Governing of Climate Change. *Environment and Planning A*, 44(10), pp. 2428–2444.
Bulkeley, H. and Betsill, M. (2005). Rethinking Sustainable Cities: Multilevel Governance and the 'Urban' Politics of Climate Change. *Environmental Politics*, 14, pp. 42–63.
Burton, I., Bizikova, L., Dickinson, T. and Howard, Y. (2007). Integrating Adaptation into Policy: Upscaling Evidence from Local to Global. *Climate Policy*, 7(4), pp. 371–376.
Carraro, C., Edenhofer, O., Flachsland, C., Kolstad, C., Stavins, R. and Stowe, R. (2015). The IPCC at a Crossroads: Opportunities for Reform. *Science*, 350(6256), pp. 34–35.
Castells, M. (1998). The Education of City Planners in the Information Age. *Berkeley Planning Journal*, 12(1), pp. 25–31.
Cochrane, A. (1999). Redefining Urban Politics for the Twenty-first Century. In: A.E.G. Jonas and D. Wilson, eds. *The Urban Growth Machine: Critical Perspectives Two Decades Later. Urban Public Policy*. Albany, NY: State University of New York Press, pp. 109–124.

Cochrane, A. (2011). Urban Politics Beyond the Urban. *International Journal of Urban and Regional Research*, 35(4), pp. 862–863.
Corfee-Morlot, J., Cochran, I., Hallegatte, S. and Teasdale, P-J. (2011). Multilevel Risk Governance and Urban Adaptation Policy. *Climatic Change*, 104(1), pp. 169–197.
Desouza, K.C. and Flanery, T.H. (2013). Designing, Planning, and Managing Resilient Cities: A Conceptual Framework. *Cities*, 35, pp. 89–99.
Dessai, S., Hulme, M., Lempert, R. and Pielke, Jr, R. (2009). Climate Prediction: A Limit to Adaptation? In: W.N. Adger, I. Lorenzoni and K.L. O'Brien, eds. *Adapting to Climate Change: Thresholds, Values, Governance*. Cambridge: Cambridge University Press, pp. 64–78.
Downs, A. (1972). Up and Down with Ecology: The Issue-Attention Cycle. *Public Interest*, pp. 38–50.
Edwards, P.N. (2001). Representing the Global Atmosphere: Computer Models, Data and Knowledge about Climate Change. In: C.A. Miller and P.N. Edwards, eds. *Changing the Atmosphere: Expert Knowledge and Environmental Governance*. Cambridge, MA: MIT Press, pp. 31–66.
Edwards, P.N. and Schneider, S.H. (2001). Self-Governance and Peer Review in Science-for-Policy: The Case of the IPCC Second Assessment Report. In: C.A. Miller and P.N. Edwards, eds. *Changing the Atmosphere: Expert Knowledge and Environmental Governance*. Cambridge, MA: MIT Press, pp. 219–246.
Fraser, A. (2010). The Craft of Scalar Practices. *Environment and Planning A*, 42(2), pp. 332–346.
Granberg, M. and Elander, I. (2007). Local Governance and Climate Change: Reflections on the Swedish Experience. *Local Environment*, 12(5), pp. 537–548.
Howitt, R. (1998). Scale as Relation: Musical Metaphors of Geographical Scale. *Area*, 30(1), pp. 49–58.
Hulme, M. (2011). Reducing the Future to Climate: A Story of Climate Determinism and Reductionism. *Osiris*, 26(1), pp. 245–266.
Hulme, M. and Mahony, M. (2010). Climate Change: What Do We Know About the IPCC? *Progress in Physical Geography*, 34(5), pp. 705–718.
Hunt, A. and Watkiss, P. (2011). Climate Change Impacts and Adaptation in Cities: A Review of the Literature. *Climatic Change*, 104(1), pp. 13–49.
Jacobs, J. (1958). Downtown is for People. *Fortune*, April.
Jasanoff, S. and Wynne, B. (1998). Science and Decisionmaking: Human Choice and Climate Change. In: S. Rayner and E.L. Malone, eds. *Human Choice and Climate Change 1: The Societal Framework*. Columbus, OH: Batelle Press, pp. 1–87.
Kahn, M.E. (2010). *Climatopolis: How Our Cities Will Thrive in the Hotter Future*. New York: Basic Books.
Kythreotis, A.P. and Bristow, G.I. (2017). The 'Resilience Trap': Exploring the Practical Utility of Resilience for Climate Change Adaptation in UK City-regions. *Regional Studies*, 51(10), pp. 1530–1541.
Kythreotis, A.P. and Jonas, A.E.G. (2012). Scaling Sustainable Development? How Voluntary Groups Negotiate Spaces of Sustainability Governance in the United Kingdom. *Environment and Planning D: Society and Space*, 30(3), pp. 381–399.
Kythreotis, A.P., Mercer, T.G. and Frostick, L.E. (2013). Adapting to Extreme Events Related to Natural Variability and Climate Change: The Imperative of Coupling Technology with Strong Regulation and Governance. *Environmental Science & Technology*, 47(17), pp. 9560–9566.
Luers, A.L. and Sklar, L.S. (2014). The Difficult, the Dangerous, and the Catastrophic: Managing the Spectrum of Climate Risks. *Earth's Future*, 2(2), pp. 114–118.
McCann, E. and Ward, K. (2011). *Mobile Urbanism: Cities and Policymaking in the Global Age*. Minneapolis: University of Minnesota Press.
Norton, S.D. and Suppe, F. (2001). Why Atmospheric Modeling is Good Science. In: C.A. Miller and P.N. Edwards, eds. *Changing the Atmosphere: Expert Knowledge and Environmental Governance*. Cambridge, MA: MIT Press, pp. 67–105.
Petersen, A., Blackstock, J. and Morisetti, N. (2015). New Leadership for a User-friendly IPCC. *Nature Climate Change*, 5(10), pp. 909–911.
Porter, J.J., Demeritt, D. and Dessai, S. (2015). The Right Stuff? Informing Adaptation to Climate Change in British Local Government. *Global Environmental Change*, 35, pp. 411–422.
Rockefeller Foundation. (2016). *100 Resilient Cities*. New York: Rockefeller Foundation. Available at: www.100resilientcities.org/#/- [accessed 31 May 2017].
Rosenzweig, C., Major, D.C., Demong, K., Stanton, C., Horton, R. and Stultset, M. (2007). Managing Climate Change Risks in New York City's Water System: Assessment and Adaptation Planning. *Mitig. Adapt. Strategies Glob. Change*, 12, pp. 1391–1409.

Smith, C.M. and Kurtz, H.E. (2010). Community Gardens and Politics of Scale in New York City. *Geographical Review*, 93(2), pp. 193–212.

Solecki, W. (2012). Urban Environmental Challenges and Climate Change Action in New York City. *Environment and Urbanization*, 24(2), pp. 557–573.

Tol, R.S.J. (2011). Regulating Knowledge Monopolies: The Case of the IPCC. *Climatic Change*, 108(4), pp. 827–839.

Tompkins, E.L., Adger, W.N., Boyd, E., Nicholson-Cole, S., Weatherhead, K. and Arnell, N. (2010). Observed Adaptation to Climate Change: UK Evidence of Transition to a Well-adapting Society. *Global Environmental Change*, 20(4), pp. 627–635.

Turnhout, E., Dewulf, A. and Hulme, M. (2016). What Does Policy-relevant Global Environmental Knowledge Do? The Cases of Climate and Biodiversity. *Current Opinion in Environmental Sustainability*, 18, pp. 65–72.

Ward, K. (2011). Urban Politics as a Politics of Comparison. *International Journal of Urban and Regional Research*, 35, pp. 864–865.

Wilson, E. (2006). Adapting to Climate Change at the Local Level: The Spatial Planning Response. *Local Environment*, 11(6), pp. 609–625.

INDEX

Page numbers in **bold** denote figures, those in *italics* denote tables.

$15 minimum wage: Los Angeles campaign 300, 451, 455, 459–461; *see also* Fight for $15
5th Street garden 328
21 Tech 510

Aboutaleb, Ahmed 477
Abrams, M. 489
Abu Dhabi 188, 193
access, relationship to allocation 529
access to water, environmental restructuring and 159
accumulation, capitalist xxxi, 37, 39, 247, 254, 529
accumulation by dispossession 127, 139; definition xxiii; resistance against 38
ACT UP 523, 529
active citizenship 42, 138, 354, 357, 559
activism: AIDS activism 529; anti-sweatshop activism 148; day labour activism in the USA 293–302 (*see also* informal labour in the USA); political activism in Copenhagen 150; relationship with the regulation of urban landscapes 27; and the roots of community gardening 135–136
Adamovsky, E. 493
adaptation to climate change: Seattle's approach 117; security as central justification for 118
Affordable Care Act 263
affordable housing 169, 176, 202, 236; treatment of in right-sizing plans 202
Africa 17
African-American neighbourhoods: changes in 53; characteristics of 320; impact of de-industrialization 320, 323
Agamben, G. xxiv, xxxviii, 565–566, 571
age-friendly cities, definition xxiii

agglomeration xxxiv, 19–23, 73, 406, 408, 542, 557
AIDS activism 529
Ailanthus altissima 534
air: commodification of 90; selling of rights in exchange for building height 85
air pollution, British policy 112–113
Airbnb 43, 504, 506
Alice's Garden, Milwaukee 323, 327, 329–330
Allen, C. 15
Allen, J. 23, 148
Allen, Will 324–325
Allende, Salvador 521
allocation, and access 529
allochtonous/authochonous, definition xxiii
Alt, J. 483
alt-labour groups, definition 293
alt-labour organizing, definition xxiii
alt-labour organizing strategies in the US Midwest 259–268; anti-union legislation/policies and 260, 263, 297; building the base 263–267; churches' role 264–265; conventions 266–267; development and testing 259–260; future expectations 267–268; post-industrial context 262–263; promise and limits of alt-labour organizing 260–262; schools' role 265–266
Alterman, R. 91
Amazon.com 404–414; anchor home of 52–53; construction and politics of spaces of circulation 411–414; countries established in 406; distribution strategy 411; founding 406; GHG emissions 117; implications of the business model 407–410; improvement through time-space compression 410–411; path to expansion 406–407; rise and performance 406–407; role

Amazon.com continued
   of ICT 405, 407; Rugeley fulfilment centre 410; space of circulation 407–408; treatment of employees 410
ambivalence (capitalist), definition xxiii
American pragmatism: characteristics 147; definition xxiv; and the normative approach to analysis 151; revival 147; and urban political ecology 147–149
Amin, A. 353, 358
Amsterdam 98, 472, 476–477; anti-immigrant mobilization 473; diversity branding 476; smart systems 98
ANC (African National Congress) 124, 131
Andrucki, M.J. 529
Animal Farm (Orwell) 539
anti-austerity protests, Malaga 521
anti-globalization movements 126, 237
anti-sweatshop activism 148
anti-union legislation and policies 261, 263
apartheid 48, 124–125, 355–356; definition xxiv
Apple 439–440, 444
appropriation of rent 40, 480
appropriation, definition 126
Arab Spring 2, 519
architectural heritage, as barrier to development 176
ArcView 101
Argent, N. 341
Argentina: mothers of the 'disappeared' 521; *see also* urban middle class in Buenos Aires
Arkin, Lois 346–347
arts of citizenship 356
Aschoff, N. 43
Asian carp: definition xxiv; efforts to contain the invasion 432; leaping **432**; potential ecological threat 427, 430–431; purpose of importation 431; regional summit on 433
Asia-Pacific Rim, scale and scope of fast-track urbanism 184–193
Asimov, I. 22
assemblage: the concept 311–312, 427–428; definition xxiv; fields of employment 426; potential of the theory for understanding urban environmental systems 427
associational forms of economic governance 41
Astana, Kazakhstan 191
asylum seekers xxv, xxxii, 469–470, 472, 476
Athens, Georgia, place-based conflict 30–31
Atlanta, role of gentrification in erasing a gay neighbourhood 530
austerity xxxi, xxxiii, 41, 61, 201–203, 263, 330, 341, 443, 521–522; community gardening and 136–139
Austin, smart growth agenda 167–178
Austin, J. 67
autoconstrução 376

automobile-centred way of life 488

Baathism 48
Bailey, R.W. 529
Balkans war 568
Baltimore: empowerment zone review 510; foreclosure governance 212–214
banlieues: centralized governance 222; and coercive administrative domination 223; definition xxiv; local resistance and neoliberal reflexiveness 226; the meaning of 217; and the middle sizing of the urban poor 217, 219, 221; neoliberal restructuring of public housing 220–221; neoliberal urban renewal programme 218–220; planning rhetoric and euphemisms 225–226; policing goals and depoliticization 224–225; post-war promotion 220; public housing authorities as urban entrepreneurs 223–224; residential dispersal/diversification 219, 224; riots 218, 224; Sarkozy reforms 218
Barber, B. 50–51
Barcelona 469; declares itself a 'refugee city' 476; smart systems 98
bare life: definition xxiv; *vs* politically embedded life xxiv
Barnett, C. 24, 148–149, 151
Barry, J. 54
Baugruppen, definition xxiv
Baum, D. 523
Bayat, A. 376
Beck, S. 596
Becker, S. 128
Bee, B.A. 112
Belgium, organization strategy for migrants 471
Bell, D. 338
Benjamin, W. 187, 544
Bennett, J. 428, 435
Berlin: community gardening 135, 140; creative destruction of railway infrastructure 550–551; in-against-and-beyond the state 128–129; struggle for energy democracy 123
Berliner Energietisch campaign 123, 128
Bernstein, R.J. 147
Bethnal Green suburbanization study 482
betterment: historical perspective 89; meaning of 89
betterment levies 86; examples of 89; historical perspective 89–90
Bible communists (Oneida Perfectionists) 49
Big Data 50–52
Bing Live Maps 100
biopolitical labour 37
biopolitical production, definition xxiv–xxv
Black Lives Matter 472, 521–522
blackout of 2003; Canada and the USA 428
Blobaum, R. 378
Block, F. 49

Blok, A. 150
Blue Origin 404
Bodnar, J. 483
Boeing 52
Bogotá, transformation 418
Boler, M. 341
Boltanski, L. 150
Bond, P. 130
bonus zoning 89; examples of 90; *see also* density bonusing
bordering, definition xxiv
Bourdieu, P. xxx, 149–150
Bouthillette, A.M. 530
Bouzarovski, S. 164
Bowling Alone (Putnam) 342
Boyer, R.H.W. 346
Brasília 186–187
Brazil 39, 376, 406; enclosure movement 39
Brenner, N. xxxiv, 15, 21, 107–108
Brexit 474
Bridge, G. 147, 149, 151
Britain 1909 Town Planning Act 89; air pollution policy 112–113; do-it-together urban development 343–345; drivers of the Industrial Revolution 376; New Towns initiative 186; the street as public space 338–339; urban 'renaissance' 340
Bronski, M. 507
Brown, E. 76
Brown, M.R. 529
Brubeck, Dave 507
Brundtland report 167
Budds, J. 111–112
Buenos Aires: evolution of the city 493–494; gentrification 499; loss of diversity 498; *Parque Indoamericano* protests 499–500; *see also* urban middle class in Buenos Aires
Buffalo, community gardening in 138
building height: motivations for increasing 87; selling of air rights in exchange for 85; tallest residential tower in Toronto 87
Bulkeley, H. 395, 422
BürgerEnergie Berlin 128
Burgess, E. 236
burkini ban 476
Bush, George W. 110
Business Improvement Area strategies 86
Bustamante, R. 130

C40 club 50
Cairo, Tahrir Square demonstrations 519
Calais, *the Jungle* 472
California: intelligent vehicle highway system 99; renewable energy 164
Calvita, N. 89
Canada: blackout of 2003 428; expansion of temporary migration streams 246; Golden Horseshoe 88; student-led uprisings 237; treatment of non-Canadian workforce 245; women's labour organizing 234; working holiday programme 250–251
Cape Town: infrastructural citizenship in 350–359; role of rent and property in sexual politics 530
capital accumulation xxix, xxxiii, 54, 80, 136, 139, 147, 150–151, 247; the city as a source of 186
Capital in the 21st Century (Piketty) 49
capitalism: compassionate *see* compassionate capitalism; geopolitics of xxix; inability to fulfil expectations 161; reappropriation of the commons from 39; tendency towards subsumption of society 43
capitalist accumulation xxxi, 37, 39, 247, 254, 529
car ownership, contribution to family-vocational privatism 484
Carlisle, flood alleviation scheme 591
Castán Broto, V. 422
Castells, M. 15, 20, 229, 295, 339, 389, 529, 592
Centemeri, L. 150
Centeno, M. 294
chain migration 482
Chandigarh 186
charter cities, as libertarian option 189–191
Chatterjee, P. 354
Chevron Oil Terminal dispute, Nigeria 38
Chiapas Rebellion, Mexico 38
Chicago 469; bonus zoning 90; foreclosure filings 206; founding 429; receivership as field of institutional expansion 209; Troubled Buildings Initiative (TBI) 209
Chicago Area Waterways System (CAWS) 426–436; the CAWS as assemblage 429–430; Chicago Diversion 430, **431**, 434; and the Clean Water Act 430; and the concept of geographical assemblage 427–428; contamination 429–430; ecological separation plan 427, 433–434; historical perspective 429; proposed solutions to the Asian carp problem 433–435; research on urban water systems 428–429; role of the water 426; Sanitary and Ship Canal 430, 432–433; threat posed by foreign fish 431–433 (*see also* Asian carp)
Chicago School 15, 17, 147, 469
children: impact of the end of child labour on family structure 483–484; providing safe spaces for play 135, 327, 329, 338–339, 343–345, 551; resilience research 160
Children's Rights, definition xxv
cholera: CAWS contamination and outbreaks of 429; KwaZulu Natal outbreak 124
Christopherson, S. 79
Cincinnati 481, 485
circulating risks, definition xxv
circulation, definition xxv

Index

circulatory spaces 389–462; Asian carp in the Chicago area waterways system 426–436; coastal cities and the spectre of climate change risk 393–401; exploring urban and non-urban spaces of *Amazon.com* 404–414; multi-racial counter-publics 451–462 (*see also* egalitarian political spaces); tech mobility and the future of urban livability 439–449; urban experimentation and the politics of sustainability 416–423; urban experimentation and the politics of sustainability (*see also* urban living laboratories)

cities: and differences in experience 27; as generators of diversity 19; as informal site of institutional experimentation 41

citizen engagement 52

citizens' networks 42

citizenship: active 42, 138, 354, 357, 559; 'acts of citizenship' 351; definition xxv; legal practices and socio-political acts 350–352; neoliberal legitimisation of conditional citizenship 321; Painter's definition of urban citizenship 273; precarious employment and the work-citizenship matrix 273; relationship with employment conditions 248

citizenship/non-citizenship, definition xxv

the city: impact on human relationships 18; Wirth's definition 18

The City and the Grassroots (Castells) 20

'City as a Growth Machine' (Molotch) 203

city branding, diversity as instrument for 476

City of Quartz (Davis) 375

CityLab 441

city-regions: logistical capacities 77; the rise of 50

civic participation 42

civil society organizations, outsourcing of state functions to 137

Clark, J. 79

class, and the urbanisation of self-management in Poland 375–385

class and territory in the American city 479–489

class monopoly rents: definition xxv; property capital and the search for 485–488

class politics: and housing markets 485–488; lineages of 376–378; territorializing 481–485

class struggle 37, 247

Clean Development Mechanism 596

Clean Urban Transport Europe Programme 420

Cleveland: decline of 200; foreclosure governance 212–213; receivership as field of institutional expansion 209

climate adaptation 590; coastal restoration techniques 398; definition xxv; New York initiatives 593; relationship to resilience 590

Climate Alliance 50

climate change: cinematic representations 589; coastal cities and the spectre of climate change risk 393–401; definition xxv; as motivation for emergence of smart cities 97; potential impact 393; Renewable Wilhelmsburg Climate Protection Concept 420; role of new technology industries 117; STS literature 595

climate change risk in coastal cities: and the concept of capital circulation 394–396; gender and inequality 396–398; mitigation and adaptation 398–400; time-space perspectives 394–396

climate governance, definition xxv

climate knowledge, planning and politics 589–598

climate policy: adaptation planning 117–119; C40 Cities Climate Change Leadership Group 112; central questions 113; depoliticization 119–120; IPCC Fourth Assessment Report 110; measurement as the pathway to 116–117; questions left off the table 113; role of climate change science 110; science and governance in Seattle 114; the scientific consensus 114–116; scientific framing of security 117–119; and the state-science relationship 112–114; and the technocratic nature of governance 111–112; US Mayors Climate Protection Agreement 110, 115

climate reductionism, definition xxv

climate resilience: definition xxv; knowledge generation/circulation and 594–596; urban imaginaries and the politics of climate change 591–593

climate science, and the city 110–120

Cloke, P. 310

coastal cities, and the spectre of climate change risk 393–401

Cobden, Richard 48

Cochrane, A. 1, 23–24, 591, 598

Coe, N.M. 70, 72, 75, 79–80

coercive care 316–317

Coghill, K. 66

cohousing 42, 342

Cold War 48, 379–380

Coleman, M. 468

collaboration/co-production, definition xxvi

collaborative commons 43

collaborative experimentation, community-based 42

collaborative governance model 321

The Collective, co-living business 43

collective consumption 234–235, 262, 308, 379; definition xxvi; and the development of urban social movements 15, 20–21; financing struggles in blue-collar cities 53

Collier, S. 378, 428

Cologne, community gardening project 140

colonial expansion 127

The Colony (Renfroe) 589

common property resources (CPR) 36

common/commons: Dardot and Laval's conception 38; definition xxvi; dispossession of native Americans' common resources 38; as an element of political praxis 38; explanations for urbanization 40; Federici on 126; Hardt and Negri's contribution to the theory 37; Harvey's dual strategy 126; Marxist perspectives 37–39, 40; multiple valences 41–43; original rehabilitation 41; Ostrom's criticisms of Hardin's work 36; and the post-crisis high-tech boom 42; radical feminist scholarship 38; resistance campaigns 38; role of the appropriation of rent 40; and 'rurbanization' processes 39; theorising 36–39; trans-historical perspective 38–39; as unifying lexicon for a cooperative society 126; urban gardening and green urban commoning 139–141; urbanizing 39–41

Commonwealth (Hardt and Negri) 37

Commun: essai sur la revolution au XXIème siècle (Dardot/Laval) 38

communicative planning 556, 562; definition xxxiv

communism 37, 49, 382

communities, development and the breakup of 481–485

community benefit agreements (CBAs) 510, 512, 514

community benefits: definition xxvi; development offsetting function xxvii, 86, 88–90; disingenuousness of the descriptor 92; examples of 89–90; implementation approaches to 89–90

community gardening: and access to nutritious food 321, 324–325, 327–329; All People's Church community gardening initiative 328–329, 331; Berlin projects 135, 140; challenges in the neoliberal city 321–322, 330–332; Cologne project 140; community gardens as new spaces of living 320–333; counter-cultural roots 135–136; cultural perspectives 329; and gentrification and image policies 136–137, 322; German movement 140–141; internal contradictions of 134; in Milwaukee 323–327; neoliberal and austerity urbanism context 136–139; New York 135–136, 140; and the outsourcing of services to the voluntary/third sector 137–138; potential effect of community gardens on property values 322; privatization concerns 367; procuring resources 331; race and 368; reliance on volunteerism 330–332; social capital building potential of 320, 329–330; as state strategy to control space 138–139; transformative potential of community gardens 320; urban gardening and green urban commoning 139–141; what is a community garden 134

community unionism, definition xxvi

community-labour coalitions 261, 263; definition xxvi

company towns 186

compassionate capitalism: the concept 505; definition xxvi; payroll tax breaks and corporate revitalization 509–510; social segregation and the new economy 505–506; tech solutions to social problems 511–512; technology companies in San Francisco 506–507; the Tenderloin district's history and demographics 507–509

competition for land uses 19

condo boom, definition xxvi

condominium: the concept 87; definition xxvi

condominiums, Rainey, Austin 174, 176

conservation movement, dispossession of native Americans' common resources 38

contingent work, definition xxvi

contraception 483

Cooper, M. 160

Cooper Marcus, C. 338

cooperative governance mechanisms, deployment of 67

cooperative society, the commons as unifying lexicon for 126

cooperatives 126

co-ownership 42

Copenhagen, political activism 150

Coradini, Bartolomeo 48

Corburn, J. 113

corporate activity, mapping 71

corporate cities paradigm 191–192

corporate social responsibility 511

corporate welfare 509

Couche Tard convenience stores, union organizing drive 238–239

counter-publics xxvii, 453; construction role of physical and virtual spaces 454; definition xxvii; function of 453; and the re-politicization of social responsibility and social justice 453

Cox, K. 6, 74, 320

Crawford, M. 186

creative destruction xxvii, xl, 184, 247, 543–544, 551–552; definition xxvii; of urban space 542–552

creative destruction of railway infrastructure: Berlin 550–551; New York 548–550; Paris 544–548

Cresswell, T. 27

Cronon, W. 107

Cumbers, A. 126–127

Curtis, S. 422

Cutchin, M.P. 147–148

Cybriwsky, R.A. 484

Czerwiński, M. 379

Dardot, P. 38

Davidson, K. 112
Davidson, M. 22, 452
Davies, J. 223
Davis, M. 375
Davis, Miles 507
The Day after Tomorrow (Emmerich) 589
day labour, definition xxvii, xxx
day labour activism in the USA 293–302; *see also* informal labour in the USA
day labour markets, and the spread of labour standards violations 294–295
De Genova, N. 254
De La Rua, Fernando 494
Dean, M. 138
Dear, M. 309
The Death and Life of Great American Cities (Jacobs) 111
degrowth machine, the concept 197
de-industrialization 234, 262, 380–381, 409, 556; impact on African-American neighbourhoods 320, 323
DeLanda, M. 312, 427
Deleuze, Giles 427
dell'Agnese, E. 47
Delorenzi, R. 342
democracy: Barber on the crisis in 50; relationship with equality 452
democratic innovation 149, 151–152
democratic politics, role of urban space 149
democratizing urban environments 122–132
Denmark, policies on binational families 475
density bonus, definition xxvii
density bonusing: Austin's programme 176; as backroom deal 86; and development in Toronto 90–92; historical perspective 89–90; implementation approaches to community benefits 89–90; origins 86, 89; Toronto's condo boom and policy framework 87–88; use of as an enticement 86
deregulation of rental controls, impact 234
Dessai, S. 596
Detroit: city regeneration strategies 199; decline of 200; influence of the rentier class 202; planned shrinkage of neighbourhood services 201; public-private partnerships study 67; right-sizing plans 197, 201
de-unionization, in US building trades 296–298
development: and the breakup of communities 481–485; uneven xxxix
DeVerteuil, G. 307
Dewey, J. 146–148, 151
DiGaetano, A. 65
Diggers and Dreamers 342
digital divide 100–103; impact of growth in wireless and mobile broadband services 103; as obstacles to implementation of e-government 102

digital technologies: impact on urban space 96–103 (*see also* information and communications technologies); tech mobility and the future of urban livability 439 (*see also* tech mobility)
Diouf, M. 356
direct action: definition xxvii; examples of 341, 521, 523; involvement of unpaid work 340; and queer political recognition 529; women's 523
'disappeared', Argentina 521
discrimination, in *Amazon.com*'s provision of services 413–414
displacement 219; of peasant populations 39
displacement of the working poor, fast-track urbanism and 186
dispositif, definition xxvii
dispossession, accumulation by *see* accumulation by dispossession
disruption: as function of protest 27; as means of innovation 410, 413; as means of insurrection 237
distributed agency, definition xxvii
distribution/fulfilment centres, definition xxvii
disused railways, and the re-designation of space 542–552
diverse economies research 38
diversity: cities as generators of 19; as instrument for city branding 476; migration policies and 475–476
Doan, P. 530
Doha 188, 193
do-it-together urban development 341–342; LA Ecovillage 345–347; liveability and 336–348; Playing Out 343–345
Dolby 509
domestic work in Indonesian cities 271–280; global geographies of domestic work 272–274; human trafficking 275; literature on Indonesian domestic workers working abroad 272; moral contrasts in Kupang 275–277; qualitative research 274–275; recruitment agents 274; rights-based perspectives of migrant domestic workers 273, 277; socio-geographical implications of domestic worker migration globally 272–274; training centres 275; wage levels 276, 277
domestic workers, definition xxvii
Donaldson, A. 149
Dorman, S. 356
Downs, A. 593
Downtown Eastside Neighbourhood House 361
Driscoll Derickson, K. 162
driverless cars *see* self-driving cars
Dropbox 509
Dubai 188, 191, 193
Dukes, T. 475

Index

Durban: in-against-and-beyond a rights-based politics in Durban 130–131; struggle for the right to water 123–124
Duyvendak, J.W. 475
dystopia, definition xxvii, 539
dystopian perspectives: climate knowledge, planning and politics 589–598; creative destruction of urban space 542–552; spaces of dystopia and utopia 539–598
dystopian spaces, and young Roma in Slovenia and Romania 565–574

Eaton, E. 139
eco-gentrification 364–365; definition xxvii
ecological modernization: definition xxviii; and socio-ecological injustice 163–164; unintended consequences 159
e-commerce, definition xxvii
economic development, spaces of 59–103; digital technologies' impact on urban space 96–103 (*see also* information and communications technologies); pro-growth politics and public-private partnerships 62–67; strategic coupling in global production networks 70–82; Toronto's condominium boom 85–93 (*see also* density bonusing; planning)
economic governance, associational forms of 41
economic informality, drivers of 294
ecovillage, definition xxviii
Edensor, T. 428
egalitarian political spaces: importance of physical spaces for the construction of 454; the making of 452–454; *see also* worker centres
e-government, concerns over 100
Eizenberg, E. 135, 139–140
Elder, G.S. 529–530
employment protections, breakdown in the USA 293
empowerment zones 509
enclosure xxiii, 38–40, 139–140
Energietisch 123, 128–129
energy, the right to access 128
energy poverty 159; definition xxviii
energy rights in Germany 123
England's Economic Heartland 16
Enke, A. 530
enterprise society 39
enterprise zones 192, 505, 509–510
entrepreneurial governance, definition xxviii
'entrepreneurial slum' 41
entrepreneurialism 62, 193, 217–218, 220, 307–308, 321
entrepreneurship, resilience and 161
environment and nature, spaces of 107–178; Austin's smart growth agenda 167–178; climate science and the city 110–120; democratizing urban environments 122–132; green spaces of urban politics 146–154; hidden costs of socio-environmental change in the Global North 157–165; politics of urban gardening 134–142 (*see also* community gardening)
environmental gentrification, meaning of 167
environmental justice, advocacy role of worker centres 455
environmental perspectives, creative destruction of urban space 542–552
Epstein, R. 222
equality: Rancière's view 452; spatio-temporality 454
erasure: creation of the political process 569; definition xxviii, 567; impact 568–570
Erlich, M. 298
Ete, H. 152
ethnic minority, definition xxviii
ethnicity, definition xxviii
'Ethnography of Infrastructure' (Star) 352
Études sur Paris (Sauvage) 545
Euromaidan protests, Kiev 519
Europe: energy poverty 164; post-war public housing policies 220; proliferation of living labs 419, 421 (*see also* urban living laboratories)
European City model: characteristics of the European City 557; and the chimera of utopia 562–563; definition xxviii; the European city as a modern utopia 557; friction and 562–563; as a political approach 557–558; potential to produce more sustainable urban areas/cities 555–563; urban development in Freiburg 558–562
European Commission 419, 475–476
European fascism 49, 378
European Union (EU): energy consumption 163; Lisbon Strategy 419, 557; smart cities 98
European Youth Capital 573
Evans, D. 159
eventful subjectivity, definition xxviii
everyday environmentalism, the concept 125
eviction, definition xxviii
exclusion, definition xxviii
expats, definition xxviii
experimental city: the concept 416 (*see also* urban living laboratories); definition xxix
extraction of natural resources 39

Facebook 100, 117, 406, 439–440, 444–445, 506
the factory, residential equivalent of as a place of organisation 14
Fainstein, S. xxxi, 562
familial-vocational privatism 481, 483, 485; contribution of car- and home-ownership 484; definition xxix
family, structural transformation 483–484
Fannie Mae 212
Farías, M. 493

fascism 49, 153, 499
fast-food workers, strikes 293 (*see also* Fight for $15)
fast-track urbanism: definition xxix; and the displacement of the working poor 186; main objective 185–186; scale and scope of 184–193; *see also* instant cities
Federal Housing Finance Agency 212
Federici, S. 37–38, 126, 129
Femen 523
Ferguson, J. 111
Fight for $15 xxix, 237, 239–240, 261, 263, 267, 293, 300
figuration, definition xxix
Filmus, Daniel 495–496
financial crisis of 2008: and Canada's condo boom 87; and the high-tech boom 42; labour organizing post-crisis 232–241; social impact 263; urban roots 40
Fine, J. 456
FIRE (finance, insurance and real estate) sector 199, 236
fissured workplace 295–296
Fitzpatrick, S. 316
Flint: decline of 200; right-sizing plans 201
flood risk management, research into 149
flooding 395, 431, 433, 589, 591
Florida, R. 238, 376, 443, 476
fLux (LAb[au]) 584
Focus E15 521–522
Follmann, A. 140
food insecurity xxix, 320, 323, 369; in US cities 320, 327
food justice: activism for in Milwaukee 327–329; definition xxix
Forbes, J. 510
Fordism 234, 262
foreclosure governance 206–215; construction of legal complexes 211–212; jurisdictional politics 208–212; knowledge of property 209–211; lacuna in academic attention 206–207; legal interpretation perspective 207–208; legal limbos 209–211; long-term effects of the foreclosure crisis 206; mass nuisance and *parens patriae* 212–214; nuisance law perspective 208–209; *parens patriae* and the city as quasi-sovereign 213–214; partnership strategies 211–212; racial discrimination suits 213–214; subprime lending as mass nuisance 212–213
Forest, B. 529
Foucault, M. xxx, xxxv, 221, 312
France: organization strategy for migrants 471; public housing in 217–226 (*see also* banlieues)
Fraser, N. 452–454, 465
Freddie Mac 212
Fredericks, R. 356
free markets, Polanyi's conclusions 49

freeways 487
Freiburg, urban development in 558–562
Freiburg Charter for Sustainable Urbanism xxix, 559
French pragmatic sociology, definition xxix
French pragmatism 146, 149–152
Friends of the High Line 549
fuel poverty, in Europe 164
Fukuyama, F. 48
Fuller, C. 150

Gandy, M. 352, 533
Gans, H. 483
Garden City model 52, 186, 188
Geddes, P. 48, 52
Gehl, J. 339, 341
Gehry, Frank 87
Geltmaker, T. 529
gemeinschaft (close-knit community) 340
gender inequality, definition xxix
gender protests, and politics of solidarity and intersection 519
gendered division of labour, and access to potable water 125, 131
gendered perspectives: disproportionate impact of climate change on women 396–398; gendering protest 518–525
gentrification: Airbnb and 43; Austin 169; Buenos Aires 499; the commodification of gay culture and 530, 534; community gardening and 136–137, 322; community impact 484; the concept 336; eco-gentrification xxvii; green 158–159; Manhattan meatpacking district 550; New Orleans 530; new-build gentrification 92; racial impact 368; relationship to livability 336; relationship with green infrastructure 364; relationship with urban agriculture 362, 364–365, 368–369; smart cities and 97; social impact 234, 236, 484; tech mobility's contribution 442, 445, 447
geopolitics of capitalism, definition xxix
Germany: community gardening movement 140–141; re-municipalization of energy supplies 123
geselligkeit ('play-form' of association) 340, 345
Gezi Park demonstrations 519
Gezi Park/Taksim Square demonstrations, as example of pragmatist urban political ecology 151–153
Ghaziani, A. 530
GHG emissions: and new technology industries 117; role of measurement in addressing climate change 116; Seattle's policies 116–117
'ghost acres' 376
Gibson-Graham, J.K. 38
Giddens, A. 41

Gilbert, L. 468
Gleeson, B. 112
global cities: and economic governance 76; function of in global production networks 76
global production networks: attributes of GPN theory 71–72; definition xxix; mapping corporate activity 71; Singapore's 'total package' offering 77; strategic coupling in 70–82; *see also* strategic coupling in global production networks
Global South, social fabric of cities in 41
global warming 110, 115
Golden Horseshoe, Ontario 88
Goldring, L. 252, 273
Gómez-Baggethun, E. 157
Goodchild, M. 100
Google 406, 439–440, 444–445, 506–507; headquarters 444
Google buses 440; definition xxix
Google Maps 100
Gordon, I. 16
governance, definition xxix
governing and planning spaces 181–226; foreclosure crisis in the US 206–215; public housing in France 217–226 (*see also* banlieues); 'right-sizing' 197–204; scale and scope of fast-track urbanism 184–193
governmentality, definition xxx
Gowan, T. 508
GPS (Global Positioning Systems) xxxviii–xxxix, 100, 439–440
Grabelsky, J. 298
Graham, S. 96, 352–353
Gramsci, Antonio xxiv
Grand Central Station, New York City 90
Great Lakes, Marquette and Joliet's exploration 429
Great Recession xxxvi, 298, 300
The Great Transformation (Polanyi) 49
green gentrification 158–159
green governmentality, and the 'science effect' in urban climate governance 112–114
Green Guerillas 135
Green Paper on the Insurance of Natural and Manmade Disasters (European Commission 2013) 162
green spaces, politics of urban gardening 134–142
green spaces of urban politics 146–154; Gezi Park example 151–153; pragmatist urban political ecology 147–151
Gregory, J. 52
Gross, J. 378
Groundwork Trusts 326
group homes, definition xxx
Grow and Play community garden 327, 329, 332
Growing Power 324–325, 329–331
growth, the role of land and development 19

growth machine, Molotch's concept 197, 202–203
Guatemala City, construction of a private city on the outskirts 190–191
Guattari, Felix 427

Habermas, J. xxix, 452, 481–483, 485–486
habitus xxx, 150, 511; definition xxx
Hackworth, J. 223
Hall, P. 16, 556
Hamburg, Renewable Wilhelmsburg Climate Protection Concept 420
Hammond, Robert 549
Hanseatic League 51
Happy City movement 418
Harassmap 523
Hardin, G. 36, 40
Hardt, M. 37, 40, 126
Harmony Society 49
Harris, C. 482, 489
Hartman, Chester xxxvi
Harvey, D. xxix, 1, 38, 59, 107, 125–126, 240, 248, 305, 342, 375, 389, 480, 485–488
Hayek, F. 49
health and safety, decline in enforcement of standards 296
Hegel, Georg 48
Henri Lefebvre 15
Herod, A. 262
Hess, M. 79
Hewitt, K. 162
Hidalgo, Anne 477
Higgins, H. 530
High Line, New York: creative destruction of railway infrastructure 548–550; Friends of the High Line 549; park development 534; photographs 552
Hinchliffe, S. 149
hipsters 52, 368
historical materialism 2, 531
Hobsbawm, E. 237
Hodson, M. 420
Hoekstra, M. 471
Holden, M. 151
Holifield, R. 149
Hollaback! 523
Holloway, J. 126–127, 129
Holocaust 377
Holston, J. 351–353, 358, 376
home sharing economy 43
homeless: technological tools for 511; treatment of in Buenos Aires 498
homeownership: as behaviour modifier 219; contribution to family-vocational privatism 484; post-war growth in 486; as primary means of investment in the US 32; social class and 377; vertical multiplication of 87
homo sacer 565, 571

homophobia, worker centres as spaces of negotiation and learning about 458
Honduras, free market cities 190
Hong Kong: migrant work opportunities 273; Umbrella Movement protests 519
hope, definition xxx
Horizon 2020 419
housing, post-war transformation 486
Housing Cooperative 42
housing markets, class politics and 485–488
Houston, residents' mobilization against highway construction 355
Howard, Ebenezer 52, 186
Howard, P. 102
Hudson, W. 507–508
Hudson-Smith, A. 96
Hughes, R. 187
Hughes, S. 113
Hulme, M. 596
Huron, A. 39
Hurricane Sandy 593
Hyman, K. 355

ICLIE Charter 50
identity, spaces of: class and territory in the American city 479–489; gendering urban protest 518–525; migration and identity politics 468–477; queering urban politics and ecologies 528–535; shifting sensibilities of middle-class identities in Buenos Aires 492–500; tech companies and the transformation of San Francisco 504–515 (*see also* compassionate capitalism; Tenderloin District, San Francisco)
identity politics: the concept 465; definition xxx
If Mayors Rule the World: Dysfunctional Nations, Rising Cities (Barber) 50
immigrants: advocacy role of worker centres 455; cities as gateways for 469; day labourer organizing in the early 1990s 300; disproportionate occupation of the informal sector in the USA 295; immigrant workforce and patterns of industrial change within the residential construction industry 298–299
immigration, privatizing effect of employer-driven systems 253
Implosions/Explosions: Towards a Study of Planetary Urbanization (Brenner) 108
Inceoglu, I. 152
inclusion, definition xxx
income inequality, in the world's cities 229
India, anti-colonial uprising 38
individual responsibility, neoliberal emphasis 452
Indonesia, domestic work in Indonesian cities *see* domestic work in Indonesian cities
Industrial Revolution 376
industrial revolution, third 43
inequality: in *Amazon.com*'s provision of services 413–414; causes and social impact 232; and the crisis of 2008 39; the digital divide 101–103; drivers of 295
informal economy, definition xxx
informal labour in the USA 293–302; disproportionate occupation by immigrant workers and members of racial minority groups 295; informalization in the construction sector 296–299; labour standards enforcement failures 295–296; and the politics of precarity 293–294; regulating informal labour markets from the bottom-up 300–301; role of the state in the spread of labour-market informality 294–295
informal settlements, and access to water 131
informal spaces of urban politics 27
informal urbanism, definition xxx
information and communications technologies (ICTs): definition xxx; the digital divide 101–103; geographic information systems and participatory planning 101; historical perspective 97; role of in *Amazon.com* 405, 407; smart cities 97–99; and urban e-government 99–100; urban planning and the growth of Web 2.0, 100–101
infrastructural citizenship: in Cape Town 350–359; the concept 354–355; the concept of citizenship 350–352; the concept of infrastructure 352–354; definition xxx; protest-based forms of 354–355
infrastructure: literature review 353–354; and non-citizen construction labour 249
inner cities: characteristics of African-American neighbourhoods 320; community gardens 320, 329 (*see also* community gardening); global revival of interest 87; property tax revenue capturing abilities 198; revival in development interest 87; 'service-dependent ghettos' 309; social impact of gentrification 484
innovation, disruption as means of 410, 413
instant cities: building cities from scratch 188–189; charter cities as libertarian option 189–191; and the conflation of history and chronology 189; corporate cities paradigm 191–192; definition xxx; development model 185; early prototypes of planned urban living 186–187; examples of 188; historical comparisons 187; master-planned cities built from scratch 191; trends and possibilities 192–193; utopian vision 187–188
institutional experimentation 41, 151–152
Insurgent Citizenship (Holston) 351, 353
insurrection, disruption as means of 237
insurrections and cities, Hobsbawm's survey 237
intelligent cities: definition xxxi; *see also* smart cities
intermediate goods and services, percentage of global trade accounted for by exchanges of 70
international relations, cities as new sites for 50

Internet: collaborative commons 43; the digital divide 101–103; internet of things 51, 98
intersectionality, definition xxxi
Intifada 520
IPCC (Intergovernmental Panel on Climate Change) 110, 114, 393, 395–396, 399, 595–596
Iran 376
Ireland, care workers campaign 240
irrigation, smart systems 98
ISIL (Islamic State of Iraq and the Levant) 476
Isin, E.F. 351, 358
Iskander, N. 299
Israel 376
issue articulation 148
Istanbul 469, 519
Italy, and the slow food movement 42
Iveson, K. 22, 452

Jacobs, J. 18–19, 111, 339, 375–376, 593
Jacobs, W. 77
James, William 147
Jasanoff, S. 595
Jayne, M. 338
Jerram, L. 378
Jerusalem, Women in Black movement 520
Jessop, B. 127
Jobs with Justice 237
Johnsen, S. 316
Joliet, Louis 429
jurisdictional fragmentation, definition xxxi
The Just City (Fainstein) 556, 562
Just City, definition xxxi
just green spaces of urban politics 146–154
Just Us! coffee shops, worker campaign 238
Justesen, R. 354–355
Justice for Janitors 260

Kahn, M.E. 589
Kaika, M. 124–125, 579
Kallipolis (Plato) xxxiii
Kasinitz, P. 509
Katznelson, I. 376
Kazakhstan 191
Keil, R. 232
Kellerman, A. 405
Kelly, P.F. 81
Kenney, P. 378
Kenworthy, J. 443
Khan, Sadiq 477
Kielce Pogrom, Poland 378
Kiev, Euromaidan protests 519
Kimeldorf, H. 233, 241
Kinderbauernhof 135–136
King, Rodney 346
Kirchner, Nestor 495
Kitchin, R. 52, 112
Kittay, E.F. 340

Knigge, L. 138
Knopp, L. 529
knowledge co-creation, definition xxxi
Koch, N. 189
Kolbitsch, J. 101
Kommando Rhino squatter collective 561–563
Krahulik, K.C. 529
Krugman, P. 375
Kurczewski, J. 380
Kuymulu, M.B. 151, 153
Kuznets, S. 50
Kuznets Curve 50
KwaZulu Natal, cholera outbreak 124
Kyoto Protocol 589, 595–596

LA Ecovillage 345–347
LAb[au] 584
labour, social division of 483
labour, spaces of 229–302; domestic work in Indonesian cities 271–280; informality and the politics of precarity 293–302; institutional geography of labour in the US Midwest 259–268; non-citizen labour and precarious construction work in Toronto 245–255; roll-against neoliberalism and labour organizing post-crisis 232–241; street work as key site of urban politics 282–290
labour organizations, as shapers of economic developmental processes 80
labour standards: definition xxxi; violations as feature of low-wage industries 295; widespread violations 293
Laclau, E. 452–453
Lagendijk, A. 77
laissez-faire economics 48
land and development 15
land and property values, as driving force of urban politics 19
land bank movement 201–202, 203
land encroachment 39
land uses, competition for 19
land value, enhancing 86 (*see also* density bonusing)
land value capture 86, 89; *see also* density bonusing
Landolt, P. 252, 273
Landscapes of Despair (Dear/Wolch) 309
Larreta, Horacio Rodriguez 499
late neoliberalism, definition xxxi
Laval, C. 38
Lawhon, M. 356
Layne, K. 100
LCD industry, growth of 75–76
Le Corbusier 186
Lebow, K. 382
Lebuhn, H. 249
Lee, Ed 505, 507

Lee, J. 100
Lee, Y.S. 75, 81
Lee, Z. 189
Lefebvre, H. xxxvi, 15, 40, 107, 125, 181, 224, 340, 379
left coast formula, of urban politics 52
Leipzig Charter on Sustainable European Cities xxviii, xxxi, 556–559
Lesbian Avengers 523
Levittown 486
LEWRG (London Edinburgh Weekend Return Group) 127–128
LG Display 75
LGBTIQ, definition xxxi
Li, T.M. 312
liberalism, and the state 48
The Life and Death of Great American Cities (Jacobs) 339, 375
Life Between Buildings: Using Public Space (Gehl) 339
Linders, D. 100
Linebaugh, P. 38
Linklater, A. 376
Linn, K. 140
Lippmann, Walter 148
Lisbon Strategy 419, 557
liveability: the concept 336; definition xxxi
liveable cities: examples of resident-led initiatives 343–347; public space and ordinary streets 338–340; from self-help to cooperative collaboration 341–342; the social practice of place-making 340–341; utopian methods of envisioning livability 342–343
Liverpool 469
living spaces 305–385; community gardens as new spaces of living 320–333; infrastructural citizenship in Cape Town 350–359; liveability and do-it-together urban development 336–348; poverty management spaces 307–317; retroactive utopia 375–385; urban agricultural politics 361–371
Living Streets 339
living wage movement 237, 239–240, 260, 293, 297, 455
Lloyd, R. 186
the local, Harvey on the dangers of fetishizing 126
Local Enterprise Partnerships 17
local government, as institutional framework for the exploration of politics 14
localism 337, 341
Lofland, L.H. 339
Loftus, A. 125, 131
Logan, J. 15, 19
The Logic of Collective Action (Olson) 36
logistics, definition xxxii
London 2012 Olympics 16, 521; for-profit co-living spaces 43; as a World city 23

London and the South East: governance challenges 16; growth region strategy 16
Los Angeles: ACT UP 529; do-it-together urban development 345–347; Ecovillage project 338; ethnic and racial diversity 454–455; history of cross-cultural and cross-racial organizing 455; minimum wage campaign 459; place-based conflict 29, 31; port closures 354; role of worker centres in 454–455; sanctuary city 476; Skid Row 307–317
Los Angeles River 429
low carbon urbanism, definition xxxii
Lowe, N. 299
low-road business practices 293, 297, 299
Lumsden, F. 131
Lyft 98, 439, 441, 446

MacKinnon, D. 162
Macri, Mauricio 495–499
Madres de Plaza Mayo 521
Magna Carta Manifesto (Linebaugh) 38
Malaga, anti-austerity protests 521
Malczewski, J. 101
Manchester, St Ann's Church 428
Manquele, Christina 130
Maoism 48
MapInfo GIS 101
Maple Spring 237
mapping websites 100
Maputo, street work as key site of urban politics 282–290
March of the Empty Pots 521
Marcuse, P. 225
Marquette, Jacques (Père Marquette) 429
Marres, N. 148–149
Martin, S. 507
Marvin, S. 96, 352–353, 420
Marx, Karl 39, 49, 124
Mason, P. 378
mass social housing policy, definition xxxii
Massachusetts, place-based conflict 31
Massey, D.S. xxxviii, 1, 22–23, 337, 389
Massuh, G. 498
material participation 149
Maurer, B. 101
mayors, global parliament of 51
Mayors Climate Protection Agreement 53
McCaffrey, A. 65–67
McCann, E. 593–594
McCarthy, J. 141
McClintock, N. 142
McFarlane, C. 312, 353
McKeithen, W. 534
Mead, George Herbert 147
Medicaid 263
megacities 41, 272–273, 379, 469
Meilvang, M.L. 150

Menem, Carlos 494
Merrifield, A. 21–22
MERS (Mortgage Electronic Recording System) 207, 210–211
method of equality, definition xxxii
metro art: definition 579; imagining networked utopias 578–580; mobilizing aesthetics of the global city 580–583; in Toronto 577–585
metromobility, definition xxxii
Mexico, Chiapas Rebellion 38
Meyer, R. 233, 241
microelectronics revolution, definition xxxii
middle sizing of the urban poor, French housing policy and 217, 219, 221
middle-class activism, hostility towards in urban studies literature 375
Midlands Connect 16
Midlands Engine, launch 16
migrant labour, definition xxxii–xxxiii
migrants: definition xxxii; and diversity as instrument for city branding 476; European migration crisis 476; shift from national to religious labels 472; voting rights 471, 473
migration, Canadian expansion of temporary migration streams 246
migration and identity politics 468–477; anti-immigrant mobilizations in cities 473–474; definition xxxii; migrants and the urban grassroots 469–472; migration policies and diversity 475–476; relationship between cities and 468–469
migration-urbanization nexus, definition xxxiii
Milan 469
Milk, Harvey 529
Miller, Z. 481–482
Million Man March, Washington 520
Million Woman March, Philadelphia 520
Milton Keynes 186
Milwaukee: community gardening in 323–327; food justice activism in 327–329; networks of association for urban agriculture **328**
Milwaukee Urban Gardens 326, 329–330
minimum wage 240, 277, 300, 412, 447, 459; setting and enforcement by worker centres 300
minimum-wage 293
Miraftab, F. 226
Mitchell, D. 246
Mitchell, K. 233
Mitchell, T. 111
Mitchell, William 418
mitigation of climate change, Seattle's approach 115
mobility, definition xxxiii
Mobypark 98
Molotch, H. 15, 19, 197, 203
monarchism, Nozick's brand 48
Monk, Thelonious 507

Montréal, immigrant worker initiatives 239
Moon, M. 101
Mooney, G. 571
moral geography, definition xxxiii
More, Thomas xl, 539
Mormons 49
Morocco, Women Shoufouch 523
Moscow metro 579
Moser, S.C. 393
Moses, Robert 375
mothers of the 'disappeared', Argentina 521
Mouffe, C. 148, 452–453
Muller, T.K. 530
multiculturalism 476
multi-racial counter-publics 451–462; see also egalitarian political spaces; worker centres
multi-racial counter publics, role of worker centres in Los Angeles 454–455
multi-racial organizing, definition xxxiii
multi-racial publics, importance of informal relations of care in the making of 461
Mumford, Lewis 48–49
Must Labour Lose? (Abrams et al) 489
My-Poznaniacy 381, 384–385

Namaste, K. 530
Naredo, J.M. 157
Nash, C.J. 530
Nation of Islam 520
National African American Leadership Summit 520
national parks 125
native Americans, dispossession of common resources 38
natural disasters, neoliberal resilience approach 162
Nazism 48
Negri, A. 37, 40, 126
neighbourhood services, planned shrinkage proposals 201
neighbourhoods, definition xxxiii
neogeography 100–101
neoliberal perspectives: community gardening in a context of austerity 136–139; drivers of infrastructural division of the city 353; neoliberal legitimisation of conditional citizenship 321; neoliberalism and the notion of publics 452; proliferation of community gardens as localized strategies to manage impacts of neoliberalization) 321–322, 330–332; remaking of cities 217; roll-against neoliberalism and labour organizing post-crisis 232–241; shifting sensibilities of middle-class identities in Buenos Aires 492–500
neoliberal privatization policies, Berlin's campaign for reversal of 123
neoliberal redevelopment governance, definition xxxiii

neoliberalism: definition xxxiii; emergence of 21; resistance against 38; the role of technocracy 111; in science fiction 22; and the undermining of redistributive functions of the state 126; as urban phenomenon 40
Netherlands: differences in local politics 471; organization strategy for migrants 471
network participation, politics of 79
networked utopia, definition xxxiii
networks of information and reciprocity 36
NeuLand, Cologne 140
'new capitalism' 41
New Deal 49
New Labour 41–42
New Orleans, gentrification 530
New Urbanism, definition xxxiii, 167
New York: bonus zoning toolkit 90; climate adaptation initiatives 593; fining of street vendors 282; for-profit co-living spaces 43; High Line Park development 534 (*see also* High Line, New York); planned shrinkage of neighbourhood services 201; response to Sandy 593; Stonewall riots 507
new-build gentrification 92
Newman, P. 443
Newton, K. 482
Nice, terrorist attack 476
Nicholls, R.J. 237
Nicholls, W. 461, 472
Nichols, Greg 53
Nichols Clark, T. 186
Nickels, Greg 110, 115
Nielsen, G.M. 351, 358
Nigella Commons 328–332
Nigeria, Chevron Oil Terminal dispute 38
NIMBY, definition xxxiii
NIMBY activism, place framing and 29
NIMBY politics: basis of the imaginaries that drive 29; homeownership and 32; motivations 32
Nobel Prize, Ostrom's receipt 42
non-citizen construction labour: and precarious construction work in Toronto 245–255; role of in the urbanization process 249 (*see also* urbanization and non-citizen labour in Toronto)
non-urban politics, identification challenges 15
'Normaltown' (Athens, Georgia) 30–31
Noterman, E. 38
novel city-networks 50
Nozick, R. 48
nuisance law, definition xxxiv

obsolessencialization, definition xxxiv
occupational communities 482–483
Occupy movement 2, 28, 126, 237, 339, 341, 519
O'Connor, S. 409–410
official decisions, spaces for making 26

official spaces 26
Olmsted, Frederick Law 533
Olson, M. 36
Olympics, Toronto's bid 585
One Kings Lane 510
Oneida Perfectionists 49
O'Neill, Thomas P. 'Tip' 26
Ong, A. 428, 566, 571, 574
Ontario: Golden Horseshoe 88; local food projects 139; primary drivers of construction markets 249; shortage of construction workers 249–250
ontology, definition xxxiv
OpenStreetMap 100
Operation Green Thumb 135
orders of worth 150
organized labour and the city: examples of organizing against neoliberalism 237–240; historical background 234–236
Orwell, G. 539
Ostrom, E. 35–37, 39, 42–43
Our City (Maria Tarantino) 477
Our Common Future (Brundtland) 167
outdoor spaces, categories of 338
outsourcing 134, 137, 296, 445
Oxford to Cambridge Arc 16

Pahl, R.E. 528
Pain, K. 16
Painter, J. 273
Pantalone, P. 92
Paradiso, M. 405
parens patriae, and quasi-sovereignty of the city 213–214
Paris 472; creative destruction of railway infrastructure 544–548; *Promenade Plantée* 549
Park, R. 236
Park am Gleisdreieck, Berlin 550–551
parking, smart strategies 98
parks, smart irrigation systems 98
Parnreiter, C. 76
participation, definition 126
participation and the culture of communicative planning, definition xxxiv
participatory planning, geographic information systems and 101
Paseo Cayala, Guatemala 190
Patrick, D.J. 534
payroll tax breaks, and corporate revitalization 509–510
Peace of Westphalia 47
Peck, J. 85, 137
Peirce, Charles Sanders 147
Pendras, M. 53
Perón, Juan 497
Persian Gulf, scale and scope of fast-track urbanism 184–193

Petite Ceinture, continuing transformation 544–548
pet-keeping, urban politics of 534
Philadelphia, Million Woman March 520
philanthropy xxvi, 43, 186, 326–327, 330, 483, 505, 510; *see also* compassionate capitalism
Philips Liveable Cities Think Tank 336
Pierce, J.L. 30
Pieterse, E. 355
Piketty, T. 50, 54
Pinterest 506
Piot, C. 188
place, thinking differently about 22–24
place framing 26, 29, 31, 33; the concept 29; definition xxxiv; as mobilizing tool 29; and NIMBY activism 29; place frames and social norms 31–33; urban politics and place frames 29–31
place imaginaries 29–32; exclusionary aspects 32
place positioning, definition xxxiv
place-based conflicts, examples of 28–32
place-making, as social practice 340
places, defining 23
places of urban politics 27–29
planetary urbanism 107
planetary urbanization 15, 21–23, 40–41, 413; critiques 22; definition xxxiv; development of the concept 21
'planned shrinkage' 201
planning: Business Improvement Area strategies 86; and the growth of Web 2.0, 100–101; research into sustainable city planning 149; securing community benefits 89–90
Plato 48
Playing Out 339, 343–345
pluralism 147
Poland: birth of contentious politics in 378; class and the urbanisation of self-management in 375–385; EU accession 380–381; Kielce Pogrom 378; post-war housing shortage 377; shift of white goods manufacturing to 381; Solidarity movement 380
Polanyi, K. 49, 54
policing, differences in experiences of 27
politically embedded life, bare life *vs* xxiv
politics of precarity, and regulatory intent 294
politics of street work 283; *see also* street work as key site of urban politics
politics of the streets 16, 27
Pollard, Ingrid 477
Pol-Potism 48
polycentric metropolitan region 16
Popper, Frank and Deborah 200
Portes, A. 294–295
Portland, urban agriculture 365–369
post-apartheid, definition xxxiv
post-growth: definition xxxiv; planning and the possibility of being 197–204

post-industrial cities, impact on communities 484
post-welfare state 21
poverty management spaces 307–317
Pow, C.P. 77
pragmatic sociology, French xxix
pragmatism: American pragmatism and urban political ecology 147–149; French pragmatist sociology and urban political ecology 149–151
precarious employment, and the work-citizenship matrix 273
precarious employment conditions, unequal citizenship and 248
precarious legal status, definition xxxiv
precariousness, tech mobility and 446
precarity, informal labour and the politics of 293–302
Pred, A. 97
Pretoria, Women's March in 520
Priestland, D. 379
privatization: of public infrastructure 248; role of enclosure 40
privatization of infrastructure, exclusionary effects 353
privatized urbanism, definition xxxv
problematization, definition xxxv
pro-growth coalitions, and public-private partnerships 62–67
pro-growth urban politics: definition xxxv; and public-private partnerships 54, 62–64
Promenade Plantée, Paris 549
propertied citizenship 354
property capital 207, 481; definition xxxv; and the search for class monopoly rents 485–488
property taxes 210, 480, 487, 550
property values, potential effect of community gardens 322
property-led development partnerships, definition xxxv
protest: disruption as function of 27; gendered perspective 519–522 (*see also* women's protests); infrastructure as a tool of 354–355 (*see also* insurgent citizenship)
protest politics, spaces of 27
protests, manifesting spaces of urban politics through 518
public benefits zoning 89
public formation: role of non-humans 149; spatiality of 148–149
public housing: authorities as urban entrepreneurs 223–224; in France 217–226 (*see also* banlieues); French subsidies 221; neoliberal restructuring of 220–221; New Towns initiative in Britain 186; redefinition of public housing policies 221
public life, contested definitions 341
public mobilization, masculinization of histories 518

public nuisance: foreclosure governance and the law of 208–209; subprime lending as mass nuisance 212–213
public services: privatization 40; role of ICTs 99–100; use of the Internet to provide 96
public spaces: categories of outdoor spaces 338; historical perspective 339; public life and 341; the role of public green spaces 339; shared 342–343; the street 338–339
public transit, and organized labour 445
public-private partnerships: accountability concerns 63; and attitude to risk 67; basic principles 65; central mechanism of pro-growth governance of urban development 62–64; the concept 452; coordination and collaboration 65–66; definition xxxv; deployment of cooperative governance mechanisms 67; entrepreneurial nature 65; evaluation criteria 64; flexibility and adaptability 66–67; and French public housing policy 220; impact of the model on civil society groups 322; inner workings of 64–67; and levels of trust 65–66; model 322, 327; pro-growth politics and 54, 62–67; regime theory perspective 65
publics: definition xxxv; the making of 452–454
Pudup, M. 137–139
Purcell, M. 29–30, 126, 452
Pussy Riot 523
Putnam 342

Quebec, student-led uprisings 237
queer ecology, definition xxxv
Queer Nation 523, 529
queering urban politics: and the emphasis on rent and property 530; heteronormative formations of natural space 533; incorporation of sexual minorities into urban political geography 529–530; and the institutionalization of heterosexuality 533; intersectional approaches 532; the 'making of gay territory' 529; pet ownership 534; 'political ecology' 533–535; power of direct action 529; queering urban political ontologies 530–532; role of activists 529–530

Rabinow, P. 312
Rabourn, M. 298
racial/ethnic perspectives: access to nutritious food 320; advocacy role of worker centres 455; changes in African-American neighbourhoods in Seattle 53; characteristics of African-American neighbourhoods 320; collaborative governance rhetoric and 321; community gardening 368; construction industry in Canada 255; criticisms of alternative food networks 363; definition xxxv; development of front and back of the house workforces 238;
discrimination in *Amazon.com*'s provision of services 413–414; discrimination suit against Wells Fargo 213–214; dispossession of native Americans' common resources 38; ethnic and racial diversity in Los Angeles 454–455; impact of de-industrialization on African-American neighbourhoods 320, 323; impact of urban renewal and planned shrinkage 201; modernization of the urban landscape 201; potential of community gardens for creating cultural and racial understanding 330–331; racial impact of gentrification 368; social mix rhetoric 225–226; urban agriculture and 368; worker centres as spaces of negotiation and learning about 457
radicalism 52
railway infrastructure: creative destruction of 542–552; definition xxxv
Rainnie, A. 80
Rancière, J. xxxii, 452, 454
Ratto, M. 341
re-appropriation of the commons 39, 135
receivership, as tool for foreclosure governance 209
Reclaim the Night 523
Reese, E. 315–316
refugee crisis 236
regeneration: plans for in London 16; property-led 199
Reid, J. 159
Reid, S. 577–578
relational approach to urban politics 23
relational thinking, sources of inspiration 1
remittances 272, 470; and domestic work migration 272
re-municipalization 123; definition xxxvi; EU rules 129; in Germany 123
re-municipalization of energy supplies, Berlin campaign 123
renewable energy, imposition of costs on the urban poor 164
Renewable Wilhelmsburg Climate Protection Concept 420
rent, appropriation of 40, 480
rentier class 199, 202; definition xxxvi
rent-seeking behaviour 19
REO properties: definition xxxvi; extension of land use and property regulatory powers to 208
reproductive rights 520, 522–523
Republic (Plato) 48
residential development, paying for the social costs of 480
residualization: definition 221; French public housing policy and 220–221
resilience: the concept 159–160, 590 (*see also* climate resilience; sustainability and resilience); definition xxxvi; development of a spatial

interpretation 162; examples of ascendancy and hegemony 161; shortcomings 162; *see also* sustainability and resilience
resilient cities, definition xxxvi
Resilient Cities Network 590
resistance movements, Hobsbawm's critique 237
resurgent radicalism 52–53
revanchism 138, 246, 310, 475, 492, 498–500, 566, 579; definition xl
re-zoning 89, 92, 482, 487–488
Rifkin, Jeremy 42
right to stay put, definition xxxvi
right to the city: the commons and the state 125–127; community gardening and 140–141; definition xxxvi
rights-based perspectives: access to energy 128; energy rights in Germany 123; of migrant domestic workers 273, 277; migrants' voting rights 471, 473; reproductive rights 520, 522–523; the right to the city, the commons and the state 125–127; Roma youth 566–568; selling of air rights in exchange for building height 85; spatial rights xxxvii; squatter rights 546; transfer of development rights 89, 90; water rights in South Africa 123–124, 130–131
right-sizing 197–204; actualized 202–203; in an atmosphere of austerity 203; the concept 197–198, 200–203; and the concept of urban growth 198–199; declining city limits and 203–204; ideational 200–201; integral measures 201–202; ontological levels 200; plans for by US and European cities 197; principles 200; statutory 201–202
right-to-work legislation, definition xxxvi
rioting, Hobsbawm on 237
riots 14, 16, 218, 224, 345, 475
The Road to Serfdom (Hayek) 49
Robinson, J. 337, 395
robots, *Amazon.com*'s use of 408
Rochester, right-sizing plans 201
Rockefeller Foundation 590
Rodgers, S. 6
roll-against neoliberalism 232–233, 237, 239; definition xxxvi; and labour organizing post-crisis 232–241
roll-back neoliberalism 39, 137, 233
roll-out neoliberalism 39, 233, 310
roll-with-it neoliberalism, the concept 232–233
Roma: definition xxxvi; origins 565
Roma youth: best-interests imaginary for 573–574; issues of rights and mobility 566–568; in Maribor 568–571; in Pata Rat 571–573; in Slovenia and Romania 565–574
Romer, P. 190
Romero-Lankao, P. 113
Roosevelt, Franklin D. 49
Rorty, Richard 147

Rosenberg, J. 509
Rosenman, E. 211
Rosser, C. 482, 489
Roth, Joseph 550
Rothbard, M. 48
Rotterdam 77; Ahmed Aboutaleb appointed mayor 477; electoral success of a populist party 475; port expansions 77; position in the global economy 77–78; Summer Carnival 476
Roy, A. 22
Rudnyckyj, D. 274
ruin, definition xxxvi
ruins, literature on 544
rural-to-urban migration xxxii–xxxiii, 247
rurbanization 39
Rust Belt 201, 365
Rutherford, S. 112

Saginaw, right-sizing plans 201
Salazarism 48
Salesforce 506
Samsung Electronics 75
San Francisco: Bay Area community gardening projects 139; earthquake and fires 507; inequality 508–509; payroll expense tax 509; tech companies and the transformation of 504–515 (*see also* compassionate capitalism; Tenderloin District, San Francisco); tech mobility 441–442; tech related job growth 508; technology companies 506–507; *see also* Tenderloin District
sanctuary cities 476
Sandilands, C. 533
Sanitary and Ship Canal, Chicago 430, 432–433 (*see also* Chicago Area Waterways System)
Santander, Spain, smart systems 98
Santiago de Chile, women's protests 521
São Paulo, Holston's exploration of democratic citizenship 351–352
Sassen, S. 87, 428
satellite cities: definition xxxvi; exemplary prototypes 185; origination and spread 185; strategic location 185
Saudi Arabia, Right-to-Drive campaigns 523
Sauvage, André 545
Savage, V.R. 77
Say Her Name campaign 521
Sbragia, A. 208
Schindler, S. 197
Schmid, C. xxxiv 15, 21
schools, impact of the digital divide on 102
Schuelke, N. 149
Schumpeter, J. 543
science effect: key features 113; in modern environmental governance 113; in urban climate governance 112–114
Scott, A. xl, 15, 19–20, 22, 50, 72, 337, 395

Scott, J. 48, 111
sea level rise, Seattle's policies 117
Seattle: anti-globalization protests 237; approach to mitigation of climate change 115; changes in African-American neighbourhoods in 53; climate science and the city 110–120; climate science and urban governance 114; median household income 53; pet-keeping 534; place-based conflict 28; smart state strategies 52–53
Sebald, W.G. 544
Section 37; definition xxxvi
Seeking Spatial Justice (Soja) 505
segregation 171, 218, 556
self-construction, definition xxxvii
self-driving cars 439, 441–444, 447; definition xxxvii; rebound effect 443; *see also* tech mobility
self-governing systems, role of social capital 37
self-management in Poland: birth of urban movements 380–385; lineages of class politics 376–378; the socialist middle class 378–380
Selle, K. 556
Sennett, R. 342
Seoul Metropolitan Area, and the growth of the LCD industry 75–76
service industry, and the growing labour movement 238
service-dependent ghettos 309–310
sexual minorities: increased visibility and urban political ecology 528; *see also* queering urban politics
sexuality, definition xxxvii
Shakers 49
sharing economies xxvi, 42–43
Shatkin, G. 185
Shaw, K. 22
Shelton, K. 355
Shenzhen 188
Shi, Y. 549
shrinking cities: most commonly used metric in the literature 198; pathway to accepting the reality 200
Siebel, W. 557
Sieber, R. 101
Silicon Valley 439, 441, 444, 506, 510
Silver, Carol Ruth 529
Simmel, G. 340, 345, 544
Simone, A. 353
Singapore: global production network 77; as international business hub 77; migrant work opportunities 273
Skid Row, Los Angeles 307–317
slack 506
slow movement 340; slow food 42
slumlord housing 201, 234
SlutWalk protests 523
smart cities: definition xxxvii, 97; examples 98; and gentrification 97; and resilience 161; savings opportunities 98; and technocratic forms of governance 112; telework and transportation systems 97–99; traffic management systems 99
Smart Cities: Big Data, Civic Hackers, and the Quest for a New Utopia (Townsend) 50
smart growth: the concept 167; definition xxxvii
smart phones xxix, xxxix, 103, 439–441, 445; and tech mobility 439
smart state: definition xxxvii; novel city-networks 50; as utopian space for urban politics 47–55
smart states: as urban political spaces 53–55; urban politics in city-regional Seattle 52–53
smart technology, tech mobility and the future of urban livability 439, 449 (*see also* tech mobility)
smart urbanism, Kitchin's argument 52
Smith, N. xl 107, 124, 162, 375
social capital 36–37, 41–42, 340; building potential of community gardening 320, 329–330; role of in self-governing systems 37; the role of public space in building 339–340
social class, education and 377
social class aesthetic, definition xxxvii
social control 138, 276–277, 308, 310
social enterprises 43
social exclusion, exclusionary effects of privatization of infrastructure 353
social justice, and the 2008 crisis 39
social media xxiii, 96, 236, 405, 506, 519, 522–523
social mix 219, 221, 224–226, 481; definition 219; in urban renewal rhetoric 225–226
social mobility, impact of the digital divide on 102
social movements: new 236; role of cities 237–238; and the roots of community gardening 135–136
social neoliberalism, principles 221
social networks, employment role 510
social norms, place frames and 31–33
social reproduction: definition xxxvii; role of the state 127
social responsibility 453, 558
social segregation, and the new economy 505–506
social service agency, definition xxxvii
social services, and place-based conflict 31–32
social spaces, relationship with development projects 17
social struggles 39
social welfare: impact of the foreclosure crisis on 206; neoliberalist re-working 310; primary outlets for last-resort services 309
socialism, and the state 48
societal governance, entrepreneurialization of 40
Soja, E. 1
Sole Food Street Farms 369, **370**
Solidarity movement 380
solidarity purchase groups 42

South Asia 17
Soviet Union 378–379
Soweto 355
space: Massey's argument 23; thinking differently about 22–24
space tourism 404
spaces, types of 26–27
spaces of dependence 74, 320–321
spaces of engagement, community gardens as 321
spaces of urban politics, framing and making the 26–33 (*see also* place framing)
Spain: late payment of utility bills 163; water prices 163
Sparke, M. 233
spatial inequities, gendered perspective 518
spatial rights, definition xxxvii
spatial scale, definition xxxvii
spatiality, of democratic politics 148
special economic zones 190, 192
speculative home building industry 484, 487
Splintering Urbanism (Graham/Marvin) 352–353
squats, and the politics of the commons 126
squatter movements, *Kommando Rhino* squatter collective 561–563
squatter rights 546
squatters movements, urban gardening initiatives 135, **547**
squatting, on the disused railway in Paris 546, **548**
Staeheli, L.A. 351
Stalinism 48
Star 352; S. 352
starchitecture, definition xxxvii
the state: in-against-and-beyond in Berlin 128–129; as an instrument of corporate power/capital 128; working within-against-and-beyond 127
state-of-exception, definition xxxviii
Stockholm, smart systems 98
Storper, M. xl, 15, 19–20, 22, 395
strategic coupling, characteristics 73
strategic coupling in global production networks 70–82; analytical potential of strategic coupling 72; asymmetries 79; characteristics of strategic coupling 73–74; in Chinese city-regions 78; coffee industry 77; dimensions of regional economic growth or decline 80–81; directionality element 75; examples 75–78, 80, 81; framing economic development of city-regions using the GPN approach 71–74; functional coupling 75, 78; indigenous couplings 75; integrating city-regions into global production networks through 74–79; intersections of urban politics and global production networks 70–71; LCD industry 75; logistics hubs and 77–78; modes of strategic coupling 75; negative consequences of territorial integration 79; politics of integration 79–81; role of global cities 76; role of innovation hubs 75; role of international partnerships are of 77; role of regional assets 73–74; structural coupling 75; the urban political perspective 82; urban politics as core domain 74

the street: reclaiming 339, 341; as space of democratic possibility 339
street work as key site of urban politics 282–290; conflict between vendors and with authorities 283, 285–286; defining street work 283; diversity of backgrounds in the informal economy 284–286; examples of informal street work 282; gender perspectives 288; political influence of worker associations 286; transnational agency and local politics 287–289
street-based harassment of women and girls, documenting and challenging 523
the streets, politics of 16, 27
Streetsblog 441
Strom, E. 65
Ström, M. 579
structural coupling: definition xxxviii; *see also* strategic coupling in global production networks
StumbleUpon 506
subprime lending, as mass nuisance 212–213
suburbanization, Bethnal Green study 482
suburbs: definition xxxviii; distinctions between city and 18; reinterpretation 17
Suez Environment 161
Sui, D. 101
surge pricing 446
surplus populations: definition xxxviii; examples of 309; management in Anglo-American cities 309; role of in urbanization processes 255; and the targeting of urban poverty management 309–310
surveillance 454
sustainability: the concept 157–158; justifying 'techno-managerial' approaches to urban development through 167; social justice perspective 167–168; tensions between three pillars of 167; urban experimentation and the politics of 416–423 (*see also* urban living laboratories)
sustainability and resilience: in cities of the Global North 157–165; competing notions of resilience 159–160; implications of the rise of resilience 164–165; the relentless expansion of resilience 160–162; successes and casualties 163–164; sustainability as ecological modernization 158–159; water and energy consumption in cities 163–164
sustainability fix, definition xxxviii
sustainability in Austin: density bonus programme 176; development and racial division in Rainey

sustainability in Austin *continued*
  neighbourhood 170–175; equity concerns 169; parkland dedication 175; Safe, Mixed-Income, Accessible, Reasonably Priced, Transit-Oriented (SMART) housing programme 169; smart growth model 168–170; urban housing transition in the Rainey neighbourhood 175–177
sustainable city planning, research into 149
sustainable communities initiative 42
sustainable urban development: Copenhagen project 150; definition xxxviii
Svarre, B. 341
Swansea, study of family change 482
Swartz, J. 507
sweatshops, activism against 148
swidden forests, regeneration of 39
Swyngedouw, E. 112, 124–125, 452
symmetrical family xxix, 481, 483
The Symmetrical Family (Young/Willmott) 483
Syrian refugee crisis 236

Taft-Hartley Act 479
Tahrir Square demonstrations, Cairo 519
Tampere, Finland, municipal Web 2.0 use 101
Tarantino, M. 477
Task Rabbit 446
Taştan, C. 152
taxation, payroll tax breaks and corporate revitalization 509–510
Teamsters and Turtles 237
Teamsters Union 445–446
tech mobility: the concept 439–441; consumers of 444; contesting the political project 447–449; contribution to gentrification 442, 445, 447; definition xxxviii; and the future of urban livability 439–449 (*see also* tech mobility); implications for urban politics of mobility 439; and labour 445, 447; and neoliberal liveability 443–445; and urban liveability 441–443; working conditions 445–446
tech sector, training requirements 510
tech volunteerism in the Tenderloin 512–514
technocracy: and climate change governance 111; key problem 112; scholarly literature on 111–112
technocratic governance, definition xxxviii
Tehran, Green Movement protests 519
telegraphy, definition xxxviii
telework, definition 98
temporary migration streams, Canadian expansion of 246
Tenderloin District, San Francisco: Airbnb description 504; community benefits agreements 510; corporate revitalization 509–510; empowerment and enterprise zones 509–510; location 508; naming 507; racial demographics **508**; social services 508; social/economic/racial segregation 512; tech transformation 504; tech volunteerism 512–514
Tenderloin district, San Francisco, history and demographics 507–509
territorial boundaries, role of in day-to-day political practice 14
territorial state, definition xxxviii
Thévenot, L. 150
'Third Way' theory 41
throwntogetherness of place: definition xxxviii; livability and 337; and the lived experiences of the urban 22–23
Tickell, A. 137
Tiebout, C. 482
time-space perspectives: *Amazon.com*'s improvement through time-space compression 410–411; climate change risk in coastal cities 394–396; worker centres 454–455
Tokyo, Japan, co-housing projects 42
Toly, N. 50
Tönnies 340
Toronto: condo boom and policy framework 87–88; condominium construction 249; density bonusing and development in 90–92; metro art 577–585; migrant work opportunities 273; non-citizen labour and precarious construction work in 245–255; SlutWalk protests 523; tallest residential tower 87
Townsend, A. 50–52
traffic management: intelligent systems 99; in smart cities 98
transaction, the concept of 147
transactional rationality 147–148
transactional spaces 149, 151
transfer of development rights (TDR) 89, 90
transit-led urbanism, definition xxxix
transnational corporate network 72, 77; and the logistical capacities of city-regions 77
transportation, intelligent systems 99
transportation demand management (TDM) 445
transportation network companies (TNCs) 76–77, 439–442, 444–447; costs and liabilities 446; definition xxxix; vulnerability of drivers 446
trickle-down economics, ineffectiveness for the working class/working poor 510
Trikala, Greece, as digital city 101
Trump, Donald J. 519–520
Tryczyk, M. 377
Tsing, A. 79
Turner, L. 237
Twitter 100, 504, 506–507, 509–510
Twitter tax breaks 505, 509
Tyler, I. 522
Tyner, J. 309
typhoid, CAWS contamination and outbreaks of 430

Index

Uber xxxix, 98, 439, 441–442, 444, 446–447, 506; price surging model 446
uber-fication 246
Uitermark, J. 472, 475
Umbrella Movement protests, Hong Kong 519
undocumented workers 234, 253–255, 298
uneven development 21, 323, 362, 364, 484–485; definition xxxix
Uneven Development: Nature, Capital, and the Production of Space (Smith) 107
UNFCCC (United Nations Framework Convention on Climate Change) 595–596
unionization, weakening of the regulatory framework 296
unionization rates, fall in the private sector 295
United Nations, global governance climate regime 592
United Nations Convention on the Rights of the Child (UNCRC) xxv
United States (USA): blackout of 2003 428; class and territory in the American city 479–489; the condominium concept 87; health care and place-based conflict 30–31; high-tech boom 42; informal labour *see* informal labour in the USA; institutional geography of labour in the US Midwest 259–268; online geographical information systems 101; postrecession translation 40; post-war suburbanization 479; predominant suburban residential imaginary 29; role of homeownership 32; the suburban ideal 31; urban 'renaissance' 340; US Mayors Climate Protection Agreement 110, 115
uplift, definition xxxix
uplift capture 89
upward social mobility, as aim of Jacobs' work 376
the urban: defining 18–20; impact of the key sides and human relationships 18; relationship with suburbia 15–18; rethinking 14–24; as a site of radical forms of politics 14; as a social phenomenon 20–22; thinking differently about 22–24
urban agriculture: and access to resources 366; contestations in Portland and Vancouver 365–369; critical perspectives 362–363; criticisms 363; definition xxxix; environmental and social benefits 362; forms of 364; gentrification's relationship with 362, 364–365, 368–369; as key component of urban sustainability 362; networks of association in Milwaukee **328**; and the political economy of sustainability 363–364; resistance to 367; and social class 367–368; as a sustainability fix 369–371; and the sustainable city 364–365; *see also* community gardening
urban citizenship: definition xxxix; Painter's definition 273

urban commons/commoning: ambivalence of 35–44; definition xxxix, 139
urban experimentation: definition xl; political perspective 418; the transformation of Bogotá 418; *see also* urban living laboratories
urban food production 361–362, 366–367; *see also* urban agriculture
urban gardening, on a disused railway in Paris **547**
urban geography, uses of the concept of assemblage 428
urban growth, the concept 198–199
urban growth machine 15, 19
urban informality 41, 283, 293; and the 'entrepreneurial slum' 41
urban infrastructure, definition xl
urban intelligence industrial complex 51
urban land nexus xl, 15, 19–20, 23; agglomeration process 19; definition xl
urban livability, tech mobility and the future of 439–449
urban living laboratories (ULLs): challenges around inclusivity and inequality 422–423; the concept 418–420; definition xl; empowerment potential 416–417; forms and examples of ownership 420–421; implicit assumptions 422; largest and most comprehensive initiative in the world 419; political dimensions 421–423; in practice 420–421; rapid spread 416; Renewable Wilhelmsburg Climate Protection Concept 420; rise of urban experimentation 417–418; social inclusion strategies 420
urban middle class in Buenos Aires: attitudes towards social mixing 497, 499; attitudes towards upward social mobility 496–497; context of the 2001 economic crisis 494–495; evolution of the city 493–494; fifteen years after the 2001 crisis 496–498; identifying the middle class in Argentina 493; impacts of the 2001 crisis for the city 495; neoliberal and revanchist character of the city 498–500; post-2001 economic crisis deterioration of middle-class fortunes 493–494; and the recovery of the Argentine economy 495–496
urban networks 51
urban outcasts 14
urban planning, definition xl
urban political ecology: American pragmatism and 147–149; the concept 124–125; definition xl; diversity of the field 146; French pragmatist sociology and 149–151; pragmatist scholarship 149; and the right to the city 125
urban political process, key elements 528
urban politics: the driving force 19; emerging spaces of 7–8; locating 5–7; the meaning of 3–5, 14; typology of spatial forms 72
urban poverty management, definition xl

621

Index

urban poverty management in the Global North: application of assemblage thinking 311–313; the concept of urban poverty management 309; exemplar space 307–308; historical perspective 309; historical perspective of Skid Row, Los Angeles 314–317; neoliberal governance perspective 310–311; and NIMBY-reinforced policies of containment 315–316; policing strategies 316; problematization of poverty 313–314

urban renewal 135, 158, 171, 174, 201, 218–219, 222–225, 323, 487, 498, 581; impact on communities 483

urban revanchism, definition xl

The Urban Revolution (Lefebvre) 125

urban space: ontologically distinctive scales 20; political resonance 24

urban studies literature, hostility towards middle-class contentious politics 375

The Urban Villagers (Gans) 483

urban violence, role of infrastructure 354

urban water, political ecology of 426

urban water systems, research on 428–429 (*see also* Chicago Area Waterways System)

urban wildlife, Hinchliffe *et al.*'s "cosmopolitical experiment" 149

urbanism as a way of life 15

urbanization, planning and the possibility of being post-growth 197–204

urbanization and non-citizen labour in Toronto: crises of urbanization and temporary migration 249–251; deportability and illegality of precarious and undocumented workers 253–255; employer-adjudicated legal status and the Federal Skilled Workers Programme 251–253; inequality of the construction industry 255; urbanization-migration nexus 247–249

user engagement, definition xl

Utopia (More) xxxiii

utopia, definition xl, 539

utopian perspectives: spaces of dystopia and utopia 539–598; sustainable urbanism in the European city 555–563; urban revitalization and the aesthetics of transportation 577–585

utopianism, globalization and utopian smartness 50–52

Valentine, G. 530
Valiani, S. 253
validation of space, definition xl
Valverde, M. 214
Vancouver: planning tools 89; spatial boundaries of sexualized urban spaces 530; urban agriculture 365–369
Vattenfall 123, 128
Versailles, Women's March on 520

Vertovec, S. 477
Viehoff, V. 140
Vietnam War 508
visible minority, definition xl
voluntary sector, community gardening and the outsourcing of services to 137–138
volunteerism: community gardening's reliance on 330–332; encouragement of as neoliberal strategy 321; and the subsidization of capital 363
Vommaro, G. 499
von Mises, Ludwig 49

Wacquant, L. 14, 236
Wafer, A. 355
wage theft 300–301, 455, 459–460; definition xli
wages, decoupling of the historic link between labour productivity and 295
walkability 175–176, 443–445
Walker, J. 160
Walker, R. 22
Walker, S. 211
Walmart 293
Walnut Way Conservation Corps, Milwaukee 325–327, 330
Ward, K. 591–594
Washington: Million Man March 520; Women's March 519–520
water, gendered division of labour and access to 125, 131
water and energy consumption in cities 163–164
water poverty, definition xli
water service billing, penalizing the precariat with 199
Waters, R. 507–508
Watson, S. 337, 339
Web 2.0: participatory qualities 101; urban planning and the growth of 100–101
Weil, D. 295
Weix, G.G. 274
Weizman, E. 376
welfare retrenchment 521, 529
Welfare State 127
welfare state, dismantling 40, 217
Wells Fargo 213
West Hollywood, coalition building 529
Westphalia, Peace of 47
Whitehead, M. 112, 122
Williams, C. 53, 272–273
Williams, O. 482, 528
Williamson, Oscar 37
Willmott, P. 481–483, 489
Wilson, D. 220, 226
windfalls capture 86, 89; *see also* density bonusing
Wingo, L. 482
Wirth, L. 15, 18–19
Wolch, J. 309

Wolfe, M. 89
Wolff, R. 342
women: access to public space and free movement 522–524; direct action 523; disproportionate impact of climate change on 396–398; documenting and challenging street-based harassment 523; participation in urban protest 518–519 (*see also* gendering urban protest); and the politics of gender in space 522, 524; presence of as feature of Middle Eastern and North African mobilization 519; protesting in Latin American cities 521; vital everyday labour historically undertaken by 126
Women in Black movement, Jerusalem 520
Women Shoufouch 523
Women's Liberation 520
Women's March 519
women's protests: *Madres de Plaza Mayo* 521; Million Woman March in Philadelphia 520; Right-to-Drive campaigns in Saudi Arabia 523; Santiago de Chile 521; Say Her Name campaign 521; SlutWalk protests 523; social media as space for 523–524; Women's March in Pretoria 520; Women's March in Washington 519–520; Women's March on Versailles 520; Women's March on Washington 519–522
women's unpaid labour, importance to the functioning of capitalist cities 229
women's urban politics, importance of the production and protection of gendered spaces 518
Woodward, D. 66
work-citizenship matrix 248, 273
worker centres: definition xli; governance flexibility 455–457; innovative organizing strategies 455–457; institutional infrastructure 456; negotiation and learning spaces 457–458; regulating informal labour markets from the bottom-up 300–301; social justice campaigns 459–461; spheres of action 455; stated goals 455; time-space perspective 454–455

worker organizing, in the informal economy 293
worker protection, failure in the USA 295–296
workers 126
workers' cooperatives 126
working class spaces, and 'culture' and 'entertainment' 236
working holidays, Canadian programme 250–251
the working poor, fast-track urbanism and the displacement of 186
workplace advocacy campaigns 263
workplace fissuring, the concept and its impact 295–296
World Trade Organization riots 52
World-City Network (WCN) 71–72
Wynne, B. 595

Yahoo! Maps 100
Yammer 506, 509–510
Yang, C. 78
Yelp 506, 509
Yeung, H.W.C. 75, 80
Young, I.M. 465
Young, M. 481–483, 489
young ethnic minorities, vulnerability to revanchism 566
Youngstown, right-sizing plan 197, 200
Youngstown 2010 Citywide Plan (City of Youngstown 2005) 197
youth activism, definition xli
Ysa, T. 67
Yurchak, A. 379

Zapatista uprisings 38
Zelizer, V. 483
Zendesk 504, 506, 509–510
The Zendesk Neighbor Foundation 511
zero marginal cost society 43
Zones of Immersion (Reid) 577–578, 584–585
Zoosk 510
Zukin, S. 238
Zwischenstadt ('in between city') 17